AUTOMOTIVE
FUEL, LUBRICATING, AND COOLING SYSTEMS

OTHER BOOKS AND INSTRUCTIONAL MATERIALS BY WILLIAM H. CROUSE AND *DONALD L. ANGLIN

The Auto Book*
Auto Shop Workbook*
Auto Study Guide*
Auto Test Book*
Automotive Air Conditioning*
Workbook for Automotive Air Conditioning*
Automotive Body Repair and Refinishing*
Workbook for Automotive Body Repair and Refinishing*
Automotive Brakes, Suspension, and Steering*
Workbook for Automotive Brakes, Suspension, and Steering*
Automotive Dictionary*
Automotive Electronics and Electrical Equipment
Workbook for Automotive Electronics and Electrical Equipment
Automotive Emission Control*
Workbook for Automotive Emission Control*
Automotive Engine Design
Automotive Engines*
Workbook for Automotive Engines*
Workbook for Automotive Fuel, Lubricating, and Cooling Systems*
Automotive Mechanics
Study Guide for Automotive Mechanics*
Testbook for Automotive Mechanics*
Workbook for Automotive Mechanics*
Automotive Engines Sound Filmstrip Program
Automotive Service Business
Automotive Technician's Handbook*
Automotive Tools, Fasteners, and Measurements*
Automotive Transmissions and Power Trains*
Workbook for Automotive Transmissions and Power Trains*
Automotive Tuneup*
Workbook for Automotive Tuneup*
General Power Mechanics* (With Robert Worthington and Morton Margules)
Motor Vehicle Inspection*
Workbook for Motor Vehicle Inspection*
Small Engine Mechanics*
Workbook for Small Engine Mechanics*

AUTOMOTIVE ROOM CHART SERIES

Automotive Brake Charts
Automotive Electrical Equipment Charts
Automotive Emission Controls Charts
Automotive Engines Charts
Automotive Engine Cooling Systems, Heating, and Air Conditioning Charts
Automotive Fuel Systems Charts
Automotive Suspension, Steering, and Tires Charts
Automotive Transmissions and Power Trains Charts

AUTOMOTIVE TRANSPARENCIES BY WILLIAM H. CROUSE AND JAY D. HELSEL

Automotive Air Conditioning
Automotive Brakes
Automotive Electrical Systems
Automotive Emission Control
Automotive Engine Systems
Automotive Steering Systems
Automotive Suspension Systems
Automotive Transmissions and Power Trains
Engines and Fuel Systems

SIXTH EDITION

AUTOMOTIVE FUEL, LUBRICATING, AND COOLING SYSTEMS

WILLIAM H. CROUSE
DONALD L. ANGLIN

GREGG DIVISION/McGRAW-HILL BOOK COMPANY

New York ○ Atlanta ○ Dallas ○ St. Louis ○ San Francisco ○ Auckland ○ Bogotá ○ Guatemala
Hamburg ○ Johannesburg ○ Lisbon ○ London ○ Madrid ○ Mexico ○ Montreal ○ New Delhi
Panama ○ Paris ○ San Juan ○ São Paulo ○ Singapore ○ Sydney ○ Tokyo ○ Toronto

ABOUT THE AUTHORS

William H. Crouse

Behind William H. Crouse's clear technical writing is a background of sound mechanical engineering training as well as a variety of practical industrial experience. After finishing high school, he spent a year working in a tinplate mill. Summers, while still in school, he worked in General Motors plants, and for three years he worked in the Delco-Remy division shops. Later he became director of field education in the Delco-Remy Division of General Motors Corporation, which gave him an opportunity to develop and use his writing talent in the preparation of service bulletins and educational literature.

During the war years, he wrote a number of technical manuals for the Armed Forces. After the war, he became editor of technical education books for the McGraw-Hill Book Company. He has contributed numerous articles to automotive and engineering magazines and has written many outstanding books. He was the first editor-in-chief of McGraw-Hill's *Encyclopedia of Science and Technology*.

William H. Crouse's outstanding work in the automotive field has earned for him membership in the Society of Automotive Engineers and in the American Society of Engineering Education.

Donald L. Anglin

Trained in the automotive and diesel service field, Donald L. Anglin has worked both as a mechanic and as a service manager. He has taught automotive courses and has also worked as curriculum supervisor and school administrator for an automotive trade school. Interested in all types of vehicle performance, he has served as a racing-car mechanic and as a consultant to truck fleets on maintenance problems.

Currently he devotes full time to technical writing, teaching, and visiting automotive instructors and service shops. Together with William H. Crouse he has co-authored magazine articles on automotive education and several automotive books published by McGraw-Hill.

Donald L. Anglin is a Certified General Automotive Mechanic, a Certified General Truck Mechanic, and holds many other licenses and certificates in automotive education service and related areas. His work in the automotive service field has earned for him membership in the American Society of Mechanical Engineers and the Society of Automotive Engineers. In addition, he is a member of the Board of Trustees of the National Automotive History Collection.

Library of Congress Cataloging in Publication Data

Crouse, William Harry [DATE]
Automotive fuel, lubricating, and cooling systems.

Includes index.
1. Automobiles—Fuel systems. 2. Automobiles—Lubrication. 3. Automobiles—Motors–Cooling systems.
I. Anglin, Donald L., joint author. II. Title.
TL214.F8C76 1981 629.2'5 80-17480
ISBN 0-07-014862-7

AUTOMOTIVE FUEL, LUBRICATING, AND COOLING SYSTEMS, Sixth Edition

Copyright © 1981, 1976, 1971, 1967, 1959, 1955 by McGraw-Hill, Inc. All rights reserved. Printed in the United States of America. No part of this publication may be reproduced, stored in a retrieval system, or transmitted, in any form or by any means, electronic, mechanical, photocopying, recording, or otherwise, without the prior written permission of the publisher.

1 2 3 4 5 6 7 8 9 0 SMSM 8 9 8 7 6 5 4 3 2 1

Sponsoring Editor: D. Eugene Gilmore
Editing Supervisor: Paul Berk
Design Supervisor: Caryl Spinka
Production Supervisor: Kathleen Morrissey
Art Supervisor: George T. Resch

Text Designer: Linda Conway
Cover Designer: David Thurston
Cover Illustration: Vantage Art, Inc.
Technical Studio: Vantage Art, Inc.

ISBN 0-07-014862-7

CONTENTS

PREFACE

This is the sixth edition of *Automotive Fuel, Lubricating, and Cooling Systems*. In the five years since the publication of the previous edition, there have been many important developments in automotive engines and in automotive fuel, lubricating, and cooling systems. New federal legislation on emission controls and fuel economy has required automotive engineers to develop many new devices and systems. New engine and fuel-system controls are now in use. The automotive industry has met the challenge of these federal regulations.

This new edition of *Automotive Fuel, Lubricating, and Cooling Systems* will help automotive teachers to meet the challenges of these new regulations. It covers new developments in engine and emission controls, new carburetors (including the Ford variable-venturi carburetor), stratified charge engines, electronic air-fuel ratio controls (such as the General Motors Electronic Fuel Control (EFC) system and the Ford Feedback Carburetor Electronic Engine Control system), three-way catalytic converters, gasoline and diesel engine fuel-injection systems, and many other innovations the automotive student must understand.

This new edition continues to feature the metric equivalents of all United States customary measurements: as an example, the dimensions of this book would be shown as 8½ × 11 inches [216 × 279 mm (millimeters)].

The text has been extensively rewritten to simplify explanations, shorten sentences, and improve readability. Objectives have been added at the beginning of each chapter so that both the student and the teacher know what the learning objectives are for that chapter.

A new edition of the *Workbook for Automotive Fuel, Lubricating, and Cooling Systems* has been prepared. It includes the basic engine-service jobs as proposed in the latest recommendations of the Motor Vehicle Manufacturers Association–American Vocational Association Industry Planning Council. Used together, *Automotive Fuel, Lubricating, and Cooling Systems* and its workbook supply the student with the background information and hands on experience needed to become a qualified and certified automotive engine technician.

To assist the instructor, the *Instructor's Planning Guide* is available. This guide contains information on the automotive service industry: the growing need for properly trained mechanics, their testing and certification, and new laws affecting them. It also lists the other materials, apart from the workbook, that are available for use with *Automotive Fuel, Lubricating, and Cooling Systems*. These include the McGraw-Hill *Automotive Transparencies*, and eight sets of *Automotive Room Charts*.

Used singly or together, these instructional materials ensure congruency between the school curriculum and the future needs of students entering the automotive service field. They will help the instructor to tailor the learning experience of the student around tested and proven competency-based objectives, and will help the student to meet the minimum standards of competence demanded of entry-level employees. The student will then be able to develop locally demanded career skills while mastering the necessary job competencies and performance indicators covered in *Automotive Fuel, Lubricating, and Cooling Systems*, Sixth Edition.

William H. Crouse
Donald L. Anglin

ACKNOWLEDGMENTS

During the preparation of this sixth edition of *Automotive Fuel, Lubricating, and Cooling Systems,* the authors were given invaluable aid and inspiration by many people in the automotive industry and in the field of education. The authors gratefully acknowledge their indebtedness and offer their sincere thanks to these people. All cooperated with the aim of providing accurate and complete information that would be useful in training automotive mechanics.

Special thanks are owed to the following organizations for information and illustrations that they supplied: AC Spark Plug Division of General Motors Corporation; American Motors Corporation; ATW; Autoscan, Inc.; British Leyland, Inc.; Buick Motor Division of General Motors Corporation; Cadillac Motor Car Division of General Motors Corporation; Carter Carburetor Division of ACF Industries; CAV Ltd.; Champion Spark Plug Company; Chevrolet Motor Division of General Motors Corporation; Chrysler Corporation; Delco-Remy Division of General Motors Corporation; Federal Mogul Corporation; Ford Motor Company; General Motors Corporation; Gunk Laboratories, Inc.; Harrison Radiator Division of General Motors Corporation; Holley Carburetor Division of Colt Industries, Inc.; Honda Motor Company, Inc.; Johnson Bronze Company; Kohler Company; Mercedes-Benz of North America, Inc.; Motor Vehicle Manufacturers Association; Oldsmobile Division of General Motors Corporation; Perfect Circle Division of Dana Corporation; Perkins Engines, Inc.; Pontiac Motor Division of General Motors Corporation; Robert Bosch Corporation; Schwitzer Division of Wallace-Murray Corporation; Snap-on Tools Corporation; Sun Electric Corporation; Suzuki Motor Company, Ltd.; Toyo Kogyo Company, Limited; Toyota Motor Sales Company, Limited; Volkswagen of America, Inc.

Special thanks must go to the manuscript reviewers. They were Eugene N. Brown, Washtenaw Community College, Ann Arbor, Michigan, and Dennis Grace, Lakeshore Technical Institute, Cleveland, Wisconsin.

William H. Crouse
Donald L. Anglin

CHAPTER 1

AUTOMOTIVE GASOLINE ENGINE SYSTEMS

After studying this chapter, you should be able to:

1. Explain the various ways in which automotive engines are classified.
2. Describe how the crankshaft and connecting rod convert reciprocating motion to rotary motion.
3. Explain how the four-cycle engine works.
4. Discuss the need for the fuel, lubricating, and cooling systems in an engine.
5. Explain the purpose of the ignition system and describe the components in the system.
6. Explain the difference between the contact-point ignition system and the electronic ignition system.
7. Describe the purpose and operation of the centrifugal and vacuum advance mechanisms in the distributor.

1-1 The gasoline engine It takes more than the burning of gasoline to keep an engine running. The pistons will not move up and down very long in the engine cylinders without the help of the fuel, lubricating, and cooling systems. And, if it was not for the ignition system igniting the fuel in the cylinders, the pistons in an engine would not move up and down at all!

This book covers the construction, operation, and servicing of four engine systems—fuel, lubricating, ignition, and cooling. Three of these systems—fuel, lubricating, and cooling—are covered in detail. The fourth system—the ignition system—is reviewed briefly in this chapter.

1-2 Engine types The two major types of engine used in automobiles today are the:

1. Piston engine (Fig. 1-1) in which pistons (Fig. 1-2) move up and down or *reciprocate* in the engine cylinders. This is the engine used in all cars made in the United States.
2. Wankel rotary engine in which rotors rotate, or spin. The major producer of this engine and the Mazda cars in which it is used is Toyo Kogyo of Japan. No American automotive manufacturer is using this engine.

The piston engine can be divided into two types:

1. The spark-ignition engine which usually burns gasoline. These engines have an electric ignition system which ignites or sets fire to the gasoline vapor in the engine cylinders. Most automotive engines are of this type.
2. The compression-ignition or diesel engine in which the *heat of compression* ignites the fuel. This engine does not have a separate electric ignition system. Instead, the diesel uses the heat produced by compressing air in the cylinders to ignite the fuel. Diesels are becoming more popular. Some United States automotive manufacturers are now making diesel automobiles. Diesel engines and diesel-engine fuel systems are covered in later chapters.

1-3 Arrangement and number of cylinders Engines have been built with one, two, three, four, five, six, eight, twelve, and sixteen cylinders. The most common arrangements for automotive engines are shown in Fig. 1-3. Most cars have either a four-cylinder (Fig. 1-4), an in-line six (Fig. 1-1), or a V-6 or V-8. In the V-type engines, the cylinders are lined up in two banks which are set at angles to each other. Figure 1-5 shows a V-6 which has one bank partly cut away to show the internal construction.

Regardless of the number of cylinders, the same actions take place in each cylinder. Let us look at what happens in one engine cylinder.

1-4 Engine cylinder Basically the cylinder is a round container. It is like a tin can closed at the top and with the bottom cut out (Fig. 1-6). The piston, which is like a slightly smaller tin can, fits rather loosely into the cylinder (Fig. 1-6*b*). The piston can slide up and down in the cylinder. This up-and-down motion is called *reciprocating* motion.

In Fig. 1-6 the cylinder has been drawn as though it were transparent so the actions in the cylinder can be seen. To start with, assume the cylinder is filled with a combustible mixture of air and gasoline vapor. The

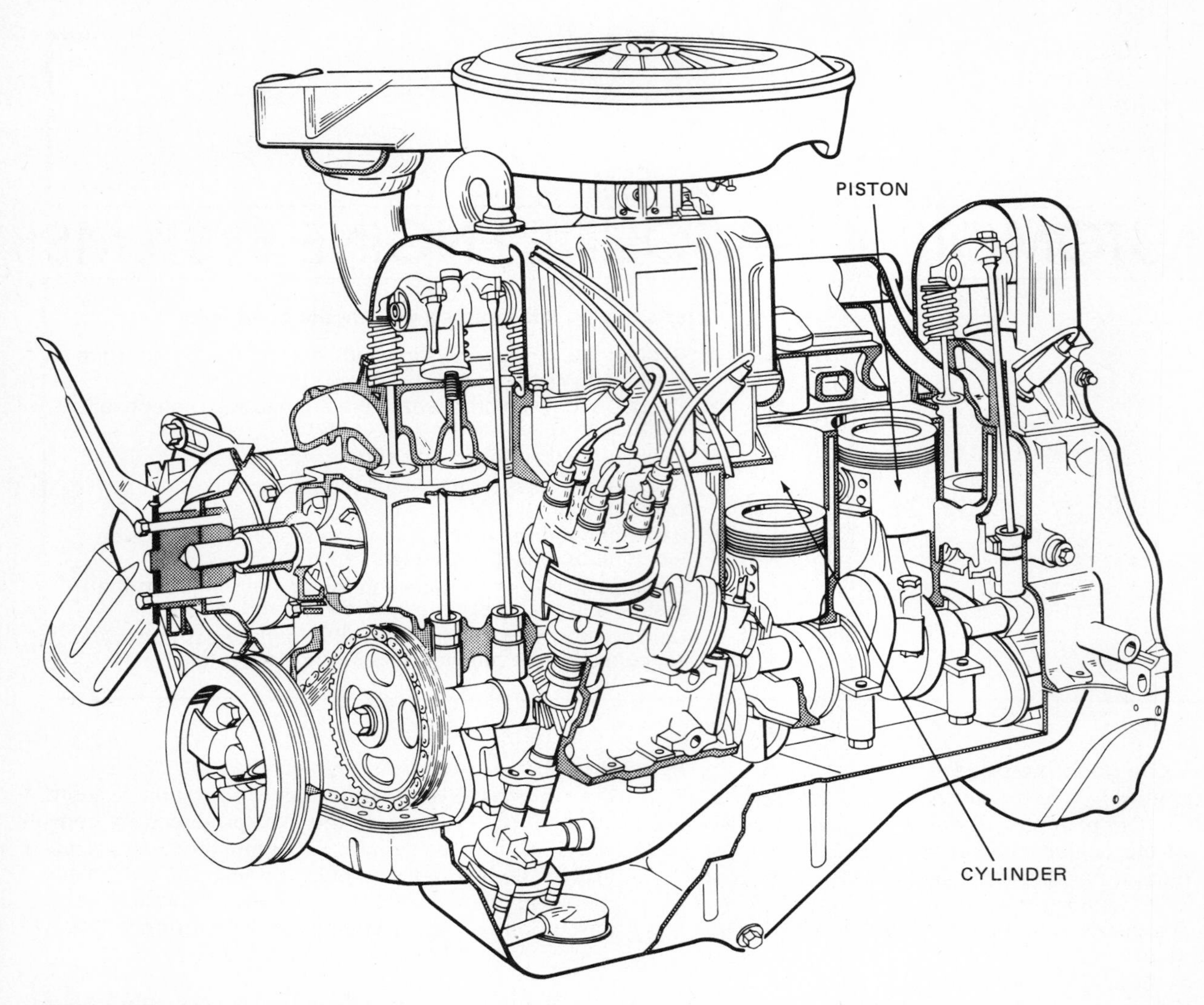

Fig. 1-1 Six-cylinder in-line engine with overhead valves, partly cut away to show internal construction.

Fig. 1-2 A piston from a piston engine.

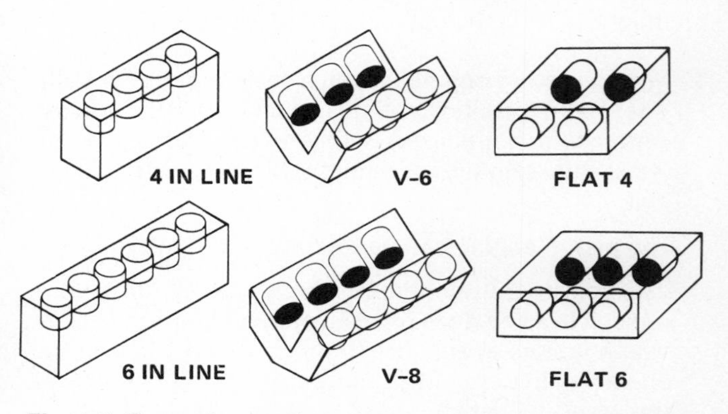

Fig. 1-3 Several ways that pistons can be arranged in an engine.

Fig. 1-4 Cutaway view of a four-cylinder engine. (*Volkswagen of America, Inc.*)

Fig. 1-5 Partial cutaway view of a V-6 overhead-valve engine. (*Chevrolet Motor Division of General Motors Corporation*)

piston pushes up into the cylinder as shown at Fig. 1-6*b*. This traps the air-fuel mixture and compresses it in the cylinder.

Now, an electric spark occurs in the top of the cylinder. This spark is produced by the ignition system. The mixture is ignited and it burns very rapidly, or "explodes." High temperatures and pressures result. The piston is driven down in the cylinder (Fig. 1-6*c*).

In the actual engine, the connecting rod and crank on the crankshaft prevent the piston from leaving the cylinder. The piston moves down. Then the rod and crankshaft push the piston up again. These actions are described later in the chapter. Now, however, let us review two things: First, how the reciprocating motion is changed to rotary motion so the car wheels are forced to rotate and move the car. Second, how the air-fuel mixture gets into the cylinder.

⚙ 1-5 Reciprocating motion to rotary motion Figure 1-7 shows a piston, connecting rod, and crank

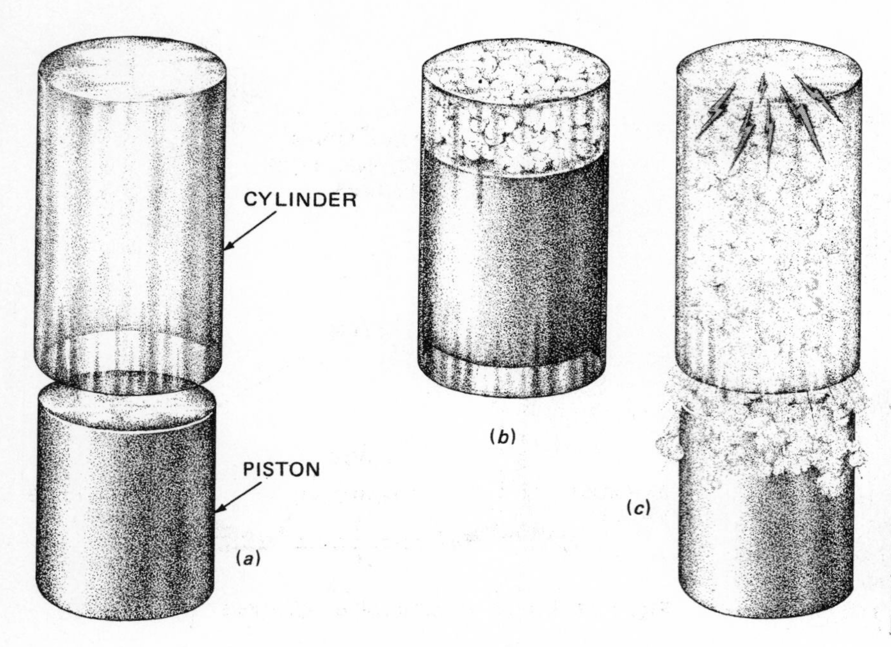

Fig. 1-6 Three steps in the actions in an engine cylinder. (*a*) The piston is a second cylinder that fits snugly into the engine cylinder. (*b*) When the piston is pushed up into the engine cylinder, air is trapped and compressed. The cylinder is drawn as though it were transparent to show the piston. (*c*) As the pressure increases because of the burning of the gasoline vapor, the piston is pushed out of the cylinder.

on a crankshaft separated so they can be clearly seen. Figure 1-8 shows them attached to each other. The connecting rod is attached to the piston by the piston pin. In Fig. 1-8, the piston is partly cut away to show how the piston pin attaches the piston and rod. The lower end of the rod is attached to a crankpin by the rod cap. The cap is held in place by cap bolts and nuts.

Figure 1-9 shows an engine crankshaft so you can see how the crankpins are offset sections of the shaft.

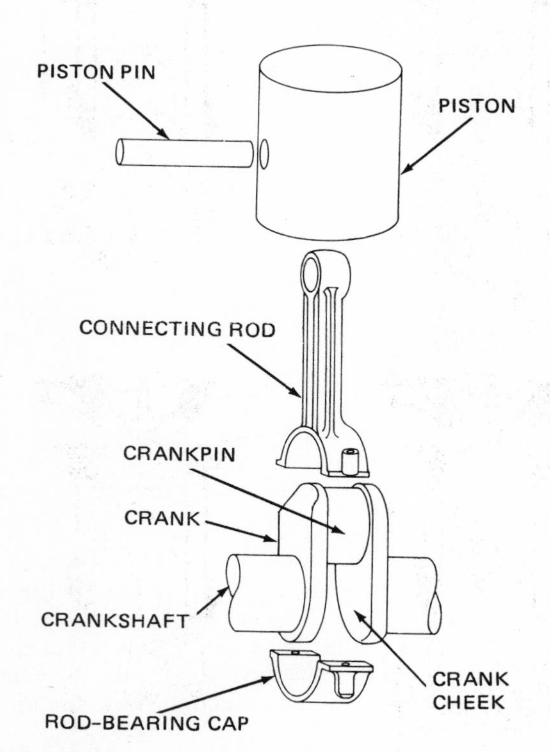

Fig. 1-7 Piston, connecting rod, piston pin, and crank of a crankshaft, disassembled. The piston rings are not shown.

As the crankshaft rotates, the crankpin moves in a circle around the crankshaft. Figure 1-10 shows the sequence of actions as the piston moves from top to bottom to top again. At the same time, the crankshaft completes one revolution. Notice that the rod tilts to one side and then the other as the lower end of the rod moves in a circle with the crankpin.

⚙ 1-6 Valves To keep the engine running, a certain amount of air-fuel mixture must repeatedly get into the cylinders. Then after the fuel is burned, the burned gas must get out of the cylinders. The valves do this job. Each cylinder has at least two valves (Fig. 1-11). The

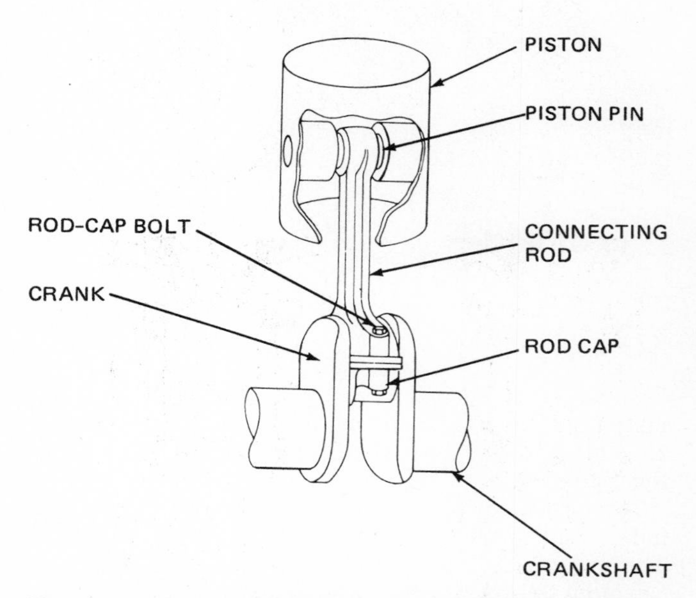

Fig. 1-8 Piston and connecting rod attached to a crankpin on a crankshaft. The piston rings are not shown. The piston is partly cut away to show how it is attached to the connecting rod.

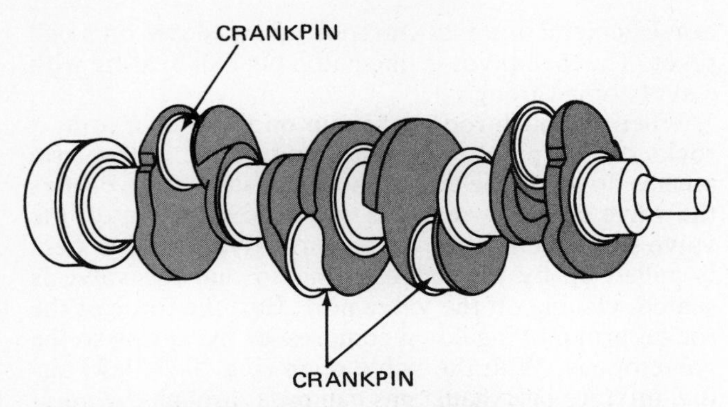

Fig. 1-9 A crankshaft for a four-cylinder engine.

valves are a type of plug on long stems. They fit into holes or *ports* at the top of the cylinder. When the valves are in the up position, they seal off or close the ports. When the valves are pushed down, they open the ports. In the open position, air-fuel mixture or exhaust gas can pass through the port.

Figure 1-12 shows an engine partly cut away so that the valve-operating mechanisms can be seen. These valve mechanisms are called the valve train. Figure 1-13 shows how the valve train works. The action starts at the camshaft, which is a long shaft with a series of cams on it. Each cam has one high spot, or lobe. The camshaft is rotated by a pair of gears or by a chain or toothed belt from the engine crankshaft.

The left part of Fig. 1-13 shows what happens when the cam lobe moves around under the valve lifter. The

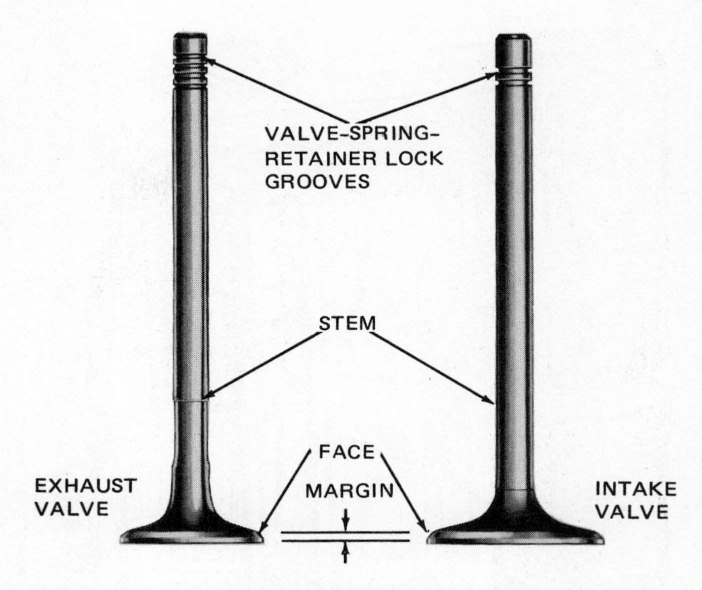

Fig. 1-11 Typical engine valves. (*Chrysler Corporation*)

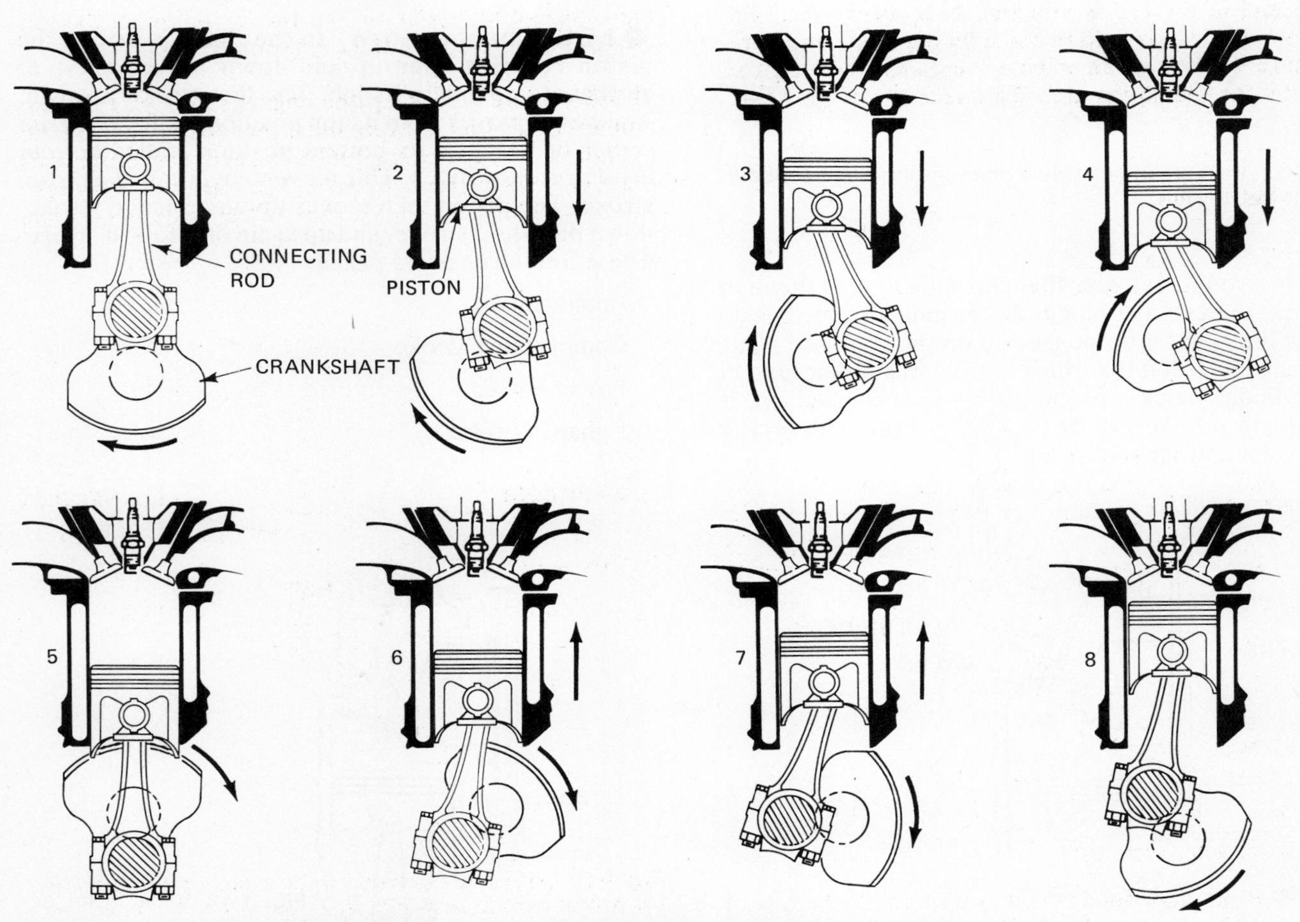

Fig. 1-10 Sequence of actions as the crankshaft completes one revolution and the piston moves from top to bottom to top again.

Fig. 1-12 Partial cutaway view of an in-line engine, showing the valve train.

lifter is a round cylinder that can slide up and down in a hole in the cylinder block. As the cam lobe pushes up on the lifter, the lifter pushes up on the push rod. This is a long metal rod that rides loosely in another hole in the cylinder block. As the push rod is pushed up, it pushes up on one end of the rocker arm. The rocker arm is a metal arm that can rock up and down on a ball pivot. This ball pivot is like half a big ball bearing with a stud going through it.

When the push rod pushes up on the rocker arm, it rocks on the pivot. The other end of the rocker arm pushes down on the end of the valve stem. This pushes the valve stem down so the valve moves down off the valve seat, opening the port. Normally, the valve stem is pulled up by the valve spring so that the valve is seated, closing off the valve port. But, the force of the rocker arm moving down compresses the spring so the valve opens. With the valve open (Fig. 1-13, left) air-fuel mixture or exhaust gas can pass through the port.

The right part of Fig. 1-13 shows what happens when the camshaft continues to rotate and the cam lobe moves out from under the valve lifter. The valve spring pushes the valve stem up, closing the valve. The rocker arm rocks back, pushing down on the push rod. The push rod pushes down on the valve lifter and the valve lifter moves down to ride on the low part of the cam.

The camshaft may be driven from the crankshaft in one of three ways. These are by gears on the crankshaft and camshaft (Fig. 1-13), and by sprockets on the crankshaft and camshaft connected by a chain (Figs. 1-1 and 1-5) or a toothed belt (Fig. 1-4). The gears are called *timing gears* because the camshaft must turn in time with the crankshaft. When a chain or belt is used, each is known as a *timing chain* or *timing belt*.

1-7 Engine operation In the running engine, the piston requires four up-and-down movements, or strokes, to complete its job. The first stroke is down. The piston moves from its top position (called *top dead center* or TDC) to its bottom position (called *bottom dead center* or BDC). This movement is called a piston stroke. The piston then moves up on a second stroke, down on a third stroke, and up again on a fourth stroke. These four strokes are called:

Intake stroke

Compression stroke

Power stroke

Exhaust stroke

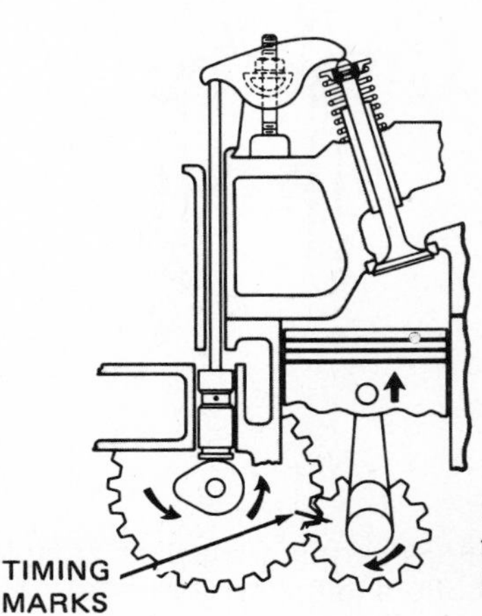

Fig. 1-13 Valve train action. Left, valve open. Right, valve closed.

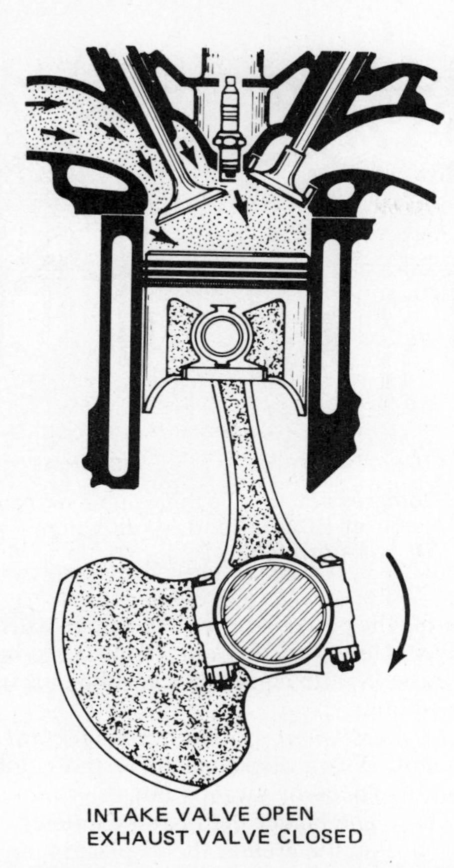

Fig. 1-14 Intake stroke. The intake valve has opened and the piston is moving down, drawing the air-fuel mixture into the cylinder.

Because each piston does its job in four strokes, the engine is known as a four-stroke-cycle engine. This is usually shortened to *four-cycle engine,* or a *four stroker*. Other engines used in some power lawn mowers and motorcycles operate on a different cycle. They need only two strokes to a complete cycle. These engines are known as *two strokers*. They are not used in automobiles today.

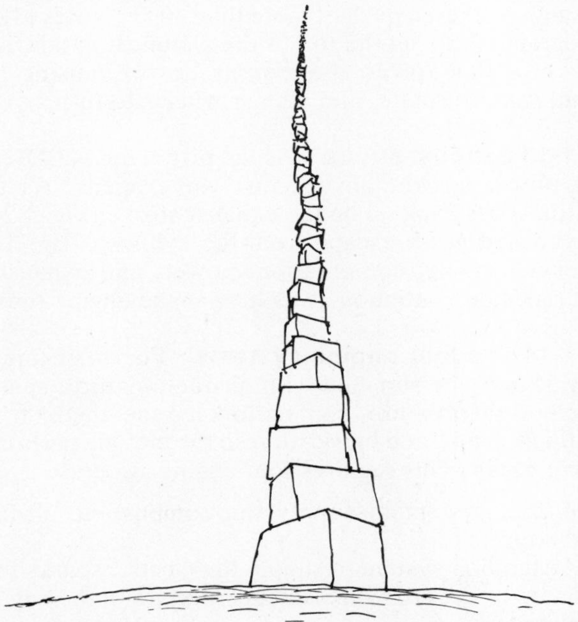

Fig. 1-15 Air pressure results from the miles of air stacked up above the earth.

⚙ 1-8 Intake stroke

Figure 1-14 shows the actions during the intake stroke. The valve train has opened the intake valve by pushing it down off its seat. The downward-moving piston produces a partial vacuum in the cylinder. Then air-fuel mixture rushes in to fill the vacuum. Atmospheric pressure (⚙ 1-9) and vacuum (⚙ 1-10) are the causes of this action.

⚙ 1-9 Atmospheric pressure

Air has weight. *Gravity* pulls down on the air just as gravity pulls down on all other objects on earth. This pull of gravity gives the air weight. At sea level and average temperature, 1 cubic foot [0.028 cubic meter (m^3)] of air weighs about 0.08 pound [36 grams (g)]. This does not seem like very much. But the blanket of air covering the earth is many miles thick. There are, in effect, thousands of cubic feet of air stacked one on top of another (Fig. 1-15).

The total weight, or downward push, amounts to about 15 pounds per square inch (psi) at sea level. In the metric system, this is 103.4 kilopascals (kPa).

⚙ 1-10 Vacuum

Any pressure less than atmospheric pressure is considered a *vacuum*. It is the absence of the normal amount of air or any other substance from a space.

At times, the automobile engine acts as a vacuum pump. Every time a piston moves down on an intake stroke (Fig. 1-14), it produces a partial vacuum in the cylinder. Atmospheric pressure then pushes air into the vacuum. The air, as it passes through the carburetor (Fig. 1-16), picks up gasoline vapor. This air-fuel mix-

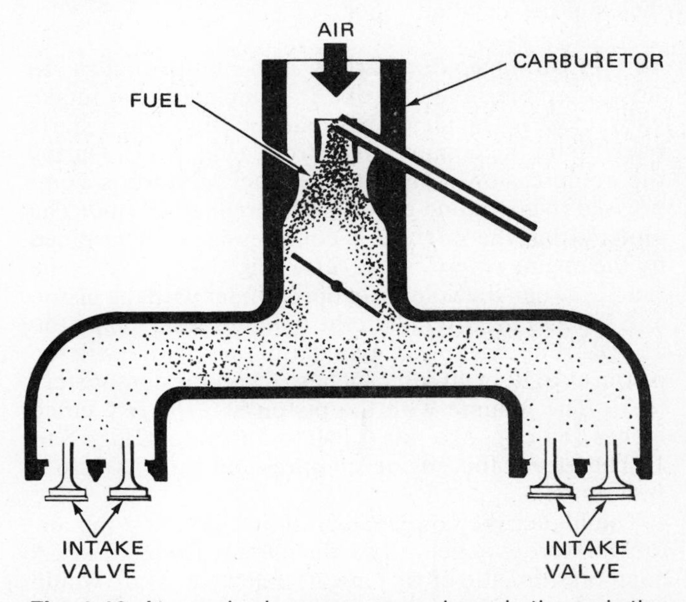

Fig. 1-16 Atmospheric pressure pushes air through the carburetor. The air picks up gasoline vapor in the carburetor and the mixture enters the cylinder past the opened intake valve.

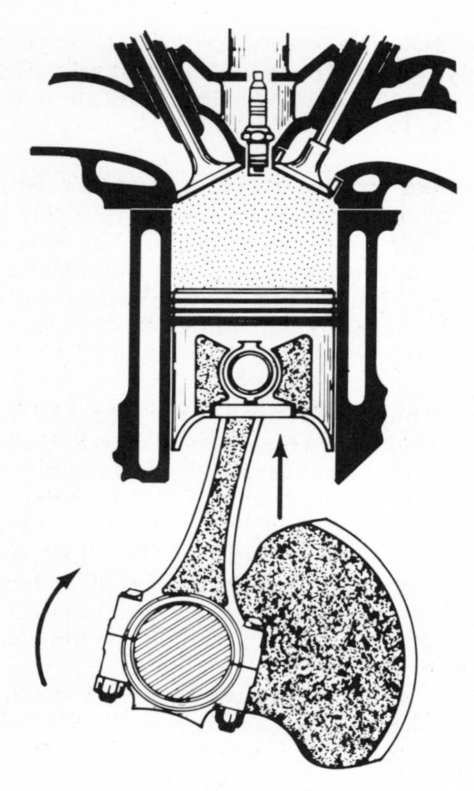

Fig. 1-17 Compression stroke. The intake valve has closed. The piston is moving upward, compressing the air-fuel mixture.

ture then flows through the intake manifold and past the open intake valve. Air pressure forces the mixture into the cylinder to fill up the vacuum. Later chapters describe how the flow of air through the carburetor picks up the gasoline vapor.

1-11 Compression stroke and compression ratio

After the intake stroke is completed, the intake valve closes. The piston has reached BDC and it starts back up on the compression stroke (Fig. 1-17). During the compression stroke, the air-fuel mixture is compressed to about one-eighth of its original volume. The amount that the mixture is compressed is determined by the engine *compression ratio* (Fig. 1-18). This is the ratio between the volume in the cylinder with the piston at BDC and the volume in the cylinder with the piston at TDC. For example, a cylinder has a volume with the piston at BDC of 40 cubic inches [656 cubic centimeters (cc)]. The volume with the piston at TDC is 5 cubic inches [82 cc]. The ratio is 40 to 5 (656 to 82), or 8 to 1. Therefore, the engine compression ratio is 8 to 1 (written 8:1).

The higher the compression ratio, the more the air-fuel mixture is squeezed by the compression stroke. A compression ratio of 10:1 means that the mixture would be compressed to one-tenth of its original volume.

Higher compression ratios mean higher initial pressures when the power stroke starts. With higher pressure at the beginning there will be more pressure on the

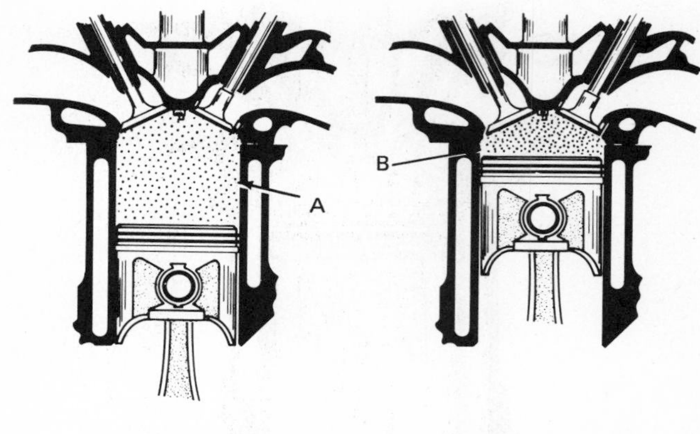

Fig. 1-18 Compression ratio is the volume in the cylinder with the piston at BDC divided by its volume with the piston at TDC, or *A* divided by *B*.

piston during the power stroke. The engine will develop more power. However, there are limits as to how high the compression ratio can be in a gasoline engine. This is discussed later.

Piston displacement is another important engine measurement. Piston displacement is the volume that the piston displaces or sweeps out as it moves from BDC to TDC. The bigger the piston diameter, and the longer the stroke, the greater the displacement. In general, engines with large piston displacement are more powerful than smaller engines.

1-12 Power stroke

The power stroke (Fig. 1-19) starts when the piston passes TDC and begins to move down again. But before the piston reaches TDC, a spark occurs at the spark plug gap. Heat from the spark ignites the compressed air-fuel mixture. High temperatures and pressures result. The pressures may go high enough to exert a push of more than 4000 pounds [1814 kilograms (kg)] on the top of the piston. It is this high pressure that forces the pistons down, causing the crankshaft to rotate, and the car wheels to turn.

1-13 Exhaust stroke

As the piston nears BDC on the power stroke, the exhaust valve opens. As the piston starts back up on the exhaust stroke (Fig. 1-20), the burned gases escape from the cylinder. The four strokes—intake, compression, power, and exhaust—are repeated continuously as long as the engine runs.

1-14 The four engine systems

For an engine to run, it must be supplied with air-fuel mixture, sparks to ignite the mixture, cooling to keep the engine from getting too hot, and lubrication so the moving parts can move easily. This requires four engine systems:

1. A fuel system to supply the combustible air-fuel mixture.
2. An ignition system to supply the electric sparks that ignite or set fire to the air-fuel mixture in the engine cylinders.
3. A cooling system to remove excess heat so the engine does not get too hot.

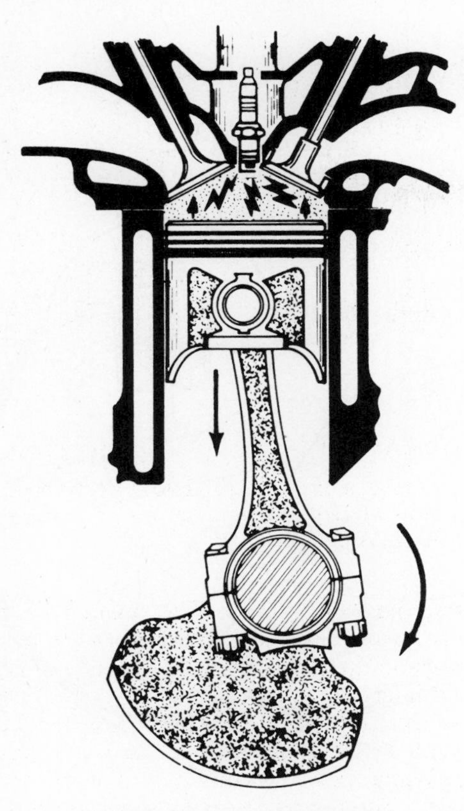

Fig. 1-19 Power stroke. The ignition system produces a spark that ignites the compressed mixture. As it burns, high pressure is produced which forces the piston down.

4. A lubricating system to keep all moving parts lubricated so they move with minimum friction and wear.

Now, let's briefly review the operation of these four systems. Then in later chapters, the construction, operation, and service of three of the systems—fuel, lubricating, and cooling—are covered in detail.

1-15 Fuel system The fuel system used in most cars includes the fuel tank, fuel lines, fuel pump, and carburetor (Fig. 1-21). The fuel tank holds a supply of gasoline. The fuel lines and fuel pump deliver gasoline to the carburetor as needed. The carburetor is a mixing valve that mixes gasoline and air to form the combustible air-fuel mixture. The mixture enters the engine where it is burned to produce power.

The carburetor varies the proportions of gasoline and air. It puts more or less gasoline into the air in response to engine needs. For example, when the engine is cold, it needs a higher proportion of gasoline in the air. The engine needs a *rich mixture*—rich in gasoline. After the engine warms up, it needs less gasoline in the air. The engine needs a *lean* mixture. Every carburetor has several built-in systems which produce these variations in mixture richness.

Most cars are equipped with carburetors. However, an increasing number of cars are now being produced with fuel-injection systems. In these systems, the gasoline is sprayed into the engine intake manifold rather than being mixed with the air in the carburetor. The fuel system for diesel engines is also a fuel-injection system, but it works differently. All of these fuel systems are covered in detail in later chapters.

1-16 Lubricating system The engine lubricating system (Fig. 1-22) includes the oil pan at the bottom of the engine which holds a supply of oil, oil pump and filter, and the oil lines to the various moving parts in the engine. When the engine is running, the oil pump picks up oil from the oil pan and sends the oil to the moving parts. The oil covers these parts so that they slide on oil, instead of on each other. This greatly reduces friction and wear. If the parts actually slid on each other, with metal-to-metal contact, they would wear very rapidly, overheat, and soon be ruined.

In the engine, the oil performs four basic jobs. It lubricates, cools, and cleans the engine. In addition, the oil helps the piston rings to seal the compression and combustion pressures into the combustion chamber.

Briefly, the oil lubricates the engine and helps prevent engine overheating by circulating between the moving parts of the engine and the oil pan. As the oil flows through the engine, the oil picks up heat. Then the oil drips down into the cooler oil pan where the oil looses heat. In this way, the circulating oil is continually removing heat from the engine.

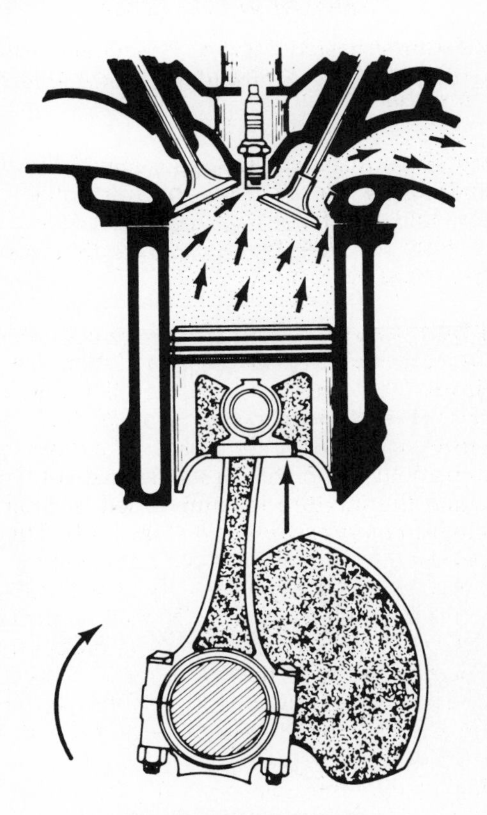

Fig. 1-20 Exhaust stroke. The exhaust valve has opened. As the piston moves up, the burned gases escape from the cylinder.

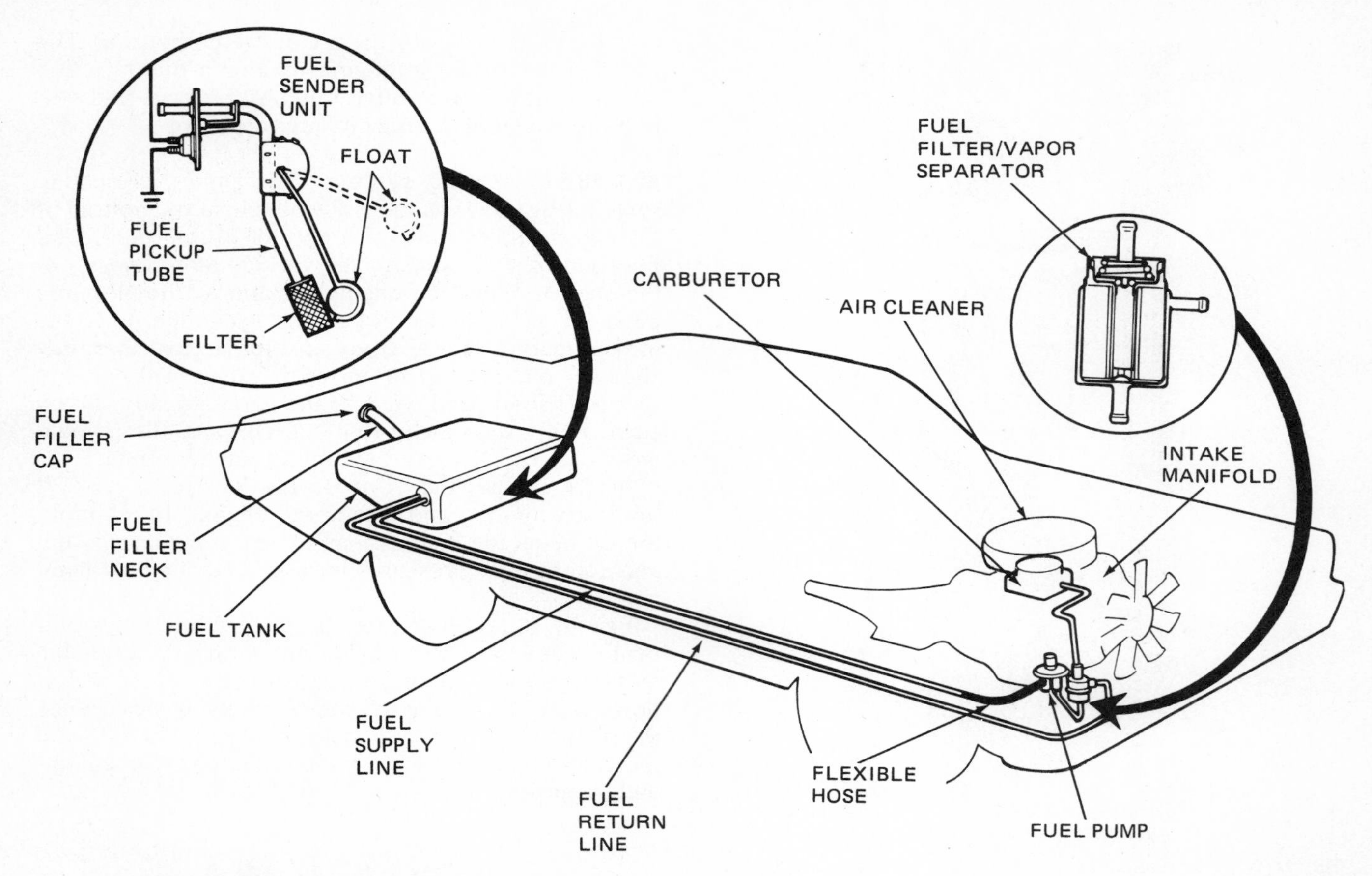

Fig. 1-21 A typical automotive fuel system. (*Chrysler Corporation*)

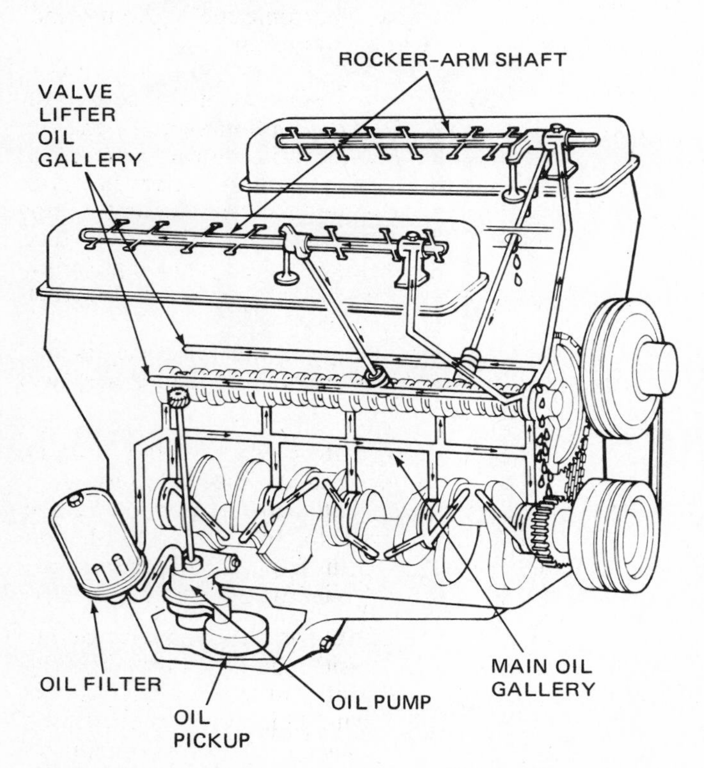

Fig. 1-22 Lubricating system for a V-8 engine. The oil dripping from the various parts runs down into the oil pan. (*Chrysler Corporation*)

The oil also helps keep the engine clean. As the oil circulates through the engine, it picks up particles of dirt, soot, and metal (from wear). These particles either fall to the bottom of the oil pan, or are filtered out as the oil passes through the oil filter. The oil goes through the filter after leaving the oil pump. After passing through the filter, the oil enters the engine.

1-17 Cooling system The burning of the air-fuel mixture in the engine cylinders produces a large amount of heat. The temperatures in the burning mixture may reach 4000°F [2204 degrees Celsius (°C)] or even higher. Part of this heat leaves with the hot exhaust gases. Part of it is removed by the circulating engine oil. The rest of the excess heat is removed by the cooling system (Fig. 1-23).

The cooling system includes a water pump, a radiator with connecting hoses, a thermostat, and water jackets in the engine. The liquid that circulates in the cooling system is part water and part antifreeze. It is called *coolant*. The coolant circulates between the engine water jackets and the radiator. As the coolant passes through the water jackets, it gets hot—almost hot enough to boil. The coolant then flows through the radiator where the coolant loses heat to the air passing through the radiator. With its temperature lowered, the coolant then reenters the engine water jackets where it gets hot again. This continuous circulation of the coolant between the engine water jackets and the ra-

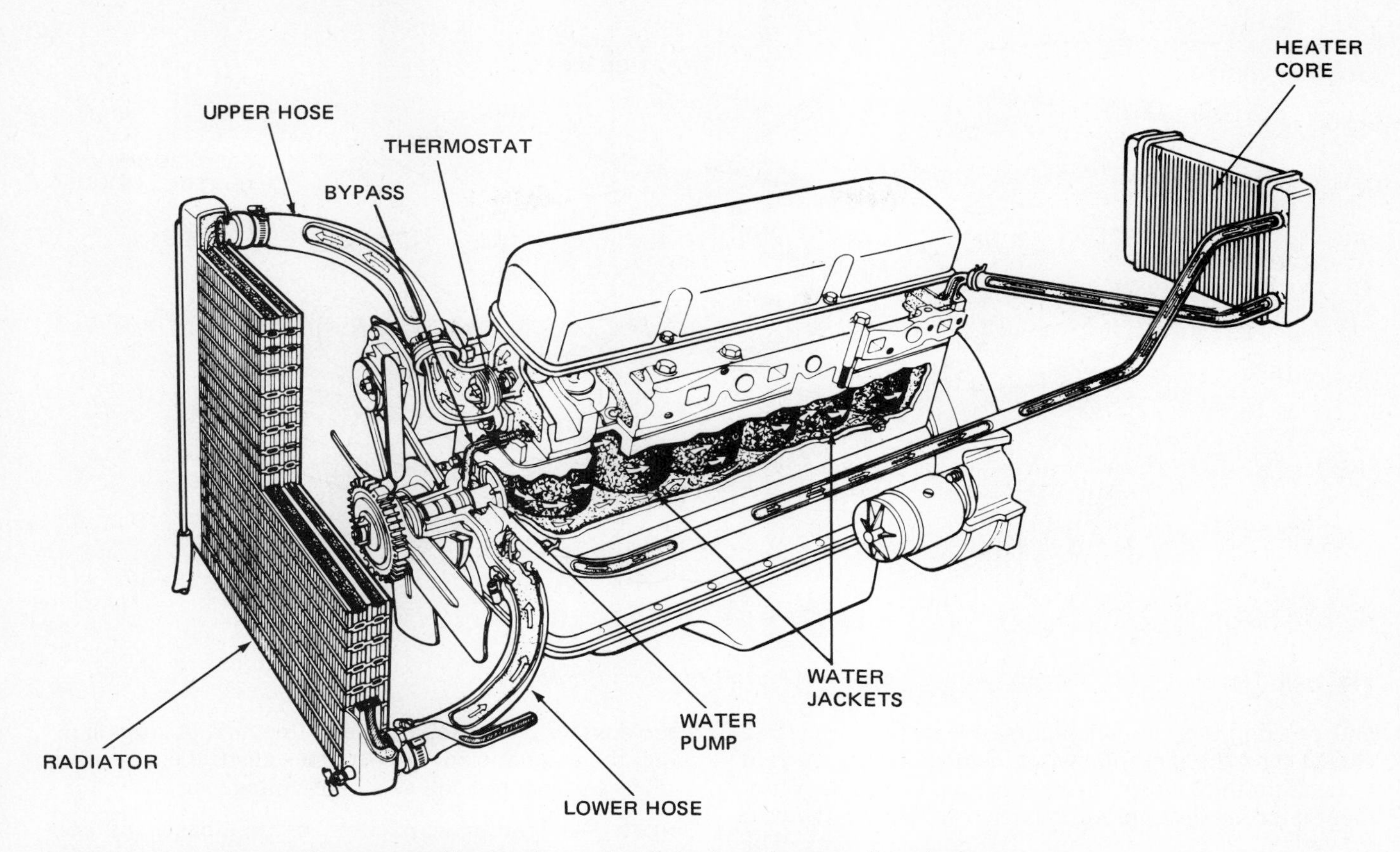

Fig. 1-23 An engine cooling system. The engine and radiator have been partly cutaway to show the coolant circulation. (*Chrysler Corporation*)

diator removes enough heat to prevent engine overheating.

The thermostat is a regulating valve that allows just the right amount of coolant to circulate. For example, when the engine is cold, the thermostat is shut and prevents any coolant flow to the radiator. This keeps all the heat in the engine, which provides rapid warm-up. Fast warm-up is important because a cold engine produces excessive pollutants in the exhaust gas and causes the engine parts to wear out rapidly. So the faster the engine warms up, the better. Then, as the engine warms up, the thermostat opens to allow coolant to circulate through the radiator and begin to take heat out of the engine.

❁ 1-18 Ignition system The ignition system is part of the automotive electric system (Fig. 1-24). The electric system does several jobs:

1. Cranks the engine for starting
2. Supplies electric sparks to ignite the compressed air-fuel mixtures in the cylinders
3. Charges the battery
4. Operates the radio, heater, air conditioner, and other accessories
5. Supplies lights for night driving
6. Operates lights or gauges on the car dash that indicate battery charging rate, oil pressure, engine temperature, and fuel in the fuel tank
7. Operates other electric devices such as window adjusters, seat adjusters, electronic fuel systems, and other electronic systems in the car

A typical ignition system consists of the source of electric current (the battery), ignition switch, ignition coil, ignition distributor, electronic control unit (ECU), spark plugs, and wiring. The ignition system has two jobs. First, it takes the low voltage from the battery (or alternator) and steps it up to the several thousand volts needed to produce the sparks at the spark plugs in the cylinders. Second, it delivers each spark to the correct spark plug at the right time.

Producing the high voltage is the job done by the ignition coil and ignition distributor. There are two basic types of distributor:

1. The type using a mechanical switch (called *contact points*) to close and open the coil primary circuit.
2. The type using a magnetic coil and an electronic switch called a *transistor* to close and open the coil primary circuit. This is the type of distributor used in electronic (*solid state*) ignition systems.

1. Contact-point distributor The contact points are mounted on a plate inside the distributor housing (Fig. 1-25). One point is stationary. The other point is mounted on a movable arm. This arm is moved by a breaker cam inside the housing. The breaker cam revolves, driven by a gear from the engine camshaft (Fig. 1-26). As the breaker cam revolves, lobes on the cam cause the movable contact-point arm to move, closing

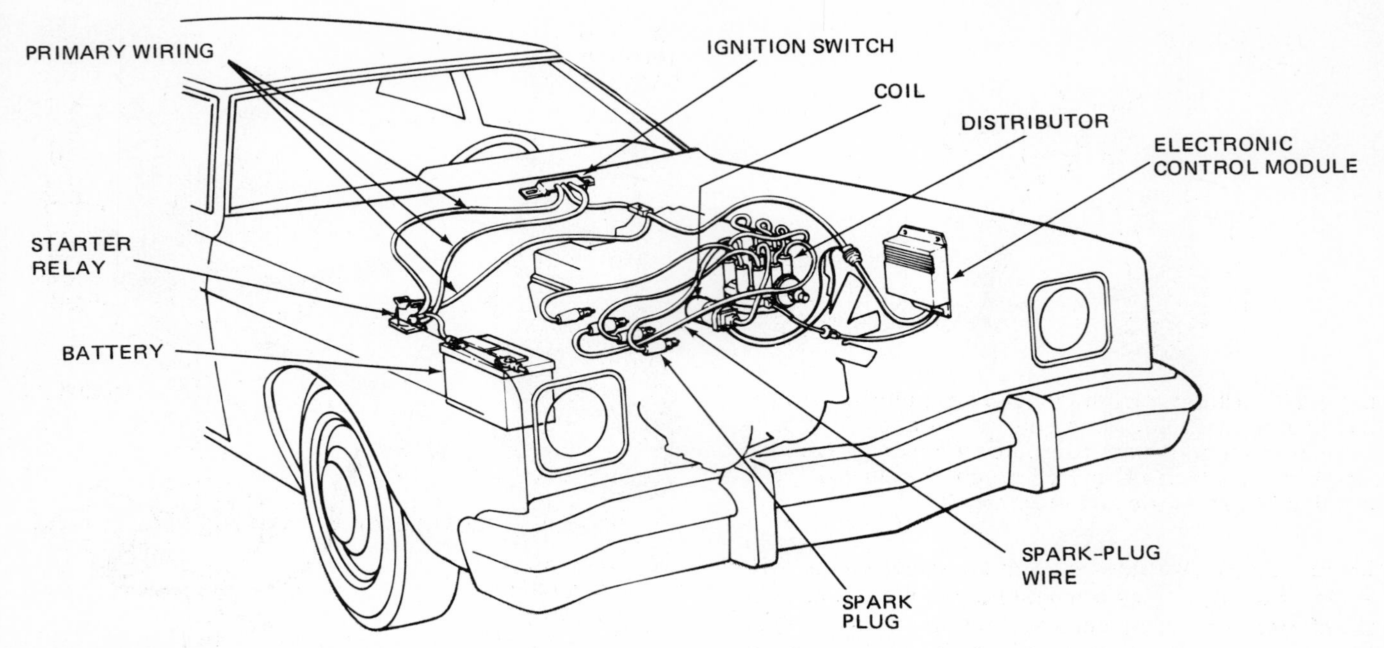

Fig. 1-24 The ignition system in a car with a V-8 engine. (*Ford Motor Company*)

and opening the contact points. When the points are closed (and the ignition switch is on), electric current flows from the battery through the ignition coil. A moment later, as the cam turns farther, a lobe on the cam moves the arm and separates the contact points. The current stops flowing.

During the time that current flows, the ignition coil becomes saturated with magnetic energy. Then, when the contact points separate and the current stops flowing, the magnetic energy becomes electric energy. It is released from the coil as a high-voltage surge.

NOTE: An ignition capacitor, or condenser, is connected across the contact points to prevent excessive arcing as they open. This prolongs point life.

2. Electronic ignition distributor The electronic

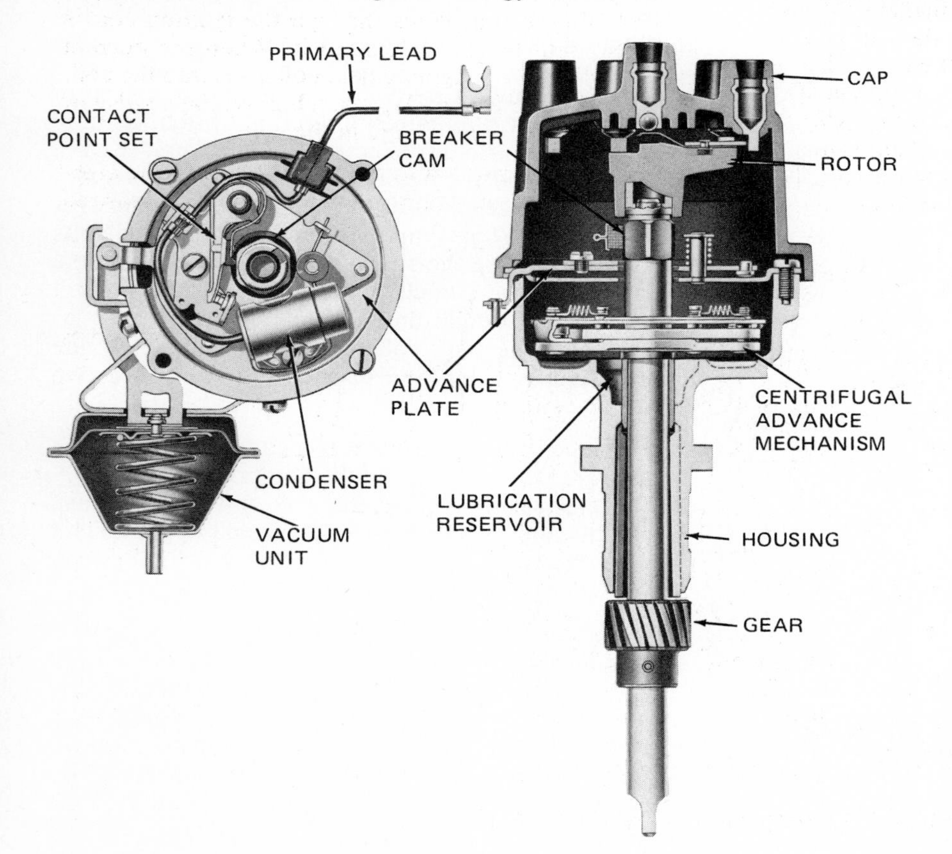

Fig. 1-25 Top and sectional views of an ignition distributor. In the top view (to left), the cap and rotor have been removed so that the advance plate can be seen. (*Delco-Remy Division of General Motors Corporation*)

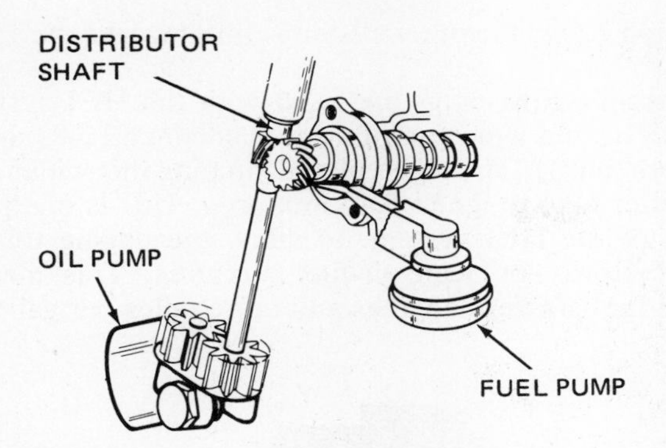

Fig. 1-26 Oil pump, distributor, and fuel pump drives. The oil pump is the gear type. A gear on the end of the camshaft drives the ignition distributor. An extension of the distributor shaft drives the oil pump. The fuel pump is driven by an eccentric on the camshaft.

ignition distributor does not have contact points (Fig. 1-27). Instead, it uses a magnetic pickup assembly in the distributor and an electronic control module which has diodes and transistors. With the cap on, many electronic distributors look about the same as contact-point distributors. However, with the caps off, the difference between the two is apparent (Figs. 1-25 and 1-27). The electronic distributor has a metal rotor, called an armature or *reluctor*, which takes the place of the breaker cam. The reluctor has the same number of tips as there are cylinders in the engine.

The distributor also has a magnetic pickup coil. As the distributor shaft turns the reluctor, the reluctor tips carry magnetism from the permanent magnet through the pickup coil. Every time a tip passes the pickup coil, the pulse of magnetism goes through the coil. This, in turn, produces a small voltage pulse or signal. This voltage signal is applied to the module (Fig. 1-28). In the module, the voltage pulse causes a transistor to turn off the current flow through the ignition coil. The effect is the same as when contact points open.

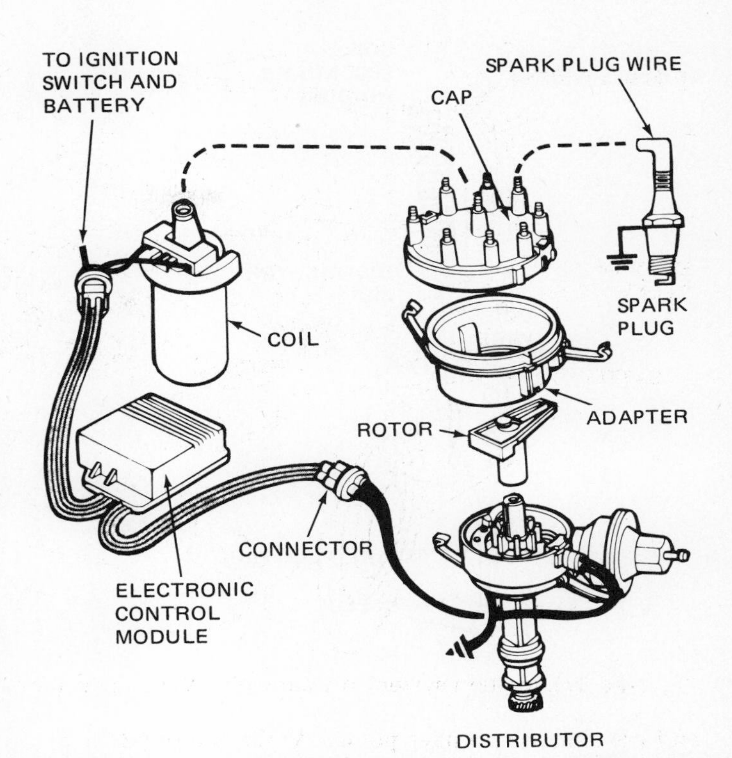

Fig. 1-28 A complete electronic ignition system. The heavy dashed line shows the secondary circuit. The secondary winding in the coil is connected to the spark plug through the distributor cap, rotor, and wiring. (*Ford Motor Company*)

While the current flows through the ignition coil, a magnetic field builds up around it. When the current stops flowing, the magnetic field collapses into the coil. As the magnetic field collapses, a high-voltage pulse or surge of current is produced in the coil.

3. Getting the spark to the spark plugs The high-voltage surge of current produced by the coil is carried

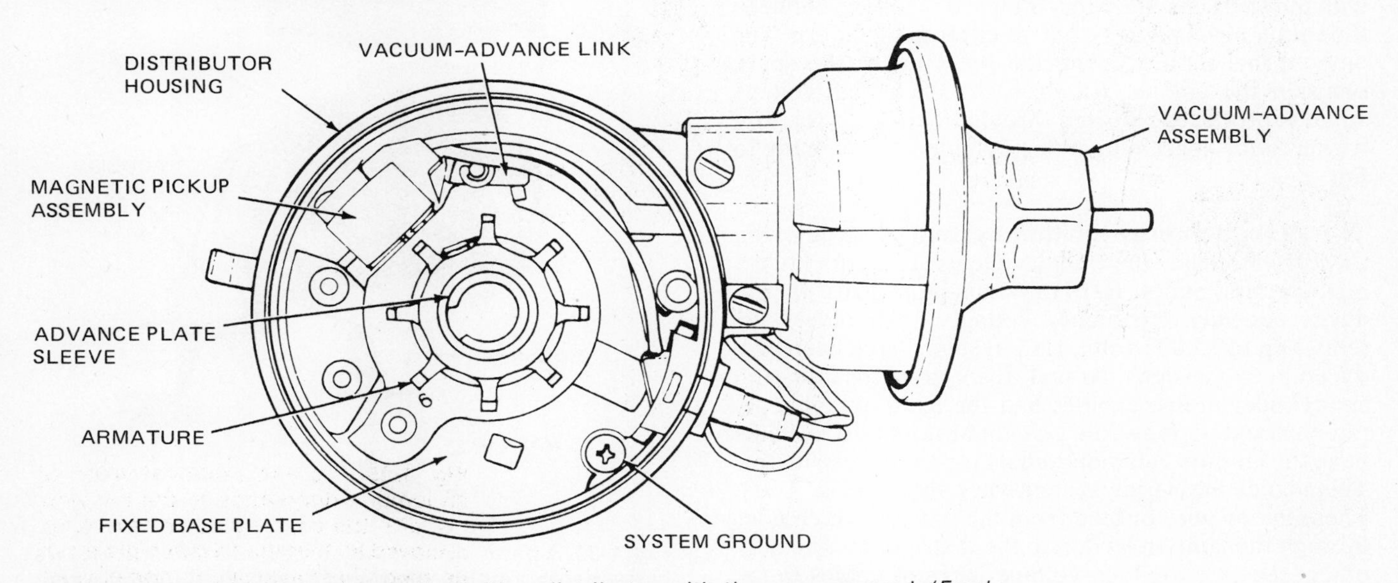

Fig. 1-27 Top view of an electronic ignition distributor with the cap removed. (*Ford Motor Company*)

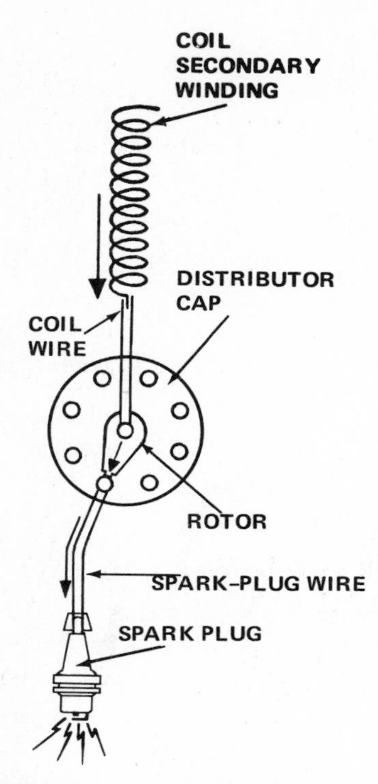

Fig. 1-29 Simplified view of the ignition-system secondary circuit.

by wires to the distributor cap and from there to the spark plug in the cylinder that is ready to fire. This is the cylinder in which the air-fuel mixture is now compressed by the piston nearing TDC on the compression stroke.

The surge of current passes through the center terminal of the distributor cap (Fig. 1-29). The center terminal is connected by a wire to the coil and inside the cap by a contact spring to the rotor (Fig. 1-25). The rotor is mounted on the distributor shaft so that it turns with the shaft. As the rotor rotates, it connects between the center terminal and each outer terminal in turn. The outer terminals are connected by wires to the spark plugs in the engine cylinders. As each high-voltage surge is produced, it is led through the cap, rotor, and wiring to the spark plug in the cylinder that is ready to fire.

⚙ 1-19 High-energy ignition system Starting with the 1975 models, cars built by General Motors have a high-energy ignition (HEI) system (Fig. 1-30). It produces considerably higher voltages than older systems—up to 35,000 volts. HEI systems have been produced in two designs. At first, the system for four- and six-cylinder in-line engines had the ignition coil separately mounted. Now, all General Motors HEI systems have the ignition coil mounted on top of the distributor. This makes the wiring system very simple (Fig. 1-31). There is one wire or lead from the battery which goes through the ignition switch to the distributor. The only other leads are the high-voltage wires or cables to the spark plugs. The ignition coil looks different from the ignition coil in the other systems, but it works the same way.

Special spark plugs are used with the HEI system which have a wide gap of up to 0.080 inch [2.03 millimeters (mm)]. The high voltage jumping this wider gap is better able to ignite lean mixtures. This is one purpose of the HEI system, to allow the engine to run satisfactorily on lean air-fuel mixtures. This means more fuel-efficient engines and more miles per gallon.

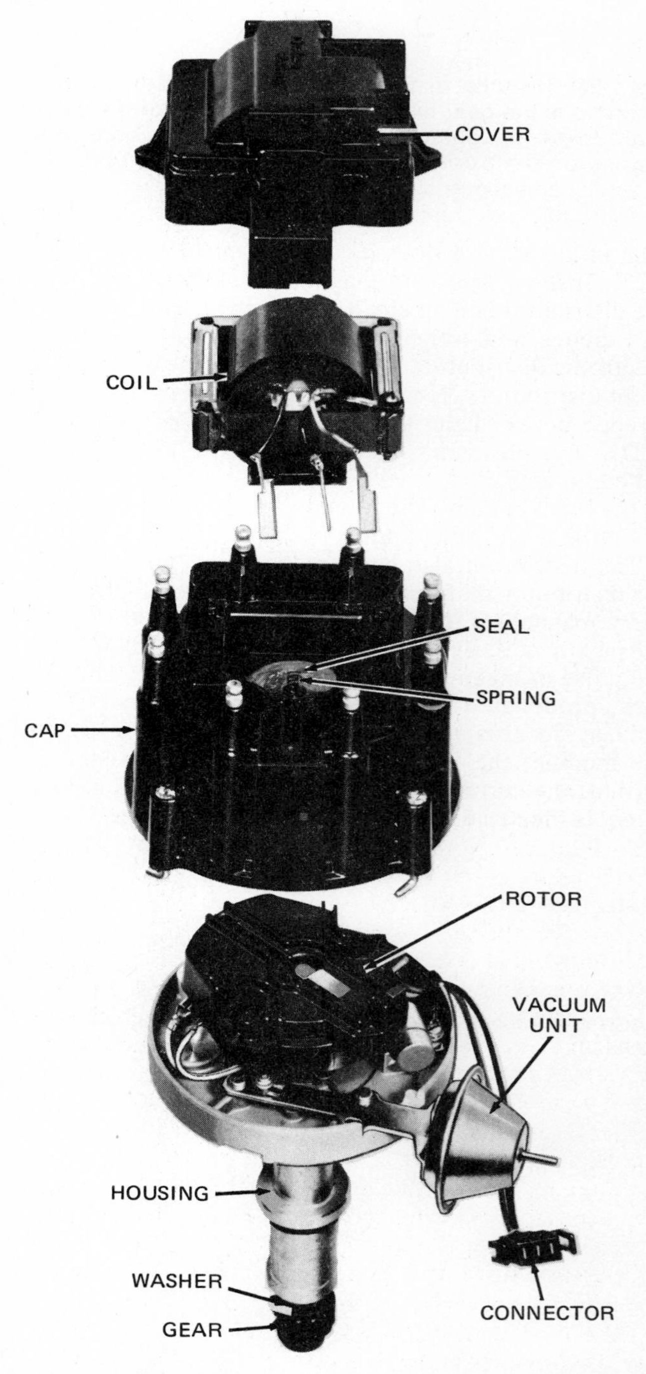

Fig. 1-30 A partly disassembled view of a General Motors high-energy ignition distributor, with coil. (*Delco-Remy Division of General Motors Corporation*)

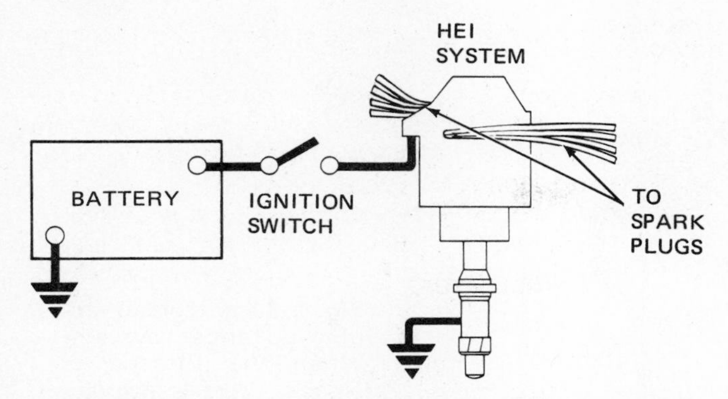

Fig. 1-31 Basic wiring diagram of the GM High-Energy Ignition system. (*Delco-Remy Division of General Motors Corporation*)

Ford also uses high-energy ignition systems on their cars. They are called "Dura-Spark" systems. These systems also use the wider-gap spark plugs, and may develop up to 47,000 volts.

1-20 Operation of the ignition-advance mechanism When the engine is idling, the sparks are timed to appear in the engine cylinders just before the pistons reach TDC on their compression strokes. But at higher speeds, the air-fuel mixture has less time to ignite and burn. If ignition still took place just before TDC on the compression stroke, the piston would be up over the top and moving down before the mixture was well ignited. This would mean that the piston would be moving away from the pressure rise. Much of the energy in the burning fuel would be wasted. However, if the mixture is ignited earlier in the compression stroke (at high engine speed), the mixture will be well ignited by the time the piston reaches TDC. Pressure will go up, and more of the fuel energy will be used.

1. Advance based on speed To ignite the mixture earlier at high speed, a spark-advance mechanism operated by centrifugal force is used (Fig. 1-32). This device usually is built into the ignition distributor. The centrifugal advance includes two weights that move out against spring tension as engine speed increases. This movement is transmitted through an advance cam to the breaker cam, or the reluctor or rotor of an electronic distributor. This causes the cam or the reluctor to advance or move ahead with respect to the distributor shaft. In contact-point distributors, this advance causes the cam to open and close the contact points earlier in the compression stroke at high speeds. In the electronic distributor, the voltage signals from the pickup coil are sent to the electronic control unit earlier. Therefore, the timing of the sparks to the cylinders varies from no advance at low speed to full advance at high speed, when the weights have reached the outer limits of their travel.

Maximum centrifugal advance may be as much as 45° of crankshaft rotation before the piston reaches TDC on the compression stroke. The shapes of the advance weights, and the tension of the advance springs, are designed so that the engine has the proper

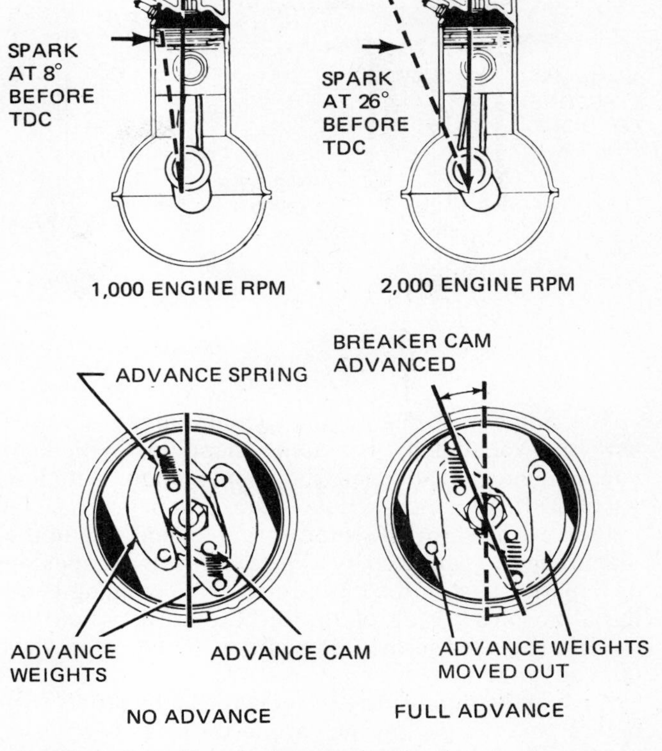

Fig. 1-32 Centrifugal-advance mechanism in the no-advance and full-advance positions. In the typical example shown, the ignition is timed at 8° before top dead center at idle. There is no centrifugal advance at 1000 engine rpm. There is 26° total advance (18° centrifugal plus 8° due to original timing) at 2000 engine rpm. (*Delco-Remy Division of General Motors Corporation*)

advance. No two makes or models of engines are exactly alike. This is why the advance curves are different for various engines.

2. Advance based on intake-manifold vacuum With a partly closed throttle valve, there is a partial vacuum in the intake manifold. Less air-fuel mixture gets into the engine cylinders. Therefore, the mixture is less highly compressed by the compression stroke. This means that the mixture, which is spread thinner throughout the combustion chamber, burns slower. Under these conditions, additional spark advance will allow the mixture more time to burn and develop pressure on the piston.

Spark advance based on intake manifold vacuum (called *vacuum advance*) is controlled by an airtight diaphragm linked to a movable advance plate in the distributor. Figure 1-33 shows the vacuum-advance mechanism. The unit is an airtight chamber that contains a spring-loaded diaphragm. When vacuum is applied to the chamber, the diaphragm compresses the spring. As the diaphragm moves, the linkage attached to it is rotating the advance plate. Mounted on the advance plate is the contact-point set or the magnetic pickup coil. Therefore, movement of the advance plate also moves the contact points or magnetic pickup coil ahead a few degrees to produce vacuum advance.

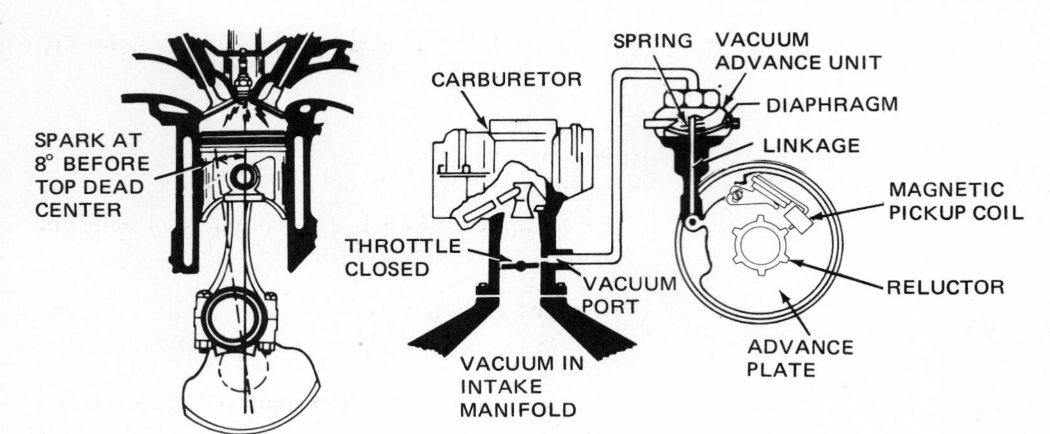

Fig. 1-33 A typical vacuum-advance system. When the throttle is closed, there is no vacuum advance.

The number of degrees of vacuum advance varies with the vacuum in the intake manifold. With high vacuum, vacuum advance is at its maximum. With low vacuum, there is no vacuum spark advance.

Figure 1-33 shows how the spring-loaded side of the diaphragm is connected by a vacuum line to an opening or "spark" port in the carburetor. This opening is on the atmospheric side of the throttle valve when the throttle is closed in the idle position. There is no vacuum advance.

As soon as the throttle is opened, the throttle valve moves past the port in the carburetor (Fig. 1-34). The intake manifold vacuum can now draw air from the vacuum line and the airtight chamber in the vacuum-advance unit. This causes the diaphragm to move against the spring. The linkage to the advance plate then rotates the plate. Moving the advance plate changes the position of the contact points with respect to the breaker cam, or the magnetic pickup coil with respect to the reluctor. As a result, current flow to the ignition coil is turned off earlier in the cycle. Now the spark appears at the spark-plug gap earlier in the compression stroke.

As the throttle is opened wider, there is less vacuum in the intake manifold and therefore, less vacuum advance. At wide-open throttle, there is no vacuum advance at all. Spark advance is provided entirely by the centrifugal-advance mechanism.

On some of the latest electronic ignition systems, spark advance is controlled by electronic means. These distributors do not have separate centrifugal and vacuum mechanisms to advance the spark. Instead, electronic devices provide the control.

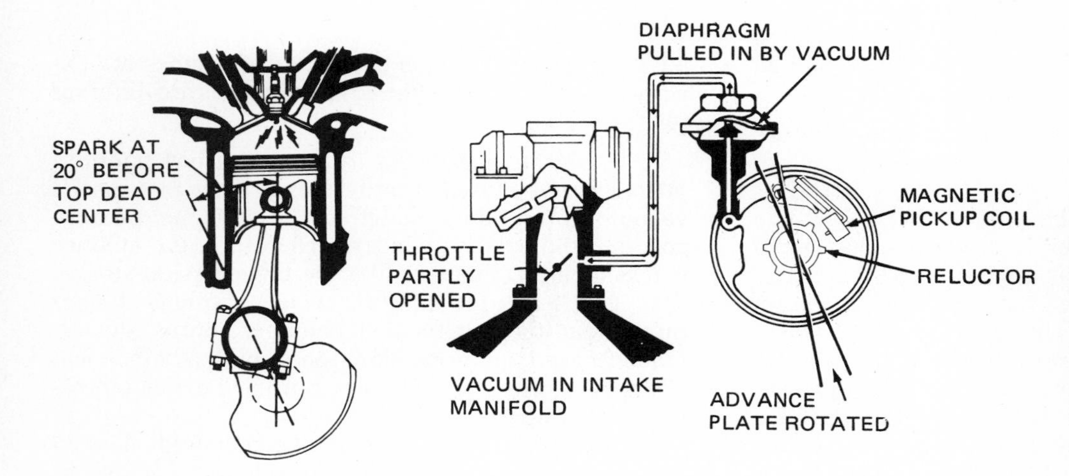

Fig. 1-34 When the throttle is partly opened so that vacuum is applied to the vacuum-advance unit, the advance plate is moved ahead so that the voltage pulses from the magnetic pickup coil occur earlier. This produces the vacuum advance.

Chapter 1 review questions

Select the *one* correct, best, or most probable answer to each question. Then check your answers against the correct answers given at the end of the book.

1. The connecting rod is attached to the piston by the:
   a. rod cap
   b. piston pin
   c. cap bolts
   d. cap bearing.
2. The connecting rod is attached to the crankpin on the crankshaft by the:
   a. piston pin
   b. crank
   c. rod cap
   d. rod boots.

3. Usually, each engine cylinder has:
   a. one valve
   b. two valves
   c. three valves
   d. four valves.
4. The two types of engine valves are:
   a. intake and port
   b. intake and inlet
   c. intake and exhaust
   d. disk and reed.
5. The four strokes in the engine, in order of their occurrence, are:
   a. intake, power, exhaust, and compression
   b. intake, exhaust, power, and compression
   c. intake, compression, power, and exhaust
   d. charging, ignition, discharging, exhaust.
6. During the power stroke, the intake and exhaust valves are:
   a. closed and opened
   b. opened and closed
   c. both closed
   d. both open.
7. The basic difference between the contact-point and the electronic ignition system is that the current flow to the ignition coil is turned on and off by:
   I. An electronic switch in the electronic distributor
   II. A mechanical switch in the contact-point distributor
   a. I only
   b. II only
   c. neither I nor II
   d. both I and II.
8. The engine camshaft has a separate cam lobe for each:
   a. engine valve
   b. engine cylinder
   c. piston
   d. crankpin.
9. The purpose of the ignition spark-advance systems is to:
   a. control engine speed
   b. advance the ignition timing as required by operating conditions
   c. better control the contact-point action
   d. control the air-fuel mixture.
10. In the ignition system the high-voltage surges produced by the coil are carried through the distributor cap and rotor to the:
    a. ignition switch
    b. spark plugs
    c. battery or source of current
    d. capacitor.

CHAPTER 2

GASOLINE AND THE COMBUSTION PROCESS

After studying this chapter, you should be able to:

1. Describe the composition of gasoline and how it burns.
2. Explain how volatility of a fuel affects engine operation.
3. Define *octane rating.*
4. Discuss the causes of detonation and preignition and how they may be prevented.
5. Explain why lead has been removed from some gasolines.
6. Describe the difference between a hemispheric combustion chamber and a wedge combustion chamber.
7. Explain the difference between conventional combustion and the combustion process in a stratified-charge chamber.
8. Discuss how changes in volumetric efficiency affect engine operation.
9. Describe the operation of the LPG fuel system.
10. Define *gasohol.*

2-1 Automotive-engine fuels Most automotive engines burn gasoline for fuel. Diesel engines use a light oil, called diesel fuel. Some engines of either type use liquefied petroleum gas (LPG). The next four chapters cover automotive fuel systems using gasoline as fuel. Following that, diesel fuel and diesel-engine fuel systems are discussed. LPG fuel systems are covered at the end of this chapter.

2-2 Gasoline and combustion Gasoline is a clear or lightly colored liquid that will evaporate quickly. It burns violently in the open air. But gasoline is not a simple liquid. It is a complex mixture of several compounds, each of which adds its own characteristics to the mixture.

Gasoline is a hydrocarbon (HC) made up largely of hydrogen and carbon compounds. These two elements unite readily with oxygen, a common element which makes up about 20 percent of our air. The chemical symbols for hydrogen, carbon, and oxygen are H, C, and O. When hydrogen and oxygen unite, a compound called *hydrogen oxide* (H_2O) is formed. We commonly refer to it as *water*. When carbon and oxygen unite, either CO (carbon monoxide) or CO_2 (carbon dioxide) are formed. These chemical reactions—the uniting of these elements—is called *combustion*. It produces high temperatures. That is the whole purpose of burning fuel in the engine—to produce high temperatures. The high temperatures in the confined space of the cylinder produce high pressure. Then the high pressure pushes the piston down so that the crankshaft turns and the car moves.

If all the gasoline (HC) burned with oxygen (O) trapped in the cylinder, only water (H_2O) and carbon dioxide (CO_2) would come out the tailpipe (Fig. 2-1). However, perfect combustion never occurs in the engine. Some of the carbon does not get enough oxygen. Therefore, instead of CO_2, carbon monoxide (CO) is formed. Some of the hydrogen does not unite with oxygen, so some HC (gasoline) remains unburned. As a result, the exhaust gas coming out the tailpipe contains some of these two pollutants, unburned gasoline vapor (HC)

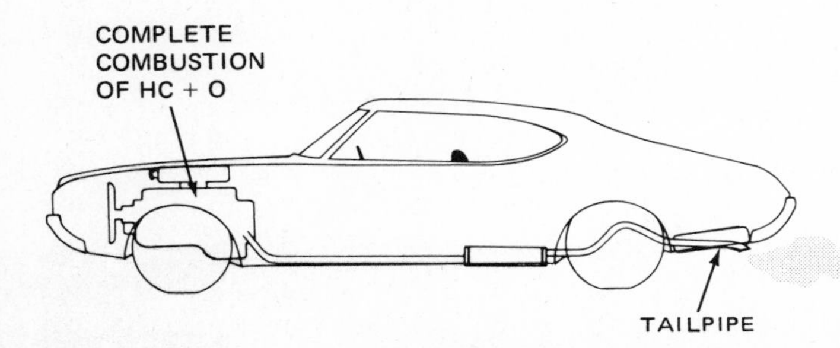

Fig. 2-1 Perfect combustion produces only carbon dioxide (CO_2) and water (H_2O) from the tail pipe.

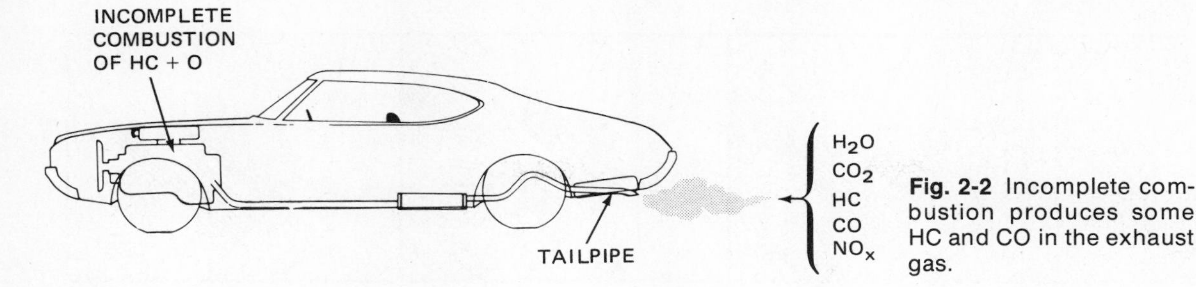

Fig. 2-2 Incomplete combustion produces some HC and CO in the exhaust gas.

and carbon monoxide (CO) (Fig. 2-2). These two compounds plus a third compound called *nitrogen oxides* (NO_x) are the pollutants from automobile engines which are regulated by law. Emission control devices are installed on engines to reduce the amount of these pollutants.

Carbon monoxide is a dangerous gas. It is colorless, odorless, and tasteless. Yet a mixture of only 15 parts carbon monoxide to 10,000 parts of air is dangerous to breathe and can make you drowsy. Greater concentrations of CO can be fatal. An engine running in a closed one-car garage for only 3 minutes can put enough carbon monoxide in the air to kill you. This is the reason you are cautioned never to run an engine in a closed garage without ventilation for the engine exhaust gases.

NOTE: Gasoline is often referred to as "gas," which can be confusing. The sort of gas that you burn in a gas stove or to heat a house is actually a vapor or gas that is delivered through gas lines or pipes. So there is a gas that is a *gas* and "gas" which is a slang expression for the liquid fuel gasoline.

CAUTION: Gasoline is a very dangerous liquid if it is not handled properly. Vapor rising from an open pan of gasoline could ignite from a spark, a flame, or a lighted cigarette. The result could be an explosion or fire. Gasoline from a leaky fuel pump could be ignited by a hot engine or any electric spark.

⚙ 2-3 Origin of gasoline Gasoline, diesel fuel oil, LPG, and many other compounds come from petroleum, or crude oil. It is found in pools under the ground. It was probably formed over a period of many millions of years from animal and vegetable matter. The petroleum usually is under considerable pressure. When a well is drilled down to a pool or reservoir, the petroleum gushes up out of the earth. However, in many wells, the natural pressure is not great enough to force the crude oil out. With these wells, pressure is applied from above the ground by artificial means to force the crude oil out.

Petroleum is a mixture of many compounds. The oil refinery separates the petroleum into various products. It alters many of the original compounds and forms new compounds in the refining process. From the refinery come many types and grades of lubricating oil, fuel oil of various types for diesel engines and for heating, gasoline of many grades and types, kerosene, LPG, and other products.

Gasoline is blended from many different basic hydrocarbons, each with its own set of characteristics. By this blending procedure, a gasoline is obtained that provides satisfactory engine operation under many different operating conditions. Factors that must be considered in blending gasoline include volatility, antiknock value, and freedom from harmful chemicals and gum. These factors are discussed in detail in following sections.

⚙ 2-4 Volatility Volatility refers to the ease with which a liquid vaporizes. The volatility of a simple compound like water is found by increasing its temperature until it boils, or vaporizes. A liquid that vaporizes at a low temperature has a high volatility: if its boiling point is high, its volatility is low. A heavy oil with a boiling point of 600°F (315.5°C) has a very low volatility. Water has a relatively high volatility; it boils at 212°F (100°C) at atmospheric pressure. Gasoline is even more volatile.

A highly volatile liquid evaporates much faster at a low temperature than liquid with a low volatility. Therefore, at room temperature, alcohol and gasoline evaporate more rapidly than water.

Gasoline is blended from different compounds. Each has a different volatility, or boiling point, which can vary from 85°F [30°C] to 437°F [225°C]. The proportions of high- and low-volatility hydrocarbons in gasoline must be correct for the following engine-operating requirements. Otherwise, the volatility will affect the *drivability* of the car. Drivability includes how well the car performs during starting, warm-up, and acceleration. It also includes any tendency toward vapor lock and stalling.

1. Easy starting For easy starting with a cold engine, gasoline must be highly volatile so that it will vaporize readily at a low temperature. Therefore, a percentage of the gasoline must have highly volatile hydrocarbons. This percentage must be higher in the cold regions than in warm climates.

2. Freedom from vapor lock If the gasoline is too volatile, engine heat will cause it to vaporize in the fuel pump. This can cause vapor lock. Vapor lock prevents normal fuel delivery to the carburetor and would probably produce stalling of the engine. Therefore, the percentage of gasoline with highly volatile hydrocarbons must be kept low to prevent vapor lock. The use of a vapor-return line to return vaporized fuel from the fuel pump to the fuel tank is discussed in ⚙ 3-13.

3. Quick warm-up The speed with which the engine warms up depends in part on how much gasoline

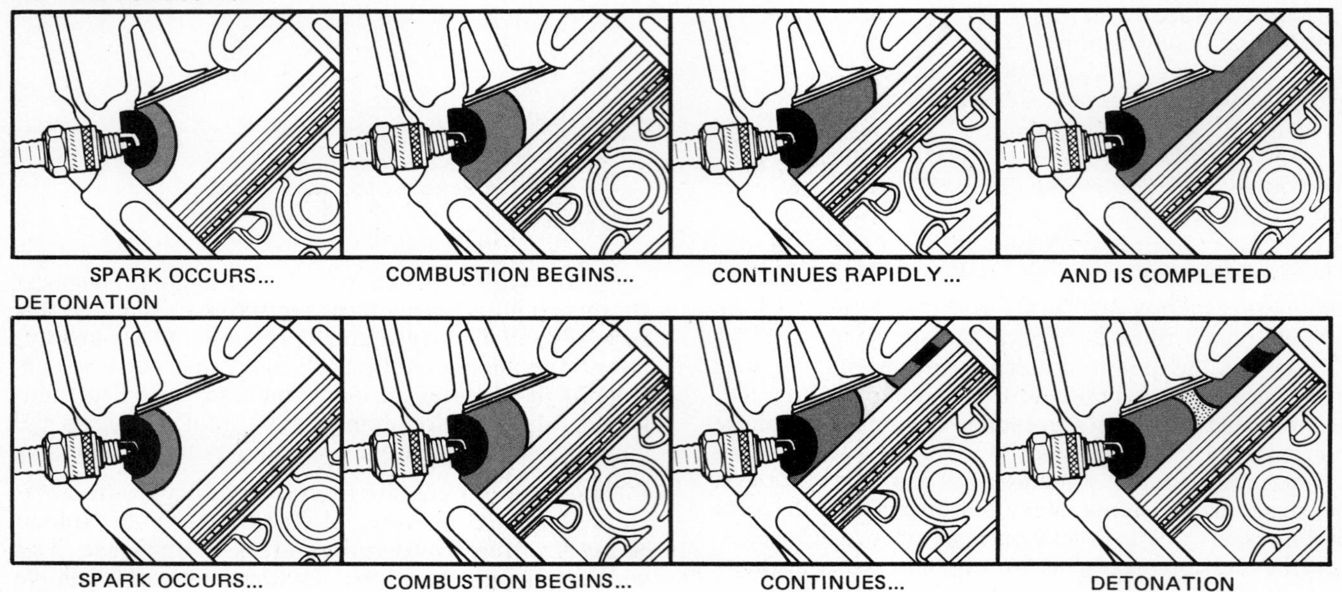

Fig. 2-3 Normal combustion without detonation is shown in the top row. The fuel charge burns smoothly from beginning to end, providing an even, powerful force to the top of the piston. Detonation is shown in the bottom row. The last part of the fuel explodes or burns almost instantly to produce detonation or spark knock. (*Champion Spark Plug Company*)

vaporizes immediately after the engine starts. Volatility for this purpose does not have to be quite so high as for easy starting.

4. Smooth acceleration When the throttle is opened for acceleration, there is a sudden increase in the amount of air passing through the throttle valve. At the same time, the carburetor accelerator pump or fuel-injection system delivers extra gasoline. If this gasoline does not vaporize quickly, there will be a momentary time during which the air-fuel mixture will be too lean. This causes the engine to hesitate, or stumble. Immediately after, as the gasoline begins to evaporate, the mixture will become temporarily too rich. Again, there is poor combustion and a tendency for the engine to hesitate. Enough of the gasoline must be sufficiently volatile to ensure adequate vaporization for smooth acceleration.

5. Good fuel economy For good fuel economy, or maximum miles per gallon (kilometers per liter), gasoline must have a high heat (or energy) content, and low volatility. Gasoline with high overall volatility tends to reduce economy. It may produce an overrich mixture under many operating conditions. Gasoline with volatility that is too low increases starting difficulty. Such gasoline lengthens warm-up time and does not give quite as good acceleration. Therefore, a blend of high- and low-volatility hydrocarbons is used in gasoline.

6. Freedom from crankcase dilution Crankcase dilution results when part of the gasoline is not vaporized when it enters the engine cylinders. The liquid gasoline does not burn. It runs down the cylinder walls and enters the oil pan, where it dilutes the oil. This process washes lubricating oil from the cylinder walls (increasing the wear of walls, rings, and pistons). Also, diluted oil is less able to lubricate other engine parts, such as the bearings. To avoid damage from crankcase dilution, the gasoline must be volatile enough so that little of it enters the cylinders in liquid form.

7. Volatility blend No one volatility of gasoline will satisfy all engine operating requirements. Gasoline must be of high volatility for easy starting and good acceleration. But it must also be of low volatility to give good fuel economy and combat vapor lock. Therefore, gasoline is blended from various amounts of different hydrocarbons having different volatilities. Such a blend satisfies the various operating requirements.

The actual gasoline blend depends on the particular part of the country and the climate in which the gasoline is to be sold. Different blends are provided by the refiners for cold weather, for hot weather, and for high and low altitudes. As a result, some gasolines may perform better in one engine than in others. However, the gasoline sold in an area is blended, and its volatility balanced to meet local conditions.

⚙ 2-5 Antiknock value During normal combustion in the engine cylinder, the pressure increases evenly (Fig. 2-3). Under some conditions, the last part of the compressed air-fuel mixture explodes, or detonates. This produces a sudden and sharp pressure increase. This detonation may cause a rattling or pinging noise that sounds almost as though the piston head had been struck by a hammer. Actually, the sudden pressure increase does hit the piston almost like a hammer blow.

This sudden shock can shatter the top ring land of the piston. In severe cases, detonation can even punch a hole through the piston head. Also, detonation wastes some of the energy in the gasoline. The sudden pressure increase does not permit the maximum force to be applied to the top of the piston.

Some gasolines detonate more easily than others. Because severe detonation is undesirable, refiners improve their fuels to reduce the tendency to detonate. Certain chemical additives, called *antiknock compounds,* have been found that reduce detonation when a small amount is added to the gasoline. The actual rating of the antiknock tendencies of a gasoline is given in terms of *octane number,* or *octane rating* (⚙ 2-10).

NOTE: For many years, detonation has been called *spark knock,* or simply *knock.* However, this makes it very easy to confuse with knock, the word also used to refer to the sound of a mechanical problem such as a *rod knock.* In this book, we will try to refer to the form of abnormal combustion as *spark knock* or *detonation.* Its characteristic sound is that of pinging.

⚙ 2-6 Compression ratio The compression ratio of an engine is the ratio between the cylinder volume with the piston at BDC divided by its volume with the piston at TDC (Fig. 1-19). The higher the compression ratio, the more the air-fuel mixture is squeezed on the compression stroke. This means there is a higher initial pressure at the beginning of the power stroke. Then there will be more pressure on the piston throughout the power stroke. The higher pressure means the engine will produce more power. This is the reason that until recently, year after year, engine designers and manufacturers increased the compression ratios of their engines. More powerful engines could be built without a comparable increase in engine weight.

The increase of compression ratio brought about certain problems. For example, detonation occurs more easily in high-compression engines. Therefore, it has been necessary to develop fuels that resist detonation in these higher-compression engines.

⚙ 2-7 Heat of compression To understand why detonation occurs, let us review what happens to air or any other gas when it is compressed. The diesel engine compresses air to less than one-fifteenth of its original volume. This compression increases the air temperature to about 1000°F [537.8°C]. The more a gas is compressed, the higher its temperature rises. This temperature rise is called *heat of compression.* Let us see how heat of compression affects detonation.

⚙ 2-8 Cause of detonation Normally the spark at the spark plug starts the air-fuel mixture burning in the combustion chamber. A wall of flame spreads out in all directions from the spark, almost like a rubber balloon being blown up (Fig. 2-4). The flame travels rapidly outward through the compressed air-fuel mixture until all the charge is burned. The speed with which the flame travels is called the *flame-propagation rate.* The movement of the flame wall during normal combustion

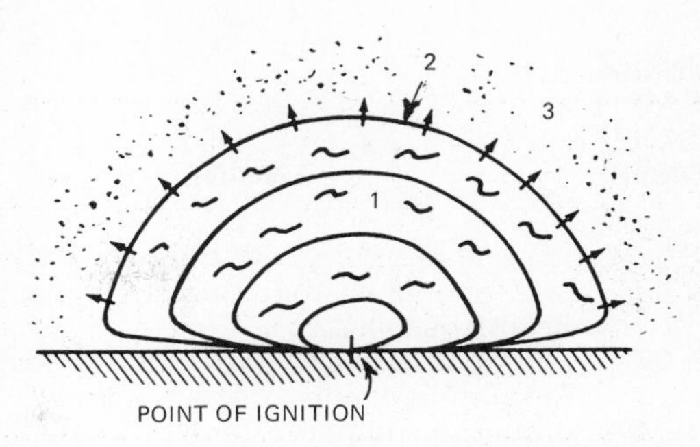

Fig. 2-4 Spherical flame travel or propagation without turbulence. (1) Flame kernel consisting of combustion products; (2) flame front area; (3) unburned mixture. *(General Motors Corporation)*

is shown in Fig. 2-3, top. During combustion, the pressure may exceed 1000 psi [6895 kPa].

Under certain conditions, the last part of the compressed air-fuel mixture, or "end gas," will explode before the flame front reaches it (Fig. 2-3, bottom). The end gas is subjected to increasing pressure as the flame progresses through the combustion chamber. This increases the temperature of the end gas (because of heat of compression and also radiated heat from the combustion process). If the temperature goes high enough, this end gas will explode before the flame front arrives. The effect is almost the same as if the piston head had been tapped rapidly with a hammer. In fact, it sounds as though this had happened.

⚙ 2-9 Compression ratio vs. detonation As compression ratios of engines have increased, so has the tendency for engines to experience detonation. Here is the reason. With a higher compression ratio, the air-fuel mixture is more highly compressed at TDC. Therefore, it is at a higher initial temperature. With higher initial pressure and temperature, the temperature at which detonation occurs is reached sooner. As a result, high-compression engines have a greater tendency towards detonation. However, higher-octane fuels have been developed for higher-compression engines. These gasolines have a greater resistance to being set off suddenly by heat of compression. They are less likely to explode suddenly. For ignition they depend only on the wall of flame traveling through the air-fuel mixture.

⚙ 2-10 Octane ratings The octane rating of a gasoline is a measure of its resistance to detonation in the engine. An octane rating does not measure the energy in a gasoline, the quality of a gasoline, or whether a gasoline contains lead.

There are three ways of designating the octane rating of gasoline. The *research octane number* (RON) is determined by testing gasoline in laboratory engines running at low speed with wide-open throttle. This tends to give a high octane rating to a gasoline, such as 100 for premium and 94 for regular. *Motor octane number* (MON) is determined by testing gasoline in engines running at full throttle with high engine speed. This gives the gasoline a lower octane rating, such as 92 for

MINIMUM ROAD OCTANE NUMBER	ENGINE
97	For engines designed to run on premium, and for engines requiring the highest octane gasoline.
93–95	For most 1970 and prior engines designed for premium, and for later engines requiring a higher octane gasoline.
91	For engines designed for premium that can run a lower octane fuel, and for engines designed for regular that require a higher octane gasoline.
89	For most 1970 and prior engines designed for regular, and for 1971 and later engines that require a higher octane gasoline.
87	For most 1971 and later engines.
85	For engines with low antiknock requirements.

Fig. 2-5 Minimum road octane numbers for various models and years of automobile engines.

premium and 86 for regular. In general, the RON for any gasoline is about eight numbers higher than the MON for the same gasoline.

Another octane rating is called the *antiknock index*, or *road octane number*. It relates reasonably well to the actual performance of gasolines in cars. This is the newest rating system, and is the number that you are most likely to see on the pumps at service stations. Road octane number is found by adding together the RON and MON for a certain gasoline, and dividing by 2. The formula is

$$\begin{matrix}\text{Road octane number} \\ \text{(antiknock index)}\end{matrix} = \frac{\text{RON}+\text{MON}}{2}$$

The typical road octane for gasoline is about 96 for leaded premium, 90 for leaded regular, and 87 for unleaded regular. Figure 2-5 shows the minimum road octane number for various models and years of automobile engines. The refiners are allowed to reduce the minimum road octane number according to the altitude in which the gasoline will be sold. Also, gasolines with a road octane number of less than 89 must have a MON of 82.0 or higher.

Notice in Fig. 2-5 that some engines require a higher-octane gasoline than that for which they were designed. This is because all engines, even of the same identical model, cannot be manufactured exactly alike. For example, most engines of one model will run satisfactorily on a gasoline with an octane rating of 87. However, a few of the engines will run satisfactorily on 85, and a few will require 89.

After the engines are in use, the variation in their octane requirements grows even wider. This is because cars are driven and maintained in different ways. Many engines need a higher-octane gasoline as they grow older. This results primarily from the deposits that build up in the combustion chamber. These increase the compression ratio and therefore can cause severe and damaging detonation. By using a gasoline with a higher octane, the engine can continue to be operated without detonation.

⚙ 2-11 Octane requirements The required octane rating of the gasoline for use in an engine is determined basically by the engine design. However, these requirements change with weather and driving conditions, and with the mechanical condition of the engine. Changing temperature and humidity change the octane requirements of the engine. Engine deposits, reduced cooling-system efficiency, carburetor or ignition troubles, and emission-control device failures may also change octane requirements.

In addition to all these, how the driver operates the car has an effect on octane requirements. If the driver does not demand rapid acceleration and high-speed wide-open-throttle operation, the engine will be less likely to detonate. Therefore, the engine will have a lower octane requirement. However, this type of operation tends to hasten engine deposits. Deposits, in turn, increase octane requirement.

Automatic transmissions make a difference in the octane rating of the fuel needed by the engine. With an automatic transmission, the engine usually operates at part to full throttle at a fairly high engine speed. There is very little lugging or low-engine-speed full-throttle operation possible with an automatic transmission. This is because of the way the engine is coupled to the transmission.

The manual transmission uses a mechanical clutch that connects the engine and drive wheels rigidly. But the automatic transmission uses a torque converter that allows slippage. On acceleration the engine may turn at high speed while the car is moving at low speed. Therefore, with an automatic transmission detonation is less likely to occur and the octane requirement of the engine is lower. However, fuels differ in how easily they detonate at high and at low speed.

⚙ 2-12 Tetraethyl lead Tetraethyl lead raises the octane rating of gasoline and reduces its tendency to detonate. The lead also improves valve-seat life. However, lead in gasoline has a bad effect on the catalysts used in catalytic converters. These devices are connected into the exhaust system to convert certain pollutants in the exhaust gases into harmless compounds. Lead from the gasoline deposits on the catalysts and stops them from doing their job. That is one reason why most gasoline produced today has no lead added to it. Without lead, the octane number of the gasoline drops. Therefore, to prevent detonation, the compression ratios of the engines must be lowered. This causes some reduction in engine power.

There is another potential problem with lead in gasoline. When the gasoline is burned, the lead is not. It leaves the engine in the exhaust gas and some gets into the air. Lead is a known poison. Some people believe that breathing air containing lead from the engine ex-

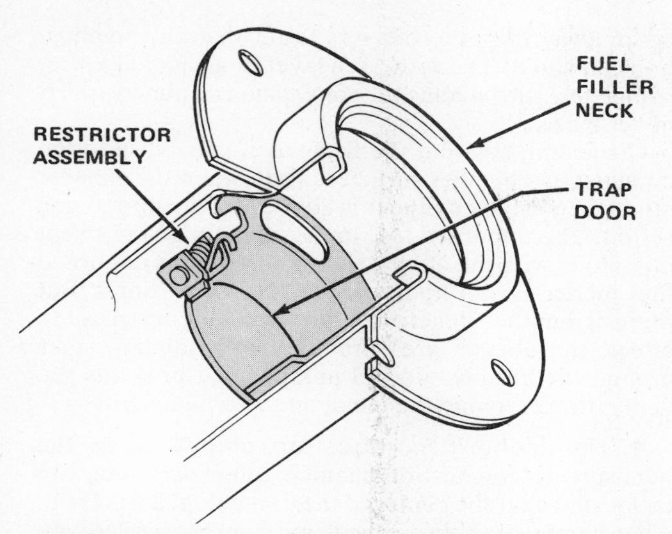

Fig. 2-6 Restrictor in the fuel tank filler neck. (*American Motors Corporation*)

haust could possibly cause lead poisoning. Severe cases of lead poisoning result in illness and death. This is the second reason why gasoline without lead is now required for most new cars.

Tetraethyl lead is not the only antiknock compound that has been used in gasoline. Others are tetramethyl lead (TML), and MMT, which is a manganese compound.

⚙ 2-13 Types of gasoline Today, there are two basic types of gasoline. These are *leaded* and *unleaded*. Leaded gasoline has had a small amount of liquid lead antiknock compound added to it. Unleaded gasoline has had no lead antiknock compound added to it. If an engine with a catalytic converter uses leaded gasoline, the lead will coat the catalyst in the converter. Then the converter will stop working. This is the reason most service stations have pumps labeled *unleaded*. The filler neck on the car requires a special small nozzle on the dispensing hose (Fig. 2-6). This combination is designed to prevent leaded gasoline from being pumped into a car requiring unleaded gasoline.

The addition of 1 teaspoon of liquid lead antiknock compound to 1 gallon of gasoline will raise the octane rating of the gasoline by about 10 numbers. When lead, or some other antiknock compound, is not added to gasoline, then the octane rating must be raised during the refining process. This requires more crude oil to produce 1 gallon of unleaded gasoline than it does to produce one gallon of leaded gasoline. As a result, the refining process takes longer and costs more. This is why unleaded regular gasoline is more expensive than leaded regular.

⚙ 2-14 Gasoline additives Gasoline is more than simply a blend of various liquid hydrocarbons. To get the needed characteristics as an engine fuel, several chemicals called *additives* are mixed into the gasoline. For example, antiknock compounds are added to some gasoline to increase its octane rating. As a result, chemical scavengers, or combustion-deposit modifiers, must then be added to the gasoline along with the antiknock compound. The scavengers minimize preignition and spark plug fouling. Other additives used in gasoline include:

1. Oxidation inhibitors to help prevent the formation of gum, especially while the gasoline is in storage.
2. Metal deactivators to protect the gasoline from the harmful effects of certain metals such as copper which can be picked up in the refining process or in the vehicle fuel system.
3. Corrosion or rust inhibitors to minimize corrosion and rusting of the vehicle fuel system.
4. Anti-icers to minimize ice formation in the carburetor around the throttle valve, and to prevent fuel-line freezing.
5. Detergents to prevent the accumulation of deposits in the carburetor around the throttle valve.
6. Dispersants to minimize engine sludge and the accumulation of deposits in the positive crankcase ventilating (PCV) valve and on engine parts.
7. Dyes for identification of various gasoline blends.

In addition, the refining process is very carefully controlled to keep sulfur compounds to a minimum. Sulfur compounds, in excess, form sulfur acids which promote rusting and corrosion of engine parts and piston ring and cylinder wall wear. These compounds can also contribute to air pollution as they burn with the fuel. Proper refining techniques minimize the amount of sulfur in gasoline.

⚙ 2-15 Mechanical factors affecting detonation The shape of the combustion chamber has a great effect on detonation. The cylinder head, intake and exhaust valves, and spark plug form the top of the combustion chamber. The piston head and top compression ring form the bottom of the combustion chamber (Fig. 2-7). Combustion chambers have two general shapes: wedge and hemispheric (Fig. 2-8). The shape determines turbulence, squish, and quench. These three factors affect detonation.

1. Turbulence When you stir coffee, you swirl it, or give it turbulence, so that the cream and sugar mix with the coffee. In the same way, giving turbulence to the air-fuel mixture entering the combustion chamber ensures more even mixing. This makes combustion

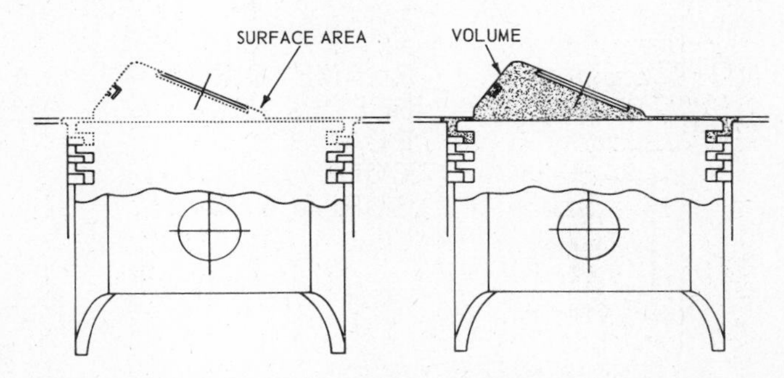

Fig. 2-7 Combustion chamber. Surface area is shown in dotted line.

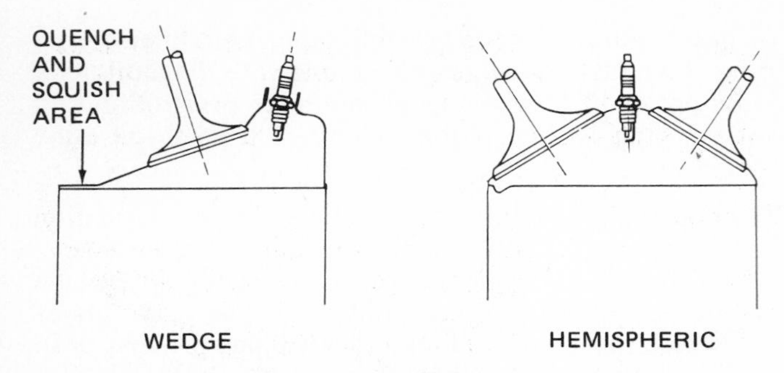

Fig. 2-8 Wedge and hemispheric combustion chambers.

more even. Turbulence also reduces the time required for the flame front to sweep through the compressed mixture. If the flame front reaches the end gases quick enough, they will not have time to detonate.

2. Squish In some combustion chambers, the piston squishes, or squeezes, a part of the air-fuel mixture at the end of the compression stroke. Figure 2-8 (left) shows the squish area in a combustion chamber, As the piston nears TDC, the mixture is squished, or pushed, out of the area. As it squirts out, it promotes turbulence and further mixing of the air-fuel mixture.

3. Quench Detonation results when the end gas temperature goes too high. The end gas explodes before the flame front reaches it. However, if some heat is taken from the end gas, then its temperature will not reach the detonation point. In the cylinder shown to the left in Fig. 2-8, the squish area is also a quench area. The cylinder head is close to the piston, and these metallic surfaces are cooler than the end gas. This causes heat to be removed from the end gas. Therefore, the end gas becomes too cool to detonate.

In addition, as the wall of flame approaches the cooler metal surfaces that make up the combustion chamber, the layers of air-fuel mixture next to the metal surfaces become too cool to burn (Fig. 2-9). Heat is taken away from these layers faster than the combustion can add it. The result is a layer of air-fuel mixture, completely surrounding the combustion chamber, which never burns.

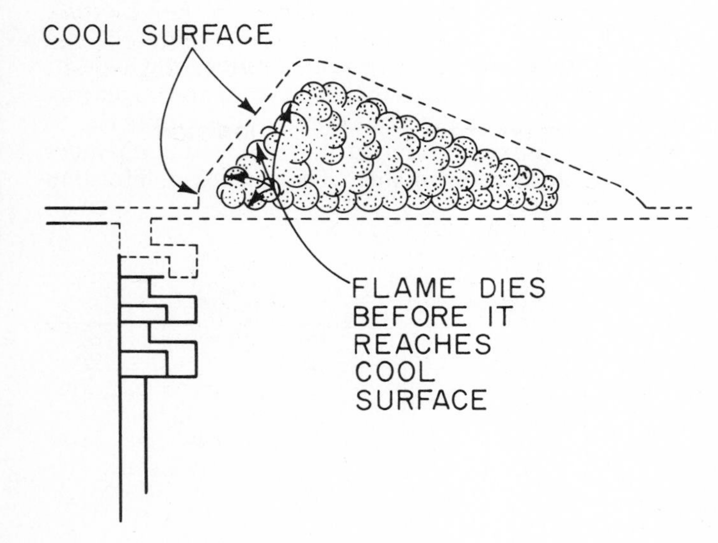

Fig. 2-9 The flame dies before it reaches the cool surface. This prevents complete combustion of the fuel.

Quenching puts out the flame as it gets close to the combustion chamber surfaces. Then when the exhaust stroke begins, the cylinder is cleared by the up-moving piston. The unburned fuel in the quench area is swept out, along with the other exhaust gases. This unburned fuel increases the hydrocarbon (HC) emissions. But there is another benefit to the quenching process. In effect, the quench area provides an insulating layer around the burning air-fuel mixture that prevents the flame from reaching the metal and overheating it.

4. Hemispheric combustion chamber In the hemispheric combustion chamber, the spark plug can be located near the center of the dome (Fig. 2-8). Then, when combustion starts, the flame front has a relatively short distance to travel. There are no distant pockets of end gas to detonate. The chamber has no squish or quench area. However, there is relatively little turbulence.

5. Wedge combustion chamber In the wedge combustion chamber, the spark plug is located to one side. The flame front must travel a greater distance to reach the end of the wedge (Fig. 2-8, left). The end of the wedge has a squish and quench area which cools the end gas. This prevents detonation and imparts turbulence to the mixture at the same time.

6. Combustion chamber shape and emissions The shape of the combustion chamber also affects the amount of pollutants in the exhaust gases. The cooler metal surfaces of the cylinder head and piston head slow combustion. Therefore, the layers of air-fuel mixture next to these metal surfaces do not burn completely. Incomplete burning means more exhaust emissions as the unburned fuel leaves with the exhaust gas. The wedge combustion chamber has a larger surface area. Therefore, it produces a greater amount of pollutants than the hemispheric combustion chamber. However, the hemispheric combustion chamber creates relatively little turbulence. As a result, combustion may not be complete, especially at low speeds.

7. Special combustion chambers Several special combustion chamber designs have been developed which improve the combustion process. In one, the intake-manifold passages are curved and the intake valve is masked (Fig. 2-10). This gives the ingoing air-fuel mixture a rapid swirling motion. The mixture burns better and is less likely to detonate. Another design uses a turbulence-generating pot (TGP). This is a small additional combustion chamber (a "precombustion" chamber) off the main combustion chamber, as shown in Fig. 2-11. The TGP has the spark plug. Therefore, ignition takes place in the TGP. Then the burning mixture streams out, producing high turbulence so that more complete combustion is achieved.

Another system also uses a precombustion chamber (Fig. 2-12). However, this precombustion chamber has its own intake valve in addition to the spark plug. The

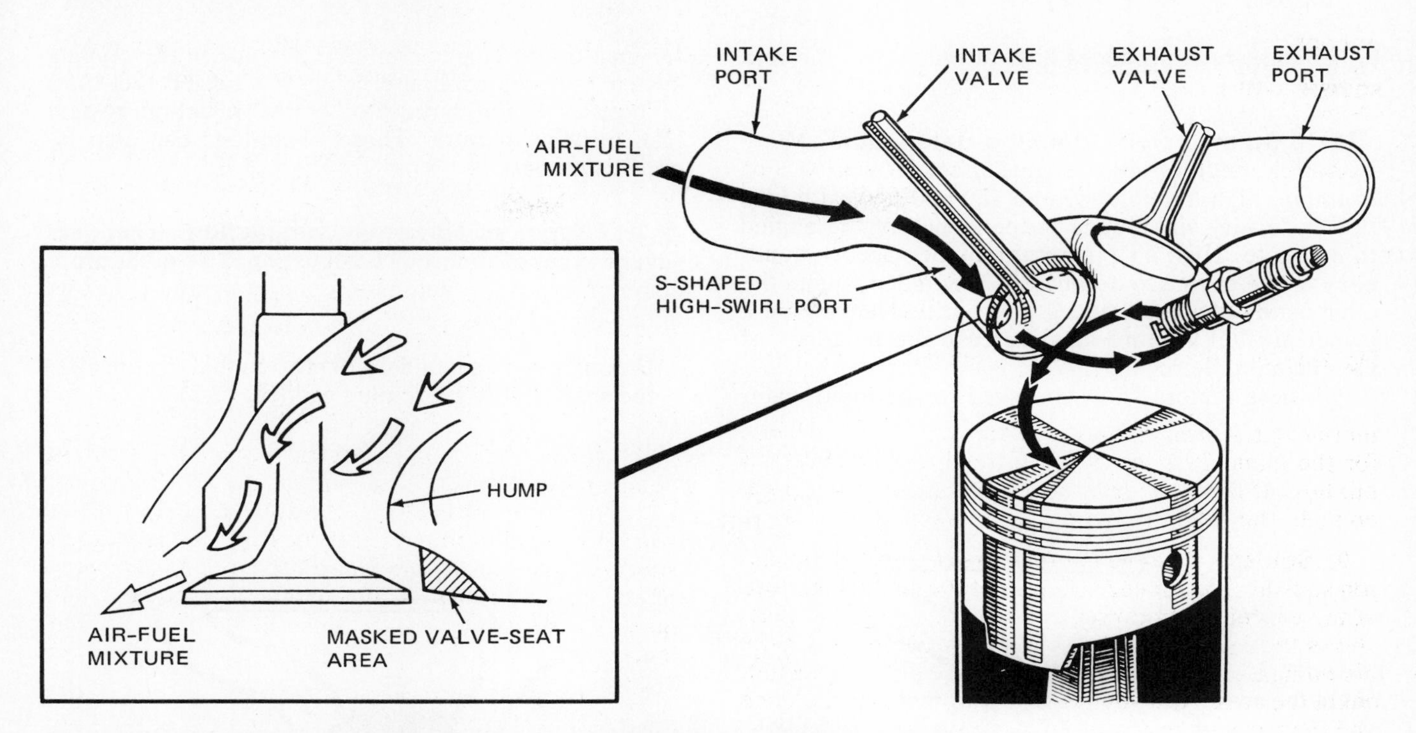

Fig. 2-10 The location of the intake valve in an S-shaped high-swirl intake port works with a masked valve seat area to swirl and direct the air-fuel mixture into the combustion chamber. This allows the engine to run on a lean mixture, which reduces exhaust emissions. (*Mazda Motors of America, Inc.*)

precombustion chamber is supplied with a rich air-fuel mixture while the main chamber fills with a lean mixture. Figure 2-12 shows the action. To the left, both intake valves are open and the piston is moving down on the intake stroke. To the right, ignition takes place and the burning mixture streams out, igniting the lean

SPARK PLUG
TURBULENCE GENERATING POT
ORIFICE
MAIN COMBUSTION CHAMBER
INTAKE MANIFOLD

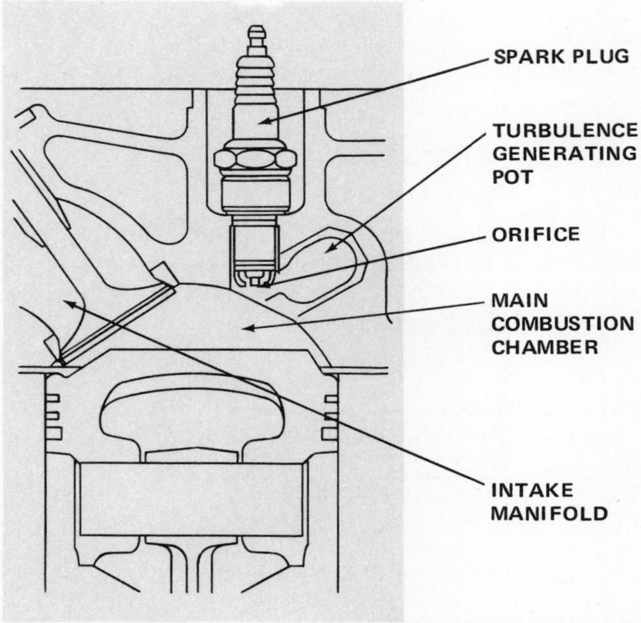

Fig. 2-11 Sectional view of cylinder head showing the location of the turbulence-generating pot. (*Toyota Motor Sales Company, Ltd.*)

mixture in the main chamber. This causes high turbulence and good combustion. An engine with this type of combustion chamber is called a *stratified-charge engine*. Stratified means *layered*.

In a stratified-charge engine, a small amount or layer of rich mixture is ignited first. Then ignition spreads to the leaner mixture which fills the rest of the combustion chamber.

A still different system of producing high turbulence is shown in Fig. 2-13. This system is called the *jet-controlled lean-combustion system*. It also uses a second intake valve. This valve admits air-fuel mixture from above the throttle valve in the carburetor. The mixture enters the combustion chamber as a high-speed

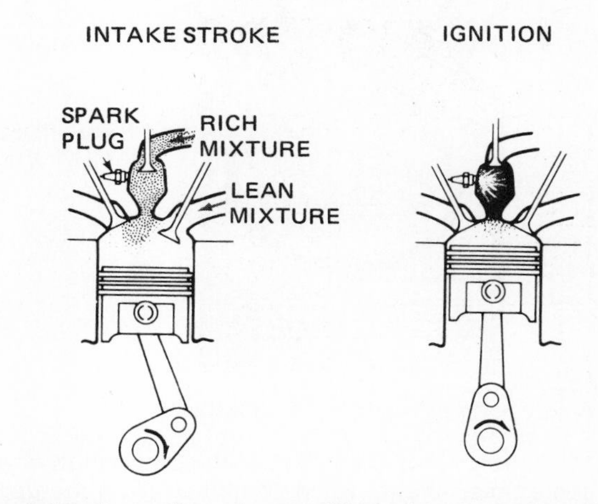

Fig. 2-12 Operation of a stratified-charge engine. (*Honda Motor Company, Inc.*)

jet stream which produces high turbulence in the rest of the mixture.

⚙ 2-16 Other factors affecting detonation Many operating conditions in an engine affect detonation. For example, high humidity (damp air) and high altitudes (lower-density air) reduce the tendency of the engine to detonate. High air temperatures increase the tendency of the engine to detonate. Carbon deposits in the combustion chamber, excessive spark advance, and a leaner air-fuel mixture also increase the tendency of the engine to detonate.

All these factors show the need for periodic maintenance of the engine. Buildup of scale in the cooling system reduces cooling efficiency. Clogged fuel lines or nozzles in the carburetor or fuel-injection system lean out the mixture. These conditions can also increase detonation.

⚙ 2-17 Types of abnormal combustion There are several types of abnormal combustion. Two of the more common types are detonation and preignition. Let us define these terms.

Detonation: A secondary explosion that occurs after the spark at the spark plug gap (Fig. 2-3).

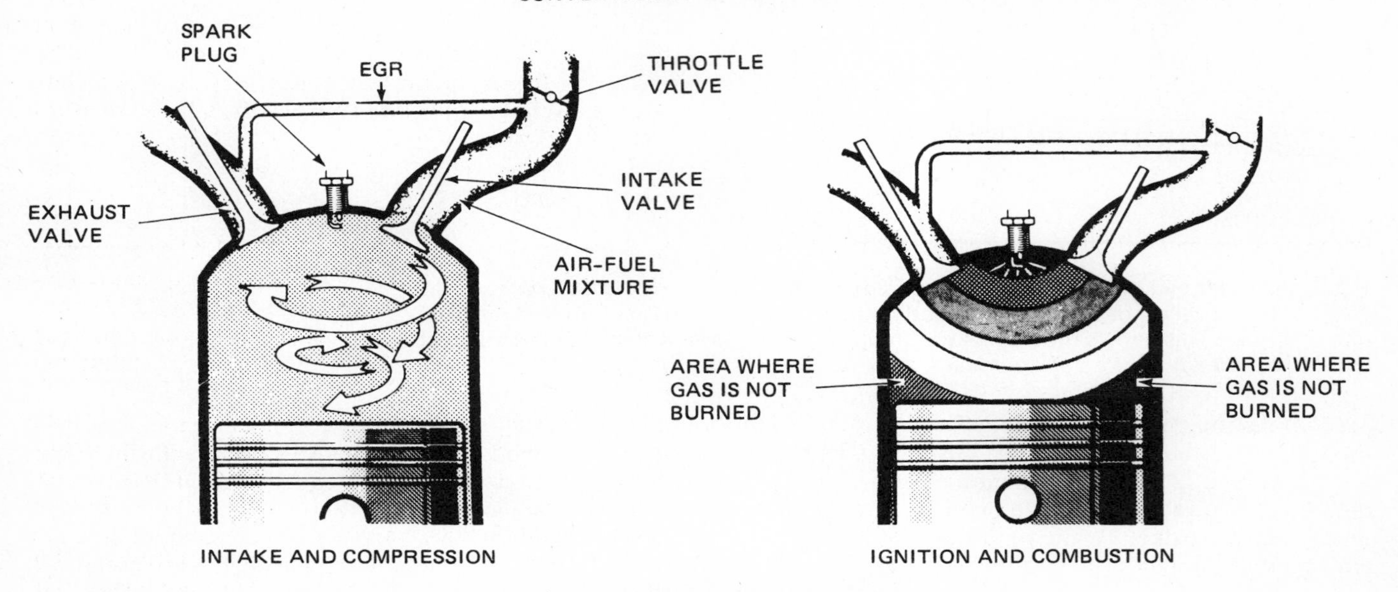

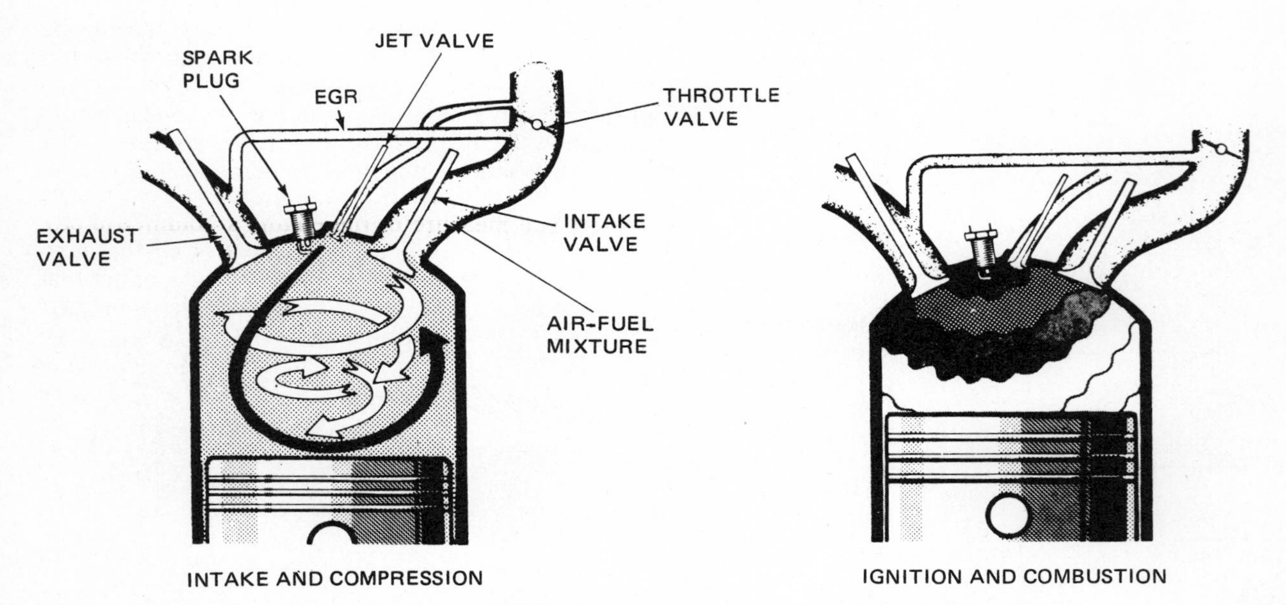

Fig. 2-13 Action of the jet valve in the combustion chamber. Both intake valves open at the same time. The jet valve admits air from the upper part of the carburetor. This air is not restricted by the partly closed throttle valve. It enters at high speed, setting up the whirling motion as shown to the left. (*Chrysler Corporation*)

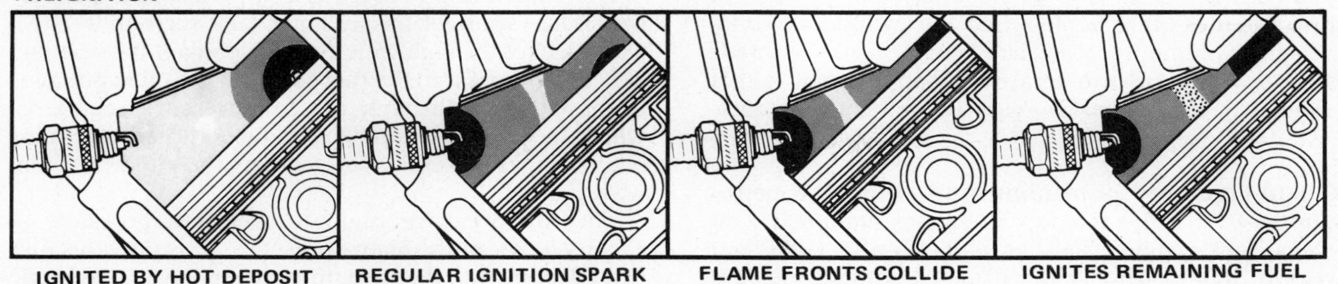

Fig. 2-14 One cause of preignition. The hot spot ignites the compressed mixture before the spark occurs at the spark-plug gap. (*Champion Spark Plug Company*)

Preignition: Ignition of the air-fuel mixture before the spark occurs at the spark plug gap (Fig. 2-14).

Detonation has a pinging sound. It is most noticeable when the engine is accelerated or is under heavy load, as when climbing a hill. Under these conditions, the throttle valve is nearly or fully wide open. The engine is taking in a full charge of air-fuel mixture on every intake stroke. This means the compression pressures reached are at the maximum. Detonation pressures and temperatures are more likely to be reached after the air-fuel mixture is ignited.

There are various types of preignition (Fig. 2-14), including surface ignition and rumble. Surface ignition can start from hot spots in the combustion chamber, such as a hot exhaust valve, spark plug, or carbon deposits. In some engines the deposits may break loose so that particles float free and become hot enough to produce ignition.

Surface ignition can occur before or after the spark occurs at the spark plug. Also, it can cause engine rumble and rough operation or mild to severe detonation. In some engines, the hot spots act as substitutes for the spark plugs. Then the engine continues to run even after the ignition switch is turned off. This condition can cause serious engine damage. Preignition can be caused by any source of heat in the combustion chamber that ignites the air-fuel mixture before the spark occurs at the spark plug. The characteristic sound of preignition is a dull "thud" (not the pinging sound that is characteristic of detonation). Preignition that happens during the intake stroke may cause a "pop-back" or backfire through the intake manifold and carburetor. For this reason, never crank an engine with the air cleaner removed. Also, never have your hands or face over the carburetor while the engine is cranked.

Surface ignition, preignition, and rumble are usually considered service problems. They result from inadequate servicing of the engine, installation of the wrong spark plugs (which run too hot), and use of incorrect fuels and lubricating oils for the engine and type of operation. With incorrect fuel or oil, engine deposits may occur. These engine deposits can cause surface ignition and rumble. They also may increase the compression ratio so that the engine is more likely to experience detonation.

⚙ 2-18 Volumetric efficiency Another condition affecting engine detonation is volumetric efficiency. This term relates to the ease with which the engine breathes or how much air-fuel mixture it can take in on the intake stroke. When the piston moves down on the intake stroke, a vacuum is produced in the cylinder. Atmospheric pressure then pushes the air-fuel mixture into the cylinder.

If the engine is running slowly with a wide-open throttle, the cylinder can be almost filled with air-fuel mixture. This means that on the compression stroke higher pressures will be reached. With higher compression pressures, there is greater tendency for detonation. This is exactly what can happen to an engine with a manual transmission when the throttle is opened wide for rapid acceleration from low speed, or when the engine is laboring to push the car up a hill.

When the car is cruising on the highway at only part throttle, less air-fuel mixture is getting into the cylinders. The mixture does not reach as high pressures on compression. Therefore, with lower pressures, there is less tendency for detonation.

The reason why higher engine speeds reduce the amount of air-fuel mixture entering the cylinders is that the intake stroke lasts a shorter time. At an idle speed of 500 rpm, the entire intake stroke takes less than one-tenth of a second. But at high engine speed the time is less than a hundredth of a second. Therefore, less air-fuel mixture enters the cylinders at high speed. The mixture flowing in is cut off by the closing of the intake valve before the cylinder is filled.

The amount of air-fuel mixture taken into the engine cylinders is a measure of the engine's volumetric efficiency. This is the ratio of the amount of air-fuel mixture that actually enters the cylinder to the amount that could enter under ideal conditions. For example, suppose that a cylinder, if completely filled, could hold 0.034 ounce [1 g]. However, the engine is running at medium speed, so that only 0.027 ounce [0.765 g] of air can enter. Then the volumetric efficiency would be only about 80 percent (0.027 is 80 percent of 0.034). At high speed, the volumetric efficiency might drop to 50 percent. This is another way of saying that the cylinders are only "half-filled" at high speeds.

To improve volumetric efficiency, intake valves are made larger, valve lift is increased (valves open wider), and intake manifolds are made as short, straight, and smooth inside as possible. Also, carburetors with additional air-fuel passages (or barrels) can be used to allow more air-fuel mixture to flow into the intake manifold (see ⚙4-10 and 4-11). All these make it easier for the air-fuel mixture to flow into the cylinders. By increasing volumetric efficiency, engines can be made more powerful. However, the penalty is that higher volumetric efficiencies increase the tendency to detonate.

⚙2-19 Liquefied petroleum gas Liquefied petroleum gas, or LPG, is used in engines equipped with special fuel systems. LPG is made up of certain light types of hydrocarbon molecules. LPG molecules are related to gasoline molecules. Both are made up of hydrogen and carbon atoms. But LPG molecules are smaller than gasoline molecules. Therefore, LPG is a vapor at ordinary temperatures. LPG is found in the earth along with natural gases and petroleum. It is normally liquid at the high pressures in the petroleum or gas reservoirs in the earth. When the pressure is relieved, the liquid turns to gas. In the recovery and refining process, the LPG is separated from other gas or petroleum products and pressurized to hold it in liquid form. It is conveniently stored and transported as a liquid in pressurized tanks.

Two different types of LPG have been used for automotive engine fuel: propane and butane. Of these, propane is the most widely used. Sometimes, small amounts of butane are added. Propane boils at −44°F [−42.2°C] at atmospheric pressure. It can be used in any climate where temperatures below this are not reached. Butane cannot be used in temperatures below 32°F [0°C], since it is liquid below that temperature. If it remains liquid, it will not vaporize in the fuel system and will never reach the engine.

Liquefied petroleum gas is a fuel that is liquid only under pressure. When the pressure is reduced, the fuel vaporizes. Therefore, the system must have a pressure-tight fuel tank to store the fuel at high pressures. A typical LPG fuel system is shown in Fig. 2-15. Pressure forces fuel through the filter, high-pressure regulator, and vaporizer. The high-pressure regulator reduces the pressure so that the fuel starts to turn to vapor.

This vaporizing process is completed in the vaporizer. The vaporizer has an inner tank surrounded by a water jacket through which engine coolant passes. The coolant adds heat to the fuel so that it is well vaporized. It then passes through the low-pressure regulator, where the pressure is further reduced. Then the vapor enters the carburetor, which mixes the vaporized fuel with air.

The low-pressure regulator reduces pressure to slightly *below* atmospheric pressure. This prevents fuel from flowing into the carburetor when the engine is off. Fuel will flow only when the engine is running and there is a vacuum in the carburetor.

LPG fuel systems are used on some trucks, buses, tractors, and forklift trucks. Systems have also been adapted to automobiles. The LPG fuel system is well suited for engines operated inside buildings, since the fuel burns clean. The exhaust gases contain very few contaminants.

⚙2-20 Gasohol The possibility of using a blend of alcohol and gasoline as an engine fuel has been studied since World War I. When a small amount of ethyl alcohol is mixed with gasoline, a new fuel, gasohol, is formed. Gasohol usually is 10 percent ethyl alcohol and 90 percent unleaded gasoline. Ethyl alcohol also is known as *ethanol* and as *denatured alcohol*.

In its pure state, ethyl alcohol has high resistance to detonation, with a RON of 106. To run straight ethyl alcohol in an engine, the carburetor must provide an air-fuel ratio of about 9:1 for normal operation. This is a much richer mixture than required for gasoline, which has a normal air-fuel ratio of about 14.5:1. However, when gasohol is used, no rejetting of the carburetor or other changes in the fuel system are required.

With crude oil in tight supply, many people believe gasohol is the automotive engine fuel of the future. Ethyl alcohol can be made from sugar, grain, and many other substances. By adding 10 percent alcohol to gasoline, the gasoline stocks are then extended by about the same amount.

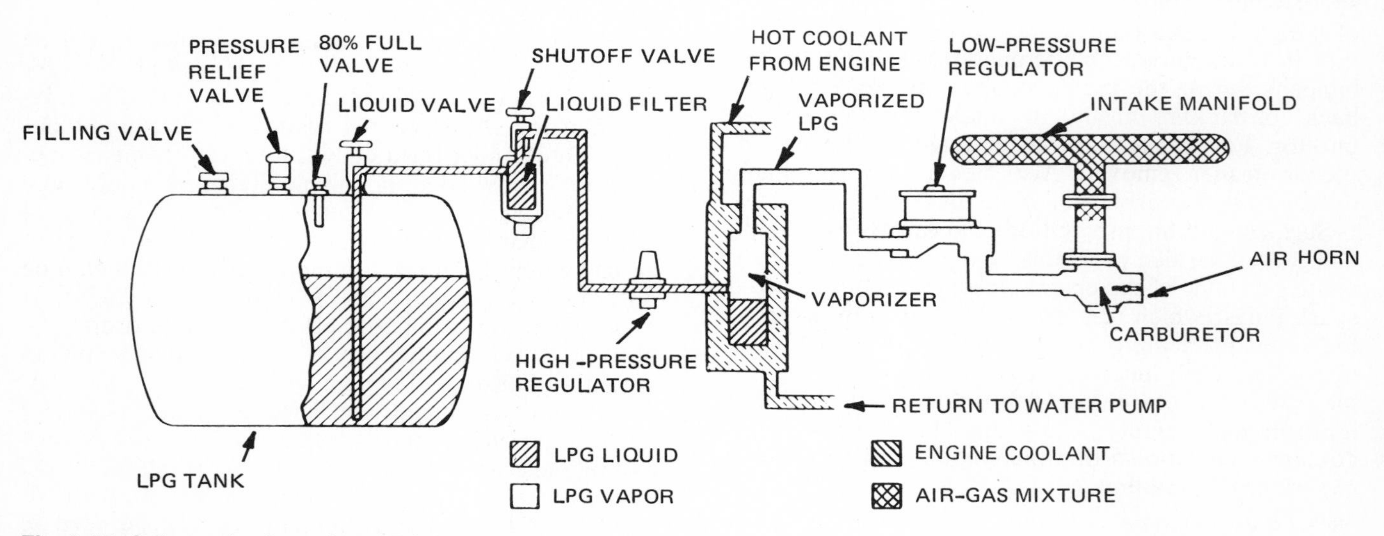

Fig. 2-15 Schematic view of an LPG fuel system.

Chapter 2 review questions

Select the *one* correct, best, or most probable answer to each question. Then check your answers against the correct answers given at the end of the book.

1. When gasoline is burned in an engine, the pollutants that come from the tail pipe are:
   a. HC, CO, and NO_x
   b. H_2O, CO_2, and O
   c. O, C, H, and N
   d. none of the above.
2. Mechanic X says that volatility refers to how well a fuel prevents detonation. Mechanic Y says the volatility refers to the ease with which a liquid vaporizes. Who is right?
   a. X only
   b. Y only
   c. both X and Y
   d. neither X nor Y.
3. Pinging is the characteristic sound of:
   a. normal combustion
   b. preignition
   c. leaded fuel
   d. detonation.
4. Which of the following statements is true?
   I. Carbon deposits in the combustion chamber raise the compression ratio.
   II. Carbon deposits in the combustion chamber increase the tendency for detonation to occur in an engine.
   a. I only
   b. II only
   c. both I and II
   d. neither I nor II.
5. Mechanic X says the octane rating of a gasoline is a measure of its energy content and quality. Mechanic Y says the octane rating of a gasoline is a measure of its resistance to detonation. Who is right?
   a. X only
   b. Y only
   c. both X and Y
   d. neither X nor Y.
6. Lead is being removed from gasoline because the lead:
   a. clogs the catalytic converter
   b. may poison the air
   c. both *a* and *b*
   d. neither *a* nor *b*.
7. Three factors in the combustion chamber which affect detonation are:
   a. intake valve size, exhaust valve size, and piston size
   b. corrosion, dispersants, and detergents
   c. engine speed, engine load, and emissions
   d. turbulence, squish, and quench.
8. Mechanic X says that in a stratified-charge engine the spark plug ignites the lean mixture. Mechanic Y says that in a stratified-charge engine the lean mixture ignites the rich mixture. Who is right?
   a. X only
   b. Y only
   c. both X and Y
   d. neither X nor Y.
9. Which of the following statements is true?
   I. Preignition occurs when the air-fuel mixture is ignited prior to the spark at the spark plug.
   II. Detonation occurs when the remaining air-fuel mixture explodes after the spark at the spark plug.
   a. I only
   b. II only
   c. both I and II
   d. neither I nor II.
10. Gasohol is a mixture of unleaded gasoline and
   a. 10 percent ethyl alcohol
   b. 90 percent ethyl alcohol
   c. 50 percent ethyl alcohol
   d. 10 percent liquefied petroleum gas.

CHAPTER 3

GASOLINE ENGINE FUEL SYSTEMS

After studying this chapter, you should be able to:

1. Explain the basic difference between the carbureted fuel system and the fuel injection system.
2. Describe the construction and operation of each major component in the carbureted fuel system.
3. Discuss the reasons for evaporative emission control systems.
4. Discuss the construction and operation of the two types of fuel gauges.
5. Explain the operation of a mechanical fuel pump.
6. Discuss the operation of the thermostatically controlled air cleaner.
7. Define PCV.
8. Discuss the difference between a single exhaust system and a dual exhaust system.
9. List the air pollutants in the exhaust gas of a running engine.
10. Explain the construction and operation of the turbocharger.

3-1 Two engines and two fuel systems Two types of engines are used in automobiles, gasoline and diesel. They have different operating cycles and burn different fuels. The gasoline engine uses an electric ignition system to ignite the air-fuel mixture in the engine cylinders. The diesel engine uses the *heat of compression* to ignite the fuel oil as it is injected into the cylinder. In the diesel engine, air alone is compressed in the engine cylinders during the compression strokes. The air gets very hot as it is compressed. Temperatures of the compressed air can exceed 1000°F [555°C]. The fuel oil is sprayed into the hot air just before the compression stroke is finished. Then the fuel oil ignites and the power stroke follows. Diesel engines and their fuel systems are covered in later chapters.

This and the next three chapters cover gasoline-engine fuel systems. These fuel systems can be divided into two types, the type using a carburetor and the type using a fuel-injection system. The various types of fuel-injection systems for gasoline engines are discussed in Chaps. 6 and 14.

3-2 Fuel systems for gasoline engines For many years, the gasoline engines used in automobiles have had carburetors in their fuel systems. The carburetor is a mixing device which mixes air and gasoline vapor in the proper proportions to produce a combustible mixture. Figure 3-1 is a simplified version of carburetor action. The combustible mixture flows from the carburetor to the engine cylinders through the intake manifold.

In recent years, some automobiles have been produced with a different kind of fuel system called a *fuel-injection system*. Figure 3-2 is a simplified drawing of one design. In this system, the carburetor is replaced by a throttle body which has only one job. It controls the amount of air flowing into the intake manifold. The

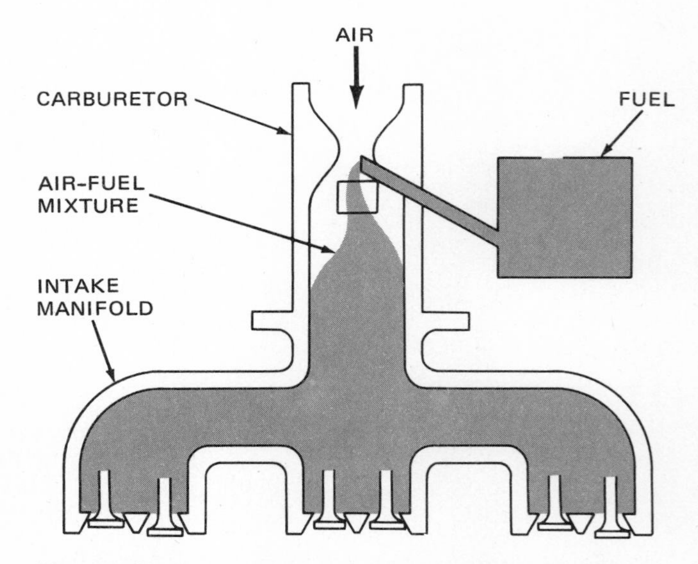

Fig. 3-1 Simplified drawing showing carburetor action.

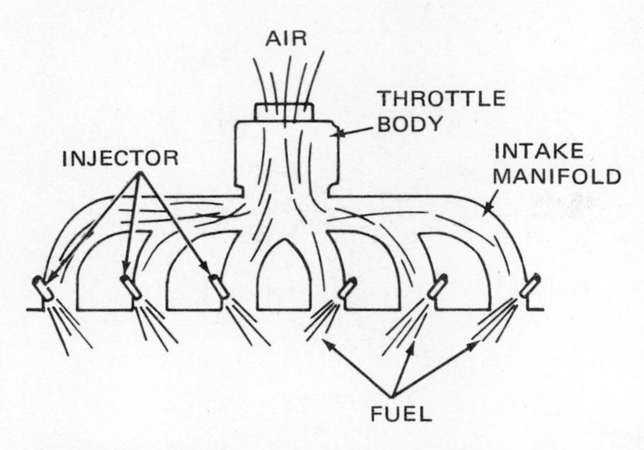

Fig. 3-2 Simplified drawing showing gasoline fuel-injection system in action.

intake manifold has a series of fuel injectors assembled to it. The injector is a fuel nozzle that sprays a metered amount of gasoline into the intake manifold, opposite the intake valve (Fig. 3-3). When the intake valve opens, this sprayed gasoline and air enters the cylinder. There is an injector for every cylinder in the system shown in Fig. 3-2. Each injector is placed opposite an intake valve.

The basic difference between the carbureted system and the fuel-injection system is how the fuel is metered. In the carburetor, the air passing through creates a vacuum which causes fuel to discharge from the main nozzle. In the fuel-injection system, the amount of gasoline discharged from the injector is determined electronically (electronic fuel injection) or mechanically (mechanical fuel injection).

NOTE: The various fuel-injection systems for gasoline engines are covered in Chap. 6. Some types of fuel-injection systems have a separate nozzle or injector for each cylinder in the engine. Other systems have the injection nozzles located in the throttle body.

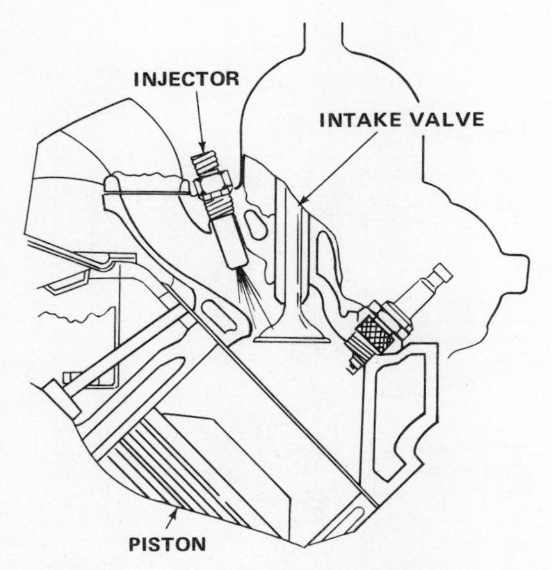

Fig. 3-3 The fuel injector sprays gasoline into the intake manifold, ahead of the intake valve.

3-3 Purpose of the fuel system Both the carbureted fuel system and the fuel-injection system do the same job. That is to supply a combustible mixture of air and fuel to the engine cylinders. The fuel system stores liquid gasoline and delivers it to the engine cylinders on the intake strokes in the form of vapor mixed with air.

The fuel system must vary the proportions of air and gasoline to handle varying operating requirements. For example, when a cold engine is started, the fuel system must deliver a very rich mixture (rich in gasoline). The proportions may be about 9 pounds [4.08 kg] of air to 1 pound [0.45 kg] of gasoline. Then, after the engine is running and has warmed up, the fuel system "leans out" the mixture (makes it less rich) to about 15 pounds [6.80 kg] of air to 1 pound [0.45 kg] of gasoline. For acceleration or high-speed operation, the mixture is again enriched.

3-4 Components of the carbureted fuel system The carbureted fuel system (Fig. 3-4) consists of the fuel tank, fuel pump, carburetor, fuel gauge, intake manifold, connecting fuel lines, and the accelerator pedal and linkage. The accelerator pedal controls the amount of air-fuel mixture entering the engine cylinders and therefore the amount of power the engine produces. The fuel tank holds a supply of gasoline for the engine. The fuel-level indicating system includes a gauge on the car instrument panel which shows the level of fuel in the fuel tank. The fuel pump draws gasoline from the fuel tank and delivers it to the carburetor. The carburetor mixes the gasoline in varying proportions and delivers it to the engine through the intake manifold.

3-5 Fuel tank The fuel tank (Fig. 3-5) is usually located at the rear of the vehicle and is attached to the frame. Figure 3-6 shows the details of installation of a fuel tank on one automobile. The fuel tank is a storage tank for fuel, made of sheet metal or plastic. It may contain baffles or metal plates attached to the inner surface of the tank, parallel to the sides. These plates have openings through which the fuel can pass. Their main purpose is to prevent the sudden surging of fuel from one side of the tank to the other when the car rounds a corner. The filler opening of the tank is closed by a cap (Fig. 3-6). The open end of the fuel line is positioned near the bottom of the tank. This prevents dirt or water that has settled to the bottom of the tank from entering the fuel line.

The fuel tank contains the tank unit of the fuel-gauge system. Figure 3-6 shows how the sender unit (view *a*) of the fuel gauge is installed in the fuel tank of one car. The fuel tank also contains a strainer that filters dirt particles from the fuel to prevent them from entering the fuel line. The strainer is located at the end of the fuel line or pickup tube (Fig. 3-6). All fuel leaving the tank must pass through the strainer. The fuel-tank cap has a vent which permits air to enter the tank as fuel is withdrawn. If this vent is plugged, a vacuum is created in the tank that prevents normal delivery of fuel to the fuel pump and carburetor. The vacuum created in the

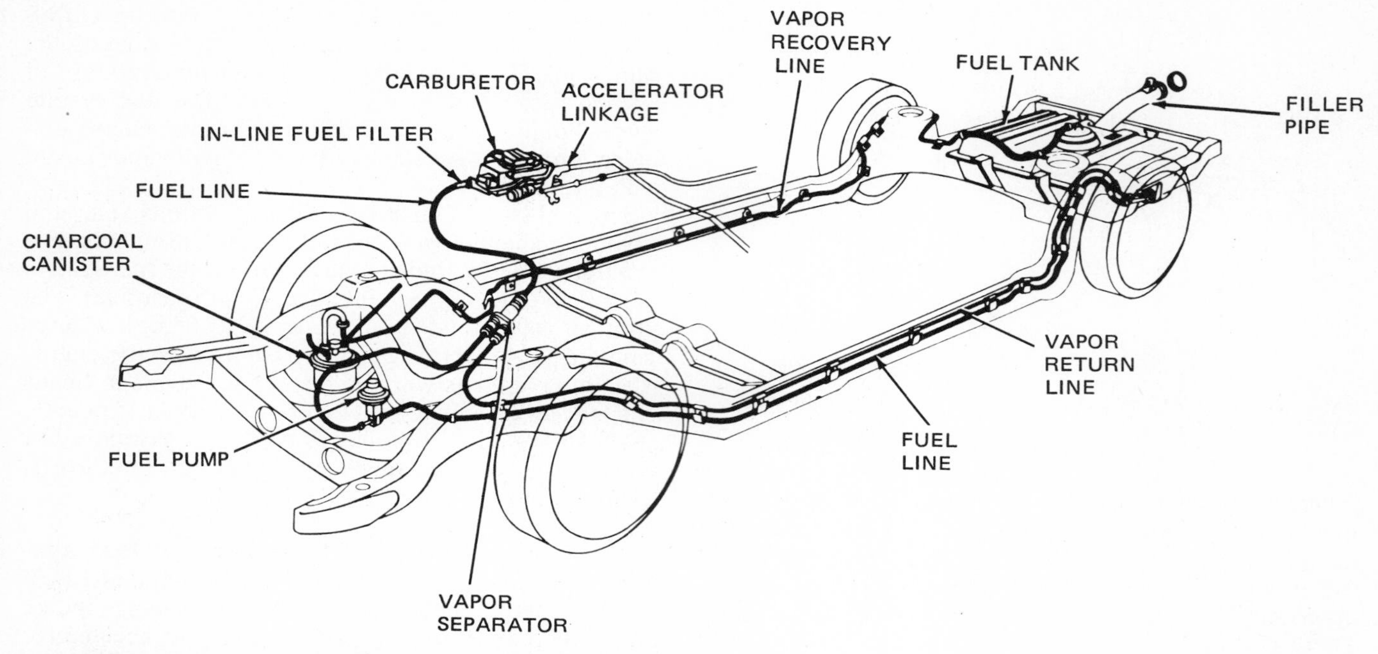

Fig. 3-4 Basic components of a carbureted fuel system.

fuel tank could cause atmospheric pressure to crush the tank. In some cars, the fuel pump also is installed in the fuel tank (⚙ 3-17).

⚙ 3-6 Fuel tanks for emission control Fuel tanks in older cars have a vent pipe which allows air to enter when fuel is withdrawn. It also allows air to escape when the tank is being filled. Gasoline vapor escaping through the vent pipe is an atmospheric pollutant that contributes to the formation of smog. Therefore, cars manufactured since 1970 have been equipped with a fuel-vapor or evaporative emission-control system. In this system, the fuel-tank vent pipe is connected to a charcoal canister. It holds the vapor and prevents its escape into the air. The fuel-vapor emission control system is covered in detail in ⚙ 3-14.

Cars with a fuel-vapor emission-control system have a special fuel-tank filler cap (Fig. 3-7). The cap has a two-way relief valve, operating on both pressure and vacuum. When gasoline is withdrawn from the tank, a slight vacuum develops. Then the vacuum valve opens to admit air. If the pressure builds up excessively, the pressure valve opens to relieve the pressure. Excessive pressure would not normally develop because any pressure is released through the fuel-vapor emission-control system. The filler cap shown in Fig. 3-7 also has a

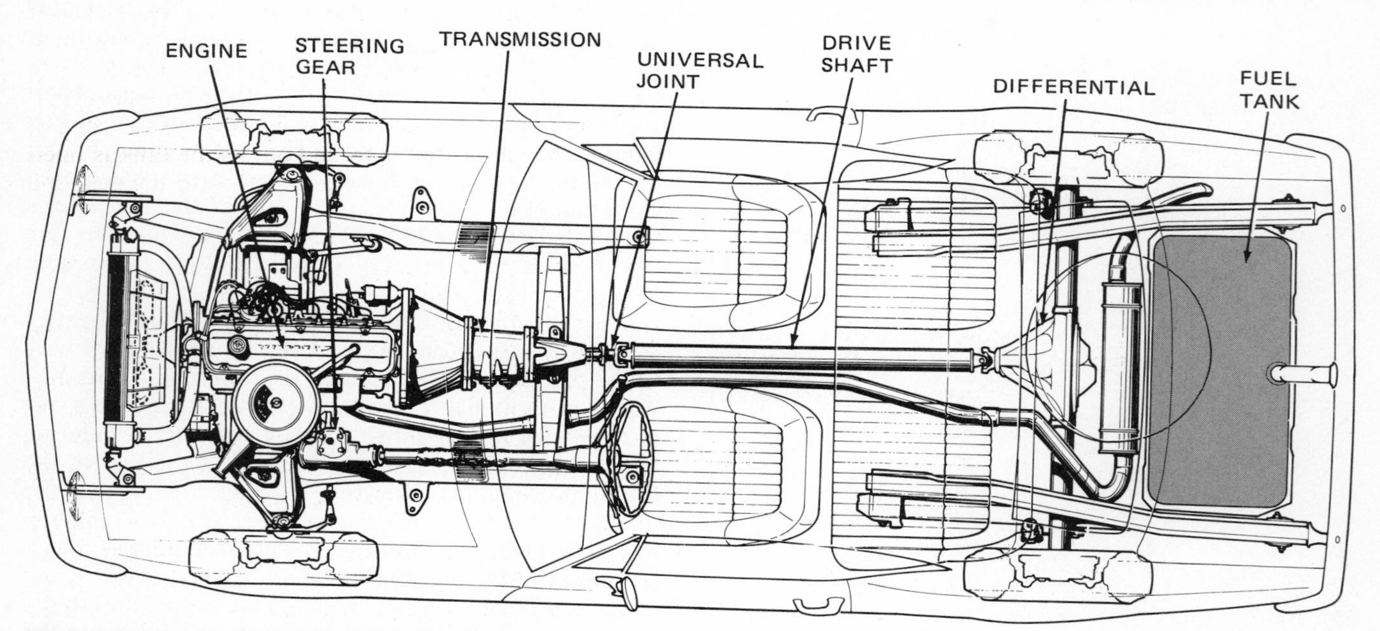

Fig. 3-5 Location of the fuel tank on the car.

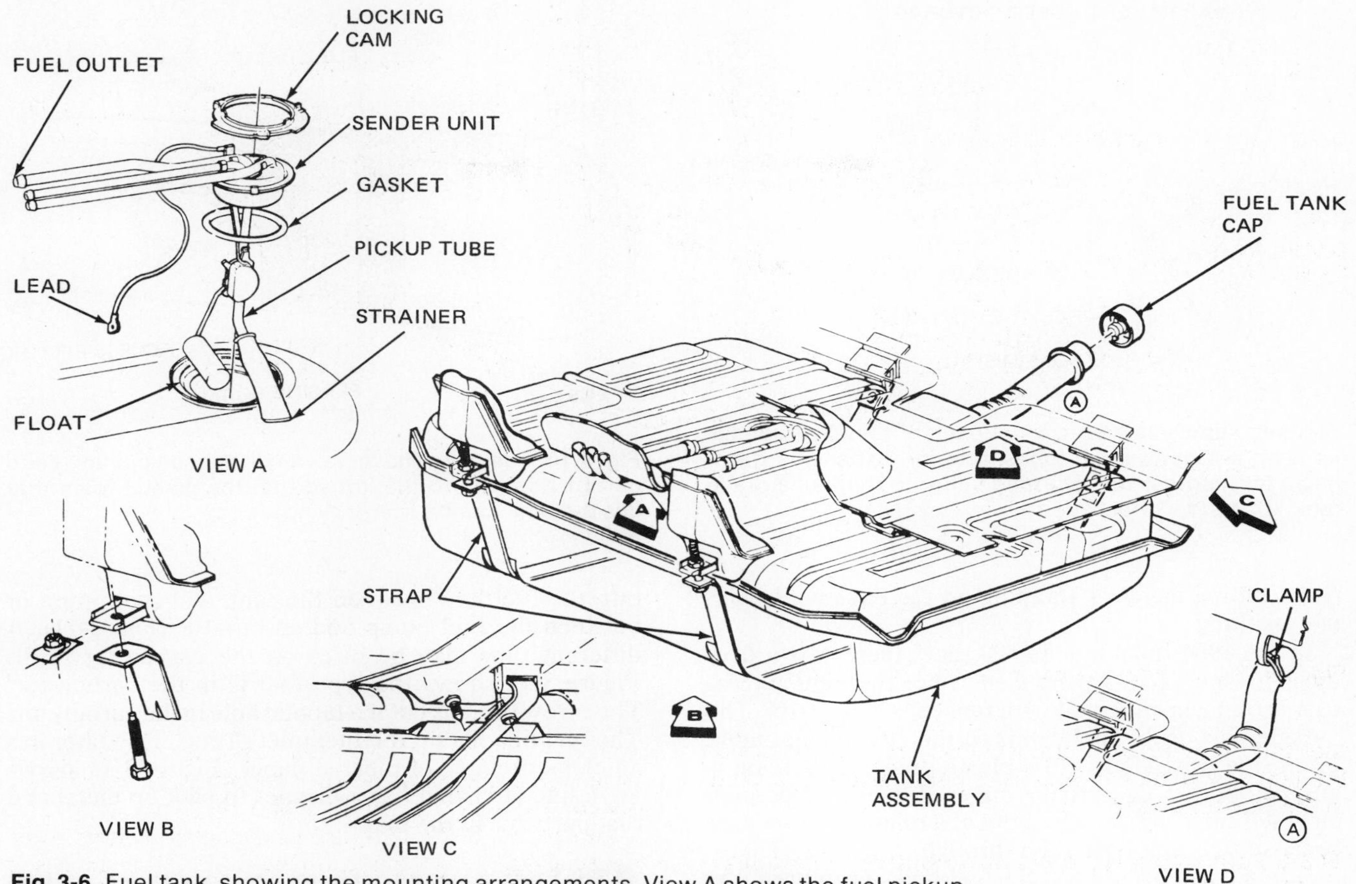

Fig. 3-6 Fuel tank, showing the mounting arrangements. View A shows the fuel pickup tube and sender unit for the fuel gauge. (*Buick Motor Division of General Motors Corporation*)

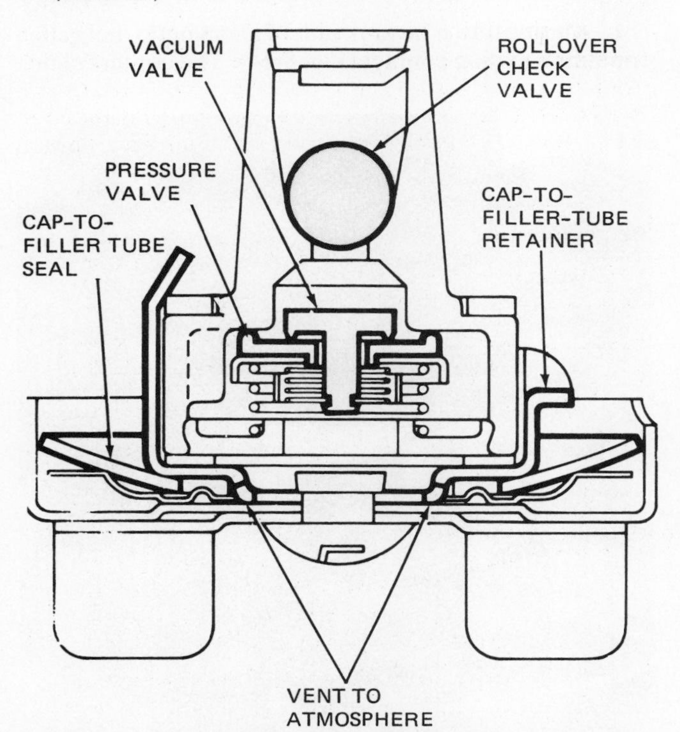

Fig. 3-7 Fuel-tank cap. It has pressure and vacuum valves, and also a rollover check valve. (*American Motors Corporation*)

rollover check valve which closes if the car rolls over. When the car and cap are upside down, the check valve closes to prevent leakage of gasoline from the tank.

Gasoline vapor (HC) escapes from the fuel tank when it is being filled, and this HC is an atmospheric pollutant. The gasoline entering the tank displaces the vapor in the tank and the vapor is forced out into the atmosphere. To prevent this, the special dispensing nozzle shown in Fig. 3-8 was developed. It has a vapor-tight seal at the tank filler pipe which prevents the escape of vapor. The vapor that is forced out as the tank is filled passes through a vent hose and goes into the gasoline storage tank.

Almost all automobiles manufactured in 1975 and later have a catalytic converter in the exhaust system. These cars must use unleaded gasoline (⚙ 2-13). To guard against putting leaded gasoline in the tank, these cars have a restrictor (Fig. 3-9). This restrictor is a small hole that may be closed by a trapdoor. The pump nozzles that dispense leaded gasoline are too large to fit. Only the smaller nozzles on unleaded gasoline pumps can enter the neck and open the trapdoor so that gasoline can flow into the tank.

⚙ 3-7 Fuel cells On race cars, where the danger of crashes is high, a special type of fuel tank is used. In a crash, an ordinary fuel tank may split, spilling fuel. This fuel could then be ignited by sparks produced by

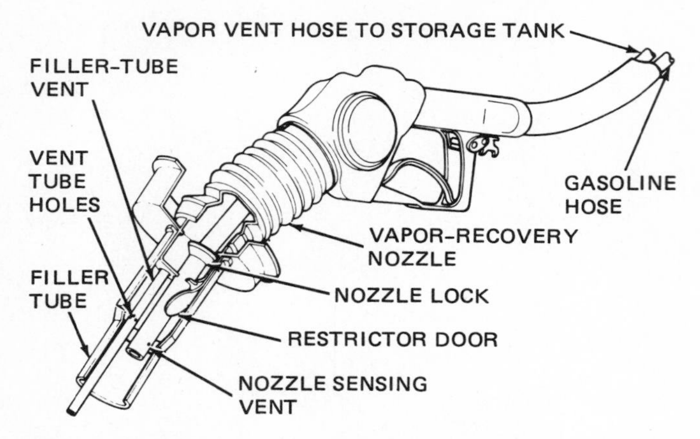

Fig. 3-8 Vapor-recovery gasoline filler tube. The fuel-dispensing nozzle has two tubes, one for dispensing gasoline and the other for returning vapor to the fuel storage tank. (*Chrysler Corporation*)

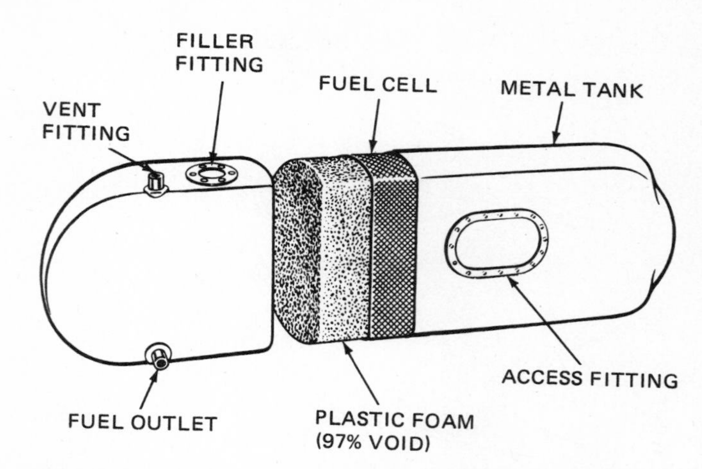

Fig. 3-10 Fuel cell. The metal tank has been cut away and part of it moved to the left so that the plastic foam that fills the tank can be seen.

metal hitting metal or scraping on the pavement, with fire resulting.

In the 1964 Indianapolis 500 race, there was a fiery crash in which two drivers died. Since that time, racing cars have been equipped with fuel cells (Fig. 3-10). The outside part of the fuel cell is formed by a metal tank, but inside it is filled with a plastic foam. This foam is an open-pore polyurethane material that is 97 percent air. Fuel can run through it, but at a relatively slow rate because the fuel must seep through the pores. Therefore, if a crash occurs and the fuel tank is ruptured, the fuel slowly oozes out.

✪ 3-8 Fuel filters and screens Fuel systems have filters and screens to prevent dirt in the fuel from entering the fuel pump or carburetor. Dirt could prevent normal operation of these units and cause poor engine performance. One older type of filter was built onto the fuel pump. Another type is a separate filter connected into the fuel line between the tank and fuel pump or between the fuel pump and carburetor (Fig. 3-11). A different type may be in or on the carburetor itself. Figure 3-12 shows the type that is in the carburetor. The screw threads enter a tapped hole in the carburetor. The fuel line fits on the fuel inlet fitting. This filter has an element made of pleated paper. Figure 3-13 shows an in-line filter that has a magnet to pick up metal and rust particles in the fuel.

✪ 3-9 Fuel gauges There are two types of fuel gauge, *magnetic* and *thermostatic*. Each of these gauges has a tank unit and an instrument-panel unit.

1. Magnetic The tank unit of a magnetic fuel gauge contains a sliding contact (Fig. 3-14). The contact slides

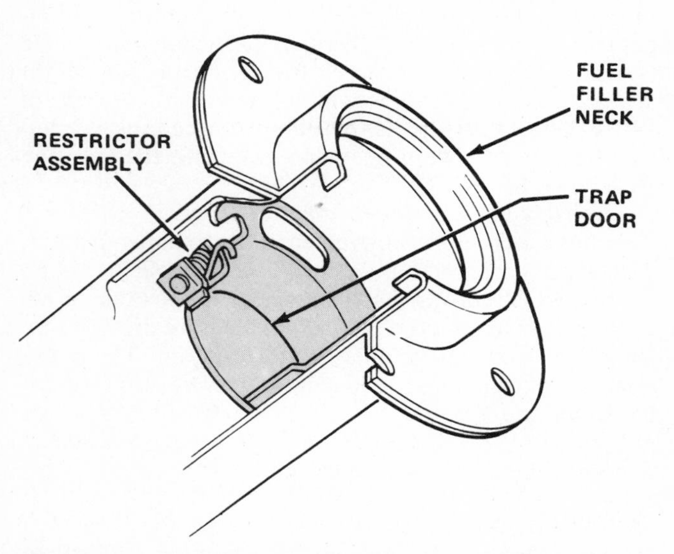

Fig. 3-9 Restrictor in the fuel tank filler neck. (*American Motors Corporation*)

Fig. 3-11 Typical in-line fuel filter installation on a six-cylinder engine. (*Chrysler Corporation*)

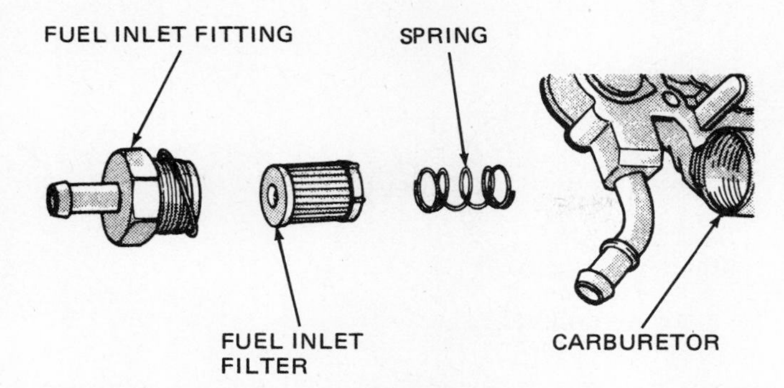

Fig. 3-12 Fuel filter that fits in the fuel inlet of the carburetor. (*Chrysler Corporation*)

back and forth on a resistor as the float moves up and down in the fuel tank. This changes the amount of electric resistance the tank unit offers. As the tank empties, the float drops and the sliding contact moves to reduce the resistance.

The instrument panel gauge contains two coils (Fig. 3-14). When the ignition switch is turned on, current from the battery flows through the two coils. This produces a magnetic field that acts on the armature to which the pointer is attached. When the resistance of the tank unit is high (tank filled and float up) most of the current flows through the F (full) coil to ground instead of through the tank unit. Therefore, the stronger magnetic field of the full coil pulls the armature to the right, so that the pointer is on the full side of the dial.

When the tank begins to empty, the resistance of the tank unit decreases. This allows more current to flow from the empty coil and through the tank unit. As the current flow through the circuit increases, the magnetic field of the empty coil becomes stronger than that of the full coil. As a result, the empty coil pulls the armature toward it. The pointer swings toward the left or E (empty) side of the dial.

2. Thermostatic Figure 3-15 is the wiring circuit of a thermostatic fuel gauge. It has a fuel-tank unit similar to the magnetic fuel gauge. The thermostatic fuel gauge also has a float and a sliding contact that moves on a resistor. Current flows from the battery through a heater wire in the fuel-gauge instrument panel unit, and through the resistance in the tank unit. When

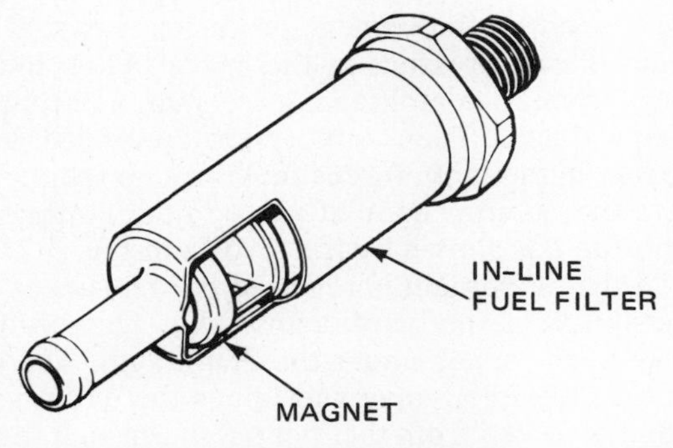

Fig. 3-13 In-line fuel filter with magnet to pick up particles of metal. (*Ford Motor Company*)

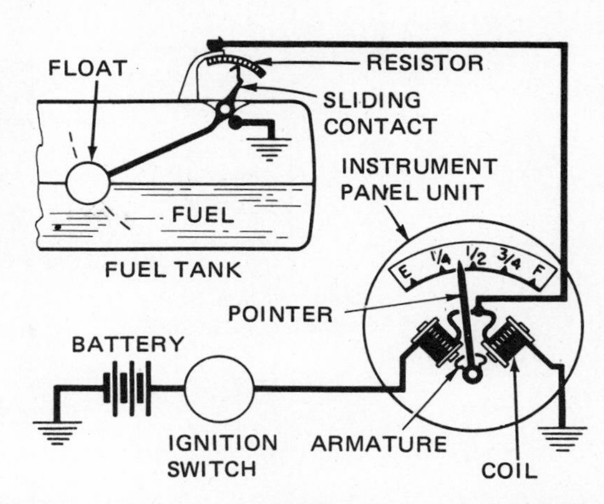

Fig. 3-14 Schematic wiring diagram of a magnetic fuel gauge indicating system, which uses balancing coils.

the fuel level in the tank is low, most of the resistance is in the circuit. Very little current can flow. When the tank is filled, the float moves up, and the sliding contact cuts most of the resistance out of the circuit. Now more current flows. As it flows through the heater coil in the fuel gauge, the current heats the thermostat. The thermostat blade bends because of the heat. This moves the needle to the right, toward the F or full mark.

Note in Fig. 3-15 that the thermostatic fuel gauge includes an instrument voltage regulator (IVR). This device also is thermostatic. Its purpose is to keep the voltage to the fuel-gauge system constant at a heating value of about 5 volts.

❁ 3-10 Low-fuel-level indicator The system shown in Fig. 3-15 also has a low-fuel-level indicator light. It includes a thermistor assembly in the fuel tank, a warning light, and a warning relay. A *thermistor* is a special sort of resistor that loses resistance as it gets hot. As long as there are more than a few gallons of fuel in the tank, the thermistor is submerged and is kept cool. However, when the fuel level is low, the thermistor is exposed to air. It gets hotter. Its resistance decreases, and more current flows. The increased current flow is sufficient to operate the warning relay. It connects the warning light to the battery. The light comes on to warn the driver that the fuel is getting low.

The system also includes a "prove-out" circuit. When the ignition switch is turned to START for cranking the engine, contacts in the ignition switch connect the warning light to the battery. Then as the switch is turned, the light comes on. When the light comes on during starting, it proves that the system is working. If the light does not come on, something is wrong. The bulb is burned out or the relay is defective.

Some late-model cars use an electronic low-fuel-level warning system (Fig. 3-16). This system also includes the fuel-level indicator. The low-fuel-level warning system operates from the voltage difference between the two terminals of the fuel gauge. As the needle moves toward E (empty), the voltage increases. When the tank is about one-eighth full, the voltage is great enough to trigger the low-fuel-level switch. It then turns the low-fuel-level warning light on.

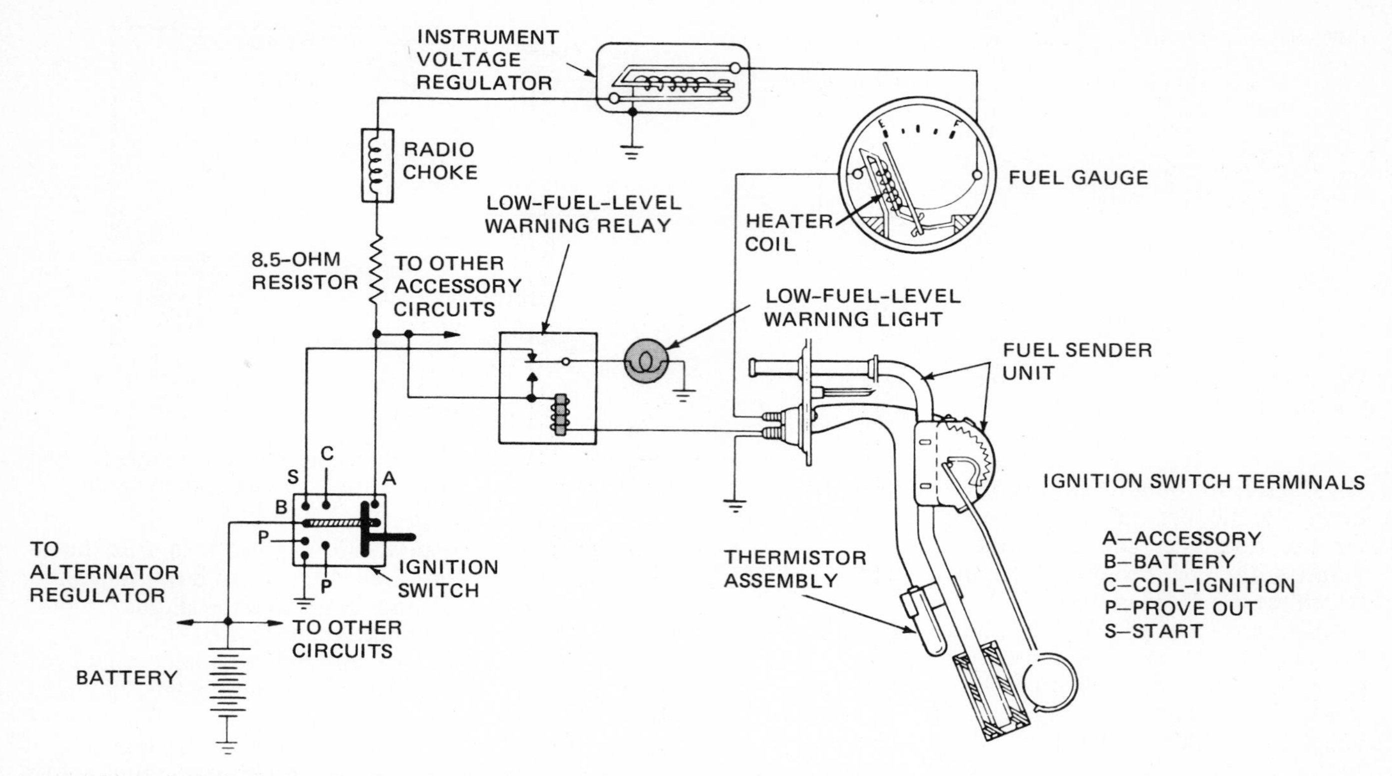

Fig. 3-15 Schematic wiring diagram of a thermostatic fuel gauge system using a variable-resistance tank unit. The system also includes a low-fuel-level warning system. (*Ford Motor Company*)

⚙ 3-11 Miles-to-empty fuel indicator This system shows the driver how many more miles the car can be driven before it runs out of gasoline. The indicator is usually located just under the fuel gauge on the dash (Fig. 3-17). When the button is pressed, the MILES TO EMPTY will flash on the indicator for a few seconds. If there is less than 50 miles of fuel remaining, the mileage reading will remain on. This system is used instead of the low-fuel-level warning system (⚙ 3-10). Figure 3-18 shows the components of the system.

The system continuously adjusts for changes in driving conditions and the way the car is used. If the car is driven around town, where gasoline mileage is relatively poor, the system will adjust to this mileage. However, if the car is generally used on the highway, where the gasoline mileage is better, the system will adjust to this type of operation. The system measures car speed, distance traveled, and the rate at which fuel is pumped from the fuel tank.

⚙ 3-12 Fuel pumps The fuel system uses a fuel pump to deliver fuel from the tank to the carburetor. There are two types of fuel pump, mechanical and electric. Electric fuel pumps are discussed in ⚙ 3-15 to 3-17. Mechanical fuel pumps are operated by an eccentric or special lobe on the engine camshaft, as explained below. The mechanical fuel pump is mounted on the side of the cylinder block in in-line engines (Fig. 3-19). In some V-type engines, the pump is mounted between the two cylinder banks. However, most V-type engines have the fuel pump on the side of the cylinder block, at the front of the engine (Fig. 3-20).

The mechanical fuel pump has a rocker arm whose end rests on the camshaft eccentric (Fig. 3-21). Many V-type engines use a push rod from the eccentric to the rocker arm (Fig. 3-20).

As the camshaft rotates, the eccentric rocks the rocker arm back and forth. The inner end of the rocker arm is linked to a flexible diaphragm. The diaphragm is clamped between the upper and lower pump housings. A spring over the diaphragm keeps tension on it. As the rocker arm rocks, it pulls the diaphragm up and then releases it. When the diaphragm is released, the spring then forces the diaphragm down. Therefore, the diaphragm moves up and down as long as the rocker arm rocks.

This diaphragm movement produces a partial vacuum and then a pressure in the space below the diaphragm. When the diaphragm moves up, a partial vacuum is produced. Then, atmospheric pressure, acting on the fuel in the tank, forces fuel through the fuel line and into the pump. The inlet valve in the pump opens to admit fuel, as shown by the arrows in Fig. 3-21.

When the diaphragm is released by the rocker arm, the spring forces the diaphragm down. This produces pressure in the space under the diaphragm. The pressure closes the inlet valve and opens the outlet valve. Now fuel is forced from the fuel pump through the fuel line to the carburetor, as shown by the arrows in Fig. 3-22.

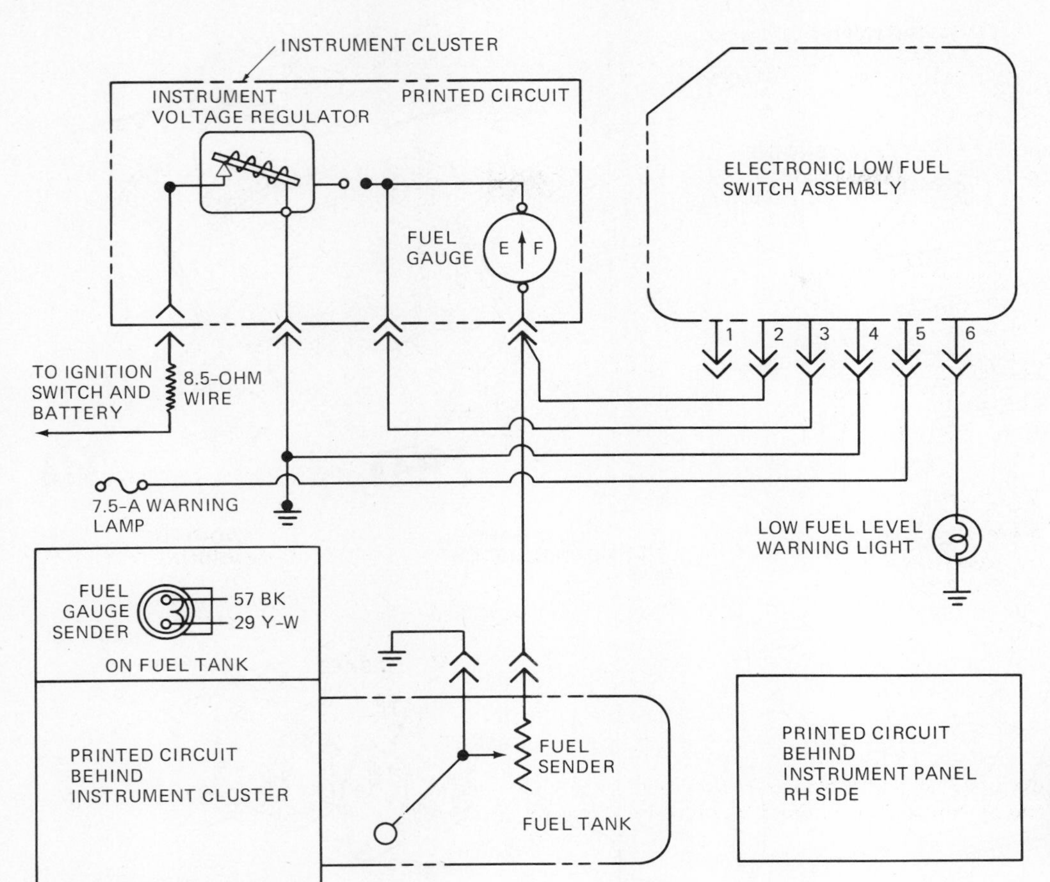

Fig. 3-16 Wiring diagram for the electronic low-fuel-level warning system. (*Ford Motor Company*)

The fuel pump works because atmospheric pressure attempts to push air toward any space where there is a vacuum. In the fuel system the only place the air can act is in the fuel tank. An air vent in the fuel tank admits atmospheric pressure. Therefore, the atmosphere pushes toward the vacuum in the fuel pump, pushing fuel ahead of it.

As the fuel is pushed toward the vacuum created by the upward movement of the diaphragm, the inlet valve is forced up off its seat. Therefore, fuel is pushed into the pump chamber (just below the diaphragm). Then, as the camshaft revolves, the high area of the eccentric moves away from the rocker arm. The rocker arm

Fig. 3-17 Fuel gauge and miles-to-empty indicator system. (*Ford Motor Company*)

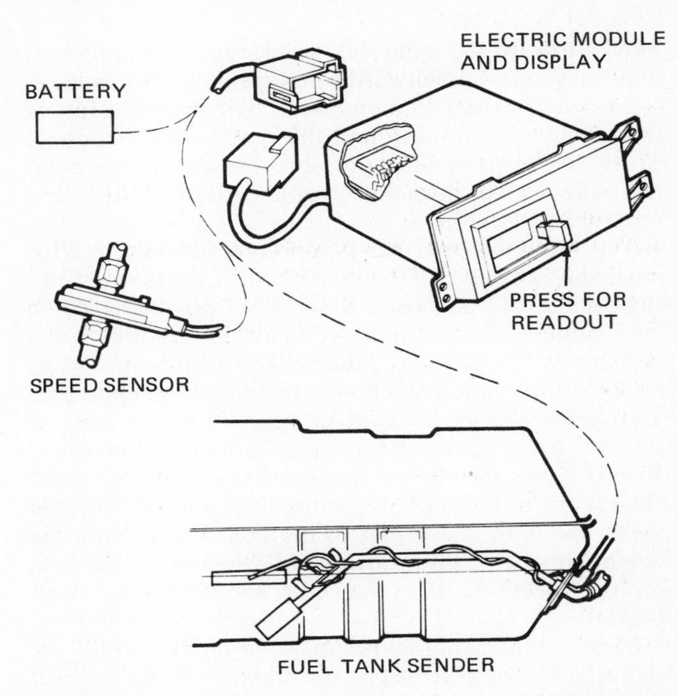

Fig. 3-18 Components in the miles-to-empty indicator system. (*Ford Motor Company*)

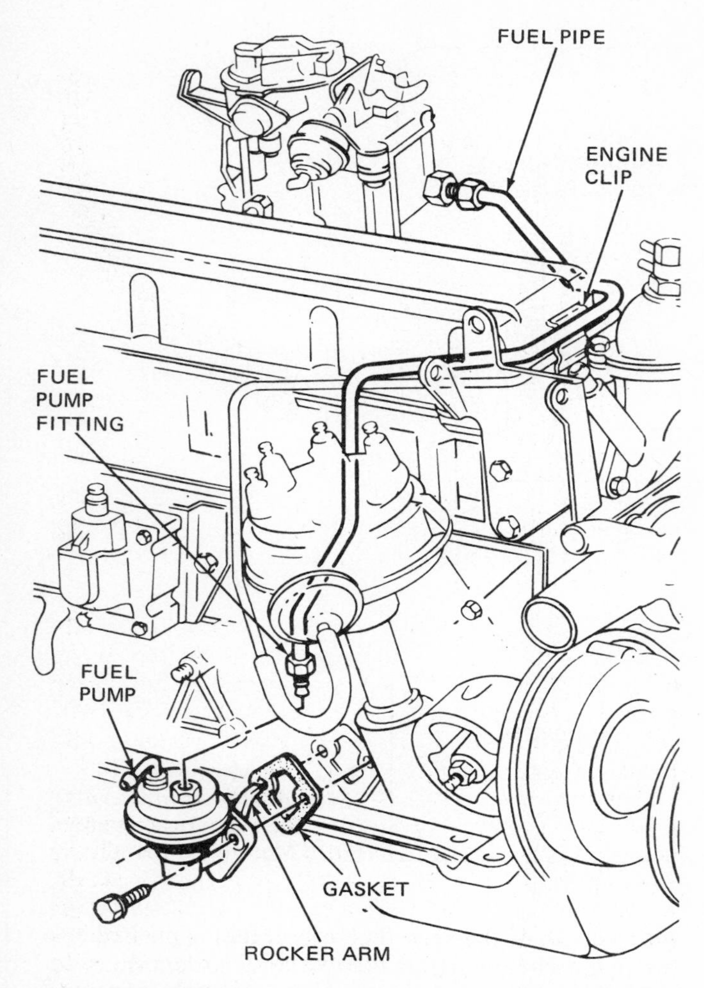

Fig. 3-19 Installation of a fuel pump on an in-line engine. (*Chevrolet Motor Division of General Motors Corporation*)

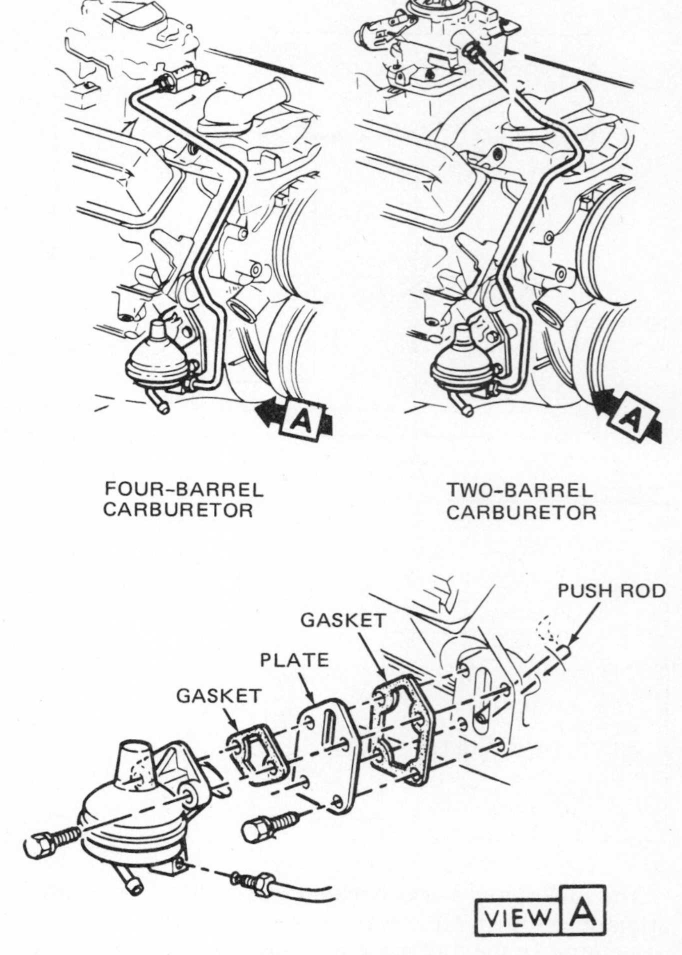

Fig. 3-20 Mounting of the fuel pump on a V-type engine which has a push rod between the fuel pump rocker arm and the camshaft eccentric. (*Chevrolet Motor Division of General Motors Corporation*)

moves toward the camshaft, releasing the diaphragm, which is pushed downward by its spring. This produces a pressure in the pump chamber. The pressure forces the inlet valve to close and the outlet valve to open. Then diaphragm-spring pressure forces fuel through the outlet valve, through the connecting fuel line, and into the carburetor.

The fuel from the fuel pump enters the carburetor past a needle valve in the float bowl. If the bowl is full, the needle valve closes so that no fuel can enter. When this happens, the fuel pump cannot deliver fuel to the carburetor. In this case, the rocker arm continues to rock within a slotted section of the diaphragm pull rod. The diaphragm remains at or near its upper limit of travel. Its spring cannot force the diaphragm downward so long as the carburetor float bowl will not accept fuel. However, as the carburetor uses up fuel, the needle valve opens to admit fuel to the float bowl. Now the diaphragm can move downward (on the rocker-arm return stroke) to force fuel into the carburetor float bowl.

Most mechanical fuel pumps used on cars today are serviced by complete replacement only. Fuel pumps of this type (shown in Figs. 3-19 and 3-20) are put together at the factory with crimped-over covers so that they cannot be disassembled. No service parts are available for these pumps. However, some fuel pumps are assembled with screws (Fig. 3-23). This type of fuel pump often can be rebuilt by the mechanic if new diaphragms and check valves are available.

❁ 3-13 Vapor-return line Many cars have a vapor-return line running from the fuel pump or the fuel filter to the fuel tank (see Fig. 3-4). Figure 3-22 shows the connection at the fuel pump (to the lower left) for the vapor return line. The purpose of this line is to return to the fuel tank any vapors that form in the fuel pump. The fuel pump can handle liquid only. Pumping stops if vapor forms in the fuel pump and is not removed.

Here's how vapor can form in the fuel pump. In operation, the pump alternately produces vacuum and pressure. During the vacuum phase, the boiling or vaporizing temperature of the fuel goes down. The lower the pressure, the lower the temperature at which any liquid vaporizes. For example, water boils at 212°F [100°C] at sea level. This is an atmospheric pressure of 14.7 psi [101 kPa]. But at 16,000 feet [4877 m] above

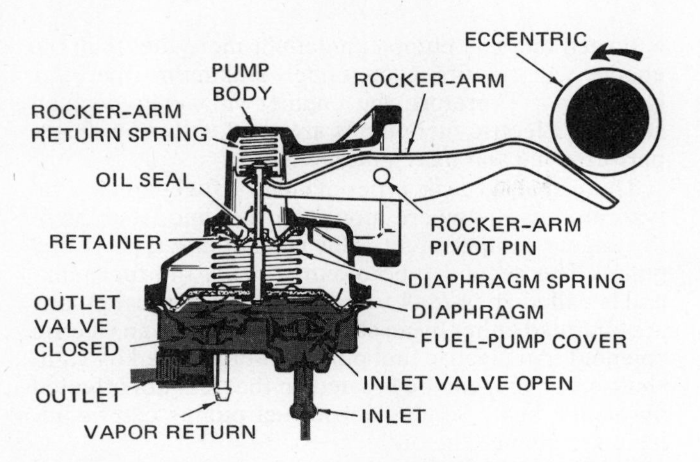

Fig. 3-21 When the eccentric rotates so as to push the rocker arm down, the arm pulls the diaphragm up. The inlet valve opens to admit fuel into the space under the diaphragm.

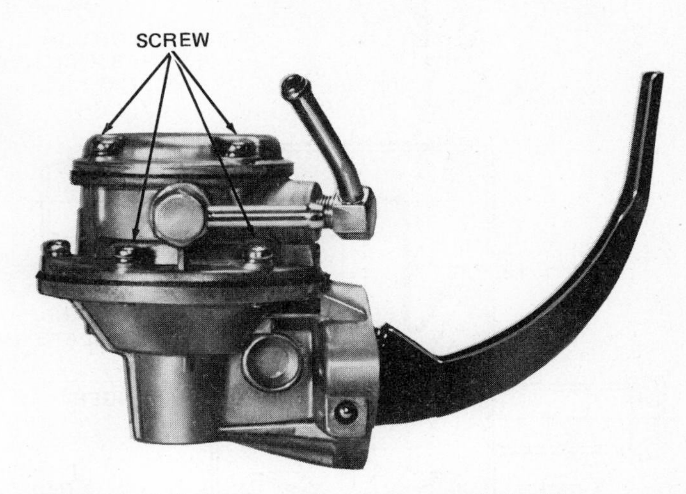

Fig. 3-23 A fuel pump that is assembled with screws, allowing diaphragm and check valve replacement. (*AC Spark Plug Division of General Motors Corporation*)

sea level, where the pressure is about 7 psi [48 kPa], water boils at 185°F [85°C].

The combination of increased temperature and partial vacuum in the fuel pump can cause fuel to vaporize. This produces a vapor lock, a condition that prevents normal delivery of fuel to the carburetor. When this happens, the engine stalls.

The vapor-return line is connected to a special outlet in the fuel pump (Fig. 3-22). It allows the vapor to return to the fuel tank. The vapor-return line also permits excess fuel being pumped by the fuel pump to return to the fuel tank. This excess fuel, in constant circulation, helps keep the fuel lines and pump cool to prevent vapor from forming.

Some cars have a vapor separator connected between the fuel pump and the carburetor (Figs. 1-21 and 3-4). It consists of a sealed can, a filter screen, an inlet and outlet fitting, and a metered orifice, or outlet, for the return line to the fuel tank. Any fuel vapor that the fuel pump produces enters the vapor separator as bubbles in the fuel. These bubbles rise to the top of the vapor separator. Then the vapor is forced by fuel-pump pressure to pass through the fuel-return line and back to the fuel tank. In the tank it condenses back into liquid fuel.

Some vapor-return lines have an in-line check valve (Fig. 3-24). This check valve prevents fuel from feeding back to the carburetor from the fuel tank, through the vapor-return line. If fuel does attempt to feed back, the pressure of the fuel forces the ball to seat, blocking the line. Normally, the pressure of the fuel vapor from the fuel pump unseats the ball and allows the fuel vapor to flow to the fuel tank.

NOTE: Fuel systems using a tank-mounted electric fuel pump do not require a vapor-return line. This type of pump maintains pressure on the fuel all the way from the tank to the carburetor float bowl. Since there is no vacuum in the line anywhere, there is little chance for vapor lock to form.

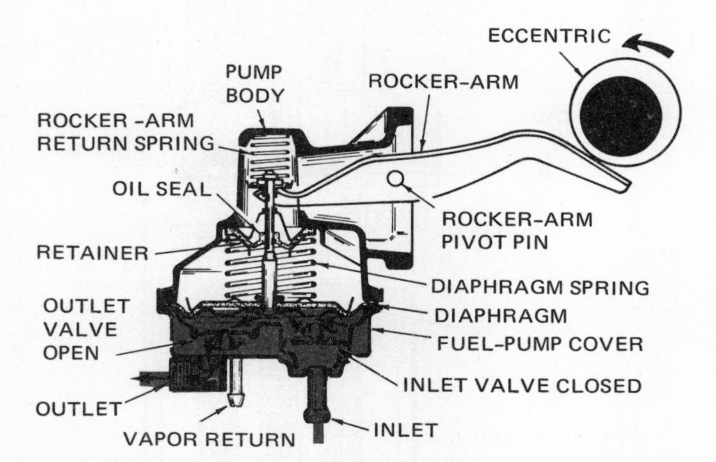

Fig. 3-22 When the eccentric rotates so as to allow the rocker arm to move up under it, the diaphragm is released so that it can move down, producing pressure under it. This pressure closes the inlet valve and opens the outlet valve so that fuel flows to the carburetor.

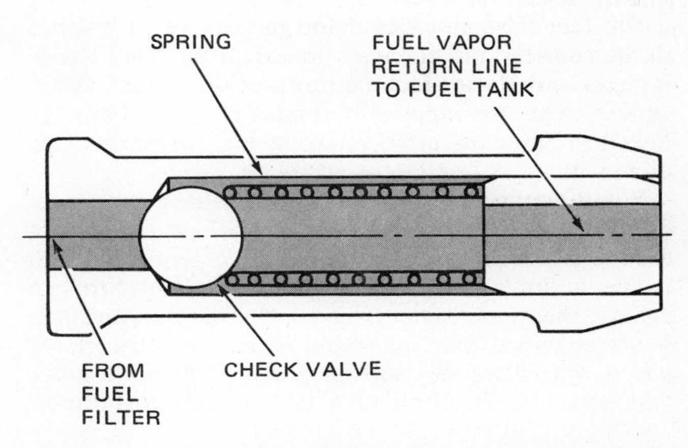

Fig. 3-24 In-line check valve which is located between the fuel filter and the fuel tank. (*American Motors Corporation*)

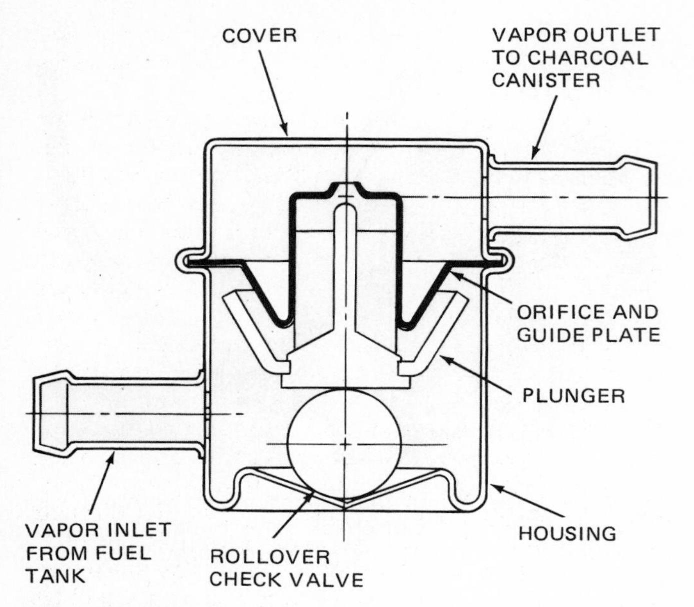

Fig. 3-25 Rollover check valve in vapor line between the fuel tank and the charcoal canister. (*American Motors Corporation*)

3-14 Fuel-vapor emission control system

Gasoline vapor can escape from the fuel tanks and carburetors of older cars. This happens when the car is parked with the engine off. At other times, when the engine is running, gasoline is being pumped from the fuel tank and carburetor, so gasoline vapor is not likely to escape. Large quantities of gasoline vapor (HC) can escape from parked cars without vapor controls and can add to atmospheric pollution.

To prevent this loss of gasoline vapor, some 1970 and later model cars are equipped with fuel-vapor emission control systems. Figure 3-4 shows one system. The fuel tank is sealed. Escaping gasoline vapor has to flow through the vapor recovery line to the charcoal canister. Gasoline vapor from the carburetor float bowl also flows to the charcoal canister. There, the charcoal particles pick up the gasoline vapor and hold it. Then, when the engine is started and runs, air flows through the charcoal canister on the way to the carburetor. This air picks up the gasoline vapor trapped in the canister and carries it to the carburetor. There, it mixes with the air-fuel mixture and enters the engine. In this way, the vapor is burned, instead of being allowed to enter the atmosphere as HC (unburned gasoline vapor).

Some vapor-emission control lines have a rollover check valve (Fig. 3-25). This valve blocks the fuel vapor recovery line in case of a rollover accident. If the car is upside down, the ball will force the plunger to seat against the guide plate. This blocks the line and prevents gasoline from flowing out of the tank through the line to the charcoal canister or carburetor. On some cars, the rollover check valve is part of the carburetor.

3-15 Electric fuel pumps

Electric fuel pumps have certain advantages over mechanical fuel pumps. The fuel system is filled with fuel as soon as the ignition is turned on. The pump can deliver more fuel than the engine will require even under maximum operating conditions. Therefore the engine will never be fuel-starved. Electric fuel pumps are used with many high-performance and heavy-duty engines.

There are two basic types of electric fuel pumps. One type mounts anywhere along the fuel line, such as in the engine compartment. This is an *in-line* type of fuel pump. The second type mounts inside the fuel tank, and is called an *in-tank* fuel pump. Electric fuel pumps are operated either by an electric motor or by an electric solenoid. An electric fuel pump is not affected by camshaft wear, and can be located so that it is not affected by engine heat. Some electric fuel pumps can be adjusted to change the outlet pressure.

3-16 In-line electric fuel pump

An electric solenoid fuel pump is shown in sectional view in Fig. 3-26. The solenoid operates a bellows. The bellows does the same job as the diaphragm in the mechanical fuel pump. As the diaphragm expands or collapses, it pulls fuel in or forces it out. The expansion or contraction is produced by an electromagnet which is repeatedly connected to or disconnected from the battery.

The electromagnet becomes connected to the battery when the ignition switch is turned on. When this happens, the electromagnet is energized. This draws the electromagnet armature downward so that the bellows expands and produces a vacuum that causes fuel to pass from the fuel tank to the fuel pump. The inlet valve opens, permitting the fuel to enter. As the armature reaches the lower limit of its travel, it opens contact points which open the circuit to the battery. Therefore,

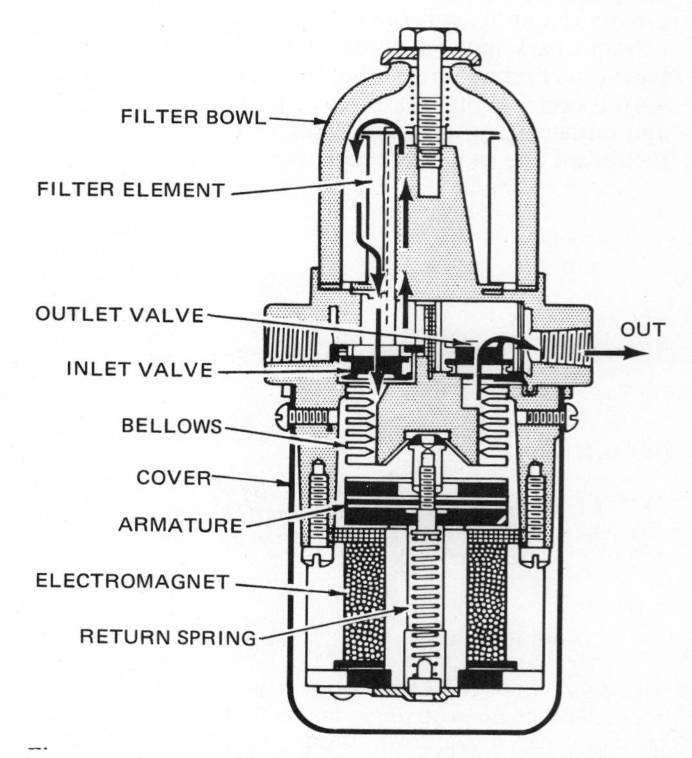

Fig. 3-26 Sectional view of an electric fuel pump. The bellows and electromagnet are in the lower part of the pump.

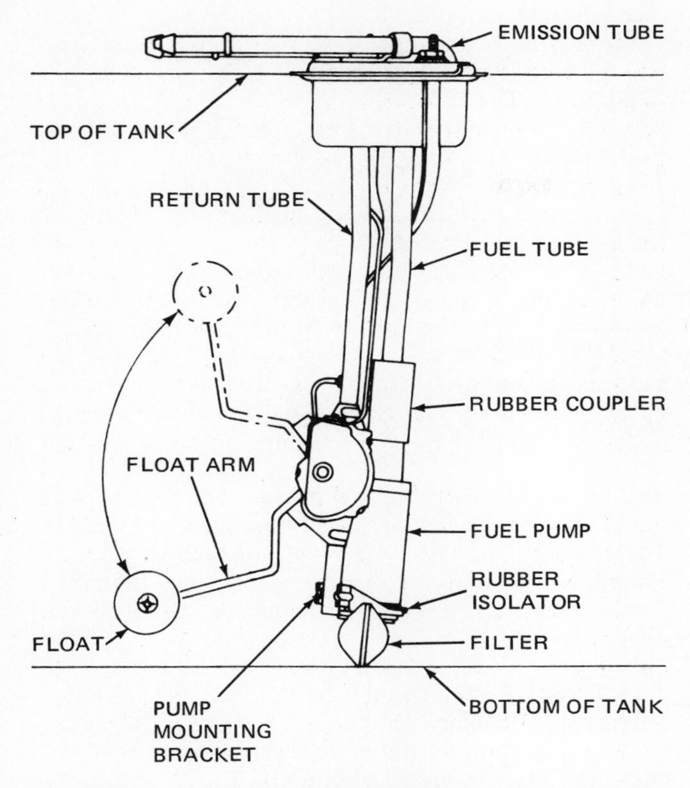

Fig. 3-27 The tank-type electric fuel pump is combined with the same support as the fuel-gauge tank unit. (*Cadillac Motor Car Division of General Motors Corporation*)

the electromagnet becomes disconnected and deenergized.

Now the pump return spring pushes upward on the armature and bellows, collapsing the bellows and forcing the fuel in the bellows out through the outlet valve. As this happens, the inlet valve is forced closed. The fuel passes from the fuel pump to the carburetor or fuel-injection system. As soon as the armature reaches the upper limit of its travel, it closes the contact points. The electromagnet is reconnected to the battery, and the above cycle is repeated. This action continues as long as the ignition system is turned on.

The frequency with which the delivery stroke of the armature is repeated depends upon the amount of fuel the engine requires. When the engine is using little fuel, the return spring collapses the bellows slowly. This is because the needle valve in the carburetor float bowl is preventing rapid delivery of fuel. But, when larger amounts of fuel are required, the bellows collapses more rapidly. Therefore the delivery stroke is repeated more often, keeping the fuel system supplied with fuel. These actions are repeated as long as the ignition is on.

❂ 3-17 In-tank electric fuel pump One type of tank-mounted electric fuel pump is shown in Fig. 3-27. This pump has an electric motor that drives a small impeller. As the impeller spins, it sends fuel through the outlet pipe and fuel line to the carburetor or fuel-injection system. An advantage of having the fuel pump in the fuel tank is that there is pressure on the fuel all the way from the tank to the engine. Fuel pumps mounted in the fuel line between the tank and the engine depend on partial vacuum for their operation. The partial vacuum can cause the gasoline to vaporize in the fuel pump under certain conditions (❂ 3-13). This can produce vapor lock, a condition that prevents delivery of fuel to the engine. Another in-tank fuel pump is shown in Fig. 3-28.

The complete wiring system for an electric fuel pump is shown in Fig. 3-29. In this system, the electric pump is located inside the pedestal filter (Fig. 3-30). The wiring circuit (Fig. 3-29) connects the fuel pump to the battery during starting through contacts in the starting-motor relay. This enables the fuel pump to start delivering fuel as soon as the starting motor begins to crank the engine. Then the circuit maintains the connection between the battery and the fuel pump through the oil pressure switch. The purpose of this is to shut off the fuel pump whenever the engine stops and the oil pressure drops.

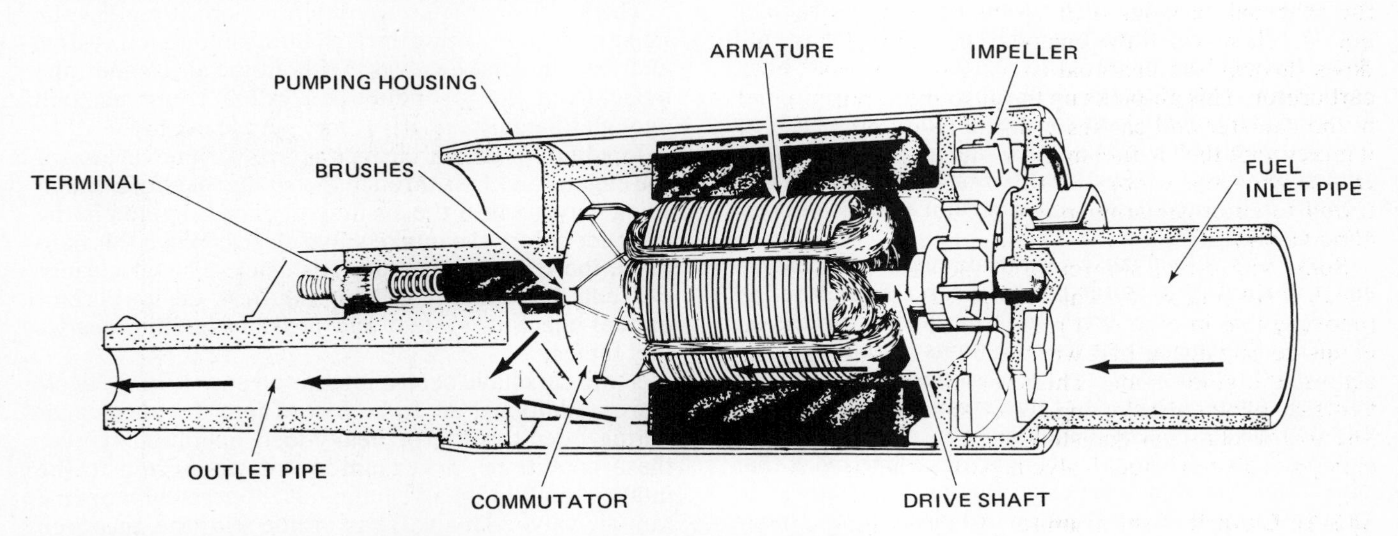

Fig. 3-28 Cutaway view of a tank-mounted electric fuel pump. (*Buick Motor Division of General Motors Corporation*)

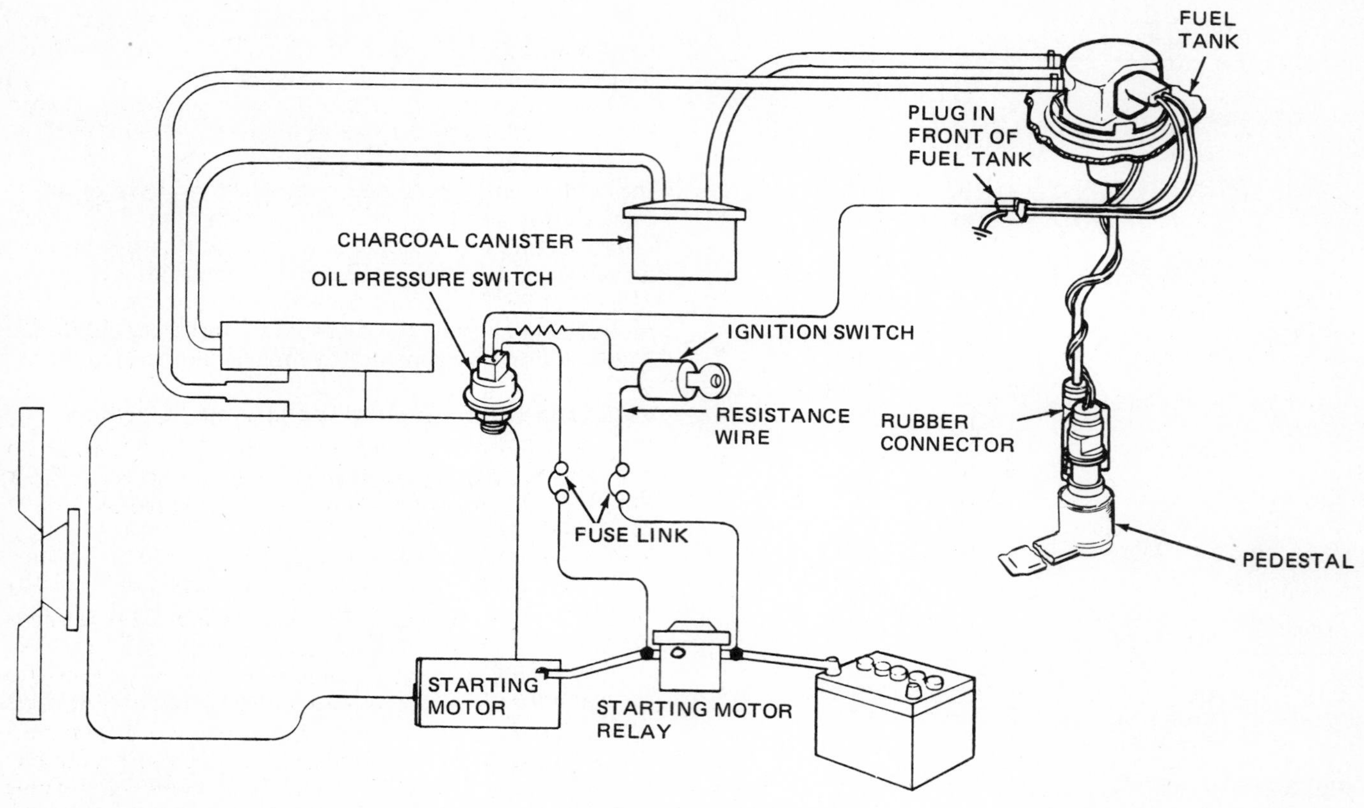

Fig. 3-29 Electric fuel-pump wiring diagram. (*Ford Motor Company*)

In addition, the wiring system for many late-model cars includes an inertia switch in the electric fuel-pump circuit. If the car rolls over, the inertia switch opens. This shuts off any fuel flow from the pump and reduces the possibility of fire.

3-18 Air cleaner A great deal of air passes through an engine when it is operating. The fuel is mixed with air, and the mixture passes on into the engine cylinders, where it is ignited and burns. During normal running of the engine, the carburetor supplies a mixture ratio of about 15:1 or 15 pounds [6.80 kg] of air for each 1 pound [0.45 kg] of gasoline. Each gallon [3.8 liters (L)] of gasoline requires 1200 cubic feet [34 m^3] of air for normal combustion in the engine. As much as 100,000 cubic feet [2832 m^3] of air may pass through the engine every 1000 miles [1609 km]. This is a large volume of air, and it contains a certain amount of floating dust and grit.

Dirt and grit can cause serious damage to the engine parts if allowed to enter the cylinders. Therefore, an air cleaner is used to filter these particles from the air (Fig. 3-31). The air cleaner contains a ring of filter material through which the air must pass. Air filters, or filter elements, are made of fine-mesh metal threads or ribbons, pleated paper, cellulose fiber, or polyurethane (plastic foam). This material provides a fine maze that filters out the dust particles. Figure 3-32 shows, in cutaway view, the construction of a pleated-paper filter element. This type is used on many cars.

An oil-bath air cleaner is shown in Fig. 3-33. The oil-bath air cleaner is for use on engines and equipment operated in very dusty conditions, such as in construction equipment and on off-road vehicles. Figure 3-33 shows a disassembled view of an oil-bath air cleaner. It contains a reservoir of oil past which the air flows. The moving air picks up particles of oil and carries them into the filter. There the oil washes accumulated dust and dirt back down into the oil reservoir. In addition to this washing action, the oiliness of the filter material improves the filtering action.

The air cleaner has a second function. It muffles the noise of the air rushing through the air-induction system and past the intake valves. Without the air cleaner, the noise from the air-induction system could be loud enough to annoy the driver and people nearby.

In addition, the air cleaner acts as a flame arrester if the engine backfires through the intake manifold. Backfiring may occur if the air-fuel mixture is ignited in the cylinder before the intake valve closes. When this happens, there is a momentary flashback. The air cleaner prevents the flame from erupting from the carburetor or fuel-injection system and possibly igniting nearby fuel fumes.

Some cars have used a ram-air cleaner (Fig. 3-34). It allows additional air to be forced into the air cleaner during open-throttle or heavy-load operation. Under these conditions, a vacuum motor, connected to the intake manifold by a vacuum hose, operates to open a ram-air valve. This valve is in line with the air scoop on the engine hood. When the valve opens, extra air from the air scoop is forced into the carburetor. This

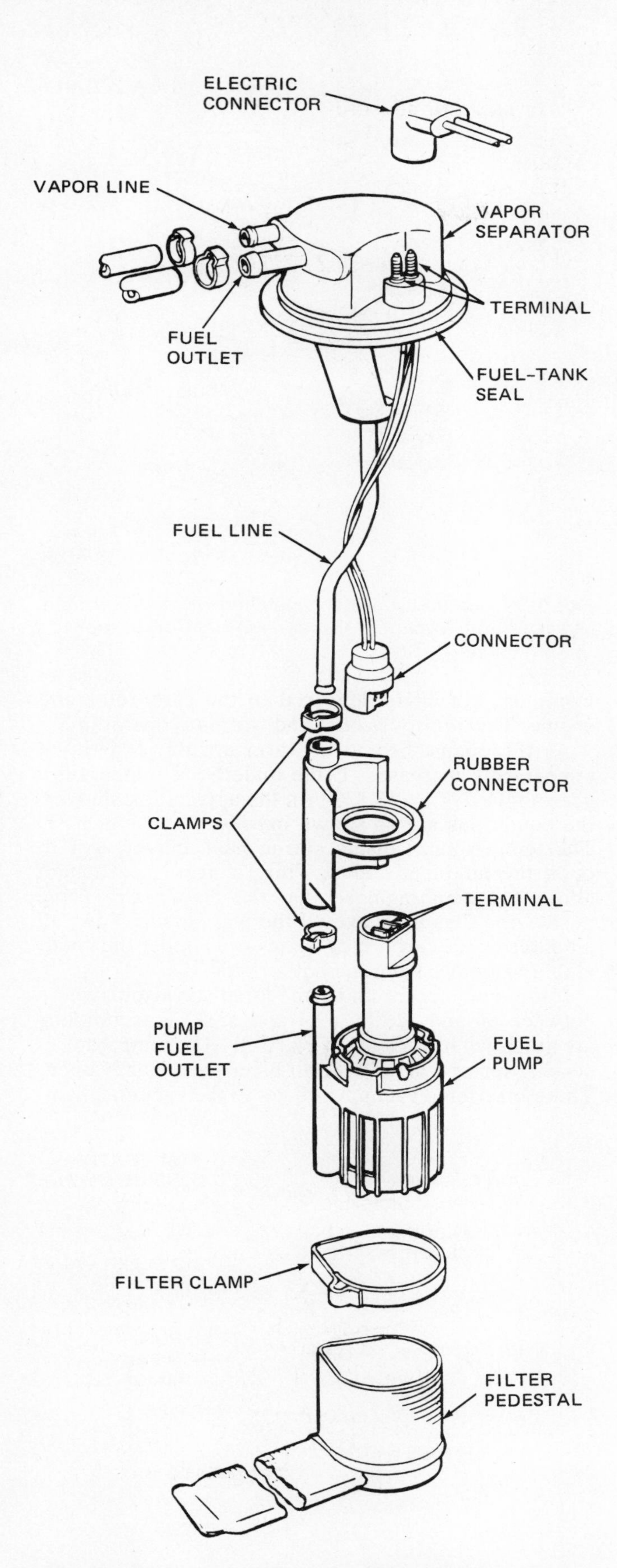

Fig. 3-30 Electric fuel pump and related parts shown removed from the fuel tank. (*Ford Motor Company*)

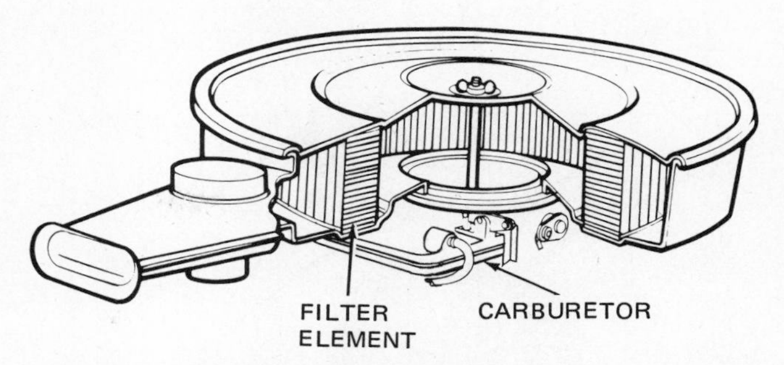

Fig. 3-31 Typical air cleaner, partly cut away to show the filter element. (*Chrysler Corporation*)

improves engine performance under these conditions. At other times, the manifold vacuum is high enough to keep the ram-air valve closed. Then air enters the filter through the snorkel tube or duct-and-valve assembly in the normal manner.

⚙ 3-19 Heated-air system with thermostatic air cleaner The heated-air system which uses a thermostatic air cleaner is installed on most cars today. It is one part of the exhaust emission-control system. To reduce engine emissions from the tail pipe, carburetors are adjusted to provide leaner mixtures at idle and part throttle. These leaner mixtures ensure a more complete burning of the fuel. This means there is less HC coming out of the tail pipe.

However, these lean mixtures can reduce engine performance when the engine is cold. To correct this, a thermostatically controlled air cleaner is used. It is also called the heated-air system (HAS) because it injects heat into the air going into the carburetor during cold weather when the engine is cold (Fig. 3-35). This improves engine performance after a cold start and during

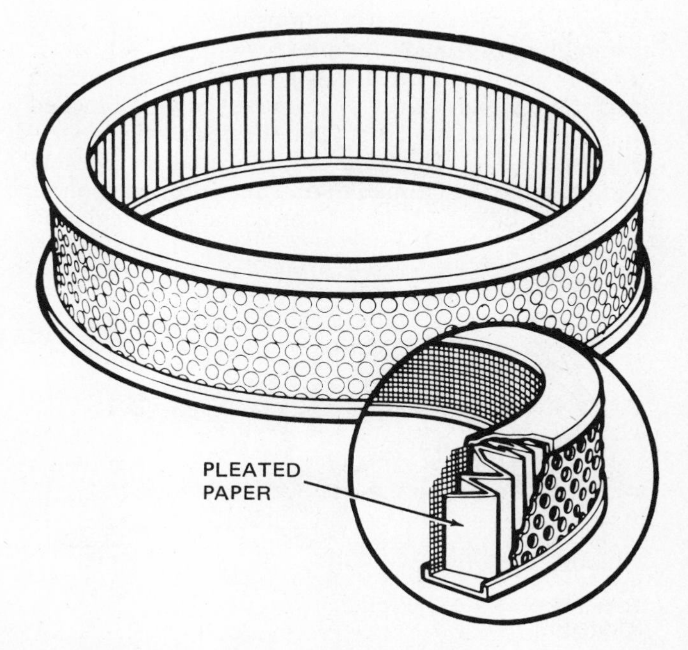

Fig. 3-32 Construction of a pleated-paper type of filter element. (*Chrysler Corporation*)

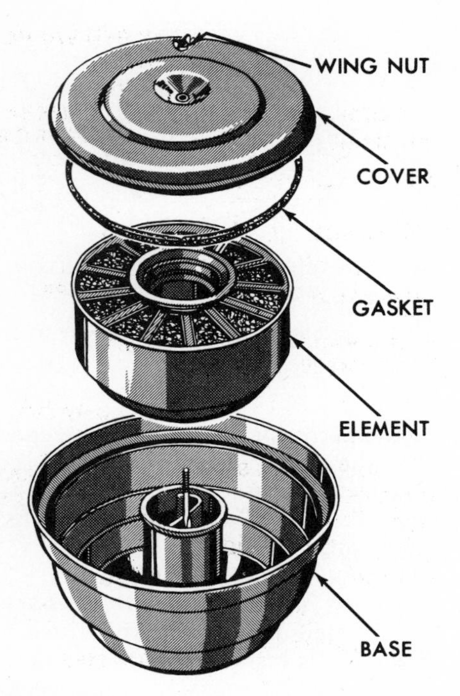

Fig. 3-33 Typical oil-bath air cleaner. (*Chevrolet Motor Division of General Motors Corporation*)

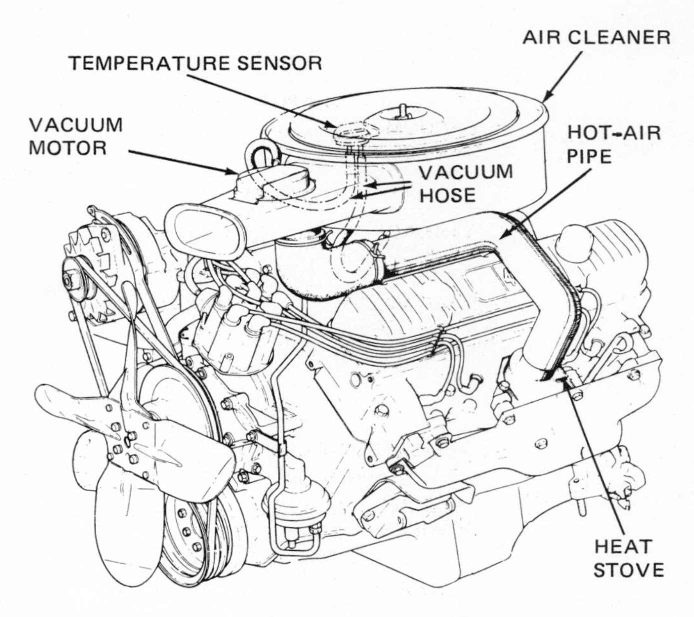

Fig. 3-35 Heated-air system installed on a V-8 engine. (*Buick Motor Division of General Motors Corporation*)

engine warm-up. Then leaner mixtures can be used to reduce emissions without affecting cold-engine performance.

One air cleaner of this type is shown in Fig. 3-36. It contains a sensing spring which reacts to the temperature of the air entering the carburetor through the air cleaner. This spring controls an air-bleed valve (Fig. 3-37). When the entering air is cold, the sensing spring holds the bleed valve closed. Now, intake manifold vacuum is applied to the vacuum chamber. The diaphragm is pushed upward by atmospheric pressure and the diaphragm spring is compressed.

In this position, the linkage from the diaphragm raises the damper valve. The snorkel tube is blocked off. All air now has to enter from the hot-air pipe [view (*b*) in Fig. 3-37]. This pipe is connected to the heat stove on the exhaust manifold. Therefore, as soon as the engine starts and the exhaust manifold begins to warm up, hot air is delivered to the carburetor and engine. This improves cold and warm-up operation.

As the engine begins to warm up, the underhood temperature increases. If the underhood temperature goes above 128°F [53.3°C] (in the application shown), the conditions are as shown in view (*c*) in Fig. 3-37. The temperature-sensing spring has bent enough to open the air bleed valve. This reduces the vacuum above the diaphragm so that the diaphragm spring pushes the damper valve all the way down. Now, all air entering the carburetor comes from under the hood, and none comes from the hot-air pipe.

If the temperature under the hood stays somewhere between 85 and 128°F [29.4 and 53.3°C], conditions are as shown in view (*d*) in Fig. 3-37. The temperature-sensing spring will hold the air bleed valve partly open. Therefore, some vacuum will get to the vacuum cham-

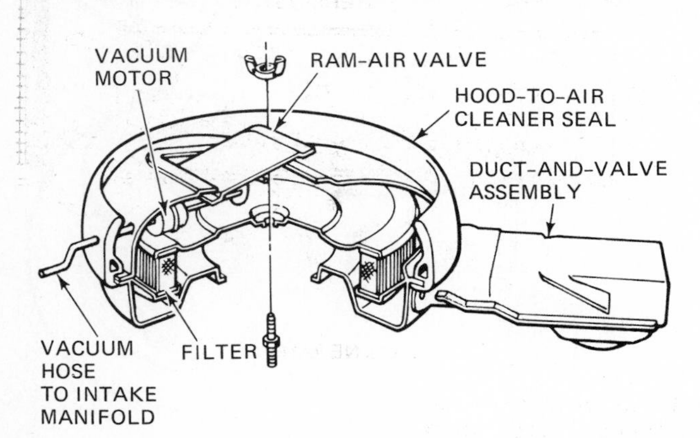

Fig. 3-34 Ram-air type of air cleaner, with the ram-air valve shown open. (*Ford Motor Company*)

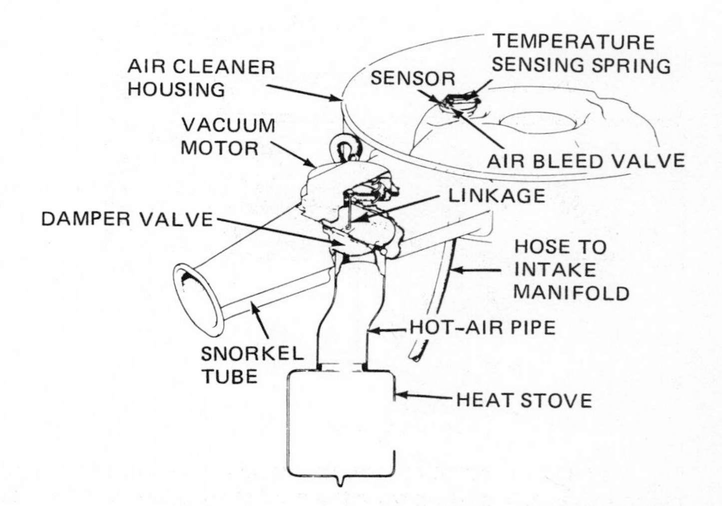

Fig. 3-36 Air cleaner with a thermostatic control. (*Chevrolet Motor Division of General Motors Corporation*)

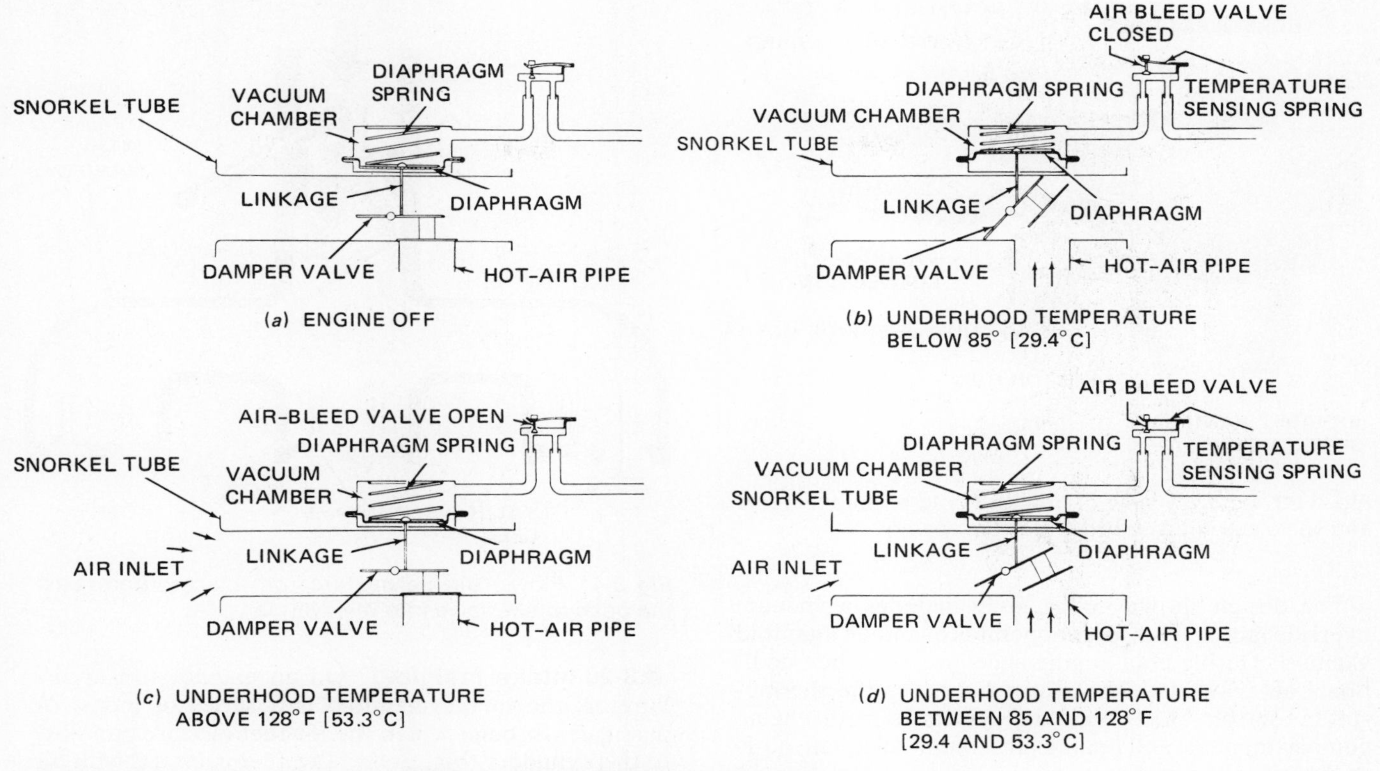

Fig. 3-37 Four modes of operation of the thermostatic air cleaner. (*Chevrolet Motor Division of General Motors Corporation*)

ber above the diaphragm. This vacuum holds the damper valve partly open. In this position, some air enters from under the hood. Some air also comes up through the hot-air pipe from the heat stove around the exhaust manifold.

A similar thermostatically controlled air cleaner is shown in Fig. 3-38. This design has a thermostatic bulb that acts directly on the valve. When the engine is cold, the thermostatic bulb positions the valve as shown in Fig. 3-38, left. All ingoing air must come from the hot-air duct, which is connected to a shroud around the exhaust manifold called a heat stove. As the engine warms up, the hotter air from the shroud causes the thermostatic bulb to move the valve. Some air begins to enter from the engine compartment. With further increases in temperature, the valve moves further so that more air enters from the engine compartment. When the engine compartment becomes hot, then most or all of the ingoing air comes from the engine compartment.

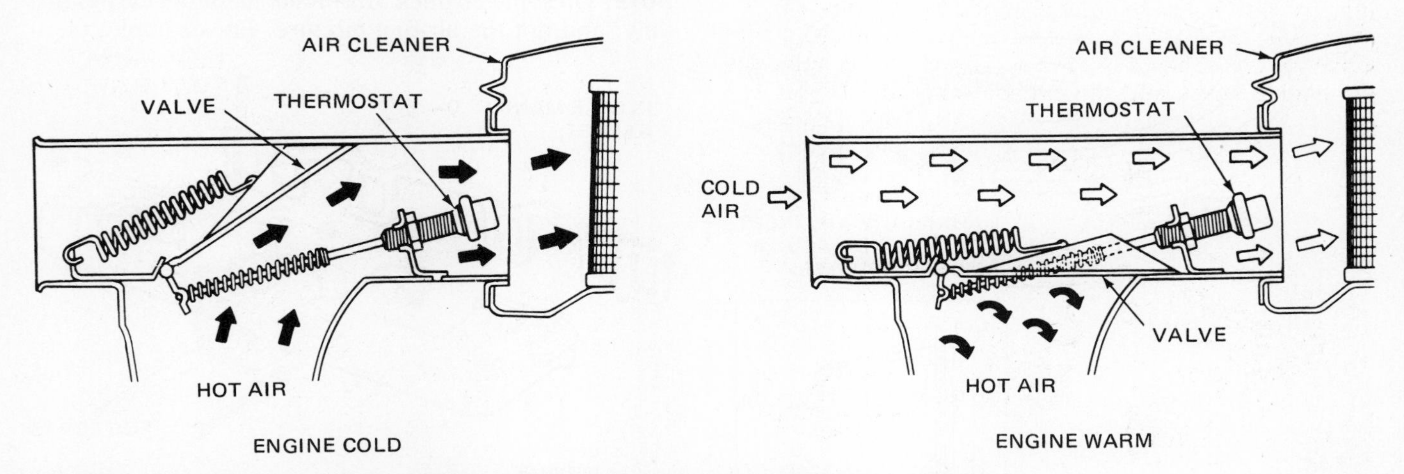

Fig. 3-38 Operation of the thermostatic air cleaner; left, when the engine is cold and heated air is being taken from the heat stove around the exhaust manifold; right, when the engine is warm and cooler air is being taken in from under the hood. (*Ford Motor Company*)

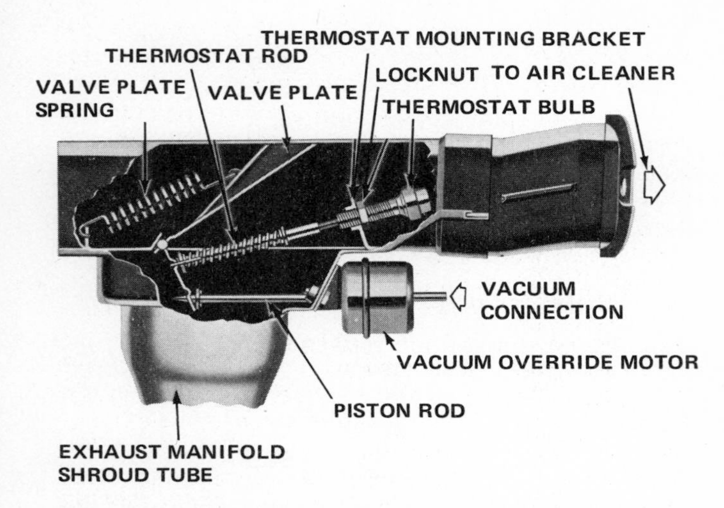

Fig. 3-39 Cutaway view of hot- and cold-air intake ducts and valve assembly. (*Ford Motor Company*)

The design shown in Fig. 3-39 includes a vacuum override motor. This motor operates on intake manifold vacuum. During cold-engine acceleration, when additional air is needed, the motor overrides the thermostatic control. This opens the system to both engine compartment air and heated air, so that adequate air is delivered to the carburetor.

Some air cleaners have a vacuum motor that opens and closes an auxiliary air passage in the air-cleaner housing. This type of air cleaner is shown in Fig. 3-40. The vacuum motor operates if a partial vacuum develops in the air cleaner. During cold-weather acceleration, not enough air may get through the heat stove to satisfy engine requirements. When this happens, the partial vacuum in the air cleaner operates the vacuum motor. It then opens the auxiliary air-inlet passage. Now, the extra air needed by the engine can enter the air cleaner.

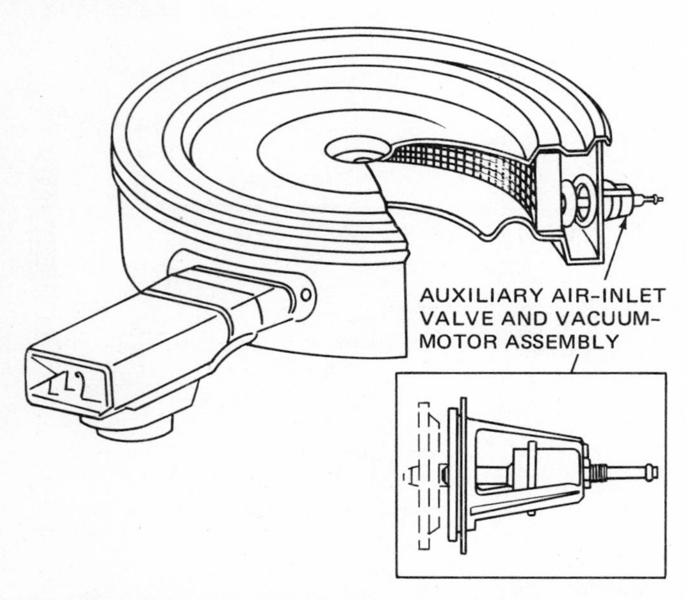

Fig. 3-40 Air cleaner with auxiliary air inlet valve and vacuum motor. (*Ford Motor Company*)

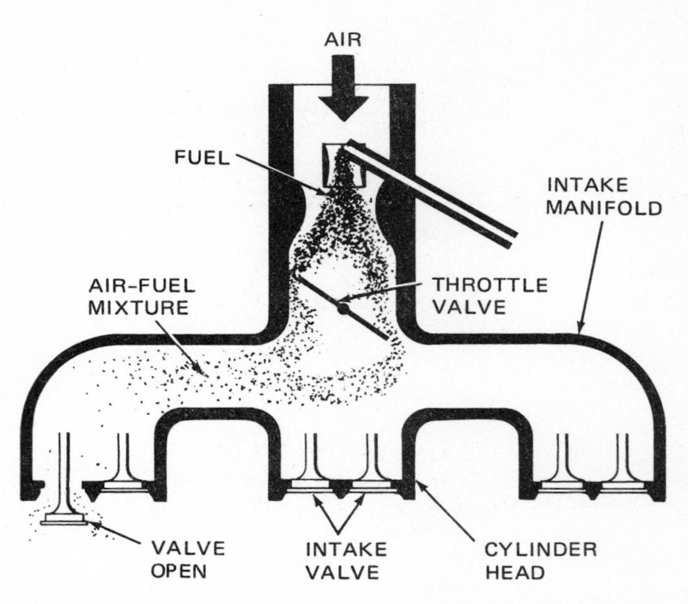

Fig. 3-41 Flow of air-fuel mixture from the carburetor past the open intake valve into the cylinder.

⚙3-20 Intake manifold On an engine with a carburetor, the intake manifold is a series of pipes, or passages, through which the air-fuel mixture can flow to the cylinders (Fig. 3-41). The fuel is mixed with air in the carburetor to form the combustible air-fuel mixture. Then this mixture flows through the intake manifold to the intake valve that is open.

Figure 3-42 shows an intake manifold for a six-cylinder engine. The intake manifold is designed to avoid sharp corners and to make the passages to the intake valves as short and as straight as possible. Also, the walls of the passages are smooth. Sharp corners and rough surfaces tend to obstruct the flow of the air-fuel mixture. The size and shape of the passages are designed to supply all cylinders with equal amounts of air-fuel mixture. There is more information on intake manifolds in ⚙4-5.

NOTE: On some engines, the intake manifold carries air only, and not the air-fuel mixture. For example, on a

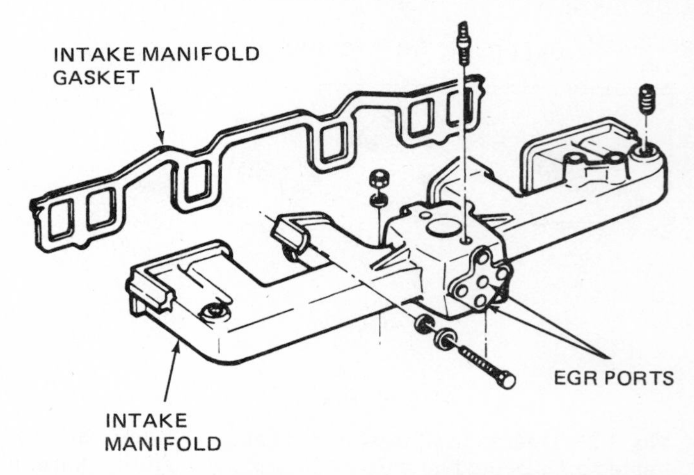

Fig. 3-42 Intake manifold for an in-line six-cylinder engine. (*American Motors Corporation*)

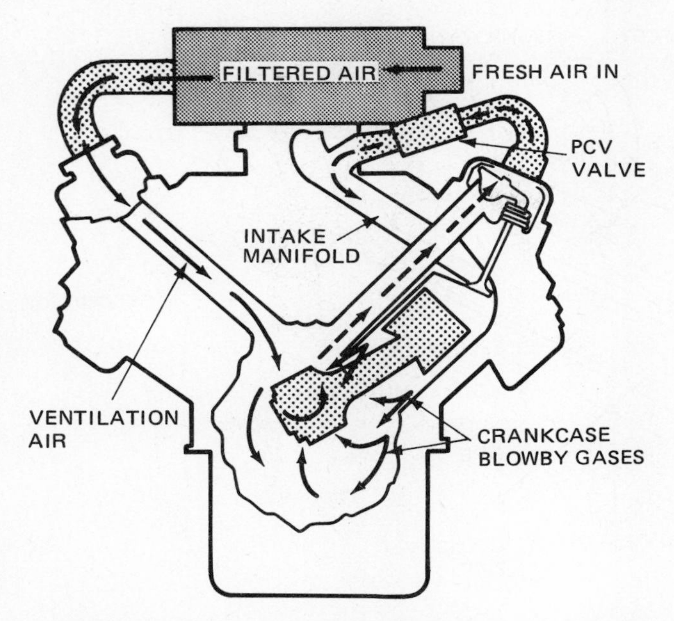

Fig. 3-43 Positive crankcase ventilating system on a V-type engine. This system permits no crankcase emissions to enter the atmosphere. (*Motor Vehicle Manufacturers Association*)

gasoline engine that has fuel injection into the intake ports, only air flows through the intake manifold. Also, on diesel engines, the intake manifold carries only air.

⚙ 3-21 Crankcase ventilation Crankcases must be ventilated. Air must flow through the crankcase to remove the blowby gases. These include the unburned and partly burned gases that get past, or blow by, the pistons and piston rings. Also entering the crankcase is any liquid fuel that seeps down past the pistons and rings, especially when the engine is cold. Water also condenses on the cylinder walls when the engine is cold. All of these by-products of combustion must be removed from the crankcase. If they are not, they can cause rusting, corrosion, and formation of gum or varnish that prevent normal engine operation.

In earlier engines, the crankcase was ventilated by an opening at the front of the engine and a vent tube at the back. The forward motion of the car, plus the rotation of the crankshaft, caused air to flow through and remove the blowby with the water and fuel vapors. However, today closed positive crankcase ventilating (PCV) systems are used on all car engines including diesels (Fig. 3-43). The air circulates through the crankcase, as before, but then enters the intake manifold and engine cylinders. Any blowby, water vapor, and fuel vapor are sent back through the engine to be burned instead of being emitted into the air.

⚙ 3-22 Exhaust system After the air-fuel mixture has been burned in the engine, the burned gases leave the cylinders when the exhaust valves open during the exhaust strokes of the pistons. The burned gases pass into the exhaust manifold and from there into the exhaust pipe, catalytic converter, muffler, resonator (on some cars), and the tail pipe (Fig. 3-44).

On V-type engines, there are two exhaust manifolds, one on each bank. The exhaust manifolds are mounted on the outsides of the cylinder banks. On some cars, each exhaust manifold connects to a separate exhaust pipe. However, on most cars, the two exhaust pipes

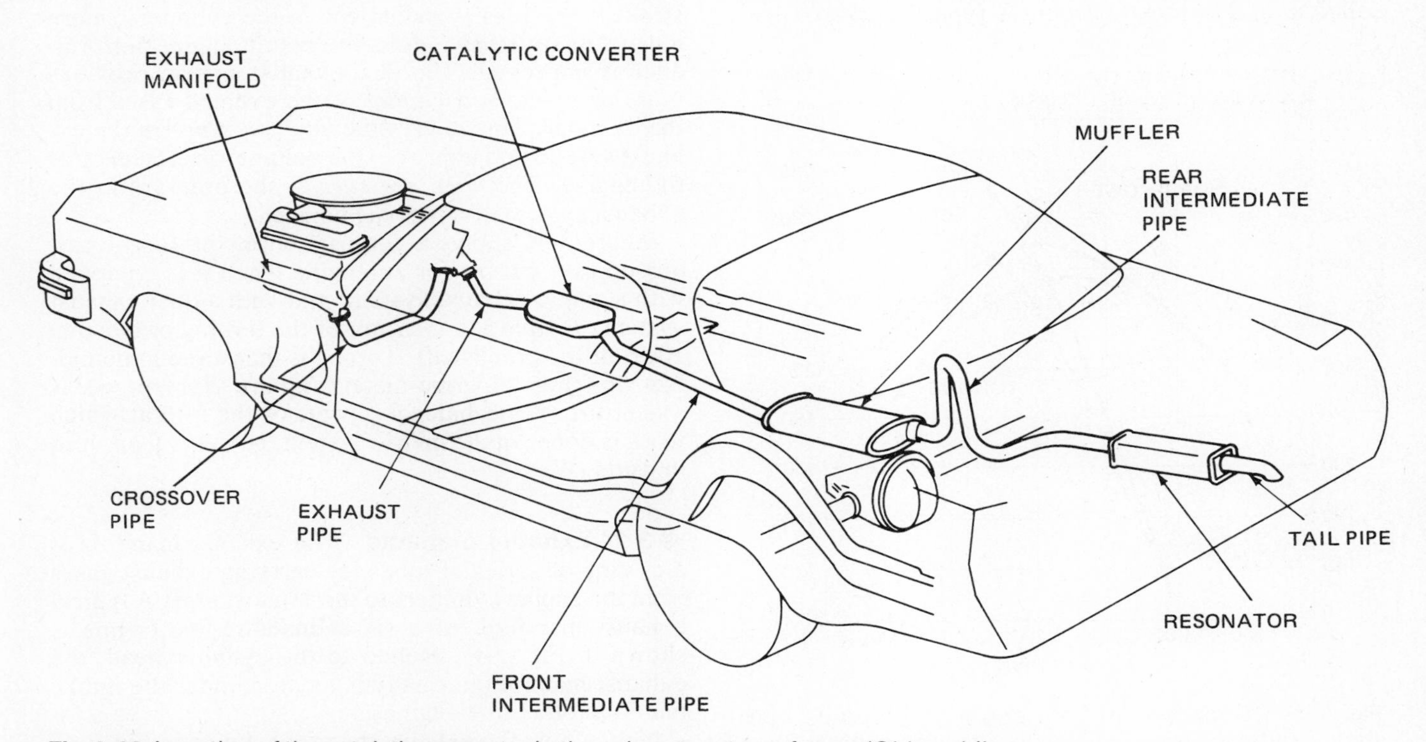

Fig. 3-44 Location of the catalytic converter in the exhaust system of a car. (*Oldsmobile Division of General Motors Corporation*)

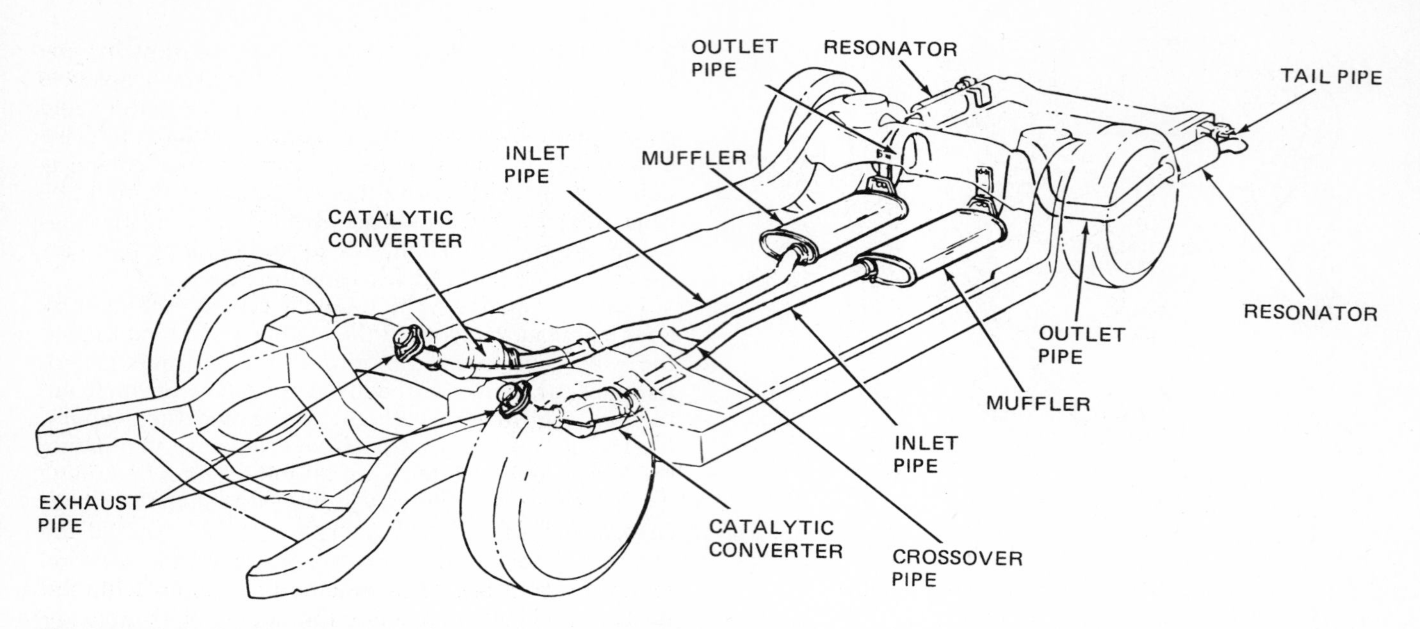

Fig. 3-45 A dual exhaust system for a V-type engine. Each bank of cylinders has its own exhaust system. (*Ford Motor Company*)

are connected together by a crossover pipe (Fig. 3-44). The crossover pipe may be connected to a single catalytic converter, muffler, and tail pipe. Then exhaust gases from the two manifolds combine and exhaust through the same muffler. Other exhaust systems have two mufflers and tail pipes connected to a single catalytic converter.

❁ 3-23 Dual-exhaust systems Some engines use two separate exhaust systems. This arrangement is called a *dual-exhaust system*. A typical dual-exhaust system for a V-type engine is shown in Fig. 3-45. Each exhaust manifold is connected to a separate exhaust pipe. Then each exhaust pipe is connected to its own catalytic converter, muffler, and tail pipe. This provides, in effect, a separate exhaust system for each bank of cylinders.

The use of two separate exhaust systems improves the ability of the engine to "breathe." This means that the engine may exhaust more freely. Less exhaust gas remains in the cylinders at the ends of the exhaust strokes. With less exhaust gas in the cylinders, more air-fuel mixture can enter. As a result, engine performance is improved. The dual-exhaust system provides more complete *scavenging* of the exhaust gases from the cylinder. This lowers the *back pressure* in the exhaust system and improves the volumetric efficiency of the engine. The *back pressure* is the pressure in the exhaust system of a running engine.

Figure 3-46 shows a comparison of the torque and horsepower curves for an engine when it is equipped with a single-exhaust system, and with a dual-exhaust system. *Torque* is a measure of the turning or twisting effort of the crankshaft. Torque is measured in pound-feet (lb-ft) or kilogram-meters (kg-m). Horsepower is a measure of mechanical power, or the rate at which work is done. In the metric system, power is measured in watts (W).

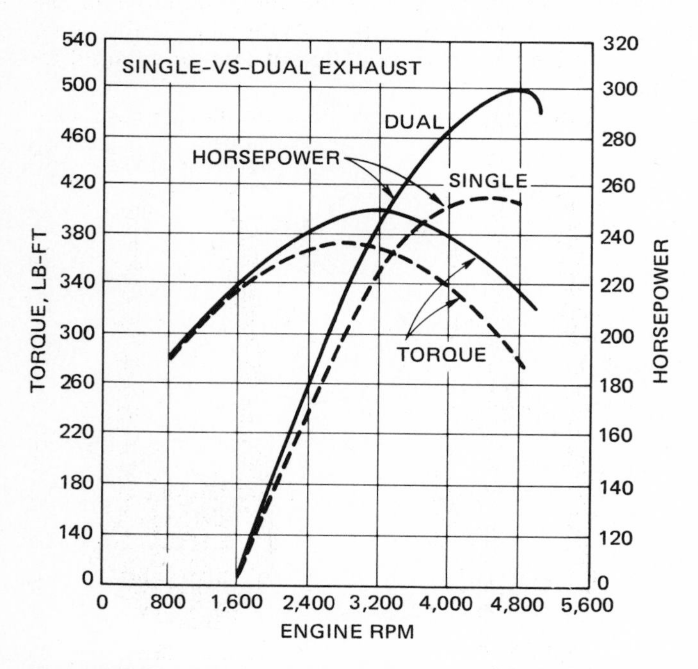

Fig. 3-46 Comparison of horsepower and torque of an engine with single and dual exhausts. (*Oldsmobile Division of General Motors Corporation*)

❁ 3-24 Exhaust manifold The exhaust manifold is a casting or series of tubes for carrying exhaust gases from the engine cylinders to the exhaust pipe. A typical exhaust manifold for a six-cylinder in-line engine is shown in Fig. 3-47. Bolted to the cylinder head, the exhaust manifold normally is located under the intake manifold on in-line engines.

Notice how the exhaust manifold shown in Fig. 3-47 bolts under the intake manifold shown in Fig. 3-42.

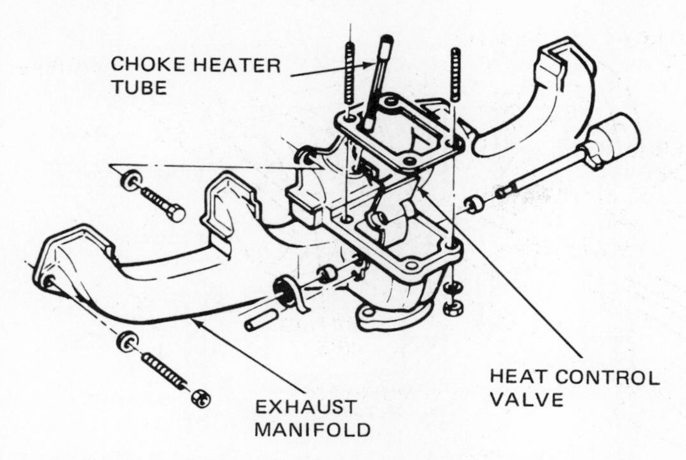

Fig. 3-47 Exhaust manifold for an in-line six-cylinder engine. (*American Motors Corporation*)

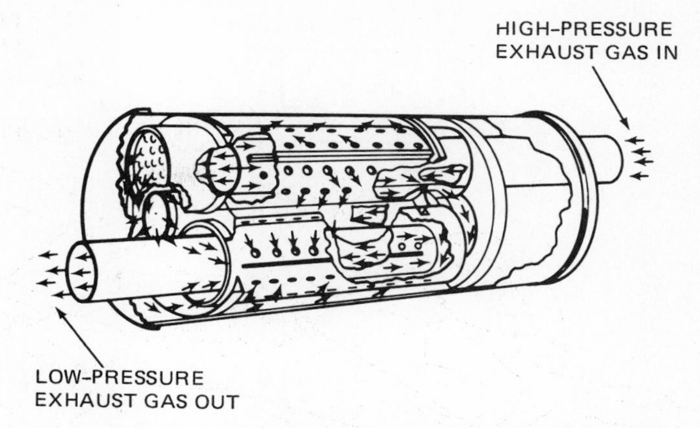

Fig. 3-49 Exhaust muffler in cutaway view. The arrows show the path of exhaust-gas flow through the muffler. (*Chevrolet Motor Division of General Motors Corporation*)

There is a connection between the two manifolds. The purpose of this connection is to supply heat to the intake manifold (from the hot exhaust gases) when the engine is first started. This improves vaporization of the gasoline entering the engine through the intake manifold. As a result, cold-engine operation is improved. The amount of heat directed around the intake manifold is determined by the position of the manifold *heat control valve* (⚙4-36).

⚙3-25 Muffler The muffler (Fig. 3-48) is located under the car body. It is connected into the exhaust system between the exhaust pipe or catalytic converter and the tail pipe (Fig. 3-44). A muffler is designed to quiet the noise of the engine exhaust by gradually reducing the pressure of the exhaust gases as they leave the engine cylinders. To do this, a muffler usually contains a series of holes, passages, and resonance chambers to absorb and damp out the high-pressure surges that occur when the exhaust valves open. The inner construction of a typical muffler and the flow path of the exhaust gases through it are shown in Fig. 3-49.

A muffler alone cannot reduce the exhaust noise to the required levels on all engines. Some exhaust systems include a smaller muffler-like device called a *resonator* (Fig. 3-44). It is mounted in series with the muffler, and helps to quiet the sound of the engine exhaust.

To further reduce exhaust noise, some exhaust pipes are made of laminated pipe (Fig. 3-50). A two-ply laminated pipe is simply one layer of pipe inside the other. Three-ply laminated pipe consists of a layer of plastic sandwiched between the two metal layers. Either combination acts to damp out exhaust-pipe "ring." This noise occurs in some exhaust systems, and cannot always be damped out by a muffler.

⚙3-26 Exhaust emissions Exhaust gases contain a variety of gaseous compounds, including unburned hydrocarbons (HC), carbon monoxide (CO), and nitrogen oxides (NO_x). These compounds are emitted from the tail pipe of a running engine and pollute the air we breathe. Because air pollution is a threat to health, laws limit the amount of each pollutant that an engine can emit.

To reduce exhaust emissions, automotive manufac-

Fig. 3-48 A typical muffler, located under the body at the rear of the car. (*Chrysler Corporation*)

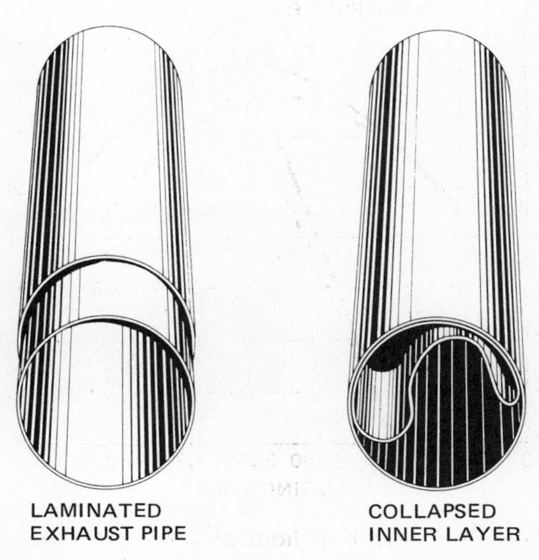

Fig. 3-50 Collapse of the inner layer of a laminated exhaust pipe will prevent normal intake and exhaust. (*ATW*)

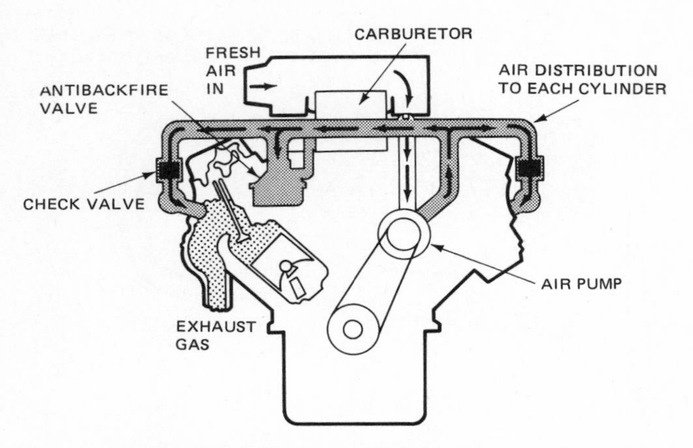

Fig. 3-51 Air-injection reactor or air pump system of exhaust emission control. With the use of the air pump, hydrocarbons and carbon monoxide in the exhaust gas are burned harmlessly in the engine exhaust manifold. (*Motor Vehicle Manufacturers Association*)

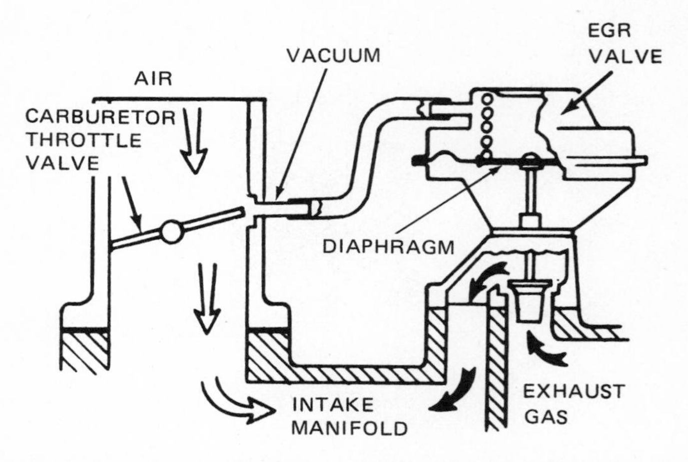

Fig. 3-52 Schematic drawing of an exhaust gas recirculation system. (*Chevrolet Motor Division of General Motors Corporation*)

turers have developed methods of converting most of these harmful compounds into harmless compounds. To do this, engines are equipped with emission control systems and devices. These include the use of air injection into the exhaust manifold (⚙ 3-27), exhaust gas recirculation (⚙ 3-28), and catalytic converters (⚙ 3-29).

⚙ 3-27 Air injection Many exhaust manifolds have air-injection systems. They include an air pump and a series of injection tubes in the exhaust manifold. In operation, the air pump sends a flow of air into the exhaust manifold, opposite the exhaust valves (Fig. 3-51). The oxygen in this extra air helps to burn any HC or CO remaining in the exhaust gas.

Some cars have a *pulse air* type of air injection system. In this system, the pulses of exhaust gas leaving the cylinder operate an air pump that injects air into the exhaust system.

⚙ 3-28 Exhaust gas recirculation The higher the combustion temperature, the more nitrogen oxides (NO_x) form during the combustion process. In the presence of sunlight, nitrogen oxides from the exhaust gas combine with hydrocarbons (HC) in the air. They form the type of haze known as photochemical *smog*. To help prevent smog, changes have been made in engines and fuel systems that reduce the amount of NO_x in the exhaust gas.

One method is to use an exhaust gas recirculation (EGR) system (Fig. 3-52). It sends a small part of the exhaust gas back through the intake manifold and into the cylinders. The exhaust gas reduces combustion temperatures. This, in turn, reduces the amount of NO_x coming out of the tail pipe. The system is operated by an EGR valve that is controlled by intake-manifold vacuum. During part-throttle operation, the vacuum causes a diaphragm in the valve to raise and open the exhaust-gas passage from the exhaust manifold. This allows some of the exhaust gas to enter the intake manifold.

Another way to get exhaust gas into the combustion chamber is to use longer valve overlap on the engine camshaft. Then the exhaust and intake valves are open

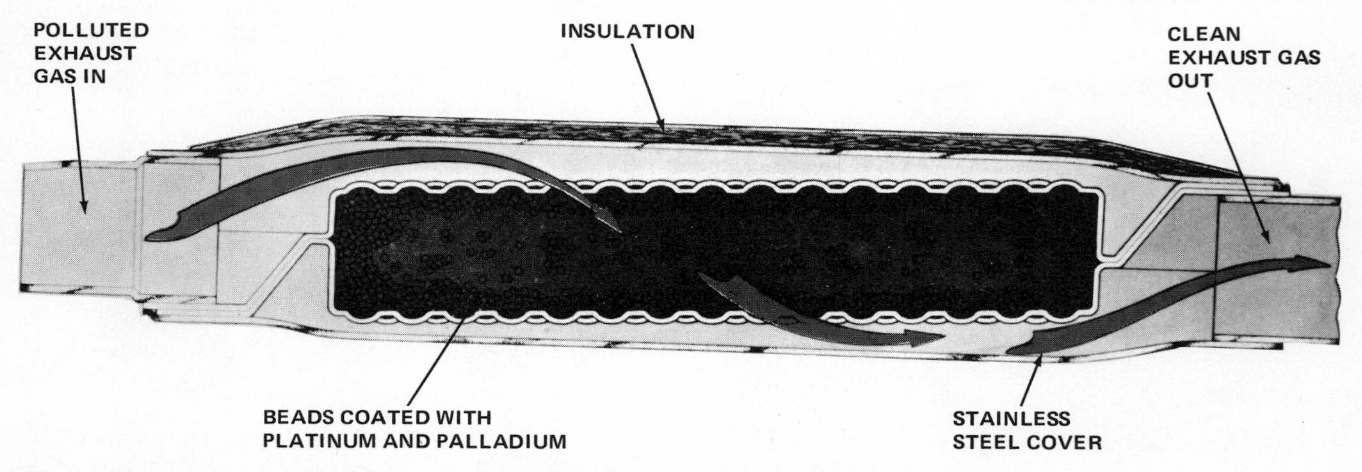

Fig. 3-53 A pellet type of catalytic converter. The flow of exhaust gas through the converter is shown by the arrows. (*General Motors Corporation*)

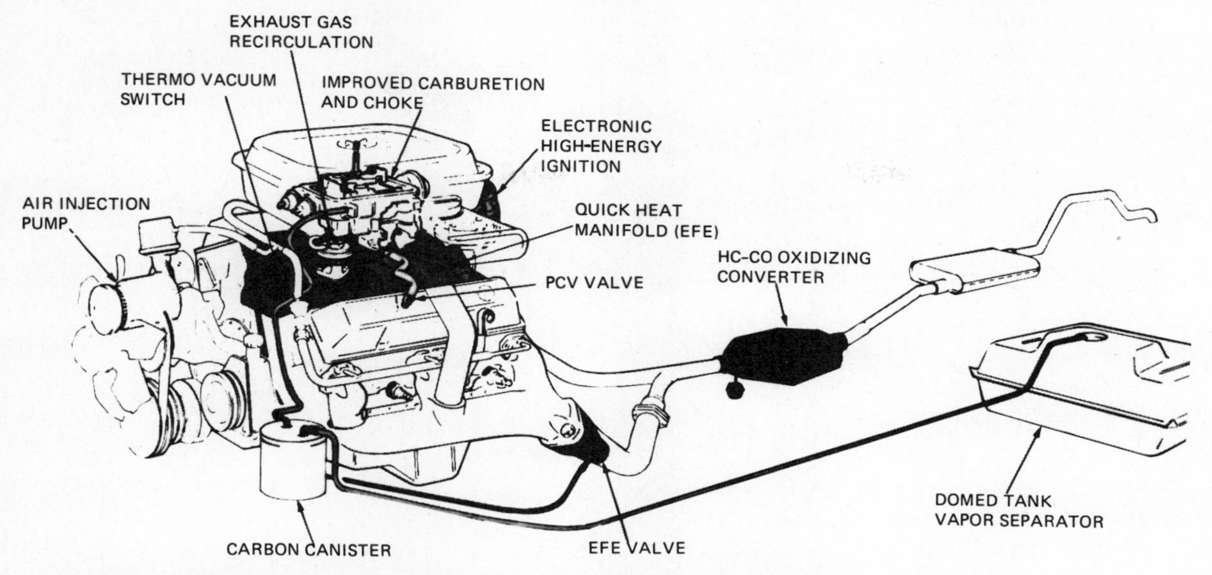

Fig. 3-54 An engine showing all of the emission control systems and devices used on it, including a catalytic converter and air injection. (*General Motors Corporation*)

longer at the same time. This allows more of the exhaust gas to remain in the cylinders and mix with the incoming air-fuel mixture. As a result, the exhaust gas lowers the combustion temperature. Less NO_x is formed during the combustion process.

❁ 3-29 Catalytic converter Catalytic converters provide another way to reduce the emissions in the exhaust gas (Figs. 3-44 and 3-45). A catalyst is a material that causes a chemical reaction without actually becoming a part of the reaction process. The metal platinum can act as a catalyst. When exhaust gas and air are passed through platinum-coated pellets (Fig. 3-53) or a platinum-coated honeycomb core, the HC and CO react with the oxygen in the air. Harmless water (H_2O) and carbon dioxide (CO_2) are formed. Reducing exhaust emissions with a catalytic converter is a chemical process. It is not a burning process, such as the process in the exhaust manifold of the air-injection system (❁ 3-27).

Figure 3-54 shows an engine using a catalytic converter, air injection, and other emission control features. Figure 3-53 is a cutaway view of a catalytic converter that uses pellets, or beads. As the exhaust gases pass through this bed of beads, the platinum changes the HC and CO into water and carbon dioxide.

Some cars may have a catalytic converter system that handles HC, CO, and also nitrogen oxides (Fig. 3-55). This is called a *dual catalytic converter*, or a *three-way catalyst*. The front converter is coated with the metals rhodium and platinum. This combination handles the NO_x and partly handles the HC and CO. Then the partly treated exhaust gas flows into the rear converter. As it does so, it mixes with the air being pumped in from the air pump. This puts more oxygen into the exhaust gas so the two-way catalyst can take care of the CO and HC.

One of the problems with the catalysts is that the lead in leaded gasoline soon coats the catalysts. When this happens, the catalysts are no longer effective. If

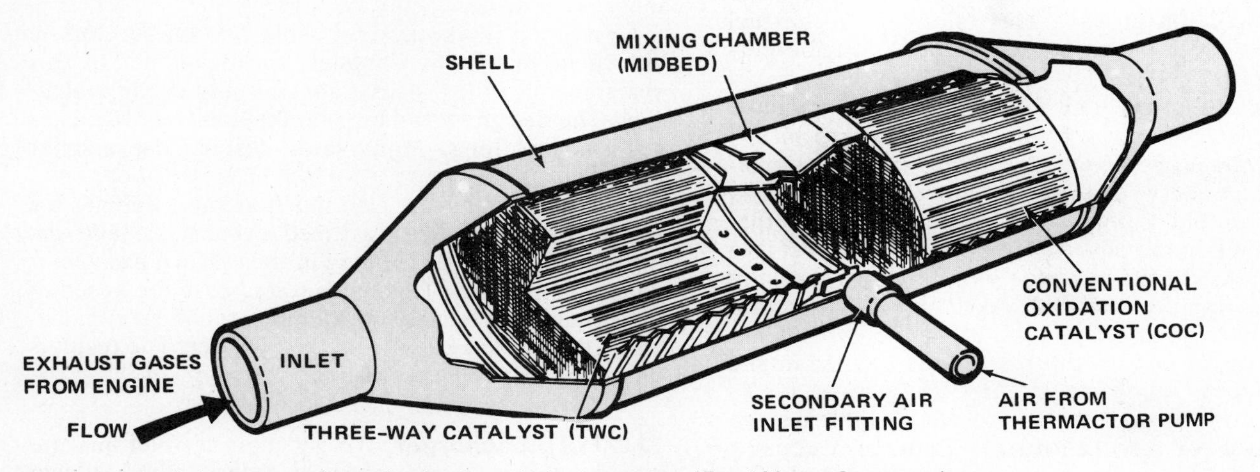

Fig. 3-55 Partial cutaway view of a dual catalytic converter. (*Ford Motor Company*)

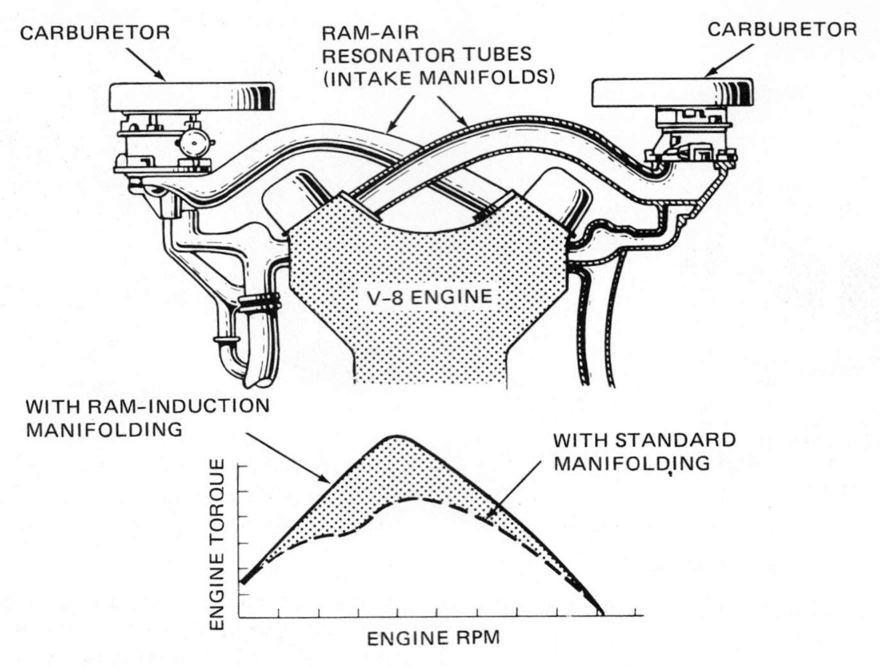

Fig. 3-56 Tuned or ram-induction system, using long intake manifold passages called *resonator tubes*. (*Chrysler Corporation*)

an engine runs on leaded gasoline, the catalysts stop doing their job in a very short time. This is one reason that cars with a catalytic converter must use unleaded gasoline.

NOTE: Removing the lead from gasoline can have a bad effect on engine valves. The lead acts as a lubricant for the exhaust-valve faces and seats. A thin coat of lead covers the valve faces and seats. This helps to reduce wear. To prevent valve troubles, manufacturers are hardening the valve seats and coating the valve faces with special metallic compounds. This overcomes the problems brought on by the loss of lead in the gasoline.

⚙ 3-30 Tuned intake and exhaust systems For improved engine breathing at high speed, valves are made as large as possible and the passages in the intake and exhaust manifolds are arranged so as to permit maximum airflow. This allows more air-fuel mixture to enter the cylinders for better engine performance.

One way to improve air intake is to tune the intake and exhaust systems. "Tuning" in this case means designing the proper length and size of the intake-manifold passages between the carburetor and intake ports. Intake manifold tuning works something like pipe-organ tuning. In the pipes of a pipe organ, air is set into vibration. As the air in a pipe vibrates, high-pressure waves pass rapidly up and down the pipe. This action produces the sound. (Sound is nothing more than high-pressure waves passing through air.) In a tuned intake manifold, the incoming air-fuel mixture is set to vibrating, not to produce sound but to increase the momentum of the mixture. This produces a "supercharging" or ramming effect. Under ideal conditions, a high-pressure wave of air-fuel mixture reaches an intake valve port just as the valve opens.

The high-pressure waves in the mixture are initiated by the sound of the intake valve closing. The waves pass back and forth in the tube, or manifold branch. If the branch is of the correct length, the waves will resonate, or pass back and forth, with little loss of energy. Under ideal conditions, when the valve opens again the high-pressure wave will hit the valve port at just the right instant to produce the ramming effect.

A manifold cannot be tuned for effective action at all speeds. The velocity of the sound waves through the mixture has little variation. However, the time intervals between valve closing and valve opening will vary greatly at different speeds. As a rule, the intake manifold is tuned to be in phase when the engine is operating near or at top speed. This is the time when volumetric efficiency begins to drop off and the ramming effect is most needed.

One tuned intake system using two carburetors is shown in Fig. 3-56. Chrysler, the developer of this system, called the process *ram charging* or *ram induction*. The design shown uses intake branches 36 inches [914.4 mm] long. Many later designs use shorter branches.

Exhaust systems are also tuned in many engines for the same reason. With a tuned exhaust system, the high-pressure waves forming in the exhaust gas give it added momentum. This tends to increase the speed of the exhaust gas and ensures a more complete exhausting of the burned gases from the cylinder. The result is improved engine performance.

⚙ 3-31 Turbocharger To get more air-fuel mixture into the engine, a *supercharger,* or *turbocharger,* can

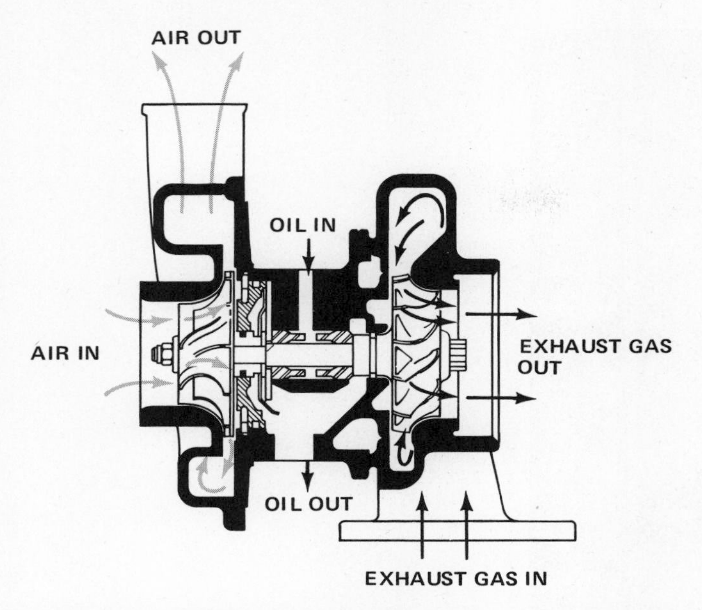

Fig. 3-57 Operation of a turbocharger. (*Schwitzer Division of Wallace-Murray Corporation*)

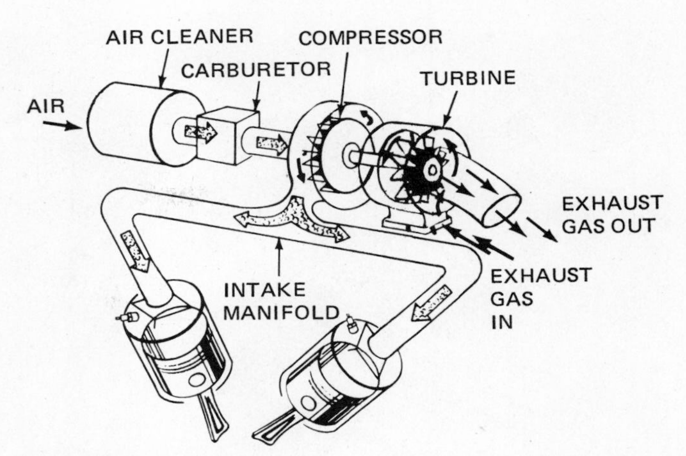

Fig. 3-58 Schematic layout of the turbocharger on a V-type engine. (*Chevrolet Motor Division of General Motors Corporation*)

be used. This is a rotary air pump that boosts engine power, sometimes up to 40 percent. It has been called a supercharger because it delivers a "super" charge of air-fuel mixture to the engine. Some early superchargers were driven by gears or chains from the engine. The supercharger contains a compressor rotor, which is a wheel with blades. The wheel is much like the impeller in a water pump. When it spins, it moves air by centrifugal force. The air between the rotor blades is pushed outward and through the outlet port of the supercharger. The air exits at relatively high pressure.

The mechanical drive gave trouble and used considerable engine power. To avoid these problems, the turbosupercharger, or turbocharger, was developed. This assembly has a turbine that drives the compressor rotor (Fig. 3-57). The turbine itself is driven by the hot exhaust gases from the engine. Figure 3-58 shows schematically the installation of a turbocharger on a V-type engine. In the installation shown, the compressor is compressing the air-fuel mixture after it leaves the carburetor. In other installations, the turbocharger compresses air alone (Fig. 3-59).

A different turbocharger installation is shown in Fig. 3-59. This shows the turbocharger mounted on a V-6 engine. The crossover pipe brings the exhaust gases from the opposing bank of cylinders to the turbine of the turbocharger. There it unites with the exhaust gases from the cylinder bank on which the turbocharger mounts. Note that the compressor is compressing air alone and sending it to the carburetor.

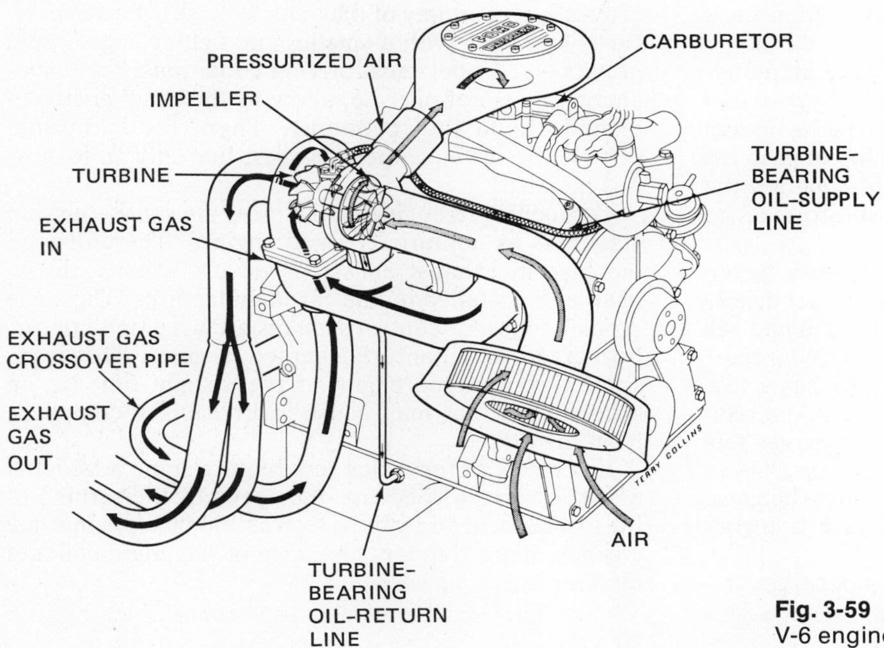

Fig. 3-59 A turbocharged Ford 182-cubic-inch (3-L) V-6 engine. (*Ford Motor Company*)

Fig. 3-60 Turbocharger installation on a fuel-injected four-cylinder engine. (*Chevrolet Motor Division of General Motors Corporation*)

Figure 3-60 shows a turbocharger installation on a four-cylinder, fuel-injected gasoline engine. Air alone is compressed and is delivered to the intake manifold. Fuel is injected into the intake manifold by the fuel injection system. The mixture enters the cylinders when the intake valves open. Fuel injection for gasoline engines is discussed in Chap. 6. On this engine, the power output is boosted almost 40 percent, from 63 hp [47 kW] to 100 hp [75 kW].

Greater power boosts have been made with turbochargers. For example, engines designed specifically for competition racing develop as much as 200 hp [149 kW] *per cylinder*. The Offenhauser (or "Offy") four-cylinder engine used in the Indianapolis 500 has a 159-cubic-inch [2605-cc] piston displacement. It is a relatively small engine. Yet with the turbocharging, this engine produces in the range of 600 to 800 hp [448 to 597 kW]. However, this is an extreme example because the engine is specifically designed to use a turbocharger.

On the passenger-car engine, the turbocharger improves fuel economy of the vehicle. This is because the car is equipped with a smaller and lighter engine than previously. Under most driving conditions, the turbocharger does not operate, allowing the small engine to provide good fuel economy. Then, for additional power, the turbocharger engages, but only so long as it is needed.

Turbocharged engines must meet the same emission standards as nonturbocharged engines. Therefore, no increase in exhaust emissions occurs when a turbocharger is installed by the car manufacturer. The same type of emission-control systems are used on both turbocharged and nonturbocharged engines. However, some turbocharged engines have a slight time-lag on acceleration that may cause a driveability complaint from the driver.

Engines with turbochargers require some additional periodic service. Because the speed of the turbine in the turbocharger may be as high as 100,000 rpm in some engines, more frequent changing of the engine oil and oil filter are required.

Chapter 3 review questions

Select the *one* correct, best, or most probable answer to each question. Then check your answers against the correct answers given at the end of the book.

1. Cars manufactured since 1970 have an evaporative emission control system to:
   a. prevent the escape of gasoline vapors from the fuel tank
   b. prevent odor in the exhaust gas
   c. prevent vapor lock
   d. all of the above.
2. The mechanical fuel pump is operated by:
   a. an eccentric on the camshaft
   b. a crankpin on the crankshaft
   c. a rocker arm on the crankshaft
   d. none of the above.
3. To prevent vapor lock, many cars are equipped with:
   a. an evaporative emission control system
   b. a vapor return line
   c. a combination fuel pump
   d. a mechanical fuel pump.
4. In the gasoline engine, the basic difference between the carbureted fuel system and the fuel injection system is:
   a. in the type of fuel used
   b. how the fuel is metered
   c. the pressure applied to the fuel
   d. all of the above.
5. In the fuel-vapor recovery system, fuel vapor escaping from the fuel tank is:
   a. vented to the intake manifold
   b. adsorbed in the charcoal canister
   c. vented to the carburetor float bowl
   d. none of the above.
6. When the engine is cold, all air entering the thermostatically controlled air cleaner comes from the:
   a. intake manifold
   b. exhaust manifold
   c. heat stove on the exhaust manifold
   d. none of the above.
7. The purpose of the EGR system is to:
   a. reduce engine temperatures
   b. reduce NO_x in the exhaust gas
   c. improve combustion of HC
   d. all of the above.
8. The purpose of the PCV system is to:
   a. remove exhaust gases from the crankcase
   b. remove blowby gases from the crankcase
   c. improve engine cooling
   d. none of the above.
9. The purpose of the catalytic converter in the exhaust system is to:
   a. convert pollutants into harmless gases
   b. reduce engine temperatures
   c. improve engine efficiency
   d. none of the above.
10. The purpose of the turbocharger is to:
   a. push more air-fuel mixture into the engine
   b. keep the battery charged
   c. increase the compression ratio
   d. none of the above.

CHAPTER 4

CARBURETOR FUNDAMENTALS

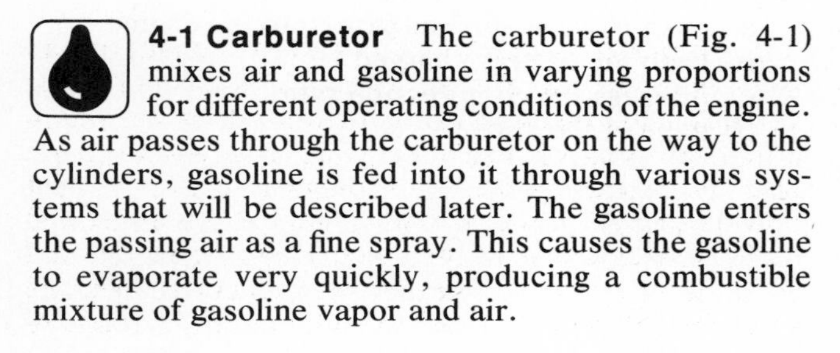

After studying this chapter, you should be able to:

1. Explain the difference between the two basic types of carburetors.
2. Describe the difference between intake manifold vacuum and venturi vacuum.
3. List the six basic systems in the automotive carburetor.
4. Describe the operation of each system in a fixed-venturi carburetor.
5. Define *air-fuel ratio.*
6. Discuss the various devices used to provide additional heat to the air-fuel mixture.
7. Explain the difference between a two-barrel and a four-barrel carburetor.
8. Describe the operation of a staged two-barrel carburetor.
9. Explain the operation of an electric choke.
10. Discuss the reason for using an idle-stop solenoid.

4-1 Carburetor The carburetor (Fig. 4-1) mixes air and gasoline in varying proportions for different operating conditions of the engine. As air passes through the carburetor on the way to the cylinders, gasoline is fed into it through various systems that will be described later. The gasoline enters the passing air as a fine spray. This causes the gasoline to evaporate very quickly, producing a combustible mixture of gasoline vapor and air.

4-2 Vaporization When a liquid changes to a vapor, or undergoes a change of state, it is said to evaporate. Water placed in an open pan eventually disappears. It changes from a liquid to a vapor. Clothes are hung on a line to dry. The water in the clothes changes to a vapor. When the clothes are spread out, they dry more rapidly than when they are bunched together. This illustrates an important fact about evaporation. The greater the surface exposed to the air, the more rapidly evaporation takes place. Water in a tall glass evaporates slowly. But water in a shallow pan evaporates much more quickly.

4-3 Atomization To produce very quick vaporization of the liquid gasoline, it is sprayed into the air passing through the carburetor. Spraying the liquid turns it into many fine droplets. This effect is called *atomization* because the liquid is broken up into small droplets (but not actually into atoms, as the name implies). Each droplet is exposed to air on all sides so that it vaporizes very quickly. Therefore, during normal engine operation, the gasoline sprayed into the air passing through the carburetor turns to vapor almost instantly.

4-4 Carburetor fundamentals A simple carburetor can be made from a round cylinder with a constricted section, a fuel nozzle, and a round disk, or valve (Fig. 4-2). The round cylinder is called the air horn, the constricted section the venturi, and the valve the throttle valve. The throttle valve can be tilted more or less to open and close the air passage (Fig. 4-3). In the horizontal position, it shuts off, or throttles, the airflow through the air horn. When the throttle valve is turned away from this position, air can flow through the air horn.

1. Venturi effect As air flows through the constriction, or venturi, a partial vacuum is produced in the venturi. This vacuum causes the fuel nozzle to deliver a fine spray of gasoline into the passing air stream. The venturi effect (of producing a vacuum) can be illustrated as shown in Fig. 4-4. Three dishes of mercury (a very heavy metallic liquid) are connected by tubes to an air horn with a venturi. The greater the vacuum, the higher the mercury is pushed up the tube by atmospheric pressure. Note that the greatest vacuum is in the venturi. This vacuum increases with the speed of the air flowing through the venturi.

One explanation of why there is a partial vacuum in the venturi is that the air is made up of countless molecules. As air moves into the top of the air horn, all the air molecules move at the same speed. But if all are to get through the venturi, they must speed up and move through faster. For example, let us consider what hap-

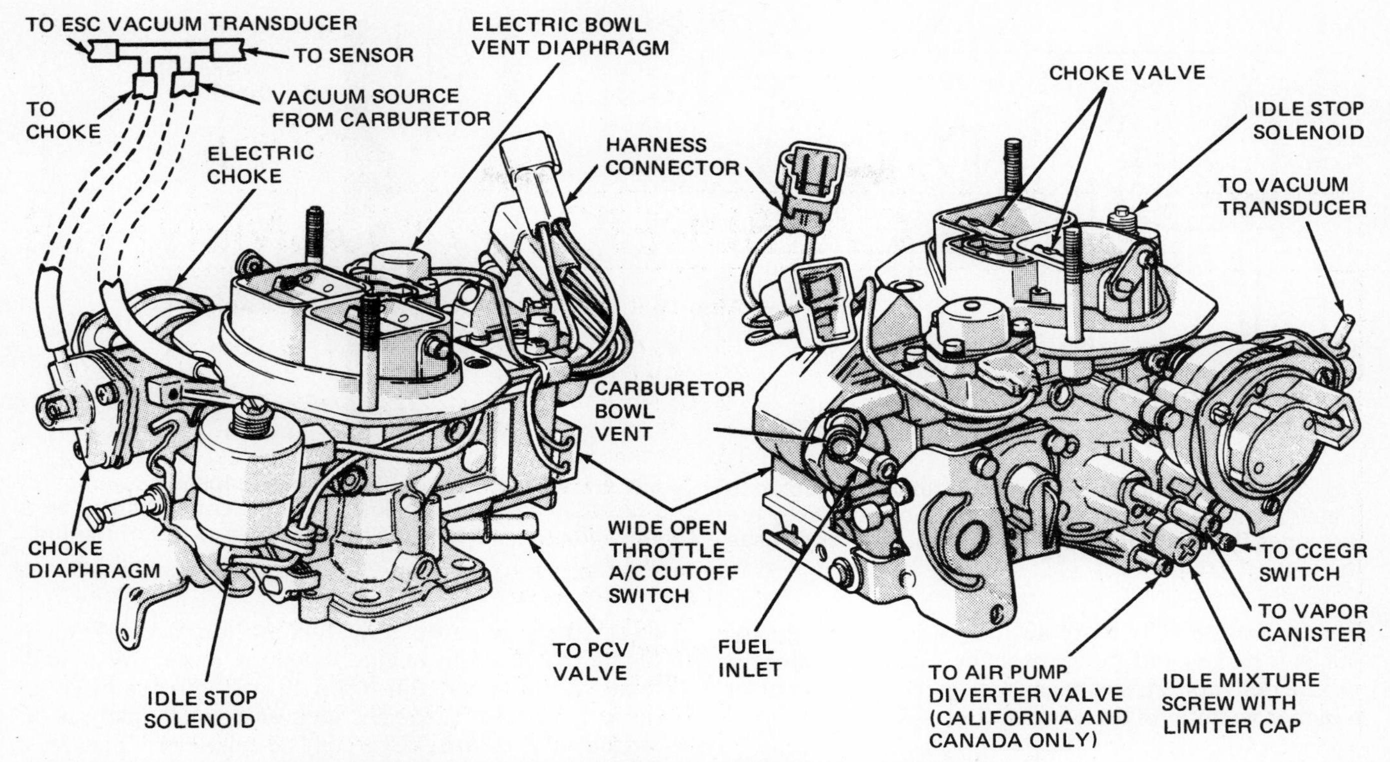

Fig. 4-1 A late-model carburetor for a car with air conditioning. (*Chrysler Corporation*)

pens to two molecules, one behind the other. As the first molecule enters the venturi, it speeds up, tending to leave the second molecule behind. The second molecule also speeds up as it enters the venturi. But the first molecule has, in effect, a head start. Therefore, the two molecules are farther apart in the venturi than they were before they entered it.

Now imagine a great number of molecules going through the same action. As they pass through the venturi, they are farther apart than before they entered. This is another way of saying that a partial vacuum exists in the venturi. A partial vacuum is a thinning out of the air, where the distance between air molecules is greater than normal (⚙ 1-10).

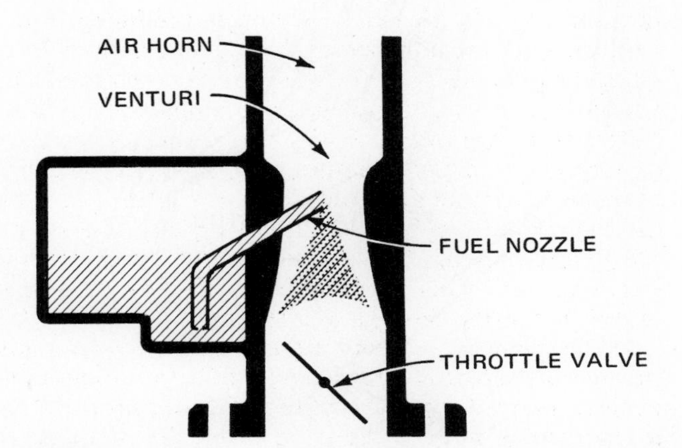

Fig. 4-2 Basic carburetor, consisting of an air horn, fuel nozzle, and throttle valve.

2. Fuel-nozzle action The partial vacuum occurs just where the end of the fuel nozzle is located in the venturi. The other end of the fuel nozzle is located in a fuel reservoir or float bowl (Fig. 4-5). Atmospheric pressure pushes on the fuel through a vent in the float bowl cover. With the partial vacuum at the upper end of the nozzle, fuel is pushed up through the nozzle and enters the passing air stream. The fuel enters as a fine spray, which quickly turns to vapor as the droplets of

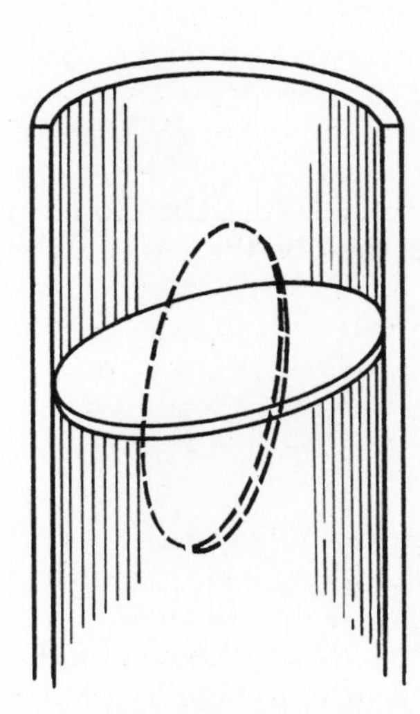

Fig. 4-3 Throttle valve in the air passage of a carburetor. When the throttle valve is closed, as shown, little air can pass through. But when the throttle valve is opened, as shown dashed, there is little throttling effect.

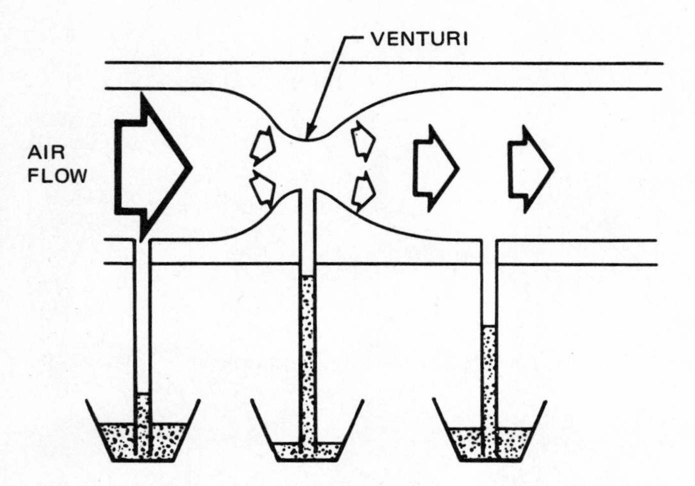

Fig. 4-4 Three dishes of mercury and tubes connected to the air passage. They show differences in vacuum by the distances the mercury rises in the tubes. The venturi has the highest vacuum.

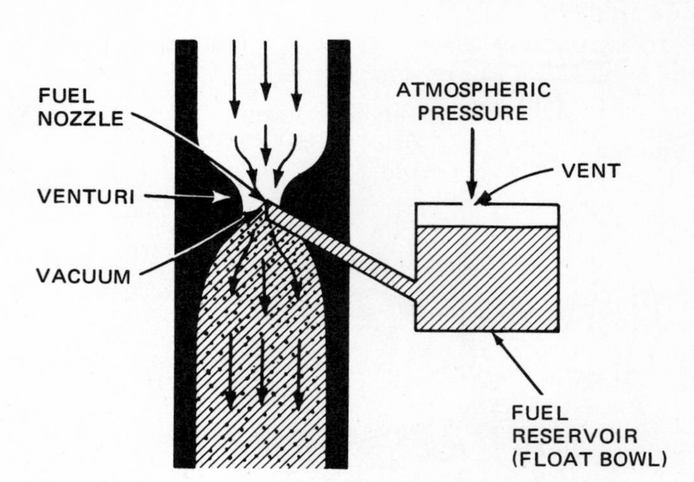

Fig. 4-5 The venturi, or constriction, causes a vacuum to develop in the airstream just below the constriction. Then atmospheric pressure pushes fuel up and out of the fuel nozzle.

fuel evaporate. The more air that moves through, the faster it moves and the greater the amount of fuel the nozzle delivers. This is because higher airspeed causes a higher vacuum in the venturi (Fig. 4-4).

3. Throttle-valve action The throttle valve can be tilted in the air horn to allow more or less air to flow through (Fig. 4-3). When it is tilted to allow more air to flow, larger amounts of air-fuel mixture are delivered to the engine. The engine develops more power and tends to run faster. But if the throttle valve is tilted to throttle off most of the air, then only small amounts of air-fuel mixture are delivered. The engine produces less power and tends to slow down.

The throttle valve is linked to an accelerator pedal in the passenger compartment (Fig. 4-6). The linkage may

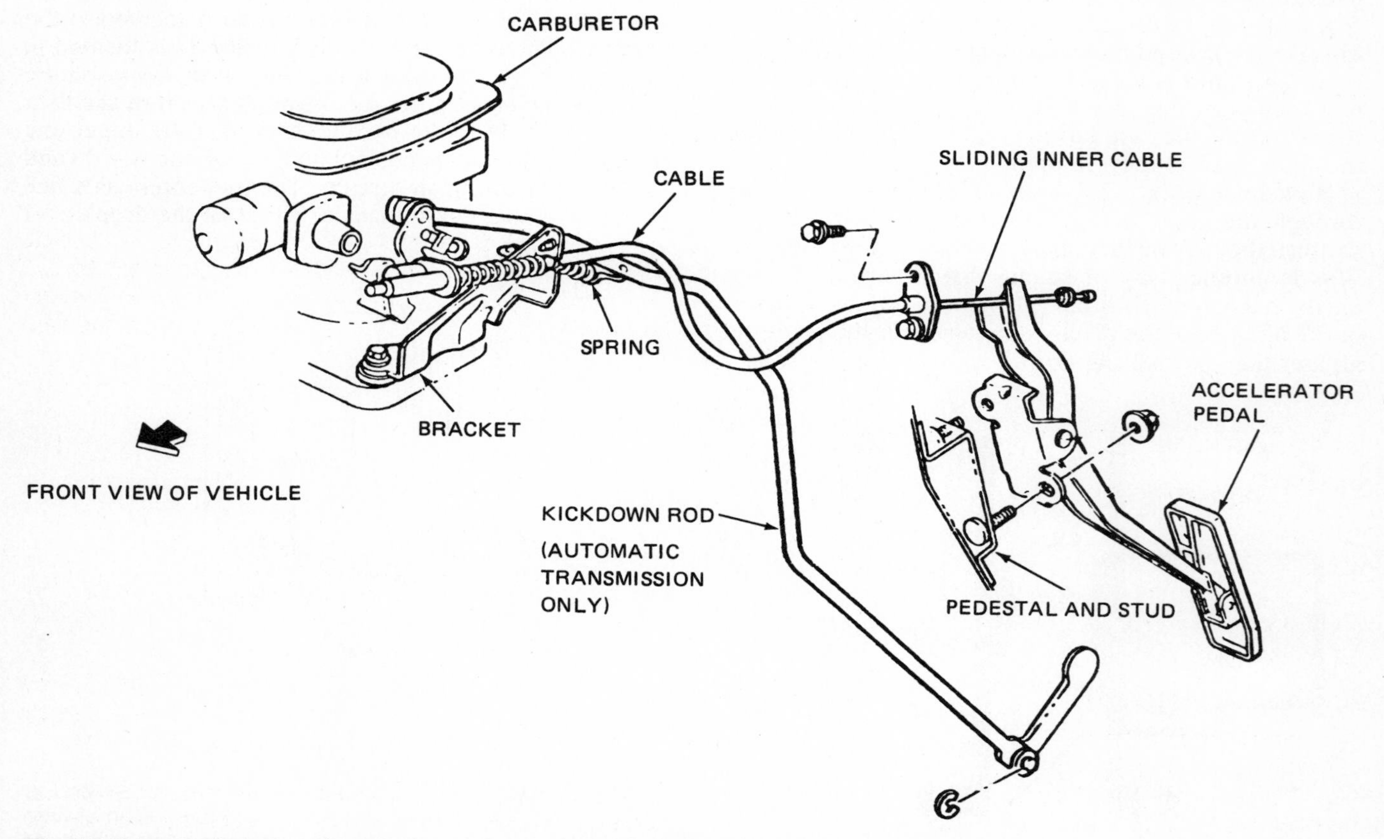

Fig. 4-6 Linkage between the accelerator pedal and the carburetor. (*Ford Motor Company*)

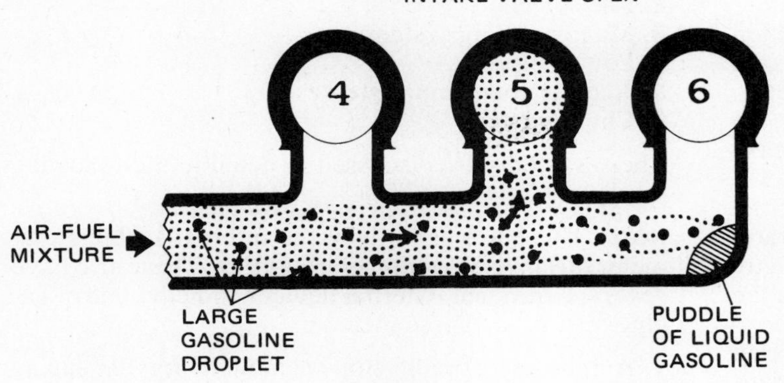

Fig. 4-7 Distribution pattern in an intake manifold. The gasoline particles tend to continue to the end of the manifold, enriching the mixture going to the end cylinders. (*Chevrolet Motor Division of General Motors Corporation*)

consist of a series of interconnected levers and rods, or levers and a cable. The cable has a flexible outer covering and an inner sliding wire. With either arrangement, movement of the accelerator pedal causes the throttle valve to change its position in the throttle body. Therefore, the driver can position the throttle valve to suit operating requirements. On some cars, the throttle linkage performs other functions such as producing a down-shift of the automatic transmission under the right operating conditions. The kickdown rod for the automatic transmission is shown in Fig. 4-6.

✪ 4-5 Distribution of the air-fuel mixture The air-fuel mixture passes from the carburetor through the intake manifold to the intake valve ports in the cylinder head. The ideal arrangement would be for precisely the same amount of air-fuel mixture, with all fuel vaporized, to be delivered to each cylinder. Also, the mixture would be identical in richness as it enters the cylinders.

Richness refers to the proportion of vaporized fuel in the mixture. A rich mixture would be approximately 12:1, or 12 parts of air to 1 part of fuel, by weight. If the proportions are changed to increase the amount of air, the mixture becomes leaner. A lean mixture would be about 16:1, or 16 parts of air to 1 part of fuel, by weight. For normal engine operation at highway speed, the best mixture would be about 15:1. There are systems in the carburetor that alter the proportions of air and fuel to suit different operating conditions (✪ 4-8).

The intake manifold can also alter the proportions of air and fuel in the mixture reaching the different cylinders. The ideal would be for all fuel to vaporize and all cylinders to receive the same amount of mixture of the same richness. Actually, the intake manifold can act as a sorting device, supplying some cylinders with a richer mixture than others. If the fuel does not completely vaporize, there will be droplets (liquid particles) in the mixture. These particles, being relatively heavy, cannot turn the corner so easily as the vaporized fuel (Fig. 4-7). Therefore, they continue in a more or less straight line until they hit the walls of the manifold.

In the example shown in Fig. 4-7, the intake valve in cylinder 5 is open, and the air-fuel mixture is flowing toward this cylinder. The vaporized fuel and air can turn the corner and enter cylinder 5. However, the fuel droplets continue on in a straight line until they strike the wall of the manifold. When the intake valve for cylinder 6 opens, air-fuel mixture begins flowing into cylinder 6. As it enters, it picks up some of the fuel on the manifold walls. This further enriches the mixture.

The result is that the center cylinders, closest to the carburetor, may receive a relatively lean mixture. At the same time the mixture entering the end cylinders may be comparatively rich. This results from the failure of all the fuel to vaporize and the sorting effect of the intake manifold. If sufficient heat is supplied to the intake manifold during engine warm-up, and the carburetor vaporizes the fuel sufficiently, the mixture will be reasonably uniform.

✪ 4-6 Air-fuel-ratio requirements The fuel system must vary the air-fuel ratio to suit different engine operating conditions. The mixture must be rich for starting, acceleration, and maximum power. It must be leaner for idle, for low-speed and low-power operation, and for medium-speed part-throttle operation.

Figure 4-8 is a graph showing typical air-fuel ratios as related to various car speeds. Air-fuel ratios and the engine speeds at which they are obtained vary with the different cars. In the example shown in Fig. 4-8, a rich mixture of about 9:1 is supplied for starting a cold engine. Then, during idle, the mixture leans out to about 12:1. At medium speeds, the mixture further

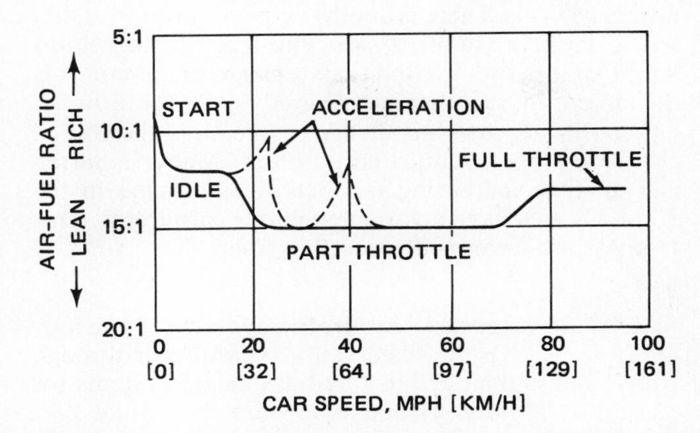

Fig. 4-8 Graph of the air-fuel ratios for different car speeds. The graph is only typical. Car speeds at which the various ratios are obtained may vary with different cars. Also, there may be some variation in the ratios.

leans out to about 15:1. But at higher speeds with wide-open throttle the mixture is enriched to about 13:1.

Opening the throttle at any speed for acceleration causes an enrichment of the mixture. This results from the operation of special systems in the carburetor, which are discussed later. In Fig. 4-8, enriching the mixture during acceleration is shown by the two peaks at about 23 mph [37 km/h] and 40 mph [64 km/h].

It would appear from a study of Fig. 4-8 that the engine itself demands a greatly varying air-fuel ratio for different operating conditions. This is not exactly true. The real problem is how to get a mixture with the needed air-fuel ratio into the cylinders.

For example, the air-fuel mixture must be very rich to start a cold engine. This is because the fuel vaporizes very poorly while the engine and carburetor are cold. Air speed is low, and much of the fuel does not vaporize. Therefore, extra fuel must be delivered by the carburetor so that enough will vaporize to provide a combustible air-fuel mixture ratio for starting.

A similar problem occurs when the throttle valve is suddenly opened for acceleration. Quick opening of the throttle valve allows a sudden in-rush of air. Again, the mixture must be enriched. Extra fuel must enter with the additional air. Following sections describe the various systems in carburetors that supply the required air-fuel mixture at the proper ratio for different operating conditions.

NOTE: Some newer cars have electronic control of the fuel system. The electronic control has a sensing device in the exhaust that measures the amount of oxygen in the exhaust gas. If there is too little oxygen, the mixture is too rich. If there is too much oxygen, the mixture is too lean. In either case, the mixture is not right. The electronic control then signals a device in the carburetor that readjusts the air-fuel mixture. The system is covered later in ⚙ 4-28 to 4-30, and in Chap. 11.

⚙ 4-7 Carburetor types So far, we have been describing the actions in a fixed-venturi (FV) carburetor. Air flowing through the fixed venturi produces a partial vacuum that causes fuel to discharge from the fuel nozzle (⚙ 4-4). There is another type of carburetor that has a variable venturi (VV). The size of the venturi varies as operating conditions change. This controls the amount of fuel that is delivered to the venturi.

Most carburetors installed in cars made in the United States are fixed-venturi carburetors. Many imported cars and an increasing number of cars made in the United States have a variable-venturi carburetor. This type of carburetor is discussed in Chap. 5.

⚙ 4-8 Fixed-venturi carburetor systems The various passages in the fixed-venturi carburetor through which fuel and air are metered are called systems (or circuits). Different systems supply fuel during idle, part throttle, full throttle, and acceleration. These systems work together or separately during certain operating conditions to supply the required air-fuel ratio. The systems in the fixed-venturi carburetor include:

1. Float system
2. Idle system
3. Main metering system
4. Power system
5. Accelerator-pump system
6. Choke system

These systems are discussed in detail in the following sections.

NOTE: These are essentially the internal systems that meter fuel in fixed-venturi carburetors. These also have several additional external devices to help control engine operation. These include:

1. Antidieseling or idle-stop solenoid to prevent engine run-on after it is turned off (Fig. 4-1).
2. Dashpot to prevent sudden closing of the throttle valve, which eliminates a sudden peak of HC in the exhaust gas.
3. Linkage to the automatic transmission which causes a downshift when the throttle is opened wide (Fig. 4-6).
4. Several openings or ports in the carburetor which send vacuum signals to other components so that they can perform a function (Fig. 4-1). For example, a vacuum passage runs from the carburetor to the ignition distributor vacuum-advance unit. Also, a vent connects the float bowl to the charcoal canister in the evaporative emission control system to trap fuel vapors.

⚙ 4-9 Single-barrel carburetor One of the main differences between carburetors is size. The carburetors described so far all have only one barrel, one main fuel nozzle, and one throttle valve. Carburetors of this type are usually called single-venturi, single-barrel, or one-barrel carburetors. They have one main passage through which air flows.

Most fixed-venturi carburetors, including single-barrel carburetors, can be divided into three basic parts (Fig. 4-9). These are the air horn, the main body or float bowl, and the throttle body. Figure 4-10 shows a one-barrel carburetor with the air horn removed. Note the location of the venturi and how the end of the fuel nozzle is centered in it.

A single-barrel carburetor is used on many four-cylinder engines. Figure 4-11*a* shows a single-barrel carburetor mounted on the intake manifold for an in-line four-cylinder engine. You can see how the intake manifold distributes the air-fuel mixture to the intake valve for each cylinder. Figure 4-11*b* shows the distribution pattern when a single-barrel carburetor is used on a typical in-line six-cylinder engine.

The single-barrel carburetor is not normally found on an engine with more than six cylinders. The intake manifold has trouble providing equal distribution of the air-fuel mixture (⚙ 4-5). In addition, a one-barrel carburetor with a large venturi is not very responsive at low speed to slight changes in throttle position. For better engine performance and drivability, a two- or four-barrel carburetor, which has smaller venturis, is installed on V-8 engines.

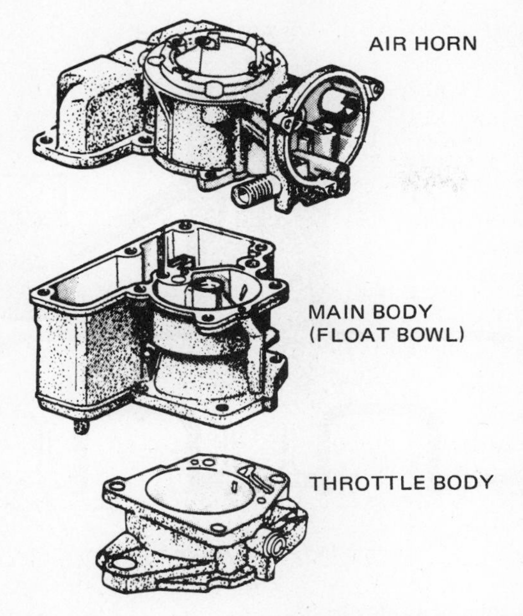

Fig. 4-9 The three main parts of a carburetor. (*Ford Motor Company*)

⚙ 4-10 Two-barrel carburetors The two-barrel carburetor, or dual carburetor, basically is two single-barrel carburetors put together into a single assembly (Figs. 4-12 and 4-13). The purpose of the additional barrel in the dual carburetor is to improve engine breathing. Engine breathing, or volumetric efficiency, is described in ⚙ 2-18. The more air-fuel mixture the engine cylinders can take in (up to a limit), the greater the power output of the engine. Adding the extra barrel permits more air-fuel mixture to enter the engine. When a single-barrel carburetor is made larger to admit more air-fuel mixture, venturi action is poor. Therefore, proper air-fuel mixture ratios for different operating conditions are difficult to achieve.

The second barrel in the dual carburetor is used in two different ways, according to carburetor design. In one design, both barrels are the same size, and both operate all the time that the engine is running. In some engines, the intake manifold is open to all cylinders. Both barrels feed into this common manifold. In other engines, the manifold is split or divided into two parts. Each barrel feeds one part which takes care of half the cylinders.

In the second design, one barrel is the primary barrel. It takes care of the air-fuel requirements for all cylinders during idle, low-speed, and intermediate-speed operation. The second, or secondary, barrel comes into operation to supply additional air-fuel mixture when the throttle is opened wide for full power.

Each design of the two-barrel carburetor is discussed below.

1. Two primary barrels When a dual carburetor has two primary barrels (Fig. 4-14*a*), each barrel provides the air-fuel mixture for half the cylinders in the engine. In Fig. 4-14*a*, notice that each barrel feeds two cylinders of a four-cylinder engine. Figure 4-14*b* shows another arrangement for a two-barrel carburetor. Both barrels feed into a common manifold which supplies all the engine cylinders. Figure 4-15 shows a divided or dual-plane manifold for a V-8 engine. Each barrel of the carburetor feeds half the engine cylinders. One barrel supplies cylinders 1, 4, 6, and 7. The other barrel supplies cylinders 2, 3, 5, and 8. The arrows show the air-fuel mixture flow from the two barrels to the cylinders.

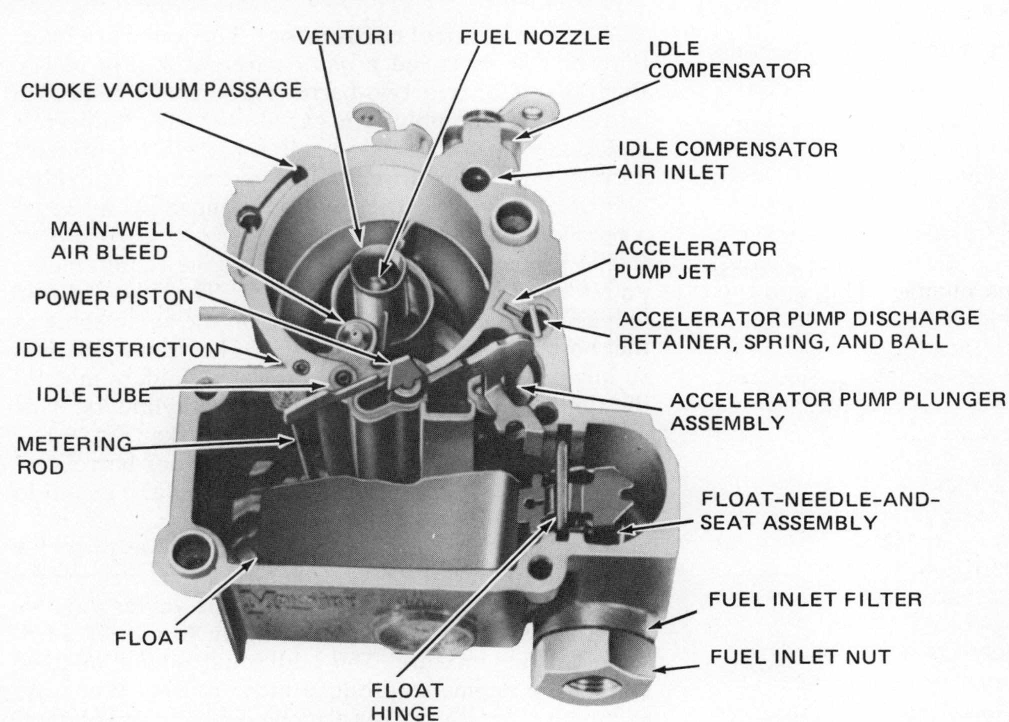

Fig. 4-10 View from above of the float bowl assembly with the air horn removed. (*Oldsmobile Division of General Motors Corporation*)

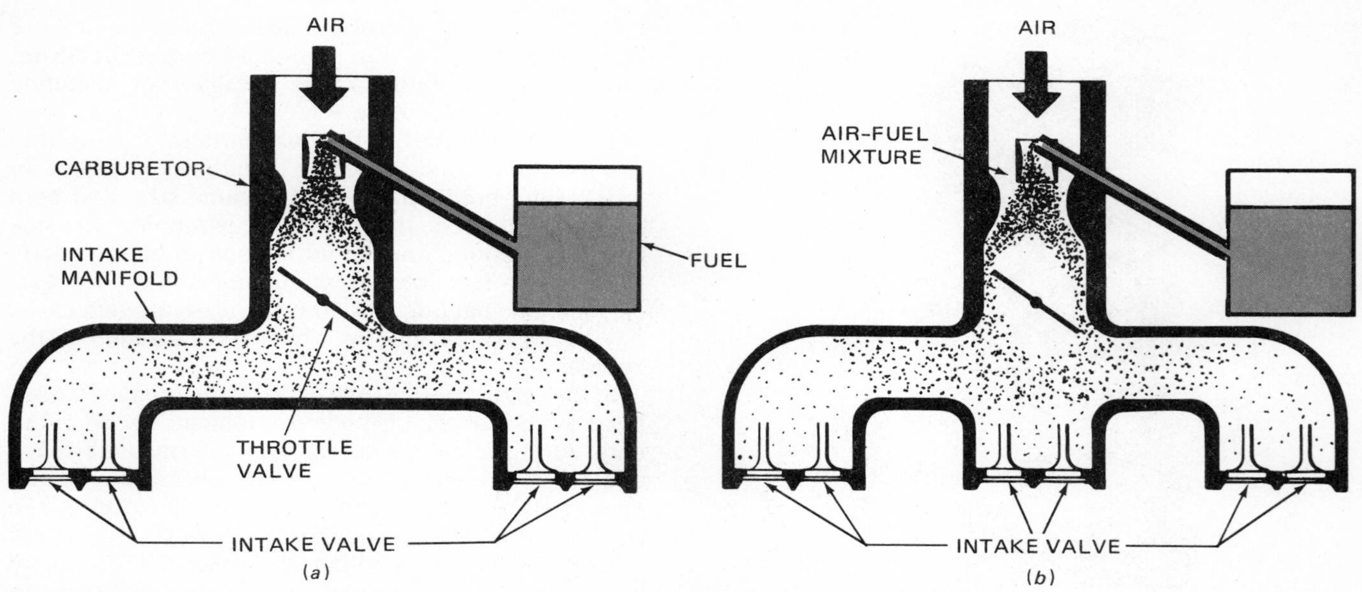

Fig. 4-11 Distribution pattern of the air-fuel mixture from a single-barrel carburetor to (*a*) a four-cylinder engine and (*b*) a six-cylinder engine.

The center passage in the intake manifold (Fig. 4-15, top) allows exhaust gas to flow around the carburetor mounting pad when the engine is cold. This adds heat to the ingoing air-fuel mixture so that the gasoline vaporizes more readily (❁ 4-36). Note also that this intake manifold has a coolant passage from each cylinder bank to a coolant outlet passage in which the cooling system thermostat is mounted. Cooling systems are covered in detail in Chaps. 18 and 19.

Each barrel of the carburetor has a complete set of fuel-metering systems. The throttle valves are fastened to a single throttle shaft (Fig. 4-13). Both throttle valves always open and close together.

2. Primary and secondary barrels The second design of two-barrel carburetor sometimes is called a *staged* carburetor (Fig. 4-16). It uses the second barrel in a different way. One barrel, which is the primary barrel, supplies the air-fuel mixture for all engine cylinders during idle, low-speed, and intermediate-speed operation. The secondary barrel comes into operation only after the primary throttle valve is about half open. When the primary throttle valve opens further, linkage to the secondary throttle valve starts to open it. Then the secondary barrel begins supplying additional air-fuel mixture to the engine. This increases engine power during medium- to high-speed operation.

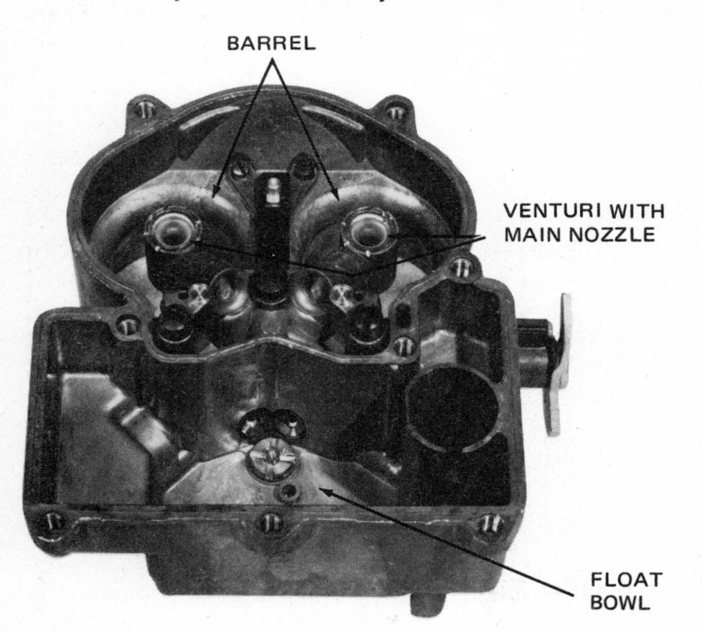

Fig. 4-12 Bowl assembly for a two-barrel carburetor. (*Chevrolet Motor Division of General Motors Corporation*)

❁ 4-11 Four-barrel carburetor The four-barrel carburetor also is called a *quad* carburetor (Fig. 4-17). Basically it is two two-barrel carburetors combined into a single assembly. The carburetor has four barrels and four main nozzles. One pair of barrels, the primary barrels, make up a primary dual carburetor. The other pair of barrels, with their main fuel nozzles, make up a secondary dual carburetor.

The primary barrels handle all engine requirements under most operating conditions. The two throttle valves in the primary barrels are on the same shaft so that both open and close together. The primary barrels supply all the air-fuel mixture for all cylinders most of the time. Each barrel feeds half of the cylinders. This is the same way the dual carburetor works (Fig. 4-15). However, each of the four barrels in a four-barrel carburetor has a separate opening into the intake manifold (Fig. 4-18).

When the throttle is moved toward wide open for acceleration or full-power operation, the secondary barrels come into action. Their throttle valves open so that the secondary barrels can begin to supply additional air-fuel mixture. This additional air-fuel mixture causes the engine to produce more power. When the secondary throttle valves are opened, engine volumetric efficiency increases.

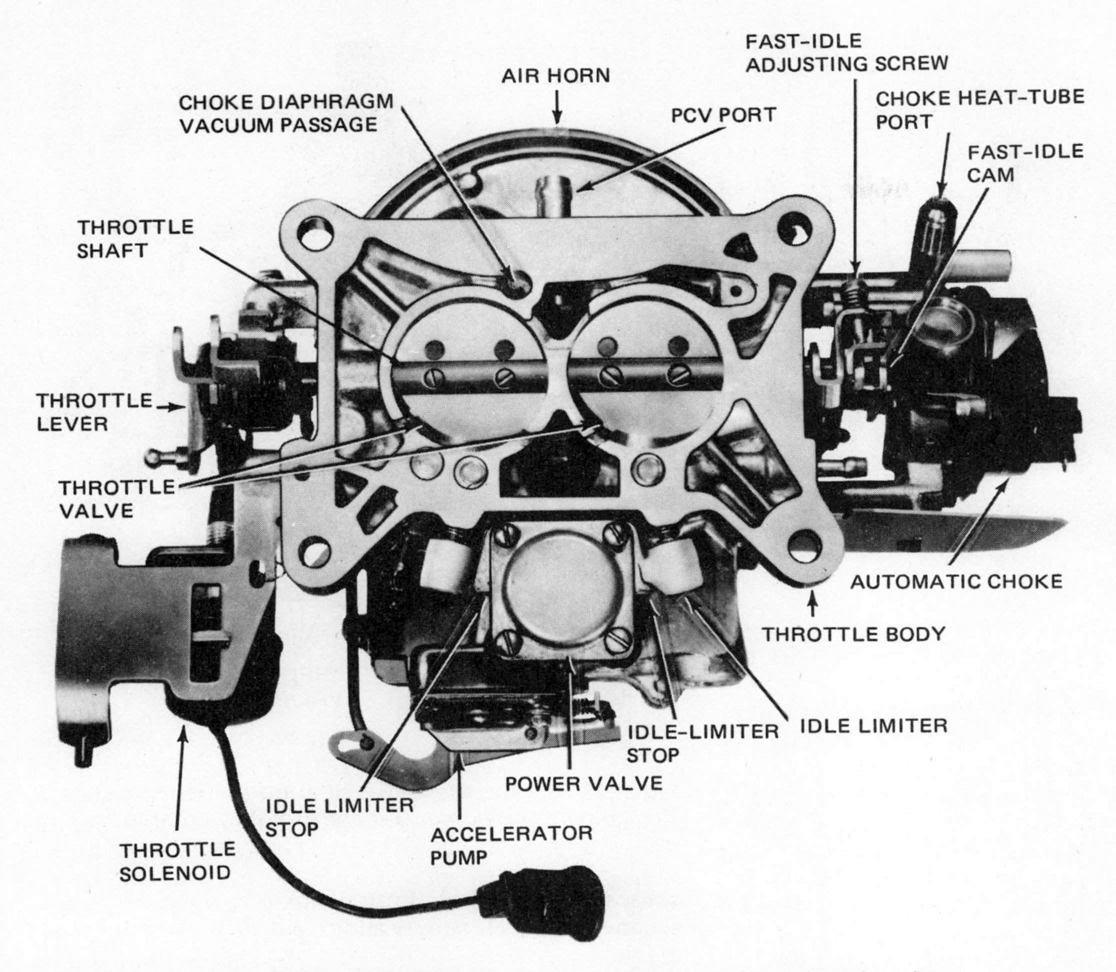

Fig. 4-13 A two-barrel carburetor, showing the locations of the two throttle valves.

There are two ways of controlling the operation of the secondary-barrels—by mechanical linkage from the primary-throttle shaft, or by a vacuum device. Figure 4-19 is a sectional view of a carburetor using a mechanical linkage. During part-throttle, only the primary throttle valves are open. However, whenever the throttle is opened wide for additional power, the linkage between the primary and secondary throttle valves

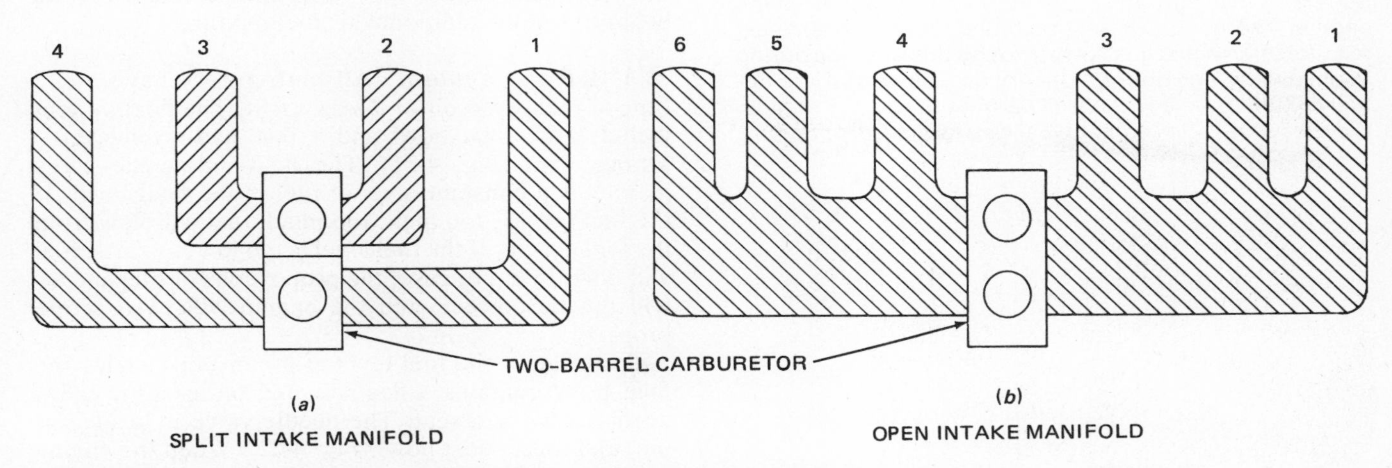

Fig. 4-14 Distribution pattern of the air-fuel mixture from a two-barrel carburetor to (*a*) a four-cylinder and (*b*) a six-cylinder engine. Note that in (*a*) the flow from each barrel of the carburetor is directed to certain cylinders, whereas (*b*) shows a manifold using a box, or open plenum chamber, without any dividers or partitions to direct the mixture flow. Many engine intake manifolds are of the open-plenum type.

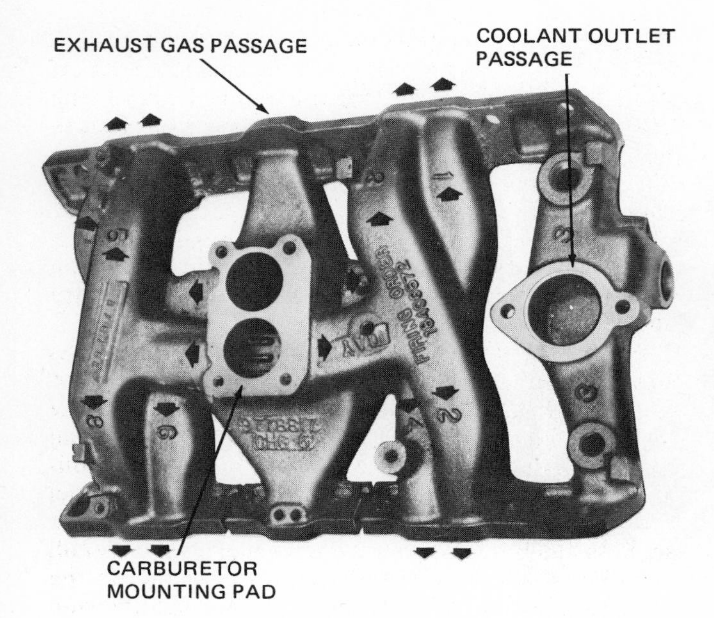

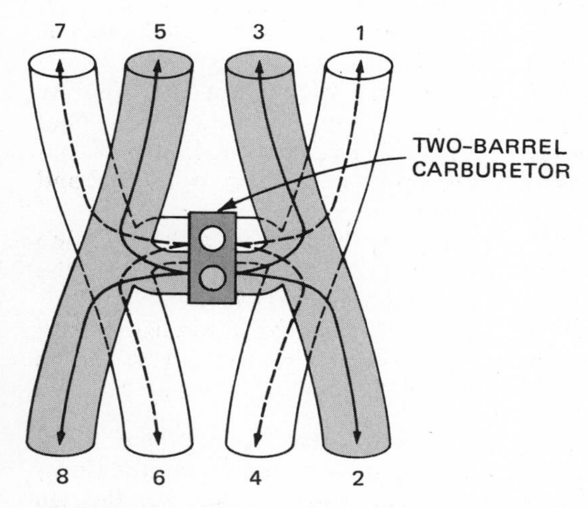

Fig.4-15 Top, intake manifold for an I-head V-8 engine. The arrows show the air-fuel mixture flow from the two barrels of the carburetor to the eight cylinders in the engine. The central passage connects the two exhaust manifolds. Exhaust gas flows through this passage during engine warm-up. Bottom, distribution pattern of the air-fuel mixture from a two-barrel carburetor to a V-8 engine. (*Pontiac Motor Division of General Motors Corporation*)

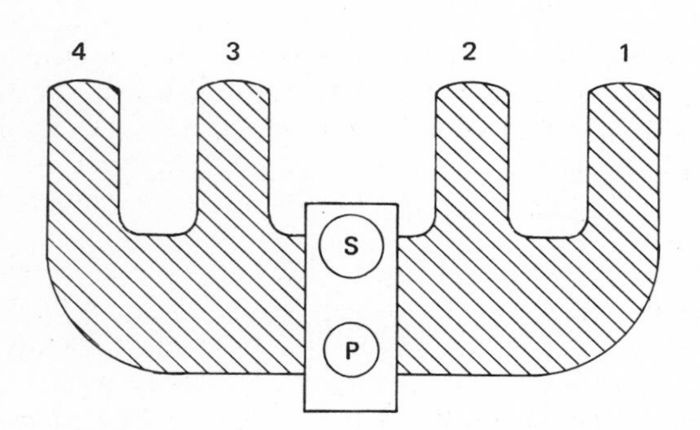

Fig. 4-16 Intake manifold for a staged two-barrel carburetor, which has a primary barrel (P) and a secondary barrel (S). (*ATW*)

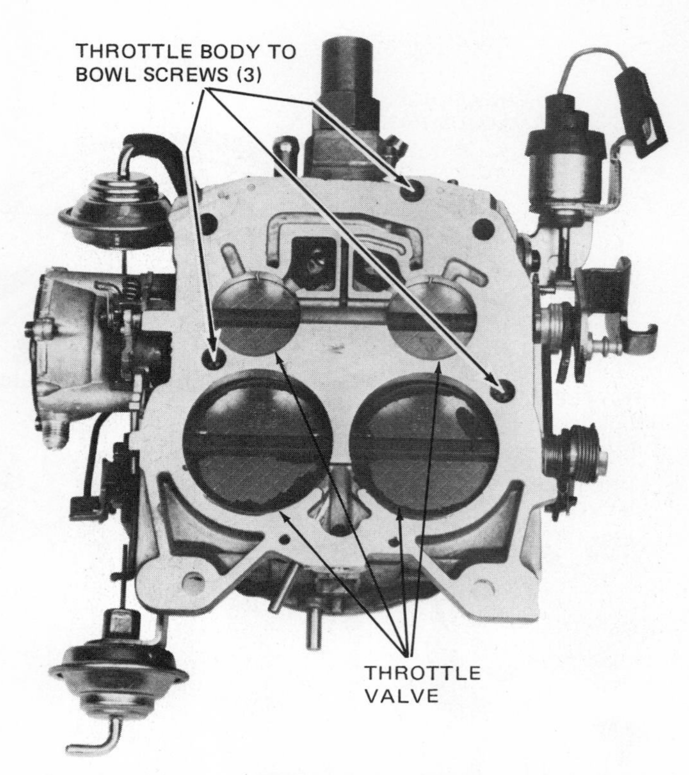

Fig. 4-17 The throttle valves of a four-barrel carburetor. (*Oldsmobile Division of General Motors Corporation*)

causes the secondary throttle valves to open. Now the secondary barrels supply additional air-fuel mixture for maximum power.

Secondary throttle valves also can be controlled by a vacuum-operated diaphragm (Fig. 4-20). The vacuum is picked up from one of the primary-barrel venturis. As airspeed through the primary barrels increases, so does this vacuum. When the vacuum reaches a predetermined amount that indicates a rather high engine rpm, the vacuum actuates the diaphragm. This opens the secondary throttle valves so the secondary barrels begin to supply additional air-fuel mixture.

⚙ 4-12 Float system All carburetors have some type of fuel reservoir or *float system*. The float system includes the float bowl and a float-and-needle valve arrangement (Fig. 4-21). The float-and-needle valve maintains a constant level of fuel in the float bowl. If the fuel level is too high, too much fuel will flow from the fuel nozzle. If the fuel level is too low, too little fuel will flow. In either case, the proportions of fuel and air will not be correct, and the engine will not operate properly.

To maintain the fuel level at a constant height, the float bowl contains a float pivoted on an arm, and a needle valve and seat. The needle valve is located at the inlet to the float bowl (Fig. 4-22). When the engine is running, the fuel pump supplies fuel to the carburetor. The fuel flows through the inlet into the float bowl. If fuel enters faster than it is being withdrawn (by the fuel nozzle), the fuel level in the float bowl rises.

As this happens, the float rises, lifting the needle

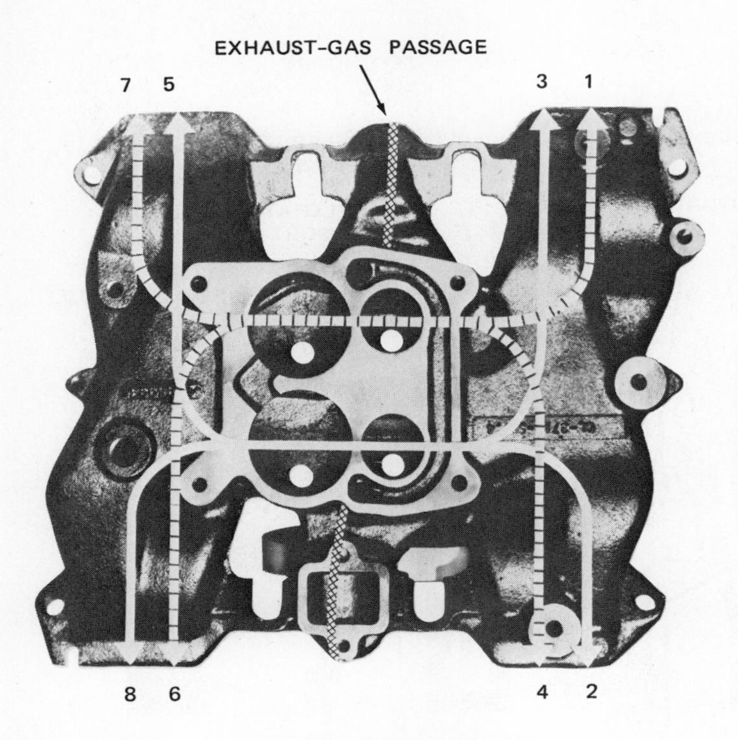

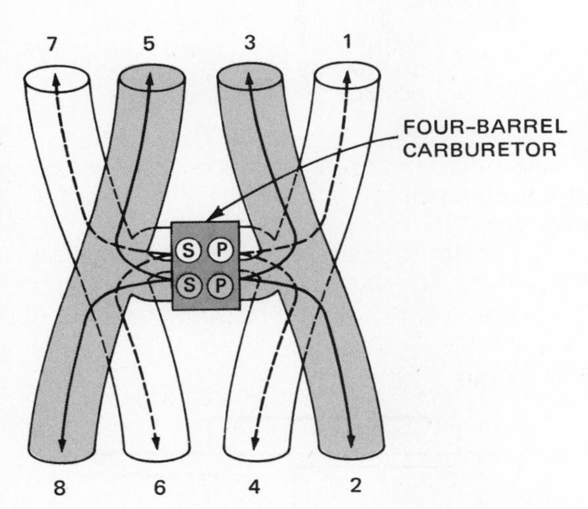

Fig. 4-18 Top, intake manifold for a V-8 engine using a four-barrel carburetor. P identifies the primary barrels. S identifies the secondary barrels. Bottom, distribution pattern of the air-fuel mixture from a four-barrel carburetor to a V-8 engine. (*Cadillac Motor Car Division of General Motors Corporation*)

valve and forcing it tightly up into the needle valve seat. This closes off the inlet, preventing further delivery of fuel. But as soon as some fuel is withdrawn, the fuel level falls, the float drops down, the needle valve is lowered off its seat, and additional fuel is delivered by the fuel pump. In actual operation, the fuel is maintained at a practically constant level in the float bowl. The float tends to hold the needle valve partly closed so that the incoming fuel just balances the fuel being withdrawn. This maintains a constant fuel level in the float bowl, which is essential to the proper operation of the carburetor.

Late-model carburetors have as small a float bowl as possible. One reason is that when the engine is shut off after having been driven long enough to heat up, the heat tends to evaporate the fuel in the float bowl. Before emission controls were installed on cars, this evaporated fuel would pass out into the air, polluting the atmosphere. All cars manufactured since 1970 have been equipped with an evaporation emission control system (⚙ 3-14). This system traps gasoline vapor from both the carburetor float bowl and the fuel tank. Figure 3-4 shows one system.

Another reason for making the float bowl as small as possible is to avoid overloading the charcoal canister with fuel vapor. The charcoal canister is part of the evaporative control system. It traps fuel vapor escaping from the float bowl. Making the float bowl small minimizes the amount of gasoline vapor that the charcoal canister has to trap. However, the float bowl must be large enough to maintain the proper fuel level under all operating conditions. If there is too little fuel in the bowl, it could be depleted under some high-demand conditions. This could result in engine hesitation or stumble just when maximum power is needed, such as when passing another vehicle on the highway.

One way of reducing the amount of fuel loss by evaporation from the float bowl is to make the float bowl of molded plastic. The plastic does not conduct heat as well as metal. So less heat gets to the fuel and less evaporation takes place.

Another way of reducing the amount of heat getting up to the float bowl is to place an insulator between the carburetor and intake manifold (Fig. 4-23).

Formerly, floats were made of brass shells. Today, they are generally made of foam plastic. These can be made smaller, and therefore, the float bowl can be made smaller.

⚙ 4-13 Typical float systems Now let us look at typical float systems used in carburetors today. Figure 4-24 shows the float bowl, float, and needle valve and seat for a carburetor. In this single-barrel carburetor, the float, as it is lifted by the rising fuel level, pivots on the float hinge pin (Fig. 4-25). This forces the lip end of the float arm to press the float needle down into the needle seat, shutting off the flow of fuel from the fuel pump.

Another float system is shown in Fig. 4-21. In this design, the end of the float arm has a tang that bears against the almost horizontal needle valve. As the tang moves up in response to a rising fuel level, the tang pushes the needle valve into its seat to shut off the flow of fuel into the float bowl. Note how the float bowl is shaped, sloping up to the needle valve, to reduce the float bowl volume.

Figure 4-26 shows a carburetor with the needle valve above the float arm. As shown in Figs. 4-21, 4-22, 4-25, and 4-26, the needle valve can be placed in any of several positions. All usually work equally well.

Two-barrel carburetors use a float system that has either a single float or a dual float. Figure 4-27 shows a two-barrel carburetor with a single float. Figure 4-28 shows a float-bowl system using two floats attached to

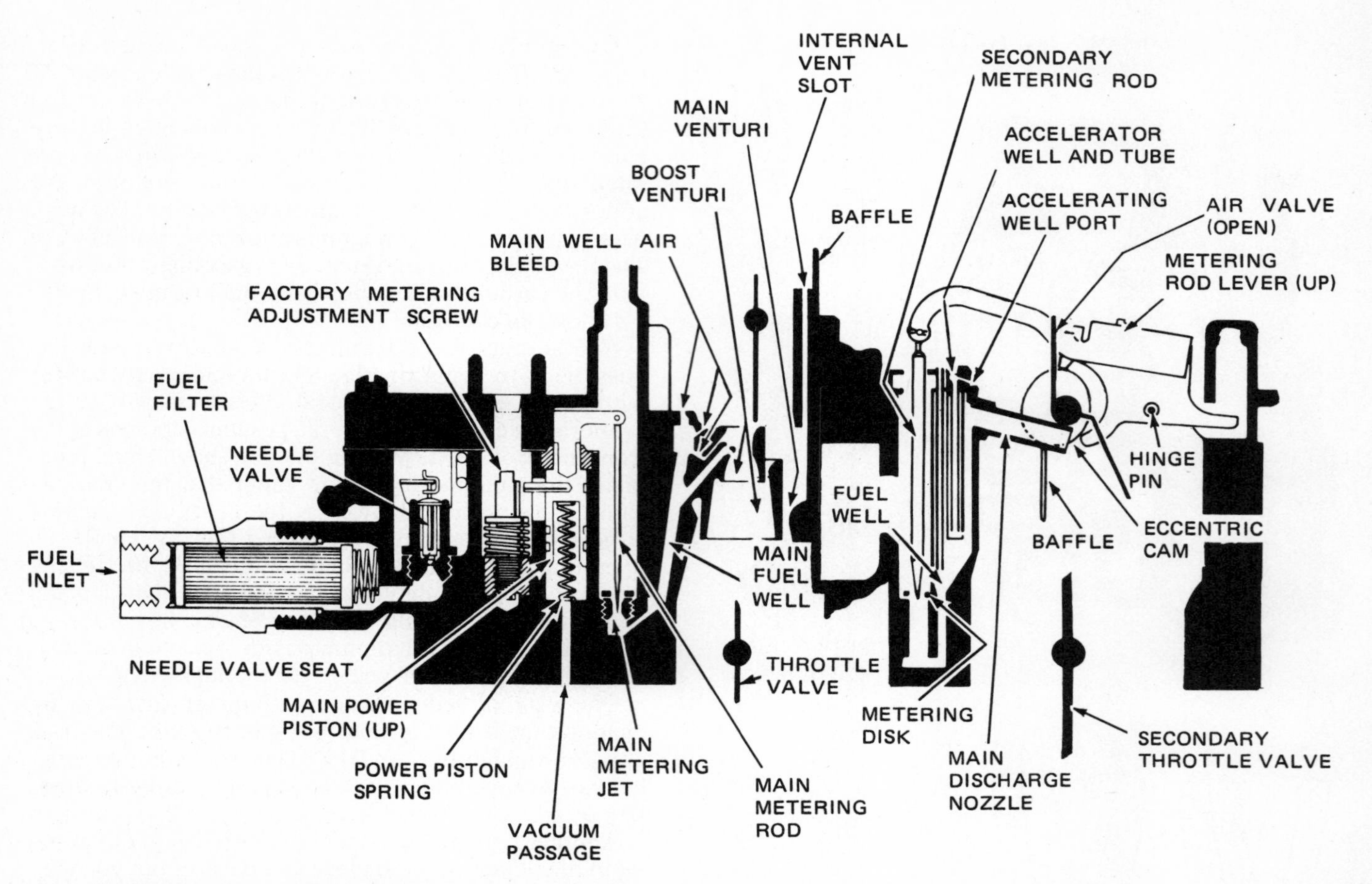

Fig. 4-19 Sectional view of a four-barrel carburetor with mechanical linkage to control the secondary throttle valves. (*Oldsmobile Division of General Motors Corporation*)

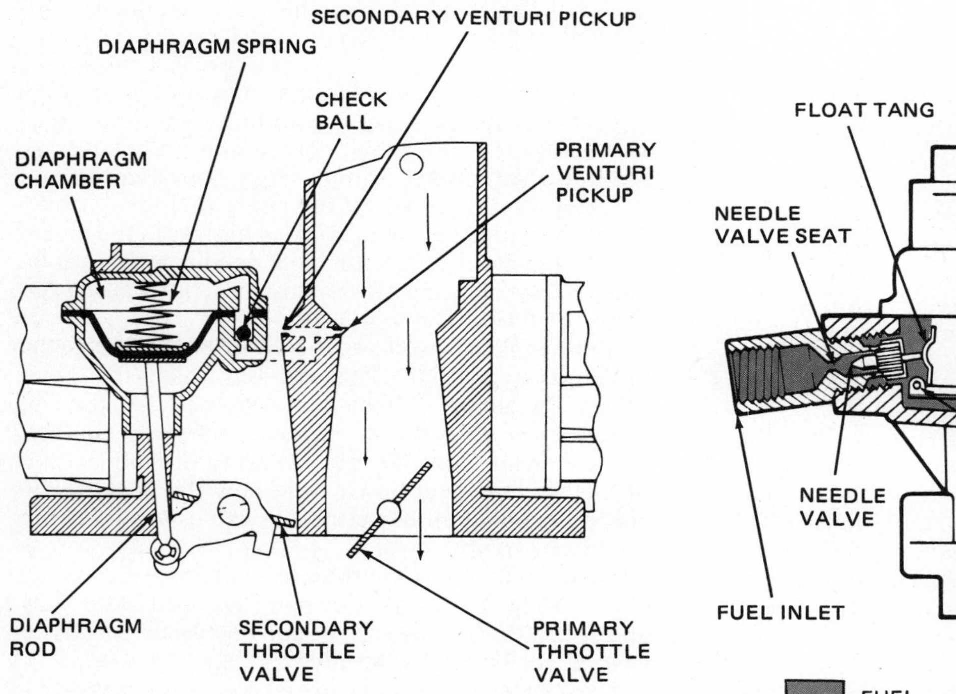

Fig. 4-20 A four-barrel carburetor with vacuum-operated secondary throttle valves. (*Holley Carburetor Division of Colt Industries, Inc.*)

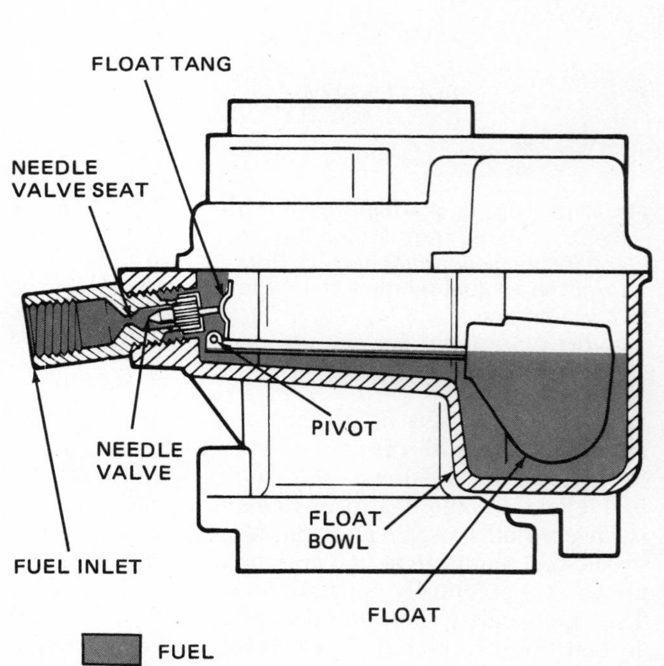

Fig. 4-21 Fuel inlet and float system of a carburetor. (*Chrysler Corporation*)

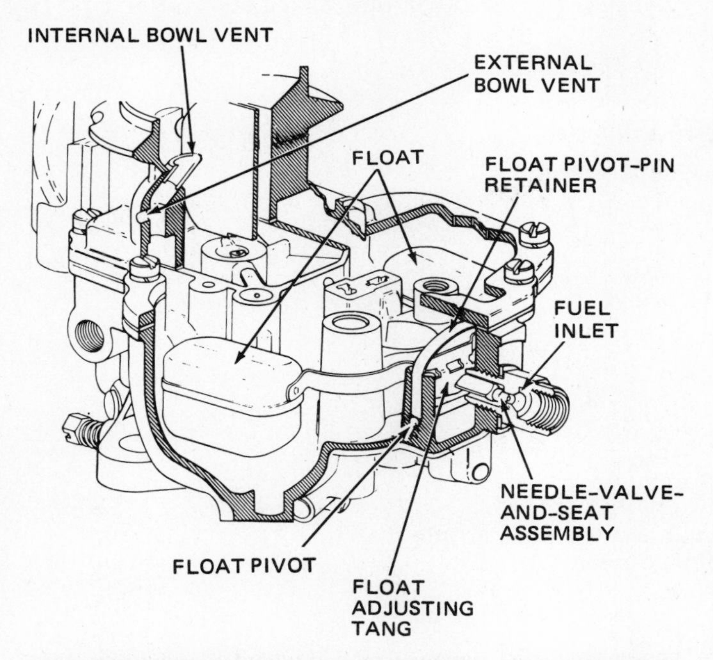

Fig. 4-22 Carburetor partly cut away to show the float system. This carburetor has a pair of floats controlling a single needle valve. (*American Motors Corporation*)

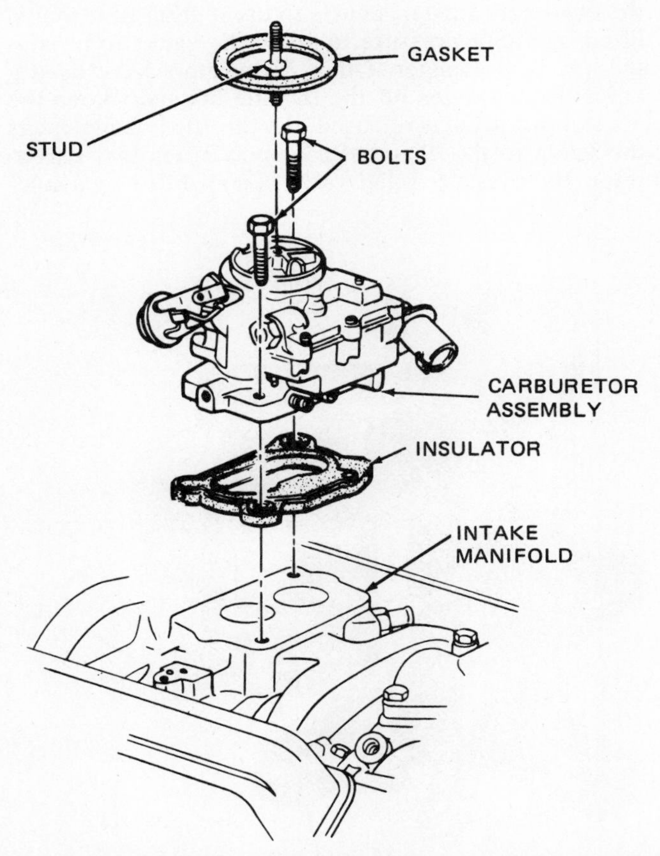

Fig. 4-23 Carburetor insulator placed between the carburetor and the intake manifold. The insulator blocks passage of heat to the carburetor. This reduces fuel evaporation from the float bowl. (*Chevrolet Motor Division of General Motors Corporation*)

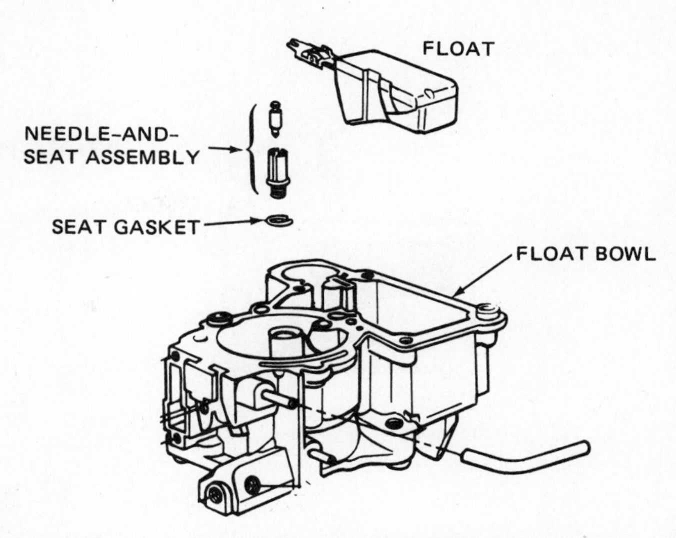

Fig. 4-24 Disassembled view of the float and related parts. (*Chevrolet Motor Division of General Motors Corporation*)

the ends of a U-shaped lever. The needle valve is closed by a tang at the center of the float lever.

The four-barrel, or quad, carburetor, uses float bowl arrangements similar to those discussed above. Figure 4-29 shows the air horn, float-and-lever assembly, and main body or float bowl of a four-barrel carburetor. Some four-barrel carburetors have two separate floats. The purpose of using two separate float systems is to assure adequate fuel to both the primary and secondary barrels under all conditions.

Another method of assuring adequate fuel flow into the float bowl under maximum-demand conditions is to use a second or auxiliary valve (Fig. 4-30). Under normal operating conditions, the main needle valve is opened to supply the fuel needs. However, when the engine is under heavy load or accelerating, it may require more fuel than the main needle valve can supply. With this condition, the float starts to drop. This causes the tab on the end of the float bowl arm to push the auxiliary valve up so that additional fuel can enter the float bowl.

⚙ 4-14 Float bowl vents There is always gasoline vapor above the gasoline in the float bowl. However, if the under-the-hood temperatures go up, and the engine gets hot enough, the gasoline vapor will produce pressure in the float bowl. The pressure has to be relieved. If it is not, it would force excessive amounts of gasoline out the main nozzle. Then the engine would receive a very rich air-fuel mixture.

Float bowls must be vented to allow the gasoline vapor pressure to escape. At one time, carburetors were designed to allow the vapor to escape into the atmosphere. These were called *unbalanced carburetors*. Emission-control laws now limit their use. Today carburetor float bowls are vented internally either into the upper air horn, or into the fuel vapor recovery system (Fig. 4-25). In this system, a charcoal canister traps the vapor and then returns it to the engine to be burned when the engine is started (⚙ 3-14). Internally vented carburetors are called *balanced carburetors*.

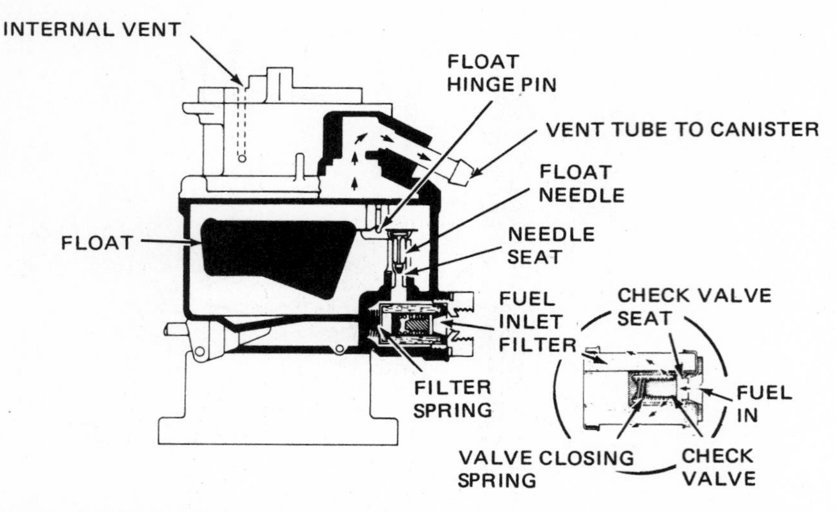

Fig. 4-25 Float system showing the two vents, one internal and the other to the charcoal canister. (*Chevrolet Motor Division of General Motors Corporation*)

In addition to polluting the air with unburned gasoline vapor (HC), an unbalanced carburetor has another disadvantage. If the air cleaner becomes clogged, it causes a partial vacuum to form in the upper part of the air horn. This adds to the vacuum produced by the airflow past the fuel nozzle. Then additional fuel is discharged from the fuel nozzle and the mixture is too rich. The balanced carburetor does not have this problem. If a vacuum forms because of a clogged air cleaner, this vacuum acts on both the fuel nozzle and the float bowl. Therefore no increase in fuel flow through the nozzle results.

There are several ways to open and close the vent to the charcoal canister. One is to use a disk valve that is lifted by vapor pressure to allow the vapor to pass to the charcoal canister. Other carburetors have used a valve that operates off the throttle linkage. When the accelerator pedal is released, the throttle linkage opens the valve to the charcoal canister. In another carburetor, the pressure relief valve is controlled by a sole-

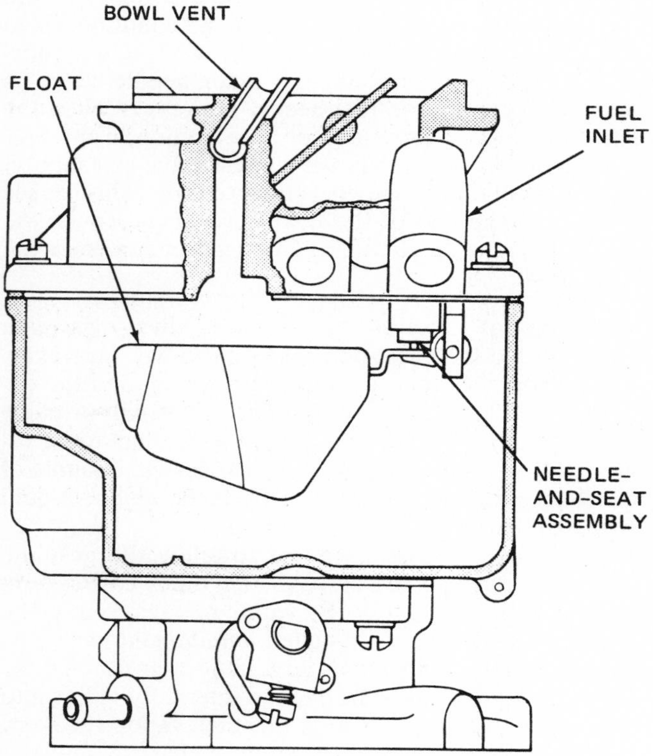

Fig. 4-26 The float system in a carburetor, with the needle valve located above the float arm. (*American Motors Corporation*)

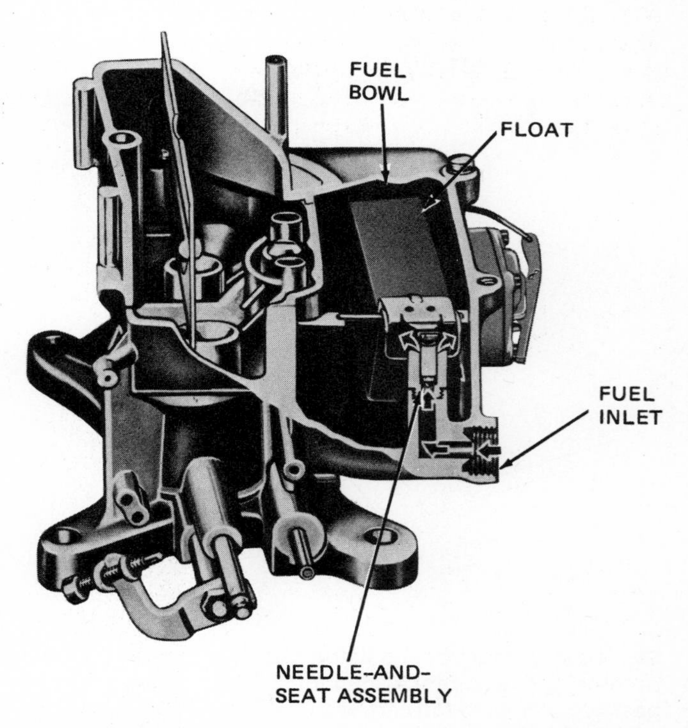

Fig. 4-27 A two-barrel carburetor using a single float. (*American Motors Corporation*)

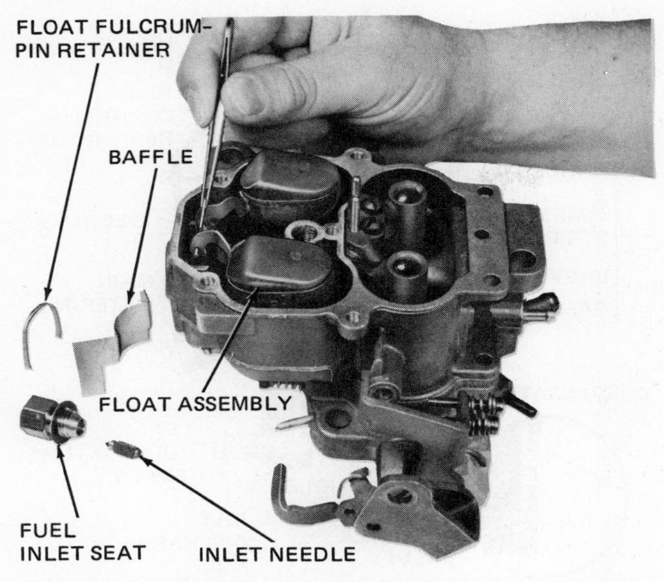

Fig. 4-28 Float bowl, showing the two floats attached to the same lever. (*Chevrolet Motor Division of General Motors Corporation*)

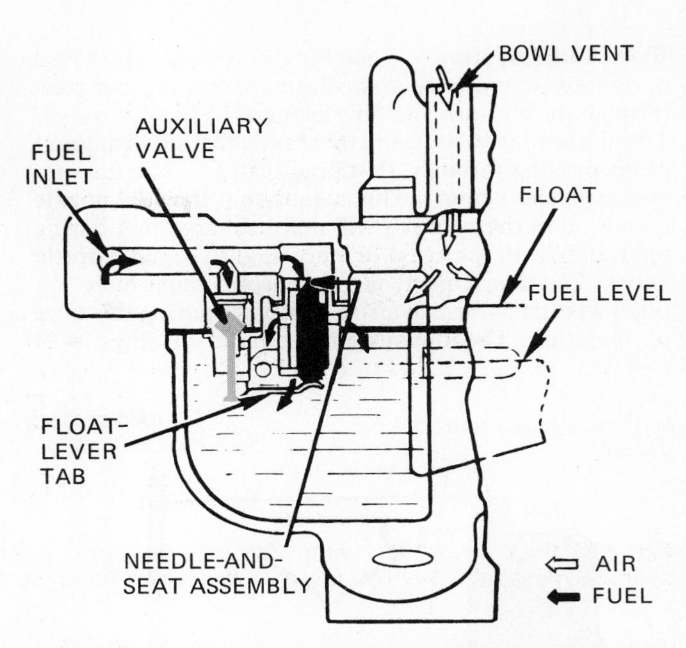

Fig. 4-30 Float system for a four-barrel carburetor. Note the auxiliary valve which opens to admit additional fuel if the float level falls. (*American Motors Corporation*)

noid (Fig. 4-31). When the ignition is turned on, the solenoid is connected to the battery. It pulls the solenoid plunger down. This blocks off the vent to the charcoal canister. Now, the float bowl is vented to the top of the air horn. When the ignition is turned off, the solenoid is disconnected from the battery. Then, the solenoid spring pushes the plunger up. This blocks the vent to the upper part of the air horn. At the same time, the vent to the charcoal canister is opened.

4-15 Hot-idle compensator One method of taking care of the excessive richness of the mixture that results from high idling temperatures is to use a thermostatic valve to admit additional air (Fig. 4-32). When under-the-hood temperatures are high, the thermostatic blade warps and lifts the valve off its seat. Now air can flow through a small air passage to an opening under the throttle plate. The added air compensates for the increased richness of the air-fuel mixture from the idle system. This valve is also called a hot-idle compensator.

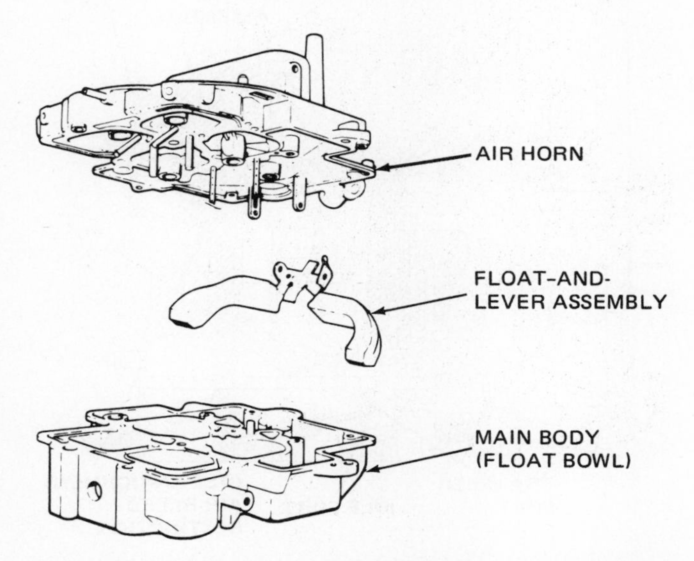

Fig. 4-29 Main parts of the float system in a four-barrel carburetor. (*American Motors Corporation*)

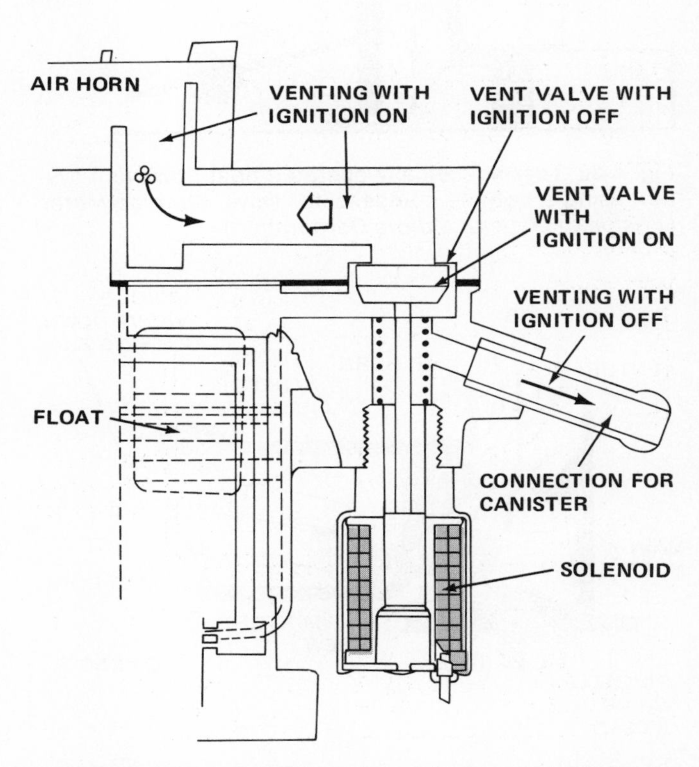

Fig. 4-31 Sectional view of a carburetor which has a solenoid-controlled vent to the charcoal canister. (*Ford Motor Company*)

⚙ 4-16 Idle system When the throttle valve is closed or slightly open, only a small amount of air can pass through the air horn and flow around the throttle valve. The airspeed is so low, and there is such a small amount of air passing through, that practically no vacuum develops in the venturi. This means that the fuel nozzle (centered in the venturi) will not discharge fuel during operation with a closed or only slightly open throttle valve. For this reason, the carburetor must have another system to furnish air-fuel mixture during this type of operation. The idle system does this job (Figs. 4-33 to 4-37).

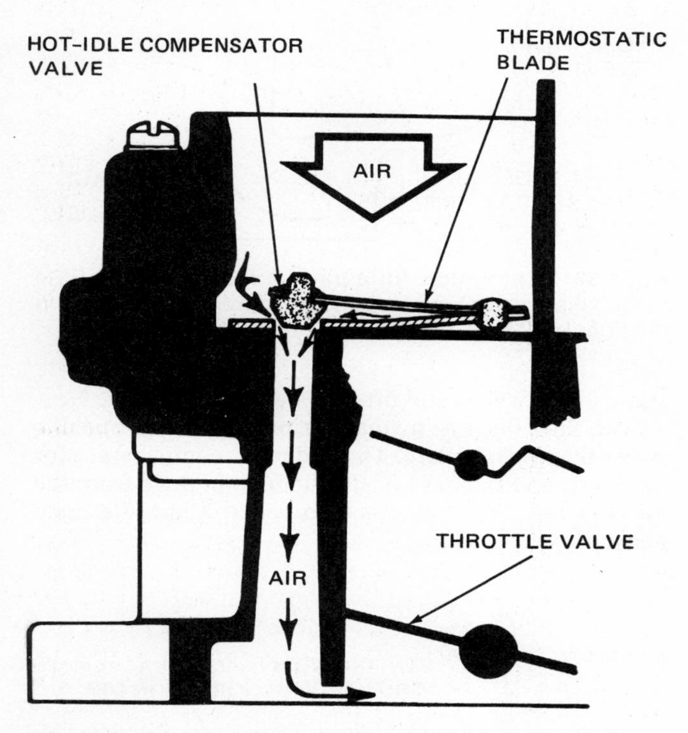

Fig. 4-32 Thermostatically operated float bowl vent system using a hot-idle compensator valve. (*Pontiac Motor Division of General Motors Corporation*)

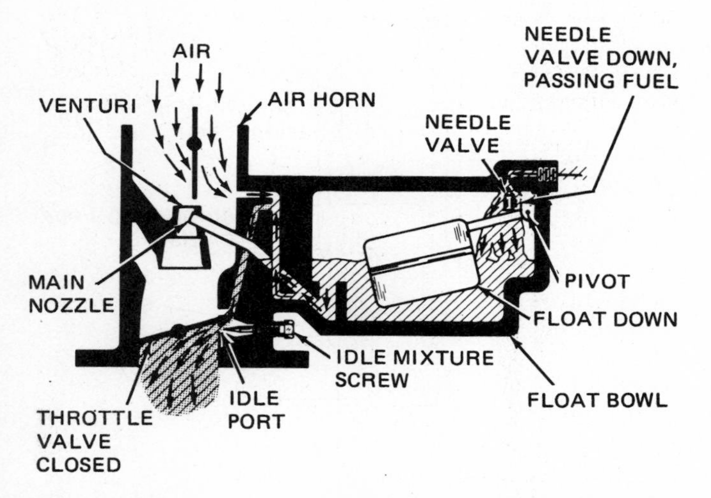

Fig. 4-33 Idle system in a carburetor. The throttle valve is closed, and all fuel is being discharged past the idle mixture screw.

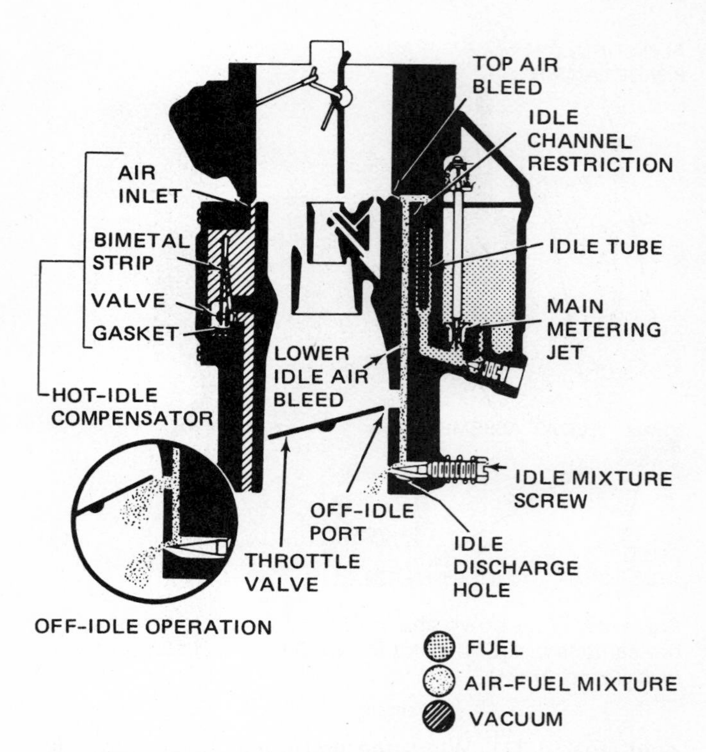

Fig. 4-34 Idle system, showing the vacuum passages, and the flow of fuel and air-fuel mixture through the carburetor. (*Buick Motor Division of General Motors Corporation*)

The idle system consists of a series of openings through which air and fuel can flow. With the throttle valve closed and the engine idling, very little air can pass between the throttle valve and the throttle bore. Therefore, a relatively high vacuum exists below the throttle valve.

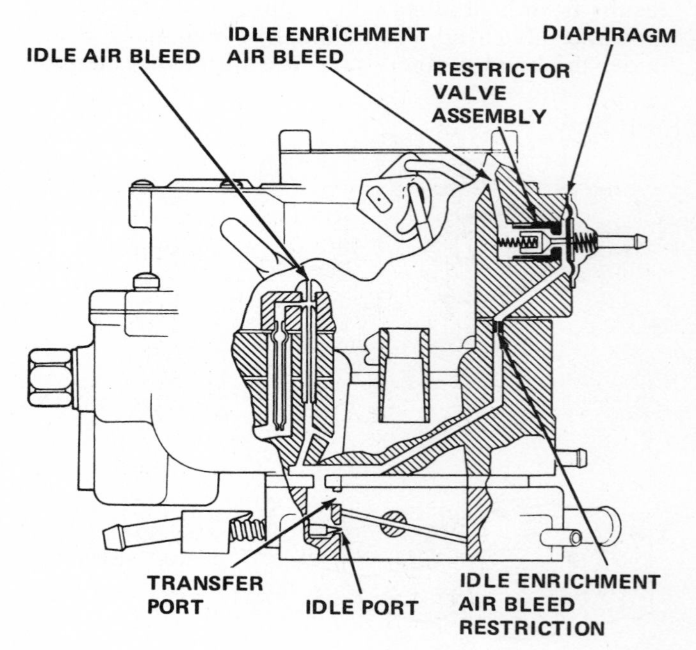

Fig. 4-35 Location of the restrictor valve in the cold-idle enrichment system. (*Chrysler Corporation*)

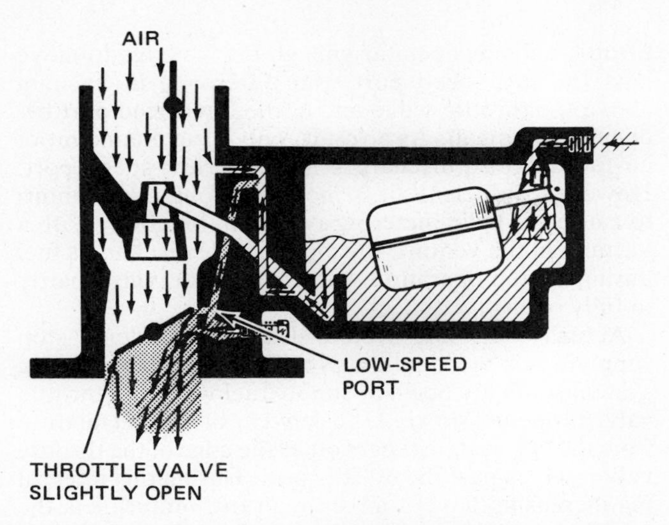

Fig. 4-36 Idle system in a carburetor. The throttle valve is slightly open, and fuel is being discharged through the low-speed port.

There is a small passage from the upper part of the air horn through the carburetor body to the idle-mixture screw. This passage is called the idle-and-low-speed passage. Atmospheric pressure forces air through this passage (as shown by the lines in Fig. 4-33). Fuel also feeds into this passage from the float bowl. Atmospheric pressure pushes the fuel toward the vacuum below the throttle valve. The air and fuel mix as they move through the passage toward the idle-mixture screw. The mixture is rich. It flows past the tapered point of the idle-mixture screw and down into the intake manifold. It mixes with the small amount of air that

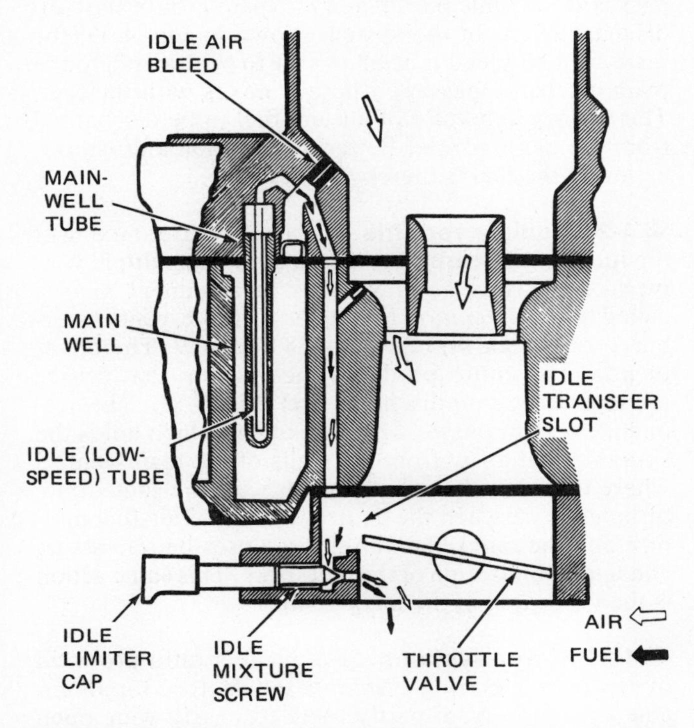

Fig. 4-37 Idle system in one primary barrel of a four-barrel carburetor. (*American Motors Corporation*)

gets past the throttle valve to form a slightly leaner mixture.

The mixture richness can be adjusted by turning the idle-mixture screw in or out. When it is turned in, less air-fuel mixture can pass through the system, and so the final mixture is leaner. When the screw is turned out, more air-fuel mixture can pass through. Therefore, the final mixture is richer. Adjustment must be made so that the engine will be supplied with the correct air-fuel mixture for minimum exhaust emissions and smooth idling.

Figure 4-34 shows a similar idle system. The circle at the left shows the position of the throttle valve after it has been opened slightly to an *off-idle* position. Notice that this carburetor has an exposed idle-mixture screw. This is no longer allowed on automotive carburetors.

NOTE: Many late-model carburetors have a locking cap on the idle-mixture screws. On others the idle-mixture screws are recessed in the throttle body and sealed with hardened-steel plugs. The locking caps permit limited adjustment of the idle mixture. The steel plugs discourage tampering with the idle-mixture screws. The only way the plugs can be removed is by disassembly of the carburetor. Servicing of carburetors is covered in later chapters.

Some carburetors have an extra feature in the idle system to provide some enrichment of the idle mixture during cold start-up. This enrichment is in addition to the choke action. The arrangement includes a restrictor valve in the air bleed to the idle system (Fig. 4-35). This restrictor valve is controlled by a vacuum diaphragm. When vacuum is applied to the vacuum diaphragm, the restrictor valve partly closes off the air flow. The air loss in the idle system then causes additional fuel to discharge from the idle system. This improves cold idle and combats the tendency for the engine to stall.

The vacuum diaphragm gets its vacuum through a thermal switch that is mounted in the cooling system. When the coolant is cold, the thermal switch opens to allow intake manifold vacuum to operate the vacuum diaphragm. Therefore, the idle air-fuel mixture is enriched. When the engine warms up, the thermal switch closes to shut off the vacuum to the vacuum diaphragm. Now, normal hot-engine idle results, with normal hot-engine air-fuel mixture discharging from the idle system.

⚙ 4-17 Low-speed operation When the throttle is opened slightly (Fig. 4-36), the edge of the throttle valve moves past the low-speed port or transfer port in the side of the throttle body (Fig. 4-37). This port may be a vertical slot or a series of small holes, one above the other. With the throttle valve only slightly open, too little air passes through to produce a vacuum in the venturi. As a result, the main fuel nozzle still does not discharge fuel. However, more fuel is needed than can be supplied through the idle port (past the idle-mixture screw). The low-speed port supplies this additional fuel.

As the edge of the throttle valve moves past the idle

port, intake-manifold vacuum is applied to the low-speed port. Now, atmospheric pressure (pushing toward the vacuum) causes this port and the idle port to start discharging air-fuel mixture. This mixture is rich, but it is leaned out by the air passing the throttle valve. As the throttle is opened more, a large part of the slotted port (or more of the drilled holes) is cleared. Therefore, more air-fuel mixture is delivered to the engine.

⚙ 4-18 Other idle systems Figure 4-37 illustrates an idle system similar to those discussed in ⚙ 4-16 and 4-17. The vacuum below the throttle valve, when it is closed, causes air to bleed into the idle air bleed. At the same time, the vacuum causes the fuel in the idle tube to discharge through the passage to the idle port, as shown by the arrows. The air entering through the idle air bleed mixes with the fuel to provide a partially atomized mixture.

The idle air bleed has another function. It prevents siphoning of fuel through the idle system during high-speed operation or engine shutdown. During these conditions, the vacuum in the intake manifold, working through the idle system, could draw out fuel from the float bowl. However, the air bleed prevents this.

There are many varieties of idle systems in addition to those shown in Figs. 4-33 to 4-37. In two-barrel carburetors, each barrel has its own system. In most four-barrel carburetors, only the primary barrels have idle systems.

NOTE: Most ignition distributors have a vacuum-advance mechanism which advances the ignition timing under part-throttle conditions (⚙ 1-20). Figures 1-33 and 1-34 show how the distributor vacuum-advance mechanism is connected to the carburetor by a vacuum line. The vacuum line opens into a hole or slot cut in the carburetor throttle body about even with the low-speed port. The two openings (vacuum and low-speed or off-idle ports) should not be confused.

⚙ 4-19 Main metering system The main metering system includes the fuel nozzle (called the *main nozzle*, or *high-speed nozzle*), the venturi, and the fuel passages from the float bowl to the nozzle (Fig. 4-38). When the throttle valve is open far enough for the edge to move past the low-speed port, the difference in vacuum above the throttle valve and at the low-speed port becomes very small. As a result, only a small amount of air-fuel mixture discharges from the low-speed port. However, sufficient air is moving through the venturi to cause the main metering system to function. With a vacuum in the venturi, the main nozzle discharges fuel during engine operation with the throttle valve partly to fully open.

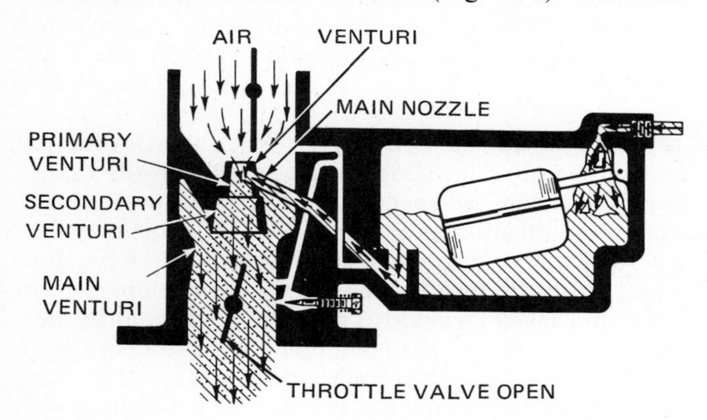

Fig. 4-38 Main metering system in a carburetor. The throttle valve is almost wide open, and fuel is being discharged through the high-speed or main nozzle.

Actually, the idle system does not suddenly stop supplying air-fuel mixture. Nor does the main metering system suddenly begin to supply fuel when the throttle valve is opened slowly. The delivery of air-fuel mixture from the idle system tapers off as the edge of the throttle valve swings past the off-idle port. During this interval the increasing flow of air through the venturi sets the main metering system into operation. Therefore, the main metering system takes over as the idle system fades out. The two systems are designed so that, as the throttle valve is moved away from idle, a nearly constant air-fuel ratio is maintained.

The wider the throttle is opened and the faster the air flows through the air horn, the greater the vacuum in the venturi. Therefore, fuel will be discharged from the main nozzle because of the greater vacuum. As a result, a nearly constant air-fuel ratio is maintained by the main metering system from part- to wide-open throttle.

Figures 4-38 and 4-39 illustrate main metering systems. Even though various fixed venturi carburetors are different in appearance and construction, all operate in basically the same way. The main metering system takes over from the idle system as the throttle passes the off-idle port. Then the main nozzle starts to discharge fuel. In the system shown in Fig. 4-39, the main-well air bleed tube allows air to be drawn into the main discharge passage where it mixes with the fuel. This causes a mixture of air and fuel to be discharged from the main nozzle. Better atomization and vaporization of the fuel is thereby achieved.

⚙ 4-20 Multiple venturis To assure better mixing of the fuel and air, carburetors usually have multiple venturis, one inside another. This arrangement also is called a *boost venturi*. Examples of triple-venturi carburetors are shown in Figs. 4-38 and 4-39. The upper or primary venturi produces the vacuum that causes the main nozzle to discharge fuel (Fig. 4-38). The secondary venturi passes a blanket of air, which holds the spraying fuel away from the walls of the main venturi where it might otherwise condense. At the same time, turbulence between the central stream of air-fuel mixture and the outer blanket of air causes better mixing and finer atomization of the fuel spray. This same action is then repeated in the main venturi.

⚙ 4-21 Power systems The air-fuel ratio provided by the main metering system is satisfactory for all engine operation from partly open to nearly wide-open throttle. However, at wide-open throttle, where full engine power is desired, an increase in mixture richness

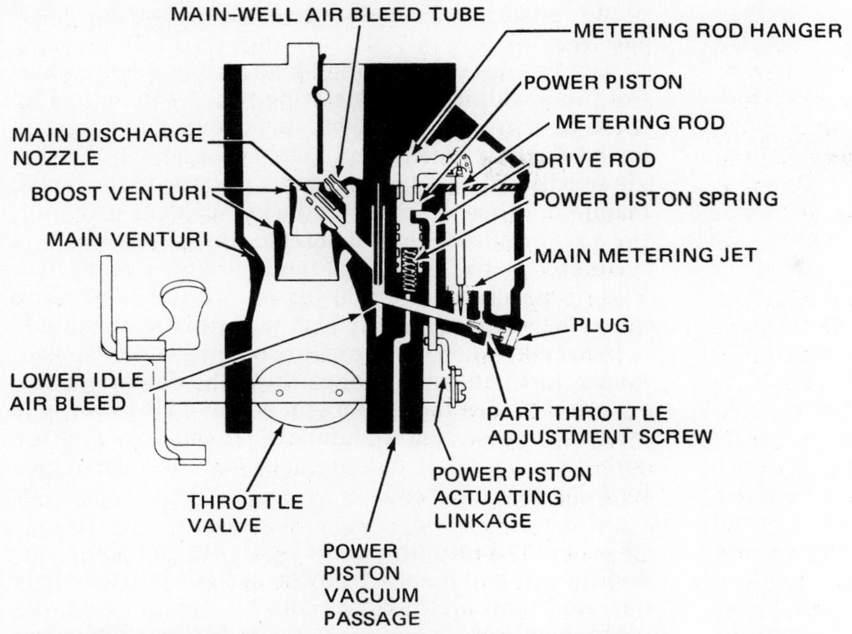

Fig. 4-39 Main metering system. (*Chevrolet Motor Division of General Motors Corporation*)

is required. To obtain a richer mixture, an additional system is incorporated in the carburetor. This is called the power system. It admits an additional flow of fuel to the main nozzle so that it discharges more fuel. Two general types of power system are in use. One is mechanically operated and the other is operated by intake-manifold vacuum.

⚙ 4-22 Mechanically operated power systems

The mechanically operated power system makes use of a metering-rod jet and a metering rod. A *jet* is an accurately drilled orifice, or small hole. The metering rod either has two or more steps of different diameters, or it is tapered at its lower end (Fig. 4-40).

At intermediate throttle, the larger diameter or step is in the metering-rod jet. This restricts the fuel flow to the main nozzle by partially blocking the jet. However, sufficient fuel does flow to provide the proper air-fuel ratio during intermediate-throttle operation.

But when the throttle is fully opened, the metering rod is raised enough to cause the smaller diameter or step to be lifted up out of the metering-rod jet (Fig. 4-41). Now the jet is less restricted, and a larger quantity of fuel can pass through it. As the fuel nozzle delivers more fuel, a richer mixture results. This is needed for full-throttle, maximum-power operation of the engine.

One reason for the use of the tapered metering rod

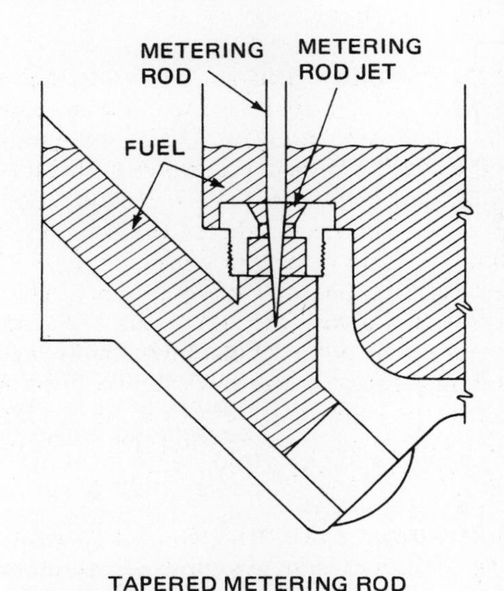

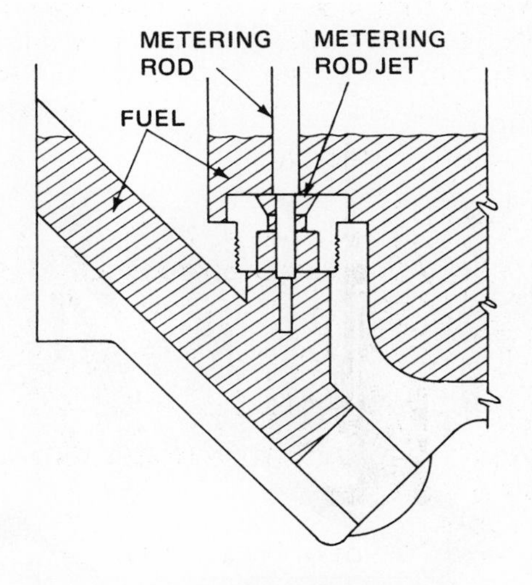

Fig. 4-40 Two types of metering rods used for controlling fuel flow through the metering rod jet into the power system.

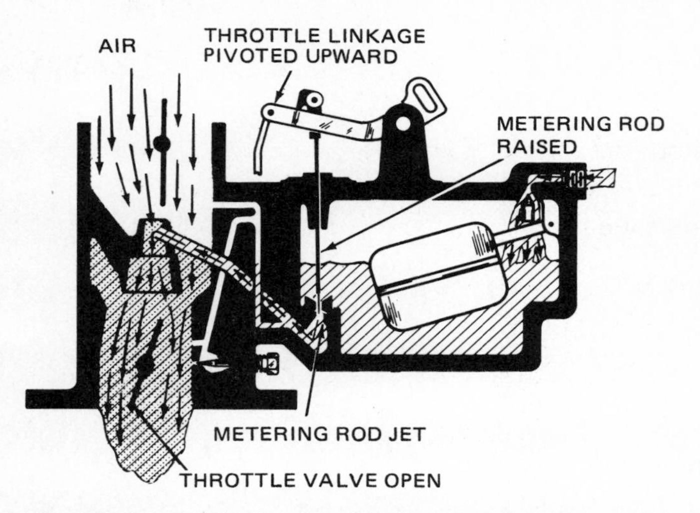

Fig. 4-41 Mechanically operated power system. When the throttle valve is open, as shown, the metering rod is raised so that the smaller diameter of the rod clears the jet. This allows additional fuel to flow through the jet.

is that it provides a more gradual increase in the amount of fuel flowing through the jet as the rod is raised. It is similar in effect to a step-type rod with many steps.

⚙ 4-23 Vacuum-operated power systems The vacuum-operated power system is operated by intake manifold vacuum. It includes a power piston or diaphragm linked to a valve or metering rod similar to the ones shown in Fig. 4-40. One design is shown in Fig. 4-42. During part-throttle operation, the piston is held in the lower position by intake manifold vacuum. However, when the throttle is opened wide, the intake manifold vacuum is reduced. This allows the spring under the power piston to push the piston upward. This motion raises the metering rod so that the smaller diameter of the rod clears the metering-rod jet. Now, more fuel can flow.

A carburetor using a spring-loaded diaphragm to control the position of the metering rod is shown in Fig. 4-43. In the system shown, a flexible diaphragm is linked to the metering rod. The space above the diaphragm is connected by a vacuum passage to the intake manifold. When there is vacuum in the intake manifold, the vacuum holds the diaphragm up. In this position the metering rod is up, and the fuel flow is restricted. The up position is normal for part-throttle operation when the vacuum is fairly high in the intake manifold.

However, when the throttle is opened wide, the vacuum is lost and can no longer hold the diaphragm up. A spring pushes the diaphragm down. This lowers the metering rod so that additional fuel can flow into the carburetor. A richer mixture is delivered to the engine for full-power operation.

⚙ 4-24 Combination power systems In some carburetors, a combination power system is used. It is operated both mechanically and by vacuum from the intake manifold. In one such carburetor, a metering rod is linked to a vacuum diaphragm and to the throttle linkage. Movement of the throttle to wide open lifts the metering rod to enrich the mixture. Or loss of intake manifold vacuum (such as during acceleration) causes the vacuum diaphragm spring to raise the metering rod for an enriched mixture.

Another type of combination power system that uses a power valve instead of a metering rod or diaphragm is shown in Figs. 4-44 and 4-45. With high intake-manifold vacuum, the power piston is held up against spring tension (Fig. 4-44). The power valve is closed. As the throttle is opened, vacuum decreases. This allows the spring in the vacuum piston to push it down, opening the power valve. Additional fuel flows through the

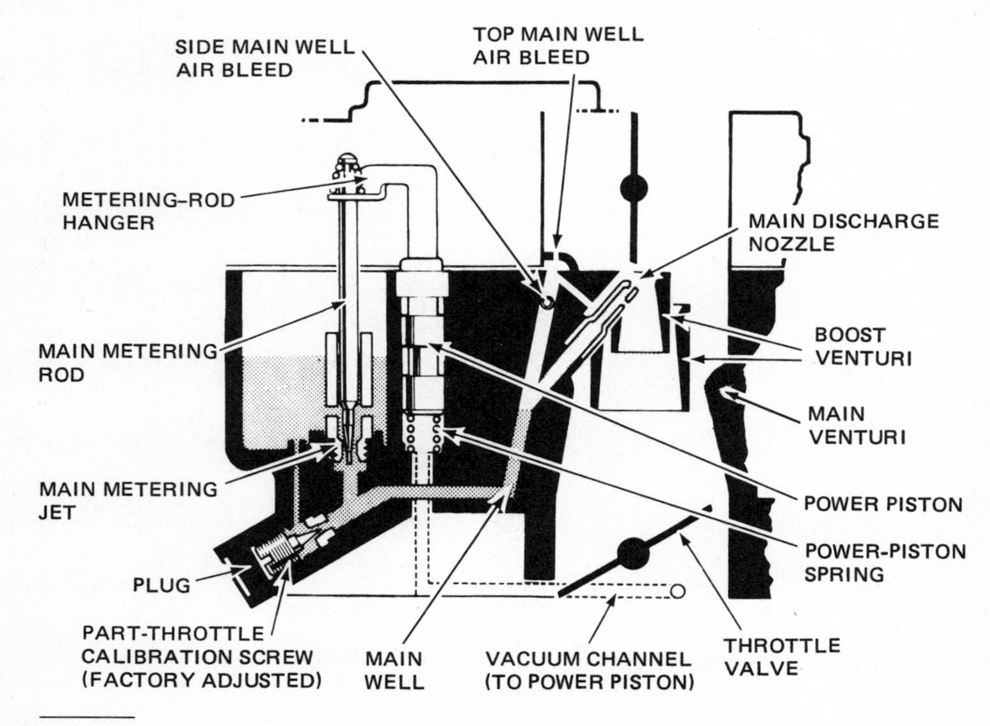

Fig. 4-42 Vacuum-operated metering rod. As the throttle valve is opened, the reduced intake manifold vacuum allows the power piston spring to raise the power piston. This, in turn, controls the position of the metering rod. (*Chevrolet Motor Division of General Motors Corporation*)

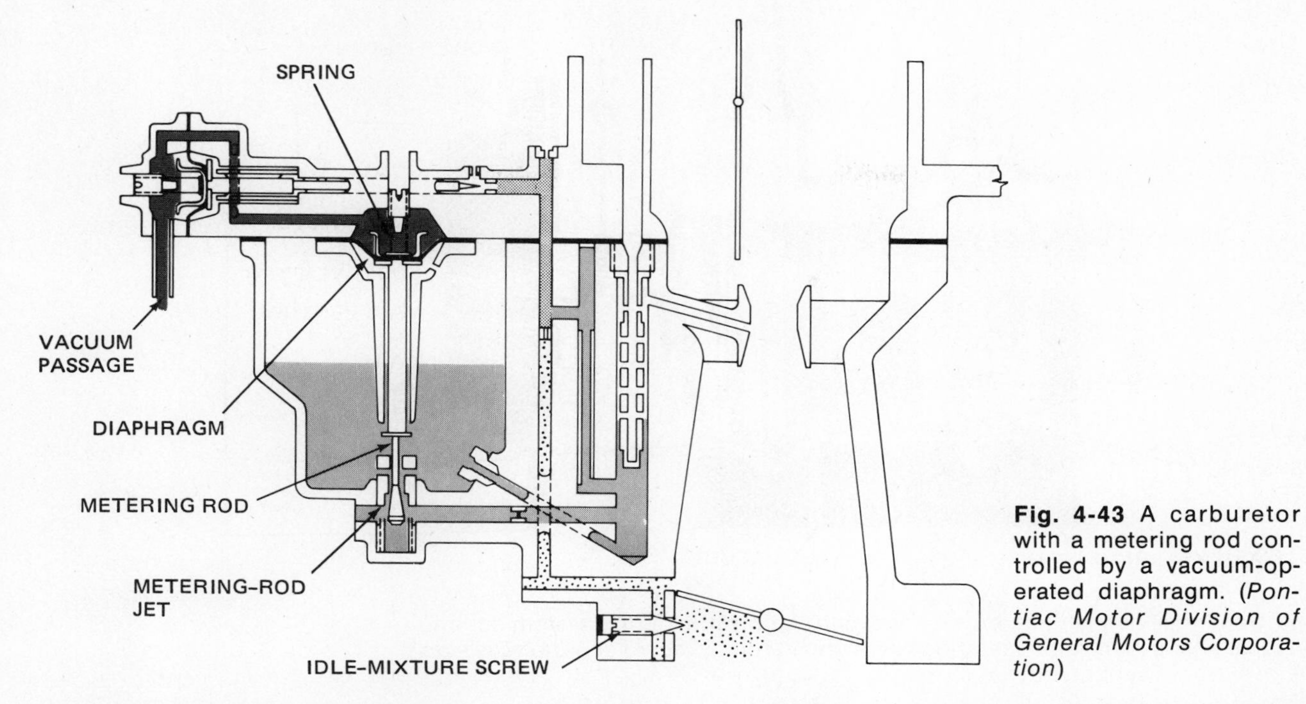

Fig. 4-43 A carburetor with a metering rod controlled by a vacuum-operated diaphragm. (*Pontiac Motor Division of General Motors Corporation*)

power valve into the power-valve restriction, and into the main well.

When the throttle valve is wide open, there is no intake-manifold vacuum. However, a lever operated by the throttle linkage pushes down on the power-valve push rod to open the power valve (Fig. 4-45). Now more fuel flows through the power valve to enrich the air-fuel mixture.

⚙ 4-25 Factory-adjustable part throttle To limit exhaust emissions during part-throttle operation, some carburetors have a factory-set fuel-metering adjustment screw (Fig. 4-19). It provides a more accurate adjustment of the fuel flow during part-throttle operation. This compensates for any slight differences resulting from manufacturing tolerances. Should replacement of the float bowl be necessary, the metering screw will already be in the new float bowl and properly adjusted.

Other adjustable part-throttle (APT) carburetors have an additional metering rod and fixed metering jet (shown to the left in Fig. 4-46). The metering rod is adjusted at the factory by turning the adjusting screw. If it is turned to lift the metering rod, more fuel can flow through the fixed metering jet. If the screw is turned to lower the metering rod, less fuel can flow.

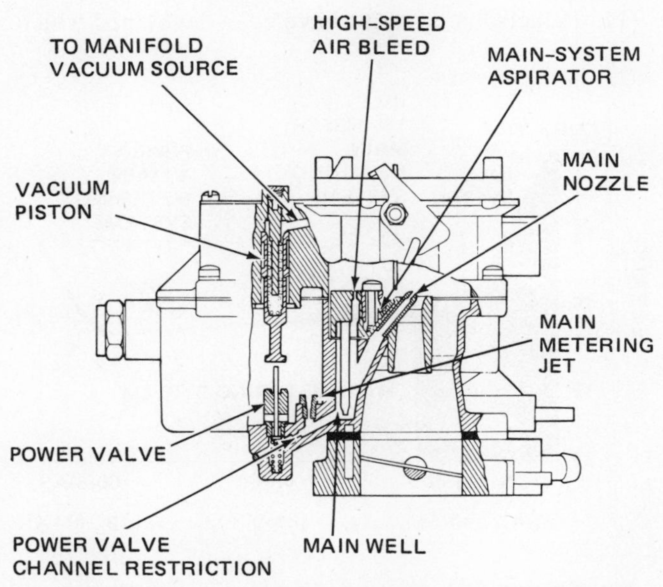

Fig. 4-44 Carburetor with a combination power system, showing the vacuum-operated power valve. (*Chrysler Corporation*)

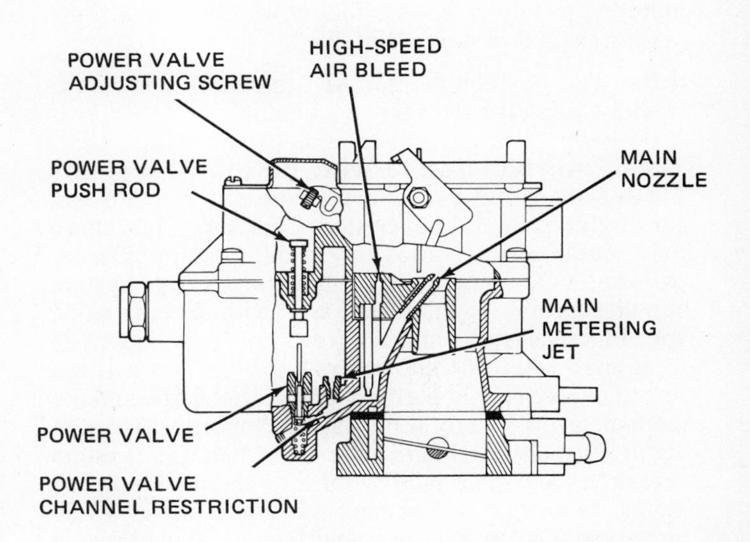

Fig. 4-45 A carburetor power system, showing the mechanically operated power valve. (*Chrysler Corporation*)

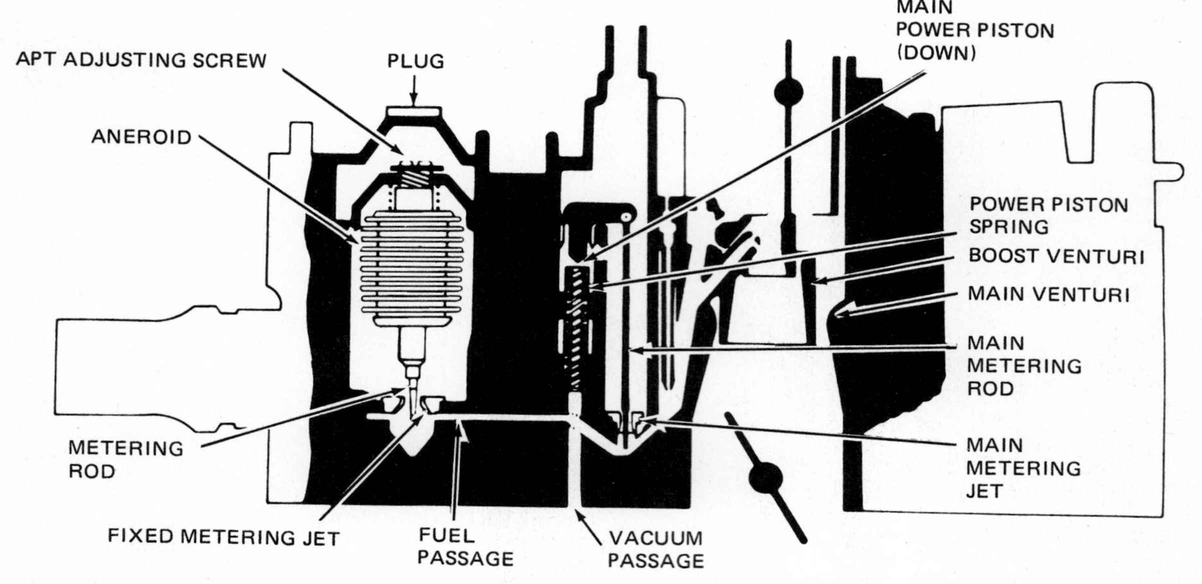

Fig. 4-46 Sectional view of a carburetor showing the main metering system and the factory-adjustable part-throttle metering rod with aneroid. (*Chevrolet Motor Division of General Motors Corporation*)

⚙ 4-26 Altitude compensation

The carburetor shown in Fig. 4-46 has an aneroid surrounding the APT metering rod. An aneroid is an enclosed bellows which is sensitive to changes in atmospheric pressure. As the pressure is increased, the aneroid is squeezed so that it is shortened. As the pressure is reduced, the aneroid expands. These actions raise or lower the metering rod as atmospheric pressure changes.

For example, suppose that the car is driven up a mountain so that the atmospheric pressure is reduced. Without any compensating device, the air-fuel mixture would become enriched. This occurs because less air would enter the carburetor. The air pressure is lower. To compensate for this, the aneroid expands as a result of the reduced pressure. This action moves the APT metering rod down so less fuel flows. As a result, the proper air-fuel ratio is maintained.

NOTE: The position of the APT metering rod is extremely critical. Never try to adjust it.

⚙ 4-27 Air-fuel ratios with different systems

Figure 4-47 shows the air-fuel ratios with the different carburetor systems in operation. This is a typical curve only. Actual air-fuel ratios may be different for different carburetors and operating conditions. The idle system supplies a very rich mixture to start with. But as engine speed increases, the mixture leans out. From about 25 to 40 mph [40 to 64 kilometers per hour (km/h)], the throttle valve is only partly open. Both the idle and the main metering system supply air-fuel mixture. Then, at about 40 mph [64 km/h], the main metering system takes over and continues by itself to about 60 mph [97 km/h]. Note that the air-fuel ratio increases as speed increases. The mixture becomes leaner. About 60 mph [97 km/h], the power system comes into operation (earlier, if the throttle valve is opened wide at a lower speed). Now, the richness of the air-fuel mixture increases with higher speeds.

⚙ 4-28 Electronic control of air-fuel ratio

The idle, low-speed, and power systems supply varying air-fuel ratios to meet different engine operating conditions. These systems depend on mechanical controls which often are not as accurate as emission laws require. Inaccurate fuel metering can sometimes cause excessively rich mixtures. These, in turn, produce excessive HC and CO in the exhaust gas. When the mixture is too lean, excessive NO_x appears in the exhaust gas and the engine tends to stall.

New electronic systems have been developed which

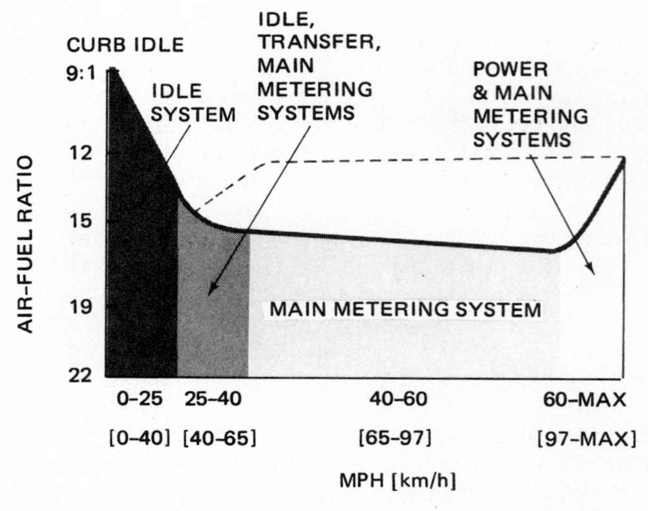

Fig. 4-47 Air-fuel ratios with different carburetor systems operating at various speeds. (*Chevrolet Motor Division of General Motors Corporation*)

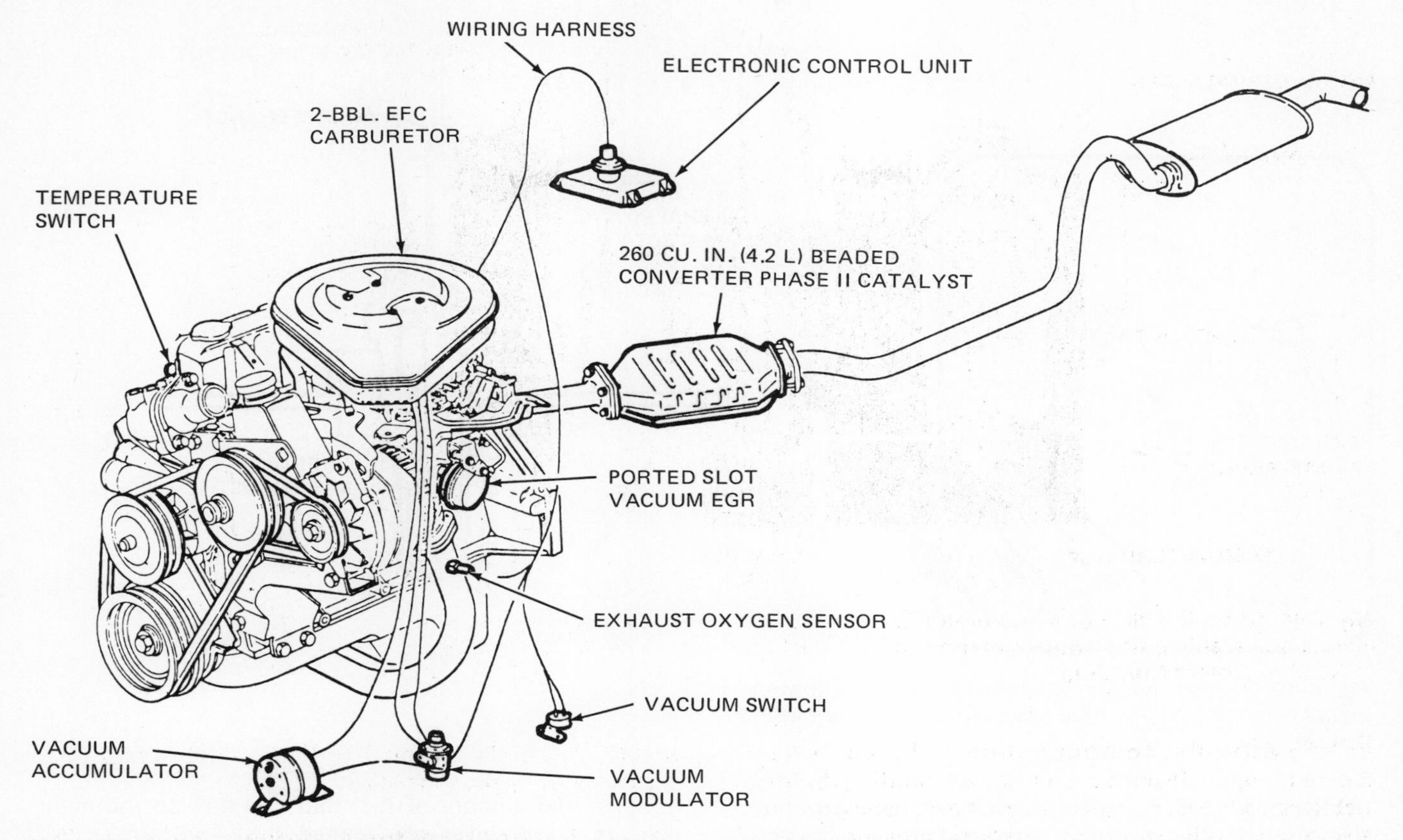

Fig. 4-48 Electronic fuel control (EFC) system on a V-8 engine. (*Chevrolet Motor Division of General Motors Corporation*)

control the air-fuel ratio of the mixture being fed to the engine from the carburetor. These new systems monitor the amount of oxygen in the exhaust gases. If the exhaust gas is low in oxygen, the air-fuel mixture is too rich. If it is high in oxygen, the air-fuel mixture is lean.

By continuously measuring the oxygen content in the exhaust gas, the system determines if the air-fuel ratio needs adjustment. It if does, and the mixture is rich, for example, the system automatically reduces the amount of fuel entering the air going through the carburetor. This leans out the mixture. If the mixture becomes too lean (higher oxygen content in the exhaust gas), the engine will not run well. It may stumble or even stall because the excessively lean mixture will not burn well in the engine cylinders. In this case, the electronic system readjusts the carburetor to enrich the mixture.

Two systems are discussed below that control the air-fuel ratio electronically, the General Motors "Electronic Fuel Control" (EFC) system (⚙ 4-29), and the Ford "Feedback-Carburetor Electronic Engine-Control" system (⚙ 4-30). All such systems reduce the pollutants in the exhaust gas.

⚙ 4-29 General Motors electronic fuel control system This system is shown in Fig. 4-48. It includes an exhaust-gas oxygen sensor or "sniffer" (Fig. 4-49), an electronic control unit (ECU), an engine temperature switch, three vacuum devices, and a three-way catalytic converter that controls NO_x as well as CO and HC. The oxygen sensor sends a voltage signal to the electronic control unit. The voltage varies as the oxygen in the exhaust gas varies. At the same time, the temperature switch is reporting the engine temperature.

If the oxygen content of the exhaust gas is low, indicating a rich mixture, the ECU signals the vacuum switch and vacuum modulator. They allow vacuum to work the carburetor feedback diaphragm. These are vacuum motors that provide a mechanical movement as the vacuum on them changes.

One of the feedback diphragms operates an idle needle that allows more or less air to bleed into the idle system. As more air bleeds in, the idle mixture is leaned out. The other feedback diaphragm meters the fuel flow from the main fuel valve. If the main metering system is feeding too much fuel so the mixture is too rich, the feedback diaphragm reduces the amount of fuel so that the mixture is leaned out. The system continually adjusts the air-fuel ratio to the lean side to minimize exhaust emissions.

The engine temperature switch sends a signal that indicates whether the engine is cold or warmed up. If the engine is cold, the signal tells the ECU not to lean out the air-fuel mixture too much. A cold engine needs a richer mixture to run satisfactorily. However, as soon as the engine approaches operating temperature, the temperature switch "tells" the ECU. The ECU then switches to the hot-engine mode and the mixture is made leaner.

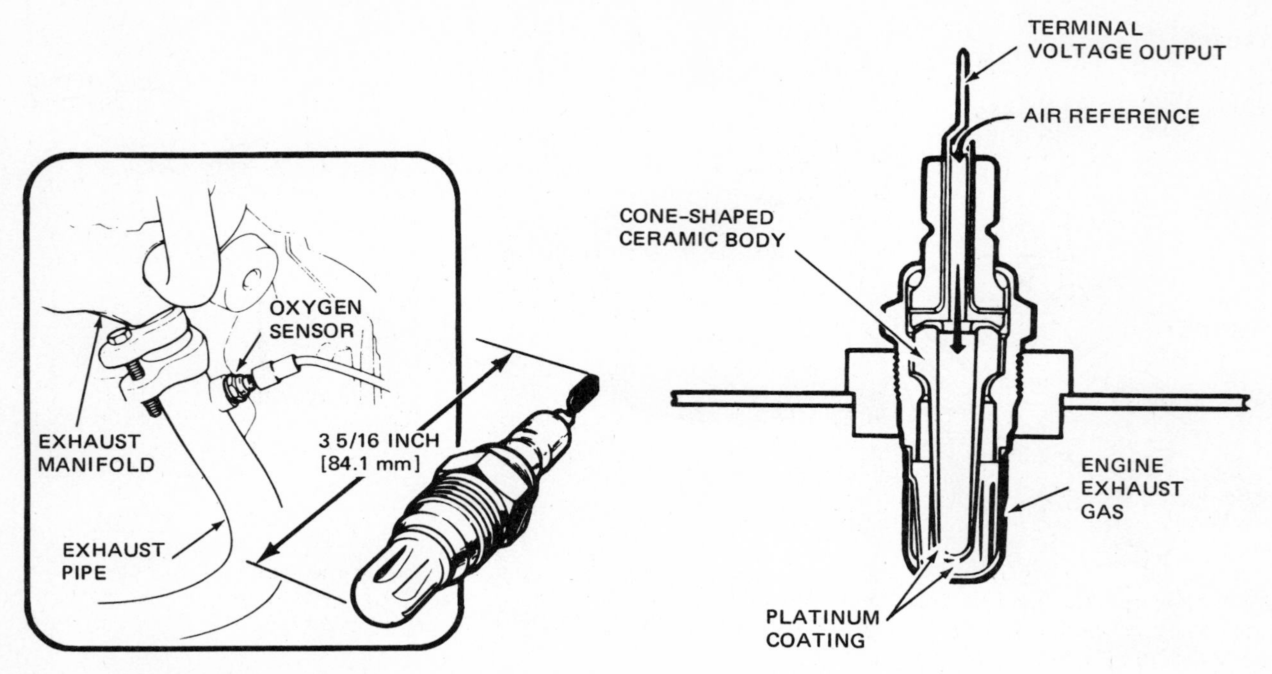

Fig. 4-49 Oxygen sensor, showing its construction and installation. (*AC Spark Plug Division of General Motors Corporation*)

Figure 4-50 is a sectional view of the carburetor showing the idle system. Vacuum from the vacuum modulator works on the idle feedback diaphragm. If the vacuum is sufficient, it causes the diaphragm to pull back the needle to the feedback-controlled idle air bleed. This allows more air to feed in so the idle mixture is leaned out.

Figure 4-51 shows the main-metering-system feedback diaphragm. It is connected to the tapered needle centered in the feedback-controlled main-metering orifice. If the mixture is too rich, there will be a sufficient vacuum on the diaphragm so it lifts the needle. This reduces the amount of fuel that can flow to the main system, so the mixture is leaned out.

NOTE: There is a detailed description of a feedback carburetor which electronically controls the idle and main metering systems in Chap. 11.

4-30 Ford feedback-carburetor electronic engine-control system The Ford system for air-fuel mixture control is tied in with the thermactor or air injection system of exhaust emission control. The air-injection system includes an air pump that blows fresh

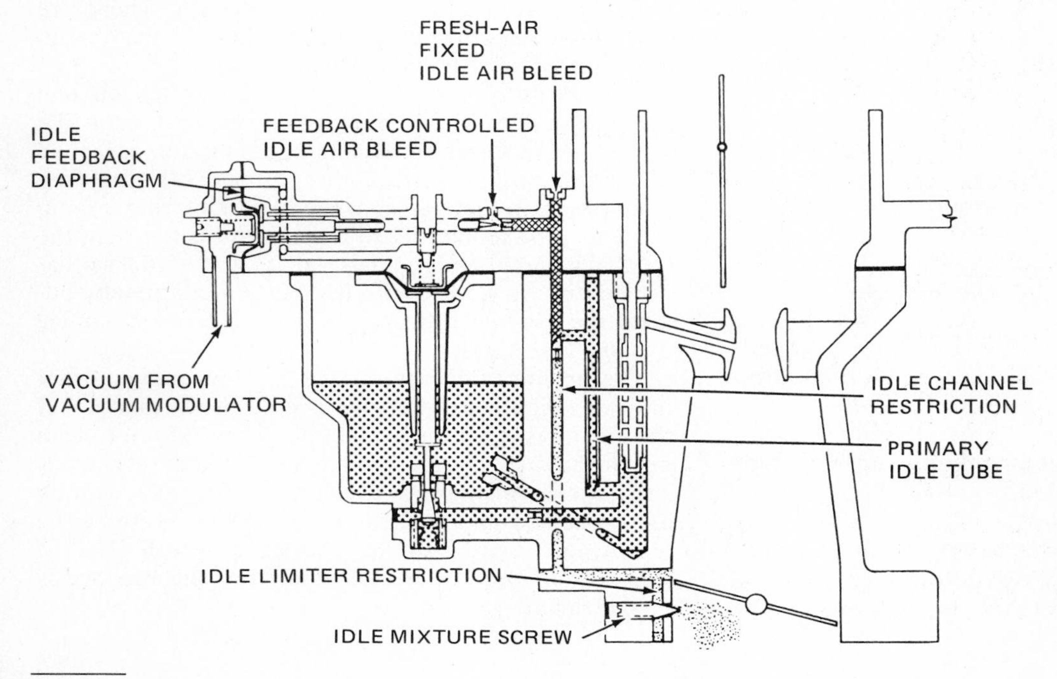

Fig. 4-50 Sectional view of the carburetor for the EFC system, showing the idle system and idle feedback diaphragm. (*Pontiac Motor Division of General Motors Corporation*)

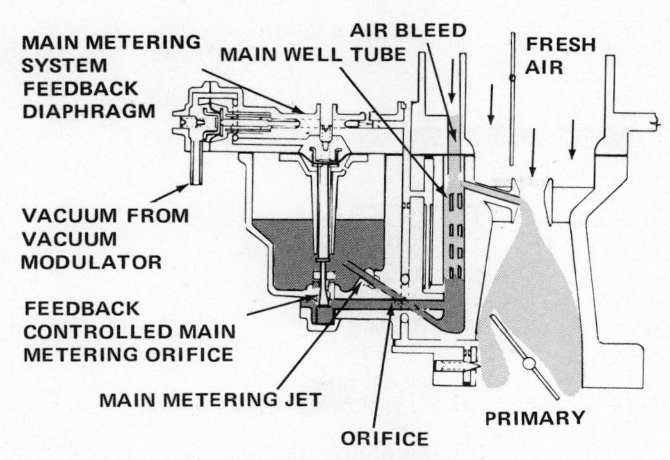

Fig. 4-51 Sectional view of the carburetor for the EFC system, showing the main metering system with feedback diaphragm. (*Pontiac Motor Division of General Motors Corporation*)

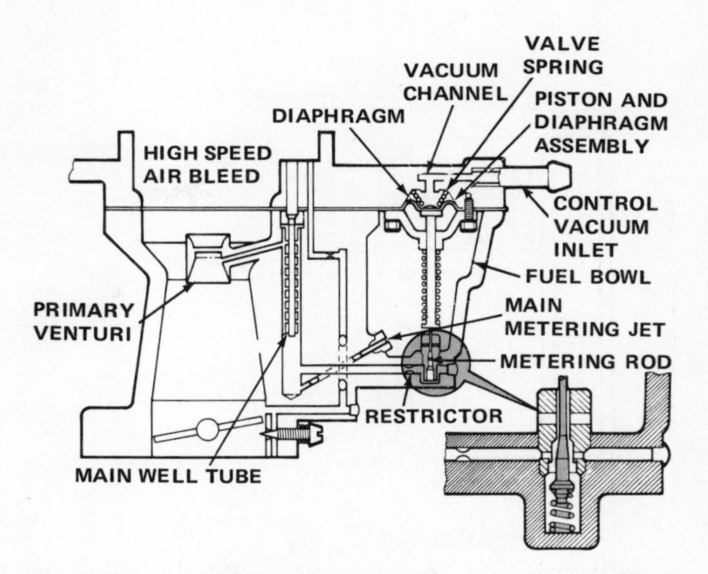

Fig. 4-52 Sectional view of the feedback carburetor for an electronic engine control system, with diaphragm control of the main metering system. (*Ford Motor Company*)

air into the exhaust manifold to help burn any unburned gasoline (HC) or partly burned gasoline (CO).

The Ford Feedback-Carburetor Electronic Engine-Control System consists of three subsystems:

1. Dual catalytic converter
2. Thermactor air control
3. Electronic feedback carburetor

The electronic feedback carburetor is very similar to the General Motors unit (Figs. 4-50 and 4-51). However, the Ford carburetor has only one feedback diaphragm and this controls the main metering system (Fig. 4-52). The system uses an oxygen sensor in the exhaust manifold. When the oxygen content of the exhaust gas goes down, indicating a rich mixture, the sensor reports this to the electronic control unit. The ECU then signals the vacuum switch system to send more vacuum to the feedback diaphragm. The diaphragm therefore lifts the metering rod to restrict the flow of fuel. Less fuel gets to the main fuel nozzle so the mixture is leaned out.

The thermactor or air-injection system has two operating modes (Figs. 4-53 and 4-54). When the engine is cold (Fig. 4-53), the fuel system is feeding a rich mixture to the engine. This is a result of the choke action, as explained later in ⚙ 4-33. Therefore, the exhaust gas is rich in unburned and partly burned fuel (HC and CO). With this condition, the air pump sends air into the exhaust manifold to help complete the combustion of these pollutants. The additional oxygen in the air helps turn the HC into H_2O and CO_2. It helps

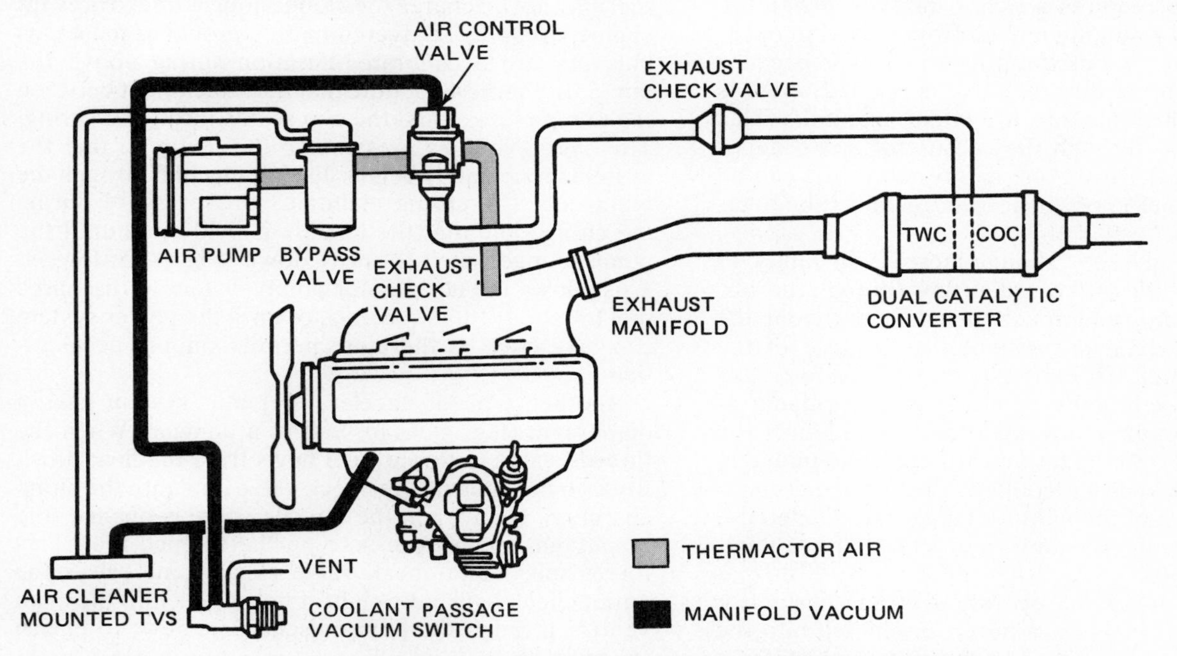

Fig. 4-53 Air injection or thermactor system for electronic engine control system, with dual catalytic converter. This shows the flow of air from the air pump with the engine cold. (*Ford Motor Company*)

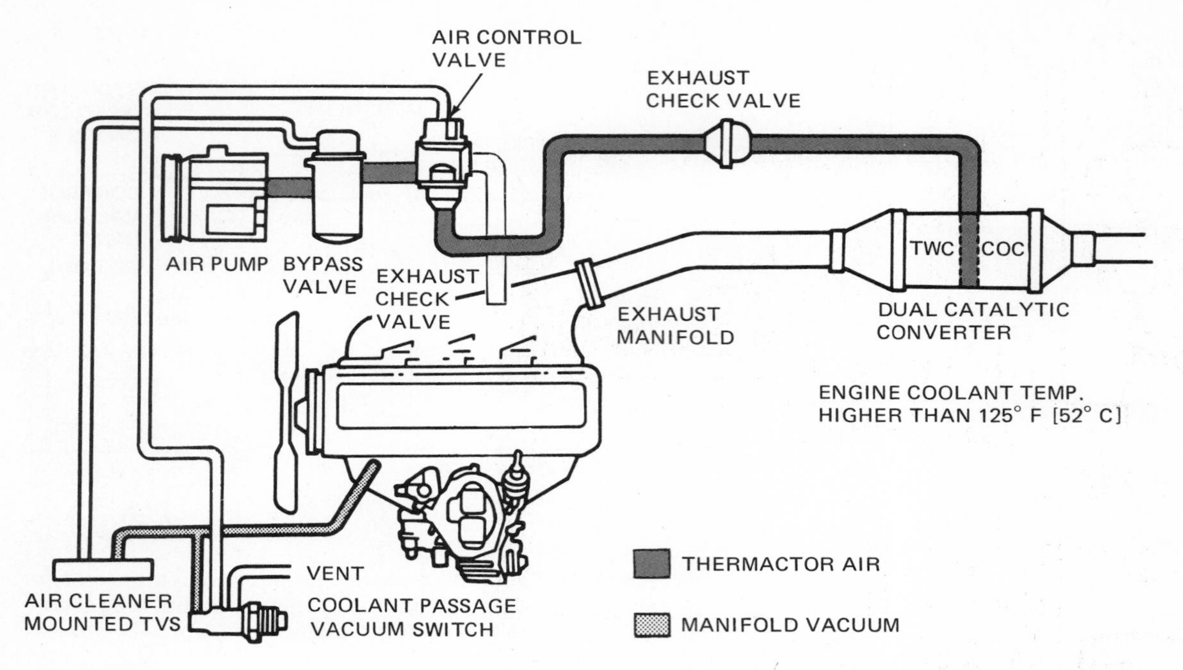

Fig. 4-54 Air injection or thermactor system with the engine hot. Air from the air pump flows to the dual catalytic converter. (*Ford Motor Company*)

turn the CO into CO_2. In doing this, the air guards the catalytic converter from overload, and possible resulting damage.

When the engine warms up (Fig. 4-54), the vacuum switch, which senses engine coolant temperature, shuts off the vacuum to the air control valve. As a result, the thermactor system sends the air it is pumping to the catalytic converter (Fig. 4-54). Here, it aids the converter in changing the pollutants to harmless gases.

⚙ 4-31 Accelerator pump system For acceleration, the engine requires a richer mixture. When the throttle valve is suddenly moved from closed to open, leaning of the air-fuel mixture occurs. This is because of inertia. When the throttle valve is opened, there is a sudden inrush of air into the intake manifold. This increases airflow through the carburetor and reduces manifold vacuum. However, before more fuel can be delivered, the inertia of the heavier fuel must be overcome.

Ideally, fuel delivery should increase instantly to correspond with the increased airflow through the carburetor. Actually, additional fuel delivery from the main nozzle is delayed momentarily because of the inertia of the fuel. To carry the carburetor over this momentary lapse in delivery of the proper amount of fuel, an accelerator pump system is included in the carburetor. This eliminates any flat spot or stumble in engine performance that could occur during acceleration as the result of the mixture going lean. Generally, the accelerator pump system is inoperative above about 30 mph [48 km/h].

Figure 4-55 shows the operation of an accelerator pump system. It includes a pump assembled into the float bowl with a fuel passage up to a jet at one side of the carburetor barrel. The pump piston is linked to the throttle so that when the throttle is opened, the piston is pushed down. This downward movement forces fuel from the pump cylinder through the fuel passage. It exits through the pump jet into the air passing through (Fig. 4-55). This momentarily enriches the mixture and causes the engine to accelerate without stall or stumble. A small check valve or check ball in the fuel passage prevents fuel from being delivered as a result of venturi vacuum in the carburetor. Fuel is delivered only when the throttle valve is opened and the pump piston is forced downward.

However, when the throttle is opened too quickly, fuel may not discharge for a long enough time to prevent engine stumble. To overcome this problem, most carburetors use a calibrated-duration spring above the pump. In Figure 4-55, note that the attachment between the pump plunger and the cup seal is through a spring. This spring applies pressure to the pump so that the system immediately begins discharging fuel through the pump jet. The spring maintains this pressure during the entire time that the throttle is held open until the pump plunger is all the way down (Fig. 4-55, lower). This allows the accelerator pump system to discharge fuel for about 1 to 3 seconds, or until the power system can take over. It therefore permits smooth acceleration.

Another type of accelerator pump system uses a diaphragm (Fig. 4-56) instead of a plunger. When the throttle valve is closed, fuel flows from the float bowl, through the open intake check valve, and into the pump chamber. Then when the throttle valve is opened, the diaphragm applies pressure on the trapped fuel. This forces the intake check valve closed, and raises the outlet check ball and weight. Fuel sprays into the main venturi through the pump discharge screw, to enrich the mixture.

An accelerator pump system for a two-barrel carburetor has two discharge nozzles, one for each barrel.

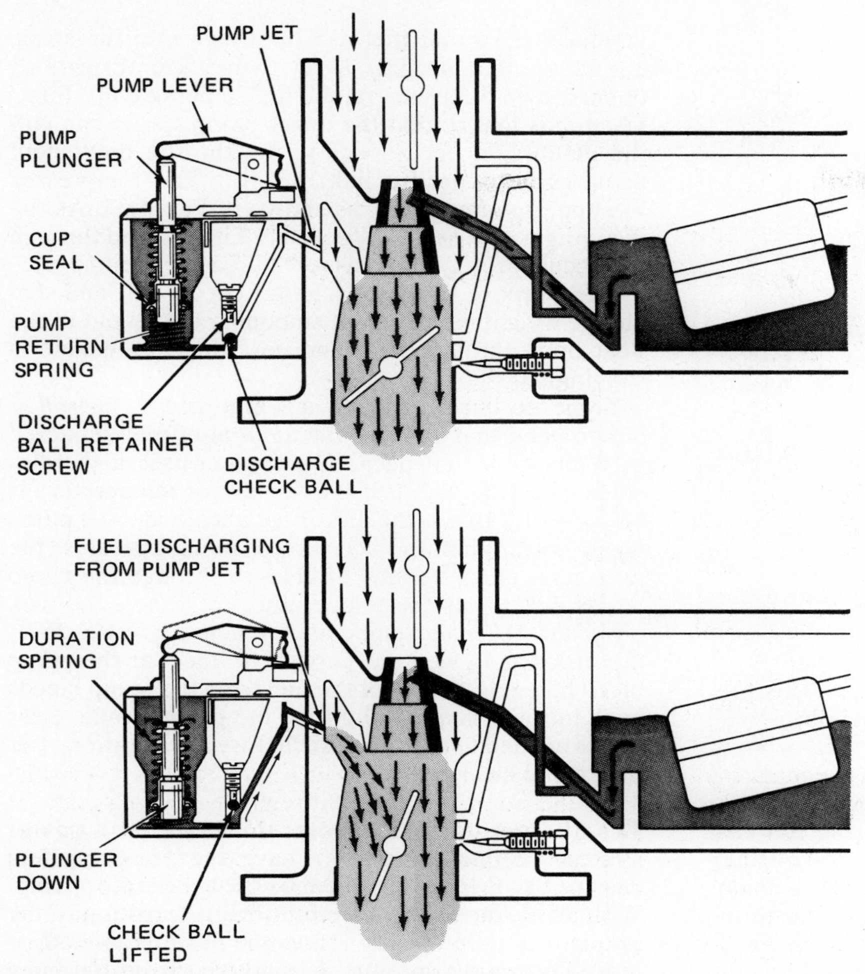

Fig. 4-55 Operation of the accelerator pump system.

The fuel flow from the accelerator pump is split between the two barrels. Regardless of the number of barrels, most carburetors use only one accelerator pump.

Some accelerator pump systems using a plunger or piston include a check ball in the pump plunger which releases vapor that might form in the pump well. If it were not for this, vapor pressure might build up enough during hot starting to cause fuel to feed from the pump well into the venturi. This, in turn, could cause hard starting from an overrich mixture. In the carburetor shown in Fig. 4-56 (which uses a pump diaphragm), an

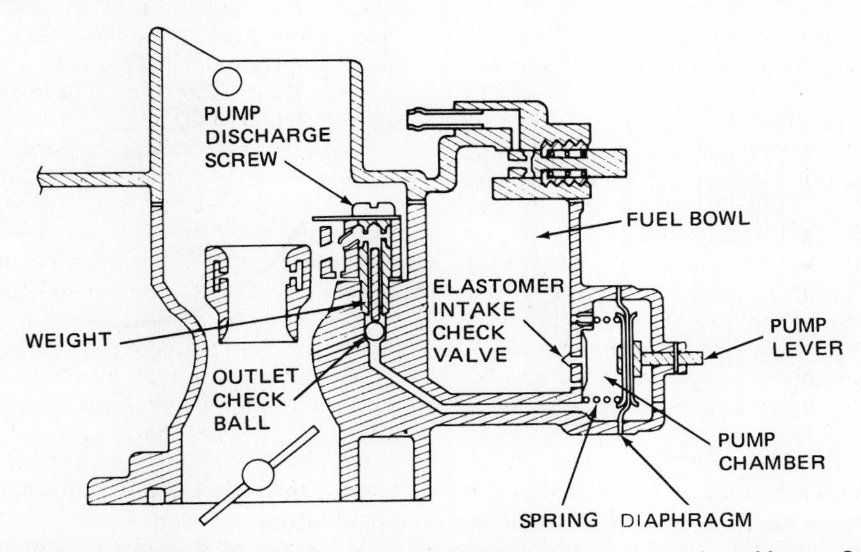

Fig. 4-56 Accelerator pump system using a diaphragm instead of a plunger. (*American Motors Corporation*)

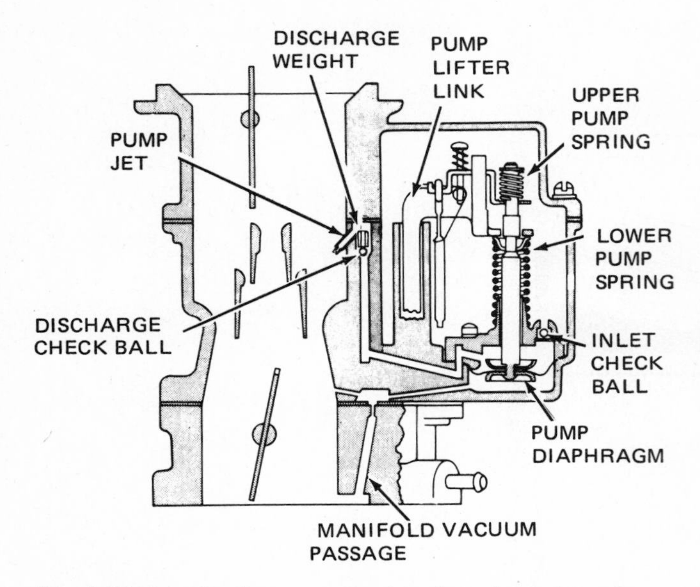

Fig. 4-57 Accelerator pump system in a single barrel carburetor. (*American Motors Corporation*)

air bleed is provided in the pump chamber. This prevents vapor accumulation and pressure buildup.

Instead of a piston, many carburetors use a pump diaphragm controlled by both manifold vacuum and throttle linkage. The same pump diaphragm that provides control of the power-system metering rod also feeds the accelerator pump system (Fig. 4-57). When the throttle closes, the pump diaphragm moves down, actuated by both the mechanical linkage from the throttle and the increased intake manifold vacuum. It is shown in the down position in Fig. 4-57.

As the diaphragm moves down, the inlet check valve is unseated so that fuel can be drawn into the space above the diaphragm. Now, when the throttle is opened, two forces act to pull the diaphragm up. First, vacuum is lost so that the lower pump spring can pull the diaphragm up. Secondly, the throttle movement actuates the pump lifter link, raising it. This compresses the upper pump spring, as shown in Fig. 4-57. As the diaphragm is raised, the fuel above it is forced through the accelerator system passage and out the pump jet.

The purpose of the discharge check ball and discharge weight is to prevent siphoning of fuel out of the float bowl through the pump jet when the engine is running.

Some carburetors have a *temperature-controlled pump bleed* that changes the amount of fuel delivered as temperature changes. The system uses a thermostatic disk (Fig. 4-58). If the carburetor temperature is below 49°F [10°C], the disk is flat and covers the pump bleed hole. When the accelerator pump operates, the maximum pump volume discharges through the pump jet. However, as the temperature increases, the disk bows out. When the temperature reaches 71°F [21.2°C], the disk has bowed out enough to uncover the pump bleed hole. Now, part of the fuel from the pump bleeds back into the float bowl during the pump stroke. Less fuel is needed to provide satisfactory acceleration after the engine has warmed up.

⚙ 4-32 Combination accelerator pump and power system Some carburetors have the power system designed so that it is operated by the accelerator pump. With this arrangement, the full-throttle position of the accelerator pump piston forces the power valve off its seat. Then additional fuel is delivered from the main nozzle. The valve operates as described in ⚙ 4-23 and

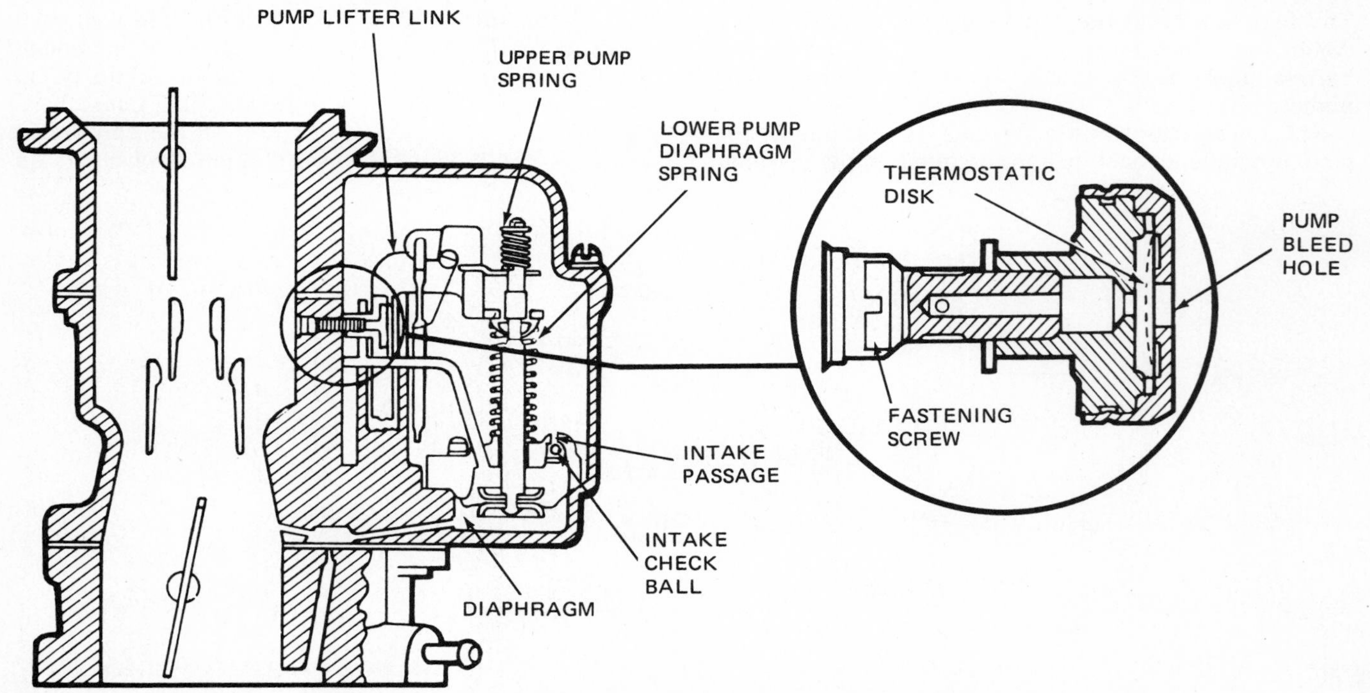

Fig. 4-58 An accelerator pump system that has a temperature-controlled pump bleed. (*Ford Motor Company*)

4-24. The only difference is that it is operated by the pump piston instead of by a vacuum piston or other throttle linkage.

The carburetor shown in Figs. 4-57 and 4-58 has the combination accelerator pump and power system. The upward movement of the pump lifter link as the throttle is opened raises the metering rod to permit mixture enrichment. At the same time, the loss of vacuum in the intake manifold releases the pump diaphragm assembly so that the accelerator pump system can further enrich the mixture.

4-33 Choke system While a cold engine is being cranked, the carburetor must deliver a very rich mixture to the intake manifold. With a cold engine and carburetor, only part of the fuel vaporizes. Therefore, extra fuel must be delivered so that enough fuel vaporizes to provide the correct air-fuel mixture for starting the engine.

During cranking, airspeed through the carburetor is very low. Venturi vacuum and intake manifold vacuum are insufficient to produce adequate fuel flow for starting. Therefore, to produce the fuel flow needed during cranking, the carburetor has a choke system with a choke valve (Fig. 4-59).

The choke valve is a round valve, shaped like the throttle valve, located in the top of the air horn. It is controlled mechanically or by an automatic device on most engines. When the choke valve is closed, it is almost horizontal, as shown in Fig. 4-59. When it is in this position, very little air gets past it. The valve has "choked off" the air flow. Then, when the engine is cranked, a fairly high vacuum develops in the air horn under the closed choke valve (Fig. 4-59). This vacuum causes the main nozzle to discharge fuel. The amount of fuel discharged is sufficient to produce the correct air-fuel mixture needed for starting the engine.

As soon as the engine starts, its speed increases from a cranking speed of about 250 to 300 rpm to over 600 rpm. Now more air and a leaner air-fuel mixture are required. One way to get more air into the engine as soon as it starts is to mount the choke valve off center on its shaft in the air horn (Fig. 4-59). Then a spring is added to the choke linkage. The spring tends to hold

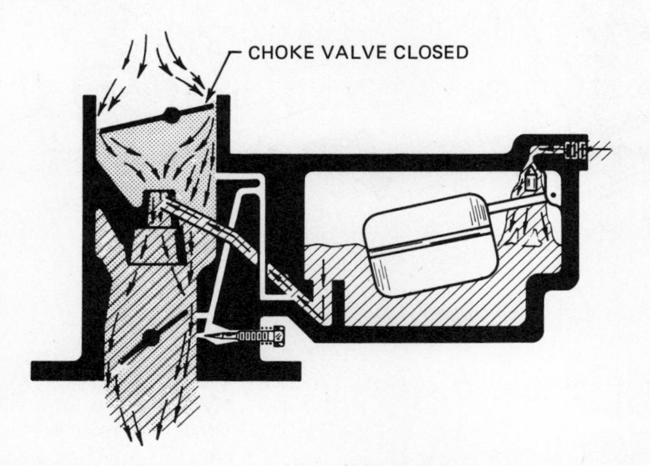

Fig. 4-59 Operation of the choke in starting the engine.

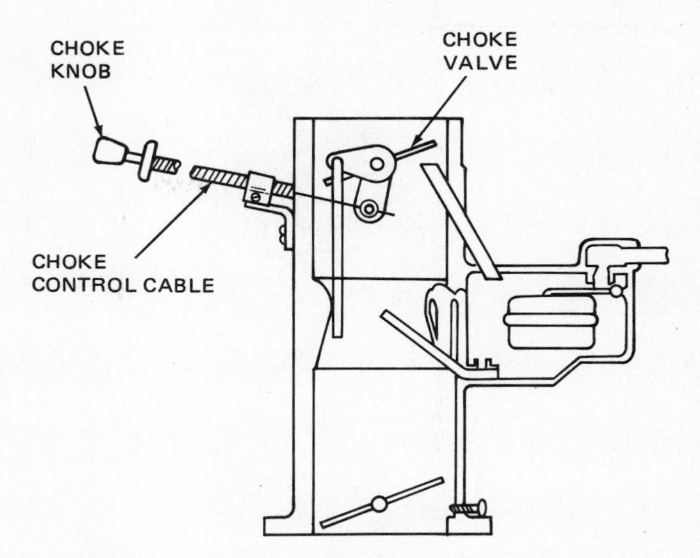

Fig. 4-60 A mechanical choke, operated by a pull knob on the instrument panel. (*Ford Motor Company*)

the valve horizontal, or in the closed position. However, when the engine starts, the vacuum produced by the running engine causes the valve to partly open against the spring pressure. More air can flow through.

Mechanically controlled chokes are operated by a pull knob on the instrument panel that is linked to the choke valve (Fig. 4-60). When the choke knob is pulled out, the choke valve is closed. The driver must remember to push in on the choke knob to open the choke valve as soon as the engine begins to warm up. If the choke knob is not pushed in, the carburetor will continue to supply a very rich air-fuel mixture to the engine. This rich mixture will cause poor engine performance, high exhaust emissions, fouled spark plugs, and poor fuel economy.

4-34 Automatic chokes To prevent the problems caused by improper driver operation of the choke, carburetors have automatic chokes. These devices close the choke valve when the engine is cold and gradually open it as the engine warms up. The automatic choke devices are all similar, although they vary in detail. Most operate on exhaust manifold temperature and intake manifold vacuum (Figs. 4-61 to 4-70).

1. Typical automatic choke In the typical automatic choke, a spiral bimetal thermostatic spring and vacuum piston are linked together to the choke valve and control its position (Fig. 4-61). The thermostatic spring winds up or unwinds with changing temperature. When the engine is cold, the spring winds up, closing the choke valve. Then the cold engine gets a rich mixture for starting. As the engine warms up, the thermostatic spring unwinds, opening the choke valve.

During warm-up, the choke piston (Fig. 4-61) comes into action. When the engine is idling, the piston is pulled down by the intake manifold vacuum, and the piston partly opens the choke. This prevents too rich an idle mixture. Then, when the throttle is opened for acceleration, the intake-manifold vacuum decreases.

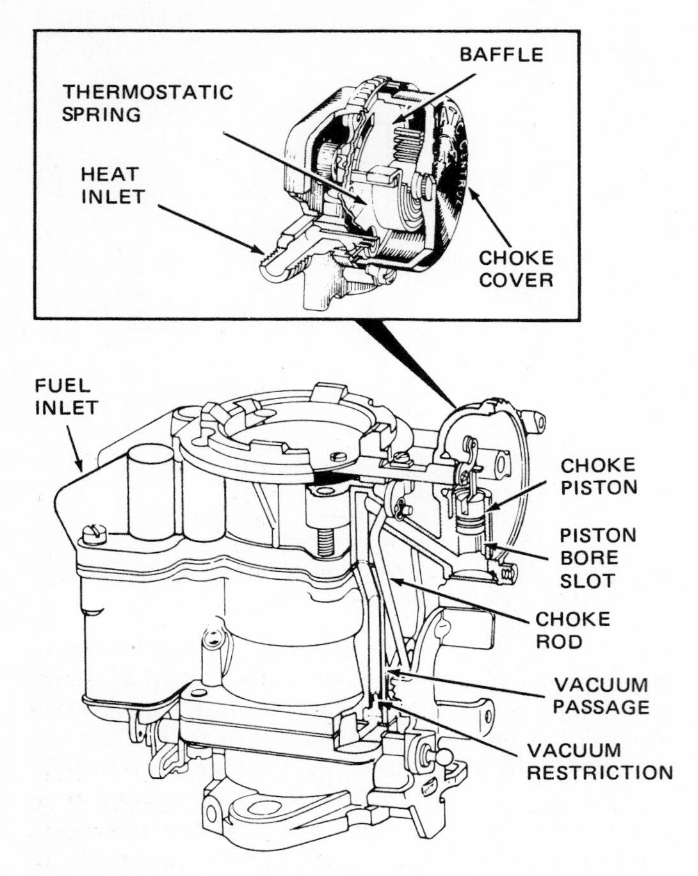

Fig. 4-61 Choke system using a choke piston and a thermostatic spring or coil mounted on the carburetor. (*American Motors Corporation*)

This releases the choke piston so that the choke valve moves toward the closed position. This action enriches the mixture for acceleration.

During closed-choke operation, the air-fuel mixture is rich. The result is that more unburned hydrocarbon goes out the tail pipe. To reduce the time the choke is closed, some automatic chokes use electric heating elements. The electric heating element speeds up the unchoking action. There is more information on other types of chokes and choke control devices later in this section.

2. Choke operation When the throttle is opened, the mixture must be enriched (⚙ 4-31). The accelerator pump provides some extra fuel. But still more fuel is needed when the engine is cold. This additional fuel is secured by the action of the vacuum piston. When the throttle is opened, intake-manifold vacuum is lost. The vacuum piston releases and is pulled inward by the thermostatic spring tension. Therefore, the choke valve moves toward the closed position, causing the mixture to be enriched.

3. Action during warm-up During the first few seconds of operation, the choke valve is controlled by the vacuum piston. However, as the engine warms up, the thermostatic spring begins to take over. The thermostatic spring is located in a housing that is connected to the exhaust manifold through a small heat tube (Fig. 4-62). Heat passes from the exhaust manifold through this tube and enters the thermostatic spring housing. Soon, the thermostatic spring begins to warm up. As the spring warms up, it unwinds. This causes the choke valve to move toward the opened position. When op-

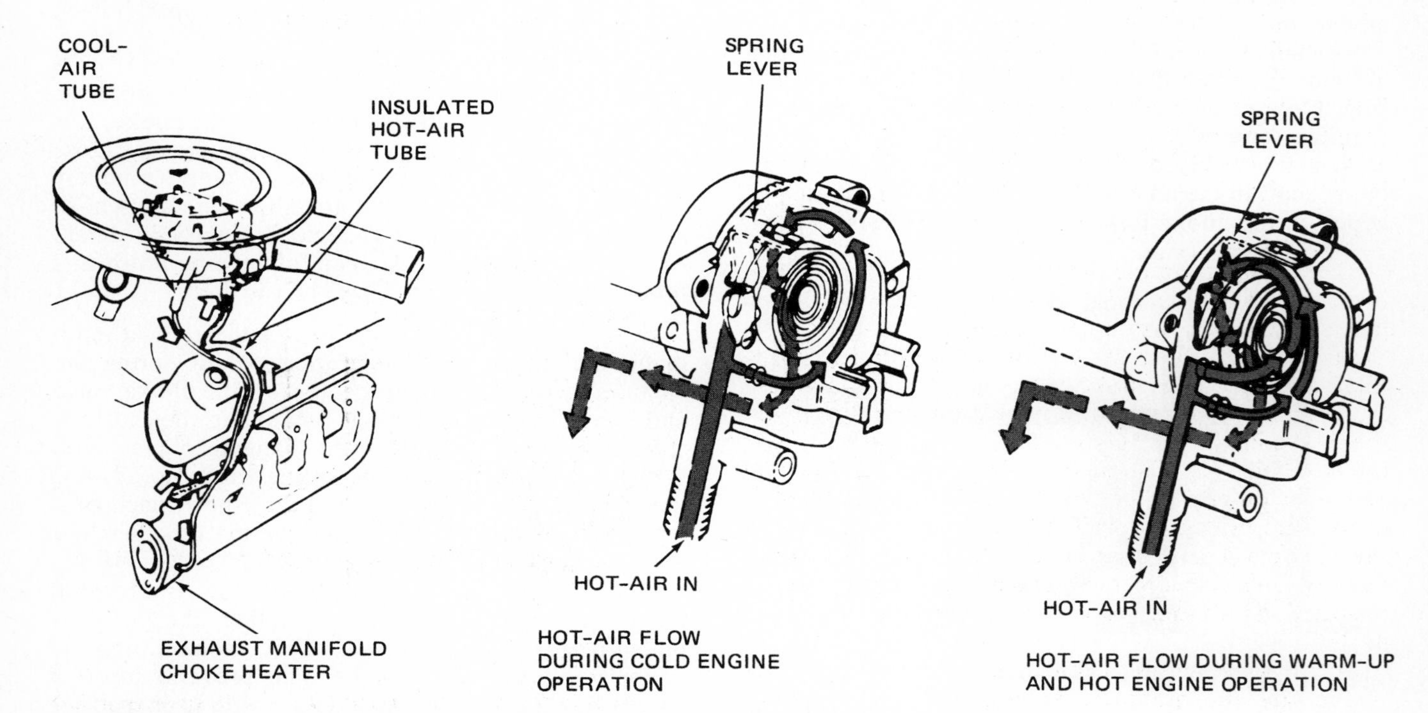

Fig. 4-62 Flow of hot air through the automatic choke. (*Ford Motor Company*)

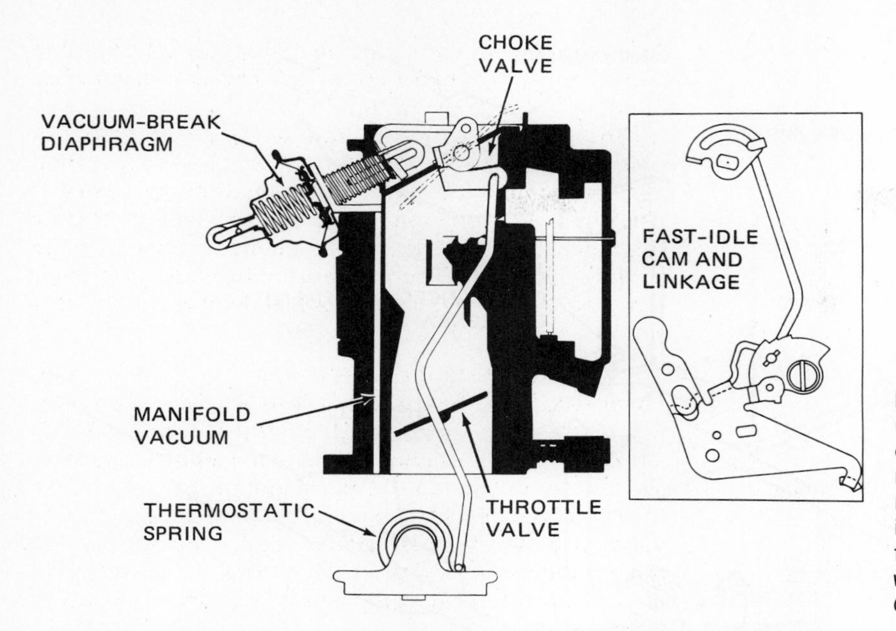

Fig. 4-63 Choke system using a vacuum-break diaphragm and a thermostatic spring or coil mounted in a well in the intake or exhaust manifold. (*Chevrolet Motor Division of General Motors Corporation*)

erating temperature is reached, the thermostatic spring has unwound enough to fully open the choke valve. No further choking takes place.

When the engine is stopped and cools, the thermostatic spring again winds up. This closes the choke valve and spring-loads it in the closed position. These chokes are often called *hot-air chokes,* since they operate when heated air from the exhaust manifold passes through them.

4. Thermostat in manifold Another type of automatic choke is shown in Fig. 4-63. The thermostatic spring is located in a well in the exhaust manifold or in the exhaust-gas crossover passage of the intake manifold. The spring is connected to the choke valve by linkage. This makes the choke faster acting because the exhaust heat is applied directly to the thermostatic spring and not through a heat tube. A vacuum-break diaphragm is used on the carburetors with the well-mounted thermostatic spring shown in Figs. 4-63 and 4-64. The vacuum diaphragm takes the place of the vacuum piston. The action is similar. However, the diaphragm provides more force to break the choke valve loose if it gets stuck.

The linkage from the diaphragm to the choke-valve lever rides freely in a slot in the lever. During certain phases of engine warm-up, the changing vacuum causes the linkage to ride to the end of the slot in the choke lever and open the choke valve. For example, when the engine is first started, a high vacuum develops in the intake manifold. This vacuum acts on the vacuum-break diaphragm. The linkage rides to the end of the slot with enough force to "break" the choke valve away from the fully closed position. This partly opens the choke valve so that an overrich mixture is prevented. The vacuum-break diaphragm exerts a more positive and stronger force than a choke piston (Fig. 4-61).

Figure 4-65 shows schematically the linkages between the thermostatic spring and the vacuum diaphragm in the pulldown motor for one type of carburetor. The thermostatic spring is in the carburetor in this design. This arrangement is similar to that shown in Fig. 4-64 except that a vacuum diaphragm is used instead of a vacuum piston.

5. Coolant-operated choke Some carburetors have a hot-coolant-operated automatic choke (Fig. 4-66). In these carburetors, hot coolant from the engine cooling system is circulated through the choke instead of hot air from the engine exhaust manifold. The hot-coolant chamber has a series of fins that help pick up heat from the coolant and carry it to the thermostatic spring or coil. Figure 4-67 shows a carburetor using both a hot-coolant heater and an electric-choke heater. Figure 4-68 is a disassembled view of the choke, showing the hot-coolant and electric heaters.

6. Electric choke Many carburetors have electric

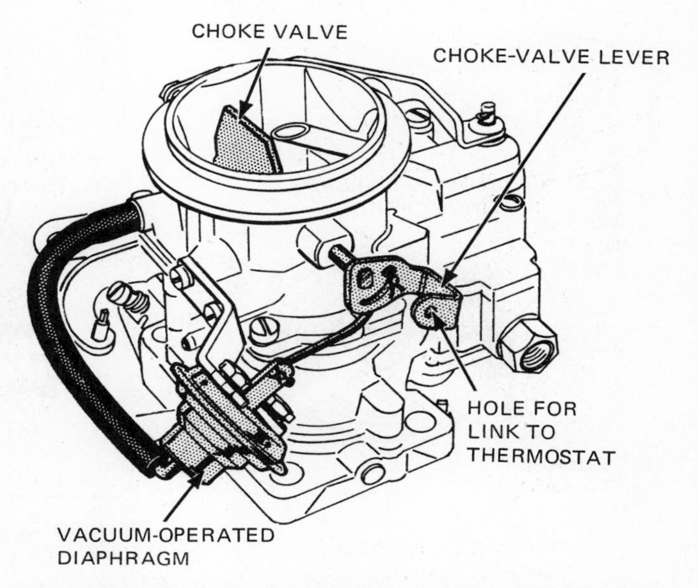

Fig. 4-64 Automatic choke using a vacuum-operated diaphragm. (*Chrysler Corporation*)

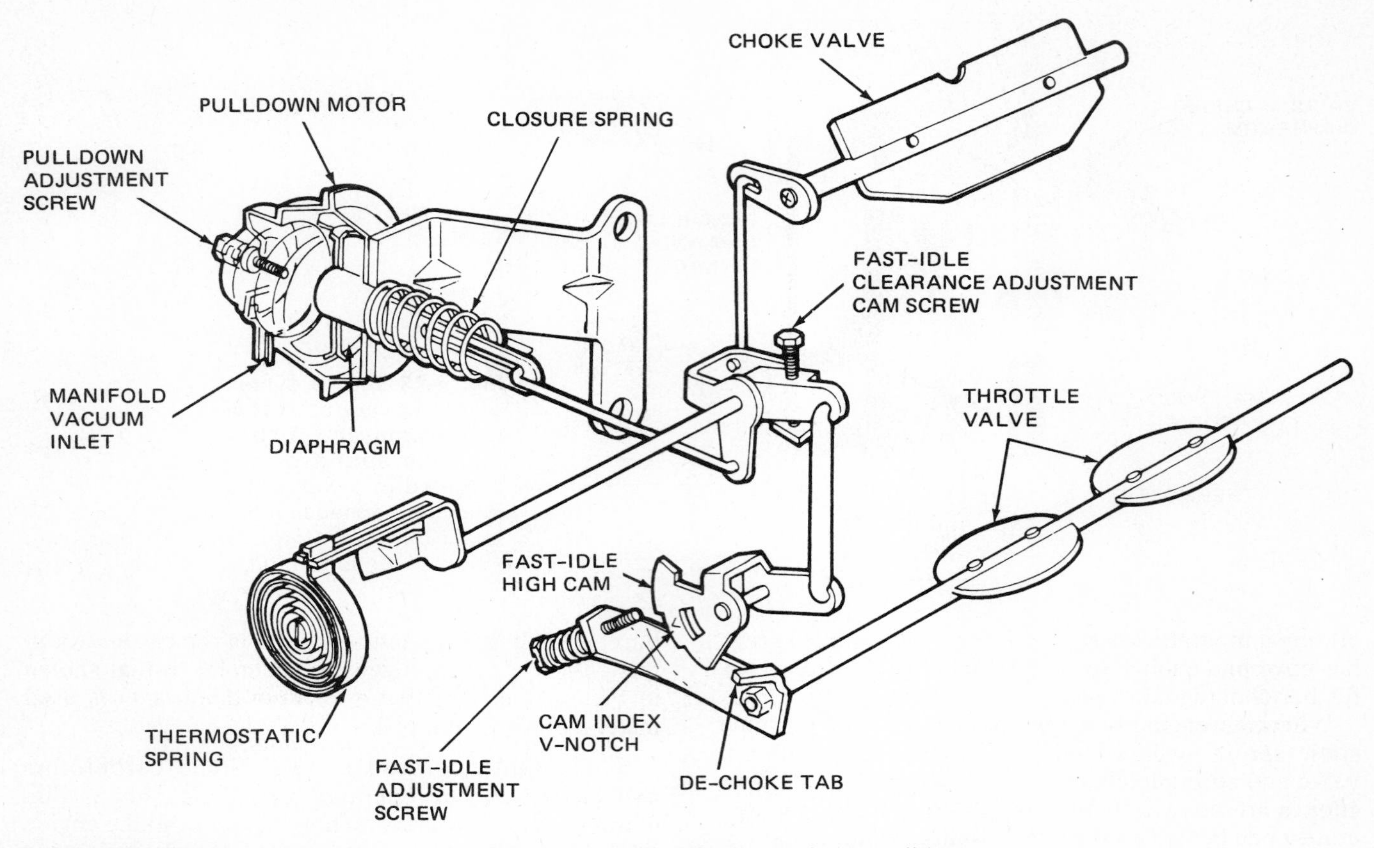

Fig. 4-65 Linkage between the thermostatic spring and the diaphragm in the pulldown motor. (*Ford Motor Company*)

automatic chokes. This type of choke includes an electric heating element (Fig. 4-69). Its purpose is to ensure faster choke opening. This helps to reduce emissions from the engine. Emissions (HC and CO) are relatively high during the early stages of engine warm-up. At low temperatures, the electric heater adds to the heat coming from the exhaust manifold. This reduces choke-opening time to as short as 1.5 minutes. Figure 4-70 shows the arrangement for a choke mounted in a well in the exhaust-gas crossover passage of the intake manifold.

7. Hot-air-operated choke Another method of speeding up the unchoking action is shown in Fig.

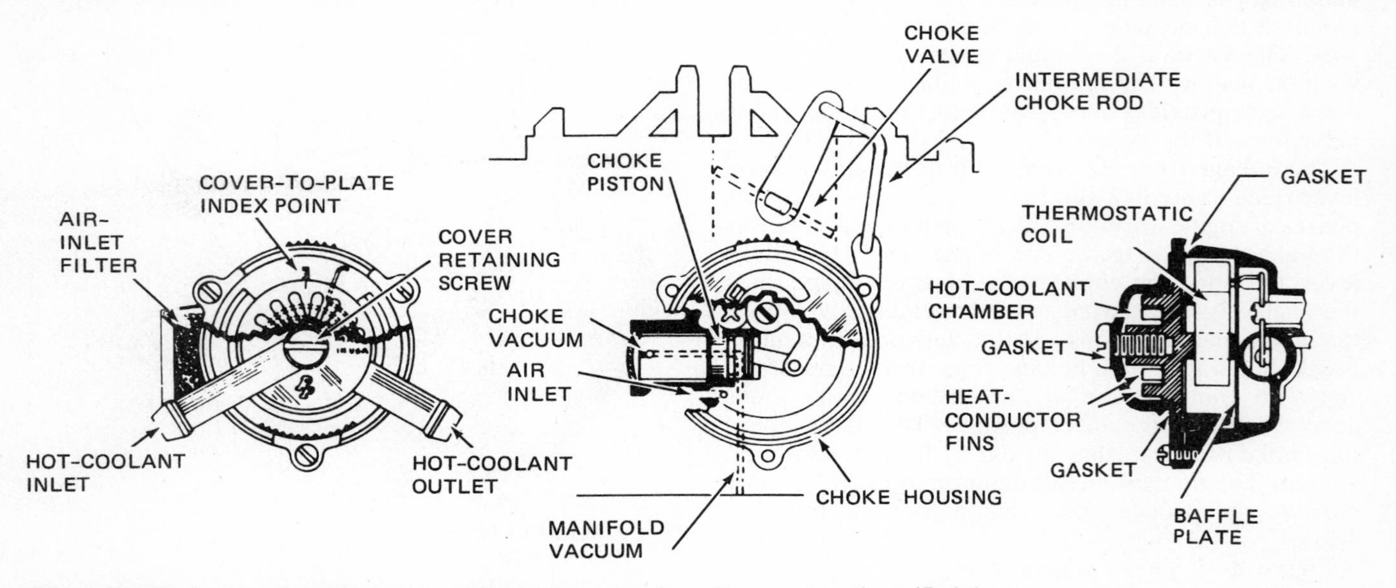

Fig. 4-66 Coolant-heated choke in cutaway views to show its construction. (*Buick Motor Division of General Motors Corporation*)

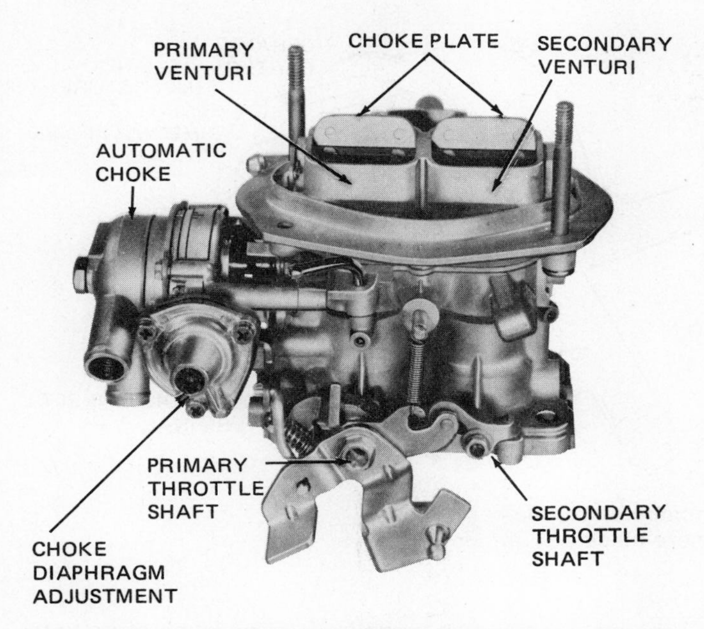

Fig. 4-67 A two-barrel carburetor using a choke with both an electric choke heater and a chamber through which coolant from the engine cooling system can flow. (*Ford Motor Company*)

4-62. Here, the hot air flowing to the thermostatic coil comes from the air cleaner. It flows through a choke heater placed in the exhaust manifold. The air is quickly heated by the hot exhaust gas. Then the hot air passes through the tube to the thermostatic spring housing. From there, the air flows into the intake manifold. Therefore, the thermostatic coil is quickly heated and rapidly opens the choke valve.

8. Counterweighted choke lever Some carburetors have a counterweighted choke lever. The counterweight provides extra closing force when the choke is closed. This assures a richer mixture to help in cold starts. When the engine starts, and the choke begins to open, the weight travels overcenter. Then it begins to apply additional opening effort to help lean out the mixture shortly after starting.

⚙ 4-35 Unloader The unloader (Fig. 4-71) opens the choke valve by linkage. When the throttle valve is opened wide, the linkage will cause the choke valve to open up (if it is not already open). The purpose of the unloader is to clear the intake manifold of an excessively rich air-fuel mixture.

For example, if the engine does not start immediately, and the choke valve is closed, the overrich mixture being delivered by the carburetor will "flood" the engine by loading the manifold and cylinders with an overrich mixture. This mixture will not ignite readily. Therefore, the engine will not start. To clear the condition, the driver pushes the throttle valve to wide open. This partly opens the choke valve. Now, with further cranking, a lean mixture is delivered which soon unloads the intake manifold so that starting can then take place.

⚙ 4-36 Manifold heat-control valve As a further means of obtaining smooth engine operation during warm-up, a manifold heat-control valve has been used on many engines. This device causes considerable heat transfer from the exhaust manifold to the intake manifold during initial operation with a cold engine. The heat transfer preheats the air-fuel mixture and assures better fuel vaporization and, therefore, better initial engine operation. Two arrangements are used, one for in-line engines and another for V-type engines.

1. In-line engines In these engines, the exhaust manifold is located under the intake manifold (Fig. 4-72). At a central point, there is an opening from the

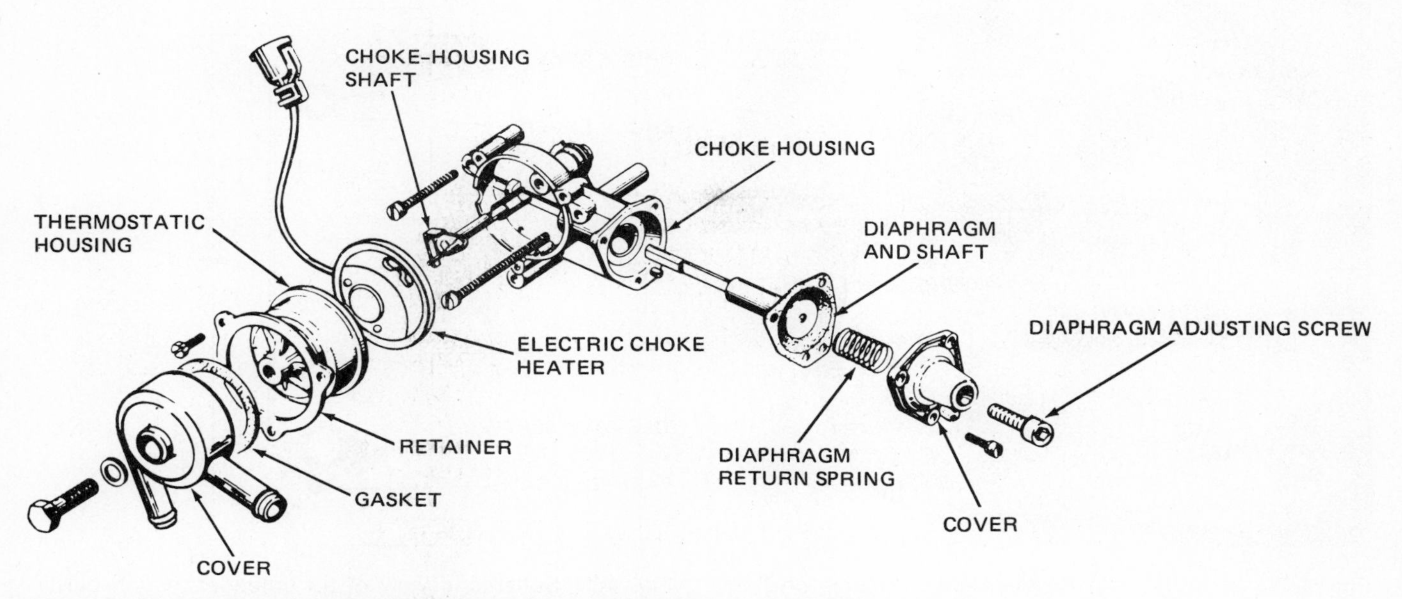

Fig. 4-68 Disassembled view of the choke system using both an electric choke heater and a chamber through which coolant can flow. (*Ford Motor Company*)

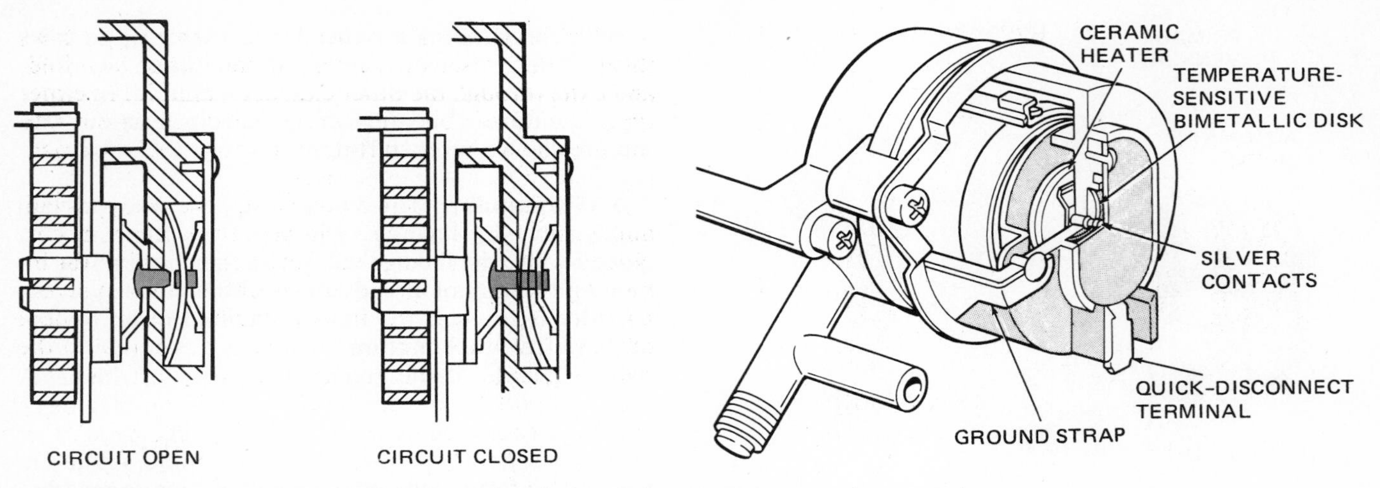

Fig. 4-69 Cutaway view of an electric-assist choke. At low temperature, the ceramic heater turns on, adding heat to the choke so that it opens more quickly. (*Ford Motor Company*)

exhaust manifold into a chamber, or oven, surrounding the intake manifold (Fig. 4-73). A heat-control valve is placed in this opening (Fig. 3-47). The position of this valve is controlled by the actions of a thermostatic spring and a counterweight (Fig. 4-72). When the engine is cold, the thermostatic spring winds up, overcoming the counterweight, and moves the heat-control valve to the closed position (Fig. 4-73, top). Now, when the engine is started, the hot exhaust gases are deflected by the closed heat-control valve. The gases circulate through the oven around the intake manifold. Heat from the exhaust gases quickly warms the intake manifold and helps the fuel to vaporize. This improves cold-engine operation. As the engine warms up, the

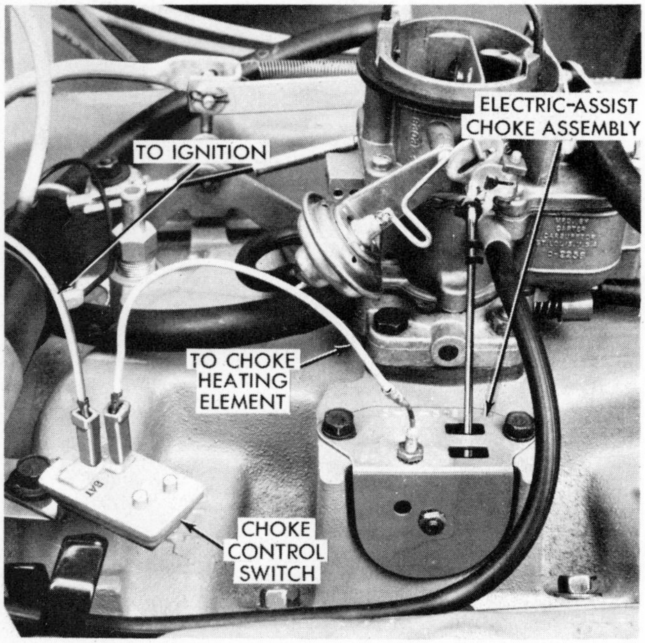

Fig. 4-70 Arrangement of an electric-assist choke mounted in a well in the intake manifold. (*Chrysler Corporation*)

thermostatic spring expands. It loses some of its tension, which was holding the counterweight up. The counterweight drops and moves the heat-control valve to the opened position (Fig. 4-73, bottom). Now, the exhaust gases no longer circulate around the intake manifold but pass directly into the exhaust pipe.

2. V-type engines In V-type engines, the intake manifold is placed between the two banks of cylinders. It has a special exhaust-gas crossover passage (Fig. 4-18, top, and Fig. 4-74) through which exhaust gases can pass. One of the exhaust manifolds has a heat-control valve that is thermostatically controlled. The heat-control valve closes when the engine is cold. This causes the exhaust gases to pass from that exhaust manifold through the passage in the intake manifold. Then the exhaust gases enter the other exhaust manifold. Therefore, heat from the exhaust gases heats the

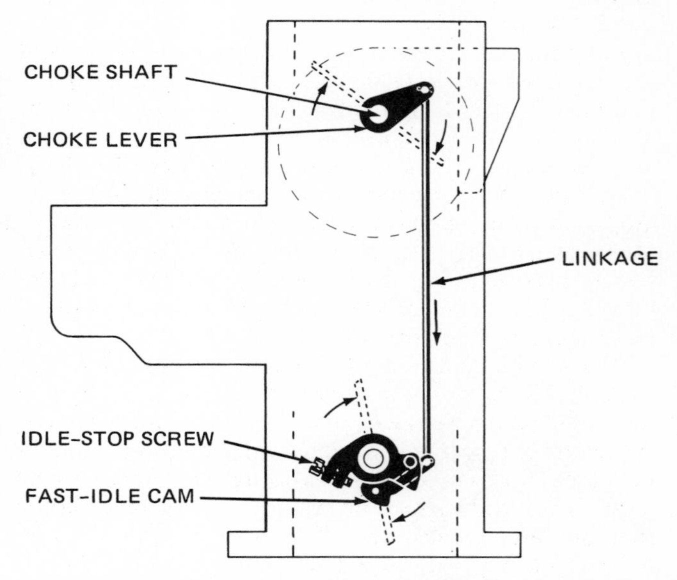

Fig. 4-71 Schematic view of an unloader, which partly opens the choke valve when the throttle valve is moved to the wide-open position.

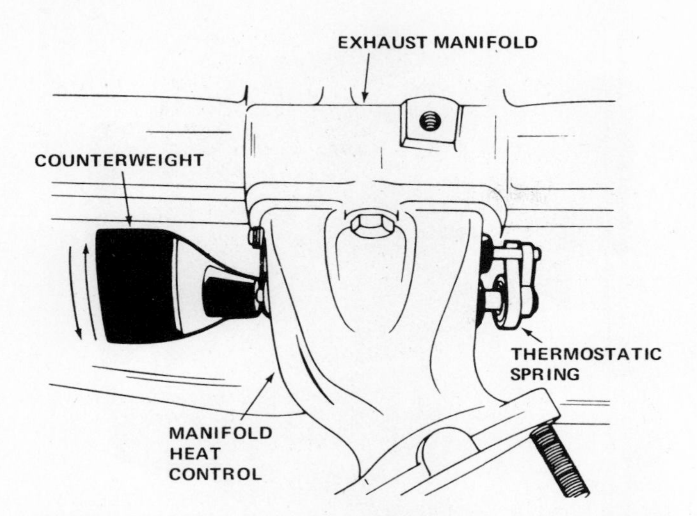

Fig. 4-72 A manifold heat control valve. (*Ford Motor Company*)

air-fuel mixture in the intake manifold for improved cold-engine operation. As the engine warms up, the heat-control valve opens. Then, the exhaust gases from both exhaust manifolds pass directly into the exhaust pipes.

NOTE: Since the introduction of the heated air system (⚙ 3-19), some engines no longer use a heat-control valve. To do so might add too much heat to the incoming air-fuel mixture. This would reduce the amount of air-fuel mixture entering and therefore reduce engine power.

3. **Early fuel evaporation system** A vacuum-controlled manifold heat control valve (Fig. 3-54) was introduced on some cars in 1975. With this arrangement, a vacuum motor instead of a thermostat controls the position of the heat control valve. The system is called an early fuel evaporation (EFE) system because of its quick action.

The heat control valve is called the EFE valve, and the vacuum to operate the vacuum motor comes from the intake manifold through a thermal-vacuum switch. You can see the locations of the EFE valve and the vacuum motor on a V-8 engine in Fig. 3-54. The EFE valve does the same job as the thermostatically operated heat-control valve described in items 1 and 2 above. However, the EFE valve does the job much faster. This reduces the time that heat is going into the intake manifold, thereby reducing the chance of overheating the air-fuel mixture.

When the engine is cold, the heat-control valve is in the cold-engine position shown in Fig. 4-73, top (for in-line engines). In V-type engines, the EFE valve is open. When the engine is started and intake manifold vacuum develops, the vacuum passes through the thermal-vacuum switch to the vacuum motor. The vacuum motor then operates to close the EFE valve as shown in Fig. 4-73, top (for in-line engines). This sends exhaust gas up through the oven around the intake manifold. In V-type engines, closing of the EFE valve shuts off one exhaust manifold from the exhaust pipe.

With the EFE valve closed, the exhaust gas flows through the crossover passage in the intake manifold, and exits through the other exhaust manifold. In either type of engine, heat from the exhaust gas passes into the ingoing air-fuel mixture to improve fuel vaporization and cold-engine operation.

As the engine begins to warm up, the thermal-vacuum switch shuts off the vacuum to the vacuum motor. Now the motor returns the heat-control valve (in in-line engines) to the hot-engine position. In V-type engines, the EFE valve opens to permit normal movement of the exhaust gases from the exhaust manifold to the exhaust pipe.

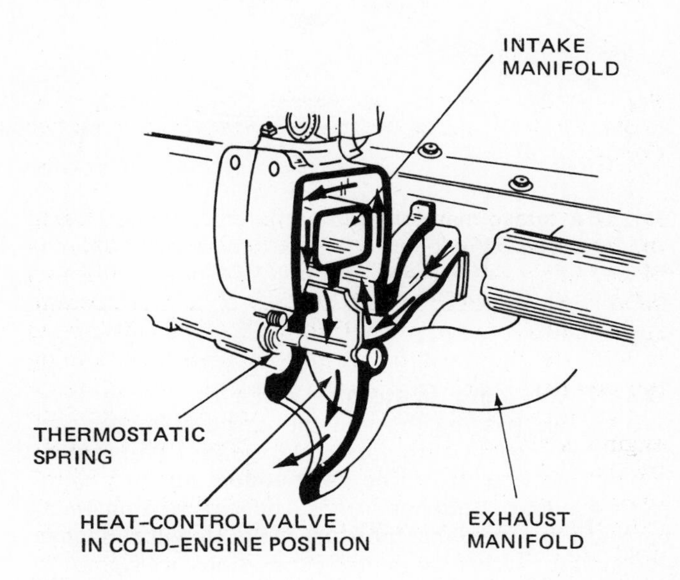

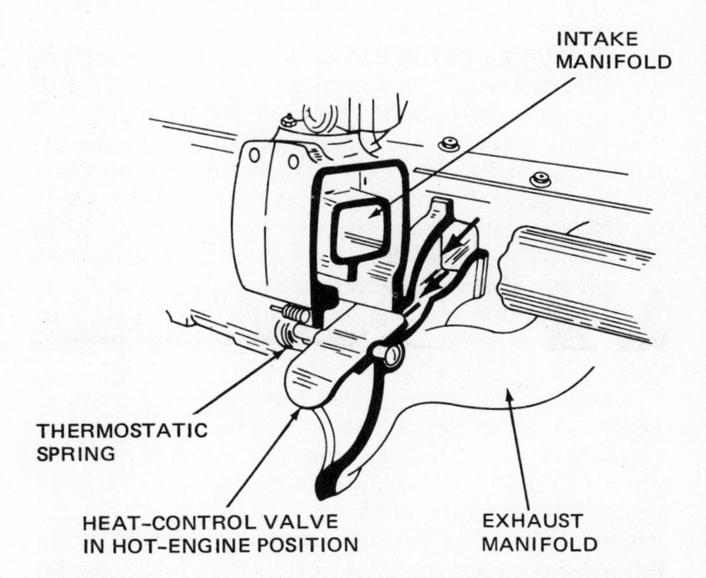

Fig. 4-73 Intake and exhaust manifolds of an in-line engine, cut away to show the location and action of the manifold heat control valve. The counterweight is not shown. At top, the heat control valve is in the HEAT ON position. It is directing hot exhaust gases up and around the intake manifold, as shown by the arrows. At bottom, the valve is in the hot-engine mode. (*Ford Motor Company*)

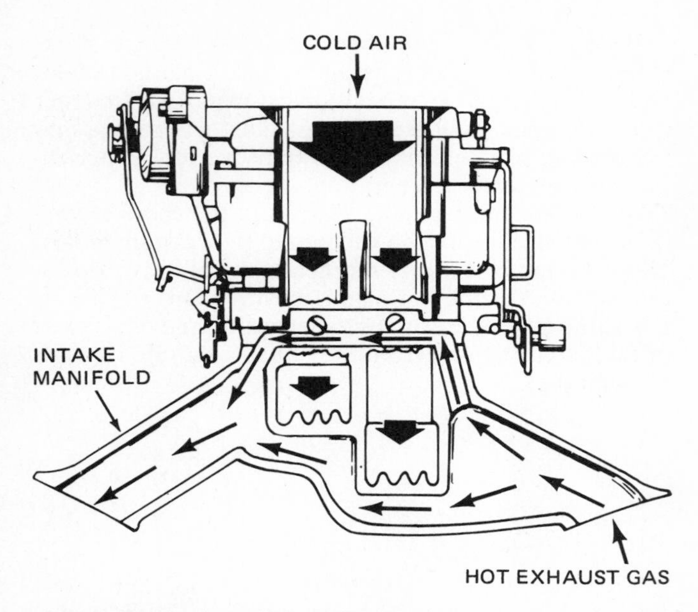

Fig. 4-74 Intake manifold and carburetor idle-port heating passages. Hot exhaust gases heat these areas as soon as the engine starts. (*Cadillac Motor Car Division of General Motors Corporation*)

Fig. 4-75 Throttle-return check or dashpot on a carburetor. (*Carter Carburetor Division of ACF Industries*)

❁ 4-37 Anti-icing systems When fuel is sprayed into the air passing through the carburetor, the fuel evaporates, or turns to vapor. As the fuel evaporates, it picks up heat from the surrounding air and metal parts. Under certain conditions, the surrounding metal parts are so cooled that moisture in the air will condense and actually freeze on them. If conditions are right, the ice can build up sufficiently to cause the engine to stall. This is most likely to occur during warm-up after the first start of the day of a cold engine. High humidity and air temperatures in the range of 40 to 60°F [4.4 to 15.6°C] are usually needed for carburetor icing.

To prevent such icing, some carburetors have special anti-icing systems. For example, during warm-up of a V-type engine, the manifold heat-control valve sends hot exhaust gases from one exhaust manifold to the other (❁ 4-36). Part of these hot exhaust gases circulate through the base of the carburetor, around the carburetor idle ports and near the throttle-valve shaft (Fig. 4-74). This circulation adds enough heat to prevent icing.

In another anti-icing system, the carburetor contains coolant passages. A small amount of the engine coolant passes through a special coolant manifold in the carburetor throttle body. This action adds enough heat to the carburetor to prevent icing.

Special anti-icing systems, as described above, are no longer required on many carburetors. With the introduction of the heated-air system using a thermostatically controlled air cleaner (❁ 3-19), carburetor icing is seldom a problem. This is because the heated-air system raises the temperature of the incoming air to above 85°F [29.4°C] very quickly. Therefore, icing cannot occur, because of the high temperature of the incoming air.

❁ 4-38 Fast idle When the engine is cold, the throttle valve must be held slightly open so that the engine will idle faster than it does when it is warm. Otherwise, slow idle and a cold engine might cause the engine to stall. With fast idle, enough air-fuel mixture gets through and air speeds are great enough to produce adequate vaporization and a sufficiently rich air-fuel mixture. Fast idle is obtained by a fast idle cam linked to the choke valve (Figs. 4-65 and 4-71).

When the engine is cold, the automatic choke holds the choke valve closed. In this position, the linkage has revolved the fast-idle cam so that the adjusting screw rests on the high point of the cam. The adjusting screw prevents the throttle valve from moving to the fully closed position. The throttle valve is held partly open for fast idle. As the engine warms up, the choke valve opens. This rotates the fast-idle cam so that the high point moves from under the adjusting screw. Now, the throttle valve closes to its normal hot-idle speed position.

1. Warm-engine operation When the engine is warm, the adjustment screw rests on the low section of the fast-idle cam (Fig. 4-65). This reduces engine speed to the normal hot-idle rpm. When the engine is turned off and cools, the thermostatic spring in the choke attempts to move the choke valve to the closed position. At the same time, the thermostatic spring attempts to turn the fast-idle cam to the fast-idle position. However, because the adjustment screw is resting on the low section of the fast-idle cam, no movement can take place.

To allow movement, which will reset the choke and return the throttle valve to the fast-idle position, the driver must push down on the accelerator pedal. This lifts the screw from the cam so that the spring can close the choke and rotate the cam to the fast-idle position. At the same time, pedal movement causes the accelerator pump to deliver a shot of fuel to the carburetor.

NOTE: The above actions occur only if the engine cools enough to cause winding of the thermostatic spring. If the engine stays warm, then no choking is needed for starting, and the above actions do not take place.

2. Cars with air conditioning Some cars with air conditioning have a device to increase idle speed when the air conditioner is running. One device consists of a vacuum-actuated diaphragm. The vacuum is directed to the diaphragm through a tube from the air-conditioner control switch.

Another device uses an electric solenoid similar to the antidieseling solenoid (⚙ 4-42). When the air conditioner is on, the solenoid increases the idle speed. This prevents stalling and engine overheating. In addition, the higher idle speed improves passenger compartment cooling.

⚙ 4-39 Antipercolator The carburetor is placed above the engine and is subject to engine heat. Under certain conditions, as when idling after a long hot drive, heat buildup may cause fuel in the main metering system to percolate. This forces fuel to dribble from the main nozzle. By using a small vent in the main metering system, vapor pressure can be relieved to prevent percolation.

The antipercolating device used on some carburetors is connected into the throttle system so that a small valve opens as the throttle is released. Another antipercolating device consists of a tube connected from the main metering system to the upper part of the air horn. This tube relieves vapor pressure enough to prevent percolation. A different design uses a small antipercolator check ball in the accelerator-pump plunger.

⚙ 4-40 Air bleed and antisiphon passages In all systems of carburetors, except some accelerator-pump systems, small openings are incorporated to permit air to enter or bleed into the systems (Figs. 4-10, 4-37, and 4-45). This action produces some premixing of the air and fuel so that better atomization and vaporization result. It also helps maintain a more uniform air-fuel ratio. For example, at higher speeds, a larger amount of fuel tends to be discharged from the main nozzle. But at the same time, the faster fuel movement through the high-speed system causes more air to bleed into the system. Therefore, the air holes tend to equalize the air-fuel ratio.

Air-bleed passages are also sometimes called antisiphon passages. They act as air vents to prevent the siphoning of fuel from the float bowl at intermediate engine speeds.

If air-bleed passages become plugged, they may cause the float bowl to be emptied after the engine shuts off. When the engine is shut off, the intake manifold cools down and a slight vacuum forms as a result. With open air bleeds, air can move through the bleeds to satisfy the vacuum. But if the air bleeds are plugged, then the vacuum will cause the float bowl to empty through the idle system.

⚙ 4-41 Vacuum vents The carburetor and intake manifold produce vacuum when the engine is running. This vacuum varies as the operating conditions change. The vacuum can be used to operate various devices (Fig. 4-1).

1. Ignition-distributor vacuum-advance mechanism This is discussed in ⚙ 1-20. At part throttle, the vacuum signal advances the spark to give the less dense mixture more time to burn.

2. Positive crankcase ventilating system This system, explained in ⚙ 3-21, removes fuel vapor and exhaust gases from the engine crankcase.

3. Vapor-recovery system This system, described in ⚙ 3-14, traps fuel vapor from the carburetor float bowl and fuel tank in a charcoal canister. When the engine is started, the vacuum port in the carburetor draws fresh air through the canister to purge it of the trapped fuel vapor.

4. Heated-air system The heated-air system, described in ⚙ 3-19, uses a thermostatically controlled air cleaner. Vacuum from the intake manifold, carried through a port in the carburetor throttle body, operates the control damper in the air-cleaner snorkel tube.

5. Exhaust-gas recirculation system The exhaust-gas recirculation system (⚙ 3-28) sends some of the exhaust gas into the ingoing air-fuel mixture. This reduces one of the atmospheric pollutants produced by the engine—nitrogen oxides (NO_x). The vacuum signal that operates the EGR valve comes from a port in the carburetor throttle body. As the throttle valve moves past the port, intake manifold vacuum is introduced through it to the EGR valve. This valve then functions to control the EGR system.

Several different types of EGR valves are in use. Some are back pressure valves, which operate only when the designed combination of back pressure in the exhaust and vacuum from the engine are available.

6. Vacuum motors for air conditioners Many air conditioners have vacuum motors to provide control of the air-conditioner doors. These doors are moved to produce the heating or cooling required. Intake manifold vacuum provides the operating force for these motors.

7. Vacuum for power brakes Power brakes use intake manifold vacuum to provide most of the braking effort. As the brake pedal is pushed down for braking, a valve admits intake-manifold vacuum into one section of the power-brake unit. Atmospheric pressure then moves a piston or diaphragm to produce the braking action.

NOTE: Diesel engines do not develop any usable vacuum in the intake manifold. Therefore, to operate the various vacuum devices on a diesel vehicle, a separate vacuum pump must be installed on the engine.

4-42 Antidieseling solenoid Some engines have a tendency to continue to run after the ignition switch is turned off. This is called *run-on* or *dieseling* (because the engine runs like a diesel engine, without electric ignition). Dieseling can be caused by high engine cylinder temperatures, particles of hot carbon, or hot spark-plug electrodes. Enough air-fuel mixture can seep around the throttle valve and through the idle system to maintain the dieseling.

To prevent dieseling, many carburetors are equipped with an antidieseling solenoid (Fig. 4-1). Its job is to control the closing of the throttle valve. When the engine is running, the solenoid is connected to the battery. It extends a plunger. The plunger serves as the idle stop and prevents complete closing of the throttle. Therefore, normal hot-idle speed operation results when the driver releases the accelerator pedal. However, when the engine is turned off, the solenoid is disconnected from the battery. It pulls in its plunger. Now, the throttle closes completely, shutting off all air flow. The engine stops running.

A second arrangement uses the antidieseling solenoid to shut off the fuel flow in the idle system when the engine is turned off. During normal operation, the solenoid pulls in its plunger so normal fuel flow can continue through the idle system. However, when the engine is shut off, the solenoid releases its plunger which blocks the idle system so no fuel can flow through it. Therefore, the engine stops.

4-43 Throttle-return check (dashpot) If the throttle valve closes too fast after the driver releases the accelerator pedal, the air-fuel mixture gets too rich. This is because the inertia of the fuel forces the fuel nozzle to dribble fuel momentarily after the airflow is largely shut off. The idle system momentarily feeds a rich mixture. This is caused by the high vacuum that results when the engine is running fairly fast with the throttle closed.

A very rich mixture will not burn properly. It can cause the engine to stall or backfire through the exhaust system. Also, the high level of HC in the exhaust gas may damage the catalytic converter. To prevent this, many carburetors are equipped with a throttle-return check or *dashpot* (Fig. 4-75). It slows the closing of the throttle valve so that the overly rich mixture is prevented.

Some carburetors on cars with catalytic converters have a throttle position solenoid which is similar to the throttle-return check. They both prevent rapid closing of the throttle valve when the driver suddenly lets up on the accelerator pedal after high-speed driving.

The throttle position solenoid is connected to the battery through an engine speed sensor. The speed sensor receives ignition pulses from the electronic ignition system. When the speed sensor senses that the engine speed exceeds 2000 rpm, it connects the throttle position solenoid to the battery. The solenoid extends a plunger that brings a throttle stop into position. Now, if the accelerator pedal is released, the throttle valve will not close completely. This prevents the sudden surge of enriched air-fuel mixture into the engine. Then, after the engine speed has dropped below 2000 rpm, the engine speed sensor opens the throttle position solenoid circuit to the battery. The solenoid withdraws its plunger and the throttle valve can return to its normal closed position.

4-44 Governors Governors control or limit top engine speed or road speed. The use of governors is largely confined to trucks and buses. They prevent overspeeding and rapid wear of the engine. One type directly controls the throttle valve. It tends to close the valve as a predetermined engine speed is reached. Another type has a throttle plate between the carburetor throttle valve and the intake mainfold. The throttle plate moves toward the closed position as a predetermined speed is reached. This prevents delivery of additional amounts of air-fuel mixture and any further increase in engine speed.

Figure 4-76 shows an electronic engine-speed governor which attaches to the carburetor. This type of governor limits engine speed to a predetermined maximum while still allowing full engine power at close to the governed speed. Intake-manifold vacuum is applied to a diaphragm which modulates or limits the opening of the throttle valves. The vacuum to the diaphragm is, in turn, modulated by the action of a solenoid. An electronic governor module (EGM) turns the solenoid on and off.

4-45 Multiple carburetors To achieve better air-fuel distribution, engine breathing, and higher volumetric efficiency, some high-performance engines are equipped with more than one carburetor. The additional carburetors supply more air and fuel to improve high-speed engine performance. They also distribute the mixture more evenly to ensure that each cylinder receives the same volume of air-fuel mixture, and that the mixture ratio received by each cylinder is the same.

Two carburetors mounted on an engine are called *dual carburetors*. Three carburetors mounted on an engine are called *triple carburetors* or sometimes *tri-power*.

Most American automobile manufacturers have used multiple carburetors in the past to provide a high-performance option to standard car engines. Foreign manufacturers have used multiple carburetors extensively to increase performance and efficiency of small-displacement engines in cars.

Figure 4-77 shows two four-barrel carburetors installed on a V-8 engine. The throttle linkage from the accelerator pedal is connected to both carburetors so that they work together. The intake manifold must have mounting pads for each of the two carburetors. Figure 4-78 shows the installation of three two-barrel carburetors on a V-8 engine. The three carburetors are linked together so that the engine idles and operates at low speed on the center or primary carburetor. Then, as the accelerator pedal is pressed down, the secondary

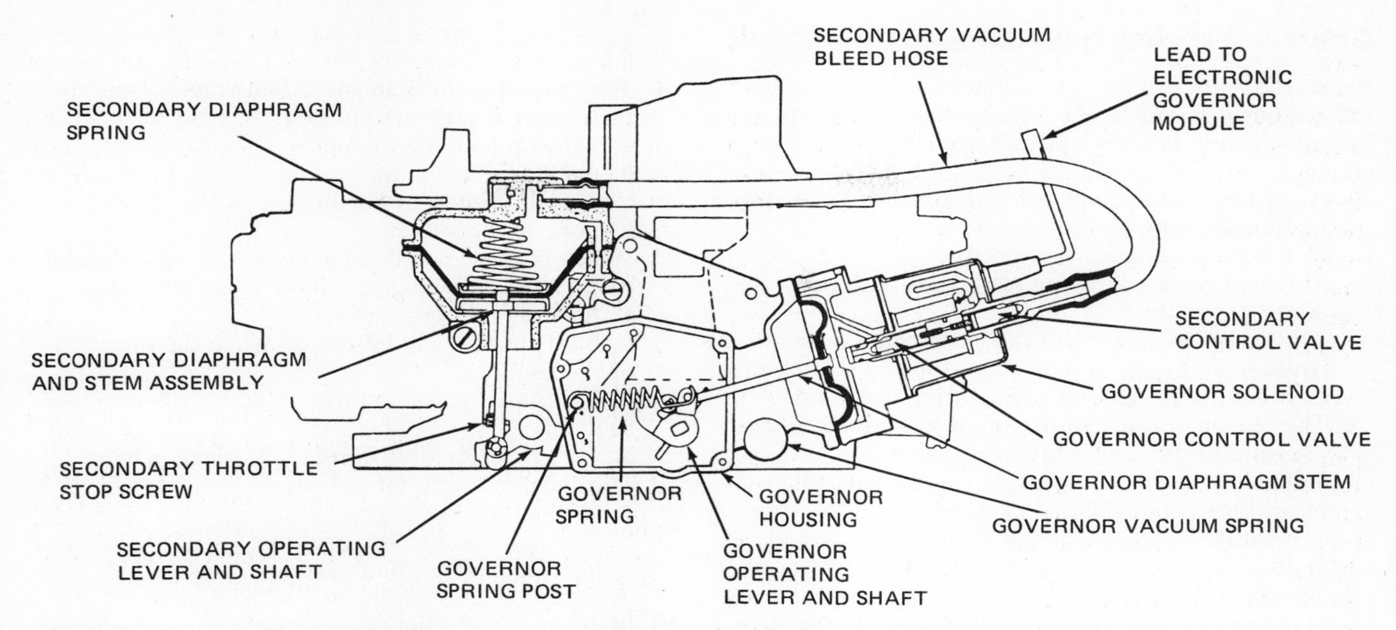

Fig. 4-76 An electronic engine-speed governor, which prevents engine overspeeding. (*Ford Motor Company*)

carburetors begin to open. The linkage is designed so that by the time the accelerator pedal is to the floorboard, the throttle valves of all three carburetors will reach the wide-open position at the same time. This type of throttle linkage is called a *progressive linkage*.

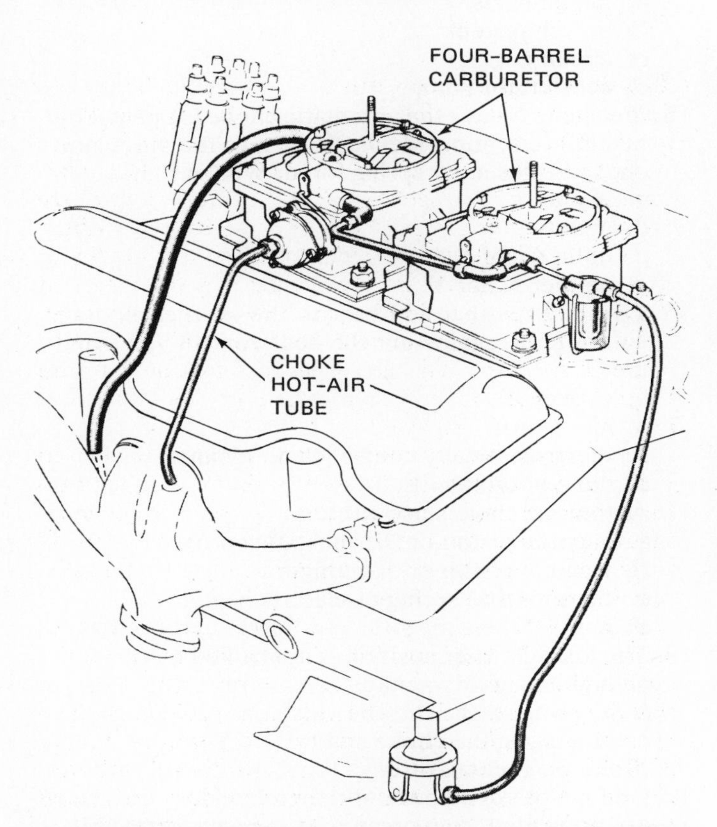

Fig. 4-77 Two four-barrel carburetors installed on a V-8 engine. (*Chevrolet Motor Division of General Motors Corporation*)

Fig. 4-78 Installation of three two-barrel carburetors on a V-8 engine. (*Ford Motor Company*)

Chapter 4 review questions

Select the *one* correct, best, or most probable answer to each question. Then check your answers against the correct answers given at the end of the book.

1. All of these statements are true about the idle-stop (antidieseling) solenoid *except*:
   a. it controls engine idle speed
   b. it can cause dieseling (afterrun) when not adjusted correctly
   c. it opens the carburetor throttle plates when the ignition is on
   d. it controls engine fast idle during warm-up.
2. The portion of the carburetor air horn that reduces pressure to cause fuel to flow is called the:
   a. throttle body
   b. air bleed
   c. venturi
   d. fuel nozzle.
3. The throttle valve:
   a. when closed, allows little or no air to flow through the air horn
   b. when open, allows air to flow freely through the air horn
   c. both of the above
   d. none of the above.
4. The reason that a richer mixture must be delivered when a cold engine is first started is that:
   a. this allows a higher cranking speed
   b. only part of the gasoline will vaporize when cold
   c. the thick engine oil must be thinned out
   d. none of the above.
5. An air-fuel ratio of 12:1 means that the mixture has:
   a. 12 pounds of gasoline to 1 pound of air
   b. 12 pounds of air to 1 pound of gasoline
   c. 1 gallon of gasoline to 12 gallons of air
   d. 12 gallons of air to 1 gallon of gasoline.
6. The carburetor system that stores the gasoline to be used by other systems is called the:
   a. power system
   b. float system
   c. main metering system
   d. choke system.
7. The purpose of the float bowl vent is to:
   a. prevent engine overheating
   b. keep the level of gasoline in the bowl constant
   c. equalize the effects of a clogged air cleaner
   d. none of the above.
8. The purpose of the hot-idle compensator is to:
   a. supply additional air for idling when the engine is hot
   b. slow down engine speed
   c. prevent a hot engine from running when the ignition is shut off
   d. increase idle speed while the engine is cold.
9. When the engine is hot and running at 600 rpm, the gasoline is supplied by the:
   a. idle system
   b. low-speed system
   c. choke system
   d. main metering system.
10. The operating mechanism of the vacuum-operated metering rod includes either a:
    a. vacuum piston or pump
    b. diaphragm or pump
    c. vacuum piston or diaphragm
    d. none of the above.
11. Which of these could happen when the accelerator pump inlet check ball is left out of a carburetor:
    a. flooding during acceleration
    b. hard starting and hesitation when the throttle is opened suddenly
    c. hesitation or stalling when the throttle is closed suddenly
    d. very rich mixtures during low-speed driving.
12. The accelerator pump operates:
    a. all the time the engine is running
    b. during initial throttle opening
    c. automatically, when vacuum drops
    d. during wide-open-throttle operation.
13. In the main metering system, the maximum amount of fuel that can flow during normal driving conditions is controlled by:
    a. main metering jets
    b. main air bleeds
    c. the choke valve
    d. the inlet check ball.
14. When the choke valve is closed, gasoline is delivered from the:
    a. main nozzle
    b. power system
    c. choke nozzle
    d. accelerator pump.
15. Mechanic X says the automatic choke is opened by manifold vacuum. Mechanic Y says the automatic choke is closed by spring force. Who is right?
    a. X only
    b. Y only
    c. both X and Y
    d. neither X nor Y.
16. Two devices that add heat to the air-fuel mixture entering the engine are the heat-control valve and the:
    a. choke
    b. air cleaner
    c. thermostatically controlled air cleaner
    d. none of the above.
17. Automatic chokes use either a:
    a. vacuum piston or thermostatic spring
    b. vacuum piston or diaphragm
    c. thermostatic spring or electric heater
    d. none of the above.
18. The fast-idle cam position is controlled by:
    a. high manifold vacuum
    b. a spring on the throttle valve
    c. an arm on the choke shaft
    d. the air-fuel mixture.
19. On four-barrel carburetors, the secondary throttles are prevented from opening at the wrong time by:
    a. secondary air valves
    b. the secondary lockout

c. a calibrated power control
d. engine temperature.

20. The idle-stop solenoid prevents dieseling by:
a. completely closing the throttle valve when the ignition switch is turned off
b. preventing excessively high idle speed
c. stopping engine idle when the engine gets hot
d. all of the above.

21. The APT carburetor has:
a. an extra throttle valve
b. an extra barrel
c. an extra metering rod adjusted at the factory
d. all of the above.

22. The exhaust manifold heat-riser valve is stuck in the open position. Mechanic X says this can cause poor gas mileage. Mechanic Y says this can cause the intake manifold vacuum to be lower than normal. Who is right?
a. X only
b. Y only
c. both X and Y
d. neither X nor Y.

23. The purpose of the unloader is to:
a. help start a flooded engine
b. help start a cold engine
c. help start a hot engine
d. none of the above.

24. The thermostatic spring on a manifold heat control valve is broken. This causes the heat control valve to remain in the:
a. cold-engine position
b. hot-engine position
c. rich-mixture position
d. none of the above.

25. A driver complains that the car is hard to start when it is cold after listening to the radio with the ignition on for several minutes before trying to start the engine. Mechanic X says this is normal operation. Mechanic Y says the electric-assist choke heater is shutting off before starting is attempted. Who is right?
a. X only
b. Y only
c. both X and Y
d. neither X nor Y.

CHAPTER 5

VARIABLE-VENTURI (VV) CARBURETORS

After studying this chapter, you should be able to:

1. Define *variable-venturi carburetor.*
2. Describe the operation of a round-piston variable-venturi carburetor.
3. Describe the operation of a Ford VV carburetor.
4. List the fuel metering systems used in the Ford VV carburetor.
5. List the systems found in fixed-venturi carburetors which are *not* used in the Ford VV carburetor.
6. Explain the operation of each fuel-metering system in the Ford VV carburetor.
7. Discuss the various types of throttle positioners used with the Ford VV carburetor.

5-1 Variable-venturi carburetors The previous chapter discussed fixed-venturi carburetors. The venturi effect and how the venturi produces a partial vacuum when the air flows through it were explained in ⚙ 4-4. This partial vacuum then causes the fuel nozzle to discharge fuel into the air passing through. The venturi in the variable-venturi (VV) carburetor does the same thing. The basic difference is that in the VV carburetor, the size of the venturi can vary. In the fixed-venturi carburetor, the venturi is located in the center of the carburetor air passage and cannot change in size.

There are two basic types of VV carburetors, the round-piston or slide-valve type, and the rectangular venturi valve type. The slide-valve type has been used on some foreign cars and many motorcycles for years. The rectangular venturi valve type is a relatively new design used by Ford. Both types of VV carburetors have float systems similar to those used in fixed-venturi carburetors.

⚙ 5-2 Round-piston VV carburetor Figures 5-1 and 5-2 show exterior and sectional views of one model of round-piston VV carburetor. The carburetor is shown in disassembled view in Fig. 5-3.

The piston (Fig. 5-2) is an assembly of two basic parts: the outer two-diameter piston and the inner oil-damper reservoir. The piston moves up and down in the piston chamber in response to the amount of vacuum between the piston and throttle valve. When the piston moves down, it reduces the size of the venturi. The venturi is formed by the end of the piston and the throttle body. Figure 5-4 shows how downward movement of the piston reduces the size of the venturi.

Movement of the piston also moves the tapered needle. The needle is fastened to the bottom of the piston. When the piston moves downward, the needle moves down into the fuel jet. This reduces the area of the jet opening, which limits the amount of fuel that can flow. At the same time, the downward movement

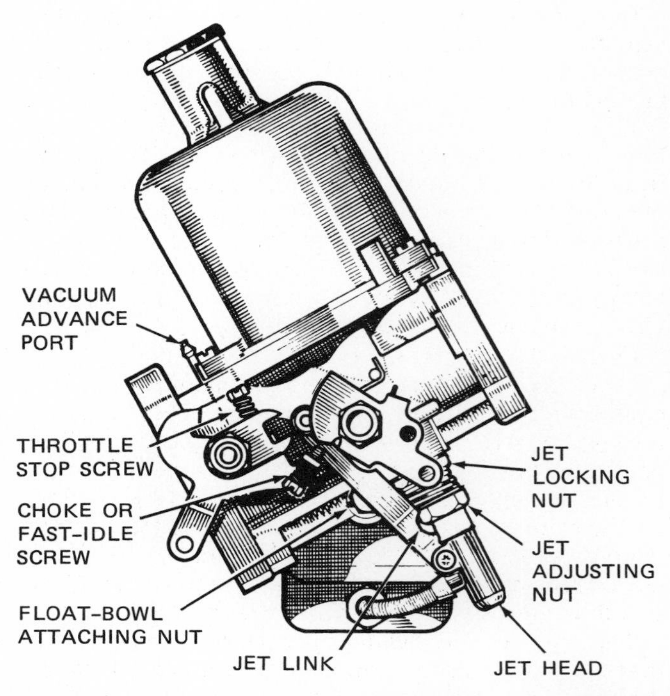

Fig. 5-1 A variable-venturi carburetor which uses a round piston, or slide valve. (*British Leyland, Inc.*)

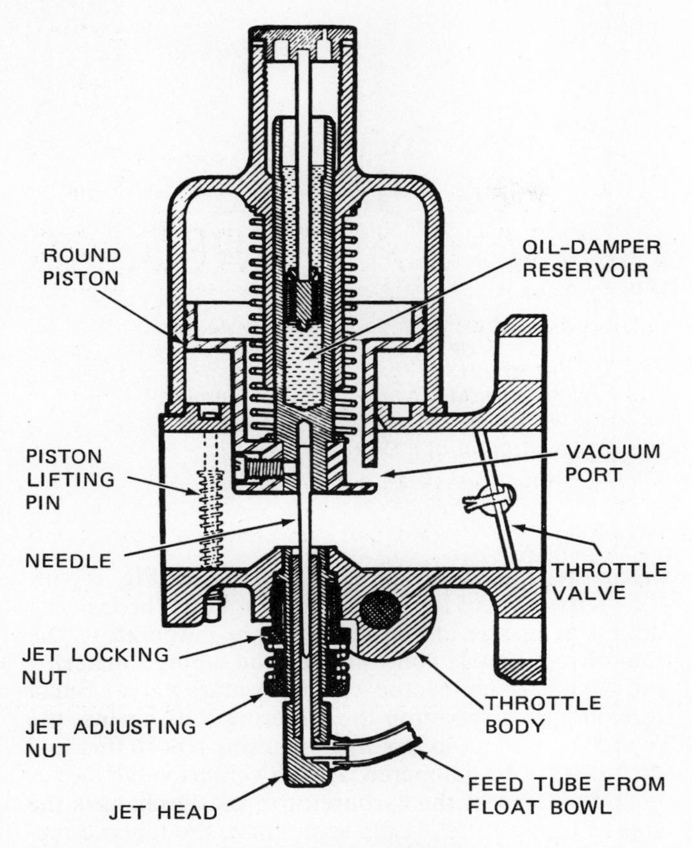

Fig. 5-2 Sectional view of a variable-venturi carburetor showing the round piston, or slide valve. (*British Leyland, Inc.*)

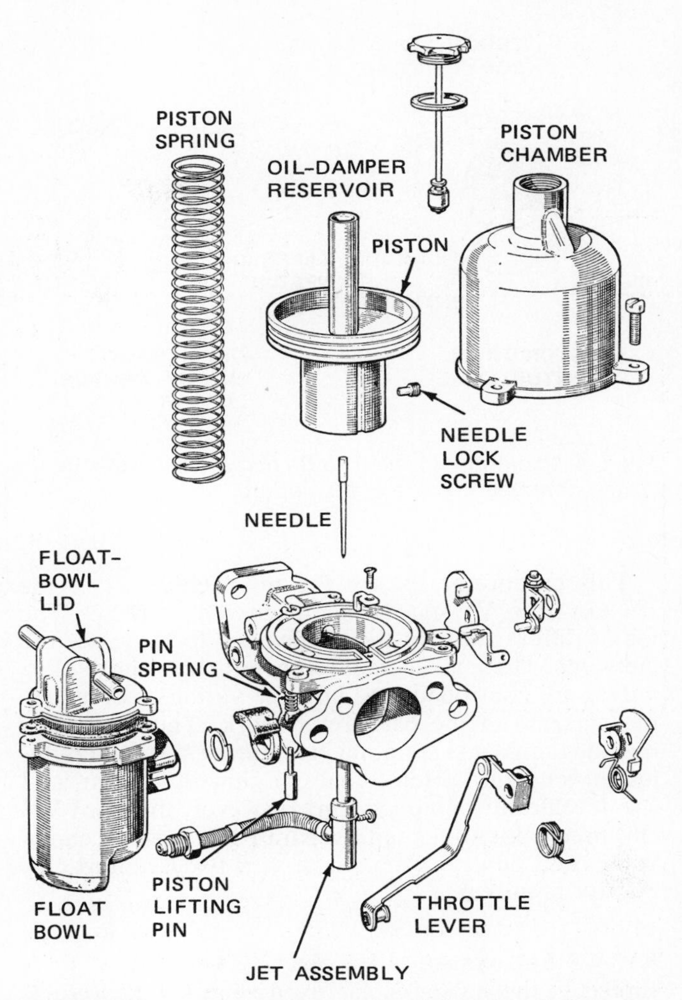

Fig. 5-3 View of a disassembled slide-valve carburetor. (*British Leyland, Inc.*)

of the piston reduces the size of the venturi. This limits the volume of air that can flow through.

In operation, as the size of the venturi changes, the size of the jet opening also changes. In this way, the proper proportions of air and fuel are maintained. The air-fuel ratio stays approximately constant.

The piston is raised or lowered in response to the movement of the throttle valve (Fig. 5-2). When the throttle valve is closed to the idling position, intake manifold vacuum is cut off from the throttle body. The piston spring pushes the piston down to its lowest position. A small amount of air flows around the throttle valve and through the venturi. It produces just enough vacuum at the venturi to cause the fuel jet to deliver enough fuel for idling.

When the throttle is opened, intake manifold vacuum enters the throttle body. This vacuum draws air from the space above the piston, acting through the vacuum port in the lower part of the piston. The piston is raised by the vacuum, partly compressing the piston spring. As the piston moves up, more air can pass through the venturi. At the same time, the needle moves up in the jet, increasing the effective size of the jet. Then more fuel flows and mixes with the air passing through.

The taper on the needle must be very accurate so that the proper air-fuel ratio will be delivered through the entire range of throttle opening. Actually, the pin taper permits additional fuel to flow at full throttle so the mixture is enriched for acceleration and full-power operation.

The oil damper reservoir (Figs. 5-2 and 5-3) that is part of the piston acts like a tiny shock absorber. It prevents excessive movements of the piston as the throttle valve is moved and vacuum conditions change. Without this shock-absorber action, the piston could bounce up and down, causing erratic fuel delivery and poor engine performance.

⚙ 5-3 Ford variable-venturi carburetor Ford began using a rectangular variable-venturi carburetor on some 1977 cars (Fig. 5-5). It is a two-barrel carburetor, which on the outside is similar in appearance to a fixed-venturi carburetor (Fig. 5-6). However, internally the Ford VV carburetor is very different.

The Ford VV carburetor has two barrels or "throats." In each throat there is a rectangular-shaped piston, or venturi valve, which slides back and forth across the throat (Figs. 5-5 and 5-7). This changes the size of the opening, or venturi, above the throttle valve. The two venturi valves are connected together. Their positions are controlled by intake manifold vacuum and throttle position. Each is connected to a tapered needle or metering rod which is positioned in a fuel jet.

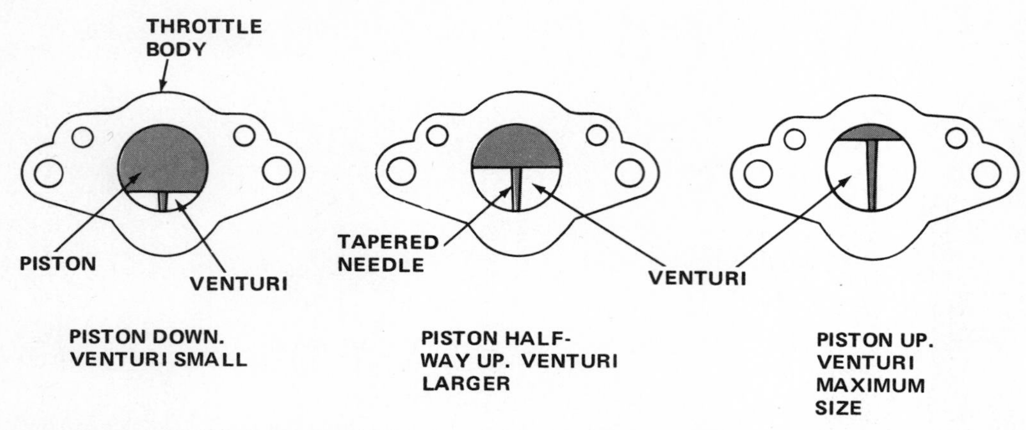

Fig. 5-4 Looking into the throttle body to see how the up-and-down movement of the piston changes the size of the venturi.

This carburetor has all the main elements of the round-piston VV carburetor discussed in ⚙ 5-2. The major difference is in the shapes and locations of the elements. There are also differences in how the venturi valves are controlled, and how the various fuel metering systems in the carburetor work. These systems include the float system, main metering system, cranking enrichment system, cold enrichment system, and the accelerator-pump system. However, the Ford VV carburetor does not need a separate idle system, choke valve, and power system such as a fixed-venturi carburetor requires.

⚙ 5-4 Float system The float system (Fig. 5-8) is similar to those used in the fixed-venturi carburetors. It uses a single float and a needle valve riding on the float lever. The fuel filter is located in the carburetor. The components of the filter are shown in disassembled view in Fig. 5-8.

⚙ 5-5 Main metering system Airflow through the venturi causes a partial vacuum. This causes fuel to discharge from the nozzle. The arrows in Fig. 5-7 indicate the flow of fuel from the float bowl through the discharge nozzle in one throat of the carburetor. The rate of fuel flow is controlled by the tapered metering rod which is connected to the venturi valve. Small torsion springs position the metering rods against the top of the orifices in the main metering jets so that fuel flow will not be hampered. As the venturi valve moves in and out across the carburetor throat, it changes the size of the venturi. At the same time, the tapered metering rod moves in and out of the main metering jet. This changes the size of the jet and therefore the amount of fuel that can flow from the discharge nozzle.

The position of the venturi valve and the metering rod is controlled by the vacuum below the venturi valve and above the throttle plate. A change in throttle plate position or in engine load changes this vacuum which changes the position of the venturi valve and metering rod.

⚙ 5-6 Control vacuum During normal warm-engine operation, the position of the venturi valve is controlled by two forces:

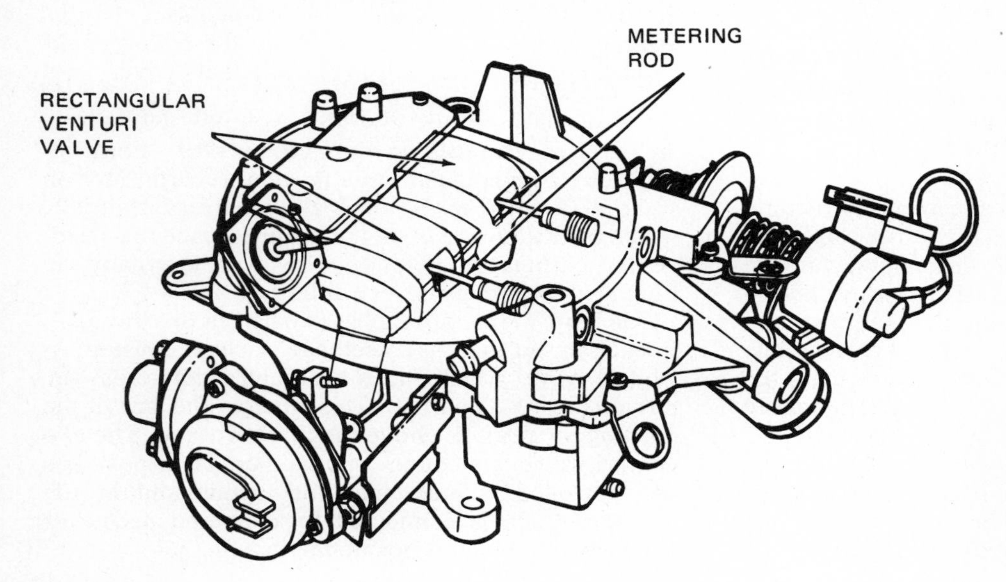

Fig. 5-5 Partial cutaway view of a two-barrel VV carburetor using rectangular-shaped venturi valves. (*Ford Motor Company*)

Fig. 5-6 Top view of a Ford VV carburetor. (*Ford Motor Company*)

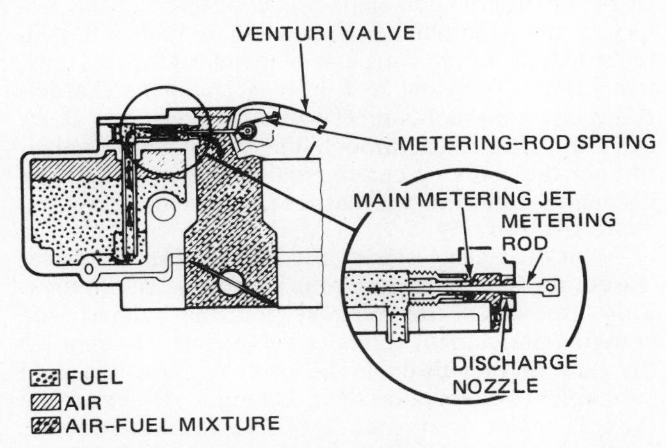

Fig. 5-7 Main metering system of the Ford VV carburetor. (*Ford Motor Company*)

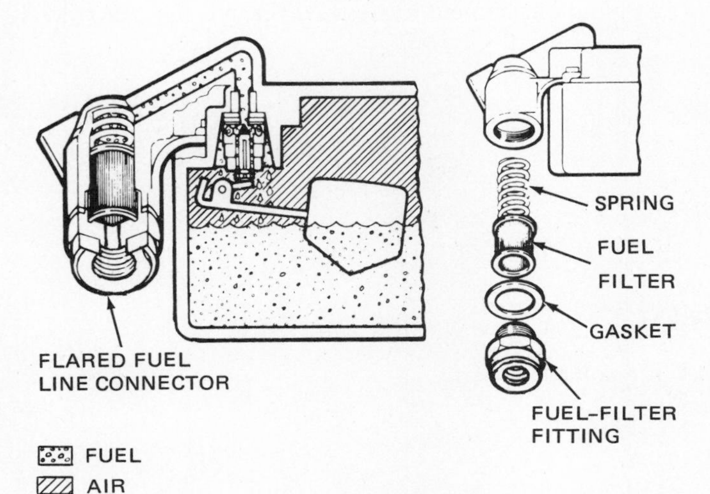

Fig. 5-8 Fuel inlet and float system, with the fuel filter shown in disassembled view. (*Ford Motor Company*)

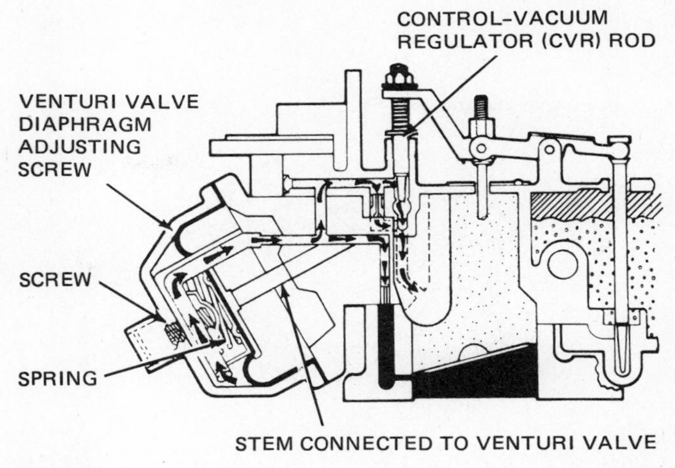

Fig. 5-9 Control vacuum system. (*Ford Motor Company*)

A spring in the diaphragm assembly that tries to close the venturi valve.

The control vacuum below the venturi and above the throttle plates.

Figure 5-9 shows the control vacuum system. Both of the forces act on the control vacuum diaphragm. The control vacuum is the vacuum below the venturi valve and above the throttle plates. It is routed to the control vacuum diaphragm through the upper body, main body, and the diaphragm housing, as shown by the arrows in Fig. 5-9. Note that the arrows are shown pointing away from the back or left end of the diaphragm. This is the way the air back of the diaphragm would flow when the vacuum is applied.

The diaphragm is connected to the venturi valve (Fig. 5-10). When the engine is running, an increase in throttle opening will normally increase the control vacuum. Increased vacuum pulls the diaphragm in against the spring. This moves the venturi valves and the main metering jets toward the open position. The venturi valves are not linked directly to the throttle. The control vacuum and spring can select the exact air-fuel ratio needed for all speed and load conditions except wide-open throttle (WOT).

⚙ 5-7 Venturi-valve limiter At wide-open throttle, under some conditions, the control vacuum will not be strong enough to override the venturi-valve diaphragm spring and open the valve fully. This could cause poor wide-open-throttle performance. To prevent this, the carburetor has a venturi-valve limiter (Fig. 5-11). The limiter has a lever on the throttle shaft. When the shaft turns to open the throttle wide, the lever moves up against the limiter-adjuster screw. This pushes the venturi valve to the wide-open position.

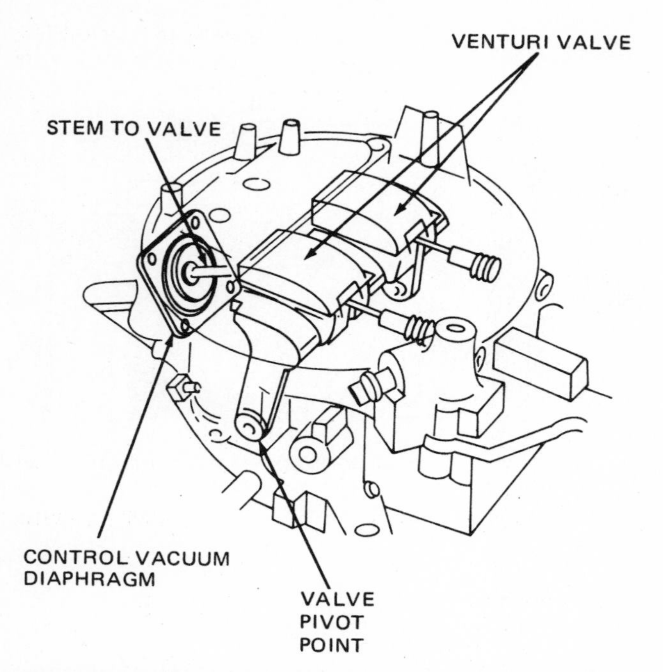

Fig. 5-10 The control vacuum diaphragm is connected by a stem to the venturi valves. As vacuum conditions in the intake manifold change, the vacuum diaphragm changes the position of the venturi valves, thereby changing the size of the venturis. (*Ford Motor Company*)

⚙ 5-8 Cold-cranking enrichment system This system is used instead of a choke. It supplies fuel for cold-engine starting below 75°F [23.9°C] while the engine is being cranked. The system includes a normally closed cranking-enrichment solenoid valve mounted at the front of the carburetor (Fig. 5-12). The system also includes a bimetal thermostatic valve (cranking fuel-control valve in Fig. 5-12) that is closed above a fuel bowl temperature of 75°F [23.9°C].

The volume of fuel flowing in the system is controlled by the position of the solenoid valve and of the bimetal thermostatic valve. Some carburetors also have a small bypass restrictor in the bottom of the fuel bowl. This bypass allows fuel to flow even if the thermostatic

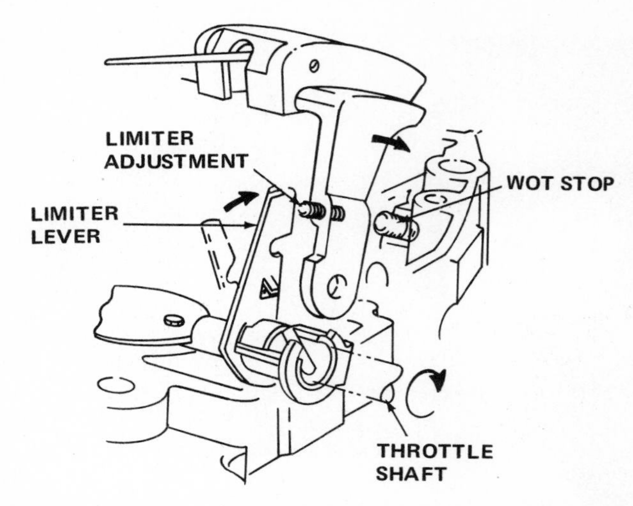

Fig. 5-11 Venturi valve limiter. (*Ford Motor Company*)

cranking-fuel-control valve is closed. This helps hot-engine starts.

When the ignition switch is turned to START, the cranking-enrichment solenoid is connected to the battery. It pulls the plunger back to allow fuel to flow in the cranking enrichment system, provided the fuel bowl temperature is below 75°F [23.9°C] and the thermostatic cranking-fuel-control valve is open. Fuel flows from the fuel bowl through the thermostatic valve, through the cranking-enrichment solenoid valve, to the discharge points in the carburetor throat below the venturi valves (Fig. 5-9).

When the engine starts, and the ignition key is released, the ignition switch returns from START to RUN. This disconnects the solenoid from the battery. The cranking-enrichment solenoid valve closes to shut off the flow of fuel through the system. Now the cold enrichment system takes over. It supplies the extra fuel the cold engine needs (⚙ 5-9). This is in addition to the fuel the main metering system is supplying (⚙ 5-5).

⚙ 5-9 Cold-running enrichment system The cold-running enrichment system takes over where the cold-cranking enrichment system leaves off. This is when

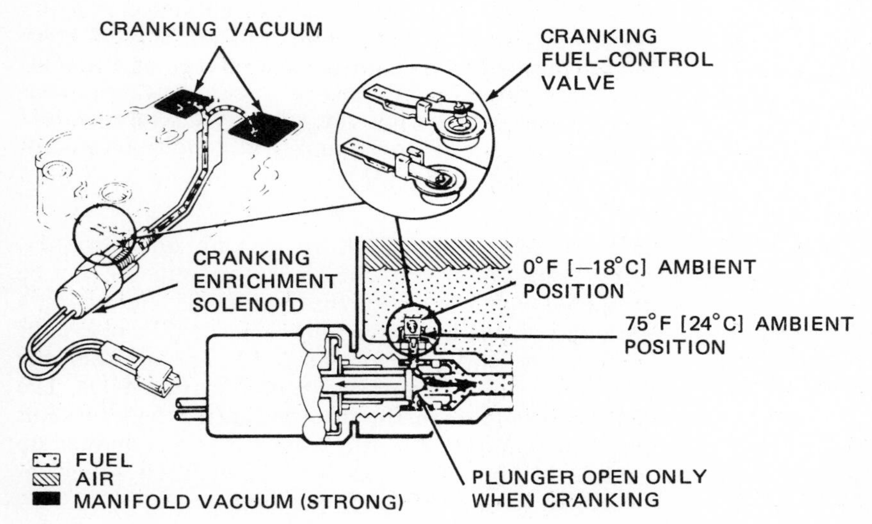

Fig. 5-12 Cold-cranking enrichment system. (*Ford Motor Company*)

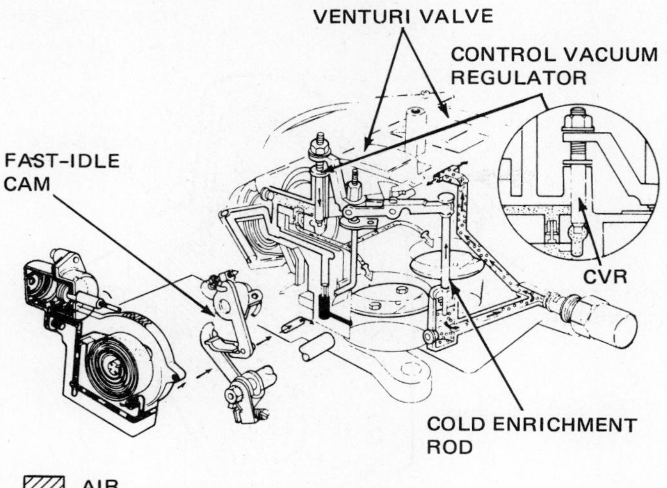

Fig. 5-13 Cold-running enrichment system. (*Ford Motor Company*)

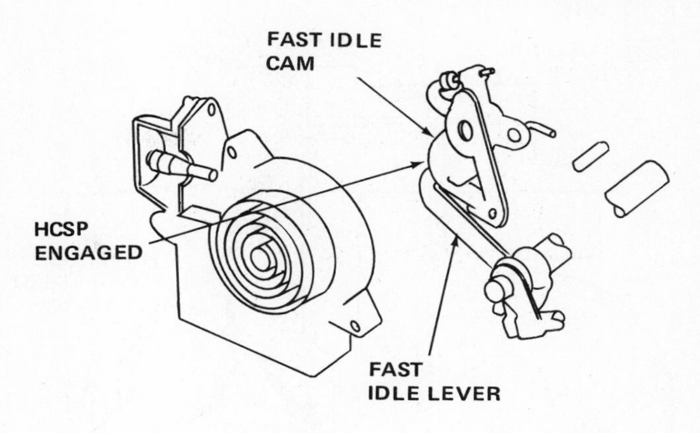

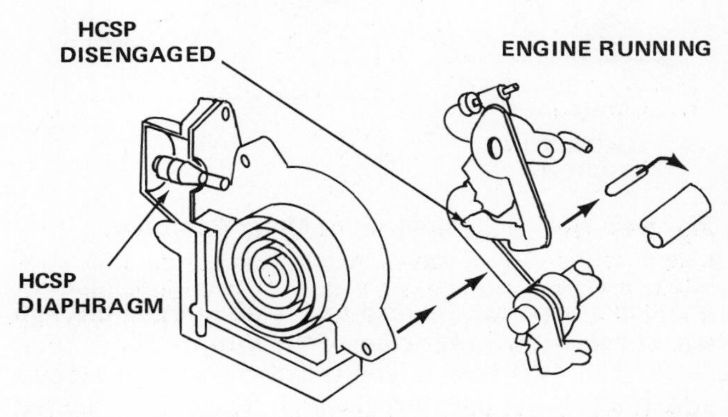

Fig. 5-14 Fast-idle cam and high-cam-speed positioner operation. (*Ford Motor Company*)

the cold engine is first running in what is called the *cold driveaway* mode. The cold-running enrichment system has an exhaust-gas-heated bimetal thermostatic coil (Fig. 5-13). This coil regulates the flow of fuel and the positioning of the venturi valve and fast-idle cam (⚙ 5-10) when the engine is running cold.

Fuel flow is regulated by a cold-enrichment metering rod suspended from the upper body of the carburetor. It hangs down through a jet or hole in the bottom of the float bowl (Fig. 5-13). The position of the metering rod, and therefore, the size of the opening, is determined by the thermostatic coil. As the engine and thermostatic coil warm up, the coil unwinds to drop the metering rod down. This reduces the flow of fuel through the same passage as the cold-cranking enrichment system.

After a cold start, the control vacuum regulator (CVR) (Figs. 5-9 and 5-13) restricts the control vacuum to the control vacuum diaphragm. This restriction allows the manifold vacuum to override the control vacuum. Then the manifold vacuum determines the position of the venturi valves. Manifold vacuum applied to the diaphragm causes the venturi opening to be slightly larger than it would normally be with a hot engine. This prevents engine loping at idle due to an overrich mixture.

When the engine is accelerated, the venturi valve assumes its normal position to supply a richer mixture for cold-driveaway.

⚙ 5-10 Fast-idle cam The fast-idle cam arrangement is similar to the systems used in the fixed-venturi carburetors (Fig. 5-14). It controls engine speed during the cold-enrichment operating mode. This is when the fast-idle cam is in position so the throttle valve is held open enough to assure fast idling. The system includes a vacuum-operated high-cam-speed positioner (HCSP).

Under cold-start conditions, a lever is inserted between the fast-idle cam and the fast-idle lever (Fig. 5-14). This provides increased throttle opening for starting. When the engine starts, manifold vacuum works on the HCSP diaphragm to retract the lever after the first throttle movement. The lever remains retracted as long as the engine is running. With the lever out of the way, the fast-idle speed is reduced for all cold-driving modes.

⚙ 5-11 Control vacuum regulator When the engine warms up, the thermostatic coil unwinds (Figs. 5-9 and 5-13). This causes the control vacuum regulator to rise. As it rises, it opens the restricted port in the control vacuum circuit. This allows the control of the control vacuum circuit to return to normal as described in ⚙ 5-6. At the same time, the cold enrichment rod is lowered until it seats in its jet. This shuts off the cold-running-enrichment fuel.

⚙ 5-12 Idle trim system This carburetor does not have an idle system. Fuel for idling is supplied by the main metering jets (⚙ 5-5). However, there is a factory-made adjustment that produces the correct idle mixture from the carburetor. The system is called the idle trim system (Fig. 5-15).

During idle, fuel is "pulled" by manifold vacuum from the main metering chamber. The fuel flows

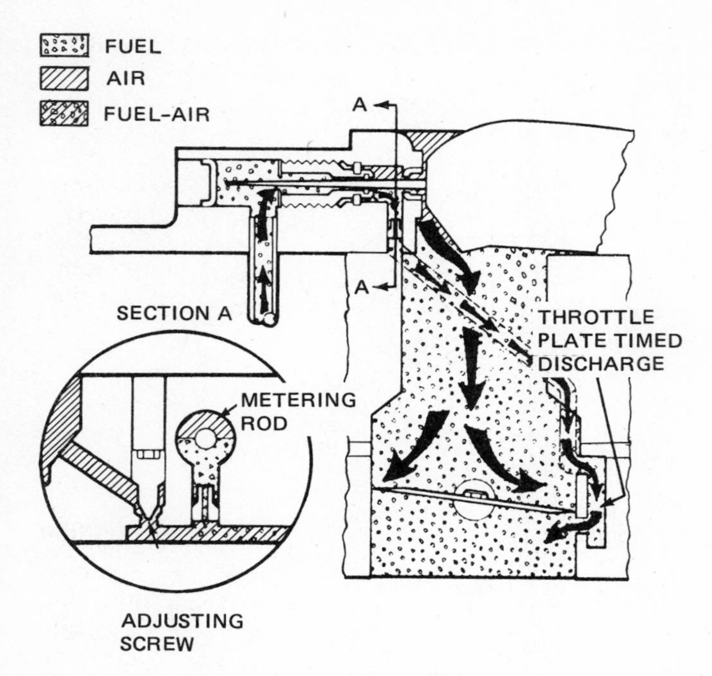

Fig. 5-15 Idle trim system. (*Ford Motor Company*)

through a separate channel and is discharged through a port below the lower edge of the throttle plates (Fig. 5-15). This fuel flow is controlled by tapered screws which admit air into the channel. There is a tapered screw and idle system for each barrel of the carburetor. The tapered screws are adjusted during original manufacture and are not adjustable by the technician.

NOTE: During servicing of this carburetor, do not remove the plug, O ring, or idle trim adjusting screw.

⚙ 5-13 Accelerator pump system The accelerator-pump system has a positive-displacement, piston-type pump. It has a piston assembly operated by an overtravel spring mounted on the throttle shaft (Fig. 5-16). The pump cavity, which is in the bottom of the fuel bowl, fills from the top. The floating cup acts as an inlet check valve.

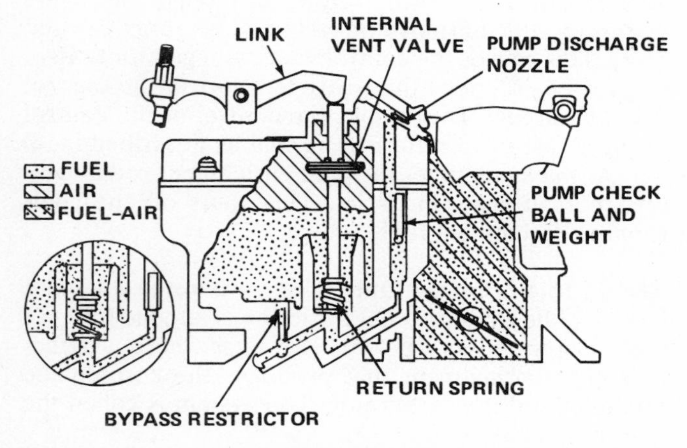

Fig. 5-16 Accelerator pump system. (*Ford Motor Company*)

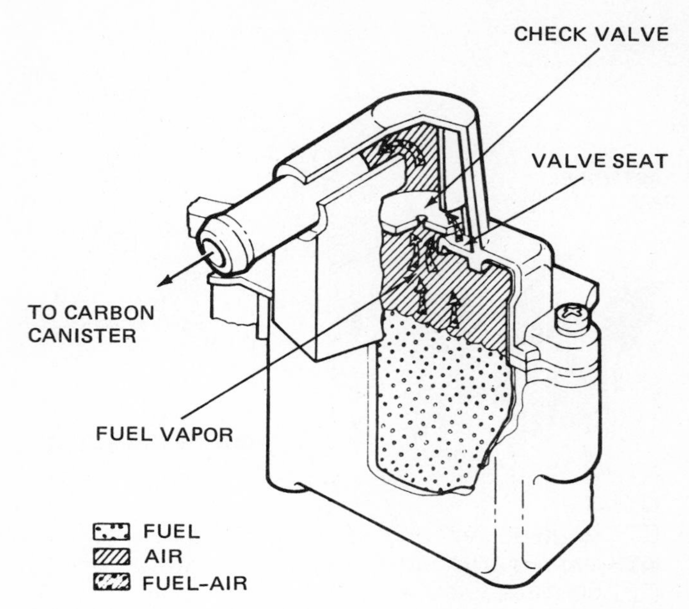

Fig. 5-17 External vent system. (*Ford Motor Company*)

When the throttle is opened, the pump piston is pushed down. Pressure develops in the cavity and this seals the cup against the piston, causing the fuel to flow. An antisiphon check ball and weight, located in the up channel, are unseated by the pressure the pump puts on the fuel. This allows fuel to flow past the check ball. The fuel flows up the channel and out through the pump discharge nozzle. Each bore has a discharge nozzle so both bores receive extra fuel for acceleration.

When no more fuel flows, the check ball and weight fall into the ball seat. This seals the channel to prevent siphoning during high-speed operation. When the throttle is returned to idle, the return spring forces the piston upward and allows the pump cavity to refill.

The system has a bypass in the main body of some carburetors which allows part of the fuel to be sent back into the fuel bowl. The size of the bypass varies with the engine the carburetor is to be used on.

The carburetor has an internal vent valve with a positive seal (Fig. 5-16). A flat disk valve is mounted on the accelerator-pump piston rod. It opens or closes the internal vent in response to the throttle position. When the throttle is closed, the vent valve is closed. This seals the internal vent so that any vapor from the fuel bowl will not flow internally into the carburetor bores. Instead, it flows through the external vent system.

⚙ 5-14 External vent system The external vent system allows any fuel-vapor pressure in the fuel bowl to be relieved through the evaporative emission control system (⚙ 3-14). When the engine is turned off after reaching normal operating temperature, the engine heat causes fuel in the fuel bowl to evaporate. As the fuel evaporates, fuel-vapor pressure in the fuel bowl increases. This pressure causes the external-vent check valve to be lifted from its seat (Fig. 5-17). With the check valve open, fuel vapor can flow to the charcoal

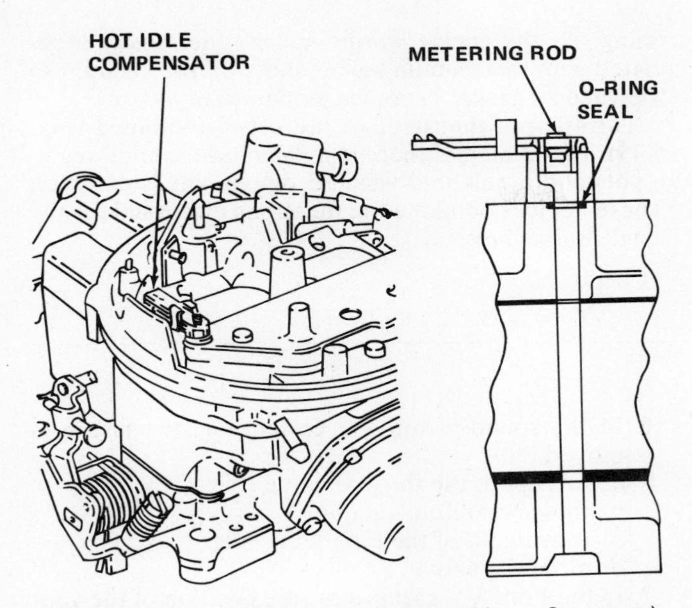

Fig. 5-18 Hot-idle compensator. (*Ford Motor Company*)

canister. The fuel vapor adheres to the charcoal. Then, when the engine is started, a reverse flow of air through the canister purges the vapor. It is carried into the carburetor and engine cylinders where it is burned.

⚙ 5-15 Hot-idle compensator The hot-idle compensator (Fig. 5-18) is located in the venturi-valve cover plate. Its purpose is to improve hot-engine idle. The compensator has a thermostatic valve which remains closed until the engine reaches operating temperature. Then the heat causes the thermostat to warp, raising the valve. This allows additional air to flow through a channel in the carburetor body. This air is discharged below the throttle plates and leans out the air-fuel mixture. A hot engine runs on a leaner air-fuel mixture than a cold engine. Leaning out the mixture allows the carburetor to supply the correct air-fuel mixture for the hot engine.

⚙ 5-16 Throttle positioners Several different throttle positioners are used on the Ford VV carburetors (Fig. 5-19). They have various purposes. The solenoids are antidieseling devices (⚙ 4-42). When the engine is running, the solenoid is connected to the battery. It extends its plunger. The end of the plunger serves as the slow-idle stop. However, when the engine is turned off, the solenoid is disconnected from the battery. The solenoid spring pulls the plunger in. Now, the throttle valve can close completely to shut off the flow of air-fuel mixture to the engine. Therefore, the engine stops running. It cannot diesel.

The dashpot prevents the throttle from closing suddenly when the driver releases the accelerator pedal. If the throttle closes suddenly, a momentarily rich air-fuel mixture is delivered which could stall the engine. The dashpot contains a spring and a diaphragm that is

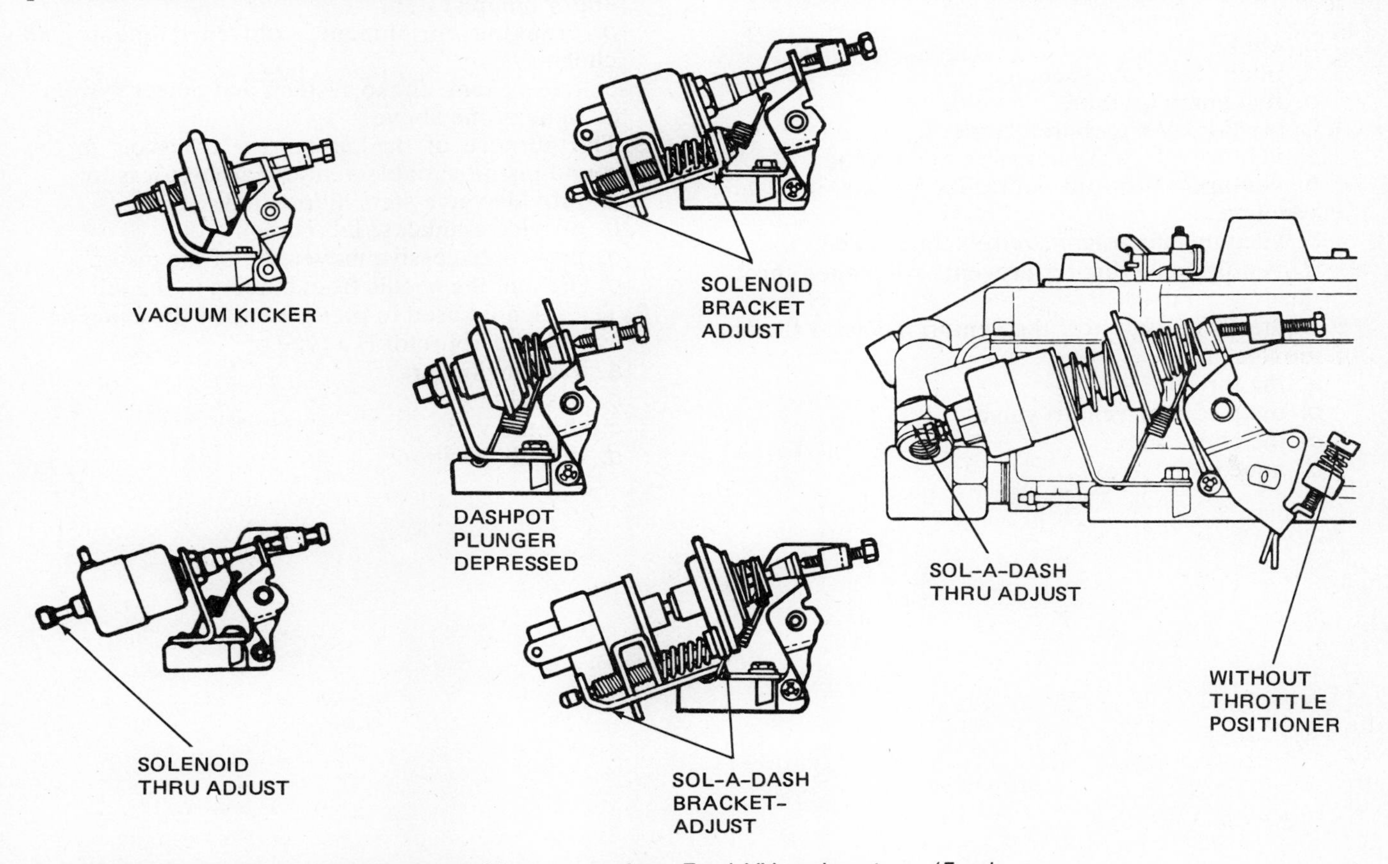

Fig. 5-19 Various types of throttle positioners used on Ford VV carburetors. (*Ford Motor Company*)

sealed at one side except for a very small hole. When the accelerator pedal is released, the air leaks out of the dashpot slowly. This prevents the throttle from instantly snapping closed. It takes a few seconds for the air to leak out and the throttle valve to close.

The vacuum kicker allows a larger throttle opening during cold-engine idle. When the engine is cold, vacuum is admitted to the kicker and this causes the throttle valve to open a little more for improved cold-engine idling. As the engine warms up, the thermostat associated with the vacuum kicker shuts off the vacuum so the vacuum kicker becomes inoperative.

Throttle-position devices are often combined (Fig. 5-19). For example, there is a solenoid-dashpot (a sol-a-pot) and a solenoid-vacuum diaphragm (sol-a-vac). These devices combine the functions discussed for the single units above.

Chapter 5 review questions

Select the *one* correct, best, or most probable answer to each question. Then check your answers against the correct answers given at the end of the book.

1. The main difference between the fixed-venturi and the variable-venturi carburetor is that the variable venturi carburetor has:
   a. a round venturi
   b. no throttle valve
   c. a choke valve
   d. vacuum control of the venturi valves.
2. The tapered needle is fastened to the:
   a. choke valve
   b. throttle linkage
   c. piston or venturi valve
   d. vacuum diaphragm.
3. The piston or venturi valve is moved by:
   a. throttle linkage
   b. choke linkage
   c. intake manifold vacuum
   d. fuel pump pressure.
4. In the Ford VV carburetor, the:
   a. venturi valves are linked to the throttle
   b. venturi valves are controlled by a spring and vacuum
   c. vacuum diaphragm controls the choke
   d. venturi-valve limiter prevents wide-open throttle.
5. In the VV carburetor, the venturi is varied by the movement of:
   a. the throttle
   b. the piston or venturi valve
   c. the choke valve
   d. none of the above.
6. In the round-piston VV carburetor, the piston is moved by:
   a. linkage to the throttle valve
   b. intake manifold vacuum
   c. movement of the tapered needle
   d. all of the above.
7. In the Ford VV carburetor, the position of the venturi valves and tapered metering rods is controlled by:
   a. intake manifold vacuum
   b. linkage to the throttle valve
   c. fast-idle cam
   d. slow-idle cam.
8. Systems *not* found in the Ford VV carburetor are the:
   a. float system, main metering system, and accelerator pump system
   b. cranking enrichment, cold enrichment, and choke system
   c. idle system, choke system, and power system
   d. none of the above.
9. The purpose of the oil-damper reservoir in the round-piston variable-venturi carburetor is to:
   a. provide valve-stem lubrication
   b. provide crankcase lubrication
   c. prevent excessive movement of the piston
   d. prevent the needle from binding in the jet.
10. The solenoid used to prevent engine dieseling on a Ford VV carburetor is a type of:
    a. vacuum kicker
    b. dashpot
    c. throttle valve
    d. throttle positioner.

CHAPTER 6 GASOLINE FUEL-INJECTION SYSTEMS

After reading this chapter, you should be able to:

1. Discuss the types of fuel injection for gasoline engines.
2. Define *injection valve.*
3. Describe the operation of the electronic fuel-injection system.
4. Explain the operation of the Bosch mechanical fuel-injection system for gasoline engines.
5. Locate the components of a gasoline fuel-injection system on various cars.
6. Explain the difference between timed injection and continuous injection.
7. Describe the basic differences between electronic fuel injection used by Cadillac and the system used by Ford.

6-1 What is fuel injection? The engine must have a continuous supply of combustible air-fuel mixture to run. In most engines, the carburetor supplies this mixture (Fig. 4-1). The carburetor is mounted on the intake manifold, which has passages connecting with the combustion chambers in the engine. The carburetor mixes air and fuel to produce the combustible mixture. Then the mixture flows through the intake manifold into the combustion chambers when the intake valves open.

In the most commonly used types of gasoline fuel-injection systems, only air enters the intake manifold. As the air approaches the intake valves, fuel injectors spray gasoline into the air (Fig. 6-1). In the carbureted fuel system, the fuel enters the air in the carburetor. With fuel injection, the fuel enters the air inside the intake manifold. There are certain advantages to fuel injection. These are discussed further in ⚙ 6-3.

NOTE: Ford uses a type of electronic fuel injection called a *fuel charging system*. Cadillac uses a similar system called "Digital Electronic Fuel Injection" (DEFI). In this system, no carburetor is used. Instead, two electronically controlled fuel injectors are mounted on the throttle body and spray fuel down the throats of the throttle body into the intake manifold. This system is discussed further in ⚙ 6-14.

⚙ 6-2 Types of fuel injection There are two basic ways to classify fuel injection systems. One way is according to where the fuel is injected into the engine. The other way is according to whether the injection of fuel is continuous or timed.

Let's look at the first classification—the point of injection. Fuel can be injected directly into the combustion chambers (Fig. 6-2), or into the intake manifold (Figs. 6-2 and 6-3). The direct injection method is currently used in diesel engines. During the compression stroke in the diesel engine, the heat of compression raises the air temperature to 1000°F [537.8°C] or above. The fuel is injected (sprayed) directly into this superheated air. The high temperature ignites the fuel as it enters.

Fuel can also be injected into the intake manifold. This is the system used for most gasoline engines. When the fuel is injected into the air approaching the intake port, the system is known as *port* injection.

The second classification of fuel-injection systems is the injection procedure—whether the fuel is injected in pulses (timed injection) or continuously. In the diesel engine, timed injection must be used because the fuel must be sprayed into the combustion chambers at the right time. This point must change as engine speed changes. Injection must take place earlier at high engine speeds. This compares with the ignition system action for gasoline engines. The spark must occur earlier at higher engine speed. In both engines, the advance gives the fuel enough time to burn and deliver its power to the piston.

Continuous injection systems (CIS) are used on some gasoline engines. In these, the gasoline is sprayed continuously into the intake manifold. The amount of fuel sprayed varies with the engine speed and power demands. The continuous injection system is simpler in many ways and less expensive. It does not require a control system that changes the length of the injection time as operating conditions change. Such a control system can be complex and expensive.

⚙ 6-3 Advantages of gasoline injection Regardless of whether the injection system is pulsed (timed) or continuous, fuel injection eliminates many carburetion

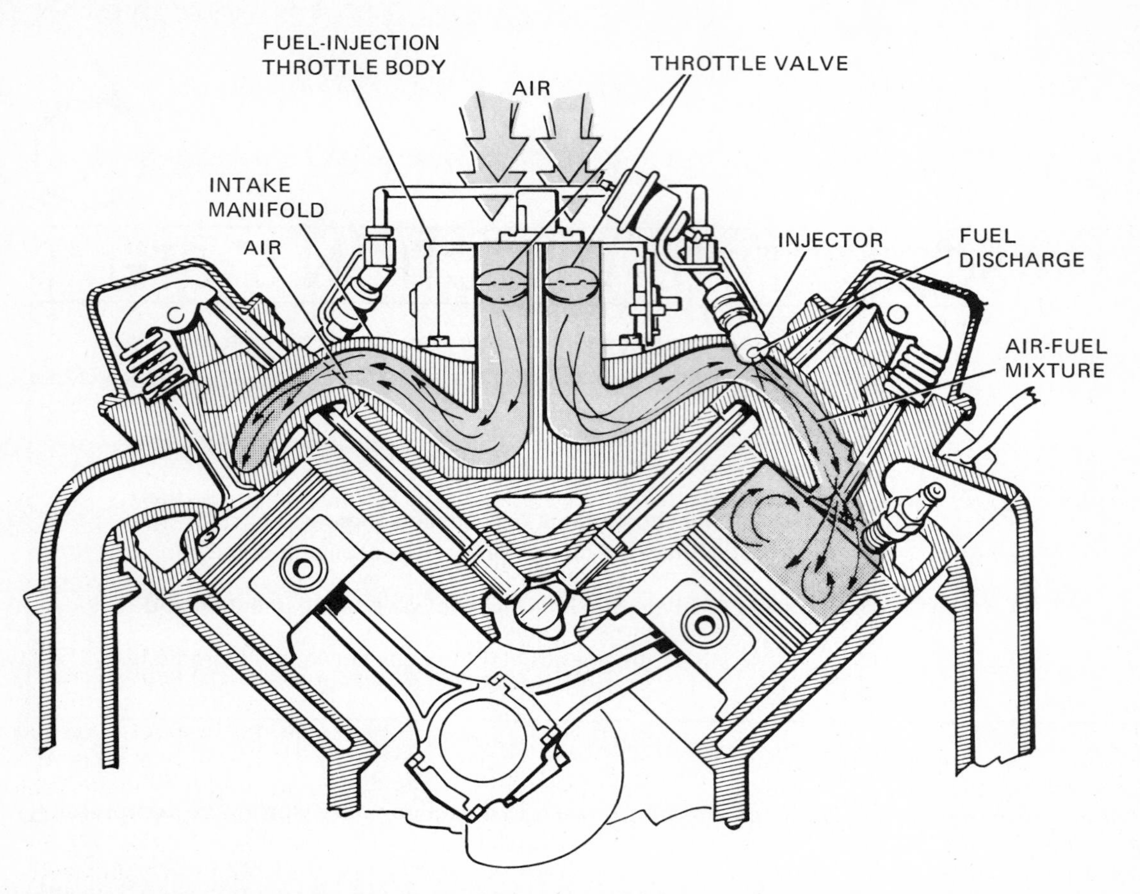

Fig. 6-1 Airflow through a fuel-injected engine. (*Cadillac Division of General Motors Corporation*)

and intake manifold distribution problems. One of the most difficult problems with a carbureted system is to get the same air-fuel mixture to each cylinder.

An intake manifold is a casting with passages of different widths and lengths. Because of this, it is difficult to design a manifold so that all cylinders receive the same amount and richness of air-fuel mixture. The air flows readily around corners and through the variously shaped passages. However, the gasoline, because it is heavier, is unable to travel as easily through the bends and around the corners in the intake manifold. As a result, some of the gasoline particles continue to move to the end of the manifold, accumulating or puddling there (Fig. 4-7). This enriches the mixture going to the end cylinders. In the example shown in Fig. 4-7, the center cylinder closest to the carburetor gets the leanest mixture.

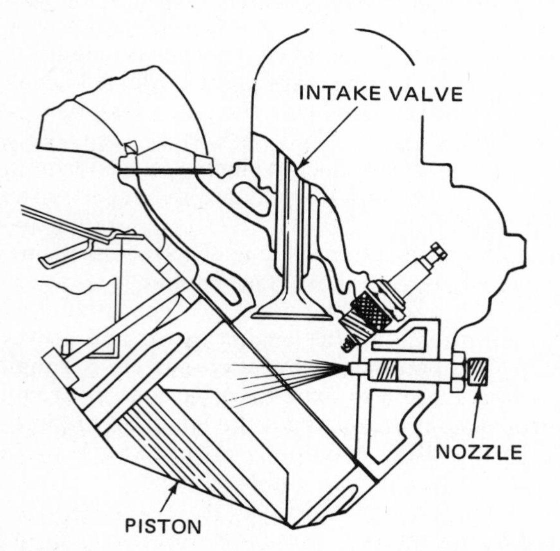

Fig. 6-2 Simplified view showing the method of injecting fuel directly into the combustion chamber of an engine.

The air-fuel mixture leaving the carburetor must be rich enough to take care of this difference and supply the center cylinders with a sufficiently rich mixture. Otherwise, these cylinders will not fire. But this requires that the end cylinders receive a richer mixture than necessary, which causes them to produce more HC and CO in the exhaust gas.

Figure 3-2 shows how a fuel-injection system solves the intake manifold distribution problem. A calibrated nozzle, or fuel injector, is located near the intake valve of each cylinder. At the right instant, fuel under pressure is sprayed out of the injector (timed injection system). Or, with the continuous injection system, the fuel sprays continuously. In either system, each injector sprays the same amount of fuel into the same amount of air entering every cylinder. As a result, each cylinder gets the same amount of air-fuel mixture having the

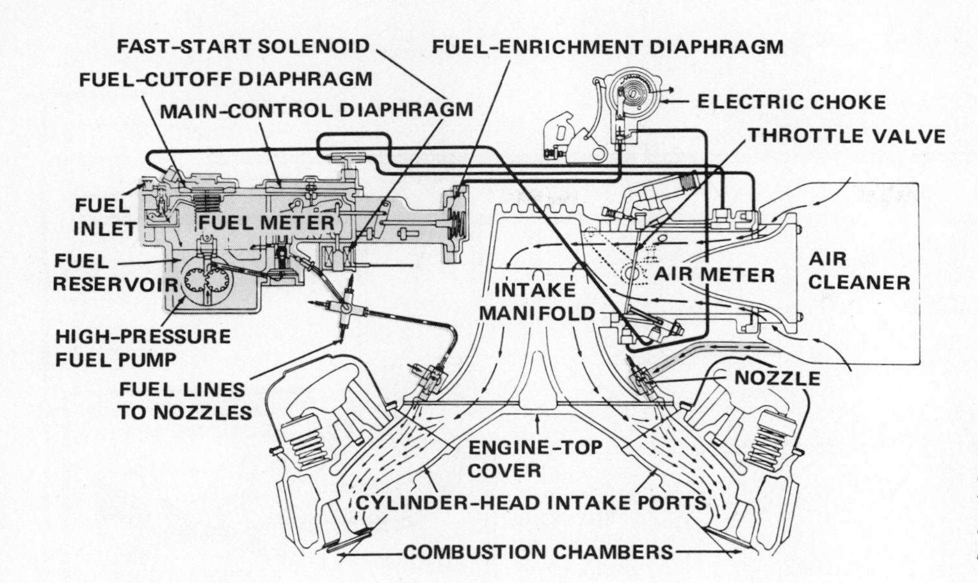

Fig. 6-3 Sectional view of a mechanical fuel-injection system. (*Chevrolet Motor Division of General Motors Corporation*)

same mixture ratio. The engine can operate on a leaner overall air-fuel ratio. This reduces HC and CO in the exhaust gas and aids fuel economy.

Another advantage of the fuel injection system is that the intake manifold can be designed for the most efficient flow of air. The manifold does not have to handle fuel, too. Also, because only a throttle body is needed instead of a full carburetor, and because of the redesigned intake manifold, the hood height of the car can be reduced.

With fuel injection, no extra heat to the fuel mixture is required during warm-up. Therefore, no manifold heat-control valve or heated-air system is required. Throttle response is faster because the fuel is under pressure at the injector at all times. It does not have to rely on the changing difference in air pressure, as in the carburetor. All that is required is to open the injector. Fuel sprays out instantly. Exhaust emissions during deceleration can be practically eliminated. The system can be designed to have complete fuel shutoff when the throttle is released and the car decelerates. This eliminates exhaust emissions during deceleration, saves gasoline, and thereby improves fuel economy.

There are two basic reasons why all engines are not equipped with fuel injection. First, there is the cost. Modern pulsed or timed electronic fuel-injection systems rely on complex sensors and controls. These are needed to control the timing of injection and the amount of fuel to be injected. They can be expensive, particularly the electronic components. However, the continuous injection system does not require a complex timing arrangement. It is less expensive, and is being used on some new cars.

Second, only recently have the electronics been available. Miniaturized electronics is a relatively new science. It has only lately produced the controlling devices at a price low enough to be acceptable. Prior to the electronic age, some completely mechanical fuel-injection systems were developed. They worked, but still had some problems. Servicing was difficult and public acceptance was limited.

6-4 Early mechanical fuel-injection systems From 1957 to 1965, Chevrolet (and briefly Pontiac) offered a continuous-flow fuel-injection system as an option. It was built by the Rochester Products Division of General Motors Corporation. The system uses vacuum and airflow signals from the engine to control the air-fuel mixture. Fuel sprays continuously from nozzles into the intake manifold opposite the intake valves. Each nozzle is simply a calibrated tube, without a valve in it.

There is always pressure behind the fuel in the nozzles. This is provided by the fuel-injection pump, which is cable-driven from the distributor. On the discharge end of the nozzle, the vacuum will vary from about 22 inches (559 mm) with the throttle closed and the engine idling, to almost zero at wide-open throttle. This varying vacuum is sensed in an air meter which sends vacuum signals to the fuel meter. Then the fuel meter regulates the amount of fuel spraying from the nozzles. Other systems in the fuel meter increase the amount of fuel as required for starting, warm-up, acceleration, and high speed.

Figure 6-3 is a schematic view of the Rochester system. It consists essentially of a special intake manifold, an air meter, and a fuel meter. The air meter controls the flow of air through the intake manifold by linkage from the accelerator pedal, which operates the throttle valve in the air meter. More air is admitted when more engine power is desired. Then the fuel meter provides corresponding amounts of fuel.

6-5 Electronic fuel injection In the late 1950s, Chrysler built a few cars with an early type of electronic fuel-injection system. This system, known as the Bendix Electrojector, was developed by the Bendix Corporation. The electronic part of this system used vacuum tubes. It was not widely produced.

With the development of solid-state electronics, such as transistors and diodes, a new type of electronic fuel injection appeared. In 1968, Volkswagen began installing a solid-state electronic fuel-injection system built

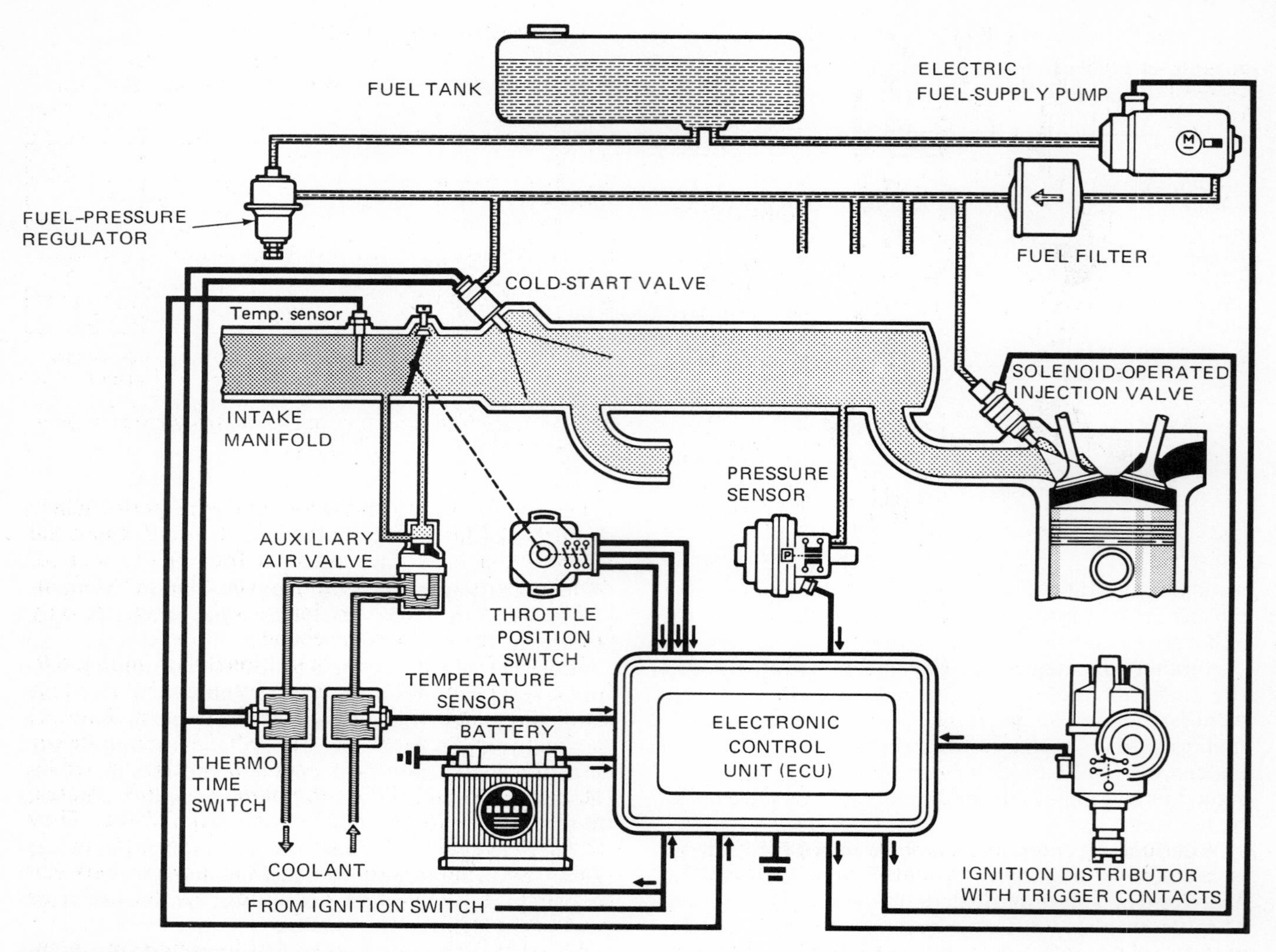

Fig. 6-4 Schematic diagram of an electronic gasoline-injection system. (*Robert Bosch Corporation*)

by Bosch on some cars imported into the United States. The Volkswagen system is discussed further in ⚙ 6-10 and in Chap. 14. Now, let us examine the basic Bosch electronic fuel-injection system for gasoline engines.

Figure 6-4 shows the system schematically. The electric fuel-supply pump maintains a high pressure in the fuel line to the injection valves in the intake manifold (Fig. 6-5).

At the proper instant, trigger contact points in the ignition distributor close. Figure 6-6 shows the distributor, cut away so that the trigger contacts can be seen. They are opened and closed by a cam on the distributor shaft. This cam is very different from the cam that opens and closes the ignition contact points. When the trigger contacts close, they send an electric signal to the electronic control unit. The ECU then connects half the solenoid injection valves to the battery. (In a four-cylinder engine, this would be two valves; in a six-cylinder engine, three valves; and in an eight-cylinder engine, four valves). The solenoid injection valves are not individually actuated. Half of them are actuated at a time. Figure 6-7 shows three of the valves in operation, spraying fuel into the intake manifold.

The fuel enters just opposite the intake valves (Fig. 6-4). Figure 6-8 is the injection timing chart for a six-cylinder engine. Note that the individual intake valves open at varying times of crankshaft degrees after injection. For example, look at the top line, which is for No. 1 cylinder. Injection takes place at 300° of crankshaft

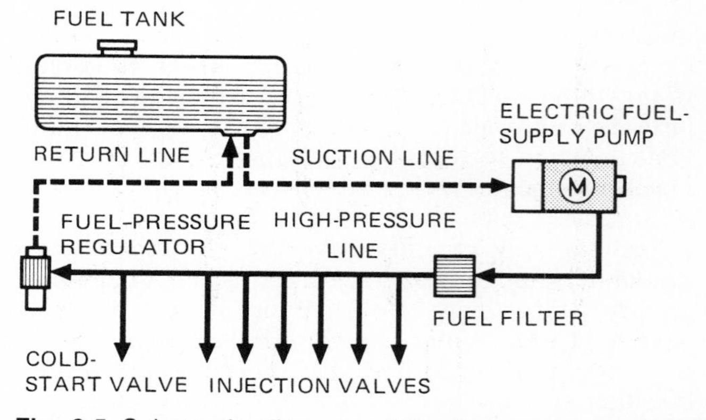

Fig. 6-5 Schematic diagram of the fuel supply system. (*Robert Bosch Corporation*)

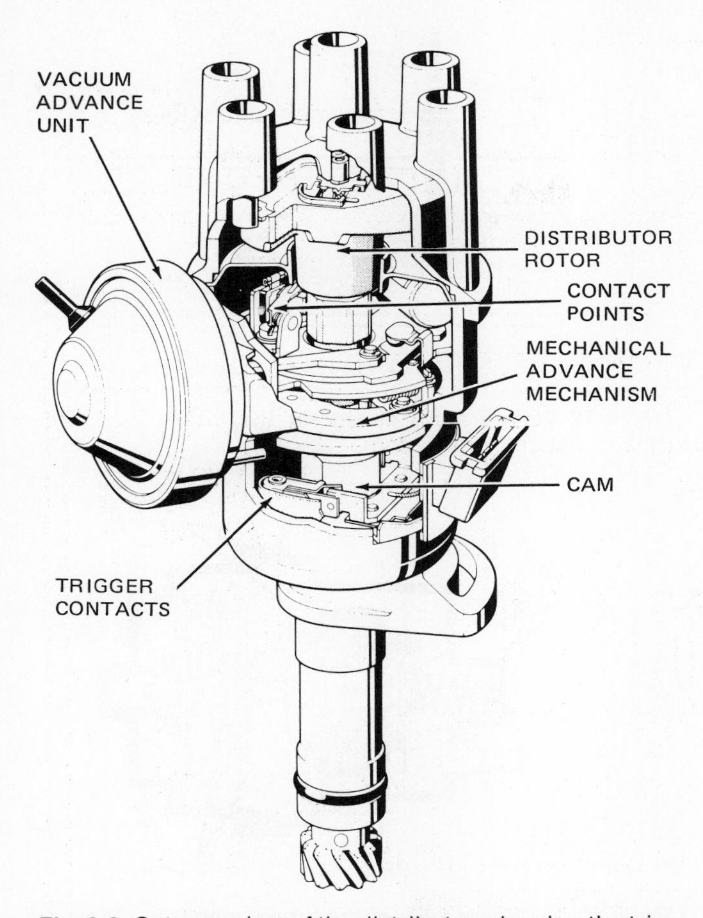

Fig. 6-6 Cutaway view of the distributor, showing the trigger contacts which activate the electronic control unit. (*Robert Bosch Corporation*)

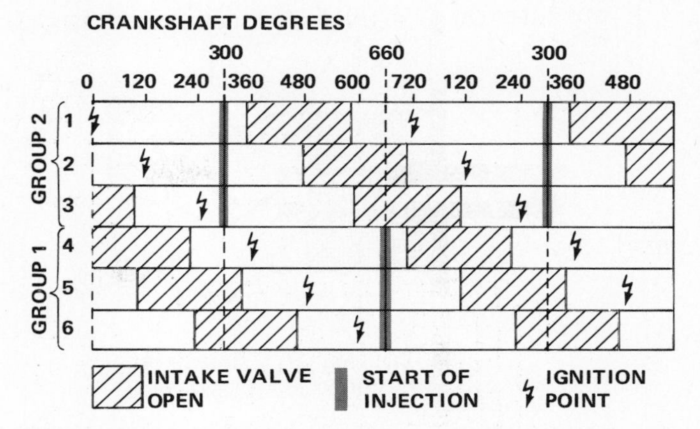

Fig. 6-8 Injection timing chart for a six-cylinder engine. (*Robert Bosch Corporation*)

rotation. Almost 60° later (near 360°), the No. 1 intake valve opens and the intake stroke starts. Cylinder No. 5 is next in the firing order. Its intake valve opens near 480°, or about 180° after injection. The intake valve for No. 3 cylinder opens about 300° of crankshaft rotation after injection. During these varying intervals between fuel injection and intake valve opening, the fuel is "stored" in the intake manifold, opposite the intake valves.

Having only two groups of injection valves simplifies the system. No appreciable loss of engine performance results from this storage of the fuel. This is because the whole action takes place in a small fraction of a second. At highway speed, for example, the time between injection and opening of the intake valve averages only about one hundredth of a second.

The fuel pump is shown in sectional view in Fig. 6-9. It is an electric motor of the "wet-pump" type. This means that the fuel flows through the pump and motor, as shown by the arrows in Fig. 6-9. The pump drives an off-center rotor with a series of notches in which rollers are located. When the pump armature rotates, the rollers are forced out by centrifugal force. They trap fuel between the rotor and the inner face of the pump. This fuel is forced out as the distance between the rotor and inner face decreases, on the outlet or pressure side.

The fuel pressure is controlled by a pressure regulator. The regulator is shown in the upper left in Fig. 6-4. It regulates by dumping some of the fuel back into the fuel tank if the pressure gets too high. Figure 6-10 is a sectional view of the pressure regulator. If the pressure exceeds a preset value, the diaphragm is pushed back against the spring. This opens the valve, allowing some of the fuel to flow out through the return line to the fuel tank. It is very important to maintain a constant pressure. The amount of fuel injected must

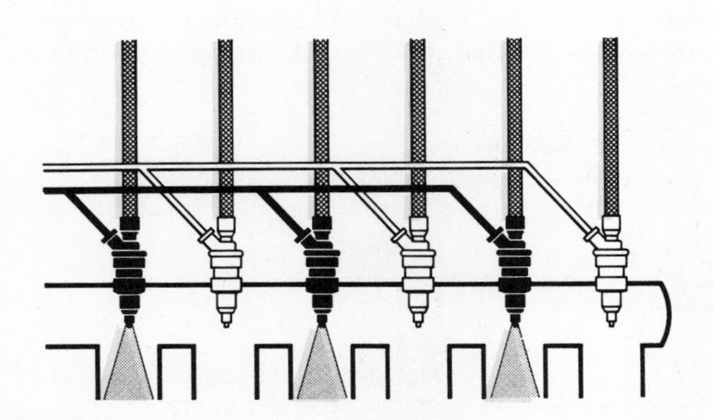

Fig. 6-7 Injection valve grouping. (*Robert Bosch Corporation*)

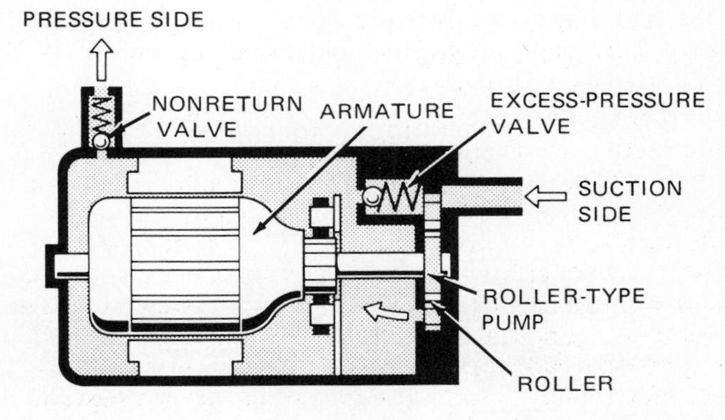

Fig. 6-9 Sectional view of the electric fuel-supply pump. (*Robert Bosch Corporation*)

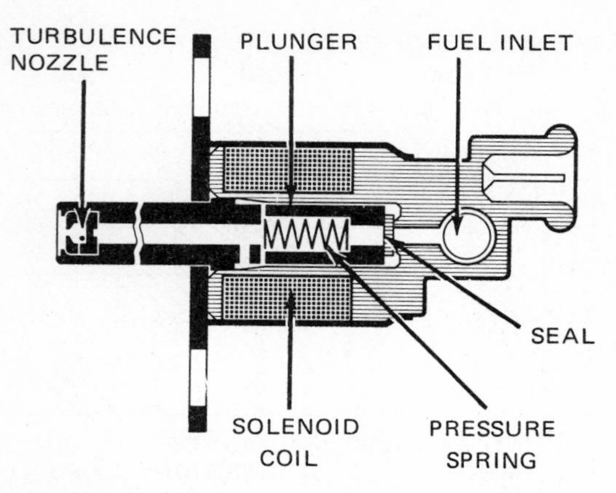

Fig. 6-10 Sectional view of the pressure regulator. (*Robert Bosch Corporation*)

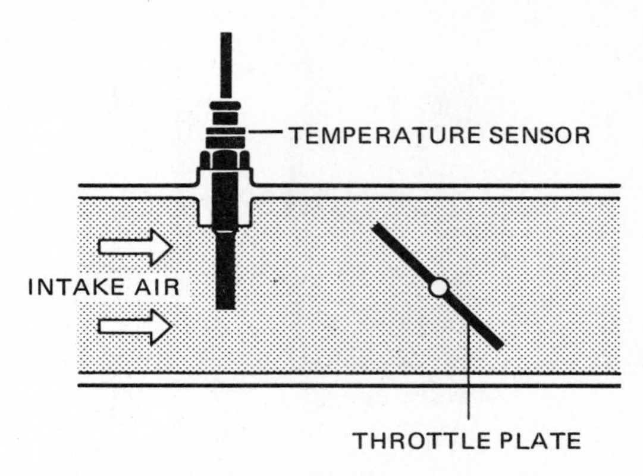

Fig. 6-12 Temperature sensor in the intake manifold. (*Robert Bosch Corporation*)

depend entirely on how long the solenoid injection valve is open, and not on the fuel pressure.

The solenoid injection valve is shown in sectional view in Fig. 6-11. It pulls the plunger and needle away from the nozzle jet when it is connected to the battery. When it is connected, the solenoid produces a magnetic field. This pulls the plunger in toward the solenoid, and the needle is lifted off the nozzle. The fuel can then spray through the nozzle into the intake manifold. The longer the needle is off the nozzle, the more fuel is sprayed.

6-6 Electronic fuel-injection controls Several factors determine how long an injection valve delivers fuel to the intake manifold. They include the throttle position, intake manifold vacuum, ingoing air temperature, and coolant (engine) temperature (Fig. 6-4). Sensing devices continuously monitor these factors and report electronically to the ECU. The ECU puts the varying signals together and determines the length of the injection cycle.

For example, consider the intake-air temperature sensor (Fig. 6-12). It constantly measures the temperature of the air entering the intake manifold. The intake-air temperature sensor contains an element that passes varying amounts of electric current as the temperature changes. At low temperatures, for example, it passes more current: it sends a stronger electric signal to the ECU. The ECU then increases the time during which the fuel-injection valves are open. More fuel is delivered to compensate for the colder and denser air.

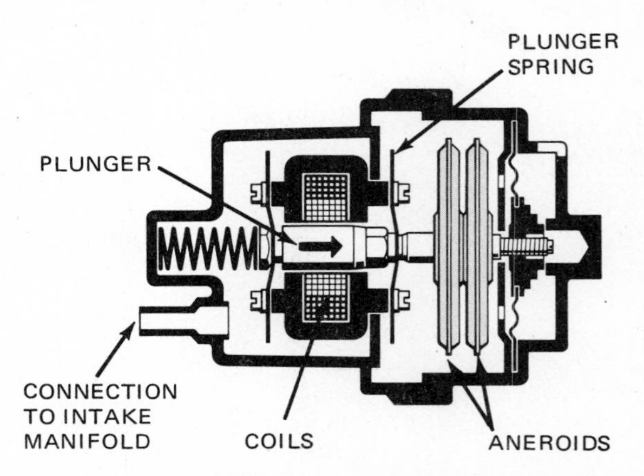

Fig. 6-13 Sectional view of the intake manifold pressure sensor with the throttle valve open. The aneroid disks are compressed. (*Robert Bosch Corporation*)

Similarly, the coolant temperature sensor sends varying amounts of current to the ECU, depending on the temperature of the engine coolant (and therefore, the engine temperature). When the engine is cold, the engine must receive more fuel, so that the mixture will be rich enough. The ECU increases the injection time when the coolant temperature sensor reports that the engine is cold. As the engine warms up, the ECU decreases the injection time.

The intake-manifold pressure sensor (Fig. 6-13) measures intake manifold pressure and compares it with atmospheric pressure. It contains a pair of *aneroids*. These are flat, hollow disks or *bellows*. In many

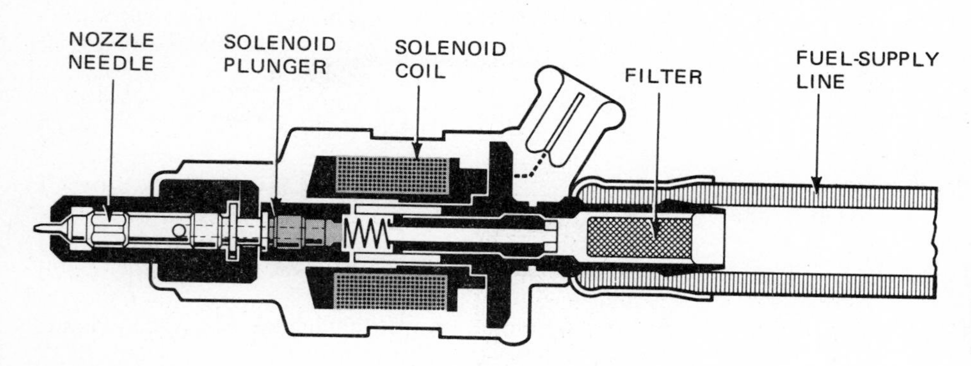

Fig. 6-11 Sectional view of the solenoid-operated injection valve. (*Robert Bosch Corporation*)

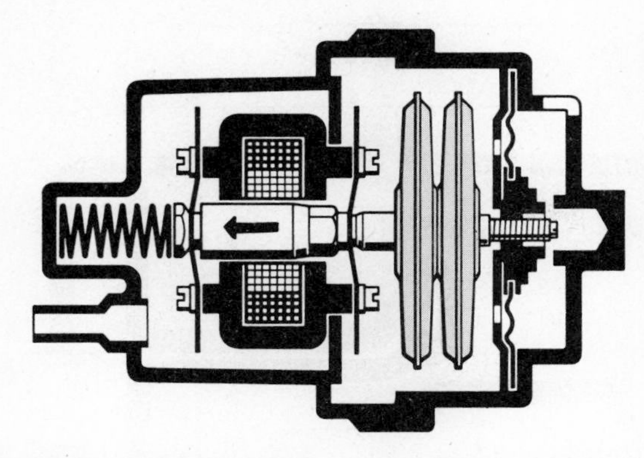

Fig. 6-14 Sectional view of the intake manifold pressure sensor with the throttle closed. Increased vacuum allows the aneroid disks to expand. (*Robert Bosch Corporation*)

models, both aneroids are *evacuated,* which means that they hold a vacuum. As the outside pressure changes, the sides of the aneroid bulge out or in, depending on whether the outside pressure is relatively low or high.

Intake manifold vacuum is introduced into the end of the pressure sensor. This vacuum acts on the aneroids. For example, if the throttle valve is open and there is little vacuum in the intake manifold, the aneroids are collapsed (Fig. 6-13). But if intake-manifold vacuum is high, the aneroids bulge out (Fig. 6-14). This repositions the plunger in the coils.

Changing the position of the plunger in the coils changes their magnetic response. The coils send a changed electric signal to the ECU. The ECU then changes the injection time so that the correct amount of fuel is injected to meet intake manifold vacuum conditions. For example, manifold vacuum is high when the throttle valve is closed or nearly closed. Only a little air is getting through to the cylinders. Therefore, only a little fuel should be injected. So the ECU shortens the injection time. This is the same as the carburetor feeding fuel to the air through the idle system.

When the throttle is opened, the vacuum in the intake manifold is reduced. The aneroids collapse slightly (Fig. 6-13) and pull the plunger into the coils. The electric signal from the coils then changes. This change causes the ECU to increase the injection time. More

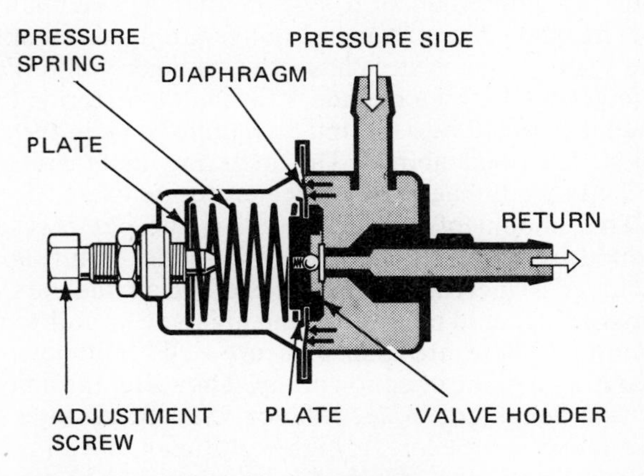

Fig. 6-15 Sectional view of the cold-start valve. (*Robert Bosch Corporation*)

fuel is injected so that the air-fuel ratio remains constant.

Other factors also affect the injection time. If the intake air is cold, the injection time is increased. Likewise, if the engine (coolant) is cold, the injection time is increased.

⚙ 6-7 Cold starts In the carburetor fuel system, the choke increases the amount of fuel delivered when the engine is being started cold. In the electronic fuel-injection system, a cold-start valve increases the amount of fuel delivered when the engine is being cold-started. Note the location of the valve in Fig. 6-4. A sectional view is shown in Fig. 6-15. The cold-start valve is triggered by the ECU. The ECU connects the cold-start valve to the battery. The solenoid then moves the plunger to allow fuel to spray into the air entering the intake manifold. This results in sufficiently rich starting mixture.

⚙ 6-8 Throttle-position switch For more exact control of the injection time, the throttle has a position switch with several contact strips and a sliding contact. As the throttle is opened, the sliding contact connects to the contact strips, one after another. As each connection is made, a different signal is sent to the ECU. This causes the ECU to modify the injection time according to throttle position. For example, on later models there is a full-load-enrichment contact strip. When the throttle is opened wide, the sliding contact connects to this contact strip. As a result, the ECU increases the injection time so that additional fuel is injected. This meets the need for a richer mixture at full power.

⚙ 6-9 ECU system operation Now let us review the operation of the system. When the engine is first started, it is cold. The air going into the intake manifold is cold. Vacuum in the intake manifold is fairly high because the throttle is nearly closed. The ECU gets signals from the air-intake temperature sensor, the coolant temperature sensor, the pressure sensor, and the throttle position switch. It puts all of these together and "decides" how long the injection valves should stay open. The ECU also decides whether to open the cold-start valve. When the trigger contacts close in the ignition distributor, the ECU opens half the injection valves. A rich mixture is delivered to the intake manifold before the intake valves open.

Then, when the engine starts, the changing intake-air temperature, coolant temperature, throttle position, and intake manifold vacuum all modify the signals going from the ECU to the injection valves.

An auxiliary air valve senses the temperature of the engine coolant. Note its location in Fig. 6-4, and in the sectional view in Fig. 6-16. When the engine is cold, the auxiliary air valve opens to allow some air to flow around the closed throttle valve. This provides the extra air needed when the cold-start valve is open and discharging fuel. However, as the engine (and the coolant) begins to warm up, the auxiliary valve closes, shutting off the flow of extra air.

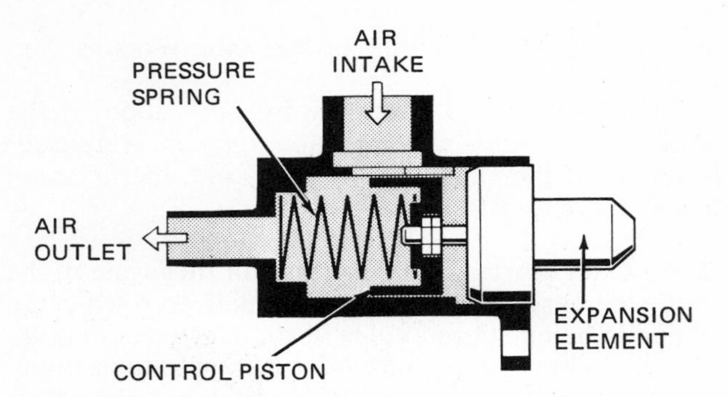

Fig. 6-16 Sectional view of the auxiliary air valve. (*Robert Bosch Corporation*)

6-10 Volkswagen electronic fuel-injection system The Volkswagen electronic fuel-injection (EFI) system is very similar to the system described above and illustrated in Fig. 6-4. This system was tailored to the Volkswagen flat-four air-cooled engine. The purpose in using EFI was to improve the combustion process and to reduce exhaust pollutants. Figure 6-17 illustrates the air supply system and its controls. Figure 6-18 illustrates the fuel supply system.

6-11 Cadillac electronic fuel-injection system The Cadillac electronic fuel-injection system (Fig. 6-19) is very similar to the Volkswagen system. Cadillac also uses a separate fuel-injector valve for each cylinder (Fig. 6-20). The eight injectors on the Cadillac V-8 engine are connected to a fuel rail. They are divided into two groups of four injectors each. Each group of injectors is alternately turned on and off by the electronic control unit. The injectors are turned on once for each two revolutions of the crankshaft. Figure 6-1 is a sectional view of the Cadillac V-8 engine with electronic fuel injection.

Figure 6-21 is a block diagram showing (on the left) the sensors that send information to the ECU. With

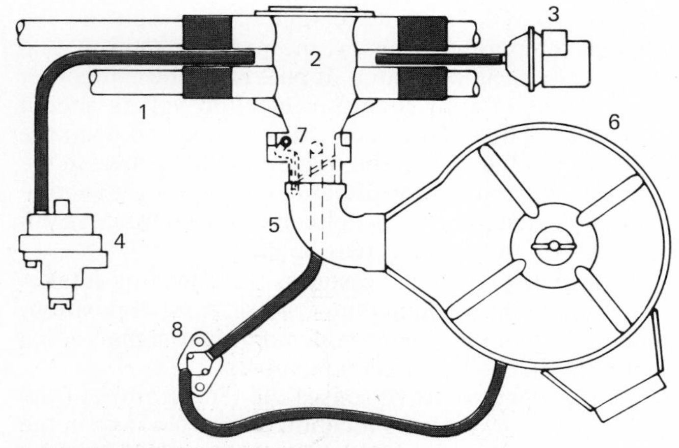

1. Air pipes to cylinders
2. Air distributor
3. Pressure switch
4. Pressure sensor
5. Idling circuit
6. Air cleaner
7. Adjusting screw
8. Auxiliary air regulator (rotary valve)

Fig. 6-17 Air supply control for the Volkswagen electronic fuel-injection system. (*Volkswagen of America, Inc.*)

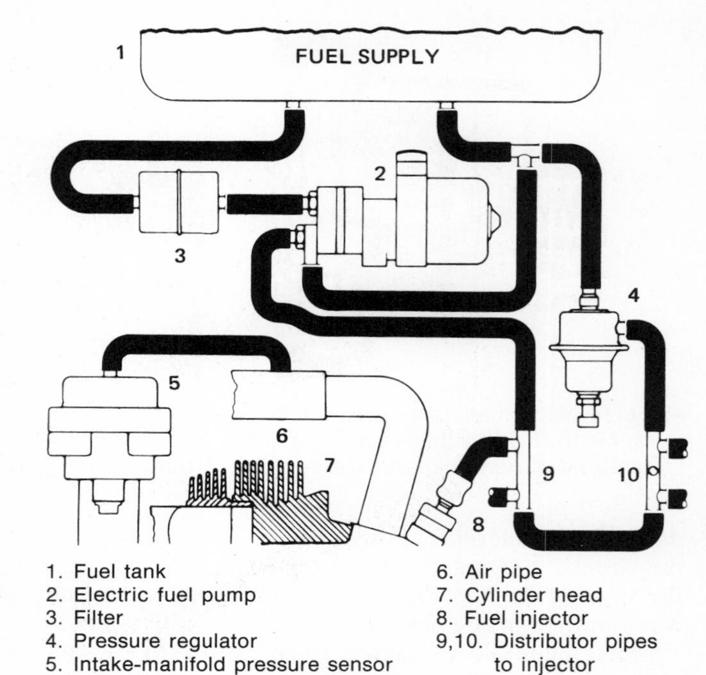

1. Fuel tank
2. Electric fuel pump
3. Filter
4. Pressure regulator
5. Intake-manifold pressure sensor
6. Air pipe
7. Cylinder head
8. Fuel injector
9,10. Distributor pipes to injector

Fig. 6-18 Fuel supply system for the Volkswagen electronic fuel-injection system. (*Volkswagen of America, Inc.*)

this and other information (such as engine displacement and volumetric efficiency), the ECU computes the amount of fuel the engine requires. It then sends control signals to the injectors and other parts of the system (on the right in Fig. 6-21).

The ECU is a preprogrammed computer installed above the glove box within the passenger compartment. It converts the input information from the sensors into an electric signal which opens the injectors for the proper duration at the proper time. The ECU cannot be adjusted or serviced. When a malfunction is traced to the ECU, it is removed from the car and a new one installed. Accurate diagnosis of ECU operation requires a special tester.

6-12 Electromechanical fuel-injection system for gasoline engines Figure 6-22 is a schematic view of a fuel-injection system that is essentially a mechanical system. Several relays are used to control the system. The system has a two-plunger pump. Each plunger feeds three cylinders through a metering unit. Fuel is injected into the intake manifold, as in the systems described above. The metering units are controlled by a linkage to the accelerator pedal.

The amount of fuel delivered by the pump is controlled by a centrifugal governor. Other controls are included to increase the richness of the air-fuel mixture for starting, cold operation, and high-speed, full-power running. There are also pressure cells built into the diaphragm of the injection pump. They alter the amount of fuel delivered, in accordance with the altitude and the density of the air. At higher altitudes, the air is less dense. Therefore, less fuel is required to achieve the normal air-fuel ratio. The pressure cells take care of this automatically.

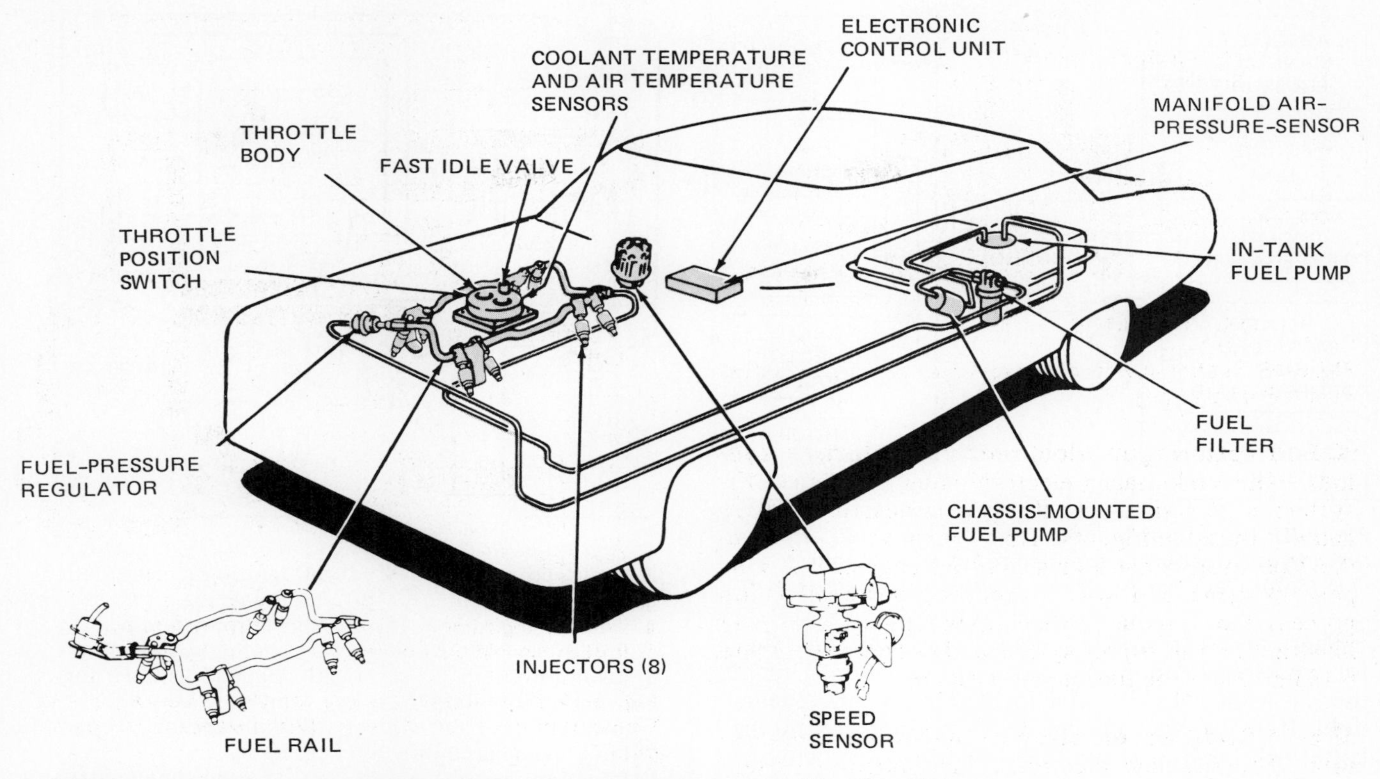

Fig. 6-19 Components of the Cadillac electronic fuel-injection system. (*Cadillac Motor Car Division of General Motors Corporation*)

⚙ 6-13 Late-model mechanical continuous fuel injection

The electronic fuel-injection system is more expensive than the simpler mechanical fuel-injection system. Also, its electric components cannot be diagnosed except with special testers. To overcome these disadvantages, Bosch has introduced a continuous-injection system that operates primarily by mechanical means (Fig. 6-23). This system uses an electric fuel pump to pressurize the fuel. To control the amount of fuel injected, the volume of intake air is measured by an airflow sensor plate. The sensor plate is located in an air funnel through which all ingoing air must pass.

As the airflow through the funnel increases, it lifts the sensor plate higher in the funnel. This causes a lever to lift the control plunger in the fuel distributor. The action increases the amount of fuel that can flow to the nozzles located in the intake manifold. A properly balanced air-fuel ratio results. The fuel is sprayed continuously from the injection valves as long as the engine is running.

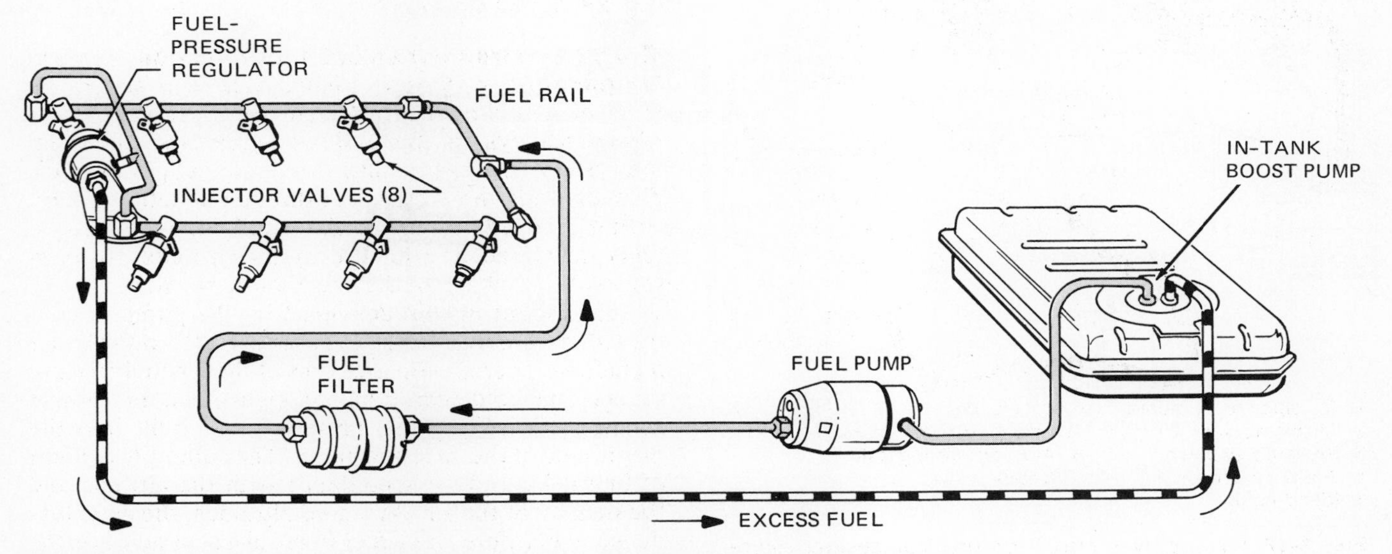

Fig. 6-20 Schematic view of the fuel system used on the Cadillac fuel-injected engine. (*Cadillac Motor Car Division of General Motors Corporation*)

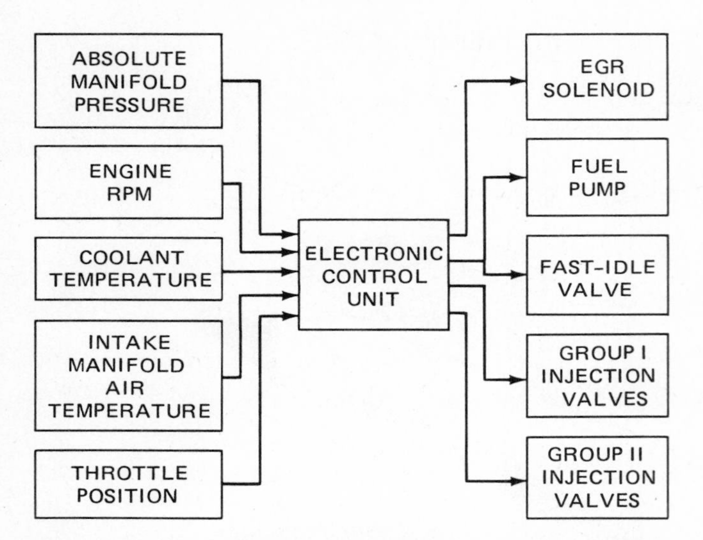

Fig. 6-21 Block diagram showing the sensors (left) that provide information to the electronic control unit. (*Cadillac Motor Car Division of General Motors Corporation*)

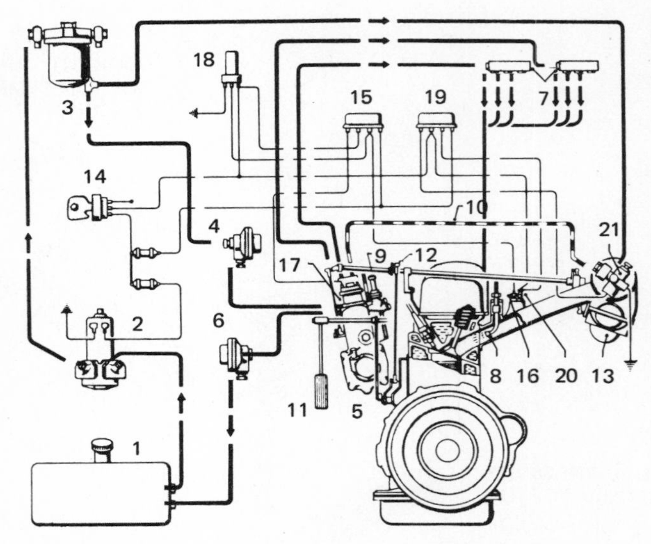

1. Fuel tank
2. Fuel-feed pump
3. Fuel filter
4. Damper container (inlet)
5. Injection pump
6. Damper container (outlet)
7. Fuel-metering units
8. Injection valves
9. Cooling-water thermostat
10. Additional air duct
11. Accelerator
12. Control linkage
13. Throttle connector
14. Ignition-starter switch
15. Relay
16. Thermo switch in cooling-water circuit
17. Magnetic switch for mixture control
18. Time switch
19. Relay
20. Thermo time switch in cooling-water circuit
21. Electromagnetic starter valve with atomizing jet

Fig. 6-22 Schematic layout of a fuel-injected six-cylinder engine. (*Mercedes-Benz of North America, Inc.*)

There are special subsystems to provide extra fuel for cold starts and during warm-up operation. One of these is a special start valve located in the intake manifold. It includes a solenoid which is connected to the battery through a thermo-time switch. This switch is mounted on the side of the engine and senses the coolant temperature. When the engine is cold, the switch is closed. However, when the engine is cranked, the start-valve solenoid is actuated, and extra gasoline is sprayed into the air going to the cylinders. This enriches the mixture for starting. As the engine warms up, the thermo-time switch opens, shutting off the start valve.

⚙ 6-14 Ford electronic fuel injection Beginning with certain 1980 cars, Ford replaced the carburetor on

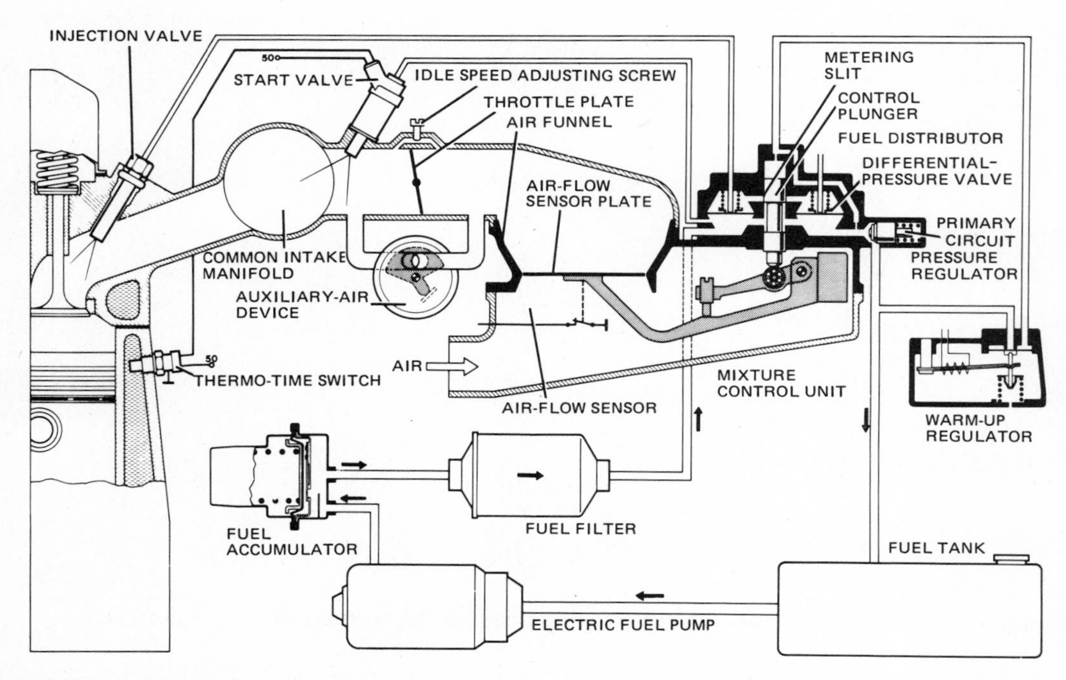

Fig. 6-23 The Bosch continuous-fuel-injection system used on some late-model imported cars. (*Robert Bosch Corporation*)

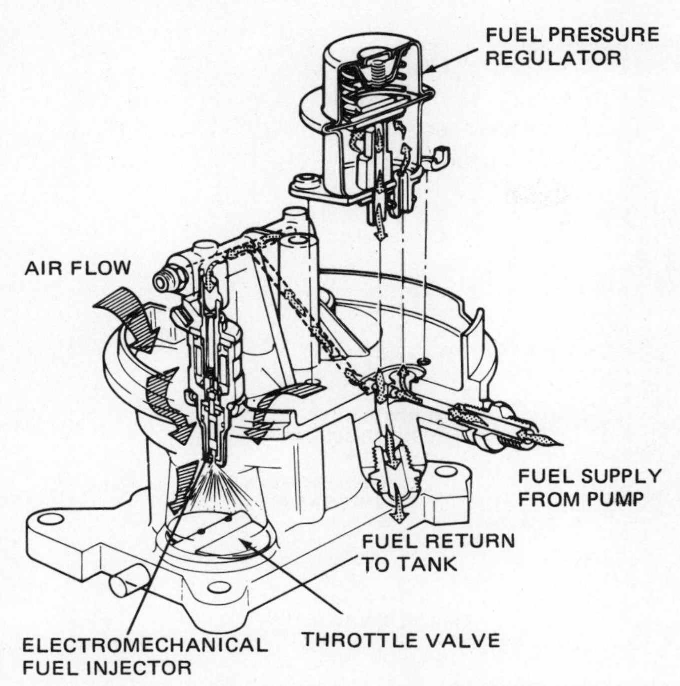

Fig. 6-24 Cutaway view of the throttle body and fuel injector in the Ford electronic fuel-injection system. (*Ford Motor Company*)

some engines with a type of electronic fuel-injection system. It sometimes is called a *fuel charging system*. The Ford system uses two pressure-actuated fuel-metering valves, or injectors. The injectors are placed in the throttle body on the intake manifold (Fig. 6-24).

The amount of fuel discharged from the injectors is controlled by Ford's "Electronic Engine Control" (EEC) III system. This system provides very accurate metering of the fuel flow to the engine. Controlled high fuel pressure in the injectors, along with the very accurate changes in fuel volume as determined by the EEC III system, improves the distribution of air-fuel mixture to each cylinder. This, in turn, provides better driveability, while keeping exhaust emissions within the legal limits.

Figure 6-25 shows the components in the Ford electronic fuel-injection system. Fuel to the injectors is provided by a high-pressure electric fuel pump located inside the fuel tank. A primary fuel filter is located in the fuel line under the passenger compartment. A smaller secondary fuel filter is mounted in the engine compartment. A fuel pressure regulator on the throttle body maintains the fuel pressure to the injectors at exactly 39 psi (269 kPa). Any excess fuel not needed by the engine is returned to the fuel tank through a fuel-return line (Fig. 6-25).

If the car is involved in a collision, contacts in the inertia switch open and stop the fuel pump. The inertia switch must be reset by pressing both buttons on it at the same time before the engine can be started. The inertia switch is located in the luggage compartment near the left wheel well on some cars.

The constant high pressure at the injectors provides a fine atomized spray of fuel when the injectors open (Fig. 6-24). The injectors are mounted vertically in the throttle body above the throttle valves so that the fuel is sprayed directly into the air stream.

Sensors monitor engine operating conditions and send this information to a microprocessor, which is a small computer (Fig. 6-25). It continuously calculates the correct injector opening time. The frequency of injection is constant at four pulses per engine crankshaft revolution, two for each injector. Fuel volume is controlled by how long each injector is open.

Figure 6-26 shows the complete electronic fuel-injection system and EEC III installed on an engine. During starting and warm-up, the EEC III system provides extra fuel on signal from a bimetal electric switch on the throttle body.

The throttle body looks similar to a conventional carburetor but is mechanically much less complicated. It includes the necessary parts and connections to cause the required downshifts in automatic transmissions, to adjust the engine idle speed, and to increase the engine idle speed when the air conditioning is on. An air cleaner fits over the throttle body exactly the same as with the carburetor.

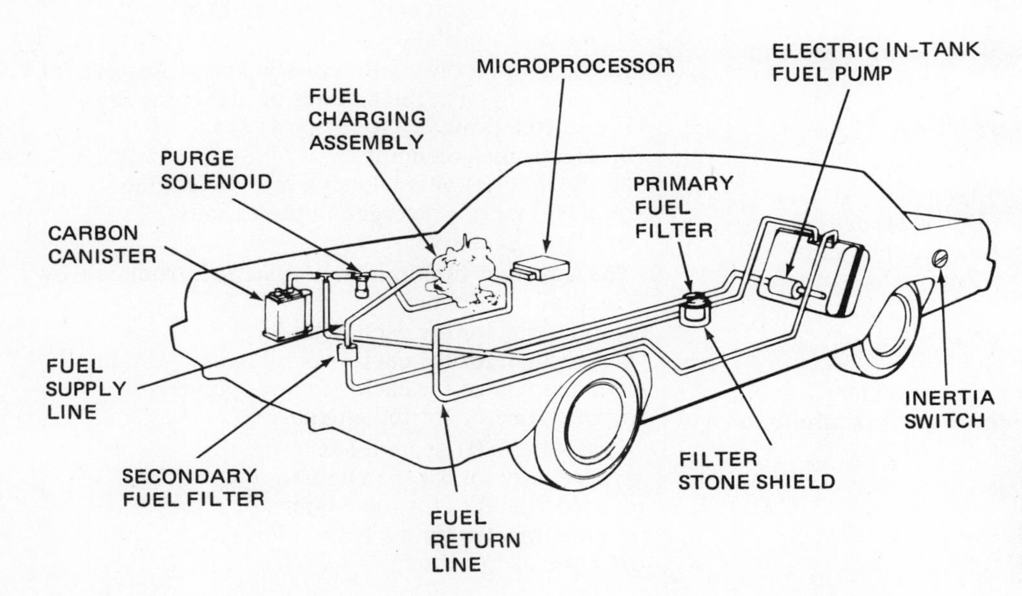

Fig. 6-25 Location of the components in the Ford electronic fuel-injection system. (*Ford Motor Company*)

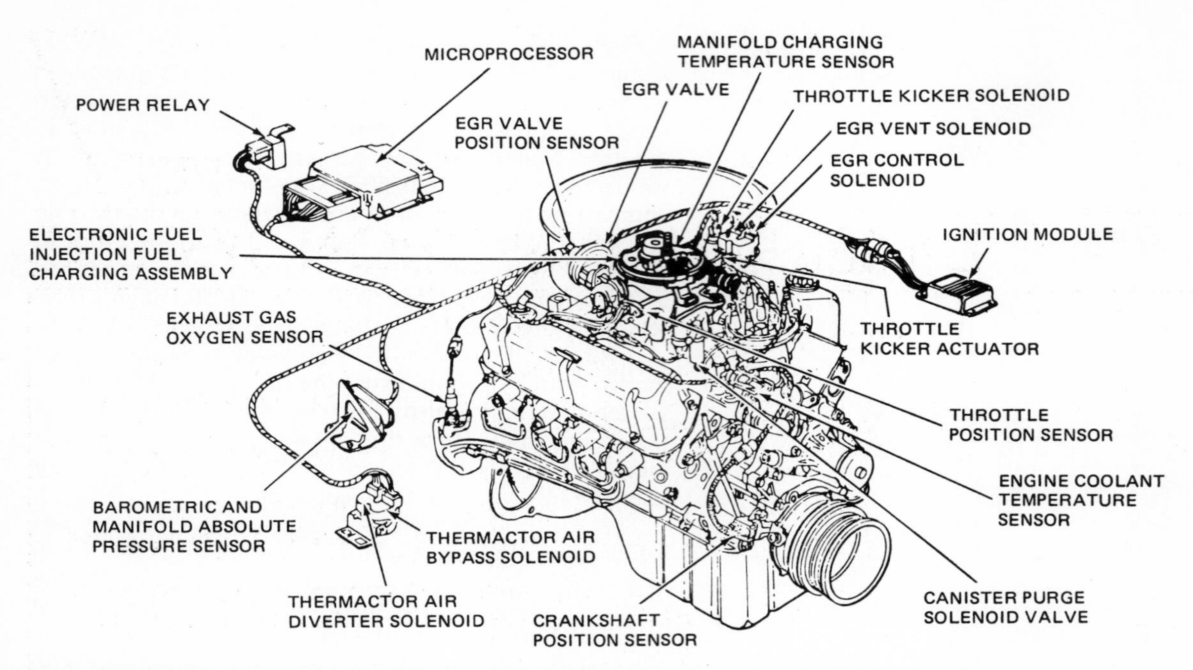

Fig. 6-26 Electronic Engine Control III system, which includes electronic throttle-body fuel injection, installed on an engine. (*Ford Motor Company*)

Chapter 6 review questions

Select the *one* correct, best, or most probable answer to each question. Then check your answers against the correct answers given at the end of the book.

1. In the gasoline fuel-injection system, the gasoline is sprayed into the air:
   a. in the combustion chambers
   b. in the intake manifold
   c. in the carburetor
   d. in the exhaust manifold.
2. Gasoline fuel-injection systems can be classified in two ways, according to whether they are:
   a. timed or pulsed
   b. continuous or controlled
   c. timed or continuous
   d. none of the above.
3. Modern timed gasoline-injection systems are controlled by:
   a. throttle position
   b. electronic devices
   c. timers
   d. glow plugs.
4. The typical gasoline fuel-injection system for a six-cylinder engine has the injectors grouped into:
   a. two groups
   b. three groups
   c. six groups
   d. none of the above.
5. The typical gasoline fuel-injection system for an eight-cylinder engine has the injectors grouped into:
   a. two groups
   b. three groups
   c. four groups
   d. none of the above.
6. The amount of fuel injected by the timed-injection system depends on:
   a. when the injection valves open
   b. how far the injection valves open
   c. how long the injection valves stay open
   d. all of the above.
7. The four factors that control the length of time the fuel-injection valves stay open are throttle position and:
   a. intake manifold vacuum, air temperature, and coolant temperature
   b. intake manifold temperature, manifold vacuum, and coolant temperature
   c. amount of oxygen in the exhaust, intake manifold vacuum, and coolant temperature
   d. none of the above.
8. In the late-model continuous-injection system for gasoline engines, the amount of fuel being injected is controlled by:
   a. electronic solenoids
   b. the amount of air flowing into the engine
   c. the amount of oxygen in the exhaust
   d. all of the above.
9. The gasoline fuel-injection system introduced by Cadillac in 1975 has:
   a. one injector for each cylinder
   b. one injector for each two cylinders
   c. one injector for each bank of cylinders
   d. one injector for the engine.
10. Ford's EEC III system has
    a. one injector for the engine
    b. two injectors for the engine
    c. one injector for each cylinder
    d. none of the above.

CHAPTER 7

DIESEL FUEL-INJECTION SYSTEMS

After studying this chapter, you should be able to:

1. Define *glow plug.*
2. Discuss the differences in the operation of diesel and gasoline engines.
3. Describe the types of fuel-injection system for diesel engines.
4. Explain the basic operation of the governor.
5. Locate the components of a diesel-engine fuel-injection system on various vehicles.
6. Explain the operation of the rotary fuel-injection pump.
7. Describe the construction and operation of a GM unit injector.

7-1 Differences between gasoline and diesel engines

NOTE: Some diesel engines in trucks operate on the two-stroke cycle. Others have both blowers and turbochargers. These engines are primarily for truck and industrial application. Diesel engine fuels are discussed in detail at the end of this chapter in ⚙ 7-13 to 7-21.

The diesel engine must be more heavily constructed because of the higher pressure in the combustion chambers and in the actions during the four strokes. Let us compare the strokes:

1. Diesel engine intake stroke The intake valve is open and the piston is moving down, producing a partial vacuum in the cylinder (Fig. 7-1). Atmospheric pressure pushes air through the air filter and intake manifold, past the open intake valve, and into the cylinder. There is no throttle valve or carburetor venturi to impede the movement of the air. The cylinder is filled completely with only air on the intake stroke. No fuel is present.

NOTE: Some diesel engines with a pneumatic governor have a throttle valve in the intake manifold. However, the throttle valve does not act to prevent more than enough air from entering the cylinder.

2. Diesel engine compression stroke During the compression stroke, both valves are closed and the upward-moving piston compresses the air (Fig. 7-2). The compression ratio of the diesel engine is much higher than that of the gasoline engine. It may be as high as 23.5:1. This contrasts with gasoline engine compression ratios which average about 9:1. The reason the diesel engine can have such high compression ratios is that air alone is compressed. Compressing air makes it hot. Compressing air to one-twentieth of its original volume (compression ratio 20:1), increases the temperature to above 1000°F [537.8°C]. This temperature is high enough to ignite almost any fuel. That is the reason why such high compression ratios are not used with gasoline engines. The air-fuel mixture would ignite before the piston reached TDC.

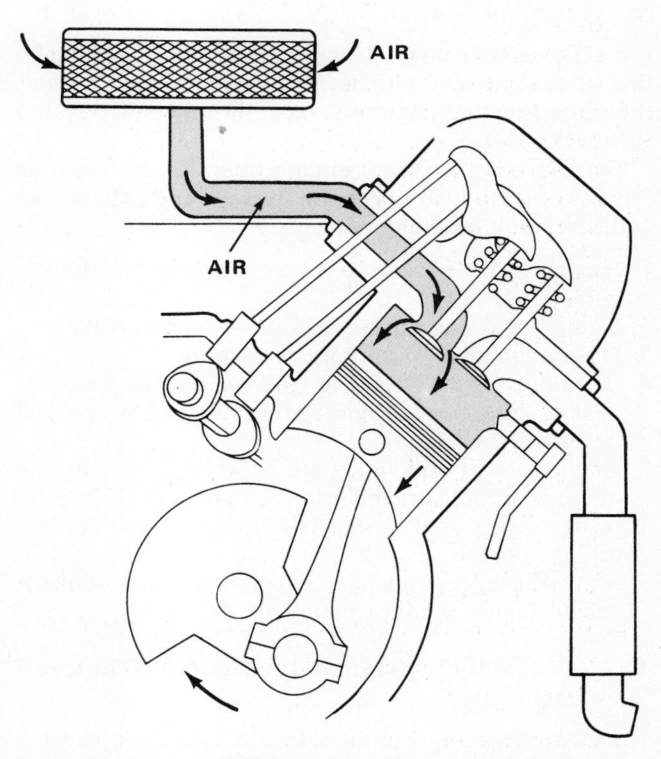

Fig. 7-1 Diesel engine intake stroke. (*Oldsmobile Division of General Motors Corporation*)

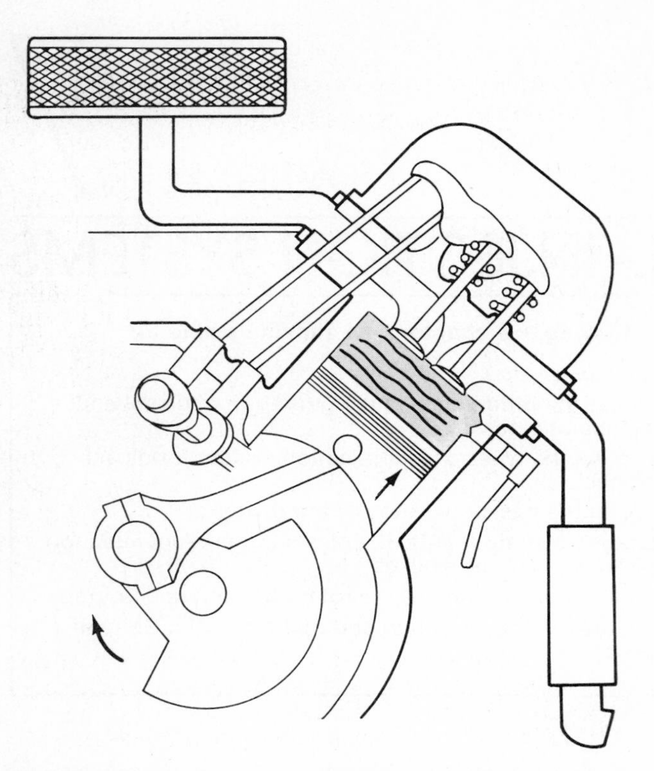

Fig. 7-2 Diesel engine compression stroke. (*Oldsmobile Division of General Motors Corporation*)

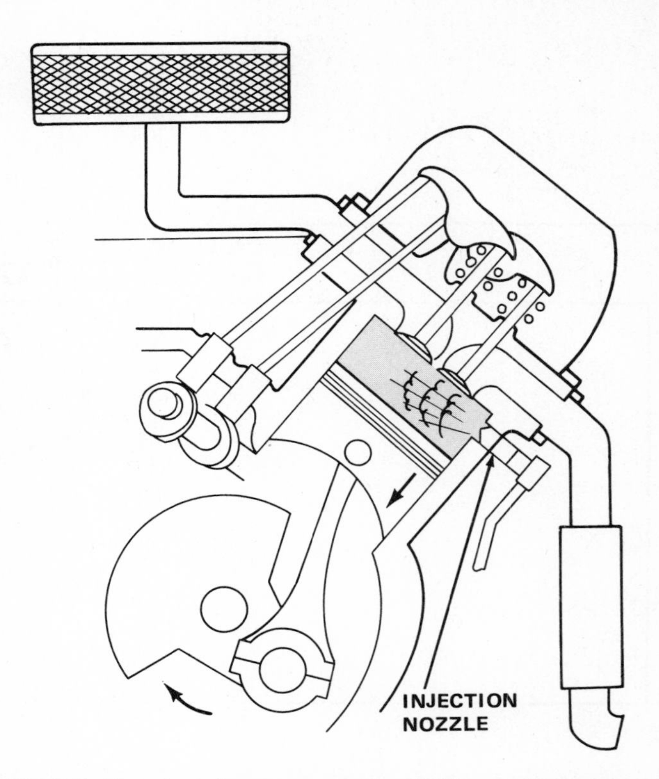

Fig. 7-3 Diesel engine power stroke. (*Oldsmobile Division of General Motors Corporation*)

3. Power stroke As the piston approaches TDC on the compression stroke, the diesel engine fuel-injection system starts to spray fuel into the cylinder (Fig. 7-3). The high temperature of the compressed air ignites the fuel and the pressure rapidly rises. The power stroke then takes place.

4. Exhaust stroke The exhaust stroke is similar to that of the gasoline engine. The piston moves up and the exhaust gases flow out past the opened exhaust valve (Fig. 7-4).

To sum up, the diesel engine operates on the four strokes of intake, compression, power, and exhaust. In addition, the diesel engine:

1. Has no throttle valve to restrict airflow into the engine.
2. Compresses only air on the compression stroke.
3. Has a much higher compression ratio.
4. Does not have an electric ignition system. Instead, heat of compression ignites the fuel as it is sprayed into the cylinders.
5. Engine power and speed are controlled only by the amount of fuel sprayed into the cylinders. For more power, more fuel is injected. For less power, less fuel is injected.
6. Many diesel engines have glow plugs which make it easier to start a cold engine (see ✲ 7-9).

✲ 7-2 Diesel fuel-system requirements The diesel fuel system must:

1. Deliver the right amount of fuel to meet the operating requirements.

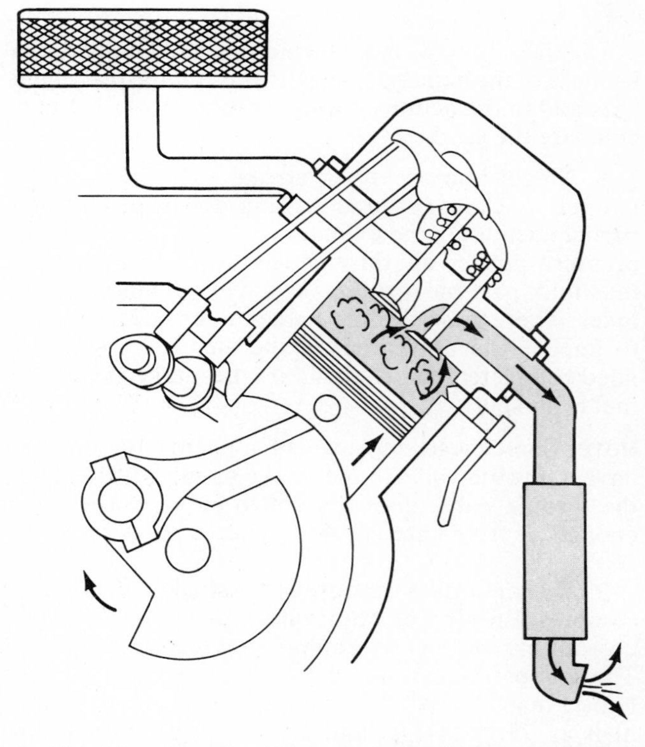

Fig. 7-4 Diesel engine exhaust stroke. (*Oldsmobile Division of General Motors Corporation*)

2. Change the timing of fuel delivery as engine speed changes. As engine speed increases, the fuel delivery must start earlier. This compares with advance of the spark in the gasoline engine as engine speed increases. The purpose is the same, to get the ignition of the fuel started before the piston reaches TDC. Without an advance, the piston would be over TDC and starting down on the power stroke before ignition was well started. The piston movement would keep ahead of the pressure rise so that most of the power in the fuel would be wasted.
3. Deliver the fuel to the cylinders under high pressure. The pressure in the cylinder at the end of the compression stroke may be more than 500 psi (pounds per square inch) [3447 kPa]. The fuel must be under pressure much higher than this in order for it to be sprayed into the compressed air.

⚙ 7-3 Types of fuel-injection systems There are basically two types of fuel-injection systems used on the diesel engine. In one, a centrally located pump pressurizes the fuel, meters it, times it, and delivers it at high pressure to the cylinders through tubes.

In the other system, the fuel is sent to the injectors under a relatively low pressure. The injectors have cam-operated plungers (like the valves), which are adjustable. At the proper instant the cams operate the plungers and they force the fuel at high pressure into the cylinders.

The centrally located pump system, used on a majority of diesel engines, can be further divided into two types, the cam-operated in-line plunger type, and the rotary distributor type. The rotary type is the most commonly used for passenger-car diesel engines.

⚙ 7-4 Cam-operated in-line plunger pump This pump has a cylinder with plunger for each engine cylinder. Figure 7-5 is a partial cutaway view of a fuel-injection pump for a six-cylinder engine. It has six plungers working in six barrels, one for each engine

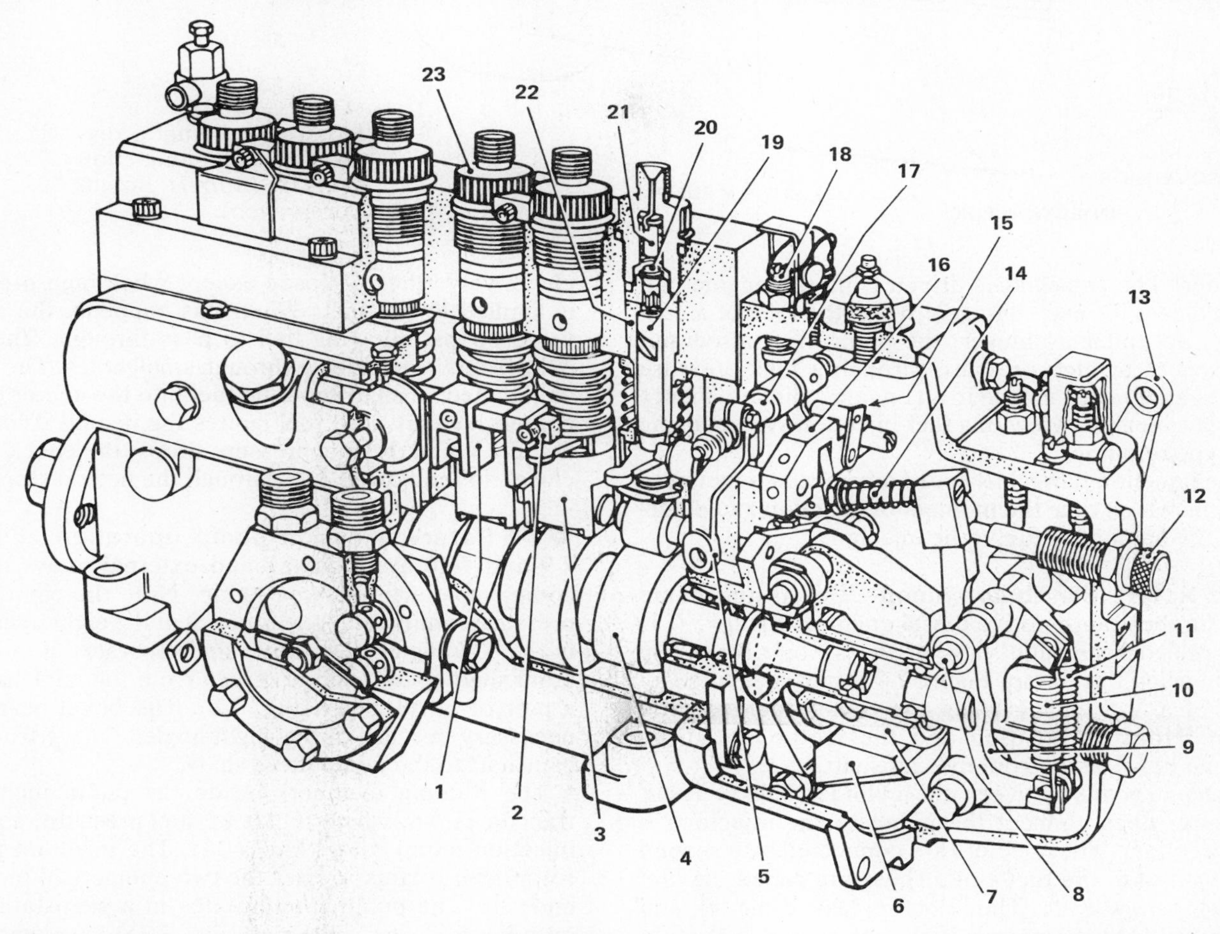

1 CONTROL FORK
2 CONTROL ROD
3 TAPPET ASSEMBLY
4 CAMSHAFT
5 STOP CONTROL LEVER
6 GOVERNOR FLYWEIGHT
7 GOVERNOR SLEEVE
8 SPEED LEVER SHAFT
9 CRANK LEVER
10 GOVERNOR MAIN SPRING
11 GOVERNOR IDLING SPRING
12 DAMPER
13 SPEED CONTROL LEVER
14 TELESCOPIC LINK
15 TRIP LEVER
16 BRIDGE LINK
17 EXCESS FUEL DEVICE
18 MAXIMUM FUEL STOP SCREW
19 PLUNGER
20 DELIVERY VALVE
21 VOLUME REDUCER
22 BARREL
23 DELIVERY VALVE HOLDER

Fig. 7-5 In-line plunger pump for six-cylinder diesel engine. (*CAV Ltd.*)

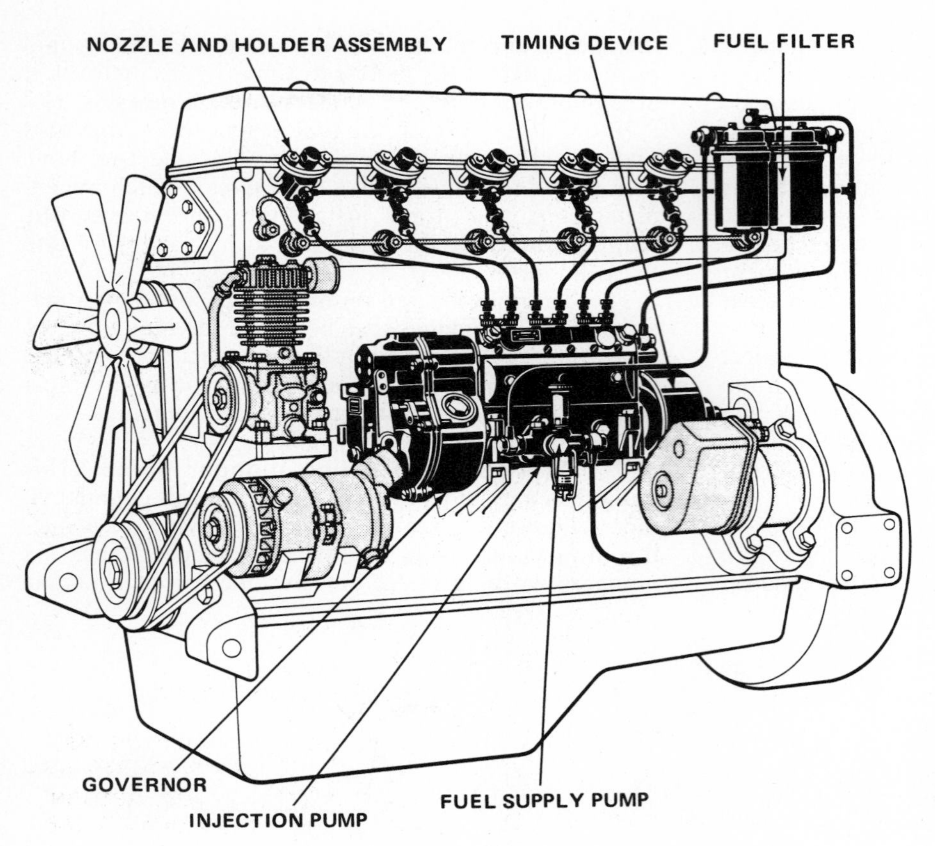

Fig. 7-6 Schematic diagram of a fuel-injection system. (*Robert Bosch Corporation*)

cylinder. The camshaft is driven from the engine and it has a cam for each plunger. When the lobe of a cam comes up under a plunger, the plunger is raised and fuel is sent at high pressure through a high-pressure tube to an injector nozzle in an engine cylinder. Figure 7-6 is a schematic view of a fuel-injection system using this type of pump.

The injection pump has speed advance and metering systems which time the moment of injection and determine the amount of fuel to be injected.

⚙ 7-5 Rotary distributor pump This is the pump used on most automotive diesel engines. Figure 7-7 is a simplified schematic view of the system for a V-6 engine using this pump. Figure 7-8 is a partial cutaway view of a V-8 engine using the rotary fuel-injection pump. Note its position and the method of drive through bevel gears from the camshaft.

The pump sits between the cylinder banks. High-pressure tubes connect the pump to the injectors in each cylinder. The injectors are connected by a second set of tubes to the fuel tank. These are called the *fuel leak-off return lines*. The injectors leak some oil, and the lines carry the excess fuel back to the fuel tank. The fuel injection pump includes a fuel supply pump which delivers the fuel to the distributor part of the pump at high pressure. The distributor then sends the fuel to the engine cylinders in the proper firing order. This compares with the electric ignition system on gasoline engines, sending high-voltage surges to the spark plugs in the proper firing order.

The nozzle of the fuel injector has a spring-loaded check valve that is closed except when high pressure is applied to the fuel. When this happens, the check valve opens, allowing fuel to pass through. The fuel exits from the nozzle tip through small holes. The holes are located so as to send the fuel into the center of the compressed air. The fuel ignites the instant it hits the hot air. When the fuel pressure drops, the check valve closes so the flow of fuel through the nozzle stops.

⚙ 7-6 Rotary injection pump operation Figures 7-9 and 7-10 are cutaway and external views of the complete rotary injection pump. Note the eight high-pressure connectors, one for each of the eight cylinders in the V-8 engine. The drive shaft operates at one-half crankshaft speed, and is driven from the camshaft by a pair of bevel gears (Fig. 7-8). The bevel gears are necessary because of the slight angle of drive from the camshaft to the pump drive shaft.

The rotating members inside the pump include a transfer pump, which builds up fuel pressure, and the injection pump rotor (Fig. 7-11). The injection pump rotor, as it rotates, causes the two plungers to move in and out. The pump rotor rotates in a semistationary internal cam ring. This ring (Fig. 7-15) has cam lobes on its inner surface.

As the rotor rotates, the rollers ride up and down on the cam lobes. When they move in this way, they cause the two plungers to move up and down in their holes. This causes the size of the chamber between the inner ends of the plungers to increase and decrease in size (Fig. 7-12). When the chamber increases in size, fuel from the transfer pump flows into the chamber. When

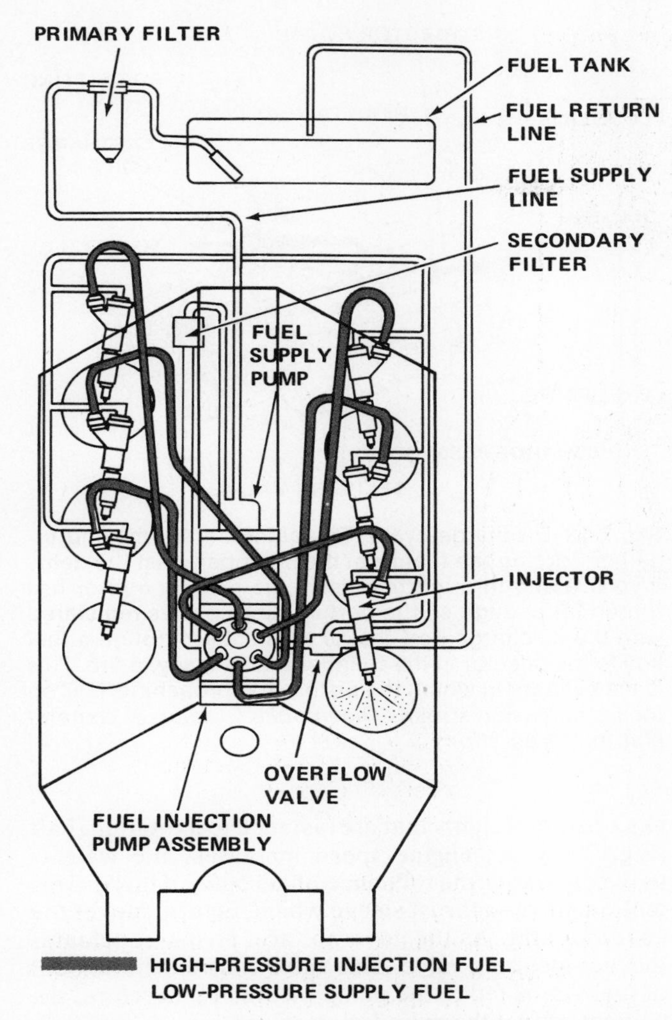

Fig. 7-7 Rotary distributor pump system for a V-6 diesel engine. (*General Motors Corporation*)

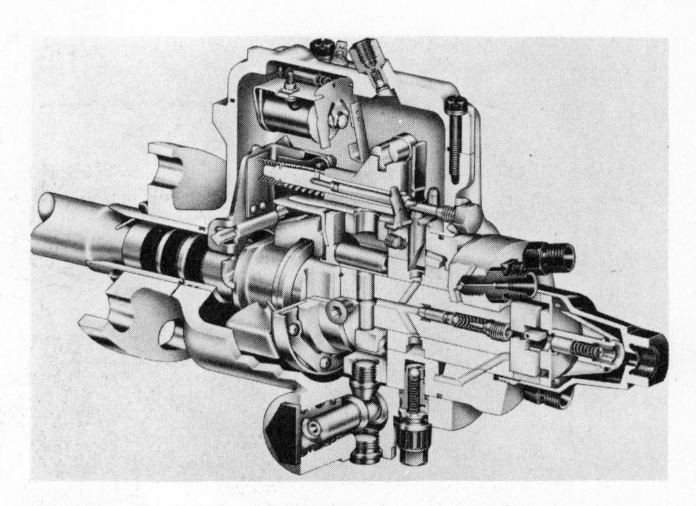

Fig. 7-9 Rotary fuel-injection pump, partly cut away to show internal construction. (*Oldsmobile Division of General Motors Corporation*)

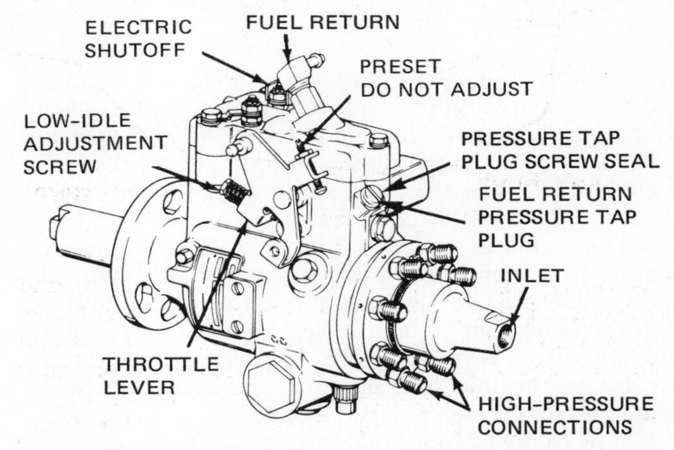

Fig. 7-10 External view of rotary fuel-injection pump. (*Chevrolet Motor Division of General Motors Corporation*)

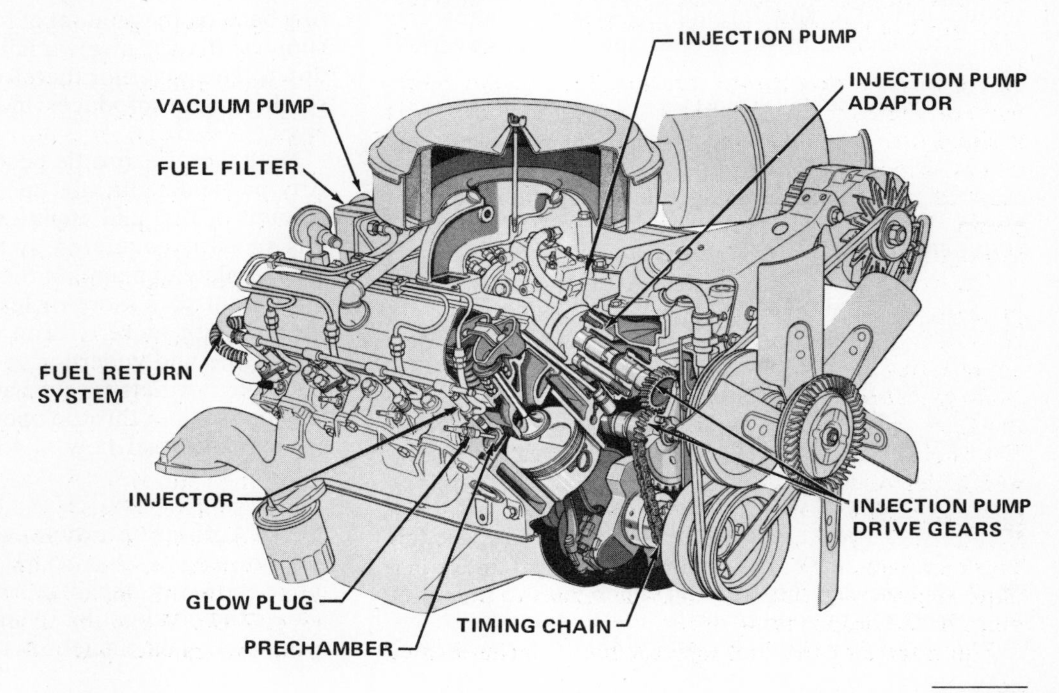

Fig. 7-8 V-8 diesel engine using rotary fuel-injection distributor pump, partly cut away to show pump drive arrangement. (*Oldsmobile Division of General Motors Corporation*)

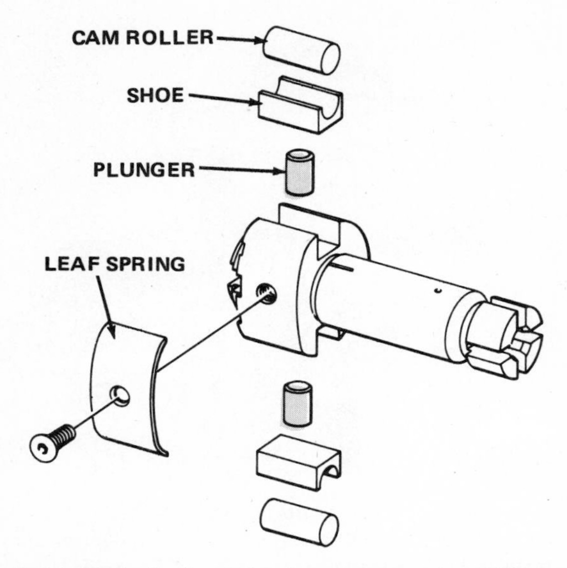

Fig. 7-11 Parts in the injection pump rotor. (*Chevrolet Motor Division of General Motors Corporation*)

the plungers move toward each other, the chamber decreases in size, forcing fuel to flow out of the chamber. The fuel flows through ports to the high-pressure line connected to the nozzle in the cylinder ready for ignition. This is the cylinder in which the piston is nearing TDC on the compression stroke.

The action is shown in Fig. 7-13. The action compares to that of the rotor in the ignition distributor. The rotor sends the high-voltage surge from the ignition coil through the high-voltage cable to the spark plug in the cylinder that is ready to fire. In a similar manner, the rotor in the injection pump aligns holes that allow the fuel under high pressure between the two plungers to flow to the cylinder that is ready for ignition.

7-7 Governor The governor allows the proper amount of fuel to flow through the system to allow the engine to operate at the proper speed. The governor

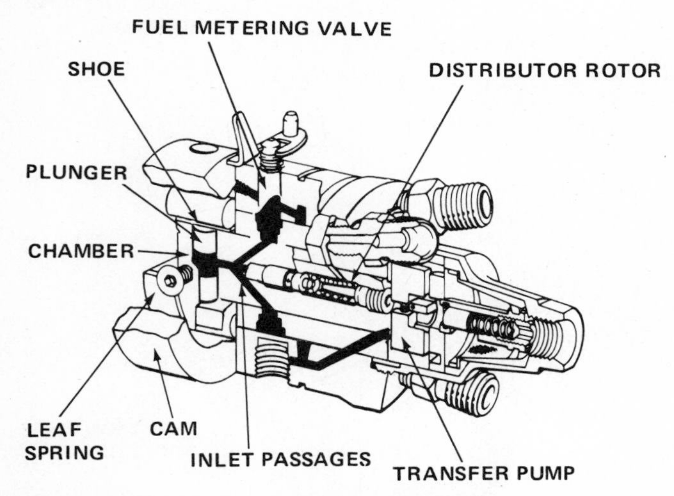

Fig. 7-12 Charging cycle during which the two plungers move apart and provide space for fuel to enter the chamber. (*Oldsmobile Division of General Motors Corporation*)

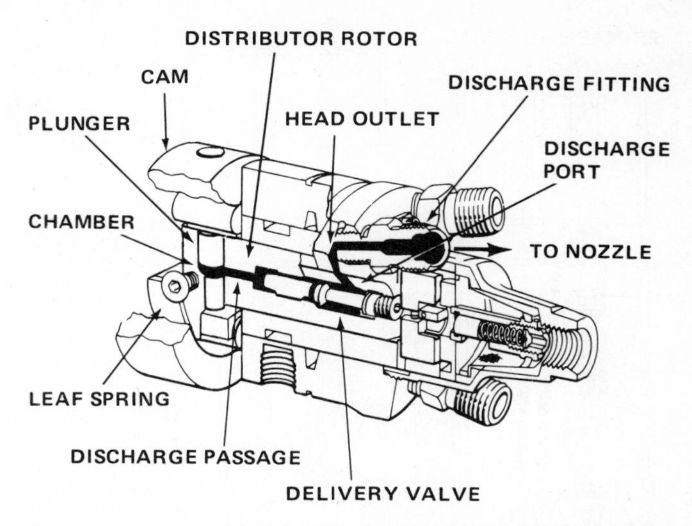

Fig. 7-13 Discharge cycle. The plungers are moving together, forcing the fuel from the chamber, past the delivery valve and through the discharge port. The rotor has turned far enough so the port in the rotor has registered with the discharge port that is connected through a fuel line to the injector in the cylinder that is ready to fire. This is the cylinder in which the piston is approaching TDC on the compression stroke. (*Oldsmobile Division of General Motors Corporation*)

has a pair of weights that are fastened to a rotating shaft (Fig. 7-14). As engine speed increases, the weights move out under the influence of centrifugal force. This motion moves a thrust sleeve which, in turn, moves the governor arm. As the governor arm pivots, it actuates the fuel-metering valve. This valve controls the amount of fuel that is fed to the pump rotor and, therefore, the amount of fuel the cylinders receive.

For example, when the car starts down a hill, the load on the engine decreases. The engine starts to speed up. Then the governor operates to prevent this by cutting back on the amount of fuel going to the engine. But suppose the car meets a hill. The engine tends to slow down. The governor therefore allows more fuel to flow so the engine produces more power. This maintains engine speed.

Note that the throttle position enters into the action. Any particular throttle opening, in effect, presets the amount of fuel and engine speed. Any variation from this speed is countered by the governor as it changes the fuel flow to maintain the preset speed. The throttle movement puts more or less tension on the governor spring. With more tension (greater throttle opening), the engine and governor speed must go higher before the governor acts to cut back on fuel flow. With less tension (smaller throttle opening), the governor can cut back on the fuel flow at lower engine and governor speed.

7-8 Automatic advance The ignition system for gasoline engines has spark advance mechanisms that move the spark ahead as engine speed increases. This allows the combustion to start earlier so it will be well advanced by the time the piston reaches TDC and starts

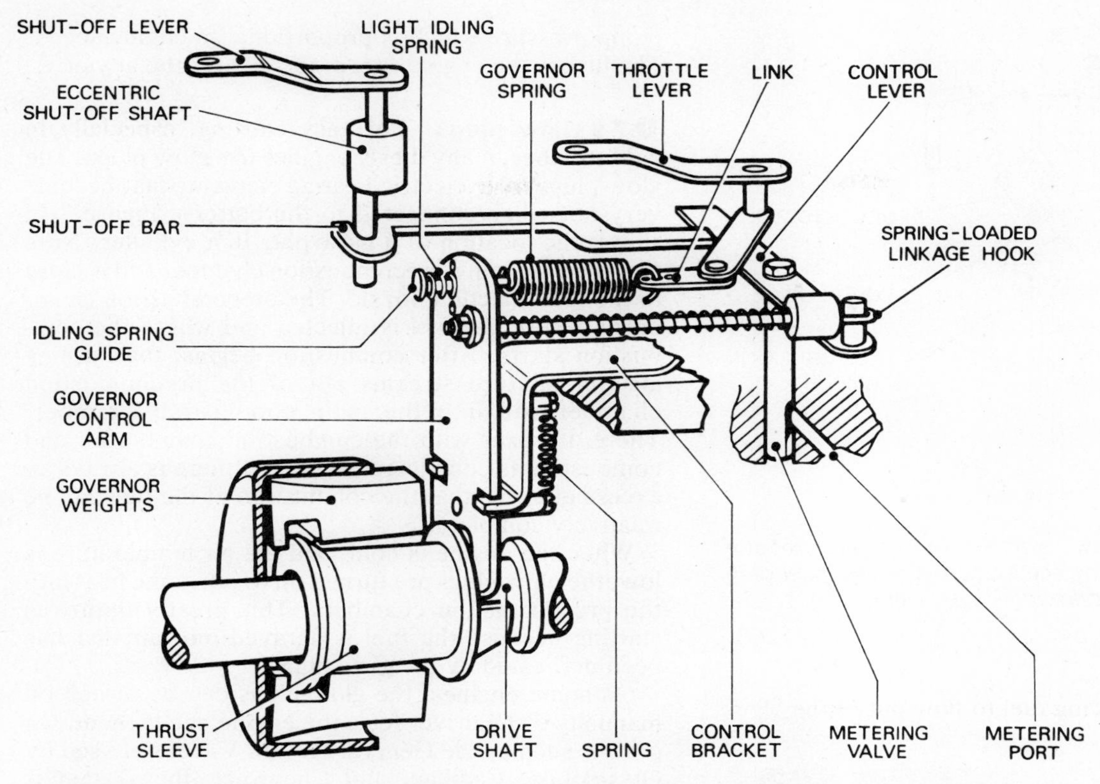

Fig. 7-14 Simplified layout of a governor. It works with the throttle to allow the correct amount of fuel to flow for the operating condition. (*CAV Ltd.*)

down on the power stroke. The diesel engine fuel-injection system also includes a speed advance. It is built into the injection pump (Fig. 7-15).

The advance mechanism uses the hydraulic pressure from the transfer pump to control the position of the cam ring. As the engine and transfer pump speed increase, the transfer-pump hydraulic pressure also increases. This increasing pressure, acting on a piston located below the cam, forces the advance pin to move (to the left in Fig. 7-15). This rotates the cam as shown by the heavy arrows. As the cam moves, the rotor rollers meet the lobes earlier and cause the plungers to be pushed together earlier. Therefore, the fuel is sent to the cylinder injection nozzles earlier. This produces the fuel-injection advance required as engine speed increases. The advance is proportional to the transfer

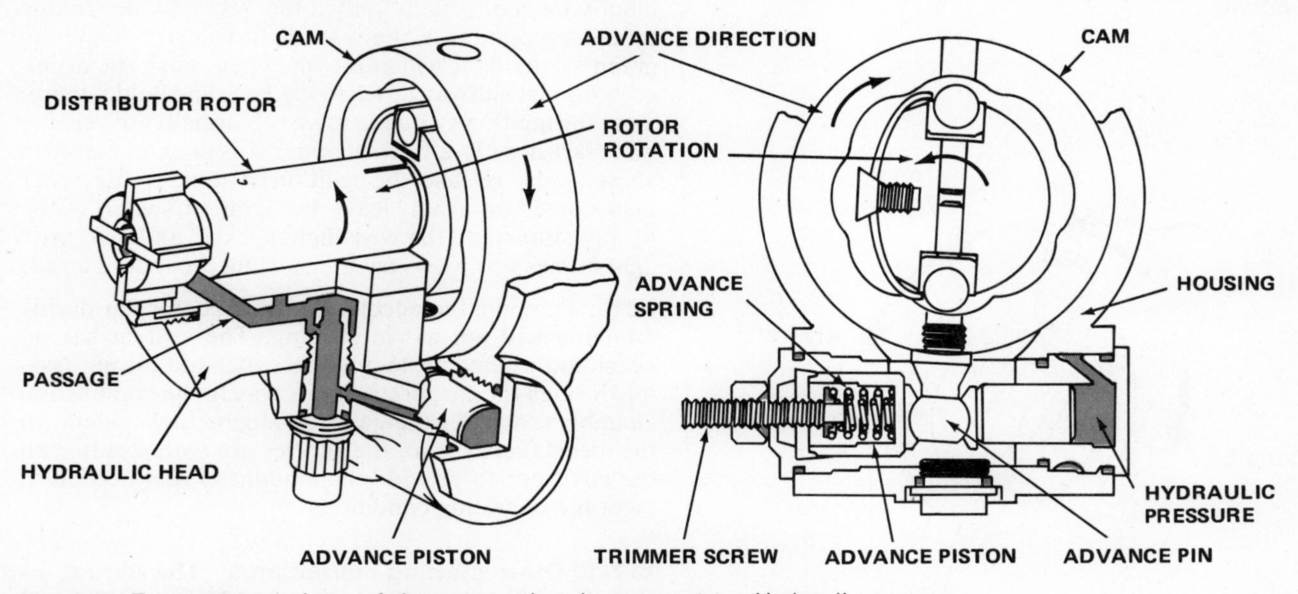

Fig. 7-15 Two cutaway views of the automatic advance system. Hydraulic pressure, which increases with speed, moves the cam ahead. (*Oldsmobile Division of General Motors Corporation*)

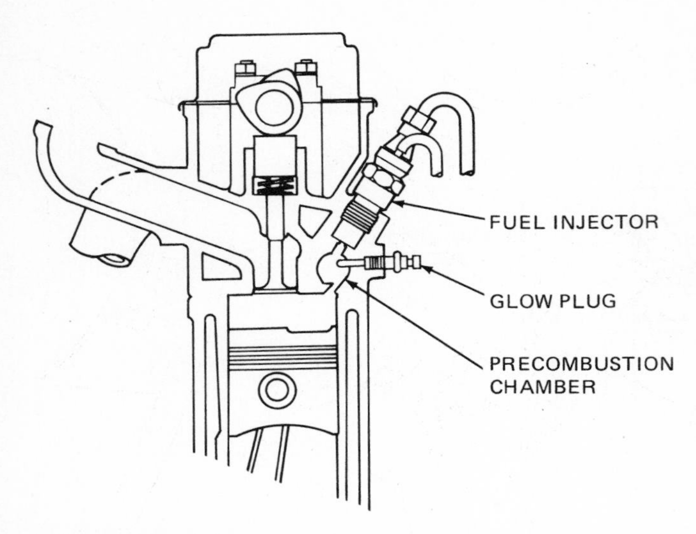

Fig. 7-16 Simplified drawing of a diesel engine precombustion chamber showing the locations of the glow plug and the fuel injector. (*Volkswagen of America, Inc.*)

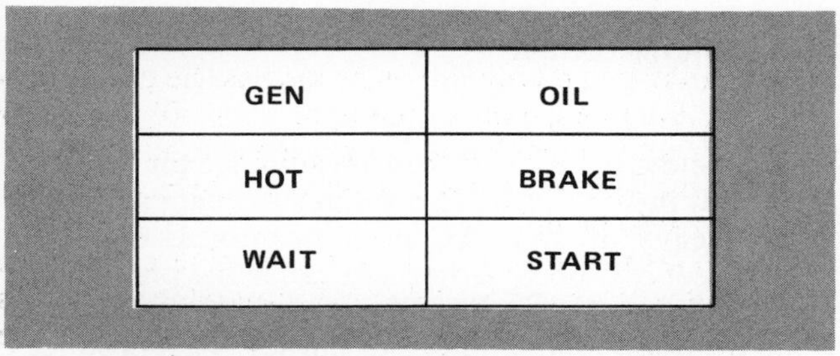

GEN	OIL
HOT	BRAKE
WAIT	START

Fig. 7-17 The instrument panel has two special lights, WAIT and START. (*Oldsmobile Division of General Motors Corporation*)

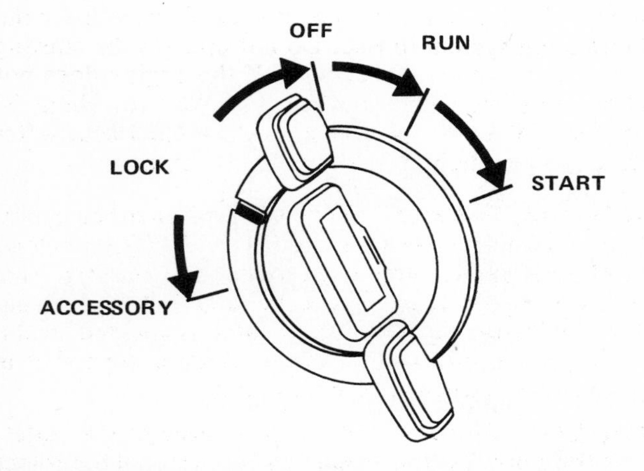

Fig. 7-18 Ignition switch positions. (*Oldsmobile Division of General Motors Corporation*)

pump pressure which is proportional to engine speed. The higher the engine speed, the greater the advance.

7-9 Glow plugs For easy starting, especially in cold weather, many diesel engines use glow plugs. The glow plugs have electric heating elements that become very hot when connected to the battery. Figure 7-16 shows the location of a glow plug in a cylinder. Note that the plug is in a precombustion chamber and is close to the fuel-injection nozzle. The precombustion chamber is where the fuel is injected and where the combustion starts. After combustion begins, the burning air-fuel mixture streams out of the precombustion chamber and into the main combustion chamber. There, it mixes with the combustion chamber air and combustion is completed. Note that there is always an excess of air so that the combustion of the fuel can be relatively complete.

When the engine is cold, and the air temperature is low, the glow plugs are turned on to put some heat into the precombustion chambers. This greatly improves starting because the fuel is sprayed into air that has been preheated by the glow plugs.

On some engines, the glow plugs can be turned on manually if the driver feels the engine needs them. On others, such as the General Motors V-8 diesel used by Oldsmobile, Cadillac, and Chevrolet, the system is semiautomatic. The instrument panel has two special lights, WAIT and START (Fig. 7-17). The starting procedure is as follows:

1. Put the transmission lever in PARK.
2. Turn the ignition switch (Fig. 7-18) to RUN. Do not turn it to START. When you turn the switch to RUN, an amber WAIT light comes on (if the engine is cold). This tells you that the glow plugs are on, heating the precombustion chambers in the engine.

After the precombustion chambers have been sufficiently heated (usually only a few seconds, depending on the temperature), the WAIT light will go out and the green START light will come on. Then, push the accelerator pedal halfway down to the floor and hold it there. Turn the ignition switch to START. Normally, the engine will start in only a few seconds. If it does not start in 15 seconds, release the ignition switch. If the WAIT light comes on again, leave the ignition switch in the RUN position until the WAIT light goes off and the START light comes on. Now, try the starting procedure again.

NOTE: Pumping the accelerator pedal before or during cranking will not aid in starting. The system has no accelerator pump system to force fuel into the air passing through the carburetor on its way to the combustion chambers, as does the gasoline-engine fuel system. In the diesel fuel system, the accelerator works only with the governor to provide the amount of fuel needed to meet the operating conditions.

7-10 Other starting instructions The starting instructions above are for the General Motors V-8 diesel engine used in some Oldsmobiles and other vehicles. Other manufacturers have slightly different instruc-

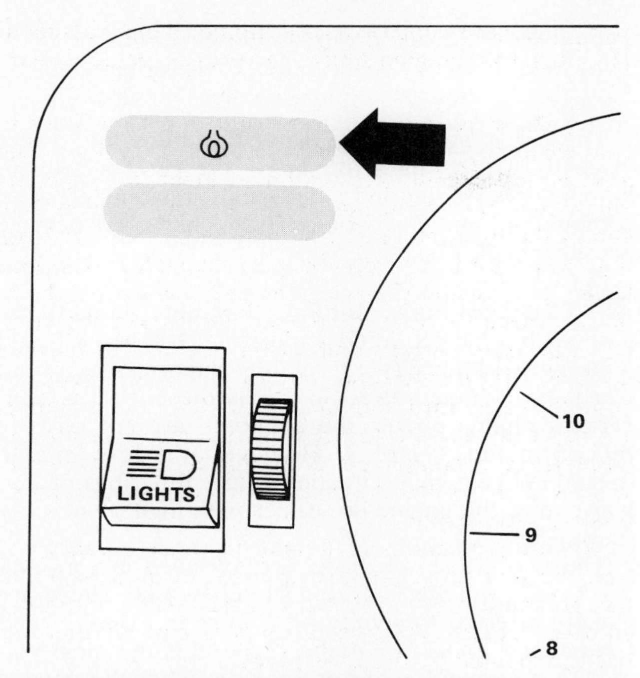

Fig. 7-19 The Volkswagen Rabbit uses a single light to indicate glow-plug action. When the plugs are on, the light comes on. (*Volkswagen of America, Inc.*)

tions and different indicating devices. The Volkswagen Rabbit with a diesel engine, for example, uses a single light to indicate glow-plug action (Fig. 7-19). Also, the ignition switch has only three positions—OFF, ON, START (Fig. 7-20). Here is the recommended starting procedure:

1. Temperatures above 32°F [0°C], engine cold—turn ignition switch to ON. The glow-plug light (Fig. 7-19)

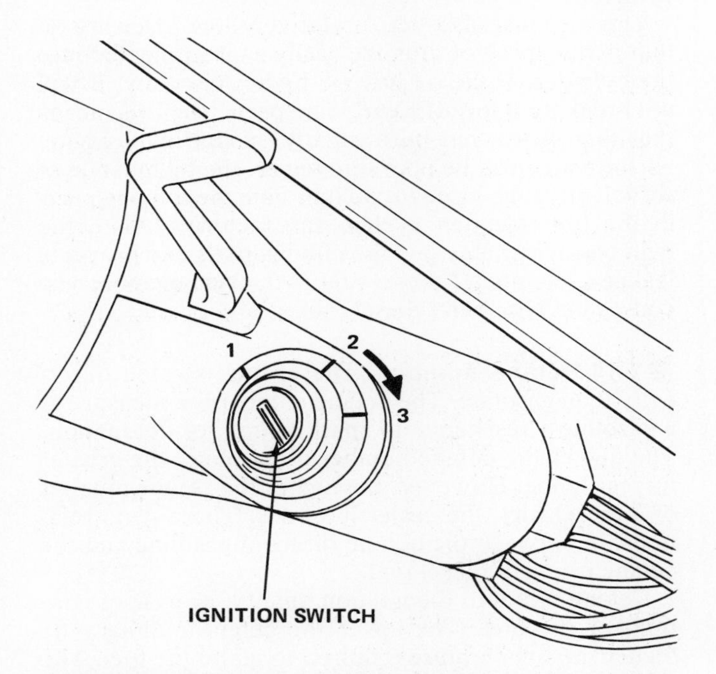

Fig. 7-20 The ignition switch on the Rabbit has three positions, off, on, and start. (*Volkswagen of America, Inc.*)

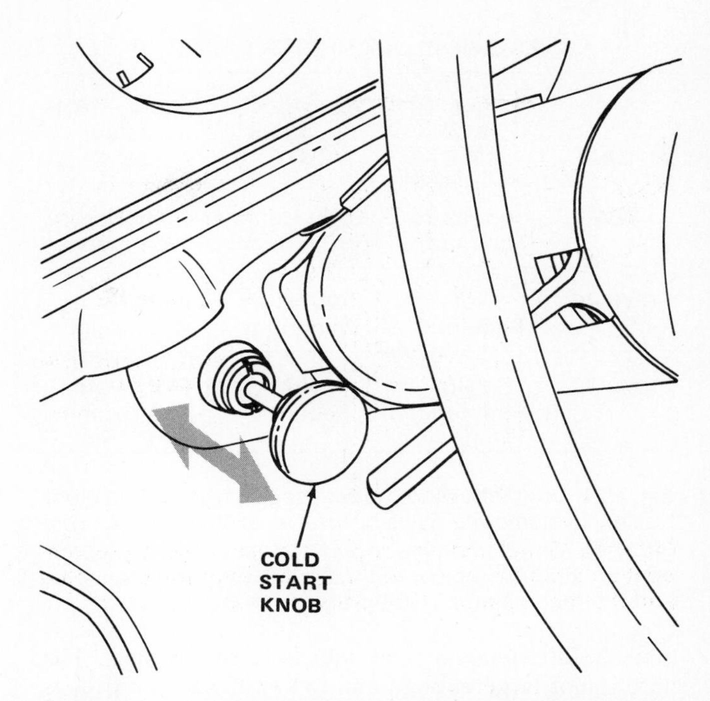

Fig. 7-21 Location of the cold-start knob. (*Volkswagen of America, Inc.*)

will come on and remain on as long as the glow plugs are heating. As soon as the light goes off, the plugs have heated enough and the engine is ready to start. Turn the ignition switch to START. Do not depress the throttle pedal.
2. Temperatures below 32°F [0°C], engine cold—turn ignition switch to ON. After the glow-plug light goes off, push the throttle all the way open. Pull out the cold-start knob under the instrument panel (Fig. 7-21). This provides extra fuel to the cylinders. Turn the switch to START. Two minutes after starting, push the cold-start knob in.
3. Starting a warm engine—do not depress the throttle. Do not use the glow plugs. Turn the ignition switch past ON to START immediately.

NOTE: Do not accelerate the engine excessively immediately after starting. Wait for the oil pressure in the lubricating system to rise. Do not operate the starting motor longer than 30 seconds. If the engine does not start, turn ignition switch to OFF. Wait for about 30 seconds. Then repeat the cold-start sequence. After the glow-plug light goes off, try another start.

7-11 Block heater For very cold weather operation, where temperatures are 0°F [−18°C] or below, block heaters are often used to assist in starting. This type of heater has an electric element that works from 115-volt house current. The heater is located in the engine block and has a special electric cord that is plugged into a regular electric outlet.

CAUTION: Be sure to plug into a three-prong outlet. The third prong is the ground and is essential to protect you from electric shock. If the electrical cord connected to the heater is not long enough, do not use a

ENGINE BLOCK HEATER USAGE

OIL	32 to 0° F [0 to −18° C]	0 to −10° F [−18 to −23° C]	Below −10° F [Below −23° C]
30W	Two Hours Minimum	Eight Hours or Overnight	Oil Not Recommended
15W–40	Not Required	Two Hours Minimum	Eight Hours or Overnight
10W–30	Oil Not Recommended	Not Required	Eight Hours or Overnight

Fig. 7-22 Chart showing proper usage of the engine block heater. (*Oldsmobile Division of General Motors Corporation*)

household extension cord with only two prongs. The wire is not large enough and the extension cord does not have the ground wire in it.

The length of time that the block heater should be plugged in depends on the type of engine oil used and the temperature. Oldsmobile recommendations are given in the chart in Fig. 7-22. The notation *oil not recommended* means that you should not use the oil indicated for the temperature shown.

NOTE: Do not use "starting aids" such as ether, gasoline, or similar materials, in the air intake. They can actually delay starting and may damage the engine.

⚙ 7-12 Vacuum pump Because there is no throttle valve or venturi in the airstream, there is no vacuum source that can be tapped on the diesel engine. Therefore, the diesel engine requires a vacuum pump to provide the vacuum to operate various devices on the car, such as power brakes and air-conditioner vacuum doors. On the General Motors diesel engine, the vacuum pump is located at the back of the engine (Fig. 7-8).

⚙ 7-13 Diesel-engine fuels Diesel engines use diesel fuel oil. The fuel oil is sprayed, or injected, into the engine cylinders toward the end of the compression stroke (Fig. 7-3). Heat of compression ignites the fuel oil, and the power stroke follows. Diesel fuel is made from crude oil with the same basic refining process by which gasoline is made. It is a light oil having a boiling point not higher than 700°F [371°C]. It is blended from several base materials so it will have the proper characteristics. These include volatility, ignition quality, viscosity, gravity, and low sulfur content.

Diesel fuels are classified according to fuel types as follows:

Type C-B: Diesel fuel oil for automobiles, city buses, etc.
Type T-T: Fuels for diesel engines in trucks, tractors, etc.
Type R-R: Fuels for railroad diesel service.
Type S-M: Heavy and residual fuels for large stationary and marine diesel engines.

The C-B type oil is the lightest and has the lowest viscosity. Other characteristics of diesel oil are cetane number, pour point and cloud point, residue left after combustion, and flash point. These characteristics are discussed below.

⚙ 7-14 Diesel-fuel volatility Volatility refers to the ease with which a liquid vaporizes (⚙ 2-4). A liquid with high volatility (very volatile) vaporizes at a relatively low temperature. So far as diesel fuel is concerned, engine power and economy are not directly related to fuel volatility. However, the less volatile fuels do have a higher heating value. But, starting and warmup of the engine are easier with higher volatility fuels.

The two grades of diesel fuel oil used in cars and trucks are Number 1-D and Number 2-D. These are usually referred to simply as "Number 1 diesel" and "Number 2 diesel." Number 1 diesel is the most volatile, and is required in some engines when the outside air temperature drops below 32°F [0°C]. Most automotive diesel engines normally operate on Number 2 diesel fuel.

⚙ 7-15 Diesel-fuel viscosity Viscosity refers to the tendency of a liquid to resist flowing. Gasoline has a very low viscosity. It flows very easily. A light oil has a moderate viscosity, since it flows easily. A heavy oil flows slowly, as it has high viscosity. The fuel oil used in a diesel engine must have a relatively low viscosity so that it will flow easily through the pumping and injection system that supplies the fuel to the engine cylinders.

The oil must also be of relatively low viscosity so that it will spray or atomize easily as it is injected into the cylinder. If the oil has too high a viscosity, it will not break up into sufficiently fine particles. This means that the oil will not burn rapidly enough, and engine performance will be poor. However, the oil must be of sufficiently high viscosity to lubricate the moving parts in the fuel-injection system satisfactorily. Also, the high viscosity helps seal the moving parts and prevents leakage. Number 2 diesel fuel is thicker, or more viscous, than Number 1 diesel fuel.

⚙ 7-16 Cetane number Diesel fuel is rated differently than gasoline. The octane number is a measure of a gasoline's resistance to spark knock, or detonation. The higher the octane number of gasoline, the greater its resistance. However, the higher the cetane number of a diesel fuel, the easier it ignites. There is no relationship between the octane rating of gasoline and the cetane rating of diesel fuel.

Cetane refers to the ignition quality, or ease of ignition, of the fuel. The lower the cetane number, the higher the temperature required to ignite the fuel. The higher the cetane number of a diesel fuel, the less tendency the fuel has to knock in the engine. To understand how the cetane number and knocking in a diesel engine are related, let's discuss what causes knocking to occur.

At the end of the compression stroke, the fuel system injects a spray of oil into the compressed air trapped in the combustion chamber. The fuel ignites throughout the combustion chamber simultaneously. This is because the fuel is sprayed into the compressed air that has reached a temperature of about 1000°F [538°C].

Although the nozzle discharges fuel under high pressure (Fig. 7-3), the oil is not delivered all at once. It takes a short time for the delivery. Fuel injection starts, continues for a fraction of a second, and then stops. Oil begins to burn almost as soon as it is injected. The delay in time between the injection of the fuel and the start of combustion is called *ignition lag*.

If the cetane number of the fuel is high (ignition temperature low), the sprayed oil will ignite almost as soon as injection begins. Now, there will be no accumulation of unburned fuel to ignite violently. Ignition continues evenly as the spray continues. An even combustion pressure results. When the cetane number of diesel fuel is too low, there is an ignition lag or delay. It takes longer for low-cetane fuel to ignite. Meanwhile, more fuel accumulates in the combustion chamber. Then, when it does ignite, there is a sudden ignition of most of the accumulated fuel. This causes a sudden pressure rise and combustion knock. Since ignition begins late, the fuel may not have sufficient time to burn completely before the exhaust stroke. Because some of the fuel fails to burn completely in the cylinder, the exhaust gas will be smoky.

The cetane number of diesel fuel must be high enough to prevent knock. However, certain conditions tend to reduce compression temperature, which encourage knock. Some of these conditions are low water-jacket temperatures, low atmospheric temperatures, low compression pressures, and light-load operation. For engines operating with these conditions, the fuel must have a sufficiently high cetane number (sufficiently low ignition point) to ignite satisfactorily. The speed of the engine also affects the cetane number of the fuel required. High-speed diesel engines require high cetane fuels. A high-speed diesel engine is any diesel that operates above 1200 rpm. Diesel engines in trucks normally do not turn faster than 3000 rpm. Automobile diesel engines sometimes turn at 5000 rpm. Engines operating at less than 1200 rpm are classed as low-speed and medium-speed diesels.

The cetane number of diesel fuel may vary, just as the allowable range of octane ratings for gasoline allow any of several octane ratings to be sold as "regular." However, the cetane rating must be equal to or above the minimum standard for that grade and season of the year. For example, this means that Number 2 diesel fuel must have a cetane number of at least 40 in summer and 45 in winter.

❂ 7-17 Pour point and cloud point These are important in very cold areas. If the oil becomes too thick to pour freely, it will not go through the fuel injection system easily. Poor engine performance will result. The cloud point is the temperature at which some components of the oil start to turn to a solid. This causes the oil to become cloudy. Cloudy oil can clog the oil filter and cause poor engine performance. In very cold areas, the oil must be specifically refined so it will flow freely and will not cloud at the lowest temperature encountered.

❂ 7-18 Residue after combustion There are two general kinds of residue that may be left after the oil is burned: carbon and ash. The carbon can clog piston rings and exhaust valves. Ash comes from suspended solids or soluble materials. They can cause wear of injector, fuel pump, and piston rings. The oil refining process normally removes these substances.

❂ 7-19 Flash point The flash point is the lowest fuel temperature at which a flame held above the oil will ignite the vapor rising from the oil. The flash point is not important so far as engine operation is concerned. However, it is important in connection with legal requirements and safety precautions involved in fuel handling and storage. An abnormally low flash point usually means the oil is contaminated with some lighter substance such as gasoline.

❂ 7-20 Other diesel oil characteristics There are several other characteristics that the oil refiners must consider when processing crude into gasoline, diesel fuel, and other products. These include sulfur content, water or sediment contamination, corrosion or rust protection, oxidation stability, and heat content.

Sulfur content must be low in diesel fuel to meet legal requirements. Also, high sulfur content can cause corrosion, engine wear, and deposits in the engine.

Water and sediment can get into the oil during fuel transfer from the refiner to the service station and fuel delivery to the vehicle fuel tank. Water and sediment can lead to engine corrosion and plugging of filters and fuel injectors. Therefore, everyone handling diesel fuel must be careful to avoid such contamination.

Oxidation stability is important when diesel oil is stored for any length of time. Oxidation in the fuel can cause gelatin-like deposits which will clog the filter, fuel pump, and injectors.

Heat content is the amount of heat the fuel oil will produce when it is burned. This varies with the composition of the oil and the refining process. The less volatile oil (higher viscosity) has a higher heating content. As a rule, oils with a higher heat content are favored but these oils must not have too low a viscosity or volatility.

❂ 7-21 Diesel fuel additives Several additives are put into diesel fuel to improve it. These include ignition-quality improvers, oxidation inhibitors, biocides, rust preventives, metal deactivators, pour-point depressors, smoke suppressants, and detergent-dispersants.

The purpose of these various additives are the same as for gasoline (❂ 2-14) with the exception of the biocides. These are chemicals that inhibit or prevent the growth of bacteria or fungi in the oil. If such life forms develop in the oil, they can produce serious contamination. This can degrade the oil to a point where it cannot be used in diesel engines.

Chapter 7 review questions

Select the *one* correct, best, or most probable answer to each question. Then check your answers against the correct answers given at the end of the book.

1. A basic difference between the gasoline and diesel engine is that:
   a. air alone is compressed in the diesel
   b. the gasoline engine is heavier
   c. the diesel has a lower compression ratio
   d. fuel alone is compressed in the gasoline engine.
2. In the diesel engine, engine power and speed are controlled by:
   a. the position of the throttle valve
   b. the amount of fuel sprayed into the cylinders
   c. the amount of air-fuel mixture the carburetor delivers
   d. the amount of air taken into the cylinders.
3. In the diesel engine, the fuel is ignited by:
   a. the ignition system
   b. the glow plugs
   c. heat of compression
   d. spark plugs.
4. The type of fuel system most common for automotive diesel engines uses a:
   a. cam-operated in-line plunger pump
   b. cam-operated plungers in the injectors
   c. rotary distributor pump
   d. carburetor.
5. In the rotary distributor pump, the fuel supply pump sends fuel at high pressure to the:
   a. fuel injectors
   b. fuel distributor
   c. fuel return line
   d. fuel filters.
6. The rotary injection pump used on many automotive diesel engines sends:
   a. air-fuel mixture to the cylinders
   b. compressed air to the cylinders
   c. diesel fuel to the cylinders
   d. gasoline to the cylinders.
7. In the rotary injection pump system, the amount of fuel delivered to the cylinders is controlled by:
   a. throttle-valve position
   b. manifold vacuum
   c. governor action
   d. none of the above.
8. The rotary injection pump system uses a speed advance mechanism which controls the position of the:
   a. cam ring
   b. injectors
   c. camshaft
   d. pistons.
9. The purpose of the glow plugs is to:
   a. spark-ignite the air-fuel mixture when it is cold
   b. heat the cylinder block when the engine is cold
   c. heat the precombustion chambers when the engine is cold
   d. none of the above.
10. The reason that pumping the accelerator pedal before or during cranking will not aid in starting a diesel engine is that:
    a. the system has no accelerator pump system
    b. this action overloads the intake manifold with fuel
    c. the accelerator pump works only when the engine is warm
    d. all of the above.

GASOLINE FUEL - SYSTEM TROUBLE DIAGNOSIS

After studying this chapter, and with proper instruction and equipment, you should be able to:

1. Connect the tachometer and measure engine speed.
2. Make a cylinder compression test on a gasoline engine.
3. Make a cylinder leakage test.
4. Connect a vacuum gauge and interpret its readings.
5. Connect the exhaust gas analyzer and measure engine emissions.
6. Connect the fuel pump tester and check the pump for pressure, vacuum, and capacity.
7. Connect the fuel mileage tester and check the fuel consumption.
8. Describe the checks and corrections for at least five fuel system troubles.
9. Explain how to make quick checks of the fuel and ignition systems.
10. Describe the causes of black, blue, and white exhaust smoke.

8-1 Cleanliness and safety The major enemy of good service work is dirt. A trace of dirt in the wrong place in a carburetor or fuel pump will cause problems. For example, dirt in the needle valve seat in the carburetor float bowl may prevent the needle valve from closing tightly. Then the float bowl will overfill and flood the engine or cause high fuel consumption. Similarly, dirt in the idle system or accelerator pump system may produce malfunctioning of the carburetor and poor engine performance. Therefore, when repairing a carburetor, be sure that your hands, the workbench, and your tools are clean. In addition, there are other safety cautions to follow.

CAUTION 1: An air hose is often used to air-dry carburetor and other parts after they have been soaked in carburetor cleaner, washed in solvent, or rinsed in water. When using an air hose, the airstream drives dirt particles at high speed. These particles could get into your eyes and injure them. Always wear safety glasses or goggles to protect your eyes while using the air hose. Also, be very careful where you point the hose to avoid particles striking someone nearby.

CAUTION 2: Gasoline vapor is highly explosive. Use care in handling fuel system parts that are covered or filled with gasoline. When removing a carburetor, fuel pump, filter, or fuel tank, drain it into a container and then wipe up all spilled gasoline with cloths. Put the cloths outside to dry. Never bring an open flame near gasoline! This could result in an explosion or fire.

CAUTION 3: Never remove a carburetor from a hot engine. Any gasoline spilled could ignite from hot engine parts and cause an explosion or fire.

CAUTION 4: Never prime an engine with the air cleaner off by pouring or squirting fuel into the carburetor. When cranked, the engine could backfire and cause an explosion or fire.

CAUTION 5: Do not run the engine with the air cleaner off unless necessary for testing. On some engines, removing the air cleaner can lean out the air-fuel mixture enough to cause backfiring. Also, a malfunctioning carburetor can cause backfire.

8-2 Need for logical procedure Always follow a logical procedure in diagnosing any fuel system trouble, since there are many conditions that could be the cause. The basic complaints that might arise from faulty operation of the fuel system are described in the trouble diagnosis chart later in this chapter. With each complaint the possible causes are listed. When a specific complaint arises, the causes for that complaint should be checked. This procedure saves time and motion which might be wasted in checking things that normally would not cause the complaint.

8-3 Testing instruments A variety of instruments are used to test the fuel system and engine performance. These include fuel mileage testers which measure fuel consumption per mile or kilometer of car

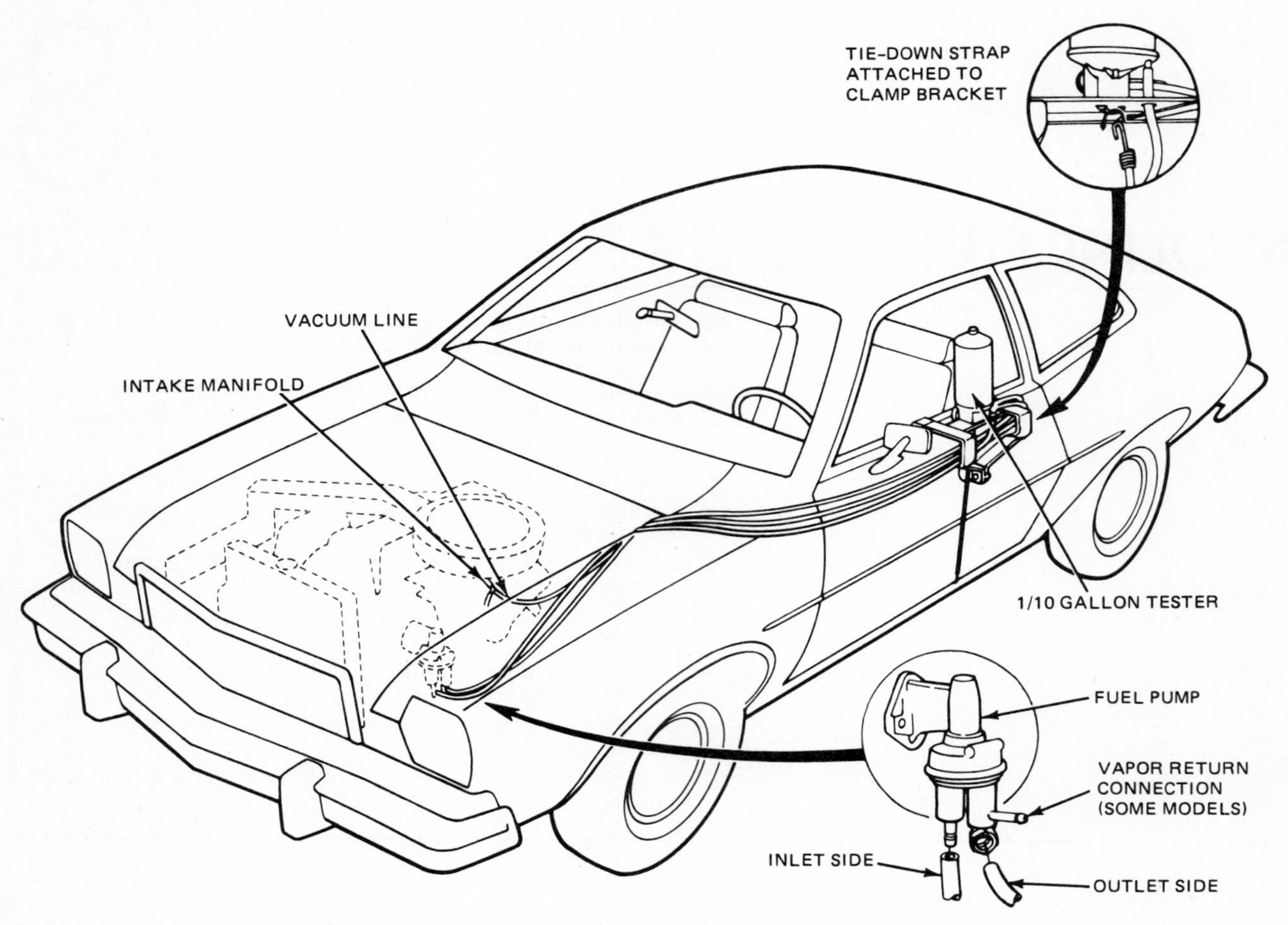

Fig. 8-1 Fuel mileage tester. A small container holding exactly 0.1 gallon (0.4 L) mounted on the driver's door. (*Ford Motor Company*)

travel, exhaust gas analyzers which check the amount of pollutants in the exhaust gas, fuel pump testers, vacuum gauges for measuring intake manifold vacuum, engine cylinder compression testers, cylinder linkage testers, tachometers for checking engine speed, and dynamometers for measuring engine power output.

8-4 Fuel mileage testers A complaint that is sometimes difficult to analyze is poor fuel economy. Many conditions can cause excessive fuel consumption. Sometimes it is necessary to make an accurate measurement of the fuel consumed if the cause of the trouble is hard to find. Fuel mileage testers vary from a fuel meter that computes fuel consumption electronically to a small container of gasoline that is connected into the fuel line. Figure 8-1 shows this type of fuel mileage tester. The test is performed by connecting the fuel mileage tester as shown in Fig. 8-1 and then recording the mileage shown on the odometer. Then the container is filled with gasoline. Operate the car until the gasoline in the container is consumed. Then, take the mileage reading again from the odometer and calculate the miles per gallon.

8-5 Exhaust gas analyzer Several emission control systems have been added to the automobile to reduce the amounts of HC, CO, and NO_x emitted. The amounts of these pollutants in the exhaust gases are measured with an exhaust gas analyzer (Fig. 8-2). This tests the operation of the emission controls, and the fuel and ignition systems.

The major use of the early type exhaust gas analyzer was to adjust the carburetor. Changing the carburetor adjustment changes the amount of HC and CO in the exhaust gas. For example, adjusting the idle-mixture screw can increase the richness of the idle mixture. This increases the amounts of HC and CO in the exhaust gas. One of the emission control steps taken in the modern automobile is to set the idle mixture as lean as possible with a satisfactory idle. This reduces the amount of HC and CO in the exhaust gas.

Today, the infrared type of exhaust gas analyzer (Fig. 8-2) is used to check how well the emission controls on the car are working, in addition to the idle-mixture adjustment. To get a reading, a probe is inserted into the tail pipe of the car. The probe draws out some of the exhaust gas and carries it through the analyzer. Two meters on the face of the analyzer (Fig. 8-3) report how much HC and CO are in the exhaust gas. Federal, state, and local laws set maximum legal limits on the amounts of HC and CO permitted in the exhaust gas.

To use the exhaust gas analyzer, the engine should be at normal operating temperature. Then check the

Fig. 8-2 Using an exhaust-gas analyzer to check the exhaust emissions from the tail pipe of an automobile engine.

vehicle exhaust system to be sure that it is free of leaks. A quick check is to block the tail pipe with the engine idling and listen for exhaust leaks anywhere in the system. If no leaks are heard, insert the exhaust gas pickup probe at least 18 inches [457 mm] into the tail pipe. Be sure the probe is securely in place. If the vehicle is equipped with a dual exhaust system, insert the probe into the side opposite the exhaust manifold heat-control valve.

To measure CO, run the engine at fast idle (1500 to 2000 rpm) for about 30 seconds. This will clear any excess fuel out of the engine. Then run the engine at the specified idle speed. Adjust the idle-speed screw as necessary to obtain the specified engine rpm. Allow 10 seconds for the meter to stabilize after each adjustment before taking a reading.

Read CO at idle on the CO meter. Be sure you are reading the correct scale on the meter. Record the reading, so you will have a record of it. Then run the engine at 2000 rpm and read the CO meter again.

A good CO reading is within specifications at idle and the same or lower at 2500 rpm. A CO reading that is higher than specified at idle or that increases at 2500 rpm is bad. In general, the higher the CO reading, the richer the air-fuel mixture. For tuneup testing, vehicles without exhaust emission controls should have less than 5 percent CO at idle. Most vehicles with exhaust emission controls (without catalytic converter) should have less than 2.5 percent CO at idle. When the engine is equipped with a catalytic converter, the reading may be in the range of 0 to 0.05 percent CO when the converter is hot.

To measure HC, run the engine at fast idle (1500 to 2000 rpm) for about 30 seconds. This will clear any excess fuel out of the engine. Then run the engine at the specified idle speed. Read HC at idle on the HC meter. Be sure you are reading the correct scale on the meter. Record the reading. Then run the engine at 2500 rpm and read the HC meter again.

A good HC reading is within specifications at idle,

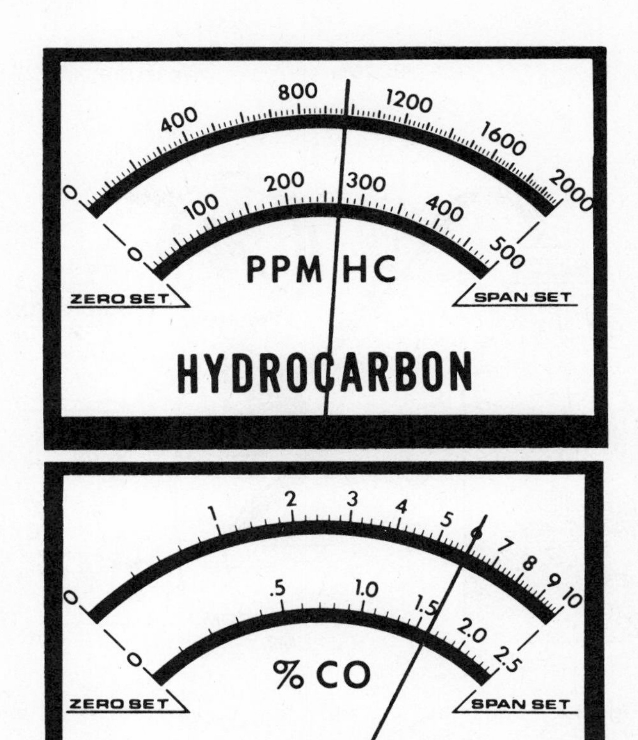

Fig. 8-3 HC and CO meter faces. (*Sun Electric Corporation*)

and the same or less at 2500 rpm. An HC reading that is higher than specified at idle or that increases at 2500 rpm is bad. In general, the higher the HC reading, the more unburned air-fuel mixture is passing out the tail pipe. For tuneup testing, most vehicles with exhaust emission controls should have fewer than 300 ppm (parts per million) HC at idle. Vehicles without exhaust emission controls should have fewer than 500 ppm HC at idle.

A different kind of tester is required for NO_x, but it works in the same general way. It draws exhaust gas from the tail pipe and runs the gas through the analyzer. The finding is reported in grams per mile. Generally, No_x testers are available only in testing laboratories. They are not used in automotive service shops.

8-6 Fuel-pump pressure test The pressure at which the fuel pump delivers fuel to the carburetor must be within definite limits. If the fuel pump pressure is too low, insufficient fuel will be delivered, and faulty engine performance will result. The air-fuel mixture will lean out excessively at high speeds and on acceleration. If the pressure is too high, flooding may result. The mixture will be too rich, causing the engine to be sluggish. An overrich mixture will also cause carbon deposits in the combustion chambers and on valves and piston rings. Also, crankcase oil dilution and rapid wear of engine parts may result.

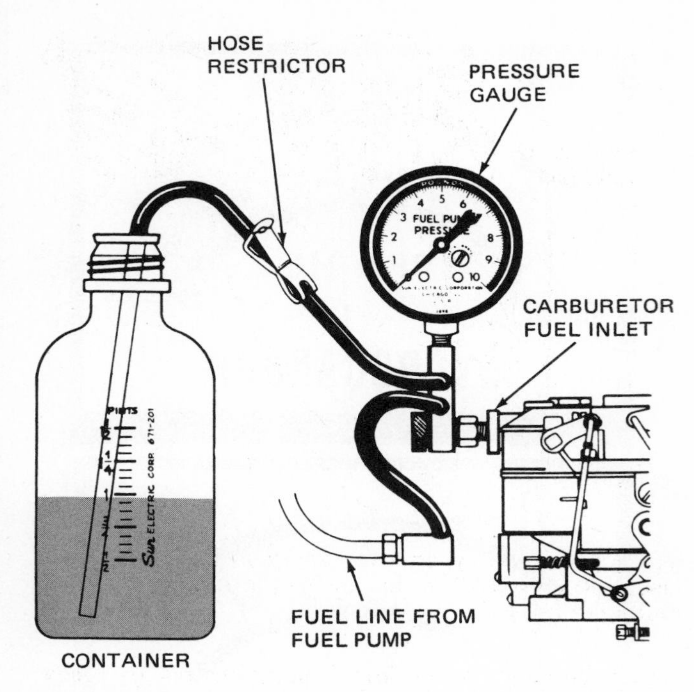

Fig. 8-4 Fuel pump pressure and capacity tests. (*Sun Electric Corporation*)

Fuel-pump static pressure is tested with a pressure gauge connected to the outlet of the fuel pump. The engine is run at approximately 2500 rpm on the fuel in the carburetor while the static fuel pressure is checked. In the flow pressure test, the gauge is connected into the fuel line between the pump and the carburetor with a T-fitting (Fig. 8-4). The engine is idled with the pump delivering fuel to the carburetor in the normal manner. Specifications vary considerably from one type of pump to another. In general, they will specify from 4 to 8 psi [28 to 55 kPa] static pressure and about 25 percent less flow pressure.

⚙ 8-7 Fuel-pump capacity test The fuel-pump capacity tester is a container that is filled to measure the amount of fuel the pump can deliver in a given time. Figure 8-4 shows the fuel container. Install a T-fitting into the fuel line at the carburetor and bleed off a portion of the fuel passing in the line. The amount that can be bled off with the engine running determines the capacity and operating condition of the pump.

A setup to test both the pressure and the capacity of a fuel pump is shown in Fig. 8-4. To make the pressure test (⚙ 8-6), start the engine, and let it idle. Open the hose restrictor momentarily, to vent any air trapped in the fuel system. Then close the hose restrictor. As soon as the pressure gauge needle is steady, read the fuel pump pressure. Some fuel pumps have a vapor-return line from the fuel pump to the fuel tank. The vapor-return line must be squeezed closed to check fuel pump capacity.

Check the fuel pump specifications in the shop manual. If fuel pump pressure is not within specifications, the pump is defective. If fuel pump pressure is within specifications, perform the fuel-pump capacity test. With the engine running at idle, open the hose restrictor (Fig. 8-4). Allow the fuel to discharge into the container for 30 seconds. On most engines in full-size American-built cars, the fuel pump should deliver at least 1 pint [0.47 L] of fuel in 30 seconds or less.

⚙ 8-8 Fuel-pump vacuum test To make a fuel-pump vacuum test, attach the fuel pump tester to the vacuum side of the fuel pump. Disconnect the fuel line at the carburetor. Idle the engine for a few seconds, while it runs on the gasoline in the float bowl. Read the fuel pump vacuum on the gauge of the fuel pump tester. A typical fuel pump should have at least 10 inches [254 mm] Hg (mercury) of vacuum while the engine is idling.

⚙ 8-9 Compression tester The cylinder compression tester measures the ability of the cylinders to hold compression. Pressure, operating on a diaphragm in the tester, causes the needle on the face of the tester to move around to indicate the pressure being applied. Figure 8-5 shows a compression tester used to measure the pressure in an engine cylinder.

To use the tester, first remove all the spark plugs. A recommended way to do this is to disconnect the wires and loosen the plugs one turn. Next, reconnect the wires and start the engine. Then, run the engine for a few seconds at 1000 rpm. The combustion gases will blow out of the plug well any dirt that could fall into the cylinder when the spark plugs are removed. The gases will also blow out of the combustion chamber any loosened carbon that is caked around the exposed threaded end of the spark plug. This procedure prevents carbon and dirt particles from lodging under a valve and holding the valve open during the compression test.

After removing the plugs, block the carburetor throttle valve wide open. This is to make sure that the maximum amount of air will get into the cylinders.

Next, screw the compression tester adapter into the spark-plug hole of No. 1 cylinder (Fig. 8-5). To protect the coil from high voltage, follow the manufacturer's recommended procedure to deactivate the ignition system. Then hold the throttle wide open and operate the starting motor to crank the engine through four compression strokes (eight crankshaft revolutions). The needle will move around to show the maximum compression pressure in the cylinder.

Record this reading. Test the other cylinders the same way. For most engines, the compression pressure of the lowest cylinder must be within 75 percent of the highest.

⚙ 8-10 Results of the compression test If everything is normal in the cylinder, the compression pressure builds up quickly and evenly. If there is leakage past the piston rings, the compression pressure is low on the first strokes but will tend to build up toward normal with later strokes. However, it does not reach normal and the pressure is lost after cranking stops.

When the results of the compression test show that compression pressure is low, there is leakage past the piston rings, valves, or cylinder head gasket. To correct

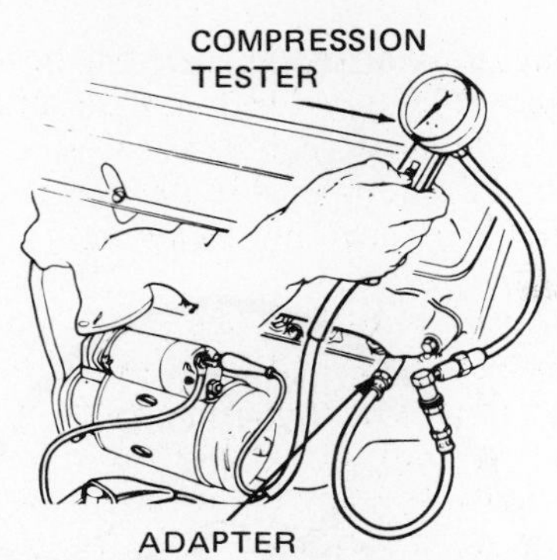

Fig. 8-5 Using a cylinder compression tester. (*Sun Electric Corporation*)

the trouble, the cylinder head must be removed so that the engine parts can be inspected.

However, before you do this, make one more test to help pinpoint the trouble. Squirt about one tablespoon [15 cc] of engine oil through the spark-plug hole into the cylinder. Then retest the compression. If the pressure increases to a more normal figure, the low compression is a result of leakage past the piston rings. Adding the oil helps seal the rings temporarily so that they can hold the compression pressure better. The trouble is caused by worn piston rings, a worn cylinder wall, or a worn piston. The trouble could also be caused by rings that are broken or stuck in the piston-ring grooves.

If adding oil does not increase the compression pressure, the leakage is probably past the valves. This could be caused by:

1. Broken valve springs
2. Incorrect valve adjustment
3. Sticking valves
4. Worn or burned valves
5. Worn or burned valve seats
6. Worn camshaft lobes
7. Dished or worn valve lifters

It may also be that the cylinder head gasket is "blown." This means the gasket has burned away so that compression pressure is leaking between the cylinder head and the cylinder block. Low compression between two adjacent cylinders is probably caused by the head gasket blowing between the cylinders.

Whatever the cause—rings, pistons, cylinder walls, valves, or gasket—the cylinder head must be removed so that the trouble can be fixed. The exception would be if the trouble is caused by incorrect valve adjustment. If adjusting the valves corrects the problem, the head would not have to be removed.

8-11 Cylinder leakage tester The cylinder leakage tester does about the same job as the compression tester but in a different way. It applies air pressure to the cylinder with the piston at top dead center on the compression stroke. In this position, both valves are closed. Very little air should escape from the combustion chamber. Figure 8-6 shows a cylinder leakage tester. Figure 8-7 shows how the tester is connected to

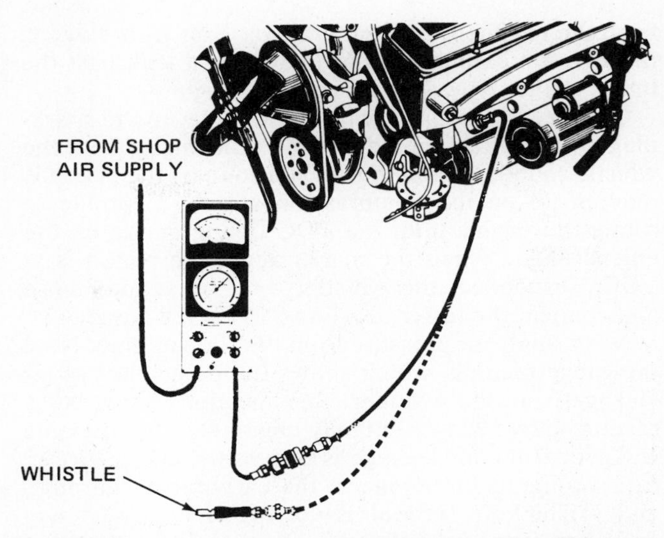

Fig. 8-6 Using a cylinder leakage tester. The whistle is used to locate TDC in the No. 1 cylinder. (*Sun Electric Corporation*)

an engine cylinder and how it pinpoints places where leakage can occur.

To use the cylinder leakage tester, first remove all spark plugs. Then remove the air cleaner, oil filler cap, and radiator cap. Set the throttle wide open, and fill the radiator to the proper level. Remove the PCV valve

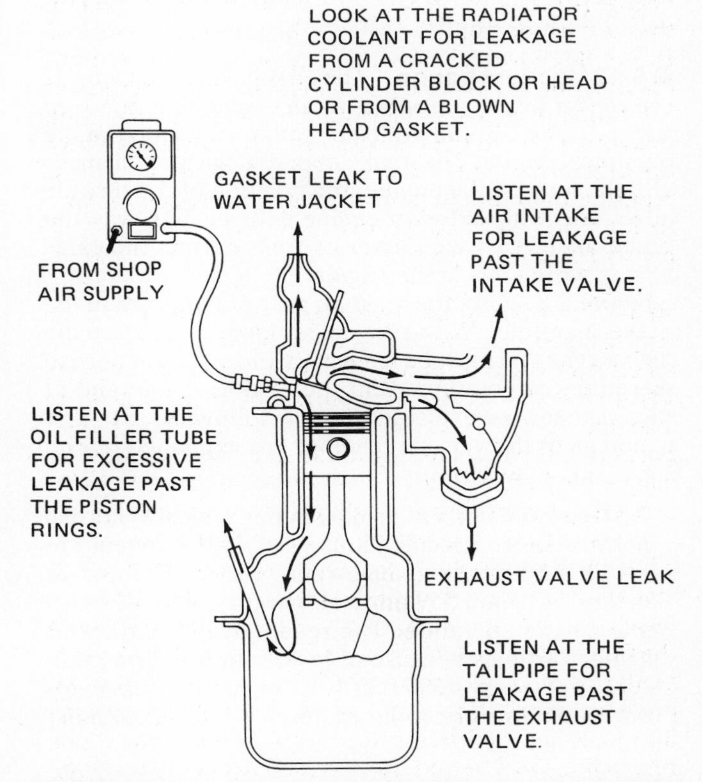

Fig. 8-7 How the cylinder leakage tester works. It applies air pressure to the cylinder through the spark-plug hole with the piston at TDC and both valves closed. Places where air is leaking can then be pinpointed. (*Sun Electric Corporation*)

from the engine to prevent crankcase air from flowing through the carburetor and indicating a leak past the intake valve. Now you are ready to begin.

Connect the adapter, with the whistle, to the spark-plug hole of No. 1 cylinder. Crank the engine until the whistle sounds. When the whistle sounds, the piston is moving up on the compression stroke. Continue to rotate the engine until the TDC timing marks on the engine align. When the marks align, the piston is at TDC. Disconnect the whistle from the adapter hose and connect the tester, as shown in Figs. 8-6 and 8-7.

Next, apply air pressure from the shop air line. Note the gauge reading, which shows the percentage of air leakage from the cylinder. Specifications vary, but a reading above 20 percent may mean there is excessive leakage. If the air leakage is excessive, check further by listening at three places: the carburetor, tail pipe, and oil filler hole. If the air is blowing out of an adjoining cylinder spark-plug hole, it means that the head gasket is blown between the cylinders.

Figure 8-7 shows what it means if you can hear air escaping at any of the three listening places. Also, if air bubbles up through the radiator, then the trouble is a blown cylinder-head gasket or a cracked cylinder head. These conditions allow leakage from the cylinder to the cooling system.

Check the other cylinders in the same manner. A test light connected across the contact points can be used to position the next cylinder in the firing order at TDC. When you use the tester, follow the instructions that explain how to use it.

8-12 Vacuum gauge The engine vacuum gauge is a tester for locating troubles in an engine that does not run as well as it should. This gauge measures intake manifold vacuum. The intake manifold vacuum changes with the load on the engine, the position of the throttle valve, and with different engine defects. The way the intake manifold vacuum varies from normal indicates what is wrong inside the engine.

Figure 8-8 shows the vacuum gauge connected to the intake manifold. With the gauge connected, start the engine. The test must be made with the engine at normal operating temperature. Run the engine at idle and at other speeds as explained in the following list. The meanings of the various readings are explained in Fig. 8-9.

1. A steady and fairly high idle reading indicates normal performance. Specifications vary with different engines, but a reading somewhere between 17 and 22 inches [432 and 559 mm] of mercury indicates normal engine operation. The reading will be lower at higher altitudes because of lower atmospheric pressure. For every 1000 feet [305 m] above sea level, the reading will be reduced about 1 inch [25.4 mm] of mercury.

NOTE: *Inches* or *millimeters of mercury* refers to the way the vacuum is measured. There is no mercury in the gauge. However, the reading compares with the changes that a vacuum would produce on a column of mercury in a barometer.

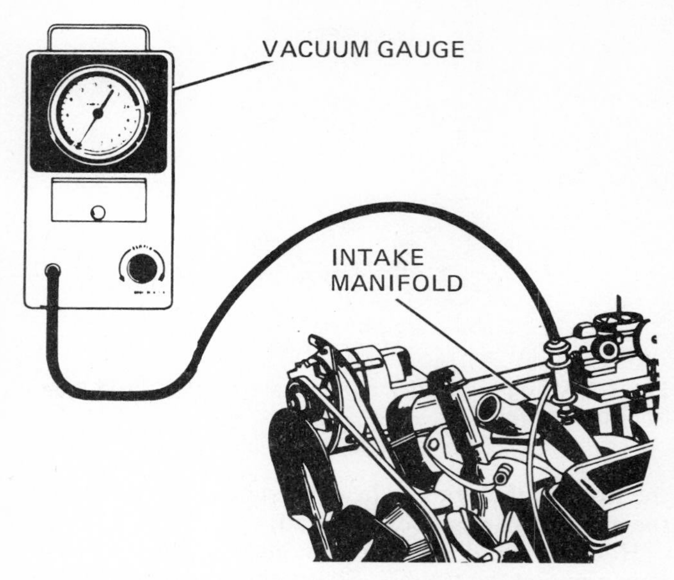

Fig. 8-8 Vacuum gauge connected to the intake manifold for a test of intake-manifold vacuum. (*Sun Electric Corporation*)

2. A steady and low idle reading indicates late ignition or valve timing, or possibly leakage around the pistons. Leakage around pistons—excessive blowby—could be caused by worn or stuck piston rings, worn cylinder walls, or worn pistons. Each of these conditions reduces engine power. With reduced power, the engine does not "pull" as much vacuum.
3. A very low idle reading indicates a leaky intake manifold or carburetor gasket, or possibly leakage around the carburetor throttle shaft. Air leakage into the manifold reduces the vacuum and engine power.

NOTE: Some engines, with high-lift cams and more valve overlap, are likely to have a lower and more uneven intake manifold vacuum. Also, certain emission control systems lower the intake manifold vacuum.

4. Back-and-forth movement of the needle that increases with engine speed indicates weak valve springs.
5. Gradual falling back of the needle toward zero with the engine idling indicates a clogged exhaust line.
6. Regular dropping back of the needle indicates that a valve is sticking open or a spark plug is not firing.
7. Irregular dropping back of the needle indicates that valves are sticking only part of the time.
8. Floating motion or slow back-and-forth movement of the needle indicates that the air-fuel mixture is too rich.

A test can be made for loss of compression resulting from leakage around the pistons. This condition would be the result of stuck or worn piston rings, worn cylinder walls, or worn pistons. Race the engine for a few seconds and then quickly release the throttle. The needle should swing around to 23 to 25 inches [584 to 635 mm] as the throttle closes, indicating good compression. If the needle fails to swing around this

	READING	DIAGNOSIS
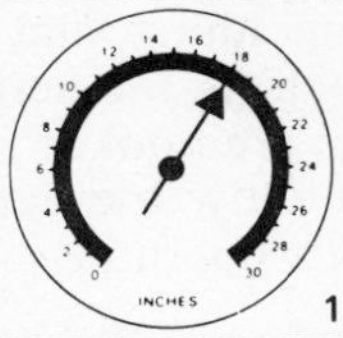 1	Average and steady at 17-21.	Everything is normal.
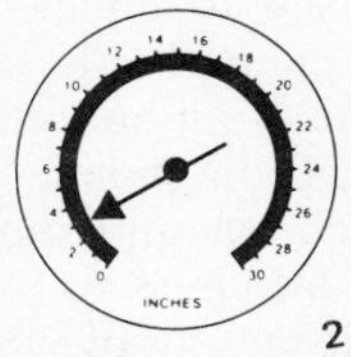 2	Extremely low reading—needle holds steady.	Air leak at the intake manifold or throttle body; incorrect timing.
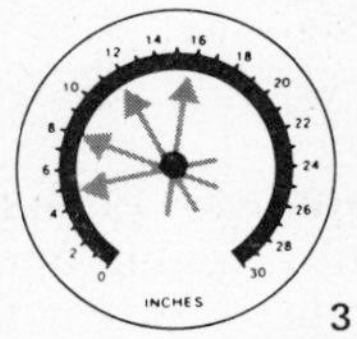 3	Needle fluctuates between high and low reading.	Blown head gasket between two side-by-side cylinders. (Check with compression test.)
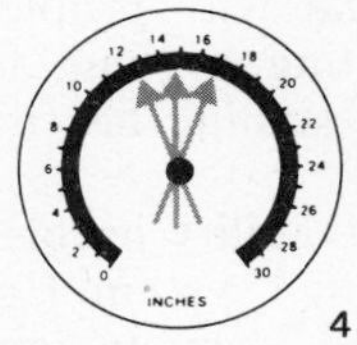 4	Needle fluctuates very slowly, ranging 4 or 5 points.	Carburetor needs adjustment, spark-plug gap too narrow, sticking valves.
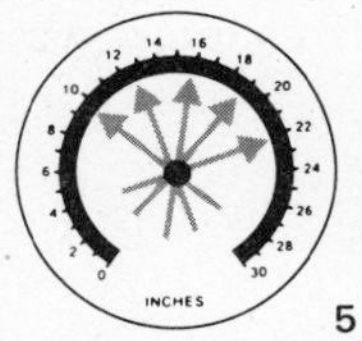 5	Needle fluctuates rapidly at idle—steadies as RPM is increased.	Worn valve guides.
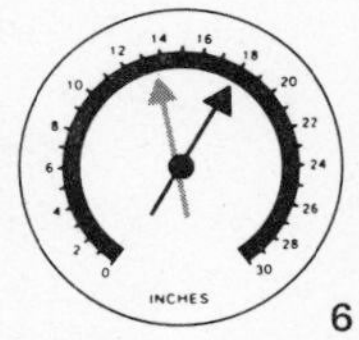 6	Needle drops to low reading, returns to normal, drops back, etc., at a regular interval.	Burned or leaking valve.
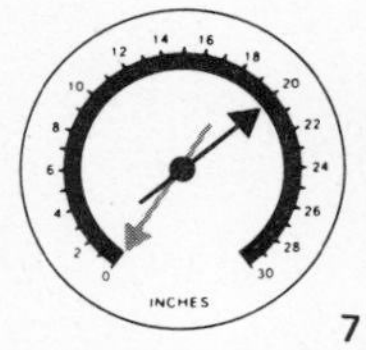 7	Needle drops to zero as engine RPM is increased.	Restricted exhaust system.
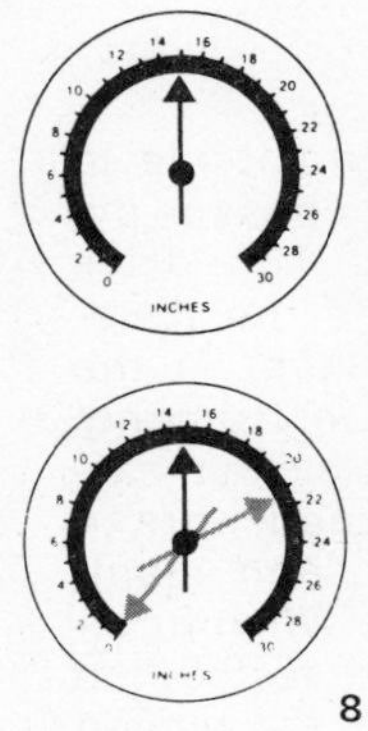 8	Needle holds steady at 12 to 16—drops to 0 and back to about 21 as you engage and release the throttle.	Leaking piston rings. (Check with compression test.)

Fig. 8-9 Vacuum-gauge readings and their meanings.

far, there is loss of compression. Further checks should be made.

NOTE: This test does not apply to engines equipped with a deceleration valve as part of the emission-control system.

8-13 Tachometer The tachometer measures engine speed in revolutions per minute of the crankshaft. It is a necessary instrument because the idle speed must be adjusted to a specific rpm. Also, many tests must be made at specified engine speeds. The tachometer is connected to the ignition system and operates electrically.

The tachometer measures the number of times the primary circuit is interrupted and translates this into engine rpm. Usually, the instrument has a selector knob that can be turned to 4, 6, or 8, according to the number of cylinders in the engine being tested. Figure 8-10 shows a tachometer connected to an engine.

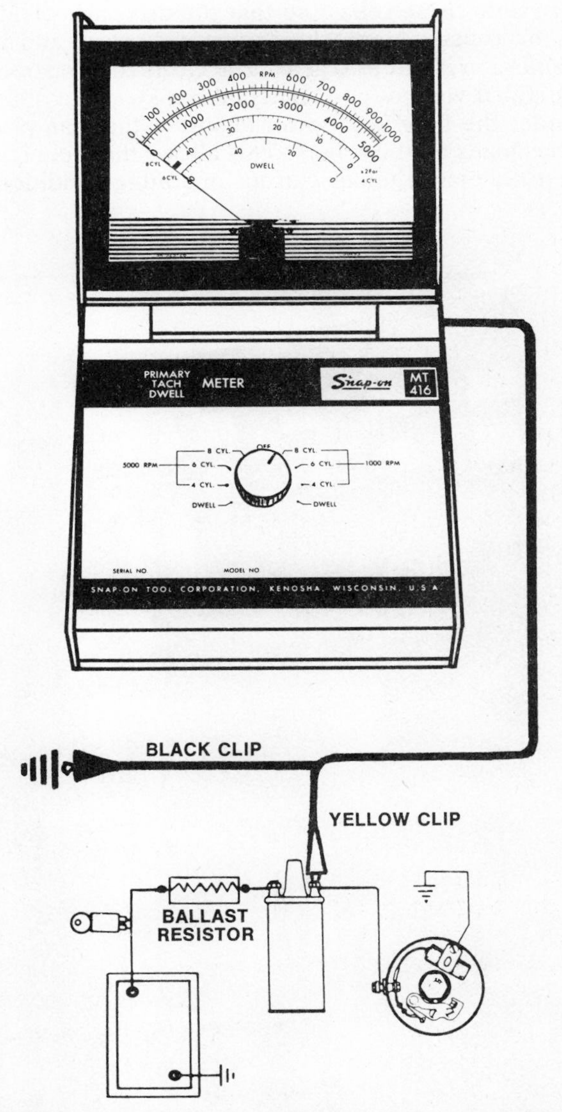

Fig. 8-10 Tachometer connected to an engine. (*Snap-on Tools Corporation*)

Some cars have tachometers mounted on the instrument panel. Their purpose is to keep the driver informed about how fast the engine is turning. Knowing this, the driver can keep the rpm within the range at which the engine develops maximum torque. This enables the driver to get the best performance from the engine. Many of these tachometers have a red line at the top rpm on the dial. The red line marks the danger point for engine speed. The driver should keep the engine below this speed.

Other in-car tachometers are mechanical instead of electrical. They are driven off a gear on the ignition-distributor shaft and operate somewhat like the speedometer.

8-14 Dynamometer The chassis dynamometer can test the engine power output under various operating conditions. It can duplicate any kind of road test at any load or speed desired. The part of the dynamometer that you can see consists of two heavy rollers mounted at about floor level (Fig. 8-11). The car is driven onto these rollers so that the drive wheels can spin the rollers. Next, the engine is started, and the transmission is put into gear. The car is then operated as though it were out on an actual road test.

Under the floor is a power absorber that can place various loads on the rollers. This allows the technician to test the engine under various operating conditions. For example, the technician can find out how the engine would do during acceleration, cruising, idling, and deceleration. The test instruments, such as the scope, dwell-tach meter, and vacuum gauge, are hooked into the engine. These instruments then show the actual condition of the engine during various operating conditions.

The dynamometer can also be used to check the transmission and the differential. For example, the shift points and other operating conditions of an automatic transmission can be checked. Special diagnostic dynamometers are available in some schools and shops. These units have many instruments attached to them and have motored rollers that permit testing of wheel alignment, suspension, brakes, and steering.

8-15 Analyzing fuel system troubles Fuel system troubles usually show up as faulty engine operation, or what is called poor "drivability." A car that runs well has good drivability. A car with problems such as poor acceleration, hard starting, missing, loss of power, stumble, hesitation, and stalling has poor drivability.

The following sections discuss drivability problems that could result from troubles in the fuel system. However, these problems could be caused by many conditions outside the fuel system. In this chapter, we are

Fig. 8-11 Automobile in place on a chassis dynamometer. The drive wheels rotate the dynamometer rollers. At the same time, instruments on the console can measure car speed, brake horsepower, and manifold vacuum. When front-drive cars are tested, the front wheels are placed on the rollers. (*Sun Electric Corporation*)

concentrating on drivability problems that might arise from faulty conditions in the fuel system.

⚙ 8-16 Troubleshooting the fuel system The tracing of trouble in the fuel system usually is a logical procedure. Fuel system troubles fall into several classifications that require specific corrections. However, there is sometimes a question as to whether the cause of complaint lies in the fuel system or in some other engine component. Therefore, the real problem is often to isolate the trouble in the improperly operating component. The trouble sometimes may be made more puzzling because it could appear as the result of any one of several conditions. For example, suppose the power valve in the carburetor is stuck open. This would produce an excessively rich mixture for all running conditions except full-power, wide-open-throttle operation. This, in turn, would cause excessive fuel consumption, spark-plug fouling, and carbon deposits that might cause defective piston ring and valve action.

The chart that follows in ⚙ 8-17 lists the various troubles that might be blamed on the fuel system, together with their possible causes, checks to be made, and corrections needed.

⚙ 8-17 Fuel system trouble diagnosis chart Most fuel system troubles can be listed under a few headings: excessive fuel consumption, poor acceleration, lack of power and high-speed performance, poor idle, engine not starting, hard starting, slow warm-up, stalling, smoky exhaust, and backfiring. The chart that follows lists possible causes of each of these troubles. Included in the chart are references to later sections which contain fuller explanations of the way to locate and eliminate the troubles.

NOTE: The troubles and possible causes are not listed in the chart in the order of frequency of occurrence. Item 1 (or item *a* under *Possible Cause*) does not necessarily occur more frequently than item 2 (item *b*). Generally, fuel system troubles and possible causes are listed first in the chart. However, in many cases, other automotive components are more likely to have caused the troubles listed.

Fuel System Trouble Diagnosis Chart

(See ⚙ 8-18 to 8-30 for detailed explanations of trouble causes and corrections listed below).

COMPLAINT	POSSIBLE CAUSE	CHECK OR CORRECTION
1. Engine cranks normally but will not start (⚙ 8-18)	a. Improper starting procedure	Use procedure outlined in owner's manual.
	b. Engine flooded	Crank with throttle open to clear engine. Use correct starting procedure.
	c. Choke valve not operating properly.	Adjust, free linkages.
	d. No fuel in carburetor	Check fuel pump, lines, filter for plugging, float for sticking.
	e. Ignition defective	Check for spark.
	f. Carburetor jets clogged	Clean.
	g. Air leaks into intake manifold or carburetor	Replace gaskets, tighten attaching nuts or screws.
2. Hard starting with engine warm (⚙ 8-19)	a. Choke valve closed	Open; adjust or repair.
	b. Manifold heat-control stuck closed	Open; free valve.
	c. Vapor lock	Use correct fuel or shield fuel line or pump.
	d. Engine parts binding	Repair engine.
3. Engine stalls cold or as it warms up (⚙ 8-20)	a. Choke valve closed	Open choke valve; free or repair automatic choke.
	b. Fuel not getting to or through carburetor	Check fuel pump, lines, filter, float-and-idle systems.
	c. Manifold heat-control valve stuck	Free valve.
	d. Engine overheats	Check cooling system (Chap. 19), ignition timing.
	e. Engine idling speed set too low	Increase idling speed to specified rpm.
	f. Malfunctioning PCV valve	Replace.
4. Engine stalls after idling or slow-speed drive (⚙ 8-20)	a. Defective fuel pump	Repair or replace fuel pump.
	b. Overheating	Check cooling system (Chap. 19), ignition timing.
	c. High float level	Adjust.
	d. Idle adjustment incorrect	Adjust.
	e. Malfunctioning PCV valve	Replace.
5. Engine stalls after high-speed drive (⚙ 8-20)	a. Vapor lock	Use different fuel or shield fuel line.
	b. Carburetor antipercolator valve defective	Check and repair.
	c. Engine overheats	Check cooling system (Chap. 19), ignition timing.
	d. Malfunctioning PCV valve	Replace.

Fuel System Trouble Diagnosis Chart *(Continued)*

(See ❋ 8-18 to 8-30 for detailed explanations of trouble causes and corrections listed below).

COMPLAINT	POSSIBLE CAUSE	CHECK OR CORRECTION
6. Engine backfires (❋ 8-21)	a. Excessively rich or lean mixture	Repair or readjust fuel pump or carburetor.
	b. Overheating of engine	Check cooling system (Chap. 19), ignition timing.
	c. Engine conditions such as excessive carbon, hot valves, overheating	Repair engine.
	d. Ignition timing incorrect	Retime.
	e. Spark plugs of wrong heat range	Install correct plugs.
7. Slow engine warm-up (❋ 8-22)	a. Choke valve open	Adjust or repair.
	b. Manifold heat-control valve stuck open	Close; free valve.
	c. Cooling system thermostat stuck open	Free; replace if necessary.
8. Poor idle (❋ 8-23)	a. Idle mixture or speed not adjusted	Readjust
	b. Automatic level control compressor operating	Check vacuum-regulator valve.
	c. PCV valve stuck	Replace.
	d. Other causes listed under item 11	
9. Engine hesitates on acceleration (❋ 8-24)	a. Loose or broken vacuum hose	Tighten, replace hose.
	b. Accelerator pump faulty or out of adjustment	Adjust, replace.
	c. Power system faulty	Clean, replace defective parts.
	d. Float level low	Adjust.
	e. EGR valve stuck open	Check, clean valve.
	f. Distributor spark advance faulty	Check, replace defective parts.
10. Engine runs but misses (❋ 8-25)	a. Fuel pump erratic in operation	Repair or replace.
	b. Carburetor jets or lines clogged or worn	Clean or replace.
	c. Fuel level not correct in float bowl	Adjust float; clean needle valve.
	d. Ignition system defects, such as incorrect timing or defective plugs, coil, points, cap, condenser, wiring	Check ignition system.
	e. Clogged exhaust	Check tail pipe, muffler; eliminate clogging.
	f. Engine overheating	Check cooling system (Chap. 19), ignition timing.
	g. Engine conditions, such as valves sticking, loss of compression, defective rings, etc.	Check engine.
11. Engine sluggish and lacks power, acceleration, or high-speed performance (❋ 8-26)	a. Accelerator pump malfunctioning	Adjust; free; repair.
	b. Power step-up on metering rod not clearing jet	Free or adjust.
	c. Power piston or valve stuck	Free.
	d. Low float level	Adjust.
	e. Dirt in filters or in line, or clogged fuel tank vent	Clean; install correct cap.
	f. Choke stuck or not operating	Adjust or repair.
	g. Air leaks around carburetor	Replace gaskets; tighten nuts or bolts.
	h. Loose or broken vacuum hose	Tighten, replace hose.
	i. Antipercolator valve stuck	Free; adjust.
	j. Manifold heat-control valve stuck	Free.
	k. Throttle valves or secondary valves not fully opening	Adjust linkage.
	l. Rich mixture because of worn jets, high float level, stuck choke, clogged air cleaner	Adjust; repair; clean; replace worn jets.
	m. Vapor lock	Use different fuel or shield fuel line or pump.
	n. Fuel pump defective	Service or replace.
	o. Clogged exhaust	Clean.
	p. Ignition defective	Check timing, coil, plugs, distributor, condenser, wiring.

Fuel System Trouble Diagnosis Chart *(Continued)*

(See ⚙ 8-18 to 8-30 for detailed explanations of trouble causes and corrections listed below).

COMPLAINT	POSSIBLE CAUSE	CHECK OR CORRECTION
	q. Loss of compression	Check engine compression; repair engine.
	r. Excessive carbon in engine	Clean out.
	s. Defective valve action	Check compression; repair engine.
	t. Heavy engine oil	Use lighter oil.
	u. Cooling system not operating properly	Check system, flush system (Chap. 19).
	v. Engine overheats	Check cooling system (Chap. 19), ignition timing.
	w. Excessive rolling resistance from low tires, dragging brakes, wheel misalignment, etc.	Correct the defect causing rolling resistance.
	x. Clutch slippage or excessive friction in power train	Adjust or repair.
	y. Transmission not downshifting or torque converter defective	Check transmission.
12. Excessive fuel consumption (⚙ 8-27)	a. Hard driving	Drive more reasonably.
	b. High speed	Drive more slowly.
	c. Short-run and start-and-stop operation	Make longer runs.
	d. Excessive fuel-pump pressure or pump leakage	Reduce pressure; repair pump.
	e. Choke not opened properly	Open, repair or replace automatic choke.
	f. Clogged air cleaner	Clean.
	g. High carburetor float level or float leaking	Adjust or replace float.
	h. Stuck or dirty float needle valve	Free and clean or replace.
	i. Worn carburetor jets	Replace.
	j. Stuck metering rod or power piston	Free.
	k. Too-rich or too-fast idle	Readjust.
	l. Stuck accelerator-pump check valve	Free.
	m. Carburetor leaks	Replace damaged parts; tighten loose couplings, jets, etc.
	n. Faulty ignition	Check coil, condenser, plugs, contact points, wiring.
	o. Loss of engine compression	Check compression; repair engine.
	p. Defective valve action	Check compression; repair engine.
	q. Excessive rolling resistance from low tires, dragging brakes, wheel misalignment, etc.	Correct cause of rolling resistance.
	r. Clutch slipping	Adjust or repair clutch.
	s. Transmission slipping or not upshifting	Adjust or repair.
13. Smoky exhaust (⚙ 8-28)	a. Blue smoke: excessive oil consumption	Oil burning in combustion chamber because of engine trouble.
	b. Black smoke: excessively rich mixture	See item 8 and ⚙ 8-28.
	c. White smoke: steam in exhaust	Tighten cylinder head bolts to eliminate coolant leakage.
14. Engine run-on or dieseling (⚙ 8-29)	a. Idle-stop solenoid adjustment incorrect	Adjust; replace solenoid.
	b. Engine overheating	Check cooling system (Chap. 19).
	c. Hot spots in cylinders	Check engine.
	d. Loose or broken vacuum hose	Tighten; replace hose.
	e. Incorrect timing	Reset.
15. Too much HC and CO in exhaust (⚙ 8-30)	a. Carburetor troubles	Check choke, float level, idle-mixture screw.
	b. Ignition troubles	Check for miss, timing.
	c. Faulty air injection	Check air-injection system.
	d. Defective TCS system or catalytic converter	Check.

⚙ 8-18 Engine cranks normally but will not start Today engines require that the starting procedure as outlined in the owner's manual be followed carefully to get normal starts. If the procedure is not followed, the air-fuel mixture could be too lean or too rich to support combustion and the engine will not start.

If the engine is flooded because of an improper starting procedure, crank the engine with the throttle wide open. This will clear the excess fuel from the engine and permit starting.

Conditions in the fuel system that could prevent starting include an improperly operating choke, no fuel, carburetor jets or passages clogged, or air leaks into the intake manifold or carburetor.

A defective ignition system that does not deliver the sparks to the spark plugs, or delivers them at the wrong time (ignition out of time) could prevent starting.

The ignition system can be quickly checked by disconnecting the lead from one spark plug (see Fig. 8-12). With insulated pliers, hold the lead clip about $^{3}/_{16}$ inch (5 mm) from the engine block and crank the engine. If a good spark occurs, the ignition system is probably operating normally, although it could be out of time.

If the ignition system operates normally, remove the carburetor air cleaner to check the accelerator pump system. With the engine off, open the throttle quickly and see whether or not the accelerator pump system is delivering fuel to the carburetor air horn. If it is not, the carburetor probably is not getting fuel from the fuel pump, because of a defective fuel pump, a clogged fuel filter or fuel line, or an empty fuel tank. Also, if the car is equipped with a fuel-vapor recovery system, installing the wrong fuel tank cap could prevent fuel delivery. The tanks of cars equipped with fuel-vapor recovery systems have special vacuum-pressure tank caps. The tanks are sealed. When fuel is withdrawn, the vacuum valve on the cap opens to admit air to the tank. If this cap is replaced with a different cap which does not have the valves, no air can enter to replace fuel withdrawn. The result is that a vacuum develops in the tank which can prevent delivery of fuel through the fuel pump to the carburetor.

If the accelerator pump system does deliver fuel to the carburetor air horn, try starting again. If the engine starts when primed by the pump operation, but then stalls, the carburetor idle or main metering system probably is clogged and not functioning normally. If the engine still does not start when primed, there is probably some problem in the engine which prevents starting.

CAUTION: Put the air cleaner back on after priming the engine with the accelerator pump before you attempt to start. If you fail to do this, the engine could backfire through the carburetor. This could burn you and also cause a fire. Any fuel in or around the carburetor could be ignited by a backfire.

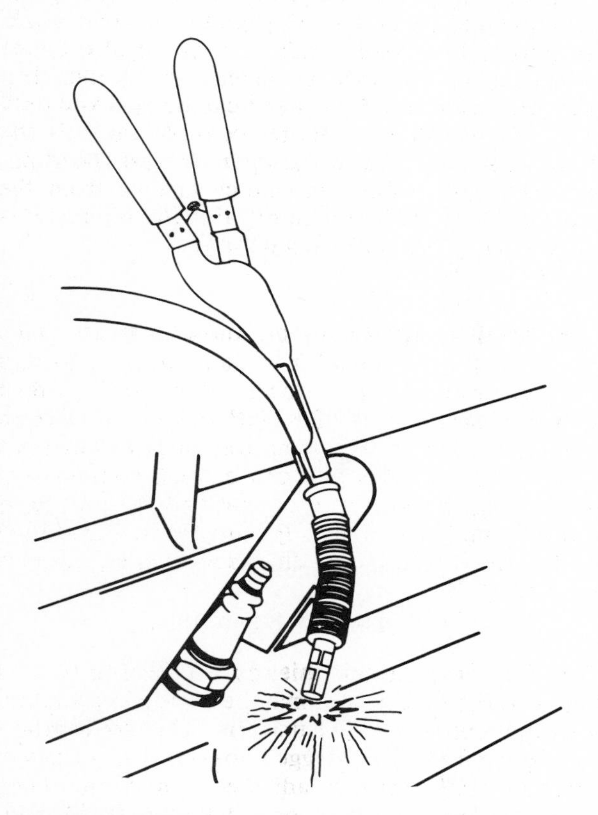

Fig. 8-12 To make a spark test, disconnect a spark-plug wire and hold the clip close to the engine block while cranking the engine. (*ATW*)

⚙ 8-19 Hard starting with engine warm If the engine starts hard when warm, it could be caused by the choke sticking closed, vapor lock, or engine binding due to overheating. Choke action can be watched with the air cleaner removed. If the choke does not open wide with the engine hot, the choke should be serviced.

⚙ 8-20 Engine stalls If the engine starts and then stalls, note whether the stalling takes place before or after the engine warms up, after idling or slow-speed driving, or after high-speed or full-load driving. Check the PCV valve (Fig. 8-13). If this valve becomes clogged or sticks, it can cause poor idling and stalling. A malfunctioning PCV valve should be replaced.

1. Engine stalls before it warms up This could be caused by an improper choke or vacuum-break setting, an improperly set fast or slow idle, or improper adjustment of the idle-mixture screw in the carburetor. Also, it could be caused by a low carburetor float setting, or to insufficient fuel entering the carburetor. This condition could result from a faulty float needle valve, dirt or water in the fuel lines or filter, a defective fuel pump, or a plugged fuel tank vent. Also, the carburetor could be icing.

In some cases, certain ignition troubles could cause stalling after starting. Generally, if ignition troubles are bad enough to cause stalling, they also would prevent starting. Burned contact points or defective spark plugs might permit starting, but could fail to keep the engine running. The other condition might be an open primary resistance. When the engine is cranked, this resistance is bypassed. Then, when the engine starts and cranking stops, the resistance is inserted into the ignition pri-

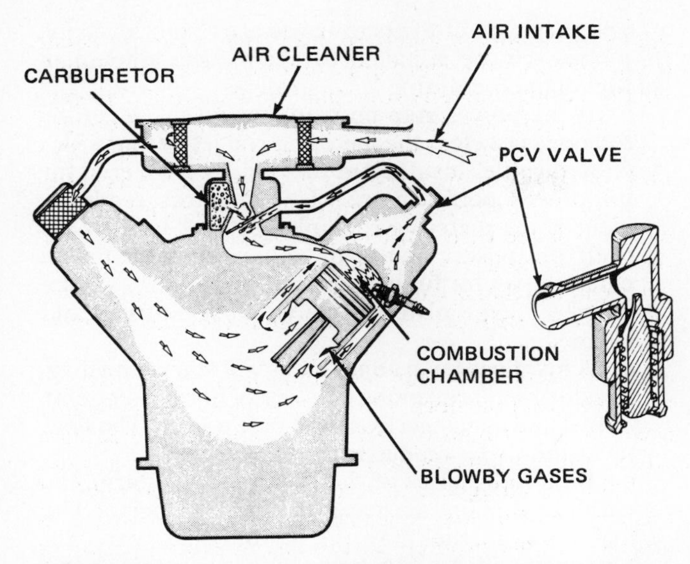

Fig. 8-13 Location of the PCV valve in an engine. (*Chrysler Corporation*)

mary circuit. If the resistor or resistance wire is open, the engine will stall when the key is released.

2. Engine stalls as it warms up This condition might result if the choke valve is stuck closed. The mixture becomes too rich for a hot engine and the engine stalls. Also, if the manifold heat-control valve is stuck, the ingoing air-fuel mixture might become overheated and too lean, causing the engine to stall. If the hot-idle speed setting is too low, the engine may stall when it warms up because its idling speed will drop too low. It is also possible that the engine is overheating, which could cause vapor lock.

3. Engine stalls after idling or slow-speed driving This may occur if the fuel pump is defective and has a cracked diaphragm, weak spring, or defective valve. The pump cannot deliver enough gasoline at low speed to replace that used by the engine. Therefore, the carburetor float bowl runs dry and the engine stops.

If the float level is set too high, or there is dirt between the needle-valve and seat, or the idle mixture is too rich, the engine may load up with an overrich mixture and stall. A lean idle adjustment also will cause stalling when the engine is hot. Overheating can cause vapor lock and engine stalling. The engine may overheat during sustained idle or slow-speed driving. Under these conditions, the air movement through the radiator may not be great enough to keep the engine temperature down. If overheating is excessive or abnormal, consider the conditions listed in Chap. 19.

4. Engine stalls after high-speed driving This may occur if sufficient heat accumulates to cause the fuel to boil in the line and produce a vapor lock. Shielding the fuel line or using a less volatile fuel (if available) reduces the tendency for vapor lock to occur. Another condition that might cause stalling after high-speed driving is failure of the antipercolator valve in the carburetor. This causes the mixture to become too rich and the engine to stall. Stalling might also result from engine overheating.

❁ 8-21 Engine backfire Backfiring may occur in a cold engine because of a temporary improper air-fuel mixture ratio or sticking intake valves. After the engine has started and is warming up, backfiring may be due to an excessively rich or lean mixture which will not ignite properly, causing backfiring through the carburetor. Backfiring may also be due to preignition caused by such engine conditions as hot valves or excessive carbon, and by such ignition-system conditions as incorrect timing or spark plugs of the wrong heat range.

❁ 8-22 Slow engine warm-up If the engine warms up slowly, the trouble could be caused by the manifold heat-control valve or the cooling system thermostat being stuck open.

❁ 8-23 Poor idle If the engine idles roughly, too slowly, or too fast, the probability is that the idle mixture and idle speed require adjustment. In addition, other conditions such as malfunctioning choke, a high or low float level, vapor lock, clogged idle system, air leaking into the intake manifold, loss of engine compression, improper valve action, overheating engine, or an improperly operating ignition system would cause poor idle. These conditions would also cause poor engine performance at speeds above idle. Improper idle mixture or idle speed can be checked only with the engine idling. Another possible cause is the PCV valve being stuck in the open or high-speed position. This could allow too much airflow from the crankcase, which would lean out the idle mixture excessively. This would cause a poor idle.

❁ 8-24 Engine hesitates on acceleration This could result from a loose or broken vacuum hose, a faulty accelerator pump or power system, or low float level in the carburetor. If the EGR valve is stuck open or opens too soon, it will admit too much exhaust gas to the engine cylinders and cause weak combustion. Also, if the ignition-distributor spark advance systems are faulty, they may not be adjusting the timing as they should to suit operating conditions when the throttle is opened.

❁ 8-25 Engine runs but misses If the engine runs but misses, the fuel system may be erratic in its action so that fuel delivery is not uniform. This could result from clogged fuel lines, clogged nozzles or passages in the carburetor, incorrectly adjusted or malfunctioning float level or needle valve, or a defective fuel pump. Other conditions that might cause missing include ignition defects, such as incorrect timing or defective

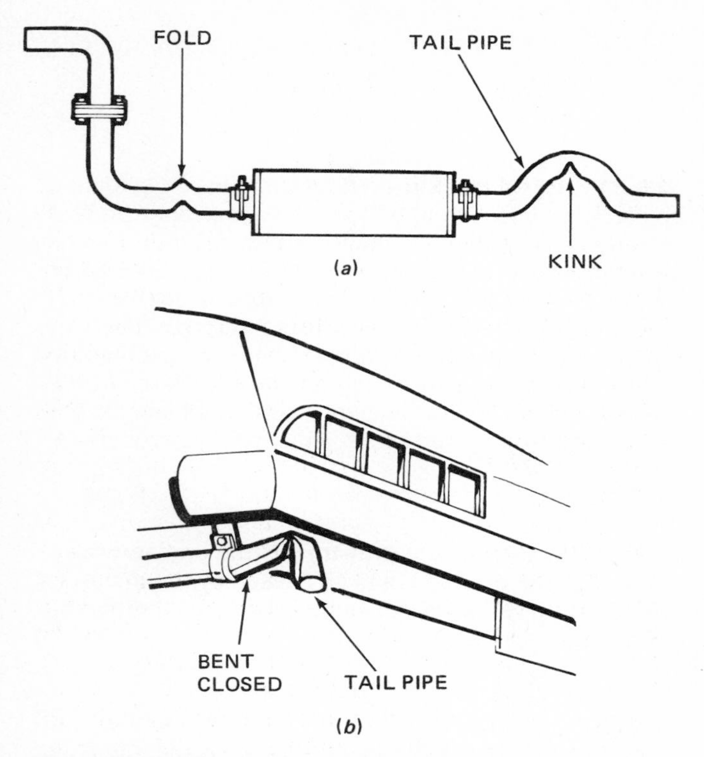

Fig. 8-14 A clogged or restricted exhaust system may cause the engine, hot or cold, to lack power. (*ATW*)

plugs, coil, points, cap, condenser, or wiring. The exhaust system might be clogged (Fig. 8-14), causing back pressure that prevents normal air-fuel mixture delivery to the cylinders. Also, the engine might be overheating, or it might have sticky valves, low compression, or stuck piston rings.

8-26 Engine lacks power, acceleration, or high-speed performance This type of complaint may be difficult to analyze. Almost any component of the engine or car, from the driver to the tires, could cause the problem. As a first step in solving this sort of complaint, some technicians road-test the car to verify the complaint. The car should be driven so as to duplicate the conditions that cause the problem. Also, the engine can be checked on the chassis dynamometer or given a tuneup.

Here are some conditions that might cause the complaint. Fuel system conditions are discussed first.

1. Many conditions in the carburetor could prevent delivery of proper amounts of fuel for good acceleration and full power. Possibilities to be considered follow.

   a. Incorrect functioning of the accelerator pump can be checked on many engines by removing the air cleaner and observing pump action. If the pump is functioning correctly, a steady stream of fuel will be discharged from the jet as the throttle is opened. The stream should continue for a few seconds after the throttle has reached the full-open position. If the pump does not operate correctly, disassembly and servicing of the carburetor are required. Some pumps can be adjusted to change the amount of fuel delivered during acceleration.

   b. If the power step-up diameter on the metering rod does not clear the metering-rod jet during wide-open throttle, insufficient fuel will be delivered for full-power performance. This requires readjustment of the metering-rod linkage.

   c. If the power piston or valve sticks so that it cannot open for full power, insufficient fuel will be delivered. The piston or valve must be freed and cleaned.

   d. A low float-level adjustment will starve the main nozzle or jet, preventing delivery of normal amounts of fuel and causing a loss of engine power. The float level should be readjusted.

   e. Dirt in the filter of the fuel line will starve the carburetor main nozzle or jet, and the engine. The dirt restricts fuel flow. Also, on older cars not equipped with a fuel-vapor recovery system, some fuel tank caps have a vent to permit air to enter as fuel is withdrawn. If this vent is clogged, no air can enter. A vacuum will develop in the fuel tank. The vacuum prevents delivery of fuel through the fuel pump to the carburetor. On cars with a fuel-vapor recovery system, installing an old-style cap could prevent fuel delivery. The fuel tanks of cars with fuel-vapor recovery systems are sealed. The required tank cap has a vacuum valve that opens when fuel is withdrawn so that air can enter. This prevents a vacuum from forming. If the wrong cap, without a vent or vacuum valve, is installed, a vacuum may form that will prevent fuel from being delivered to the carburetor.

   f. A stuck or inoperative choke will cause loss of power when the engine is cold. It may also cause loss of power when the engine is hot, if it is stuck in a partly closed position. This produces an excessively rich mixture. The choke should be serviced.

   g. If air leaks into the intake manifold around the carburetor or manifold mounting, or past worn throttle shaft bearings, the air-fuel mixture may become too lean. Gaskets should be replaced and mounting nuts or screws should be tightened as necessary. Excessively worn throttle shaft bearings require replacement of the throttle body.

   h. A stuck antipercolator valve may also cause an excessively lean mixture. The valve requires freeing or adjustment.

   i. A stuck manifold heat-control valve, if stuck in the closed position, overheats the air-fuel mixture in the intake manifold with the engine hot, so that the mixture expands excessively. This starves the engine, causing poor performance. If the valve sticks open, warm-up will be slowed. The valve should be freed.

   j. If the throttle linkage is out of adjustment, the throttle valve may not open fully, preventing delivery of full power. Throttle linkage should be correctly adjusted.

   k. Most of these conditions produce an excessively

lean mixture. However, conditions that produce an excessively rich mixture also cause poor engine performance.

2. Vapor lock also causes fuel starvation in the engine. Vaporization or boiling of the fuel in the fuel pump or fuel line prevents delivery of normal amounts of fuel to the carburetor and carburetor nozzles and jets. One way to check for this condition is by inserting a section of clear plastic tubing in the fuel line. Then watch for bubbles in the fuel to pass through the tube with the engine hot and running. To prevent vapor lock, shield the fuel line and fuel pump from engine heat. If available, use of a fuel with a lower volatility also will prevent vapor lock.
3. A defective fuel pump also can starve the engine by not delivering sufficient amounts of fuel to the carburetor. This requires servicing or replacement of the fuel pump. In addition, a clogged filter screen in the fuel tank will cause fuel starvation.
4. A clogged exhaust system caused by rust, dirt, or mud in the muffler or tail pipe, or by a pinched or damaged muffler or tail pipe could create sufficient back pressure to prevent normal exhaust-gas flow from the engine (Fig. 8-14). This would result in reduced engine performance, particularly on acceleration or at high speed. Some exhaust systems include exhaust pipes that are laminated. A laminated exhaust pipe consists of a single ordinary exhaust pipe placed tightly inside a slightly larger pipe. This combination reduces exhaust noise because the laminated exhaust pipe deadens the ringing noise from some engines that gets through the standard muffler. There have been cases where the inner layer of pipe separated from the outer layer and sagged or collapsed, partly blocking the passage. This is another point to check carefully, since collapse of the inner pipe often cannot be seen by visual inspection.
5. Defective ignition can reduce engine performance, just as it can increase fuel consumption. Conditions in the ignition system that might cause the trouble include a "weak" coil or condenser, incorrect timing, faulty advance-mechanism action, dirty or worn contact points or spark plugs, defective pickup coil or electronic control unit (electronic ignition system), and defective wiring.
6. A sluggish engine will result from loss of compression, excessive carbon in the engine cylinder, defective valve action, or heavy engine oil.
7. Failure of the cooling system to operate properly could cause the engine to overheat, with a resulting loss of power (Chap. 19). Also if the cooling-system thermostat fails to close as the engine cools, it will prolong engine warm-up the next time the engine is started. This reduces engine performance during warm-up.
8. Any condition that increases the rolling resistance reduces acceleration and top speed. These conditions include low tires, dragging brakes, and misaligned wheels.
9. Clutch slippage or excessive friction in the power train will reduce acceleration and top speed.
10. An automatic transmission that does not downshift, or a defective torque converter, can reduce performance and acceleration.

⚙ 8-27 Excessive fuel consumption The first step in analyzing a complaint of excessive fuel consumption is to make sure that the car is really having this trouble. Usually, this means taking the word of the car owner that the car is using too much fuel. A fuel mileage tester (⚙ 8-4) can be used to determine accurately how much fuel the car is using. After it has been determined that the car is using too much fuel, the cause of trouble must be found. It could be in the fuel system, ignition system, engine, or elsewhere in the car.

The compression tester and the intake manifold vacuum gauge will determine the location of the trouble.

A quick check of mixture richness that does not require any testing instruments is to install a set of new or cleaned spark plugs of the correct heat range for the engine. Then take the car out on the highway for 15 to 20 minutes. Stop the car. Remove and examine the plugs. If they are coated with a black carbon deposit, the indication is that the mixture is too rich.

If the trouble appears to be in the fuel system, the following points should be considered.

1. Drivers who pump the accelerator pedal when the engine is idling and jump away when the stoplight changes use an excessive amount of fuel. Each downward movement of the accelerator pedal causes the accelerator pump to discharge gasoline into the carburetor. This extra fuel is wasted, since it contributes nothing to the movement of the car.
2. High-speed operation requires more fuel per mile. For example, a car that will give 20 miles per gallon at 30 mph may give less than 15 miles per gallon at 60 mph. At 70 to 80 mph the mileage may drop below 10 miles per gallon. Therefore, a car operated consistently at high speed will have higher fuel consumption than a car driven consistently at intermediate speed.
3. Short-run, stop-and-start operation uses more fuel. In short-run operation, when the engine is allowed to cool off between runs, the engine is operating mostly cold or on warm-up. This means that fuel consumption is high. When the car is operated in heavy city traffic, or under conditions requiring frequent stops and starts, the engine is idling much of the time. Also, the car is accelerated to traffic speed after each stop. All this uses more fuel, and fuel economy will be poor.
4. If the fuel pump has excessive pressure, it will maintain an excessively high fuel level in the carburetor float bowl. This will cause a greater discharge at the fuel nozzle or jet, thereby producing high fuel consumption. Excessive pump pressure is not a common cause of excessive fuel consumption, however. It could result only from installation of the wrong pump or diaphragm spring or from incorrect reinstallation of the pump diaphragm during repair. Fuel pumps can develop leaks that will permit loss of gasoline to the outside or into the

crankcase. This requires replacement of the diaphragm, tightening of the assembling screws, or replacement of the pump.

5. If a manually operated choke is left partly closed, the carburetor will deliver too much fuel for a warm engine, and fuel consumption will be high. On manually operated chokes, it is possible for the choke valve linkage to get out of adjustment so that the valve will not open fully. This requires readjustment to prevent high fuel consumption. With an automatic choke, the choke valve should move from the closed to the open position during engine warm-up, reaching the full-open position when the engine reaches operating temperature. This action can be observed by removing the air cleaner and noting the changing position of the choke valve during warm-up. If the automatic choke does not open the choke valve normally, excessive fuel will be used. The choke must be serviced.
6. A clogged air cleaner acts much like a closed choke valve, since it chokes off the free flow of air through it. The element should be cleaned or replaced, or fresh oil added (on the oil-bath air cleaner).
7. In the carburetor itself, the following conditions could cause delivery of an excessively rich air-fuel mixture:

   a. A high float level or a leaking float will permit delivery of too much fuel to the float bowl and therefore through the fuel nozzle or jet. The float level must be readjusted and a leaky float must be replaced.

   b. A stuck or dirty float-needle valve will not shut off the flow of fuel from the fuel pump. Too much fuel will be delivered through the carburetor fuel nozzle or jet. The needle valve should be freed and cleaned or replaced.

   c. Worn carburetor jets pass too much fuel, causing the air-fuel mixture to be too rich. Worn jets must be replaced.

   d. If the power system operates during part-throttle operation, too much fuel will be delivered through the main fuel nozzle. This could be caused by a stuck metering rod or power piston, which must be freed.

   e. An idle mixture that is set too rich or an idle speed that is set too fast wastes fuel. Resetting of the idle mixture and idle speed is required.

   f. If the accelerator-pump check valve sticks open, it may permit discharge of fuel through the pump system into the carburetor air horn, causing excessive fuel consumption. This requires freeing and servicing of the check valve.

   g. Carburetor leaks, either internal or external, cause loss of fuel. The correction is to replace gaskets or damaged parts and tighten loose couplings on fuel lines, loose jets or nozzles, and loose mounting nuts or screws.
8. Faulty ignition can also cause excessive fuel consumption. The ignition system could cause engine miss and therefore failure of the engine to utilize all the fuel. Faulty ignition would also be associated with the loss of power, acceleration, or high-speed performance (⚙ 8-26). Conditions in the ignition system that might cause trouble include a "weak" coil or condenser, incorrect timing, faulty advance mechanism action, dirty or worn spark plugs or contact points, defective pickup coil or electronic control unit (electronic ignition system), and defective wiring.
9. Several conditions in the engine can also produce excessive fuel consumption. Loss of engine compression from worn or stuck rings, worn or stuck valves, or a loose or burned cylinder head gasket causes loss of power. This means that more fuel must be burned to achieve the same speed or power.
10. Any condition that increases rolling resistance and makes it harder for the car to move along the road increases fuel consumption. For example, low tires, dragging brakes, and misalignment of wheels increase fuel consumption. Similarly, losses in the power train such as from a slipping clutch increase fuel consumption.

⚙ 8-28 Smoky exhaust gas The color of smoky exhaust gas is a clue to what is causing it (Fig. 8-15). For example, if the exhaust gas is blue, that means the engine is burning oil. This is normally an engine problem and is caused by oil entering the combustion chambers. Oil can enter through the PCV system, through clearances between the valve stems and valve guides, and past the piston rings.

If the exhaust gas is black or sooty, it means the air-fuel mixture is too rich so that it is not burning completely. Refer to ⚙ 8-27 for causes of excessive fuel consumption.

If the smoke is white, coolant from the engine cooling system probably is leaking into the combustion chambers. The water is turned into steam by the heat of combustion and gives the exhaust gas the white color. The remedy is to tighten cylinder head bolts or replace the cylinder head gasket. This is a likely place for coolant to leak into the combustion chambers. The condition also could be caused by a cracked cylinder head or cylinder block.

NOTE: A quick check for coolant leakage is to put a drop of oil from the crankcase on a sheet of aluminum foil. Then apply heat to the foil under the drop of oil. If the oil begins to crackle and pop, there is water in it. If coolant is leaking into the combustion chambers, it will also probably be found in the engine oil.

⚙ 8-29 Engine run-on, or dieseling Engines with emission control devices require a fairly high hot-idle speed for best operation. This makes run-on, or dieseling, possible. If there is any source of ignition, the engine will continue to run even after the ignition switch is turned off. Hot spots in the combustion chamber can take the place of the spark plugs. If the throttle is slightly open, enough air-fuel mixture can be getting past it to keep the engine running. Many engines have an idle-stop solenoid to close the throttle completely when the ignition switch is turned off.

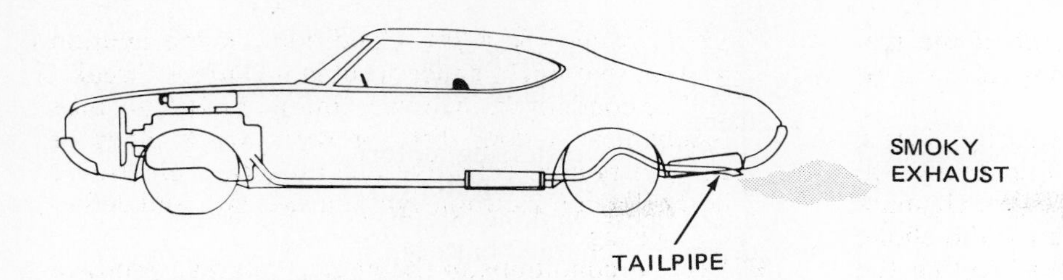

Fig. 8-15 The color of the exhaust gas may be a clue to what is causing it.

If an engine diesels, check the idle-stop solenoid. It may need adjustment so that it will allow the throttle to close completely when the ignition is turned off. Also, make sure the engine is not overheating. This can contribute to the problem of run-on.

⚙ 8-30 Too much HC and CO in exhaust If the exhaust-gas analyzer (⚙ 8-5) indicates that there is too much HC and CO in the exhaust gases, several conditions could be the cause. In the fuel system, a choke sticking closed, worn jets, high float level, and other conditions listed in ⚙ 8-27 could be the cause. In the ignition system, missing or incorrect timing could be responsible. Also, emission controls such as the air-injection system, transmission-controlled spark system, or defective catalytic converters could be causing the problem.

⚙ 8-31 Quick carburetor checks Several quick carburetor checks can be made to give an indication of whether the various carburetor systems are functioning. The results of these checks must not be considered final. They give only a preliminary indication of possible trouble. Accurate diagnosis requires an exhaust-gas analyzer and an intake manifold vacuum gauge.

1. Float-level adjustment With the engine idling, remove the air cleaner. Look for fuel dribbling from the main metering nozzle. If it is, the float level is high.

CAUTION: A backfire can occur with the air cleaner off. Removing the air cleaner can lean the air-fuel mixture enough to cause backfiring. A malfunctioning carburetor can also cause backfiring.

2. Low-speed (off-idle) and idle system If the engine does not idle smoothly, the idle system may be malfunctioning. Other possible causes of poor idling are noted in ⚙ 8-26 and ⚙ 8-19. Open the throttle slowly to increase engine speed to about 2500 rpm. If engine speed does not increase evenly and the engine runs roughly through this speed range, the low-speed system probably is malfunctioning.

3. Accelerator pump system With the engine off and the air cleaner removed, look down the carburetor. Open the throttle suddenly and note whether the accelerator pump jet discharges fuel. If no fuel discharges, and you are sure the float bowl is full, the accelerator-pump system is malfunctioning.

4. Main metering system With the engine warmed up and running at approximately 2500 rpm, slowly cover the air-cleaner intake with a piece of stiff cardboard. Do not use your hand! The engine should speed up slightly since closing the air intake should cause the main metering system to discharge more fuel.

Chapter 8 review questions

Select the *one* correct, best, or most probable answer to each question. Then check your answers against the correct answers given at the end of the book.

1. The compression readings on a four-cylinder engine were 140, 135, 5, and 140. After these compression readings were taken, a wet compression test was made. The second set of readings was almost the same as the first. Mechanic X says that a burned valve could cause these readings. Mechanic Y says that a broken piston ring could cause these readings. Who is right?
   a. X only
   b. Y only
   c. both X and Y
   d. neither X nor Y.
2. When the tachometer is connected between the distributor primary terminal and ground, it indicates:
   a. engine speed
   b. engine vacuum
   c. engine compression
   d. engine power.
3. If pouring a small amount of oil into the cylinder increases the compression presure, then the loss of compression probably is a result of leakage:
   a. past the valves
   b. past the head gasket
   c. past the piston rings
   d. past the oil seals.
4. If the vacuum gauge needle swings around to 23 to 25 inches [584 to 635 mm] of mercury as the throttle is quickly closed after the engine has been raced, it indicates:

a. stuck valves
b. low compression
c. satisfactory compression
d. leaky valves.

5. A steady but low vacuum reading with the engine idling indicates that the engine:
a. is losing power
b. has a stuck valve
c. exhaust line is plugged
d. none of the above.

6. A very low vacuum reading with the engine idling indicates:
a. stuck valves
b. air leakage into the manifold
c. loss of compression
d. faulty piston rings.

7. A valve that sticks open or a spark plug that is not firing will cause the vacuum gauge needle to:
a. oscillate slowly
b. drop back regularly
c. fall back slowly to zero
d. read too high.

8. A compression test has been made on an in-line six-cylinder engine. Cylinders 3 and 4 have readings of 10 psi [69 kPa]. The other cylinders all have readings between 130 and 135 psi [896 and 931 kPa]. Mechanic X says this could be caused by a blown head gasket. Mechanic Y says this could be caused by wrong valve timing. Who is right?
a. X only
b. Y only
c. either X or Y
d. neither X nor Y.

9. The device which can give a very close approximation of a road test in the shop is called a:
a. engine dynamometer
b. chassis dynamometer
c. tachometer
d. engine tester.

10. A compression test shows that one cylinder is too low. A leakage test on that cylinder shows that there is too much leakage. During the test, air could be heard coming out of the tail pipe. Which of these could be the cause?
a. broken piston rings
b. a bad head gasket
c. a bad exhaust gasket
d. an exhaust valve not seating.

11. Pressure and volume tests of a mechanical fuel pump are both below specs. Mechanic X says that an air leak in the fuel line between the tank and pump could be the cause. Mechanic Y says a plugged fuel tank pickup filter could be the cause. Who could be right?
a. X only
b. Y only
c. either X or Y
d. neither X nor Y.

12. Missing in one cylinder is likely to result from:
a. a clogged exhaust
b. an overheated engine
c. vapor lock
d. a defective spark plug.

13. Irregular missing in different cylinders may result from:
a. a defective starting motor
b. a defective carburetor
c. an open cranking circuit
d. an overcharged battery.

14. Loss of engine power as the engine warms up is most likely caused by:
a. vapor lock
b. excessive rolling resistance
c. the throttle valve not closing fully
d. heavy oil.

15 An engine will lose power (hot or cold) if it has:
a. incorrect idle speed
b. an automatic choke valve that is stuck open
c. worn rings and cylinder walls
d. none of the above.

16. An engine may stall as it warms up if the:
a. ignition timing is off
b. choke valve sticks closed
c. battery is run down
d. throttle valve does not open fully.

17. The most probable cause of an engine stalling after a period of idling or slow-speed driving is:
a. loss of compression
b. a defective fuel pump
c. sticking engine valves
d. all of the above.

18. Stalling of an engine after a period of high-speed driving is likely to be caused by:
a. vapor lock
b. incorrect ignition timing
c. worn carburetor jets
d. all of the above.

19. Engine backfiring may result from:
a. spark plugs of wrong heat range
b. vapor lock
c. a run-down battery
d. worn piston rings.

20. A smoky blue exhaust may be caused by:
a. an excessively rich mixture.
b. burning of oil in the combustion chamber
c. a stuck choke valve
d. incorrect valve adjustment.

21. A smoky black exhaust may be caused by:
a. worn piston rings
b. worn carburetor jets
c. spark plugs of wrong heat range
d. none of the above.

22. Which of these is *least likely* to cause a car to hesitate or stumble when the accelerator pedal is opened quickly?
a. retarded ignition timing
b. low carburetor float level
c. leaking carburetor accelerator pump check
d. leaking carburetor power valve.

23. A car has a steady, even miss *only* while climbing a hill. The cause could be:
a. vapor lock

b. a bent push rod
c. a dirty air cleaner
d. none of the above.

24. A carburetor is flooding, with liquid gasoline leaking from around the outside and also running down inside the carburetor. Which of these is the *least* likely cause of the problem?
a. high float level
b. plugged air bleeds
c. dirt in the needle-and-seat
d. high fuel pump pressure.

25. If the EGR valve is stuck open, it could cause:
a. poor idling
b. poor acceleration
c. loss of power
d. all of these.

CHAPTER 9 FUEL SYSTEM SERVICE

After studying this chapter, and with proper instruction and equipment, you should be able to:

1. Check and service the air cleaner.
2. Diagnose and repair troubles in the heated-air system.
3. Adjust the choke and check its operation.
4. Service the fuel filter.
5. Diagnose a defective fuel gauge.
6. Perform pressure, vacuum, and capacity checks on a fuel pump.
7. Remove and install a fuel pump.
8. Diagnose troubles in the PCV system, evaporative control system, air-injection system, EGR system, and the catalytic converter.
9. Service the emission control systems.

9-1 Air-cleaner service The air cleaner passes a large volume of air through its filter element. The filter element removes dirt and dust from the air. This dirt gradually accumulates in the paper element and partially clogs it. In the oil-bath air cleaner, much of this dirt is washed down to the oil, and so the oil gradually gets dirty.

At one time, car manufacturers recommended that the filter element should be periodically removed, cleaned, and reinstalled. Today, only replacement of the filter element is recommended. Because of this, and because of improved filter elements, most manufacturers recommend replacing the filter element after a certain time or mileage, whichever occurs first. Manufacturers no longer provide cleaning instructions. The typical recommendation today is to install a new filter element every 30,000 miles (48,000 km), or oftener for dusty conditions.

Figure 9-1 shows an air cleaner and its related parts. The filter element is removed from some air cleaners by taking off the wing nut and air cleaner cap. Then the element can be lifted out. On other air cleaners, the cap is part of the air cleaner cover. The assembly includes the snorkel, the connections to the PCV system, and the heat stove for the heated-air intake system. On these, the air cleaner cover can usually be lifted up enough to remove the element. If not, the hoses must be disconnected.

NOTE: Chrysler recommends that if the air cleaner must be loosened from the carburetor for any purpose, the air cleaner should be removed from under the hood. If the air cleaner rests on or hooks into linkage parts, the parts could be damaged. Cap all carburetor air fittings which could leak air if the engine is started while the air cleaner is removed.

With the filter element out, the bottom of the air cleaner, gasket surfaces, and cover should be cleaned. Check the cover seal for tears or cracks, and replace it if it is damaged. Examine the filter element for oil. If it is wet with oil over more than half of its outside surface, or circumference, check the PCV system. It is carrying oil up to the air cleaner.

Chevrolet recommends inspection of the element for dust leaks after the first 15,000 miles (24,000 km) of operation. Look for dust in the air cleaner housing on the inside of the filter element. Dust could enter through holes in the filter element, and under or over the element, past the side seals. If no dust is found, rotate the element one-half turn, and reinstall it.

CAUTION: Do not operate the engine with the air cleaner off unless it is when making tests. If the engine backfires with the air cleaner off, it could cause a fire in the engine compartment.

1. Cleaning paper element Here is the procedure that is recommended by Chrysler for cleaning the paper element on older cars. Examine the element, and if it is wet with oil over more than half of its circumference, discard it. To clean the element, use compressed air to blow through the element from the inside (Fig. 9-2). Hold the nozzle at least 2 inches [51 mm] away from the inside screen of the element. Do not blow from the outside in. This will embed the particles in the paper.

After cleaning the element, examine it for punctures. If you can see any pinholes when the element is held up to the light, discard the element. Make sure the

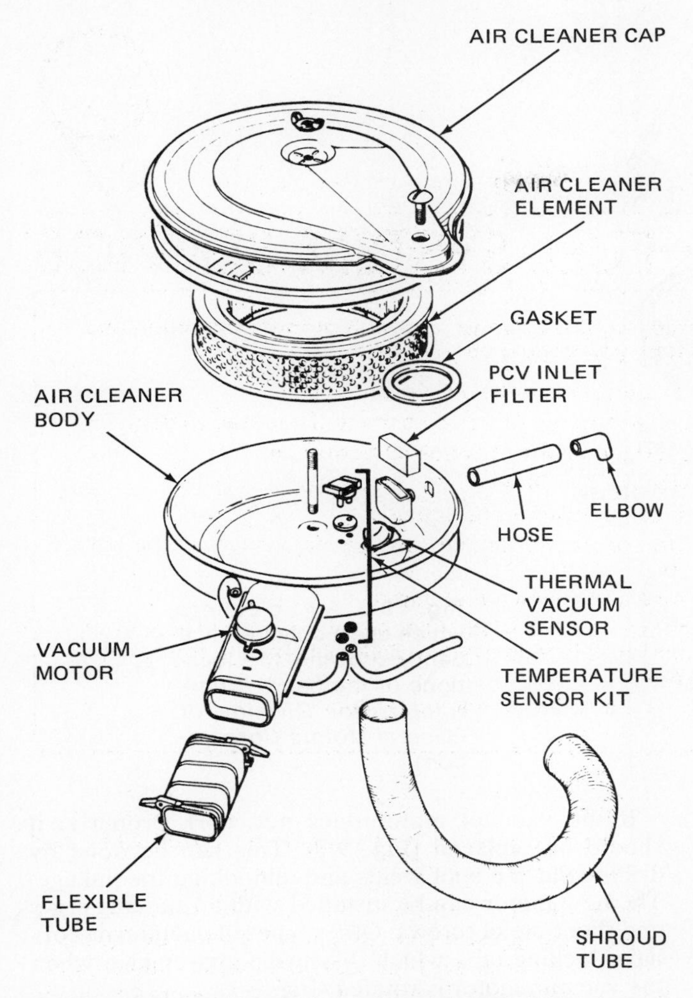

Fig. 9-1 Disassembled view of a carburetor air cleaner and related parts for a six-cylinder engine. (*American Motors Corporation*)

plastic sealing rings on both sides of the element are smooth and uniform. If these are in good condition, install the element. Be sure it seals both top and bottom when the air cleaner cover is reinstalled.

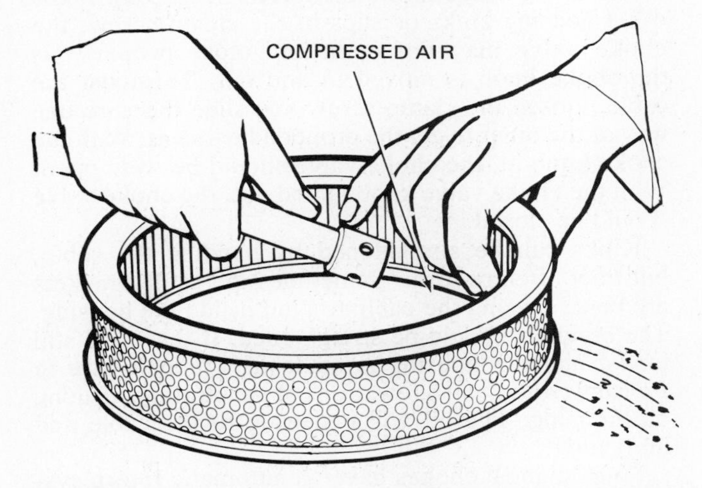

Fig. 9-2 Cleaning the filter element with compressed air blown from inside the element. (*Chrysler Corporation*)

2. Cleaning polyurethane element On some vehicles, a plastic wrapper, or polyurethane element, has been used outside the paper element (Fig. 9-3). After the polyurethane element is removed, inspect it carefully for rips or other damage. Discard it if it is damaged. Wash the element in kerosene. Then squeeze it gently to remove excess kerosene (Fig. 9-4).

NOTE: Do not use solvents containing acetone or similar compounds, since they could ruin the polyurethane element. Never wring out, shake, or swing the element. This could tear it. Instead, fold the element and then gently squeeze it.

Clean the cover and other parts of the air cleaner. Then dip the cleaned element in SAE 10W-30 engine oil, and squeeze out the excess. Reinstall the element and its support in the air cleaner. Make sure that the element is not folded or creased and that it seals all the way around. Use a new gasket when installing the air cleaner on the carburetor.

3. Oil-bath air cleaner Figure 9-5 shows an oil-bath air cleaner. After removing the filter element, clean it by sloshing it up and down in clean solvent. Dry it with compressed air. Dump the dirty oil from the air cleaner body, wash the body with solvent, and dry it. Refill the body to the full mark with clean SAE 10W-30 engine oil. Reinstall the filter element and air-cleaner body on the engine.

NOTE: Air cleaners which are not attached directly to the carburetor are connected to it by a flexible hose or tube. This hose must have an airtight connection to both the air cleaner and the carburetor. Also, the hose must have no tears or punctures which would admit unfiltered air.

⚙ 9-2 Servicing the thermostatically controlled air cleaner To check the thermostatically controlled air cleaner, first make sure the hoses are tightly connected. The hot-air tube from the heat stove to the snorkle of the air cleaner should be in good condition. Inspect the system to ensure that there are no leaks. The system can be checked with a thermometer. Failure of the thermostatic system usually results in the damper door staying open. This means that the driver probably will not notice anything wrong in warm weather. But in cold weather, the driver will notice hesitation, surge, and stalling. A typical checking procedure follows.

Remove the air cleaner cover. Install the thermometer as close to the sensor as possible. Allow the engine to cool below 80°F [29°C] if it is hot. Reinstall the air cleaner cover without the wing nut.

Start and idle the engine. When the damper begins to open, remove the air cleaner cover, and note the temperature reading. It should be between 85 and 115°F [29 and 46°C]. If it is difficult to see the damper, use a mirror.

If the damper does not open at the correct temperature, check the vacuum motor and sensor.

With the engine off, the control damper should be in the "compartment" or cold-air-delivery position (Fig. 3-37). To determine if the vacuum motor is operating,

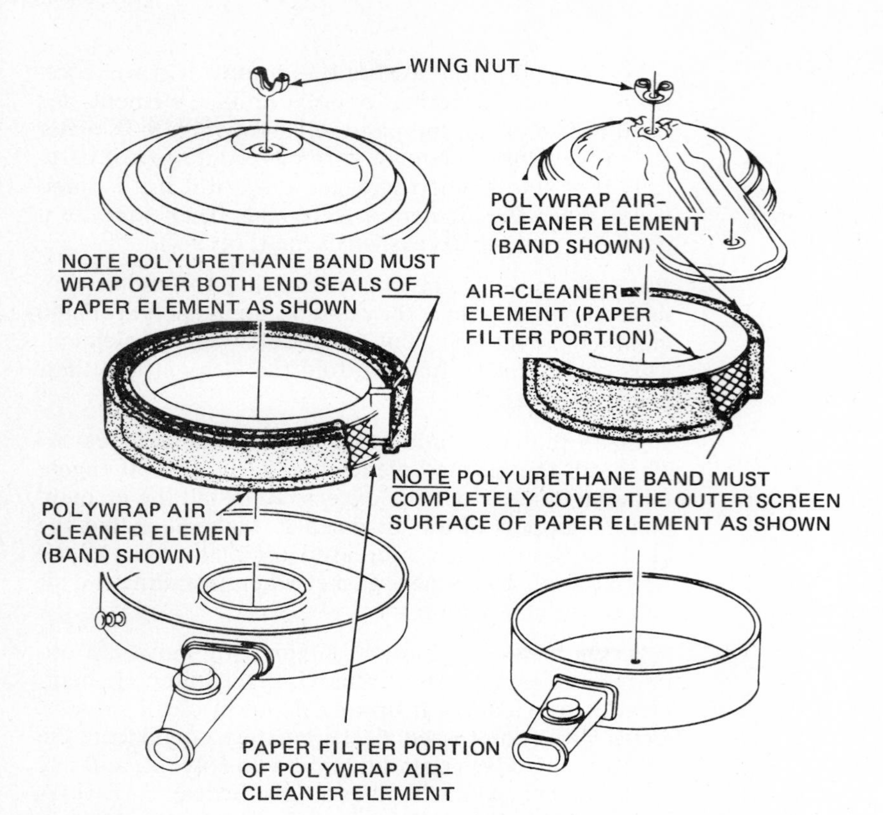

Fig. 9-3 Air cleaners which use a polyurethane band, showing the locations of the band. (*Chevrolet Motor Division of General Motors Corporation*)

apply at least 9 inches Hg [229 mm] of vacuum to the fitting on the vacuum motor (Fig. 9-6). The vacuum can be from the engine, from a distributor tester, or from a hand vacuum pump (Fig. 9-6). With vacuum applied, the damper should move to the hot-air-delivery position (Fig. 3-37).

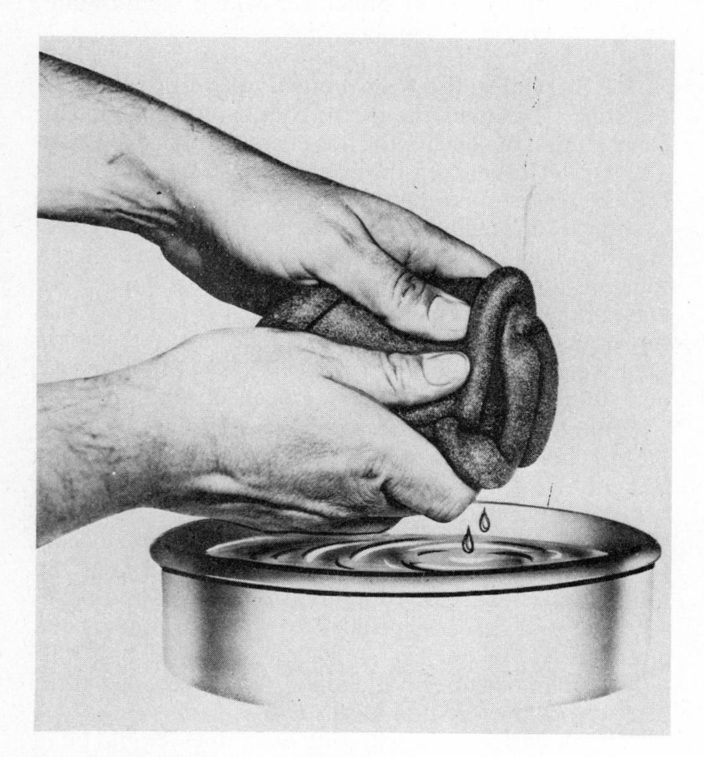

Fig. 9-4 Washing a polyurethane air-cleaner element. (*Chevrolet Motor Division of General Motors Corporation*)

If the vacuum motor does not work properly, it should be replaced (Fig. 9-7). This can be done by drilling out the spot welds and unhooking the linkage. The new motor can be installed with a retaining strap and sheet-metal screws. Other types of vacuum motors have locking tabs which disengage and engage when the vacuum motor is rotated (Fig. 9-7).

If the vacuum motor does work properly, the sensor should be replaced (Fig. 9-8). This is done by prying up the tabs on the retaining clip. The new sensor is then installed, and the tabs are bent down again.

⚙ 9-3 Adjusting the manual choke On manual chokes, a knob on the dash is linked by a wire or cable to the choke valve in the carburetor (Fig. 9-9). If the cable housing kinks or slips in the clamp screw, the choke valve may not open and close properly as the choke knob is moved in and out. To adjust the cable, loosen the clamp screw and slide the wire one way or the other to get the proper adjustment. With the choke knob in, the choke valve should be wide open. With the choke valve knob pulled out, the choke valve should be closed.

Kinks can be straightened by bending the cable, but they often re-form. Sometimes the cable brackets are bent, causing the cable to bind inside the housing. The brackets should be straightened. If the cable still binds, and the housing is made of wrapped wire (a *Bowden cable*), put a few drops of penetrating oil along the housing. The oil will penetrate to the cable and lubricate it.

Some manual chokes have an automatic return system which opens the choke valve when the engine is hot (Fig. 9-10). The choke valve is held closed by a

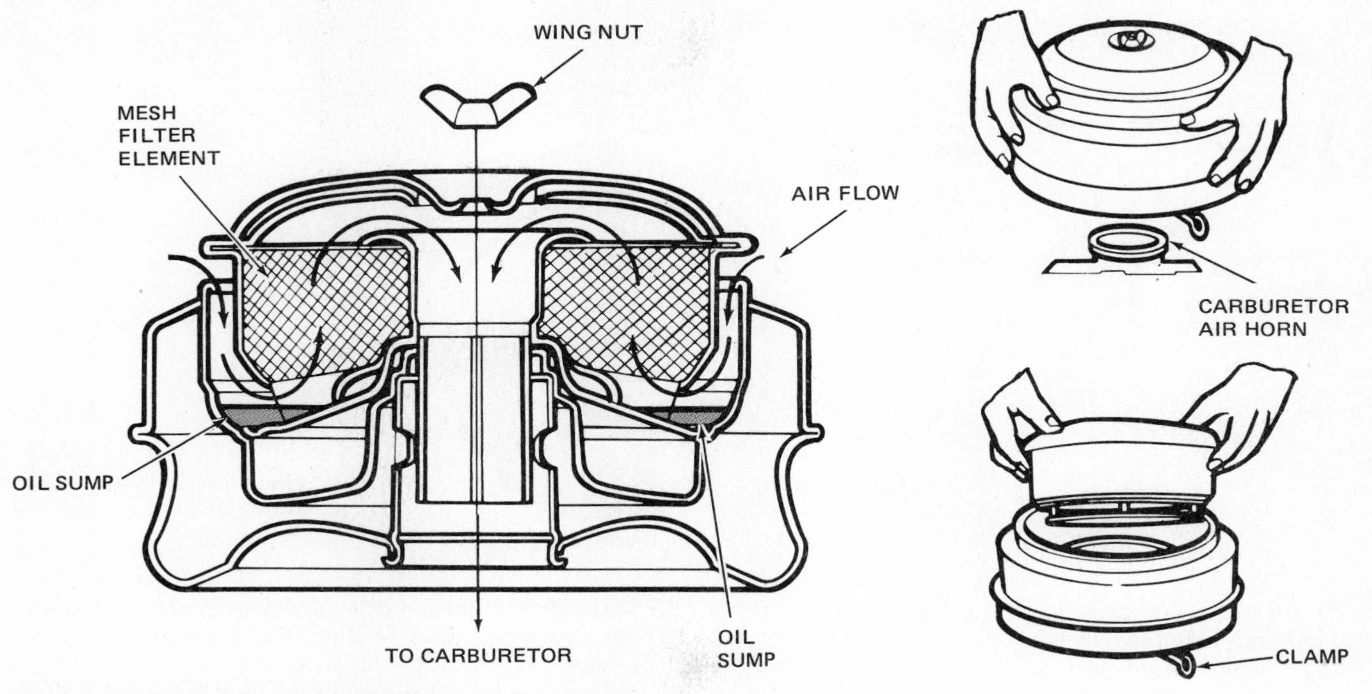

Fig. 9-5 An oil-bath air cleaner. (*Chrysler Corporation*)

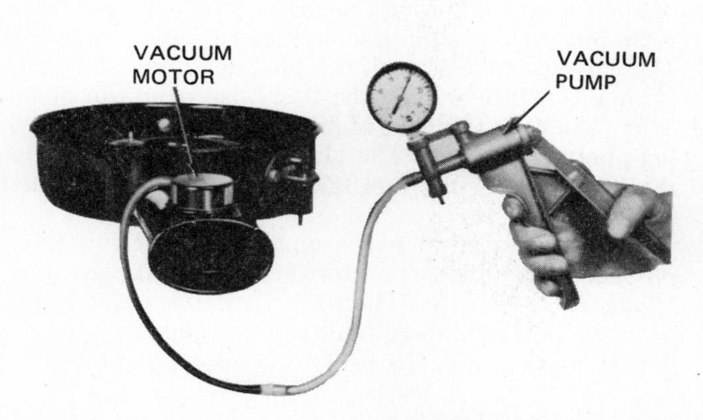

Fig. 9-6 Testing the vacuum motor on a thermostatically controlled air cleaner. (*Chrysler Corporation*)

spring, and open by a solenoid. When the system is checked with the engine cold, the solenoid should hold the choke knob in any position while the ignition switch is on. When the ignition switch is turned off, the knob should return to its in or choke-open position. After the engine has warmed up, a thermo (heat) switch should open. This disconnects the solenoid from the battery. Now when the choke knob is pulled out with the ignition on, the choke knob automatically returns to the in or choke-open position. Failure to operate properly is probably caused by a defective thermo switch or a defective solenoid.

9-4 Automatic-choke adjustment Automatic chokes are of several types, some that can be adjusted, others that require no adjustment. In normal operation, a properly adjusted choke does not usually get out of adjustment. It is possible for a choke to become stuck

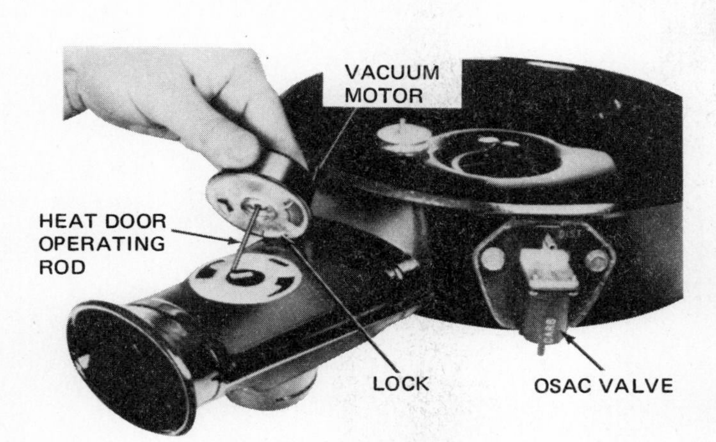

Fig. 9-7 Replacing the vacuum motor on a thermostatically controlled air cleaner. (*Chrysler Corporation*)

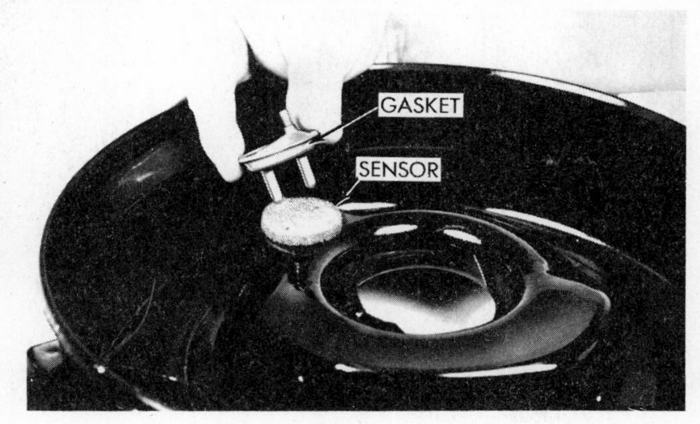

Fig. 9-8 Replacing the temperature sensor in a thermostatically controlled air cleaner. (*Chrysler Corporation*)

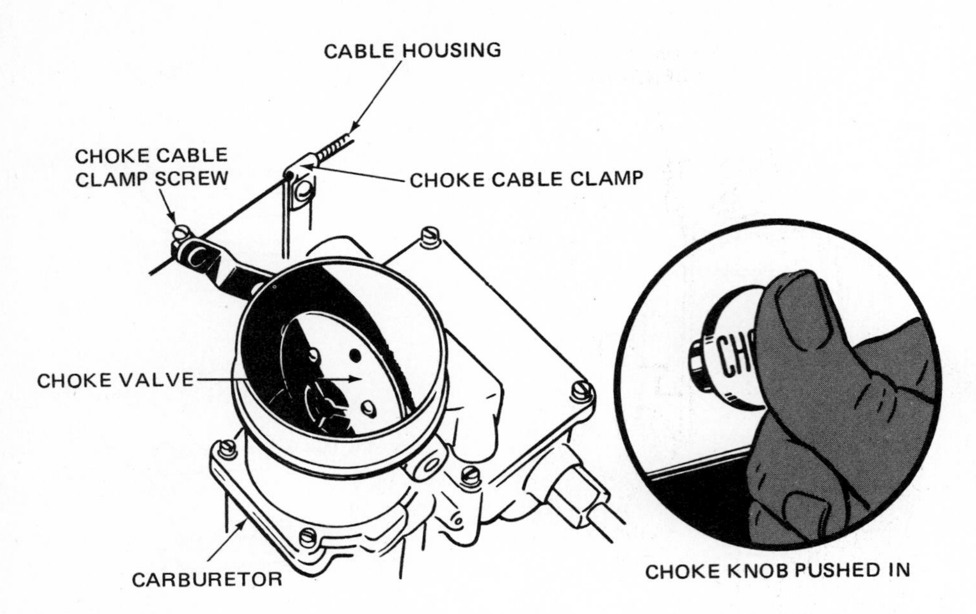

Fig. 9-9 Linkage to the choke valve in the carburetor.

because of accumulation of dirt or gum. Adjustments would not help this. The choke would require cleaning and freeing up and then readjustment. If hard starting is a problem, other causes should be eliminated before the choke is adjusted. If some other condition is causing the trouble (faulty spark plugs, for example), readjusting the choke could make the condition worse (by further fouling the plugs from an overrich mixture).

An electric-assist choke that is not adjustable is shown in Fig. 4-70. However, the carburetor-mounted choke shown in Fig. 4-61 can be adjusted. Adjustment is made by loosening the choke cover clamp screws and turning the cover one way or the other to lean out or enrich the mixture. Other chokes that have the thermostatic bimetal spring heated by the exhaust gas can be adjusted by bending the rod connecting the spring to the choke valve (Fig. 9-11). Details of automatic-choke adjustment are given in the manufacturer's service manuals.

9-5 Fuel filter service Various types of fuel filters are shown in Fig. 9-12. Fuel filters require no service except periodic checks to make sure they are not clogged, and replacement of the filter element or cleaning of the filter, according to type. On some cars, the filter is part of the fuel pump and can be removed so that the element can be replaced. Another type is the carburetor-mounted in-line fuel filter (Figs. 9-12 and 9-13). To service this type, the filter is removed by unclamping and detaching the fuel hose from the filter. When installing this type of filter, use the special crimping pliers (Fig. 9-13). The clamps must be located on the two sides of the fuel tube and filter, missing the beads.

To change a filter that is mounted in the fuel inlet passage inside the carburetor, first detach the fuel line from the carburetor. Then the carburetor nut is removed so that the dirty filter can be slipped out and a new filter installed (Fig. 9-12).

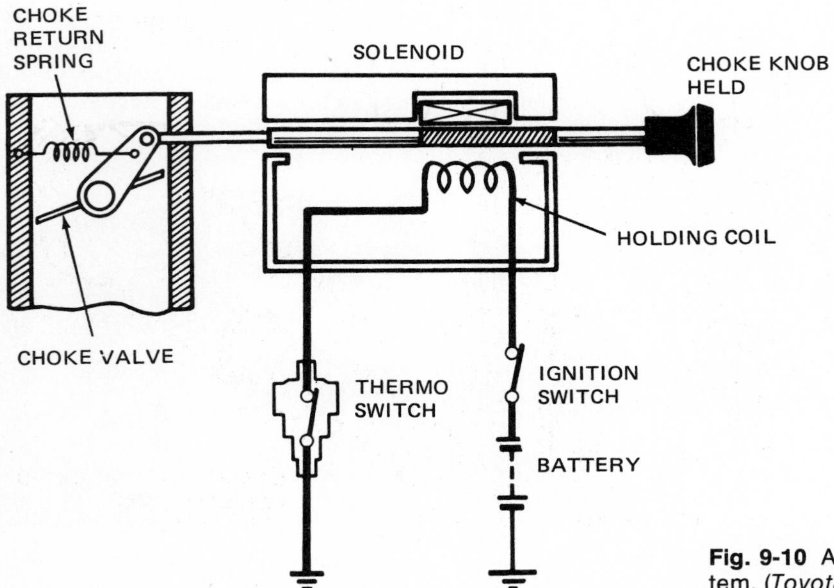

Fig. 9-10 A manual choke with an automatic-return system. (*Toyota Motor Sales Company, Ltd.*)

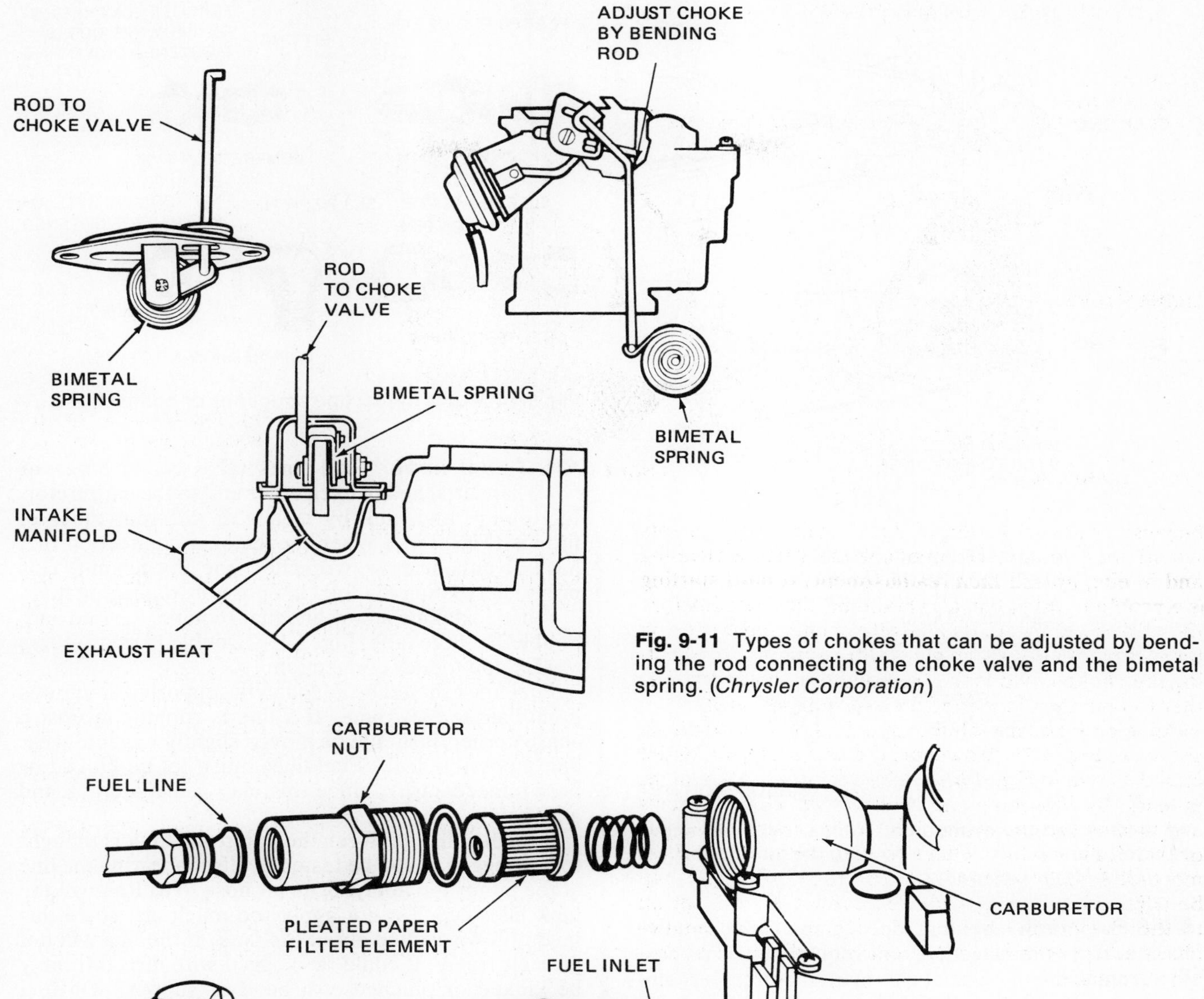

Fig. 9-11 Types of chokes that can be adjusted by bending the rod connecting the choke valve and the bimetal spring. (*Chrysler Corporation*)

Fig.9-12 Various types of fuel filters. (*Chrysler Corporation*)

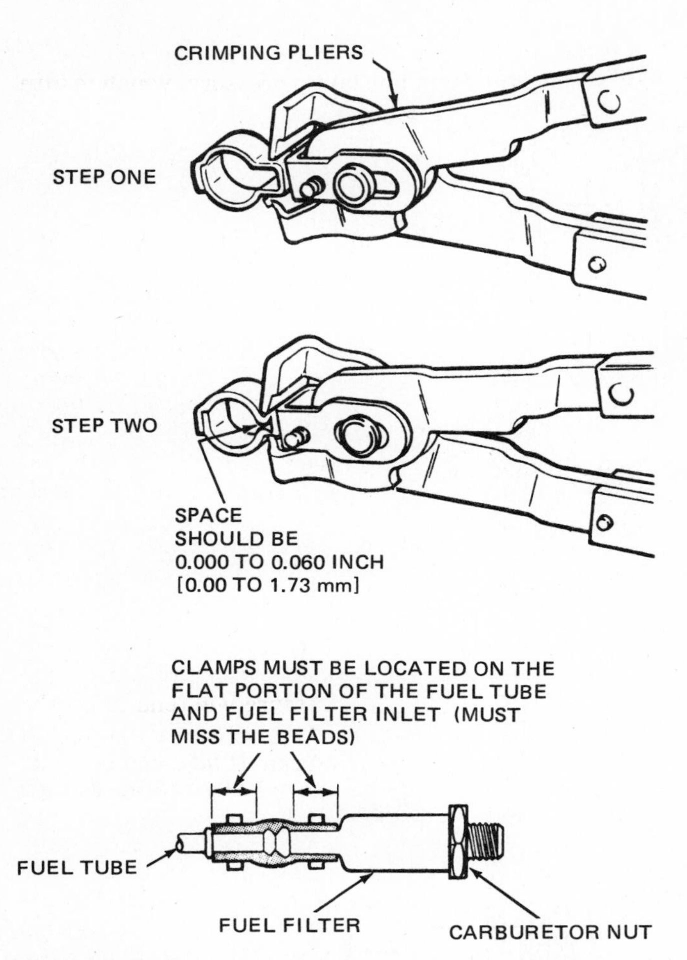

Fig. 9-13 A carburetor-mounted in-line fuel filter, showing how to crimp the clamp attaching the filter to the fuel line. (*Ford Motor Company*)

9-6 Fuel tank Seldom does a fuel tank require service. If it is damaged, it will require repair or possibly replacement. Some car manufacturers recommend that the fuel tank be drained once a year to remove accumulated dirt and water. Some tanks have a drain plug. Fuel tanks usually are supported by straps bolted to the frame (Fig. 3-6). Removal of these straps permits removal of the tank. Before the tank is removed, the fuel-gauge wire and fuel line must be detached. Fuel should be drained from the tank into a safety container.

CAUTION: If a tank is to be repaired, great care must be used to make sure that it is completely free of gasoline vapors. A spark from a hammer blow or from a torch might set off vapor remaining in the tank and cause an explosion.

The fuel filter or strainer can be cleaned, if the tank is removed, by blowing air through the strainer from an air hose. Air should be directed into the strainer from the fuel outlet (Fig. 3-6).

When installing the tank, make sure that the support straps are firmly fastened. If the fuel-gauge terminal is dirty, clean it so that good contact will be made when the wire is connected. Make sure that the sending unit is properly grounded.

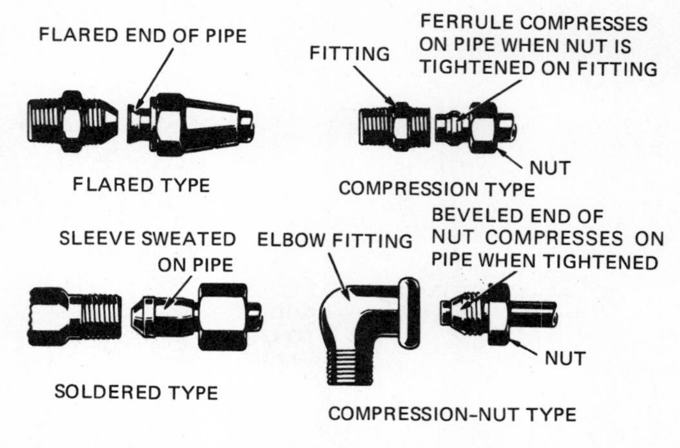

Fig. 9-14 Types of fuel line couplings or connections.

9-7 Fuel lines Fuel lines (also called *pipes* or *tubes*) are attached to each other and to the carburetor, fuel pump, and tank by means of different types of fittings (Fig. 9-14). When loosening a connection that has two nuts, use two wrenches, one on each nut. This prevents twisting the line and possibly damaging it.

When installing a new line that requires a flared end, double-flare the tube (Fig. 9-15). Double flaring assures a safer and tighter connection.

Fuel lines should be adequately supported at various points along the frame. If a line is rubbing against a sharp corner, it should be moved slightly to avoid wear and a possible leak. Fuel lines must not be kinked or bent unnecessarily, since this may cause a crack and a leak.

If the fuel line between the pump and tank is thought to be clogged, it can be tested by disconnecting the line at the pump and applying an air hose to it. Remove the tank filler cap. Do not apply too much air, since this might blow gasoline out of the tank. If the line will not pass air freely, it could be clogged with dirt, or it may be kinked or pinched at a bend or support. On fuel tanks with an internal filter (Fig. 3-6), the filter may have become clogged. Kinked or pinched fuel lines should be replaced. Even if a pinched or kinked line can be straightened, it could later crack and leak.

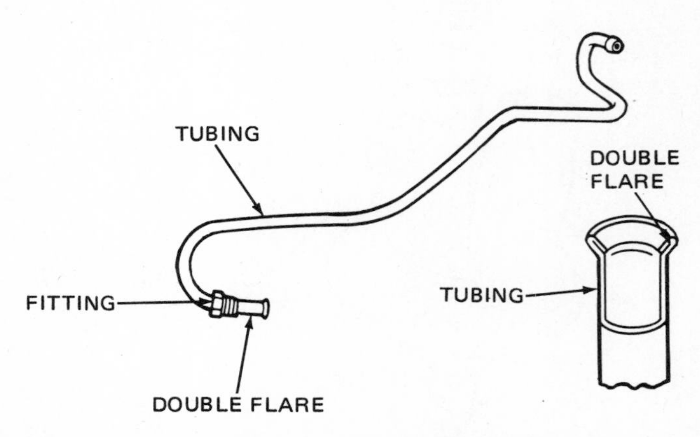

Fig. 9-15 Fuel line showing the double flare at one end. (*ATW*)

⚙ 9-8 Fuel gauges The fuel gauge provides the driver with a continuous indication of the fuel level in the fuel tank. The fuel-gauge indicating system consists of two components: the tank unit and the instrument-panel unit. There are two types of fuel gauges: the magnetic gauge and the thermostatic gauge (⚙ 3-9). There is very little in the way of service that fuel gauges require. Defects in either the dash unit or the tank unit usually require replacement of the defective unit.

If a fuel gauge is defective or if malfunctioning of the gauge is suspected, check out the gauge with a fuel-gauge tester. If a fuel-gauge tester is not available, substitute a new tank unit for the old one. This can be done without removing the old tank unit by disconnecting the tank-unit terminal lead from the old unit and connecting it to the terminal of the substitute unit. Then connect a jumper wire from the frame of the substitute unit to a ground on the car.

With these connections made, turn on the ignition switch and operate the float arm of the substitute tank unit. If the dash unit now works and indicates as the float arm is moved up and down, then the old tank unit is defective. If the dash unit still does not work, then either it is at fault or the wiring is defective.

NOTE: On the thermostatic fuel gauge, allow 2 minutes for the thermostat in the gauge to heat up. This assures that the instrument panel unit will be indicating normally.

⚙ 9-9 Check the fuel pump The fuel pump can be checked for pressure, capacity, and vacuum with the fuel pump tester (⚙ 8-6 to 8-8). Readings obtained should be compared with the specifications in the manufacturer's service manual. A quick check of fuel pump operation can be made by disconnecting the fuel line from the carburetor and sticking the fuel line in a suitable container. Then crank the engine. Ignition should be off and the distributor primary lead should be disconnected from the coil so that the engine does not start. During cranking, the fuel pump should deliver a spurt of gasoline with each rotation of the engine camshaft. Wipe up any spilled gasoline with cloths and put the cloths outside to dry.

In addition to checks of the operating action, the fuel pump should be checked for leaks. Leaks might occur at fuel line connections or around sealing gaskets. For example, leaks may occur at the joint between the fuel pump cover and the fuel pump body.

If the fuel pump pressure is too high or too low, if the pump does not deliver fuel normally to the carburetor, if leaks show up, or if the pump is noisy, then the pump should be replaced. The following section describes various pump troubles and their causes.

⚙ 9-10 Fuel pump troubles The trouble diagnosis chart in Chap. 8 lists various fuel system troubles and their causes. Some of these causes may be in the fuel pump. Many troubles are in the other fuel system components or engine components. Following is a discussion of fuel system troubles that might be caused by the fuel pump.

1. Insufficient fuel delivery Insufficient fuel delivery could result from low pump pressure, which in turn could be caused by any of the following:

1. Broken, worn-out, or cracked diaphragm
2. Improperly operating fuel pump valves
3. Broken diaphragm spring
4. Broken or damaged rocker arm
5. Clogged pump filter screen
6. Air leaks into sediment bowl through loose bowl or worn gasket

In addition to these causes of insufficient fuel delivery resulting from conditions within the pump, many other conditions outside the pump could prevent delivery of normal amounts of fuel. They include such causes as a clogged fuel tank cap vent, a clogged fuel line or filter, air leaks into the fuel line, and vapor lock. In the carburetor, an incorrect float level, a clogged inlet screen, or a malfunctioning inlet needle valve would prevent delivery of adequate amounts of fuel to the carburetor.

2. Excessive pump pressure High fuel pump pressure will cause delivery of too much fuel to the carburetor. The excessive pressure will tend to lift the needle valve off its seat so that the fuel level in the float bowl will be too high. This results in an overrich mixture and excessive fuel consumption. Usually, high pump pressure would result only after a fuel pump has been rebuilt. If a fuel pump has been operating satisfactorily, it is unlikely that its pressure would increase enough to cause trouble. High pressure could come from installation of an excessively strong diaphragm spring or from incorrect reinstallation of the diaphragm. If the diaphragm is not flexed properly when the cover and housing are reattached, it may have too much tension and produce too much pressure.

3. Fuel pump leaks The fuel pump will leak fuel from any point where screws have not been properly tightened and anywhere the gasket is damaged or incorrectly installed. If tightening the screws does not stop the leak, the pump must be serviced or replaced. Leaks also may occur at fuel line connections which are loose or improperly coupled.

4. Fuel pump noises A noisy pump is usually the result of worn or broken parts inside the pump. These include a weak or broken rocker-arm spring, a worn or broken rocker-arm pin or rocker arm, or a broken diaphragm spring. In addition, a loose fuel pump or a scored rocker arm or cam on the camshaft may cause noise. Fuel pump noises may sound something like engine valve tappet noise, since its frequency is the same as camshaft speed. If the noise is bad enough, it can actually be felt by gripping the fuel pump firmly with your hand. Also, careful listening will usually disclose that the noise is originating in the vicinity of the fuel pump. Tappet noise is usually distributed along the engine or is located distinctly in the valve compartment of the engine.

⚙ 9-11 Fuel pump removal As a first step in removing the fuel pump, wipe off any dirt or accumulated

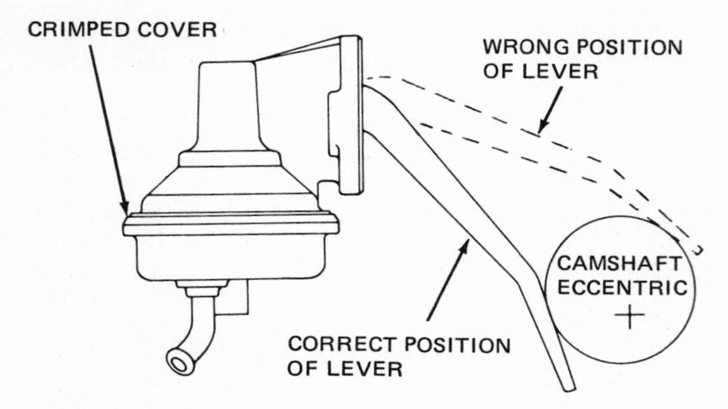

Fig. 9-16 Correct positioning of the fuel pump lever on the camshaft eccentric. (*Chevrolet Motor Division of General Motors Corporation*)

grease so that dirt will not get into the engine. Then take off the heat shield (where present), and disconnect the fuel lines (Fig. 9-16). Remove attaching nuts or bolts, and lift off the pump. If it sticks, work it gently from side to side, or pry lightly under the mounting flange with a screwdriver to loosen it. Do not damage the flange or attaching studs. On engines using a push rod to operate the fuel pump, remove the rod so that it can be examined for wear or sticking.

⚙ 9-12 Fuel pump service Today, many fuel pumps cannot be disassembled for any service. They are put together by crimping (Fig. 9-16) and are only serviced by complete replacement. Earlier fuel pumps could be disassembled. On these, a repair kit could be purchased which had all the necessary parts to rebuild the fuel pump. Now the cost of labor is high and the cost of a new fuel pump is relatively low. Usually it is cheaper to buy a new pump than it is to repair a defective pump.

⚙ 9-13 Fuel pump installation Make sure that the fuel line connections are clean and in good condition. All old gasket material must be off the mounting pad on the engine and the fuel pump flange. On some engines, connect the fuel lines to the pump before attaching it to the engine. (On other engines, the lines are attached after the pump is installed.) Then place a new gasket on the studs of the fuel pump mounting or over the opening in the crankcase.

Many automotive manufacturers recommend coating both sides of the new gasket with oil-resistant sealer before installation. The mounting surface of the engine should be clean. Insert the rocker arm of the fuel pump into the opening, making sure that the arm goes on the proper side of the camshaft (Fig. 9-16) or that it is centered over the push rod. If it is difficult to get the holes in the fuel pump flange to align with the holes in the crankcase, crank the engine until the low side of the camshaft eccentric is under the fuel pump rocker arm.

Some V-type engines have a push rod between the camshaft and the fuel pump. It may be necessary to use "mechanical fingers" or other holding tool to hold the push rod up in the cylinder block while starting the mounting bolts. Tighten the mounting bolts or nuts to the specified torque. Then check the operation of the fuel pump (⚙ 9-9).

⚙ 9-14 Servicing emission control systems Someday cars that give off excessive amounts of air pollutants from the crankcase, exhaust system, fuel tank, and carburetor may not be allowed on the streets. Stronger automotive emission standards (laws that limit air pollution from the car), plus state inspections that include testing of the exhaust gas and proposed mandatory maintenance requirements indicate that such a time is coming. Generally, these programs are referred to as *inspection and maintenance,* or I/M. Their purpose is to ensure that the cars on the road contribute the minimum amount possible to the air pollution problem.

The operation of the most widely used automotive emission control systems is covered in earlier chapters. In the following sections, possible troubles with automotive "smog devices" are discussed. Then the checks and service recommended for each are given.

NOTE: It is illegal for a mechanic to tamper with safety devices or emission controls on vehicles or vehicle engines. The penalty is a fine of $2500 for every vehicle tampered with. When servicing engines and emission controls, do only those jobs specified in the manufacturers' service manuals. If a mechanic removes, disconnects, damages, or renders inoperative any safety or emission control device, the mechanic is breaking the law and may be prosecuted.

⚙ 9-15 PCV system troubles Several engine troubles can result from defective conditions in the PCV system (⚙ 3-21). These troubles can also result from faults in other systems as explained in Chap. 8. Here, we look only at troubles arising from the PCV system.

Rough idle and frequent stalling could result from a plugged or stuck PCV valve or from a clogged PCV air filter (Fig. 9-1). In either case, the remedy is to replace the valve or filter.

Vapor flow into the air cleaner and oil in the air cleaner can result from backflow. Instead of filtered air flowing into the crankcase, vapors from the crankcase are flowing into the air cleaner. The cause is a plugged PCV valve or plugged or leaking condition somewhere in the PCV system. It could also be caused by worn piston rings or cylinder walls which allow more blowby than the PCV system can handle.

Sludge or oil dilution in the crankcase can result from a plugged PCV valve or hose. Either condition prevents normal circulation of the ventilating air and blowby gases.

⚙ 9-16 Testing the PCV system Here is a quick check to make of the PCV system and the PCV valve. With the engine running, remove the PCV valve from the valve cover (Fig. 9-17). A hissing sound should be heard from the valve. This indicates that the valve and the line to it are not plugged. Next, shake the valve. A rattle should be heard. This indicates that the valve is not stuck. Place your thumb over the opening in the

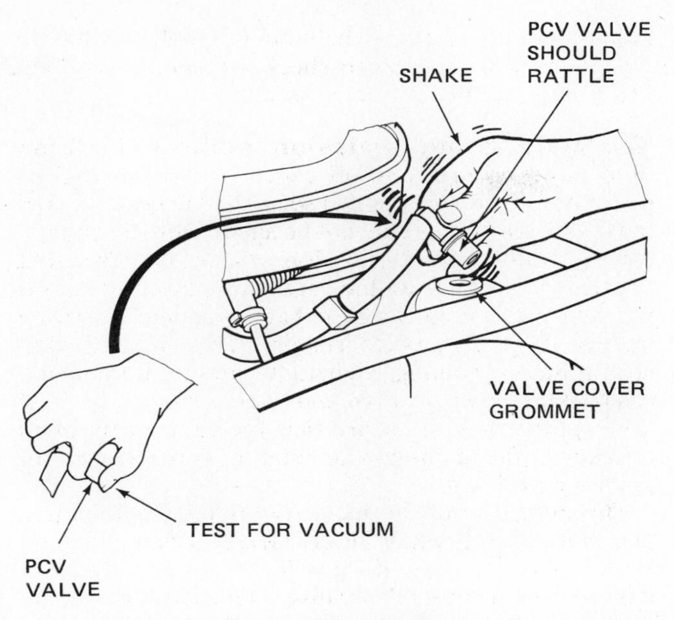

Fig. 9-17 Making a quick check of the PCV valve. (*Chrysler Corporation*)

valve and feel for vacuum (Fig. 9-17). If there is no vacuum, or if you can feel a positive pressure, then something is wrong. The PCV valve should be checked.

Toyota recommends checking the PCV valve by blowing through it from each end. Air should pass freely in the direction of the intake manifold and should be restricted (but not blocked) in the opposite direction. All hoses and connections should be checked for free flow.

Special testers are available which can be used to check the operation of the PCV valve. If the valve becomes clogged, it will cause engine loping (speeding up and slowing down) and rough idle. The PCV valve cannot be cleaned. A clogged or sticking PCV valve must be replaced. For engine loping and rough idle, install a new PCV valve. If the idling improves, leave the new PCV valve in. If the loping or rough idle persists, check for restrictions in the lines.

After the quick checks above are made, Chrysler recommends reinstalling the PCV valve in place. Then remove the oil filler cap. Hold a piece of stiff paper over the opening for the oil filler cap. In less than 1 minute, the vacuum in the crankcase should be attempting to pull the paper into the opening. If the PCV system does not pass these tests, replace the PCV valve and try the test again. If the system still does not pass the test, the hose may be clogged. It should be cleaned out or replaced. If neither the hose nor the PCV valve is the problem, the vacuum passage in the carburetor may be plugged. Clean the passage by twisting a ¼-inch (6.35 mm) drill through it with your fingers.

Chevrolet suggests another method of checking the PCV system. First, connect a tachometer. With the engine idling, remove the PCV valve from the valve cover. Be sure the hose remains attached to the PCV valve. Hold your thumb over the end of the PCV valve and note the change in engine speed on the tachometer. A decrease of less than 50 rpm indicates a plugged PCV valve.

⚙ 9-17 Servicing the PCV system The PCV valve must be replaced at regular intervals and whenever it clogs or sticks. When you install a new PCV valve, inspect and clean the system thoroughly. This includes all hoses, grommets, and connectors. Clean the insides of the hoses with a brush, and wash the outsides in solvent. Thoroughly clean all connectors, especially the elbow connection. On vented crankcase systems, wash the oil filler cap in solvent and shake it dry. Some types of oil-filler caps must not be dried with compressed air.

After all parts of the PCV system are clean, inspect them carefully. Replace any component that shows signs of damage, wear, or deterioration. Be sure the grommet that the PCV valve fits into is not damaged or torn (Fig. 9-17). Replace any cracked or brittle hose with hose of a similar type. Replace any component, hose, or fitting that does not allow a free flow of air after cleaning.

Air must be filtered before it enters the crankcase. There are two different methods in use. In one method, the carburetor air-cleaner filter does the cleaning. In this system, the hose for the crankcase ventilation air is connected to the downstream, or clean-air, side of the carburetor air filter. No special service is required. The second method of cleaning the crankcase ventilation air is to use a separate filter, called a PCV filter (Fig. 9-1). This filter mounts on the inside of the air-cleaner housing. Ventilation air comes into the air cleaner through the inlet or snorkel. It then passes through the PCV filter and into the crankcase.

Whenever the PCV system is serviced, the PCV filter must also be checked. To check the filter, remove the air-cleaner cover, and take out the PCV filter. Check it for damage, dirt buildup, and clogging. If the filter is clean, reinstall it in the air cleaner. A dirty or damaged PCV filter must be replaced.

⚙ 9-18 Evaporative emission control system troubles Here are troubles that might be caused by conditions in the fuel vapor emission-control system (⚙ 3-6 and 3-14).

Fuel odor or loss of fuel could be caused by several conditions. These include:

1. Overfilled fuel tank
2. Leaks in fuel, vapor, or vent line
3. Wrong or faulty fuel tank cap
4. Faulty liquid-vapor separator
5. Excessively high fuel volatility
6. Vapor line restrictor missing
7. Canister drain cap or hose missing

A collapsed fuel tank can result if the wrong fuel tank cap is installed or if the vacuum valve in the cap sticks. In either case, no air can enter to replace fuel being withdrawn by the fuel pump. The result could be a vacuum in the fuel tank great enough to allow atmospheric pressure to crush the tank.

Excessive pressure in the fuel tank could result from

a combination of high temperatures and a plugged vent line, liquid-vapor separator, or canister. Pressure can be released by turning the tank filler cap just enough to allow the pressure to slowly escape.

Many different engine-idling problems can result from faulty or improper connections of a hose. A plugged canister, vapor line restrictor missing, or high-volatility fuel can also cause poor idle.

9-19 Servicing evaporative control systems No testers are needed to check evaporative control systems (Fig. 3-4). Almost all problems can be found by visual inspection. Problems are also indicated by a strong odor of fuel. Some technicians use an infrared exhaust gas analyzer to quickly detect small vapor losses from around the fuel tank, canister, air cleaner, lines, or hose. Any loss will register on the HC meter of the exhaust analyzer.

Most problems with evaporative control systems can be noticed during an inspection. Typical defects are damaged lines, liquid fuel and vapor leaks, and missing parts. The filler cap can be damaged or corroded so that its valves fail to work properly. A problem with the fuel tank cap could result in deforming of the tank. This could also occur when the wrong cap is installed on the tank. Be sure that the fuel tank filler cap is the type specified by the manufacturer for the vehicle and that the cap seals the fuel tank.

To service the evaporative control system, inspect the fuel tank cap. Check the condition of the sealing gasket around the cap. If the gasket is damaged, replace the cap. Check the filler neck and tank for stains resulting from fuel leakage. Usually, you can trace a stain back to its origin. Then fix the cause of the leak. This may require replacing a gasket, clamp, or hose, or replacing the tank.

Inspect all lines and connections in the fuel and evaporative control systems for damage and leakage. Perform any necessary repairs. Check all clamps and connections for tightness.

NOTE: The hoses used in evaporative control systems are specially made to resist deterioration from contact with gasoline and gasoline vapor. When you replace a hose, make sure that the new hose is specified by the manufacturer for use in evaporative control systems. Sometimes this type of hose is marked EVAP.

Check the charcoal-canister lines for liquid gasoline. If any is present, replace the liquid-vapor separator or the liquid check valve.

Some types of charcoal canister require replacement of the canister filter (Fig. 9-18). To replace the canister filter, remove the canister, turn it upside down and remove the bottom cover. Pull out the old filter with your fingers, and put the new filter inside. If the canister itself is cracked or internally plugged, a new canister assembly should be installed.

9-20 Air-injection system troubles Troubles related to the air-injection system (Fig. 3-51) include noise, no air supply, backfire, and high HC and CO levels in the exhaust.

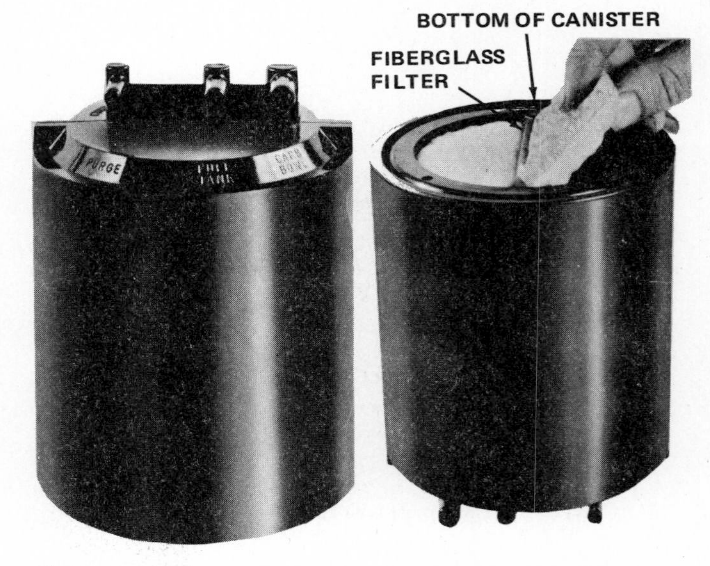

Fig. 9-18 Replacing the air filter in the charcoal canister.

Noise from the belt or air pump could result from a loose belt, loose air pump mounting bolts, worn pump bearings or other internal trouble, or air leaks from the system. On many cars, the pump is not repairable. It must be replaced if damaged. The air pump should be adjusted so the belt has the correct tension. Do not pry on the pump housing because this can ruin the pump. If you have to use a pry bar, pry as close to the pulley end as possible. Some manufacturers recommend using a belt tensioner and a belt-tension gauge (Fig. 9-19).

Air leaks should be stopped by tightening hose connections and replacing any hose that is defective.

If no air is getting to the air manifold, the exhaust gas will probably be high in HC and CO. Causes of no

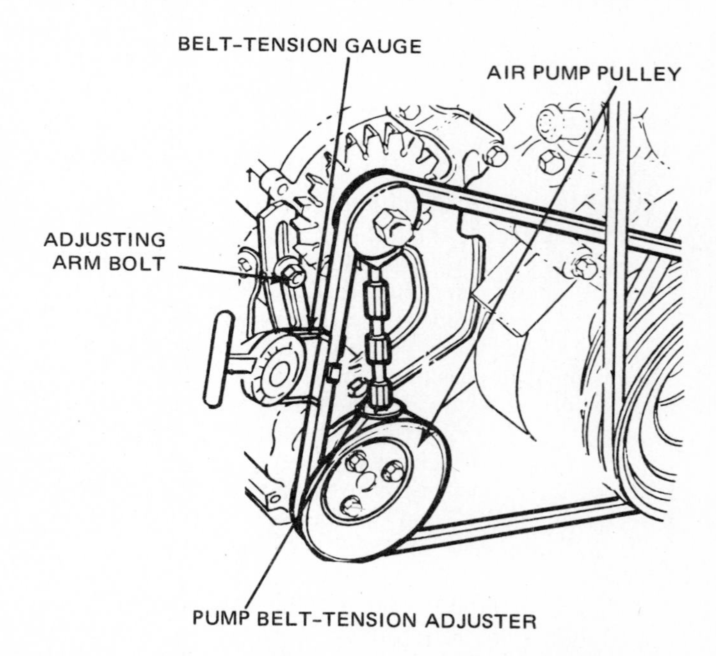

Fig. 9-19 A pump belt tension adjuster being used to hold the air pump in place while the adjustment bolts are tightened. (*Ford Motor Company*)

or inadequate air include a loose belt, frozen pump, leaks in the hoses or connections, and failure of the diverter or check valve. A defective pump or valve must be replaced. They are not repairable.

Backfire is usually caused by a defective diverter valve which fails to block off the air supply when the accelerator pedal is suddenly released. The same thing happens if the vacuum hose becomes disconnected or blocked.

⚙ 9-21 Air-injection system service No routine service is required on the air-injection system except to replace the air filter every 12,000 miles [19,312 km] on those systems using a separate air filter. All late-model air pumps use a centrifugal filter which requires no separate service. Rusted-out or corroded exhaust check valves should be replaced.

The air-pump drive belt should be checked periodically for wear and tension. Replace any belts that are worn, cracked, or brittle. Do not overtighten the air pump drive belt. Overtightening can cause early failure of the bearing in the air pump.

⚙ 9-22 Heated-air system troubles The heated-air system or thermostatically controlled air cleaner (Fig. 3-35) is designed to let the engine run better when cold, with a relatively lean mixture. It does this by adding heat to the air going into the carburetor almost as soon as the engine is started. This means the mixture is warm and the gasoline vaporizes better. If the heated-air system is not working to add heat, the control damper door has closed off the hot-air pipe. As a result, the mixture is not getting warmed. The cold engine will hesitate or stumble, or even stall, because the mixture is too cold and lean to fire consistently. If the damper does not shut off the hot-air pipe when the engine is hot, the mixture will be overheated so not enough mixture gets into the cylinders. Therefore, the engine will not develop full power.

Servicing of the thermostatically controlled air cleaner is covered in ⚙ 9-2.

⚙ 9-23 Exhaust gas recirculation troubles Trouble in the EGR system (⚙ 3-28) is reflected in poor engine performance. For example, rough engine idle and stalling could be caused by a leaky EGR valve or valve gasket that allows exhaust gas or air to enter the intake manifold during idling. A defective thermal vacuum switch (on engines having this valve) could cause vacuum to operate the EGR valve when it should not.

Poor part-throttle performance, poor fuel economy, and rough running on light acceleration could also be caused by the vacuum valve being defective. In addition, a sticking or binding EGR valve, or deposits in the EGR passages, could cause these conditions. If deposits have clogged the EGR passages, remove the manifold to clean them away.

If the engine stalls on deceleration, it could be due to a restricted vacuum line that is preventing the EGR valve from closing promptly. Detonation at part throttle could be caused by insufficient exhaust gas recirculation. This could be caused by clogged or damaged

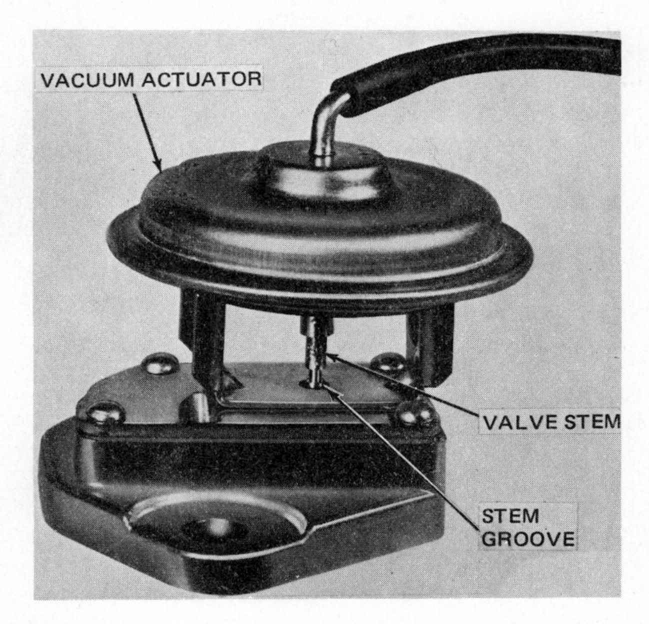

Fig. 9-20 EGR valve with exposed valve stem. (*Chrysler Corporation*)

hoses, an EGR valve that is stuck, or a defective thermal vacuum switch.

⚙ 9-24 Servicing the EGR system Some EGR valves have the stem visible under the diaphragm (Fig. 9-20). Check this type of EGR valve with the engine warmed up and idling. With the transmission in neutral, snap the throttle open to bring the engine speed up to about 2000 rpm. The EGR stem should move up, indicating the valve has opened. If it does not, connect a vacuum tester to the vacuum hose on the valve.

With the engine warmed up and idling, apply about 8 inches Hg [203 mm] to the valve. The valve should operate. If it does not, it is either defective or dirty. You can tell when the valve operates because the engine will idle roughly and may even stall. A backpressure-type valve will open too soon when there is a restriction in the exhaust system.

To test the thermal-vacuum switch connect a vacuum gauge and a vacuum pump as shown in Fig. 9-21. With the engine cold, no vacuum should pass through the switch. When the engine warms up, vacuum should pass through.

Typical service intervals for EGR systems are to check the system every 12 months or 12,000 miles [19,312 km] if the engine is running on leaded gasoline. If the engine is running on unleaded gasoline, check the system half as often, at every 24 months or 24,000 miles [38,624 km]. Some cars made by Chrysler have an EGR maintenance-reminder light that comes on at 15,000 miles [24,140 km] to remind the driver to have the system checked. However, many late-model cars do not require any regular check. Instead, if trouble develops then the system should be checked as explained above.

⚙ 9-25 Catalytic converter troubles Catalytic converter (⚙ 3-29) troubles would show up as noise, BB-

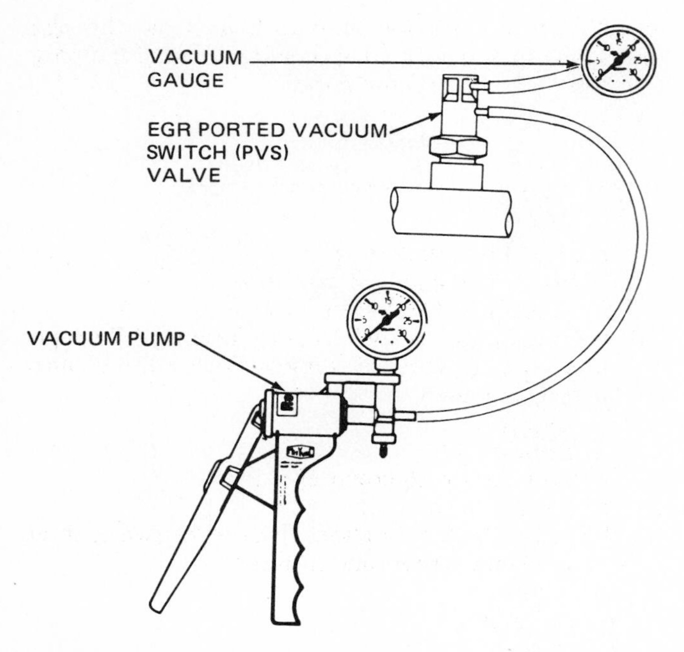

Fig. 9-21 Testing the EGR thermal-vacuum switch. (*Ford Motor Company*)

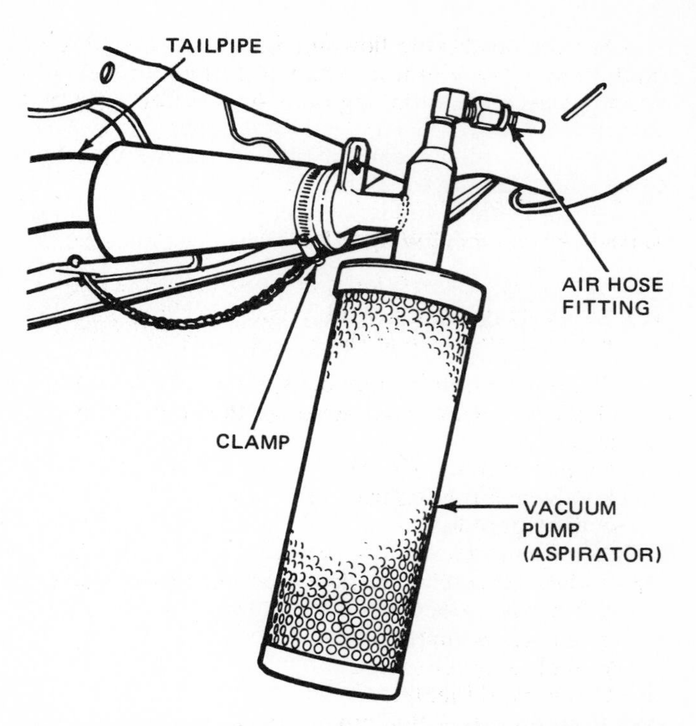

Fig. 9-22 Vacuum pump, or aspirator, mounted on the tail pipe of a car to change the beads in a pellet-type catalytic converter. (*American Motors Corporation*)

size particles coming out the tail pipe, a rotten-egg smell, or high CO and HC levels in the exhaust gas.

Noise could be caused by loose exhaust-pipe joints, a ruptured converter, or a loose or missing catalyst replacement plug (GM and AMC).

BB-size particles coming out the tail pipe mean the converter has been overheated so the catalyst support has warped. This allows the beads to be blown out by the exhaust gas. The condition can happen only on the type of converter using beads (GM and AMC). The remedy is converter replacement.

A rotten-egg smell comes from hydrogen sulfide (H_2S) that the catalytic converter is producing. The S, or sulfur, is in the gasoline. Some gasolines have more than others. A different brand of gasoline may eliminate the smell. Also, check the carburetor adjustments. The smell is more noticeable when a momentarily rich mixture enters a hot converter.

⚙ 9-26 Catalytic converter service Damaged or overheated converters must be replaced. They are not repairable. However, on the bead-type converter, the old beads can be removed and a fresh charge of beads installed. Figures 9-22 and 9-23 show the special devices required. The "aspirator" or vacuum pump is turned on while the vibrator and can are attached. It keeps the beads from falling out when the converter filler plug is removed. After the vibrator and can are attached, the vacuum is turned off and the air supply to the vibrator is turned on. Beads will now start falling in the can. It takes about 10 minutes to empty the converter.

To install new beads, dump the old beads and fill the can with new beads. Attach the can to the vibrator. Turn on the air and vacuum lines. The beads will flow from the can into the converter.

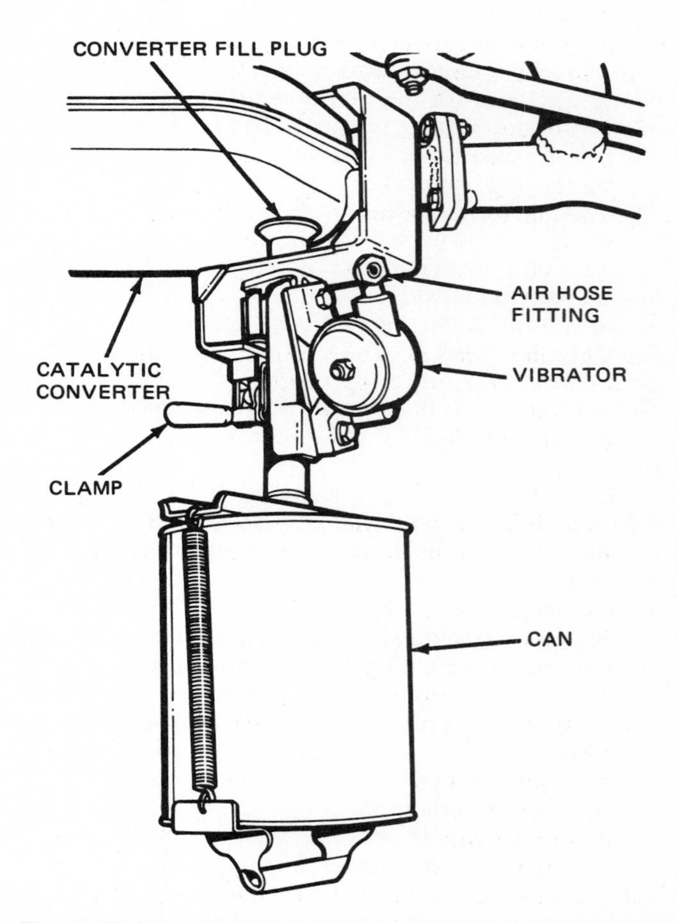

Fig. 9-23 Vibrator mounted on a catalytic converter. (*American Motors Corporation*)

After the beads stop flowing, disconnect the air hose and remove the vibrator. The vacuum pump will prevent the beads from falling out. The converter should be filled flush with the fill-plug hole. Coat the plug threads with antisieze compound and install the plug. Then remove the vacuum pump.

Chapter 9 review questions

Select the *one* correct, best, or most probable answer to each question. Then check your answers against the correct answers given at the end of the book.

1. A broken diaphragm or spring, stuck valves, clogged screen or air leaks in the fuel pump can cause:
   a. high pump pressure
   b. low pump pressure
   c. high float level
   d. a rich mixture.
2. A high fuel level in the float bowl and an excessively rich mixture could result from:
   a. a cracked diaphragm
   b. high pump pressure
   c. a kinked fuel line
   d. pump valves defective.
3. Periodically, the PCV valve should be:
   a. adjusted
   b. replaced
   c. cleaned
   d. reverse-connected.
4. The exhaust-gas analyzer tests for:
   a. HC and NO_x
   b. HC and CO
   c. NO_x and CO_2
   d. H_2O and NO_x.
5. The fine for tampering with emission controls is:
   a. $5
   b. $100
   c. $1000
   d. $2500.
6. The only service the evaporative control system requires is periodic replacement of the:
   a. canister
   b. canister filter
   c. charcoal
   d. fuel lines.
7. In addition to replacing the filter (on some cars) in the air-injection system, the only other periodic service is to:
   a. replace the air line
   b. replace the manifold nozzles
   c. replace the fan blades
   d. check the drive belts.
8. If the EGR valve remains open during idling, the result would be:
   a. engine stalling
   b. smooth idle
   c. quick warm-up
   d. engine overheating.
9. A rotten-egg smell with BB-size particles coming out the tail pipe could be caused by:
   a. worn carburetor jets
   b. a defective fuel pump
   c. a defective catalytic converter
   d. a defective EGR system.
10. To prevent loss of air-cleaner efficiency, the paper filter element must be removed from the cleaner periodically and:
    a. replaced
    b. cleaned
    c. blown out with compressed air
    d. washed in solvent.
11. To adjust the hot-air choke, loosen the two or three cover clamp screws and turn the:
    a. clamp
    b. carburetor
    c. choke valve
    d. cover.
12. If a fuel tank is to be repaired, care must be used to make sure it is completely free of:
    a. water vapor
    b. attaching studs
    c. gasoline vapor
    d. fuel-gauge wires.
13. Double-flaring the fuel line tube assures:
    a. shorter tubing
    b. a safer and tighter connection
    c. stiffer tubing
    d. longer tubing.
14. To locate trouble in a fuel-gauge system, temporarily substitute for the old unit a:
    a. new instrument panel unit
    b. new wire
    c. new tank unit
    d. new switch.
15. Mechanic X says excessive hydrocarbon (HC) emissions in the exhaust gas are caused by an overrich air-fuel ratio. Mechanic Y says excessive hydrocarbon (HC) emissions in the exhaust gas are caused by a misfiring spark plug. Who is right?
    a. X only
    b. Y only
    c. both X and Y
    d. neither X nor Y.
16. Mechanic Y says that the air bypass (diverter) valve can be tested by squeezing the hose from the valve to the air cleaner, and revving the engine to see if air is diverted during deceleration. Mechanic Y says that a catalytic converter can overheat if the air-injection system malfunctions. Who is right?
    a. X only
    b. Y only
    c. both X and Y
    d. neither X nor Y.
17. When a mechanic tests the PCV system, the tester indicates REPAIR. This means that:

a. there is a vacuum in the crankcase
b. there is a pressure in the crankcase
c. there is an air leak into the crankcase
d. none of the above.

18. Pressure and volume tests of a mechanical fuel pump are both below specs. Mechanic X says that an air leak in the fuel line between the tank and pump could be the cause. Mechanic Y says a plugged fuel tank pickup filter could be the cause. Who could be right?
a. X only
b. Y only
c. either X or Y
d. neither X nor Y.

19. While the engine is running, a mechanic pulls the PCV valve out of the valve cover and puts a thumb over the valve opening. There are no changes in engine operation. Mechanic X says the PCV valve could be stuck in the open position, Mechanic Y says the hose between the intake manifold (carburetor base) and the PCV valve could be plugged. Who could be right?
a. X only
b. Y only
c. either X or Y
d. neither X nor Y.

20. A car equipped with an air pump emission control system backfires when decelerating. Which of these should the mechanic check?
a. the operation of the exhaust manifold check valve(s)
b. the output pressure of the air pump
c. the operation of the diverter or gulp valve
d. the air manifolds for restrictions.

CHAPTER 10

SERVICING SINGLE-BARREL CARBURETORS

After studying this chapter, and with proper instruction and equipment, you should be able to:

1. Remove and install carburetors.
2. Check and reset the choke.
3. Check and adjust the idle-stop solenoid.
4. Check and adjust the idle mixture with propane.
5. Clean and inspect single-barrel carburetors.
6. Make internal adjustments on single-barrel carburetors.
7. Make external adjustments on single-barrel carburetors.

10-1 Servicing single-barrel carburetors This chapter covers the servicing of two different single-barrel carburetors. The first carburetor discussed is the Rochester model 1ME, as used by Chevrolet. The second carburetor is the Holley model 1945, which is on several versions of the Chrysler 225-cubic-inch [3.7-L] slant-six engine. First the carburetors are described. Then adjustments and servicing procedures are discussed.

When a carburetor or any other part of the fuel system is worked on, cleanliness is essential and safety cautions must be observed. Before you begin shop work on carburetors and fuel systems, review ⚙ 8-1, Cleanliness and Safety.

⚙ 10-2 Rochester model 1ME carburetor The Rochester model 1ME carburetor is a one-barrel carburetor (Fig. 10-1), which is also called a "Monojet." Figure 4-39 shows the main metering system. The float system is shown in Fig. 4-25. To prevent an excessively rich air-fuel mixture when the engine is hot, the idle system has a hot-idle compensator (⚙ 4-14). The choke has an electrically heated coil. A vacuum break is mounted externally on the air horn and connects to the choke lever through a link.

A fuel filter is located in the fuel inlet (Fig. 4-25), along with a check valve. The check valve prevents fuel draining out in case the car rolls over. An idle-stop solenoid is used (Fig. 10-1). This allows the throttle valve to close completely when the engine is turned off so the engine cannot diesel, or run on.

Following sections cover the on-car adjustments of the carburetor, and the removal, servicing, and installation of the carburetor.

⚙ 10-3 On-car service The following on-the-car adjustments can be made to the Monojet 1ME carburetor:

1. Float level (Fig. 10-2)
2. Metering rod (Fig. 10-3)
3. Fast idle (Fig. 10-4)
4. Choke coil lever (Fig. 10-5)
5. Automatic choke (Fig. 10-6)
6. Fast-idle cam (Fig. 10-7)
7. Vacuum break (Fig. 10-8)
8. Unloader (Fig. 10-9)
9. Idle speed (Fig. 10-10)

Figures 10-2 to 10-10 show the procedures for making each of these adjustments. Perform each of the numbered steps in the sequence shown in each illustration. Refer to the manufacturer's service manual for the specifications for these adjustments. The specifications

Fig. 10-1 A Rochester model 1ME single-barrel carburetor. (*Chevrolet Motor Division of General Motors Corporation*)

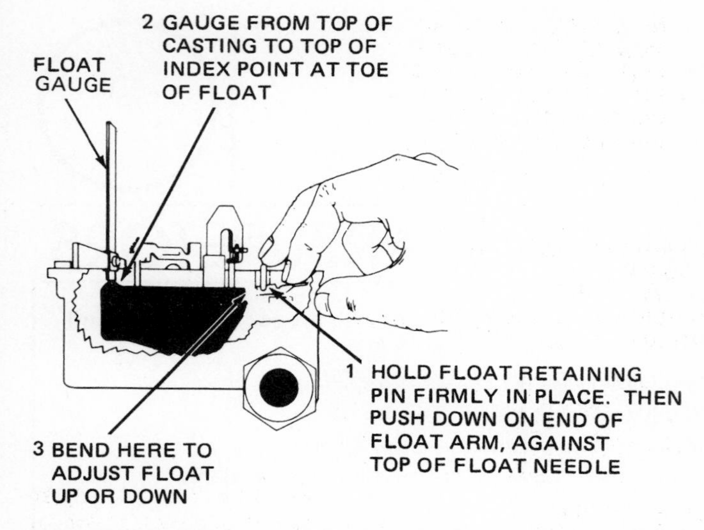

Fig. 10-2 Float level adjustment. (*Chevrolet Motor Division of General Motors Corporation*)

for the idle-speed adjustment (Fig. 10-10) are given on the vehicle emission control information label in the engine compartment. Following sections describe the procedure for checking the choke, and how to check and adjust the idle mixture using propane.

⚙ 10-4 Checking the choke Remove the air cleaner. Hold the throttle half open, and then open and close the choke valve several times (Fig. 10-11). If the linkage or the choke sticks, clean the sticking parts with a spray can of choke or carburetor cleaner (Fig. 10-11). Follow the directions on the can. If the cleaner does not free the choke and linkage, remove the carburetor for additional cleaning (⚙ 10-7).

Make sure all vacuum hoses are connected properly and in good condition. Replace or tighten hoses as necessary.

The vacuum-break diaphragm shaft should be fully extended when the engine is off. If the shaft is not fully extended, replace the vacuum-break assembly.

Check the position of the choke valve when the engine is cold. When the throttle is opened slightly, the choke valve should be closed completely. Install the air cleaner and make the following running test.

1. Start the engine. Measure the time in seconds it takes the choke valve to reach the full-open position. If the valve fails to reach wide open within 3½ minutes, make these additional checks.
2. Check the voltage at the choke heater connection with the engine running. If the voltage is between 12 and 15 volts, replace the electric choke unit.
3. If the voltage is low or zero, check all wiring and connections. If the connections at the oil-pressure switch are faulty, the oil warning light will be off with the ignition key turned to ON with the engine not running. If a fuse is blown, the radio or turn signals will not work. Repair connections or wires as required.
4. If step 3 does not correct the problem, replace the oil-pressure switch. Do not use a gasket between the choke cover and housing. This would destroy the necessary ground connection.

⚙ 10-5 Checking the idle-stop solenoid This solenoid (Figs. 10-1 and 10-10) allows the throttle valve to close completely when the engine is turned off so that the engine cannot diesel. If the solenoid is suspected of not working properly, check it as follows:

1. Turn ignition key to ON but do not start the engine.
2. Open the throttle to allow the solenoid plunger to extend.
3. Hold the throttle wide open. Feel the end of the plunger and disconnect the wire at the solenoid. The plunger should move as the spring pushes it out.
4. If the plunger does not move, back out the hex screw (Fig. 10-10) one full turn and repeat steps 2 and 3. If the plunger now moves, reconnect the wire to the solenoid and adjust idle speed (Fig. 10-10).
5. If the plunger does not move, connect a test light between the solenoid feed wire and ground. If the light glows, replace the solenoid.

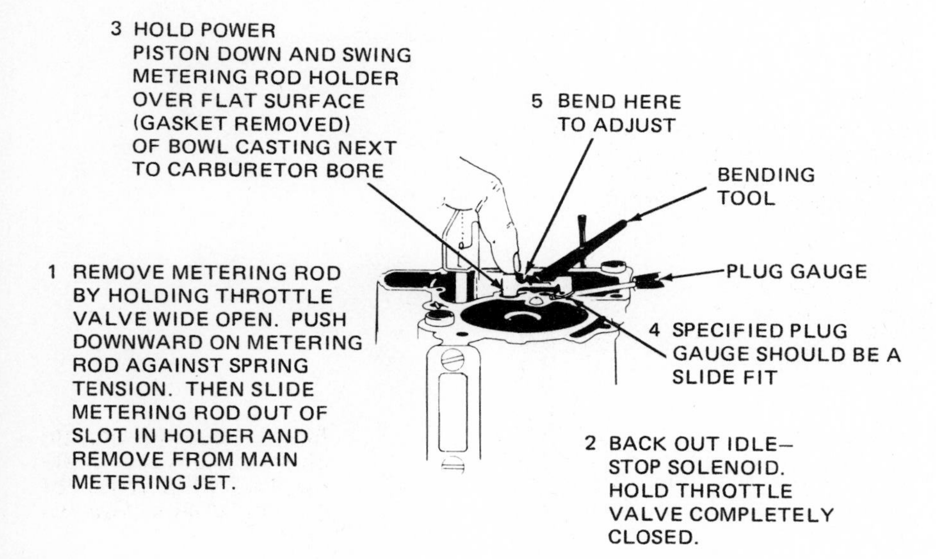

Fig. 10-3 Metering rod adjustment. (*Chevrolet Motor Division of General Motors Corporation*)

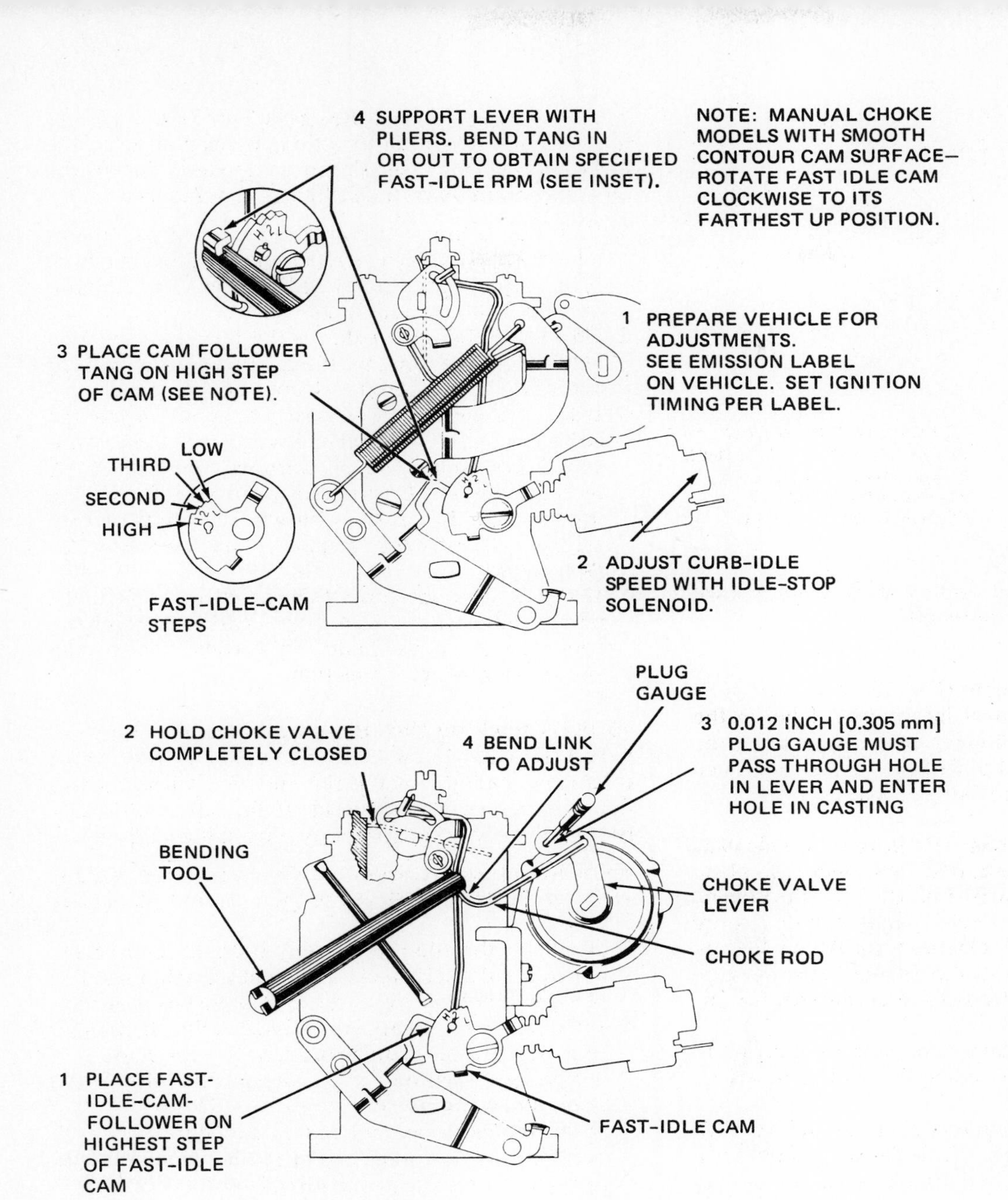

Fig. 10-4 Fast-idle adjustment. (*Chevrolet Motor Division of General Motors Corporation*)

Fig. 10-5 Choke coil lever adjustment. (*Chevrolet Motor Division of General Motors Corporation*)

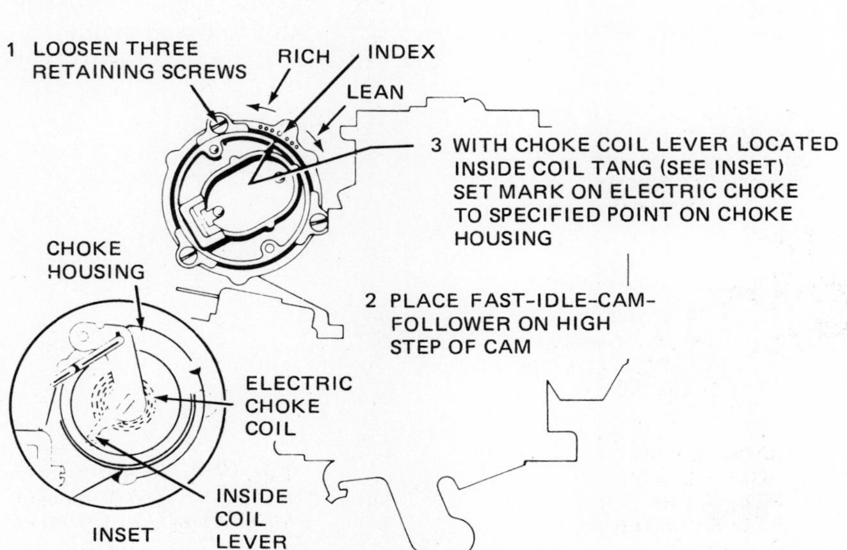

Fig. 10-6 Automatic-choke adjustment. (*Chevrolet Motor Division of General Motors Corporation*)

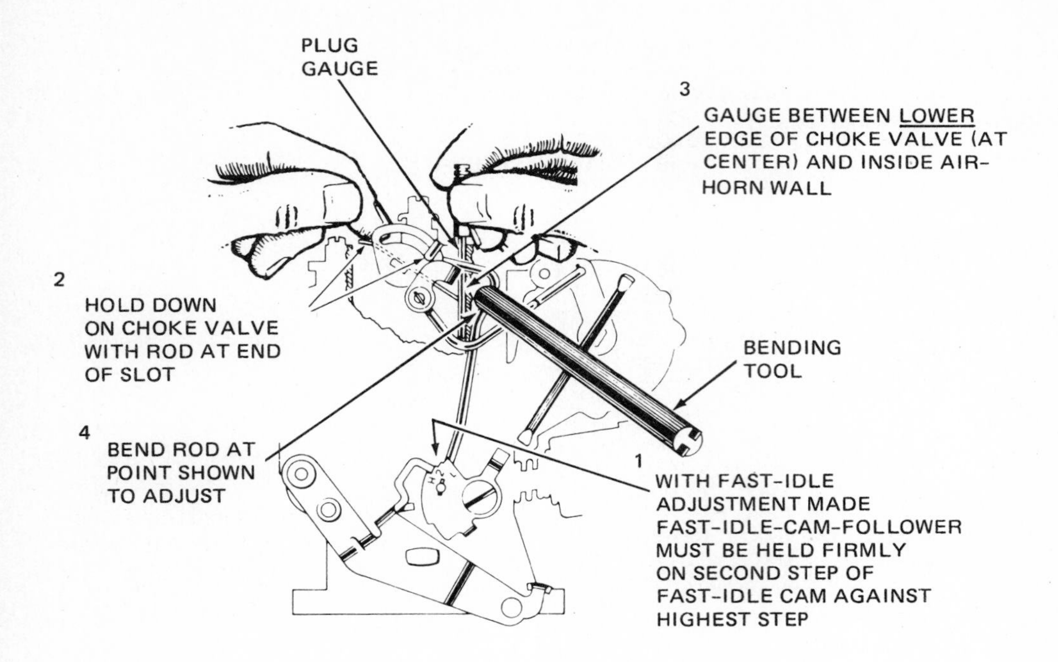

Fig. 10-7 Fast-idle cam adjustment. (*Chevrolet Motor Division of General Motors Corporation*)

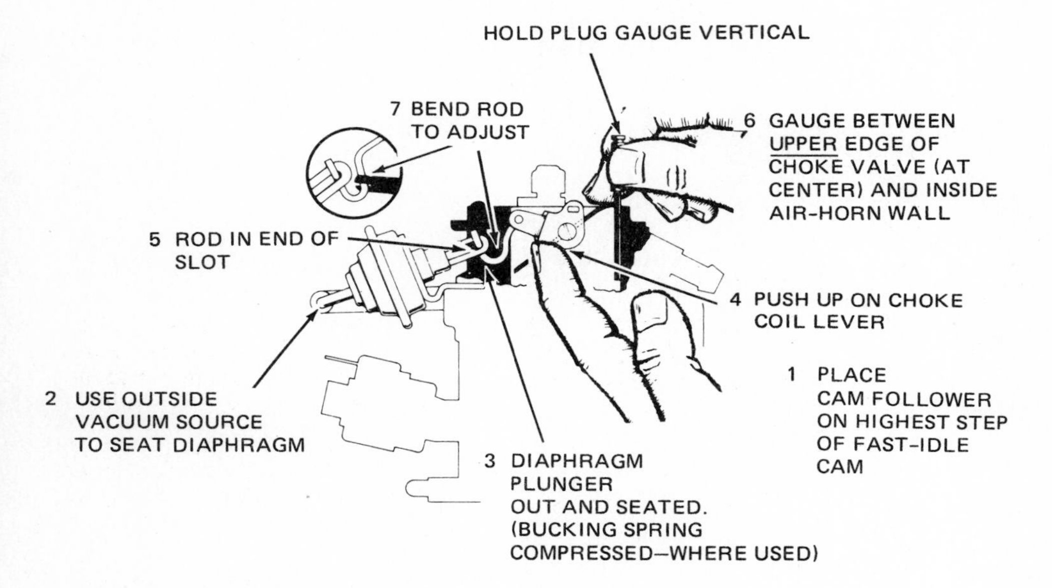

Fig. 10-8 Vacuum-break adjustment. (*Chevrolet Motor Division of General Motors Corporation*)

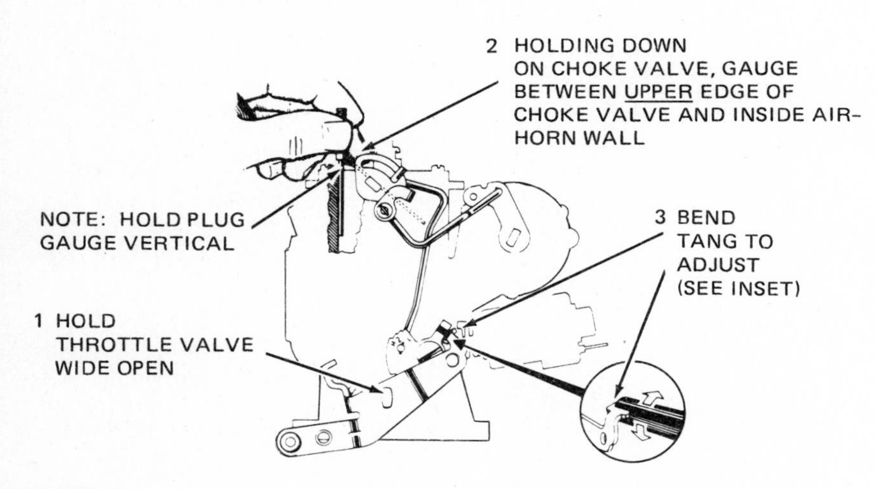

Fig. 10-9 Unloader adjustment. (*Chevrolet Motor Division of General Motors Corporation*)

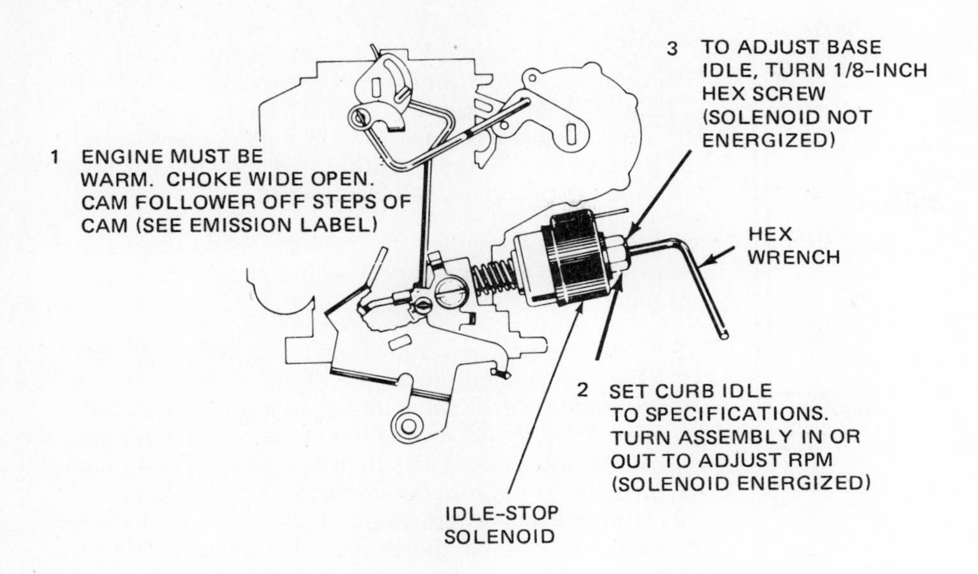

Fig. 10-10 Idle-speed adjustment. (*Chevrolet Motor Division of General Motors Corporation*)

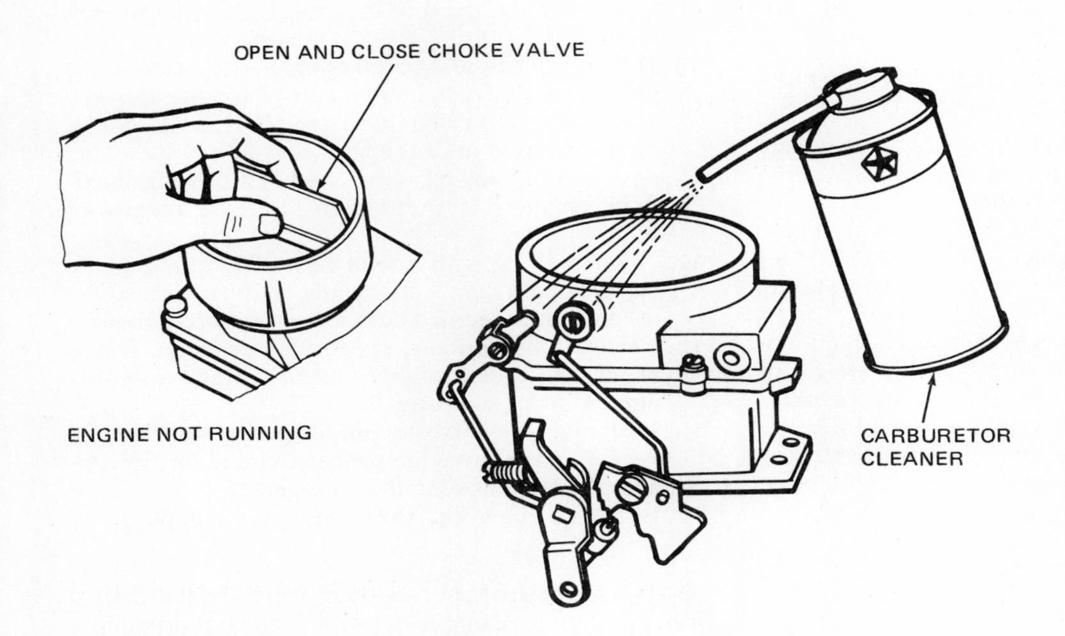

Fig. 10-11 Check the choke by opening and closing it several times. If it sticks, clean the sticking parts with spray carburetor cleaner. (*Chrysler Corporation*)

6. If the light does not come on, there is an open in the feed line. Check the wires and connections.

⚙ 10-6 Idle-mixture check and adjustment with propane The catalytic converter cleans the exhaust gas by converting the HC and CO into harmless gases. This means the idle mixture cannot be adjusted by reading the CO content of the exhaust gas with an exhaust gas analyzer. Therefore, the propane or *artificial enrichment* method of adjusting the idle mixture should be used.

The idle-mixture screws have been preset at the factory and capped. Idle mixture should be adjusted only during a major carburetor overhaul or when an inspection determines that there is excessive HC and CO in the exhaust gas.

Adjusting the mixture by any method other than that given below may violate federal, state, or local laws.

NOTE: Cars requiring the propane adjustment procedure have an emission control information label in the engine compartment (Fig. 10-12). The label explains how to disconnect and plug hoses before adjustment is made. Following is the procedure for the 1ME carburetor.

Before checking or resetting the carburetor to correct poor engine performance or idling, check:

1. Ignition system—distributor, timing, spark plugs, and wires.
2. Air cleaner.
3. Evaporative emission system.
4. EFE and PCV systems, EGR valve, and engine compression.
5. Intake manifold, vacuum hoses, and connections for leaks.

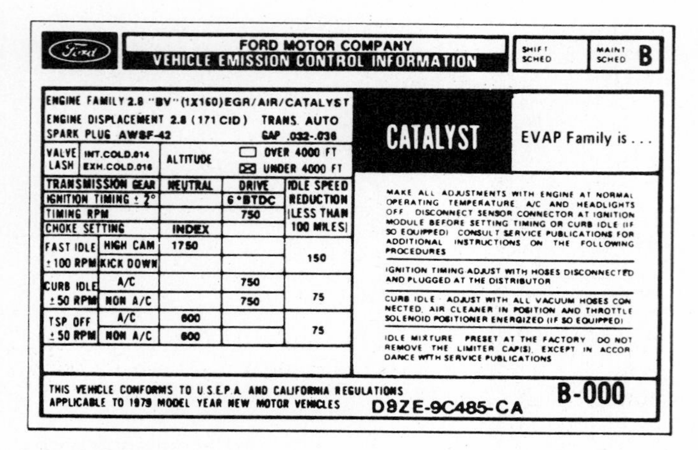

FORD MOTOR COMPANY
VEHICLE EMISSION CONTROL INFORMATION

SHIFT SCHED | MAINT SCHED B

ENGINE FAMILY 2.8 "BV" (1X160) EGR/AIR/CATALYST
ENGINE DISPLACEMENT 2.8 (171 CID) TRANS. AUTO
SPARK PLUG AWSF-42 GAP .032-.036

VALVE LASH: INT. COLD .014, EXH. COLD .016
ALTITUDE: ☐ OVER 4000 FT ☒ UNDER 4000 FT

CATALYST

EVAP Family is . . .

TRANSMISSION GEAR		NEUTRAL	DRIVE	IDLE SPEED REDUCTION (LESS THAN 100 MILES)
IGNITION TIMING ± 2°			6°BTDC	
TIMING RPM			750	
CHOKE SETTING		INDEX		
FAST IDLE ± 100 RPM	HIGH CAM	1750		150
	KICK DOWN			
CURB IDLE ± 50 RPM	A/C		750	75
	NON A/C		750	
TSP OFF ± 50 RPM	A/C	600		75
	NON A/C	600		

MAKE ALL ADJUSTMENTS WITH ENGINE AT NORMAL OPERATING TEMPERATURE. A/C AND HEADLIGHTS OFF. DISCONNECT SENSOR CONNECTOR AT IGNITION MODULE BEFORE SETTING TIMING OR CURB IDLE (IF SO EQUIPPED). CONSULT SERVICE PUBLICATIONS FOR ADDITIONAL INSTRUCTIONS ON THE FOLLOWING PROCEDURES

IGNITION TIMING: ADJUST WITH HOSES DISCONNECTED AND PLUGGED AT THE DISTRIBUTOR

CURB IDLE: ADJUST WITH ALL VACUUM HOSES CONNECTED, AIR CLEANER IN POSITION AND THROTTLE SOLENOID POSITIONER ENERGIZED (IF SO EQUIPPED)

IDLE MIXTURE: PRESET AT THE FACTORY. DO NOT REMOVE THE LIMITER CAP(S), EXCEPT IN ACCORDANCE WITH SERVICE PUBLICATIONS

THIS VEHICLE CONFORMS TO U.S.E.P.A. AND CALIFORNIA REGULATIONS APPLICABLE TO 1979 MODEL YEAR NEW MOTOR VEHICLES D8ZE-9C485-CA B-000

Fig. 10-12 Vehicle emission control information decal, with tuneup specifications, located in the engine compartment. (*Ford Motor Company*)

6. Carburetor bolts for tightness. If necessary, retorque to specifications.

After checking the above, proceed with the idle-mixture check and adjustment using propane (Fig. 10-13).

1. Set parking brake and block drive wheels. On cars equipped with a vacuum parking-brake release, disconnect and plug the hose at the brake.
2. Disconnect and plug hoses as directed on the emission control information label under the hood.
3. Connect a tachometer to the engine to measure engine rpm.
4. Disconnect vacuum advance and set the ignition timing to the specification shown in the emission control information label. Reconnect the vacuum advance.
5. Engine must be at normal operating temperature. The choke must be open and the air conditioning off.
6. Set carburetor idle speed to neutral set speed as shown on the idle speed chart.

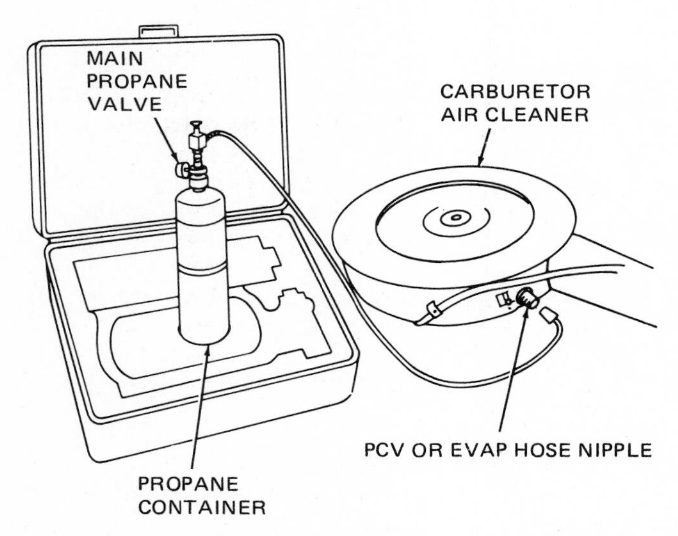

Fig. 10-13 Setup for adjusting the idle mixture with propane. (*Ford Motor Company*)

7. Disconnect the PCV hose from the air cleaner.
8. Connect hose from the propane container valve to the PCV hose nipple in the air cleaner (Fig. 10-13). Propane gas will now flow through the air cleaner, carburetor, intake manifold and into the cylinder when the propane valve is opened.

NOTE: On some engines, the procedure is to make the connection at the evaporative-hose nipple on the air cleaner.

9. With the propane container in a vertical position (Fig. 10-13), and the engine idling, slowly open the propane control valve while watching the tachometer. Engine speed will increase to a maximum (enriched idle speed) and then decrease as too much propane is fed into the engine.
10. Note the maximum rpm. Then close the propane control valve. If the maximum rpm is within the enriched idle rpm range as shown on the idle speed chart, the idle mixture is correct. Proceed to step 14 below.
11. If the enriched idle speed is incorrect, remove the idle-mixture screw limiter cap. Turn the mixture screw in until it is lightly seated. Then back out the mixture screw until the lean best-idle point at the enriched idle speed is reached. This is the maximum engine idle speed with the least amount of fuel.
12. Now, starting with a lean best-idle setting at the enriched idle rpm, lean the idle mixture (turn screw in) until the specified curb idle speed is reached.
13. Recheck the enriched speed with propane. If it is not within specifications, repeat the above procedure, starting with step 9.
14. With the idle mixture properly set, turn off the engine and remove the propane container. Reconnect the PCV hose to the air cleaner.
15. Reset the idle speed as shown in Fig. 10-10.

⚙ 10-7 Carburetor removal and installation

Flooding, stumble on acceleration, and poor drivability are often caused by dirt, water, or other foreign material in the carburetor (see the fuel system trouble diagnosis chart in ⚙ 8-17). This requires removal of the carburetor for disassembly and cleaning.

The steps in the removal procedure are as follows:

1. Remove the air cleaner.
2. Disconnect all fuel and vacuum lines from the carburetor.
3. Disconnect the electrical connector from the choke.
4. Disconnect the throttle linkage.
5. Disconnect the solenoid electrical connection.
6. Remove the attaching nuts and lift the carburetor off the manifold.
7. Take off the insulator gasket.

NOTE: If the carburetor is to be off for any length of time, cover the exposed manifold holes with masking tape (Fig. 10-15). Do not use shop cloths because threads and lint can drop into the manifold. Covering

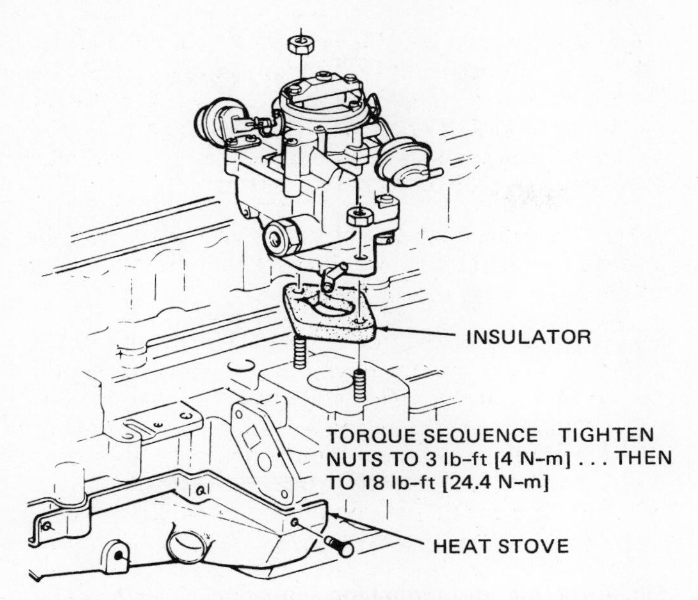

Fig. 10-14 Replacing a carburetor. (*Chevrolet Motor Division of General Motors Corporation*)

the manifold holes prevents engine damage from loose parts and dirt that might get into the manifold.

Before installing the carburetor, fill the float bowl with fuel. This reduces the strain on the starting motor and battery, and reduces the possibility of backfiring during starting. Operate the throttle lever several times to see if fuel discharges from the accelerator pump jets before installing the carburetor.

Installation procedure is the reverse of removal. Make sure the sealing surfaces of the carburetor and intake manifold are clean. Then put the insulator and carburetor in place and secure them with the nuts (Fig. 10-14).

Connect the vacuum and fuel lines, accelerator linkage, and wires to the choke and solenoid. Install the air cleaner. Then start the engine, and check and adjust idle speed.

10-8 Carburetor disassembly To disassemble the carburetor, first separate it into its three basic assemblies—air horn, float bowl, and throttle body. Then each assembly can be serviced as needed. It is not always necessary to completely disassemble the carburetor.

NOTE: Carburetor overhaul kits are supplied for many carburetors. These kits contain instructions and all necessary parts such as gaskets and washers which are required to overhaul the carburetor.

Here are the major steps in disassembling the Rochester model 1ME carburetor. Use separate trays or pans for the parts from the three basic assemblies so you do not mix the parts of the air horn, float bowl, and throttle body.

1. Put the carburetor in a holding fixture.
2. Remove the vacuum break by detaching the hose and backing out the two tapered-head screws (Fig. 10-8).

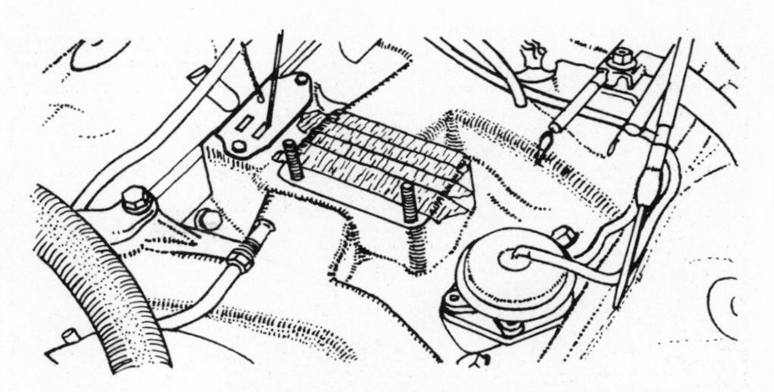

Fig. 10-15 If the carburetor is to be off the engine for any length of time, cover the manifold holes with masking tape to prevent parts from falling into the manifold.

3. Remove the fast-idle cam and detach the choke rod from the choke valve lever (Fig. 10-5). Take out three screws and remove the thermostatic coil assembly from the choke housing.
4. Remove one short and three long screws attaching the air horn to the float bowl.
5. Remove the air horn by twisting it back toward the choke housing so that the choke coil-lever link will disengage from the choke coil lever.
6. Turn the air horn upside down and put it on a clean bench.
7. The float bowl can now be disassembled. This includes removing the float and needle valve, and the metering rod and accelerator pump (Fig. 10-16). The idle tube and accelerator pump spring and ball can also be removed (Fig. 10-17).
8. Next, the throttle body can be detached by inverting the carburetor (float bowl down) on the bench and taking out two screws.

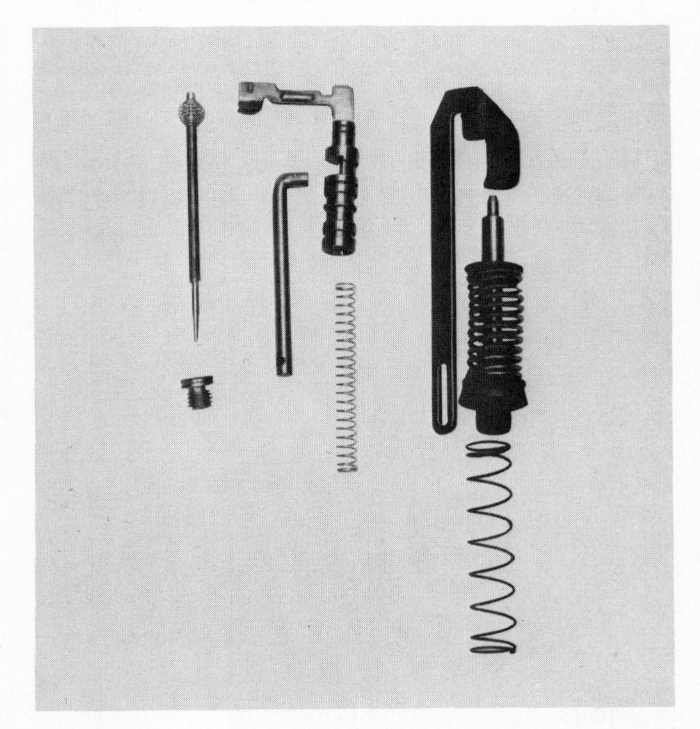

Fig. 10-16 Metering rod and accelerator pump. (*Chevrolet Motor Division of General Motors Corporation*)

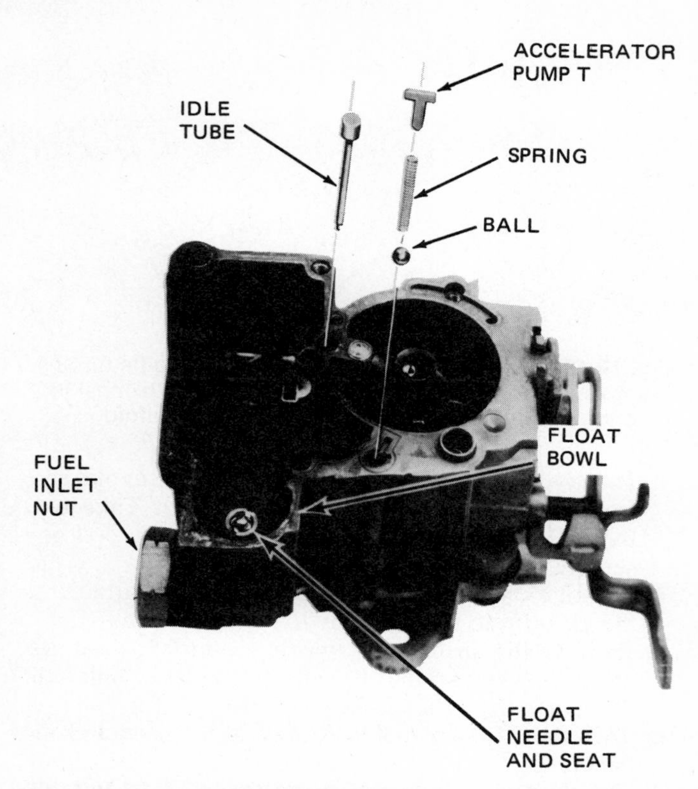

Fig. 10-17 Float bowl showing location of idle tube and accelerator pump piston. (*Chevrolet Motor Division of General Motors Corporation*)

❁ 10-9 Cleaning and inspection Clean the metal parts in an approved carburetor cleaner (Fig. 10-18). Do not clean rubber or plastic parts with the cleaner. Blow out all passages in the castings with compressed air.

NOTE: Do not use drills or wire to clean jets or passages. This could enlarge the holes and upset carburetor action.

If the float needle or seat is worn, install a new set. Check all mating surfaces on the castings for damage.

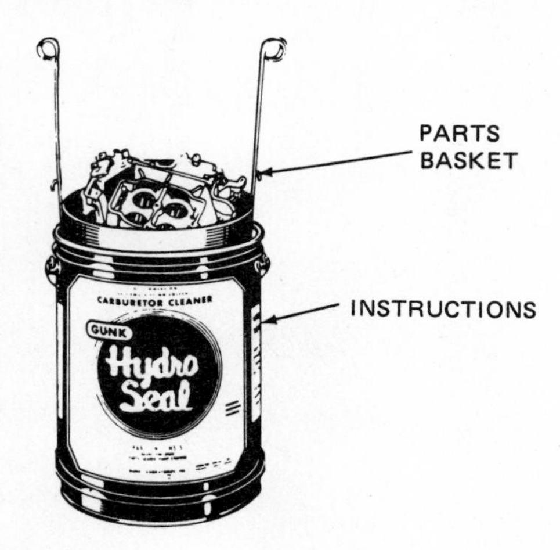

Fig. 10-18 A shop-size container of carburetor cleaner. The basket has long handles so your hands do not get in the chemical. (*Gunk Laboratories, Inc.*)

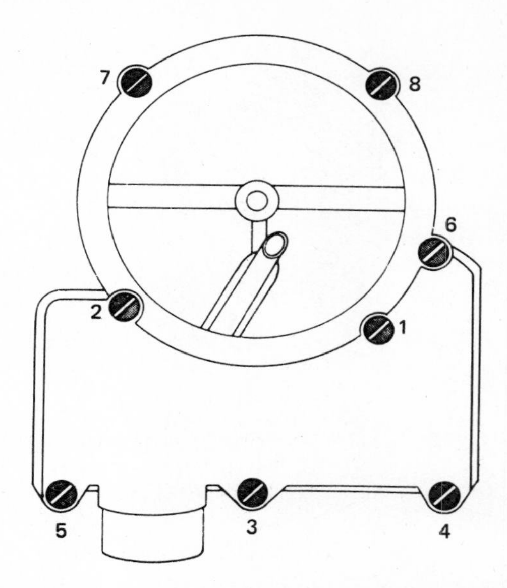

Fig. 10-19 Air horn tightening sequence. (*Chevrolet Motor Division of General Motors Corporation*)

Examine the holes in the levers and in the fast-idle cam for wear. Replace any worn parts.

❁ 10-10 Reassembly When reassembling the float bowl, adjust the float level (Fig. 10-2) and metering rod (Fig. 10-3). Figure 10-19 shows the proper attaching-screw tightening sequence for installing the air horn. After assembly is complete and the carburetor is installed on the engine, make the other adjustments shown in Figs. 10-4 to 10-10 and described in ❁ 10-3 to 10-5. Then check and adjust the idle mixture as explained in ❁ 10-6.

NOTE: As the carburetor is reassembled, install all of the new gaskets, seals, and clips that are in the carburetor kit. Also, install a new fuel filter.

❁ 10-11 Holley model 1945 carburetor The Holley model 1945 carburetor is used by Chrysler on many slant-six engines. Figure 10-20 shows an external view of the carburetor. The systems in the carburetor are similar to those in other carburetors (Chap. 4). The carburetor choke has an electric heater which comes on in cold weather to assist engine heat in opening the choke valve rapidly.

Checking and adjusting the idle mixture with propane, and removing and installing the carburetor are done the same way as for the Rochester model 1ME carburetor (❁ 10-6 and 10-7). Disassembly is also similar. Figures 10-21 to 10-28 show various steps in disassembling and reassembling the carburetor. The major steps in the procedure are discussed in following sections.

❁ 10-12 Disassembling the Holley model 1945 carburetor Remove the carburetor from the engine and put the carburetor in a repair stand. Remove the float bowl vent cover and spring. Then proceed as follows:

1. Remove the fast-idle cam retaining clip, cam, and link (Fig. 10-21).

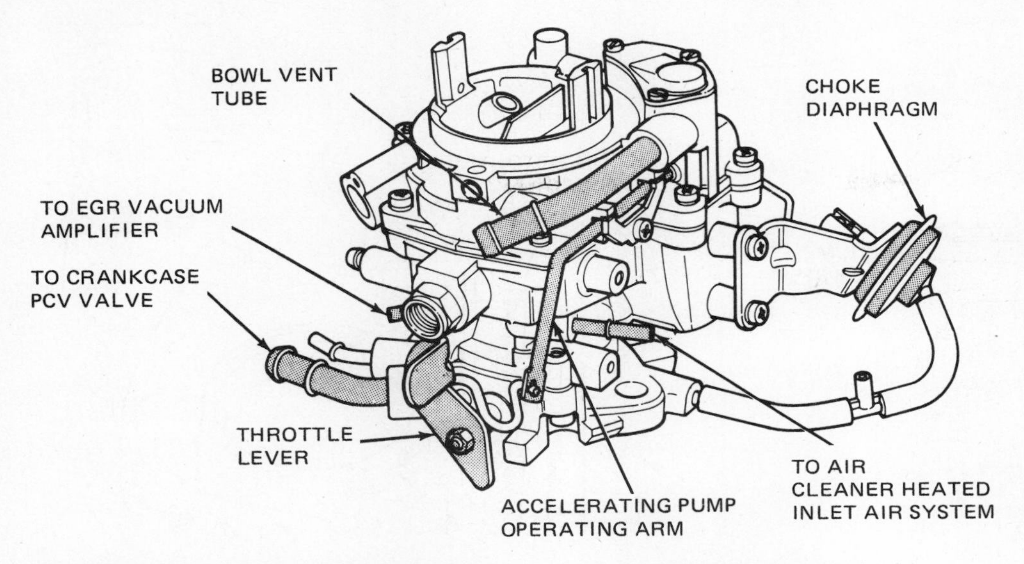

Fig. 10-20 Holley model 1945 single-barrel carburetor. (*Chrysler Corporation*)

2. Remove the choke vacuum diaphragm, link, and bracket assembly. Disengage the link from the slot in the choke lever.
3. Remove dashpot, if on the carburetor.
4. Remove cotter pin and accelerator pump link from the throttle lever. Note the hole position of the link. There are two positions for different models (Fig. 10-22). The link must be reinstalled back in the same hole when reassembling the carburetor.
5. Remove the seven bowl cover screws and lift the bowl cover from the bowl (Fig. 10-23). If the cover sticks to the bowl, loosen the cover by tapping it lightly with a plastic hammer or the handle of a screwdriver. *Do not pry the cover off!*
6. Remove the accelerator pump operating rod and the accelerator pump (Fig. 10-24).
7. To remove the vacuum piston, remove the staking around the retaining ring with a suitable tool (Fig. 10-25). Then push the piston down and allow it to snap up against the ring. It will then come loose.
8. Remove the fuel-inlet-fitting valve assembly from the main body.
9. Remove the float and associated parts (Fig. 10-26). Turn the main body upside down and the pump discharge check ball and weight will fall out.
10. If the main jet is worn or damaged, remove it with a special tool or a wide-blade screwdriver.
11. If the power valve requires replacement, use a ⅜-inch-wide [9.5-mm] screwdriver to push down on the valve until the blade can fit into the slot on top of the valve. Remove the valve. Install the new seat, needle, and spring from the carburetor overhaul kit.

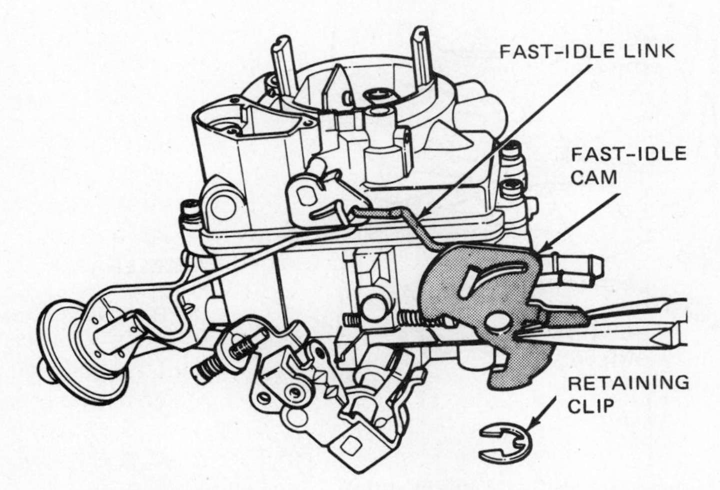

Fig. 10-21 Removing or installing the fast-idle cam and link. (*Chrysler Corporation*)

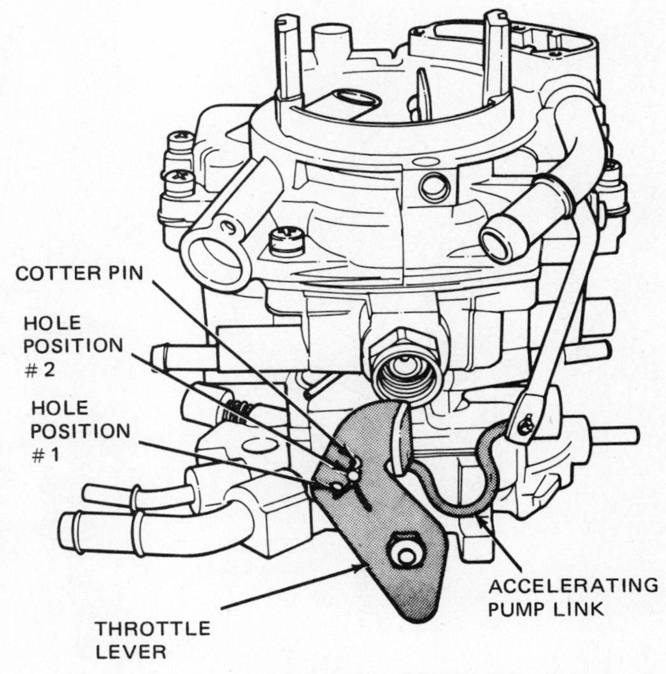

Fig. 10-22 Removing or installing the accelerator pump linkage. (*Chrysler Corporation*)

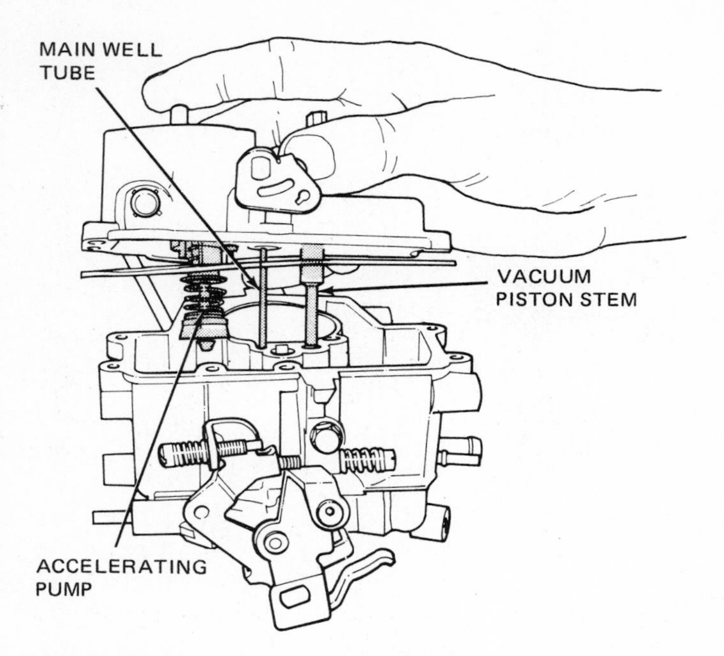

Fig. 10-23 Removing or installing bowl cover. (*Chrysler Corporation*)

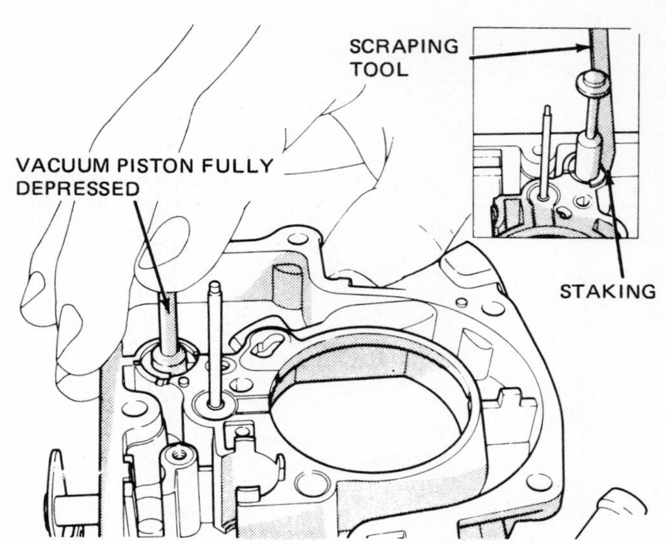

Fig. 10-25 Removing the vacuum piston. (*Chrysler Corporation*)

⚙ 10-13 Cleaning and inspection Plastic and rubber parts, and the choke diaphragm, can be damaged if they are immersed in or cleaned with carburetor cleaner (⚙ 10-9).

Check the plastic type of float for fuel absorption by lightly squeezing it between your fingers. If wetness appears on the surface, or if the float feels heavy, replace it.

Before installing the vacuum piston in the bowl cover, remove all staking from the retainer cavity. Install the spring and piston and stake the retainer lightly (Fig. 10-27).

Test the accelerator pump discharge check ball and seat before reassembly by coating the pump piston with oil, or filling the float bowl with clean fuel. Install the discharge check ball and weight. Hold the ball and weight down with a small brass rod and push the pump plunger down by hand (Fig. 10-28). If the ball is leaking, you will feel no resistance. Remove the weight. Use a

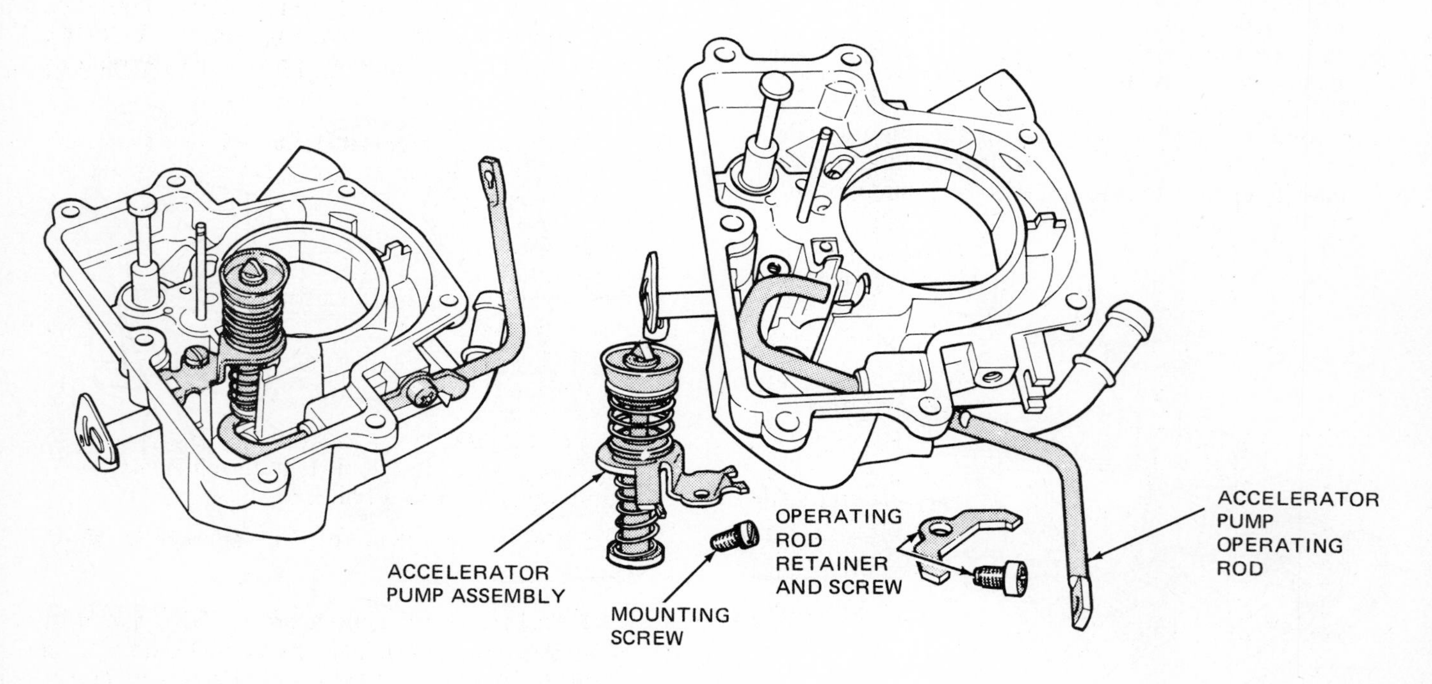

Fig. 10-24 Removing or installing accelerator pump. (*Chrysler Corporation*)

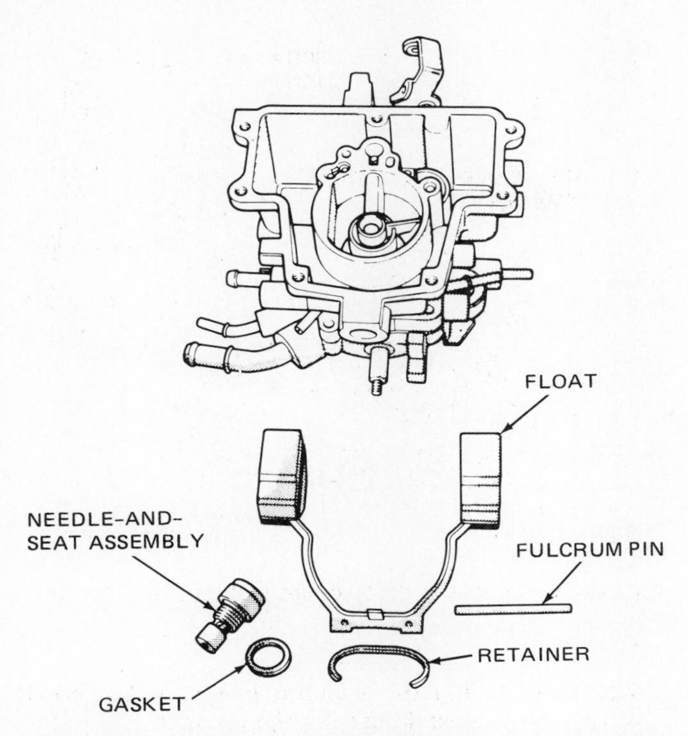

Fig. 10-26 Removing the float assembly. (*Chrysler Corporation*)

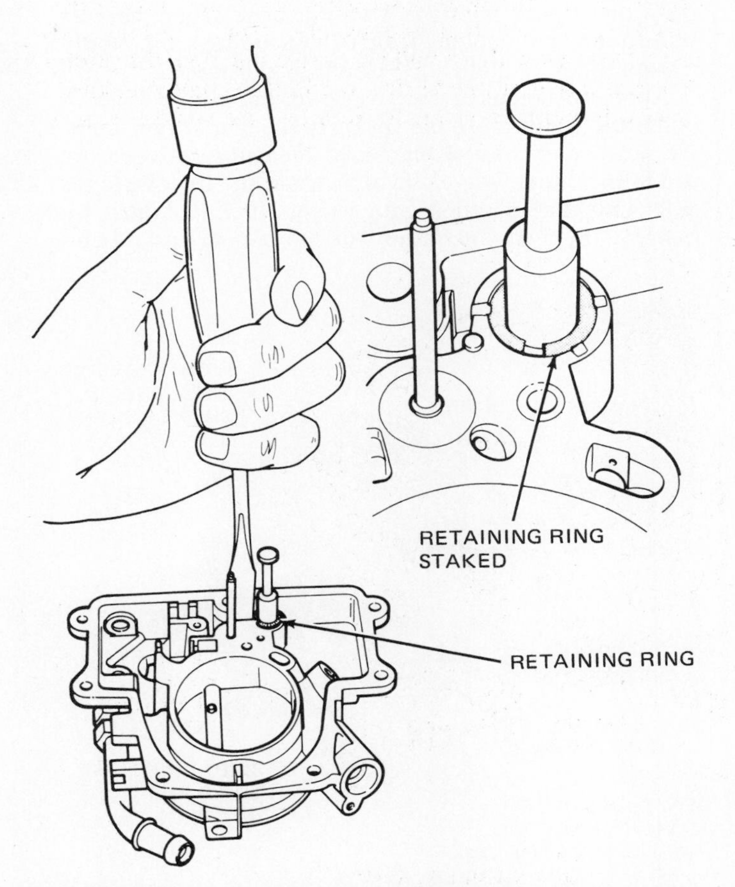

Fig. 10-27 Staking vacuum piston in place. (*Chrysler Corporation*)

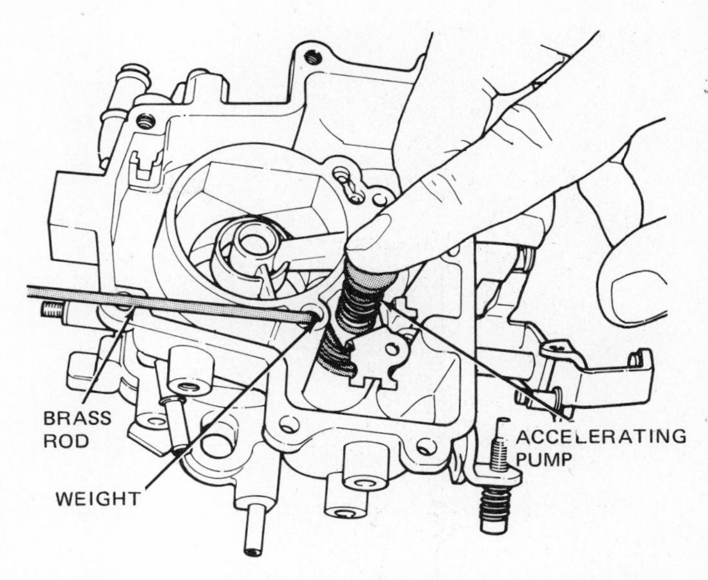

Fig. 10-28 Testing the accelerator pump discharge check ball and seat. (*Chrysler Corporation*)

brass drift punch and lightly tap on the ball to improve the seat. Then remove the old ball and install a new ball and the weight. Recheck for leaks.

New gaskets should be used on reassembly and installation of the curburetor.

⚙ 10-14 Carburetor reassembly In general, reassembly is the reverse order of disassembly. All necessary parts in the carburetor overhaul kit should be installed during reassembly. Figure 10-27 shows how to stake the vacuum piston retaining ring. During reassembly and after installation of the carburetor on the engine, several checks and adjustments should be made, as explained in the following section (⚙ 10-14). Idle mixture setting is made with propane as explained ⚙ 10-6.

⚙ 10-15 Carburetor adjustments Following are the adjustments to be made after reassembly of the Holley model 1945 carburetor:

1. Float setting
2. Idle setting
3. Accelerator pump piston stroke
4. Fast-idle-cam position
5. Choke unloader
6. Choke vacuum kick
7. Fast-idle speed
8. Dashpot adjustment (manual transmissions only)
9. Bowl vent valve (on vehicle at curb idle rpm)
10. Float adjustment on vehicle

These procedures are covered in ⚙ 10-16 to ⚙ 10-25.

⚙ 10-16 Float setting adjustment With the float and retaining spring installed in the bowl, hold the bowl cover gasket in place and turn the bowl upside down. Put a straightedge across the gasket surface (Fig. 10-29). The floats should just touch the straightedge. Adjust by bending the float tang.

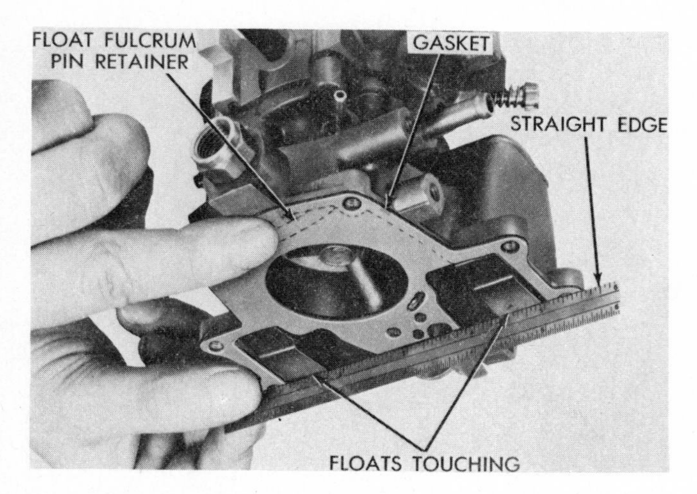

Fig. 10-29 Float setting adjustment. (*Chrysler Corporation*)

⚙ 10-17 Idle settings This adjustment is made with the carburetor installed and the engine warmed up. Put the transmission in neutral and set the parking brake. Headlights and air conditioner should be off.

Remove the vacuum hose at the distributor and plug the hose.

Check and reset, if necessary, the ignition timing. (See the emission control information label in the engine compartment.) Disconnect the PCV hose at the valve cover and the purge hose at the charcoal canister. Leave both hoses unplugged.

Adjust the idle mixture with propane as explained on the emission control label (⚙ 10-6). Adjust idle speed. Reconnect all hoses after setting idle speed.

NOTE: On new vehicles with under 300 miles [483 km], the idle-speed setting should be reduced 75 rpm for both propane rpm and idle-set rpm.

⚙ 10-18 Accelerator pump piston stroke adjustment Put the throttle in the idle-set rpm position with the accelerator pump operating link in the proper hole in the throttle lever (Fig. 10-30). Measure the pump operating link. Bend the link to adjust it. If this adjustment is made, then the bowl vent adjustment must be reset.

⚙ 10-19 Fast-idle-cam position adjustment With the fast-idle-speed adjusting screw on the second-highest-speed step of the fast-idle cam, move the choke valve toward the closed position with light pressure on the choke shaft lever (Fig. 10-31). Use the specified gauge to check between the top of the choke valve and air horn on the throttle lever side. Adjust by bending the fast-idle connector rod.

⚙ 10-20 Choke unloader adjustment Hold the throttle valve in the wide-open position (Fig. 10-32). Lightly press against the control lever to move the choke valve toward the closed position. Measure with the specified gauge between the top of the choke valve

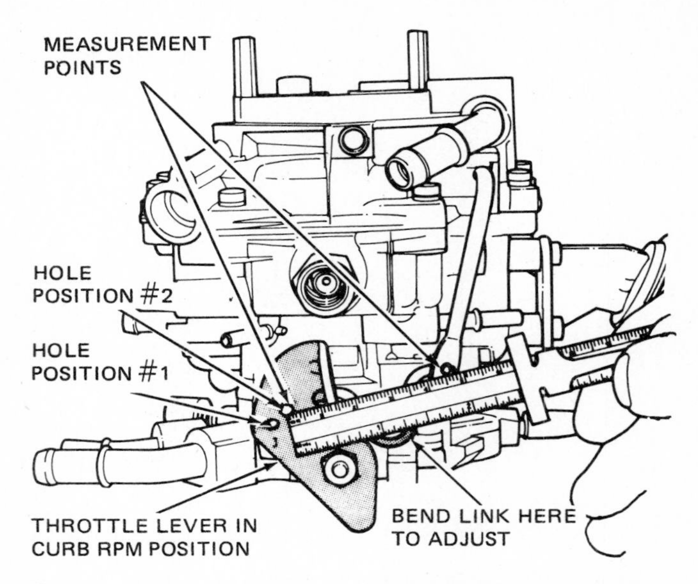

Fig. 10-30 Adjusting the accelerator pump piston stroke. (*Chrysler Corporation*)

and the air horn at the throttle lever side. Adjust the unloader by bending the tang on the throttle lever.

⚙ 10-21 Choke vacuum kick adjustment Open the throttle and close the choke (Fig. 10-33). Then close the throttle to trap the fast-idle cam at the closed-choke position. Disconnect the vacuum hose from the carburetor and connect the hose to a vacuum pump. Apply a vacuum of 15 inches [381 mm] Hg or more.

Apply sufficient closing force on the choke lever to completely compress the spring in the diaphragm stem. Do not use excessive pressure. Measure between the top of the choke valve and air horn at the throttle lever side with the specified gauge (Fig. 10-33). Adjust by bending the link. Check for free movement and rebend

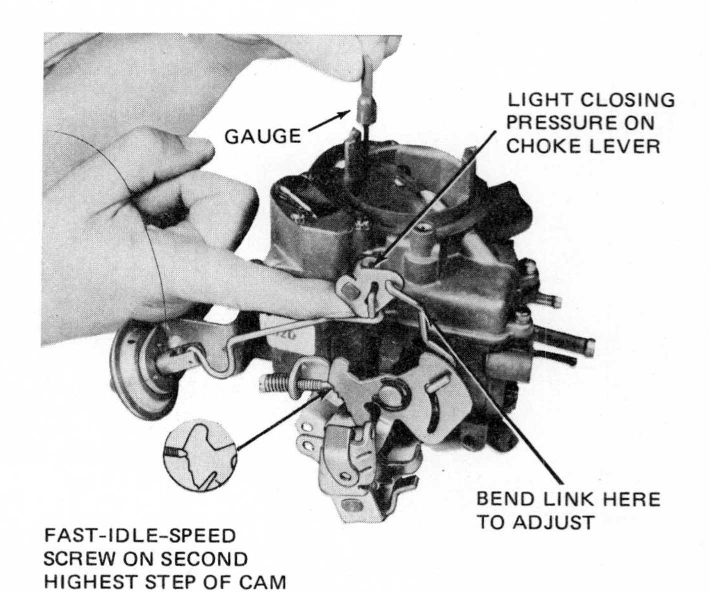

Fig. 10-31 Adjusting the fast-idle cam. (*Chrysler Corporation*)

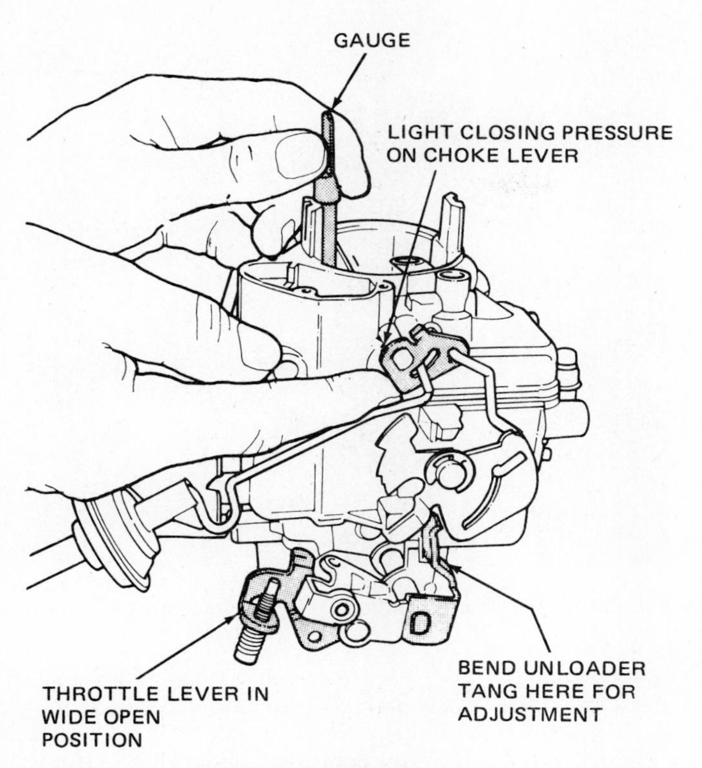

Fig. 10-32 Choke unloader adjustment. (*Chrysler Corporation*)

if necessary. Reattach the vacuum hose to the carburetor.

⚙ 10-22 Fast-idle-speed adjustment Remove the air cleaner and disconnect hoses to eliminate vacuum advance and EGR signals. Cap and plug all disconnected hoses and fittings.

Set the parking brake and put the transmission in neutral. With engine off, open the throttle and close the choke. Then close the throttle. Rotate the fast-idle cam until the speed screw drops from highest to second speed step (Fig. 10-31). Start engine and allow it to stabilize. Check idle rpm. Adjust by turning the idle-speed screw.

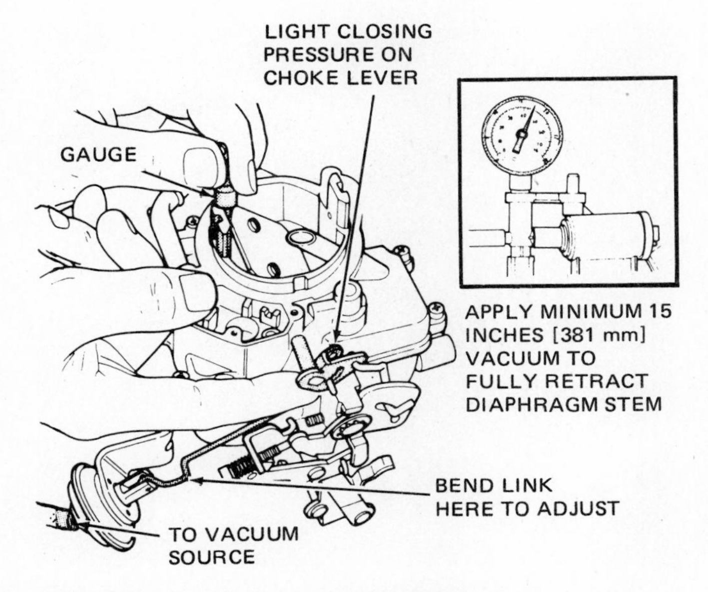

Fig. 10-33 Choke vacuum kick adjustment. (*Chrysler Corporation*)

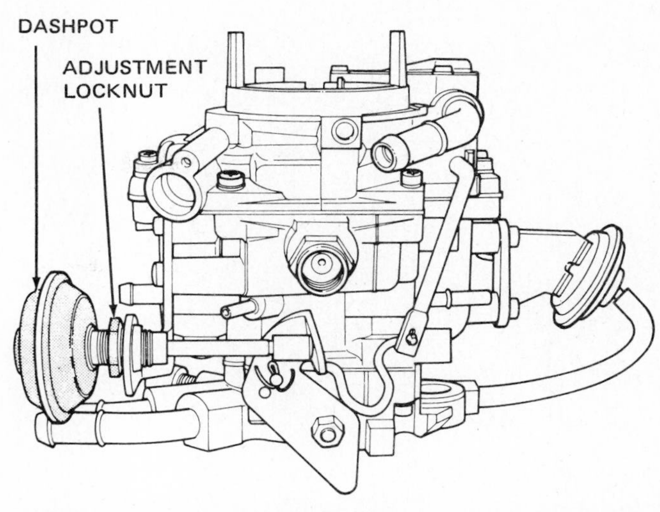

Fig. 10-34 Adjusting the dashpot. (*Chrysler Corporation*)

⚙ 10-23 Dashpot adjustment This adjustment is made only on cars with a manual transmission. Set idle speed (⚙ 10-17). Be sure that all vacuum hoses from the carburetor are connected. Remove the air cleaner and plug the disconnected vacuum fitting. Set the parking brake and put the transmission in neutral. Start the engine and position the throttle lever so it touches the dashpot stem without depressing it. Wait 30 seconds to stabilize engine speed. Adjust to specifications by turning the dashpot after loosening the locknut (Fig. 10-34). Tighten the locknut. Open and close the throttle to make sure the throttle returns to the set rpm.

⚙ 10-24 Bowl vent valve adjustment With the throttle at idle-set rpm, measure the distance from the cover support surface down to the flat on the plastic bowl vent lever (Fig. 10-35). If adjustment is necessary,

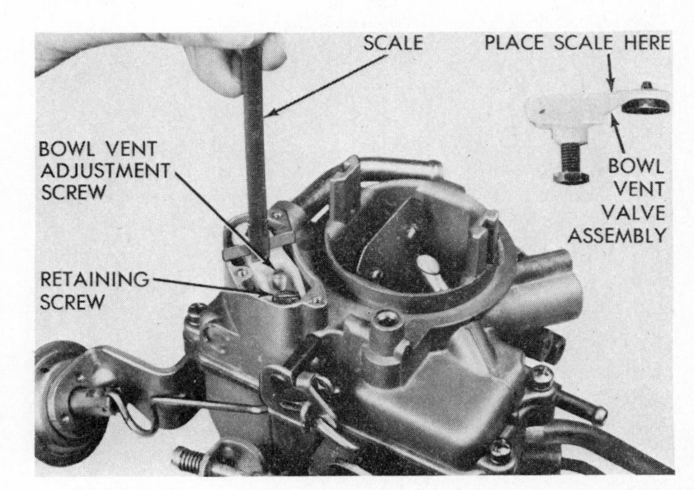

Fig. 10-35 Bowl vent valve adjustment. (*Chrysler Corporation*)

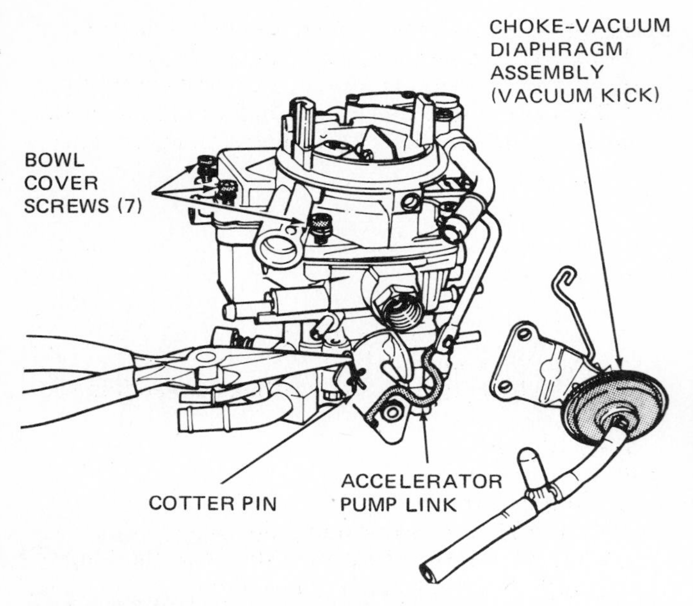

Fig. 10-36 Removing or installing the choke vacuum diaphragm and accelerator pump linkage. (*Chrysler Corporation*)

turn the bowl vent lever adjusting screw. Install bowl vent spring and cover plate.

If the accelerator pump piston stroke is changed, this adjustment must also be reset.

⚙ 10-25 Float adjustment on vehicle Remove the air cleaner, bowl vent hose, and disconnect the choke. Remove the fast-idle cam and link. Remove the vacuum choke diaphragm (Fig. 10-36). Remove dashpot if present. Remove cotter pin and accelerator pump link. Remove the bowl cover screws and lift bowl cover off.

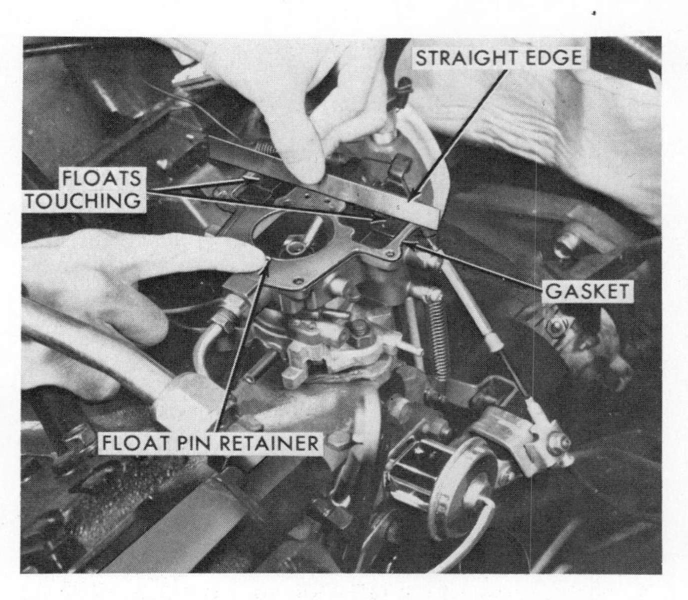

Fig. 10-38 Checking the float. (*Chrysler Corporation*)

1. Depress the float manually to allow residual line pressure to overfill the bowl to within ⅛ inch [3.2 mm] of the top of the float bowl (Fig. 10-37). If not enough fuel enters, add fuel as necessary.
2. Use two wrenches and back off the flare nut and tighten the inlet fitting to the recommended torque.
3. Put bowl cover gasket on top of the float bowl. Seat the float pin retainer by hand and lay a straightedge across the gasket surface (Fig. 10-38). The floats should just touch. Adjust if necessary by bending the float tang.
4. Remove bowl cover gasket. Empty fuel from accelerator pump well with a syringe (Fig. 10-39). This is done to prevent the discharge check ball and weight

Fig. 10-37 Overfilling the float bowl. (*Chrysler Corporation*)

Fig. 10-39 Removing fuel from the accelerator pump well. (*Chrysler Corporation*)

from leaving their position during assembly of the bowl cover.

5. Install gasket and bowl cover on main body. Be sure the leading edge of the accelerator pump cup is not damaged as it enters the pump bore and that the weight and check ball remain in place.
6. Install the seven bowl cover screws. Tighten them alternately to compress the gasket evenly.
7. Install all other parts removed from the carburetor.
8. Adjust idle speed, following instructions on the emission control label in the engine compartment.

Chapter 10 review questions

Select the *one* correct, best, or most probable answer to each question. Then check your answers against the correct answers given at the end of the book.

1. After a cold start, the choke should open within:
   a. 1½ minutes
   b. 3½ minutes
   c. 5½ minutes
   d. 10 minutes.
2. The purpose of the idle-stop solenoid is to:
   a. prevent dieseling
   b. allow the throttle to close completely when the engine is turned off
   c. shut off the flow of air-fuel mixture to the engine when the engine is turned off
   d. all of these.
3. On the Rochester model 1ME carburetor, the idle-mixture screw has been preset at the factory and:
   a. staked in place
   b. covered with a steel plate
   c. capped
   d. locked with a nut.
4. To make the idle-mixture adjustment with propane, increase the flow of propane until:
   a. the engine stalls
   b. the engine reaches intermediate speed
   c. engine speed decreases to a minimum
   d. engine idle speed increases to a maximum and then drops off.
5. The purpose of filling the carburetor float bowl before installing the carburetor is to:
   a. keep dirt from falling into the bowl
   b. check the float level
   c. assure quick initial starting
   d. check the power system.
6. On the Holley model 1945 carburetor the idle settings are made:
   a. on the bench
   b. with propane
   c. without propane
   d. with the accelerator pump disconnected.
7. The accelerator-pump piston stroke in the Holley model 1945 carburetor is adjusted by:
   a. turning the throttle screw
   b. relocating the linkage in the hole
   c. adjusting the throttle linkage
   d. bending the operating link.
8. To adjust the float setting in the Holley 1945 carburetor:
   a. bend the float rods
   b. bend the float tang
   c. turn the needle jet
   d. bend the float linkage.
9. Before the idle-mixture adjustment is made with propane:
   a. the engine must be warmed up
   b. ignition timing should be checked and adjusted
   c. transmission must be in neutral
   d. all of the above.
10. On chokes with an adjustable cover, the reference mark for making the initial setting is called the:
    a. position mark
    b. fast-idle mark
    c. cold-engine mark
    d. index mark.

CHAPTER 11

SERVICING TWO-BARREL CARBURETORS

After studying this chapter, and with proper instruction and equipment, you should be able to:

1. Explain the operation of the electronic control on a feedback carburetor.
2. List the adjustments to be made on a feedback carburetor.
3. Demonstrate how to make the required checks and adjustments on a feedback carburetor.
4. Overhaul a feedback carburetor.
5. Diagnose troubles in a feedback carburetor.

11-1 Types of two-barrel carburetors There are two types of fixed-venturi two-barrel carburetors (4-10). In one type, both barrels are the same size and both operate all the time that the engine is running. Two-barrel carburetors of this type are checked, adjusted, and serviced much like the single-barrel carburetor (Chap. 10).

The second type of two-barrel carburetor has one primary and one secondary barrel. With this arrangement, one barrel is the primary barrel. It provides the air-fuel mixture during idle, low-speed, and intermediate-speed operation. The secondary barrel comes into operation only after the primary throttle valve has opened about 45°. When the primary throttle valve moves past the 45° position, the linkage to the secondary throttle valve starts to open it. The secondary barrel then comes into operation, and starts to supply additional air-fuel mixture to the cylinders. This increases the supply of air-fuel mixture to the engine for improved medium- to high-speed operation.

Some late-model two-barrel carburetors have a feedback system that electronically adjusts the richness of the air-fuel mixture. This system is discussed in 4-27 to 4-29. Some of these carburetors control both the idle-mixture richness and the main metering system mixture richness. Other carburetors with an electronic feedback system control only the richness of the main metering system mixture.

This chapter describes the construction, operation, adjusting, and servicing of a two-barrel carburetor with a primary and a secondary barrel. This carburetor also has the electronic feedback system which automatically adjusts mixture richness to prevent excessively rich or excessively lean mixtures. The carburetor is the Rochester model E2SE, which is used on several models of General Motors cars.

11-2 Rochester model E2SE carburetor The Rochester model E2SE carburetor is a two-stage, two-barrel carburetor which has a primary and a secondary barrel (Fig. 11-1). It is used by General Motors on both four-cylinder and V-6 engines.

Aluminum die castings are used for all the major

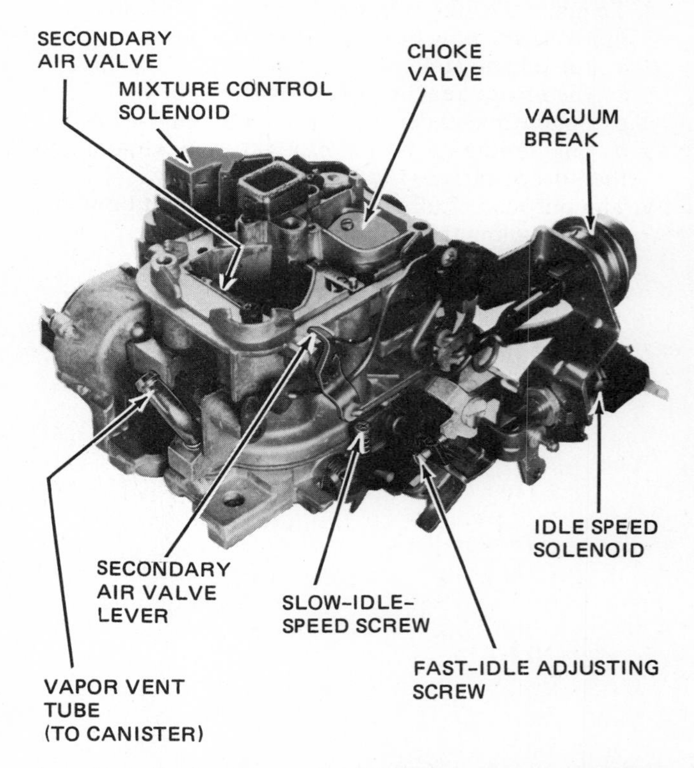

Fig. 11-1 Rochester model E2SE two-barrel two-stage carburetor. (*Chevrolet Motor Division of General Motors Corporation*)

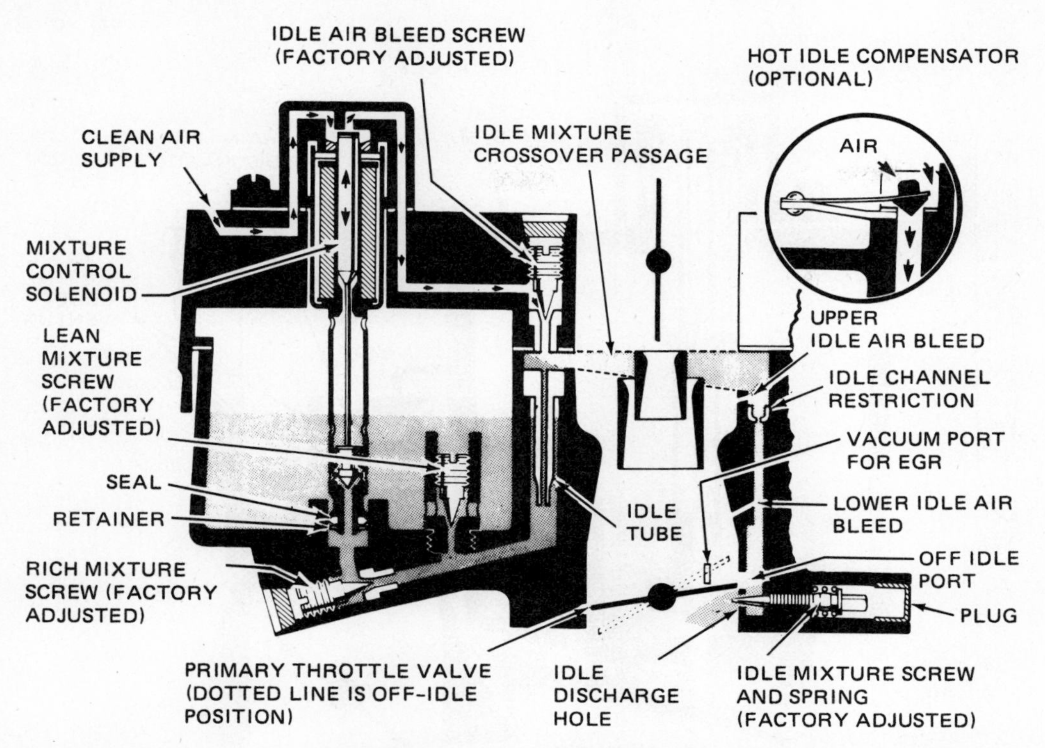

Fig. 11-2 Idle system in the Rochester model E2SE carburetor. (*Chevrolet Motor Division of General Motors Corporation*)

parts of the carburetor. Aluminum is softer than some other metals, so the parts must be handled with care to avoid their getting nicked or scratched.

⚙ 11-3 Electronic control The model E2SE carburetor is called a feedback carburetor. It has an electronically controlled solenoid which continually readjusts the idle and main metering systems to maintain the proper air-fuel ratio. Figure 11-2 shows the idle system for the carburetor that is used on four-cylinder engines. The solenoid receives pulses of electric current, which cause the solenoid plunger to be pulled down. The strength and frequency of the electrical pulses are what determine how far down the plunger will move. When the plunger is pulled down, the idle mixture is leaned out. This results from the upper end of the plunger moving away from the idle air bleed passage.

The pulses are controlled by the electronic system, which puts together signals from the oxygen sensor in the exhaust system and the engine temperature sensor. If the oxygen content of the exhaust gas is low, it means the mixture is too rich. Then the pulses will average out to move the solenoid plunger down. This allows more air to bleed into the idle system. The result is that the idle-mixture leans out.

If the oxygen content in the exhaust gas is too high, it means the mixture is too lean. The pulses average out to be weaker, which allows the plunger to move up. The upper end of the plunger reduces the air bleed passage. Less air bleeds into the idle system and the idle mixture becomes richer. There is a factory-adjusted idle air bleed screw in the air bleed system (Fig. 11-2). This screw is set to compensate for any slight variation in the carburetor due to manufacturing tolerances.

At the same time, the needle valve on the lower end of the plunger rod is moving up or down in the main metering system (Fig. 11-3). If the plunger and valve move down, it reduces the size of the opening through which fuel can feed to the main discharge nozzle. If the plunger and valve move up, the opening is increased and more fuel can feed through. The electronic system controls the pulses to the solenoid so that the valve takes the proper position to maintain the correct air-fuel mixture ratio.

There are two adjustment screws in the main metering system (Fig. 11-3). Both are factory-adjusted. The one inside the float bowl (the lean-mixture screw) is set to allow a minimum of fuel to flow into the main metering system. For satisfactory engine operation, this minimum fuel flow must be supplemented with the additional flow through the solenoid plunger valve system. The rich-mixture screw, also factory-adjusted, allows fine tuning of the main metering system. It is an additional fixed control of the fuel flow from the solenoid plunger valve system.

The model E2SE carburetor used on the V-6 engines has an additional sensor that feeds information to the electronic control unit. This is a throttle position sensor (Fig. 11-3) which electrically signals the throttle position to the electronic control unit. As the throttle is moved, a tang on the pump lever extension changes the position of the plunger inside the throttle position sensor. This changes the strength of the signal going to the

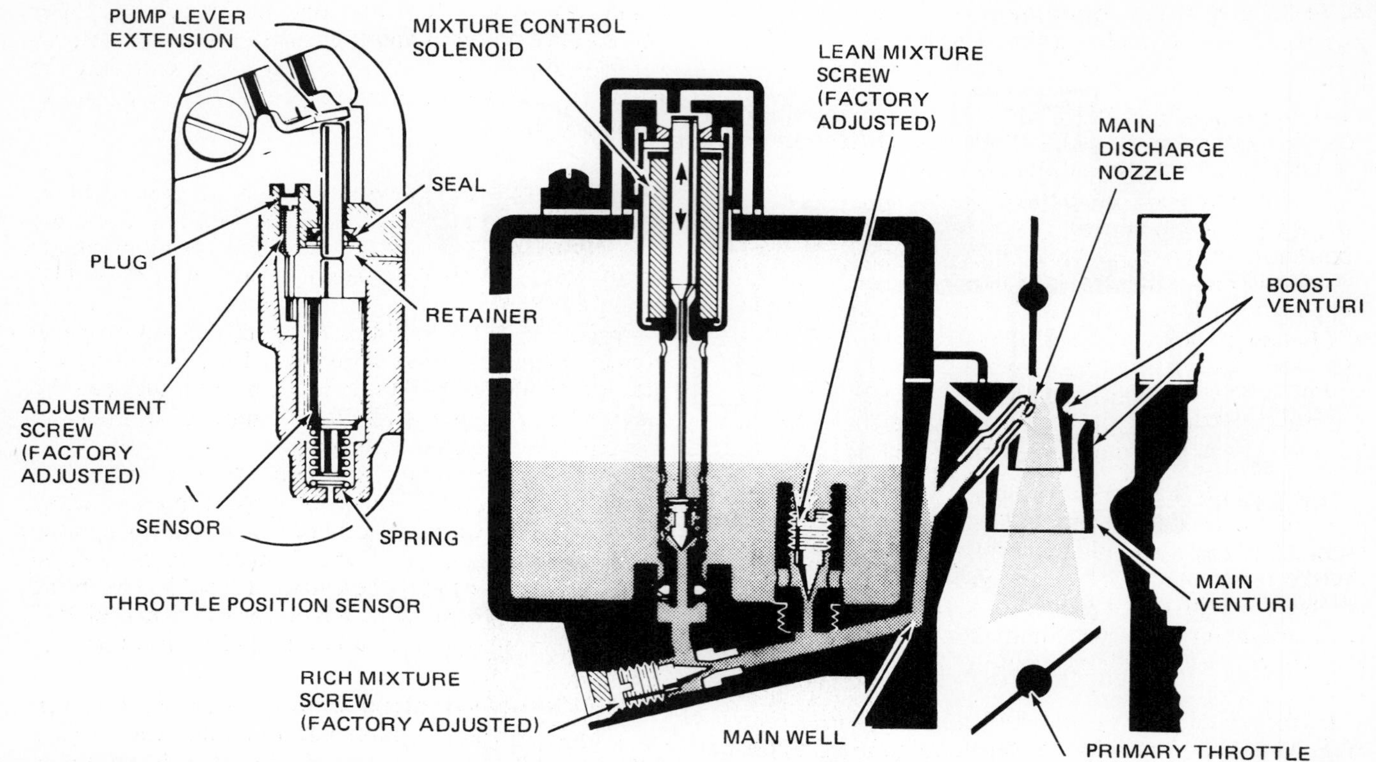

Fig. 11-3 Main metering system in carburetor used on a V-6 engine. (*Chevrolet Motor Division of General Motors Corporation*)

ECU. The ECU checks its memory bank and momentarily holds the last known air-fuel ratio for a short time. This prevents engine stumble or hesitation.

⚙ 11-4 Choke systems The choke system of the carburetor for the four-cylinder engine uses a single vacuum-break unit (Fig. 11-4). It is mounted on the idle-speed solenoid bracket on the primary side of the carburetor. The carburetor for the V-6 engine uses a dual vacuum-break system. The secondary vacuum-break unit is mounted on a bracket on the secondary side of the carburetor.

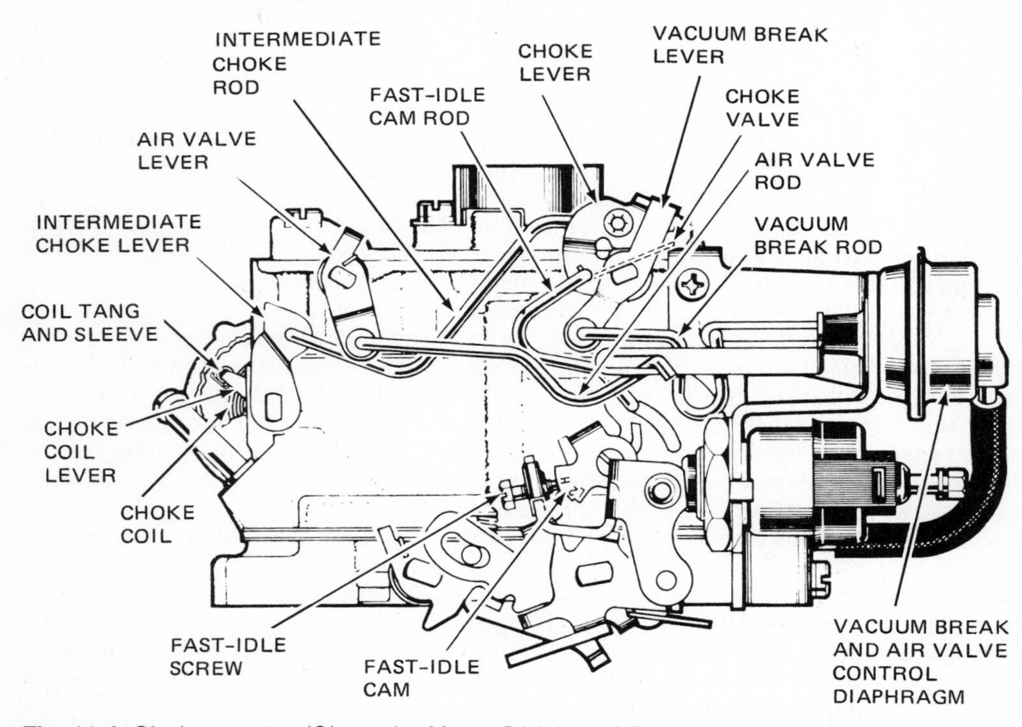

Fig. 11-4 Choke system. (*Chevrolet Motor Division of General Motors Corporation*)

⚙ 11-5 Carburetor adjustments The following adjustments can be made on the E2SE feedback carburetor:

1. Float adjustment (Fig. 11-5)
2. Accelerator pump adjustment (Fig. 11-6)
3. Off-car fast-idle adjustment (Fig. 11-7)
4. Choke coil lever adjustment (Fig. 11-8)
5. Choke rod adjustment (Fig. 11-9)
6. Air valve rod adjustment (Fig. 11-10)
7. Primary vacuum-break adjustment (Fig. 11-11)
8. Secondary vacuum-break adjustment (Fig. 11-12)
9. Unloader adjustment (Fig. 11-13)
10. Secondary lockout adjustment (Fig. 11-14)
11. On-car fast-idle adjustment (Fig. 11-15)
12. Air-conditioning idle-speed adjustment (Fig. 11-16)
13. On-car idle-speed adjustment (Fig. 11-17)

Figures 11-5 to 11-17 show the procedures for making each of these adjustments. These are typical adjustments that may not apply to all carburetors of the E2SE design. There may be some variations that could change the adjustment procedures. If possible, refer to the manufacturer's service manual that covers the carburetor you are working on before making any adjustments.

Perform each of the numbered steps in the sequence shown in each illustration. Refer to the manufacturer's service manual for the specifications for these adjustments. The specifications for the idle-speed adjustment are given on the vehicle emission control information label in the engine compartment (Fig. 10-10).

⚙ 11-6 Servicing the Rochester model E2SE carburetor Servicing of the Rochester model E2SE carburetor can be performed either on the car or off the car on the work bench. Some fuel system problems can be solved without removing the carburetor from the engine. However, there are some problems that can be solved only by removing the carburetor and overhauling it on the bench. Both procedures for servicing the Rochester model E2SE carburetor are covered in following sections.

⚙ 11-7 On-the-car service This procedure includes checks of the choke and hoses, the fuel filter, and the idle-speed solenoid. In addition, several adjustments can be made to the carburetor as shown in Figs. 11-5 to 11-14.

The hoses should be in good condition and properly connected. The choke action can be checked as explained in ⚙ 10-4. The idle-stop solenoid can be checked as outlined in ⚙ 10-5. Figures 11-15 to 11-17 explain how to set the idle speed.

Engine performance problems such as flooding, stumble, hesitation, and poor drivability are often caused by dirt, water, or other foreign materials in the carburetor. If analysis of the problem, as detailed in Chap. 8, indicates that the cause of trouble lies in the carburetor, it should be removed for overhaul. ⚙ 10-7 describes how to remove and reinstall a carburetor.

⚙ 11-8 Carburetor disassembly A carburetor overhaul kit should be available before carburetor is disassembled. On reassembly, all gaskets, diaphragms, seals, and worn or damaged parts should be replaced. As a first step, put the carburetor in a holding fixture. Following are the various servicing procedures required in disassembling the Rochester model E2SE carburetor:

1. Removing the idle-speed solenoid (⚙ 11-9)
2. Removing and disassembling the air horn (⚙ 11-10 and 11-11)
3. Disassembling the float bowl (⚙ 11-12)

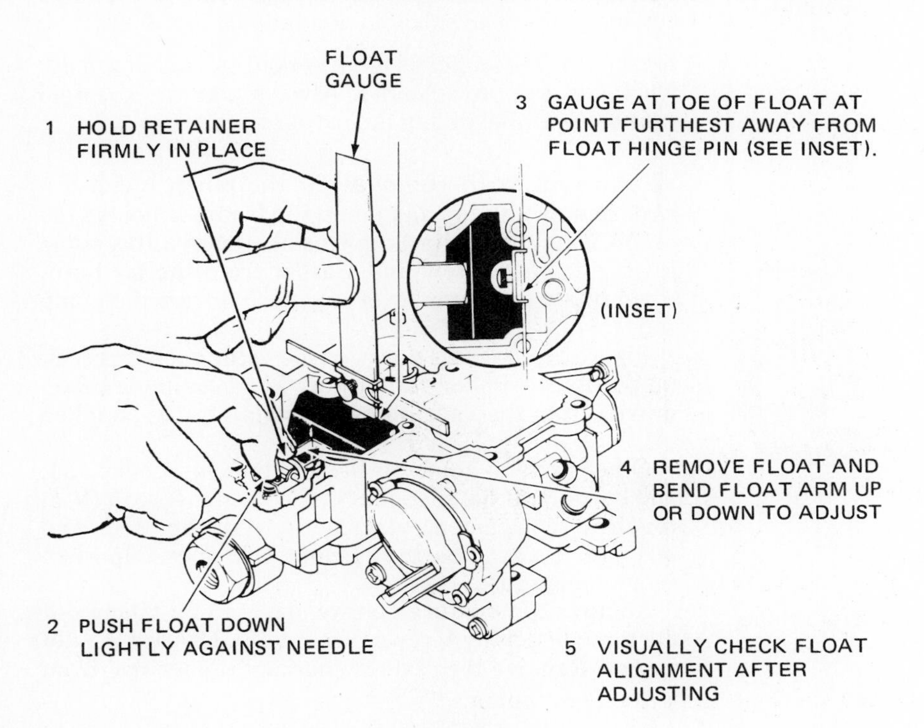

Fig. 11-5 Float adjustment. (*Chevrolet Motor Division of General Motors Corporation*)

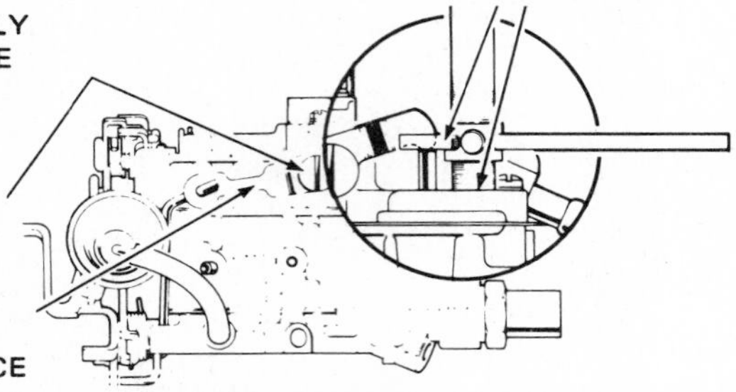

NOTE: ON MODELS USING A CLIP TO RETAIN PUMP ROD IN PUMP LEVER, NO PUMP ADJUSTMENT IS REQUIRED. ON MODELS USING THE "CLIPLESS" PUMP ROD, THE PUMP ADJUSTMENT SHOULD NOT BE CHANGED FROM ORIGINAL FACTORY SETTING UNLESS GAUGING SHOWS OUT OF SPECIFICATION. THE PUMP LEVER IS MADE FROM HEAVY DUTY, HARDENED STEEL MAKING BENDING DIFFICULT. DO NOT REMOVE PUMP LEVER FOR BENDING UNLESS ABSOLUTELY NECESSARY.

1 THROTTLE VALVES COMPLETELY CLOSED. MAKE SURE FAST-IDLE SCREW IS OFF STEPS OF FAST-IDLE CAM.

2 GAUGE FROM AIR HORN CASTING SURFACE TO TOP OF PUMP STEM. DIMENSION SHOULD BE AS SPECIFIED.

3 IF NECESSARY TO ADJUST, REMOVE PUMP LEVER RETAINING SCREW AND WASHER AND REMOVE PUMP LEVER BY ROTATING LEVER TO REMOVE FROM PUMP ROD. PLACE LEVER IN A VISE, PROTECTING LEVER FROM DAMAGE, AND BEND END OF LEVER (NEAREST NECKED DOWN SECTION).

NOTE: DO NOT BEND LEVER IN A SIDEWAYS OR TWISTING MOTION.

4 REINSTALL PUMP LEVER, WASHER AND RETAINING SCREW. RECHECK PUMP ADJUSTMENT AND 2 . TIGHTEN RETAINING SCREW SECURELY AFTER THE PUMP ADJUSTMENT IS CORRECT.

5 OPEN AND CLOSE THROTTLE VALVES CHECKING LINKAGE FOR FREEDOM OF MOVEMENT AND OBSERVING PUMP LEVER ALIGNMENT.

Fig. 11-6 Accelerator pump adjustment. (*Chevrolet Motor Division of General Motors Corporation*)

4. Disassembling the choke (⚙ 11-13)
5. Disassembling the throttle body (⚙ 11-14)

After these steps are performed, the parts should be cleaned and inspected (⚙ 10-9). Each of the above procedures is described in following sections.

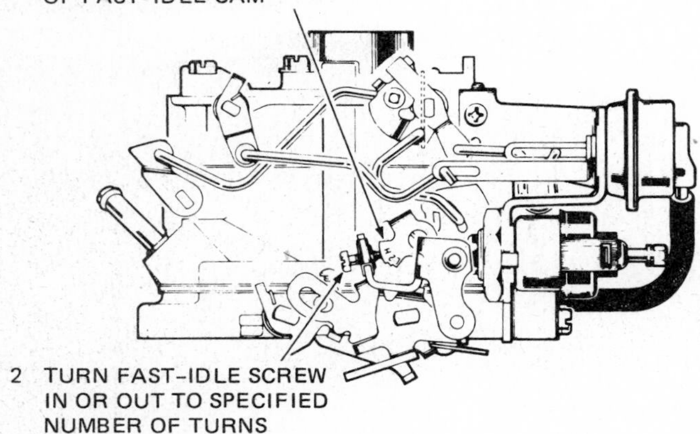

Fig. 11-7 Fast-idle adjustment with the carburetor off the car. (*Chevrolet Motor Division of General Motors Corporation*)

⚙ 11-9 Idle-speed solenoid removal To remove the idle-speed solenoid (Fig. 11-1), bend back the retaining tabs on the lockwasher and remove the large retaining nut. The solenoid can then be lifted off.

NOTE: Do not immerse the solenoid in cleaning fluid. This will ruin the solenoid. Always take the solenoid off when disassembling the carburetor.

⚙ 11-10 Air horn removal If the pump rod has a clip, remove the clip and pump rod from the hole in the pump lever. If a clip is not used, remove the pump lever retaining screw and washer from the air horn. Then rotate the pump lever to detach it from the pump rod.

Figure 11-18 shows how to remove the vacuum break on the four-cylinder carburetor. The hose must be removed from the vacuum break before it can be taken off.

Other parts to come off include the clip, choke rod, and bushing. If the secondary-side vacuum break (V-6) requires replacement, detach the rod clip, rod, and bushing. Always save the bushing. Use a new clip when reattaching rods.

Remove the mixture control solenoid by taking out the three attaching screws (Fig. 11-19). Discard the old gasket. Remove the seal retainer and rubber seal from

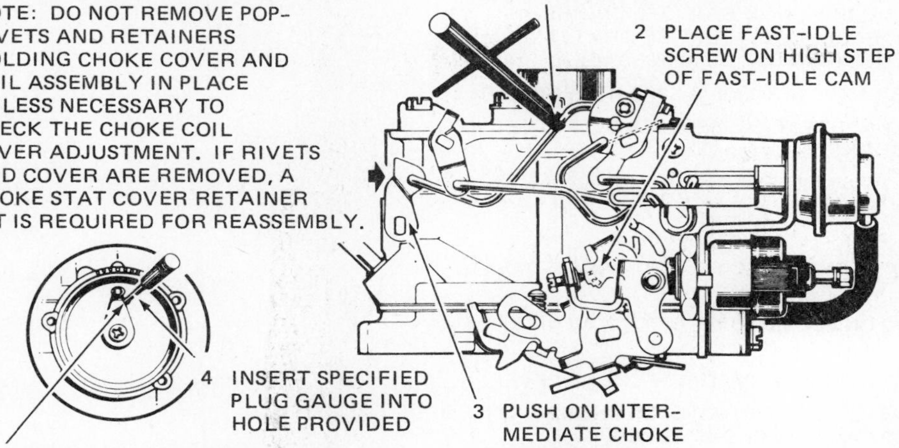

1 REMOVE RIVETS AND CHOKE COVER AND COIL ASSEMBLY FOLLOWING INSTRUCTIONS IN CHOKE STAT COVER RETAINER KIT (SEE NOTE).
NOTE: DO NOT REMOVE POP-RIVETS AND RETAINERS HOLDING CHOKE COVER AND COIL ASSEMBLY IN PLACE UNLESS NECESSARY TO CHECK THE CHOKE COIL LEVER ADJUSTMENT. IF RIVETS AND COVER ARE REMOVED, A CHOKE STAT COVER RETAINER KIT IS REQUIRED FOR REASSEMBLY.
6 BEND INTERMEDIATE CHOKE ROD AT THIS POINT TO ADJUST
2 PLACE FAST-IDLE SCREW ON HIGH STEP OF FAST-IDLE CAM
4 INSERT SPECIFIED PLUG GAUGE INTO HOLE PROVIDED
3 PUSH ON INTERMEDIATE CHOKE LEVER UNTIL CHOKE VALVE IS CLOSED
5 EDGE OF LEVER SHOULD JUST CONTACT SIDE OF PLUG GAUGE AS SHOWN

Fig. 11-8 Choke coil lever adjustment. (*Chevrolet Motor Division of General Motors Corporation*)

the end of the solenoid plunger. Discard the seal and retainer.

Remove the hot-idle compensator, if present. Remove the air-horn attaching screws and lockwashers. Rotate the fast-idle cam to full UP. Tilt the air horn to disengage the fast-idle cam rod from the slot in the cam. Lift the air horn off. Do not remove the fast-idle cam from the float bowl. It is serviced as part of the float bowl assembly.

⚙ 11-11 Air horn disassembly About the only additional part that can be removed from the air horn is the throttle position sensor plunger (where present). Discard the old seals and retainers. The air valve screws are permanently staked and should not be removed. The plugs covering the idle-air bleed screw and throttle position sensor adjustment screw must not be removed. These screws are preset at the factory and should not be changed. If the air horn or float bowl requires replacement, then special service instructions regarding their setting are supplied with the new float bowl or air horn.

If any connecting rods in the linkage require replacement, the old clips should be discarded. But the bushings can be reused.

⚙ 11-12 Float bowl disassembly Remove the air horn gasket, pump plunger, and spring from the pump well. If the carburetor has a throttle position sensor, push up from the bottom on the electrical connector to remove it and the spring under it. Be careful to avoid damaging the sensor.

Then remove the following if necessary:

1. Plastic filler block over the float bowl.
2. Float assembly and valve.
3. Extended metering jet (Fig. 11-20). Do not change the adjustment of the small calibration screw inside the metering jet. This is preset at the factory.
4. Plastic retainer holding the pump discharge spring and check ball. Use a small slide hammer. Do not remove the retainer by prying it out. This will damage the bowl casting and a new bowl assembly will be required.
5. Turn the bowl upside down and catch the spring and ball in your hand.

⚙ 11-13 Choke disassembly The choke cover is made to be tamper-resistant to discourage field adjustment. However, if it is necessary to remove the cover, use a No. 21 drill (0.159 inch) to drill out the Pop rivet heads. Then drive out the rivets with a small drift punch and hammer. Be careful to avoid damaging the cover or housing.

Once the rivets are out, the rest of the parts will come off after retainers, screws, and intermediate choke shaft and lever assembly are removed.

⚙ 11-14 Throttle body There is little in the way of disassembly that the throttle body requires. The throttle valves are permanently staked on the throttle shaft. The plug covering the idle-mixture screw should not be removed. However, if the idle system is clogged and soaking and compressed air will not clear it, the plug can be removed. Use a punch to drive against the steel plug. When the plug breaks, hold the punch at a 45° angle and break out the throttle body casting to get to the plug and drive it out.

⚙ 11-15 Trouble diagnosis Here are specific things to look for if various troubles have been reported:

1. Flooding. Check float valve and seat for damage and seating. Check float adjustment. If float is heavy and

FIGURE (A)

1. CHOKE COIL LEVER MUST BE CORRECT AND FAST-IDLE ADJUSTMENT MUST BE MADE BEFORE PROCEEDING.
2. USE CHOKE VALVE MEASURING GAUGE, WHICH MAY BE USED WITH CARBURETOR ON OR OFF ENGINE. IF OFF ENGINE, PLACE CARBURETOR ON HOLDING FIXTURE SO THAT IT WILL REMAIN IN SAME POSITION WHEN GAUGE IS IN PLACE.
3. ROTATE DEGREE SCALE UNTIL ZERO (0) IS OPPOSITE POINTER.
4. WITH CHOKE VALVE COMPLETELY CLOSED, PLACE MAGNET SQUARELY ON TOP OF CHOKE VALVE.
5. ROTATE BUBBLE UNTIL IT IS CENTERED.

FIGURE (B)

6. ROTATE SCALE SO THAT DEGREE SPECIFIED FOR ADJUSTMENT IS OPPOSITE POINTER.
7. PLACE FAST-IDLE SCREW ON SECOND STEP OF CAM AGAINST RISE OF HIGH STEP.
8. CLOSE CHOKE BY PUSHING ON INTERMEDIATE CHOKE LEVER.
9. PUSH ON VACUUM BREAK LEVER TOWARD OPEN CHOKE UNTIL LEVER IS AGAINST REAR TANG ON CHOKE LEVER.
10. TO ADJUST, BEND FAST-IDLE CAM ROD UNTIL BUBBLE IS CENTERED.
11. REMOVE GAUGE.

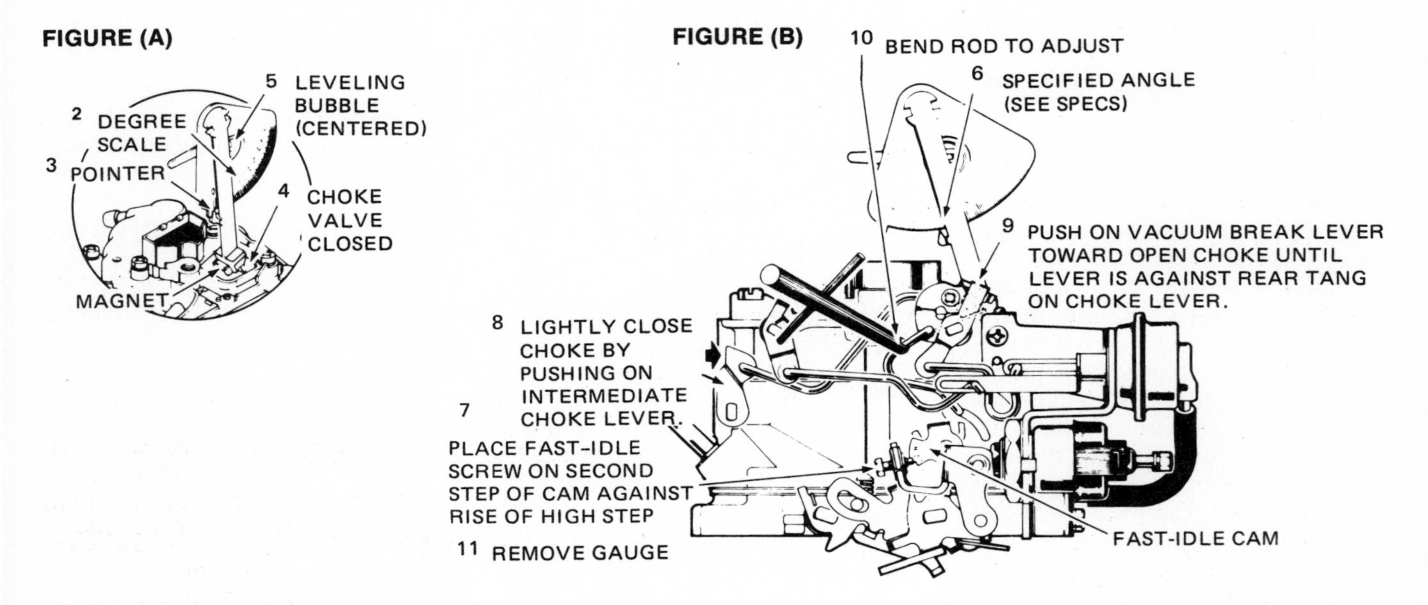

Fig. 11-9 Choke rod adjustment. (*Chevrolet Motor Division of General Motors Corporation*)

has absorbed fuel, replace it. Clean or replace filter at fuel inlet.

2. Hesitation. The trouble probably is in the accelerator pump. Check the pump and cup for fit and damage. Check return spring for weakness. Check pump passages and jets, and the discharge check ball and its seating. Check pump linkage for wear or damage.
3. Hard starting and poor cold operation. This is probably caused by trouble in the choke, which should be checked. It could also be caused by clogged inlet filter or a stuck or dirty float valve. Causes under flooding in item 1 above could also be responsible.
4. Poor performance and poor gasoline mileage. This could result from dirt, wear, or damage inside the carburetor. It could also be caused by a sticky mixture-control solenoid. The solenoid can be checked by connecting it to a 12-volt battery. Then apply vacuum and check the number of seconds it takes for the solenoid to leak down from the lean position.
5. Rough idle. This could be caused by the same conditions listed in item 4 above. It could also be caused by binding linkages and valves, and by damaged diaphragms.

NOTE: The above troubles could be caused by many problems outside the carburetor. See the trouble diagnosis chart in Chap. 8.

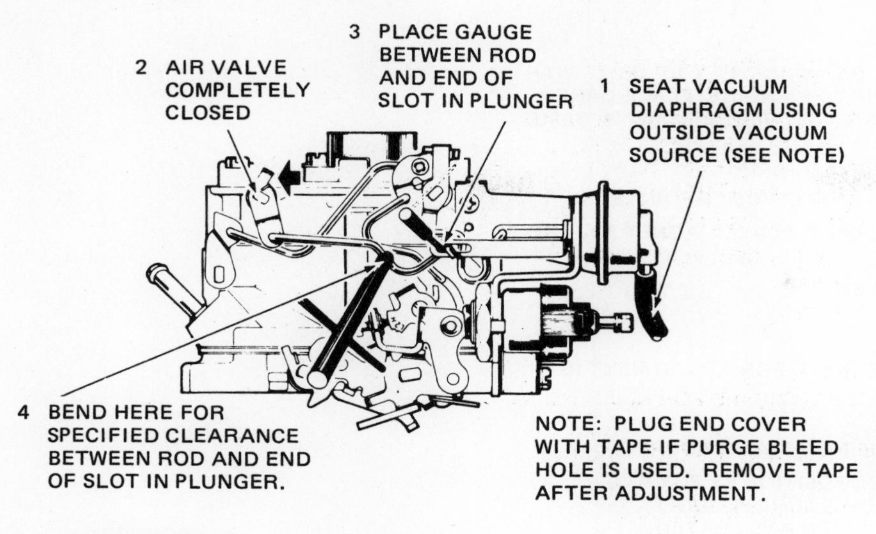

Fig. 11-10 Air valve rod adjustment. (*Chevrolet Motor Division of General Motors Corporation*)

FIGURE (A)

1. USE CHOKE VALVE MEASURING GAUGE, WHICH MAY BE USED WITH CARBURETOR ON OR OFF ENGINE. IF OFF ENGINE, PLACE CARBURETOR ON HOLDING FIXTURE SO THAT IT WILL REMAIN IN SAME POSITION WHEN GAUGE IS IN PLACE.
2. ROTATE DEGREE SCALE UNTIL ZERO (0) IS OPPOSITE POINTER.
3. WITH CHOKE VALVE COMPLETELY CLOSED, PLACE MAGNET SQUARELY ON TOP OF CHOKE VALVE.
4. ROTATE BUBBLE UNTIL IT IS CENTERED.

FIGURE (B)

5. ROTATE SCALE SO THAT DEGREE SPECIFIED FOR ADJUSTMENT IS OPPOSITE POINTER.
6. SEAT CHOKE VACUUM DIAPHRAGM USING VACUUM SOURCE.
7. HOLD CHOKE VALVE TOWARD CLOSED POSITION BY PUSHING ON INTERMEDIATE CHOKE LEVER. MAKE SURE PLUNGER BUCKING SPRING (IF USED) IS COMPRESSED AND SEATED.
8. TO ADJUST, BEND VACUUM BREAK ROD UNTIL BUBBLE IS CENTERED.
9. REMOVE GAUGE.

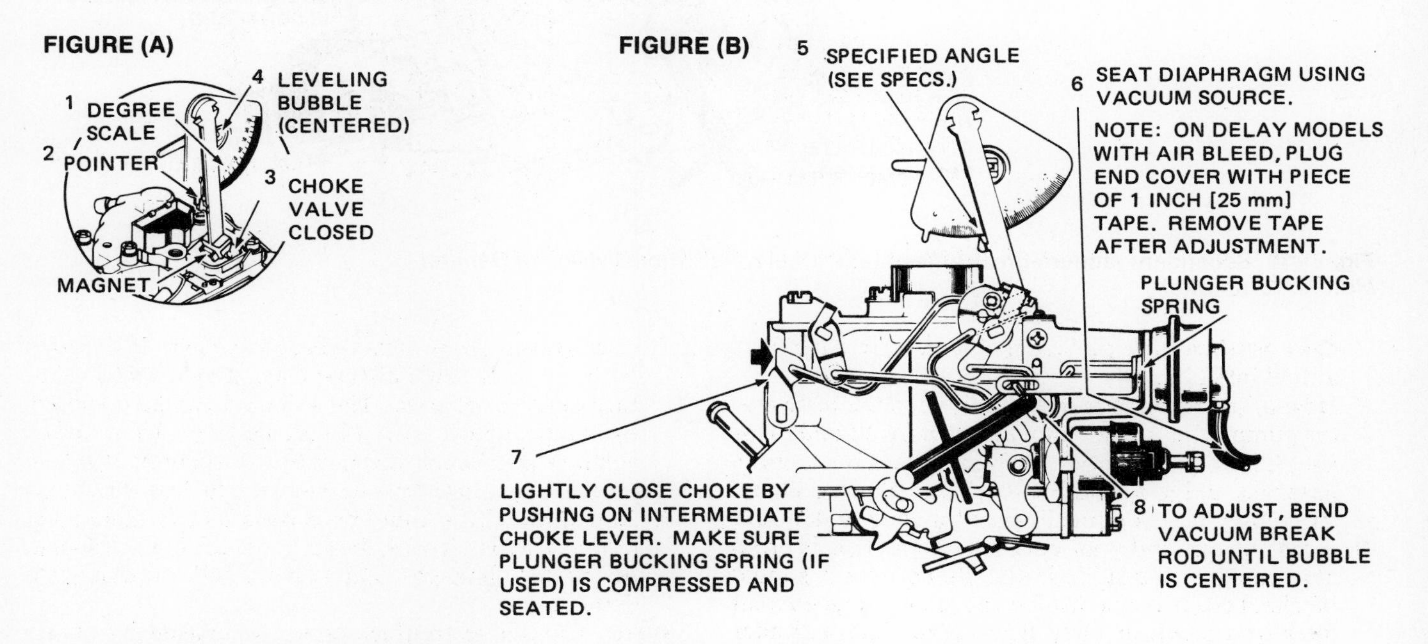

Fig. 11-11 Primary vacuum-break adjustment. (*Chevrolet Motor Division of General Motors Corporation*)

FIGURE (A)

1. USE CHOKE VALVE MEASURING GAUGE, WHICH MAY BE USED WITH CARBURETOR ON OR OFF ENGINE. IF OFF ENGINE, PLACE CARBURETOR ON HOLDING FIXTURE SO THAT IT WILL REMAIN IN SAME POSITION WHEN GAUGE IS IN PLACE.
2. ROTATE DEGREE SCALE UNTIL ZERO (0) IS OPPOSITE POINTER.
3. WITH CHOKE VALVE COMPLETELY CLOSED, PLACE MAGNET SQUARELY ON TOP OF CHOKE VALVE.
4. ROTATE BUBBLE UNTIL IT IS CENTERED.

FIGURE (B)

5. ROTATE SCALE SO THAT DEGREE SPECIFIED FOR ADJUSTMENT IS OPPOSITE POINTER.
6. SEAT CHOKE VACUUM DIAPHRAGM USING VACUUM SOURCE.
7. HOLD CHOKE VALVE TOWARD CLOSED POSITION BY PUSHING ON INTERMEDIATE CHOKE LEVER. MAKE SURE PLUNGER BUCKING SPRING (IF USED) IS COMPRESSED AND SEATED.
8. TO ADJUST, BEND VACUUM BREAK ROD UNTIL BUBBLE IS CENTERED.
9. REMOVE GAUGE.

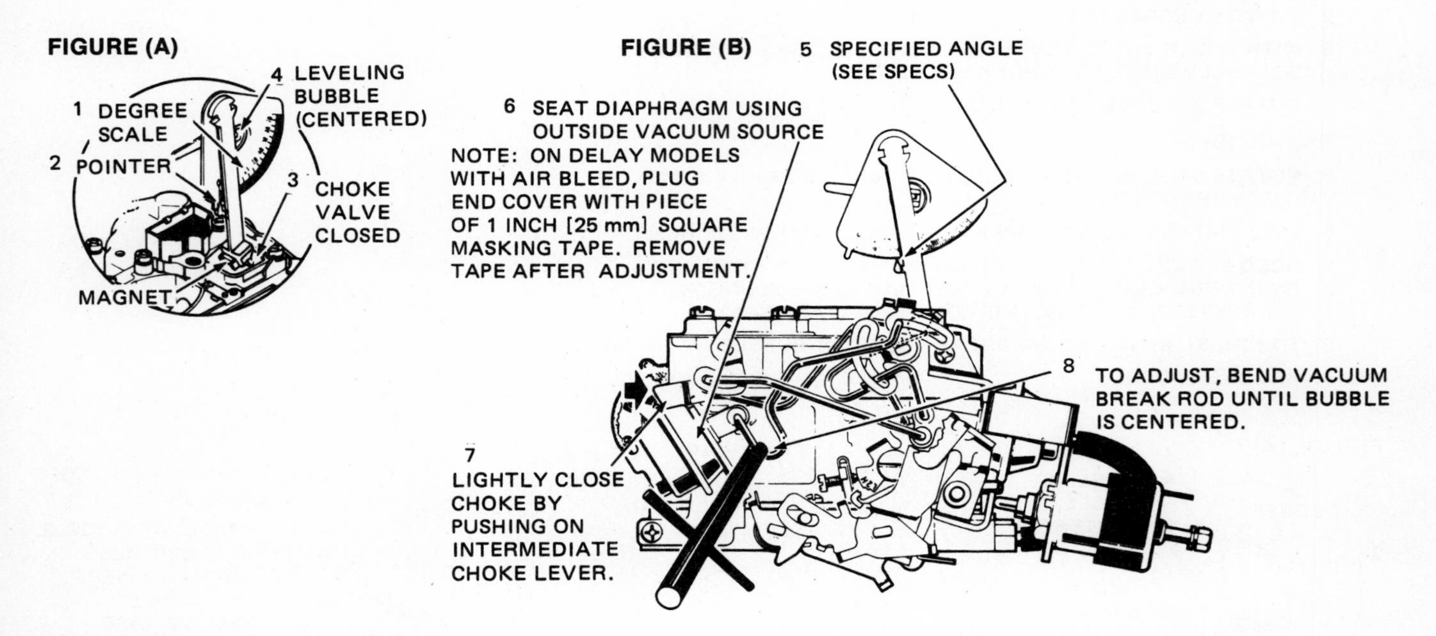

Fig. 11-12 Secondary vacuum-break adjustment. (*Chevrolet Motor Division of General Motors Corporation*)

FIGURE (A)

1. USE CHOKE VALVE MEASURING GAUGE, WHICH MAY BE USED WITH CARBURETOR ON OR OFF ENGINE. IF OFF ENGINE, PLACE CARBURETOR ON HOLDING FIXTURE SO THAT IT WILL REMAIN IN SAME POSITION WHEN GAUGE IS IN PLACE.
2. ROTATE DEGREE SCALE UNTIL ZERO (0) IS OPPOSITE POINTER.
3. WITH CHOKE VALVE COMPLETELY CLOSED, PLACE MAGNET SQUARELY ON TOP OF CHOKE VALVE.
4. ROTATE BUBBLE UNTIL IT IS CENTERED.

FIGURE (B)

5. ROTATE SCALE SO THAT DEGREE SPECIFIED FOR ADJUSTMENT IS OPPOSITE POINTER.
6. INSTALL CHOKE THERMOSTATIC COVER AND COIL ASSEMBLY IN HOUSING. ALIGN INDEX MARK WITH SPECIFIED POINT ON HOUSING.
7. HOLD PRIMARY THROTTLE VALVE WIDE OPEN.
8. ON WARM ENGINE, CLOSE CHOKE VALVE BY PUSHING CLOCKWISE ON INTERMEDIATE CHOKE LEVER (HOLD IN POSITION WITH RUBBER BAND).
9. TO ADJUST, BEND TANG ON THROTTLE LEVER UNTIL BUBBLE IS CENTERED.
10. REMOVE GAUGE.

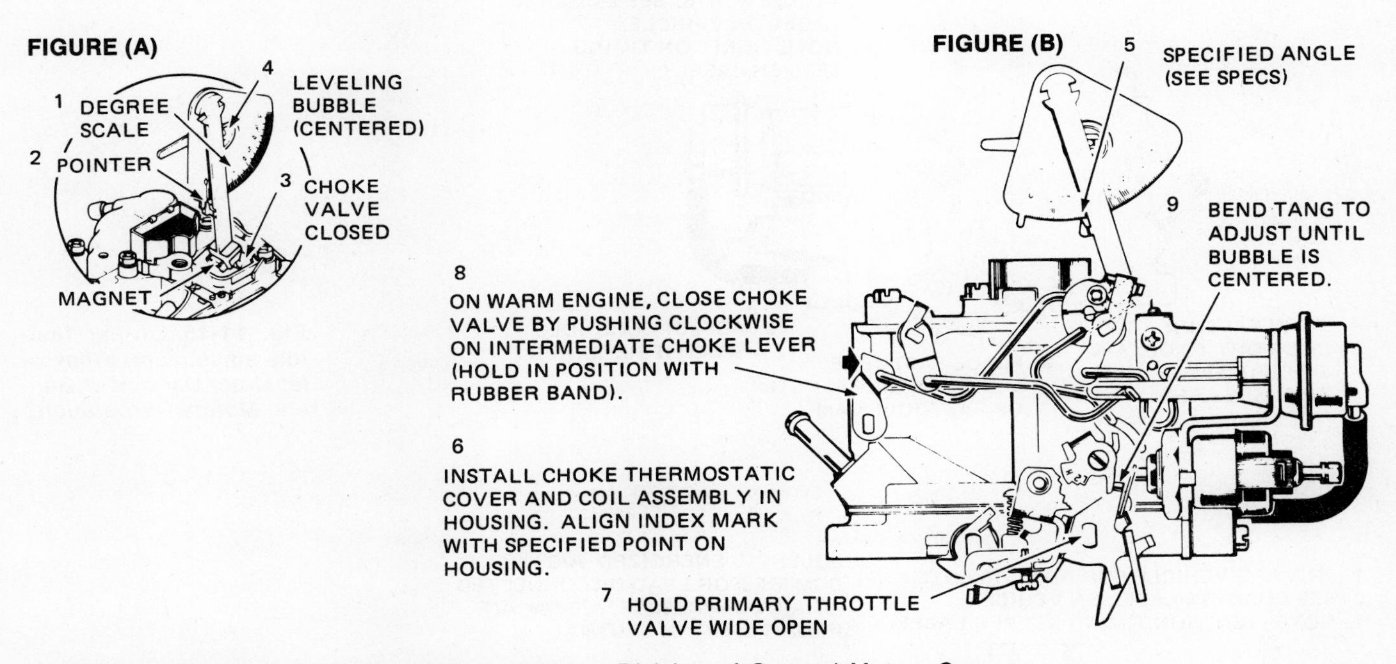

Fig. 11-13 Unloader adjustment. (*Chevrolet Motor Division of General Motors Corporation*)

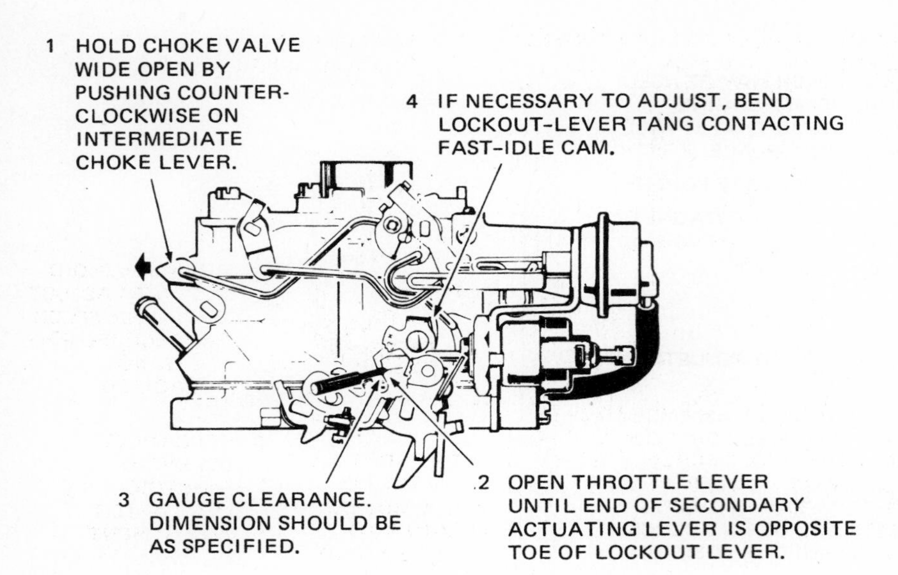

Fig. 11-14 Secondary lockout adjustment. (*Chevrolet Motor Division of General Motors Corporation*)

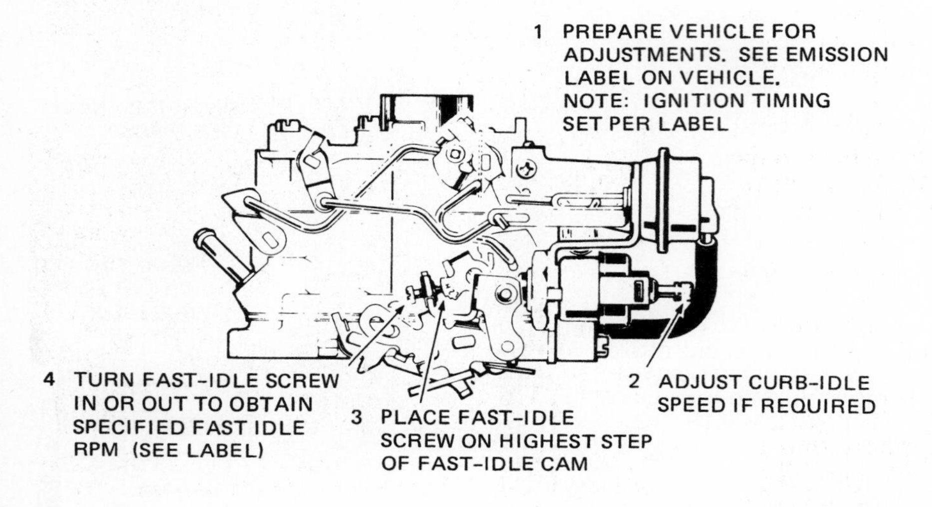

Fig. 11-15 On-car fast-idle adjustment. (*Chevrolet Motor Division of General Motors Corporation*)

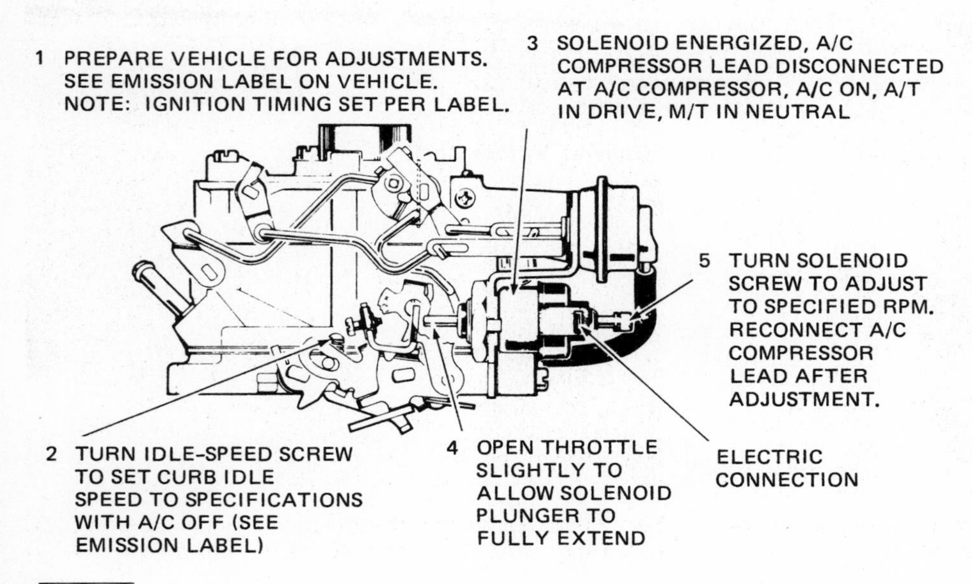

Fig. 11-16 Idle-speed adjustment on car equipped with air conditioning. (*Chevrolet Motor Division of General Motors Corporation*)

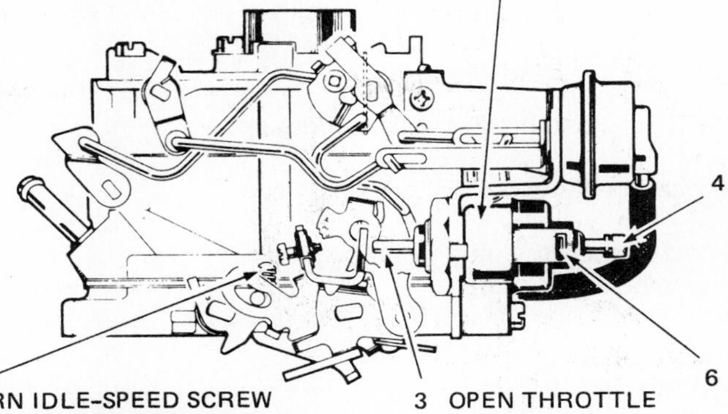

1 PREPARE VEHICLE FOR ADJUSTMENTS. SEE EMISSION LABEL ON VEHICLE. NOTE: IGNITION TIMING SET PER LABEL.

2 SOLENOID ENERGIZED, A/T IN DRIVE, M/T IN NEUTRAL

3 OPEN THROTTLE SLIGHTLY TO ALLOW SOLENOID PLUNGER TO FULLY EXTEND

4 TURN SOLENOID SCREW TO ADJUST CURB-IDLE SPEED TO SPECIFIED RPM SOLENOID ENERGIZED

5 TURN IDLE-SPEED SCREW TO SET BASIC IDLE SPEED TO SPECIFICATIONS (SOLENOID DE-ENERGIZED)

6 RECONNECT SOLENOID ELECTRIC LEAD AFTER ADJUSTMENT

Fig. 11-17 Adjusting the idle speed on a car without air conditioning. *(Chevrolet Motor Division of General Motors Corporation)*

11-16 Carburetor reassembly On reassembly, new clips and seals should be used. All new parts included in the carburetor overhaul kit should be installed. Basically, reassembly is installing the rods, throttle position sensor, accelerator pump, and other parts removed during the disassembly.

If you install an additional in-line fuel filter, it must have a built-in check valve. This is required to meet the Motor Vehicle Safety Standards (MVSS), which require that fuel delivery to the carburetor be stopped if the car rolls over.

If the choke cover was removed, reinstall it with self-tapping screws.

During and after reassembly, make the carburetor adjustments shown in Figs. 11-5 to 11-17.

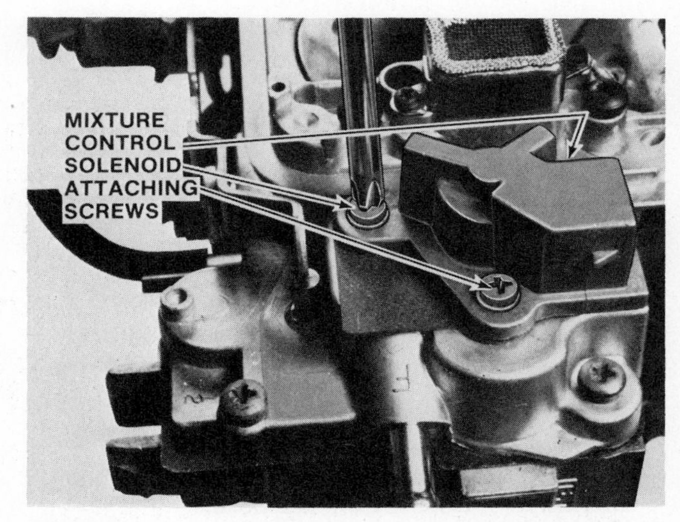

Fig. 11-19 Removing the mixture control solenoid. *(Chevrolet Motor Division of General Motors Corporation)*

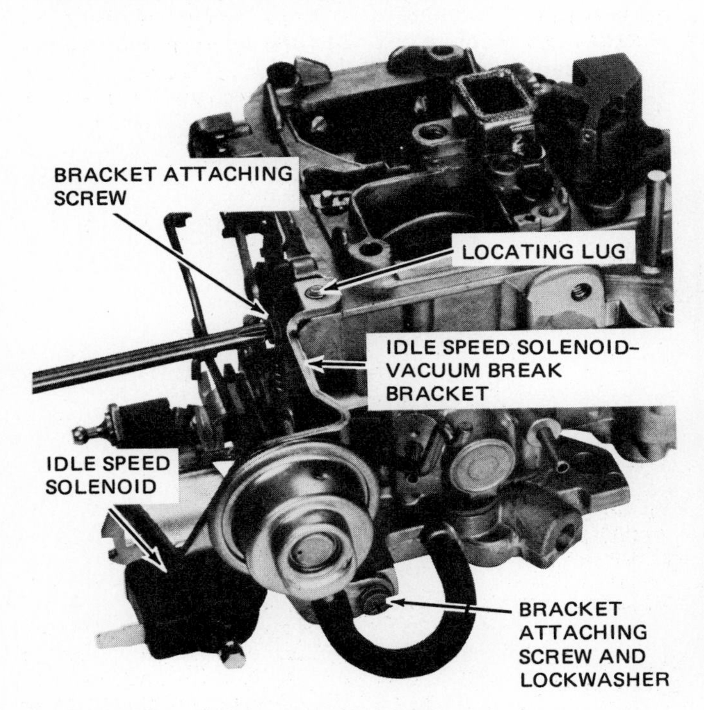

Fig. 11-18 Removing the vacuum break. *(Chevrolet Motor Division of General Motors Corporation)*

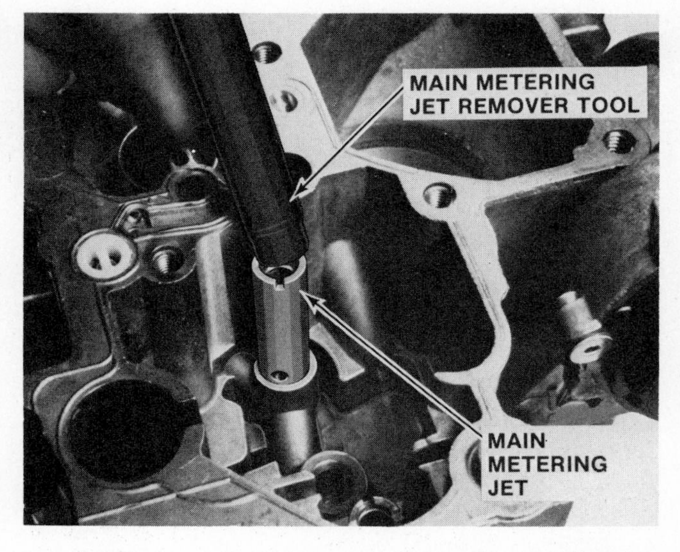

Fig. 11-20 Removing the extended main metering jet. *(Chevrolet Motor Division of General Motors Corporation)*

Chapter 11 review questions

Select the *one* correct, best, or most probable answer to each question. Then check your answers against the correct answers given at the end of the book.

1. In a carburetor with one primary and one secondary barrel:
   a. the secondary barrel handles idle and low-speed requirements
   b. the primary barrel operates only at high speed
   c. both barrels are the same size
   d. the secondary barrel supplies air-fuel mixture for medium- to high-speed operation.
2. In two-barrel carburetors in which the two barrels are the same size:
   a. both work all the time the engine is running
   b. only one has an idle system
   c. only one has an accelerator pump system
   d. only one has a choke system.
3. In some engines, the manifold is open and both barrels feed into it. In others, the manifold is divided and:
   a. each barrel feeds one cylinder
   b. each barrel feeds half the cylinders
   c. only the carburetor idle system is divided
   d. only the carburetor power system is divided.
4. The feedback carburetor:
   a. feeds back fuel to the engine when needed
   b. does not need an idle system
   c. adjusts the air-fuel mixture ratio as information is fed back to it by an electronic control unit
   d. controls the power and choke systems electronically.
5. The feedback carburetor controls the richness of the idle mixture by:
   a. admitting more or less air into the idle system
   b. raising or lowering the idle needle
   c. controlling the vacuum diaphragm attached to the needle
   d. controlling the linkage between the throttle and idle cam.
6. If the mixture is rich, the exhaust gas is:
   a. rich in oxygen
   b. rich in H_2O
   c. low in oxygen
   d. low in NO_x.
7. In the Rochester model E2SE carburetor, idle-mixture richness and main-metering-system richness are controlled by:
   a. two separate solenoids
   b. an idle needle and a main metering system air bleed
   c. an idle air bleed and a main metering system needle valve
   d. two vacuum diaphragms.
8. The Rochester model E2SE carburetor for six-cylinder engines has an additional sensor that "tells" the ECU the:
   a. richness of the NO_x in the exhaust
   b. position of the throttle
   c. temperature of the coolant
   d. location of the choke valve.
9. The two factory-adjusted screws in the main-metering system of the Rochester model E2SE carburetor are the:
   a. idle-mixture screw and idle-speed screw
   b. lean-mixture screw and rich-mixture screw
   c. idle-speed screw and idle-air-bleed screw
   d. none of the above.
10. On an engine with a Rochester model E2SE carburetor, the oxygen content of the exhaust gas is too high. Mechanic X says the air-fuel mixture is too rich. Mechanic Y says the air-fuel mixture is too lean. Who is right?
    a. X only
    b. Y only
    c. both X and Y
    d. neither X nor Y.

CHAPTER 12

SERVICING THE FORD VARIABLE-VENTURI CARBURETOR

After studying this chapter, and with proper instruction and equipment, you should be able to:

1. Perform the adjustments on a Ford VV carburetor.
2. Overhaul a Ford VV carburetor.

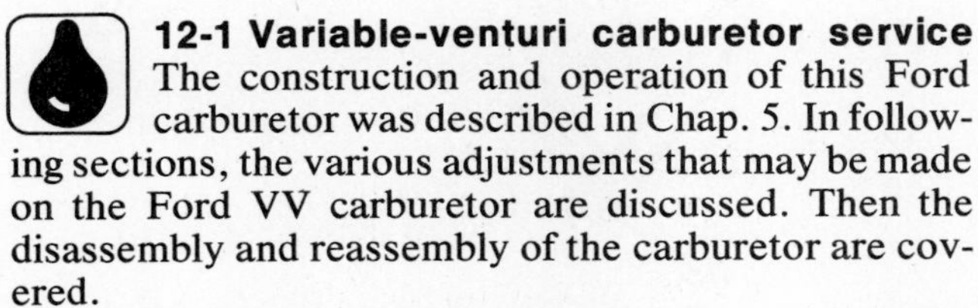

12-1 Variable-venturi carburetor service The construction and operation of this Ford carburetor was described in Chap. 5. In following sections, the various adjustments that may be made on the Ford VV carburetor are discussed. Then the disassembly and reassembly of the carburetor are covered.

The following adjustments can be made on the Ford variable-venturi carburetor:

1. Curb idle speed (⚙ 12-2)
2. Cold enrichment metering rod (⚙ 12-3)
3. Internal vent (⚙ 12-4)
4. Fast-idle speed (⚙ 12-5)
5. Fast-idle cam (⚙ 12-6)
6. Fuel level (⚙ 12-7)
7. Float drop (⚙ 12-8)
8. Venturi-valve limiter (⚙ 12-9)
9. Control vacuum regulator adjustment (⚙ 12-10)
10. High-cam-speed positioner (⚙ 12-11)

All adjustments made with the engine running require that the engine be at normal operating temperature. The parking brake should be applied and the wheels blocked. If the car has a vacuum parking-brake release, disconnect the vacuum line to the brake control and plug the line.

On vehicles with automatic transmissions, the kickdown rod must be adjusted whenever the carburetor is removed and installed. Figure 12-1 shows the procedure for one line of Ford cars. Following sections cover the other adjustments.

⚙ 12-2 Curb idle speed The procedure for the adjustment of curb idle speed depends on the type of throttle positioner on the car (Fig. 5-19). Refer to the emission control label in the engine compartment for specifications and details.

⚙ 12-3 Cold enrichment metering rod Remove the choke cap and install the stator cap tool and a dial indicator as shown in Fig. 12-2. With the tip of the indicator on the top surface of the cold enrichment rod, adjust the dial to zero. Raise the weight in the stator and release it to make sure the needle returns to zero.

Remove the stator cap and install it at the index position. The dial indicator should read as shown in the factory specifications. If it does not, adjust the adjusting nut (Fig. 12-2).

⚙ 12-4 Internal vent Curb idle speed should be set to specifications. Put the specified size thickness gauge between the accelerator pump stem and the lever pump (Fig. 12-3). Turn the nylon adjusting nut until there is just a slight drag when the gauge is pulled out.

NOTE: Make this adjustment whenever the curb idle speed is adjusted.

⚙ 12-5 Fast-idle speed With the engine at normal operating temperature, put the fast-idle lever on the specified step of the fast idle cam (Fig. 12-4). EGR must be disconnected and the vacuum line plugged.

Make sure the high-cam-speed positioner lever is disengaged. Then turn the fast-idle adjusting screw to adjust speed.

⚙ 12-6 Fast-idle cam Remove the choke cap. Put the fast-idle lever in the corner of the specified step of the fast-idle cam (counting the highest step as step 1) with the high-cam-speed positioner retracted (Fig. 12-5).

Hold the throttle lightly closed with a rubber band to maintain the cam position. (This step is not required if the carburetor is on the vehicle.)

Install the stator cap tool and rotate it clockwise until the choke lever touches the adjusting screw (Fig. 12-5). Turn the fast-idle-cam adjusting screw until the index mark on the stator cap lines up with the specified notch on the choke casting.

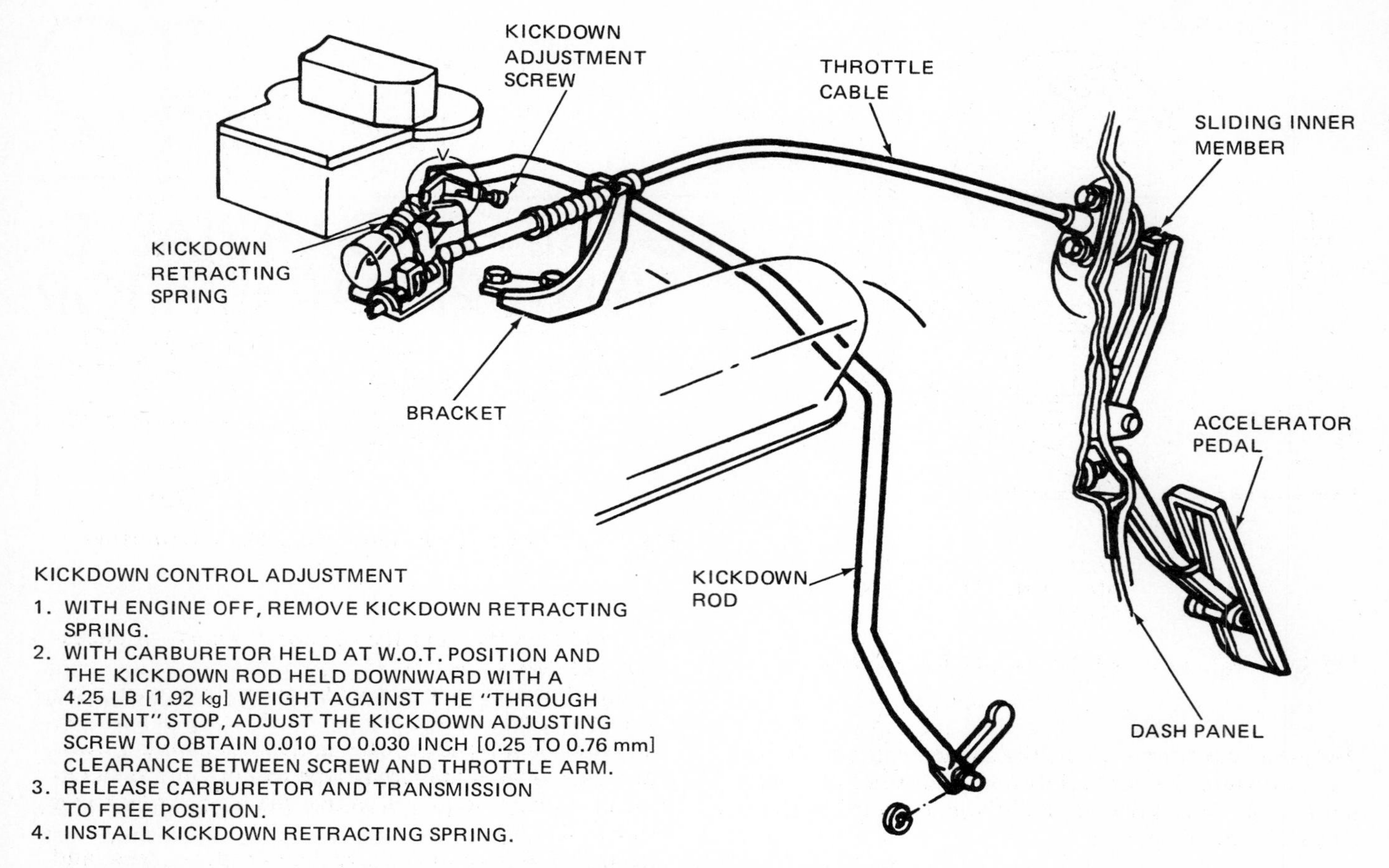

Fig. 12-1 Adjusting the throttle and downshift linkage. (*Ford Motor Company*)

Remove the stator cap and install the choke cap to the specified setting.

⚙ 12-7 Fuel level Remove the upper body assembly and gasket. Install a new gasket. Make a special tool to the specified dimensions, as shown in Fig. 12-6. Turn the upper body upside down and put the special tool on the metal surface of the upper body (not on the gasket). To adjust, bend the float operating lever away from the fuel-inlet needle to decrease the setting and toward the needle to increase the setting. Then check and adjust the float drop (⚙ 12-8).

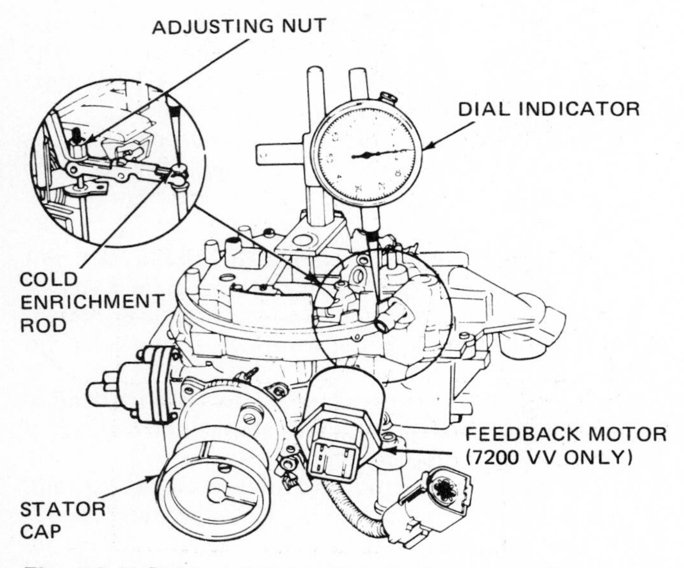

Fig. 12-2 Cold-enrichment metering rod adjustment. (*Ford Motor Company*)

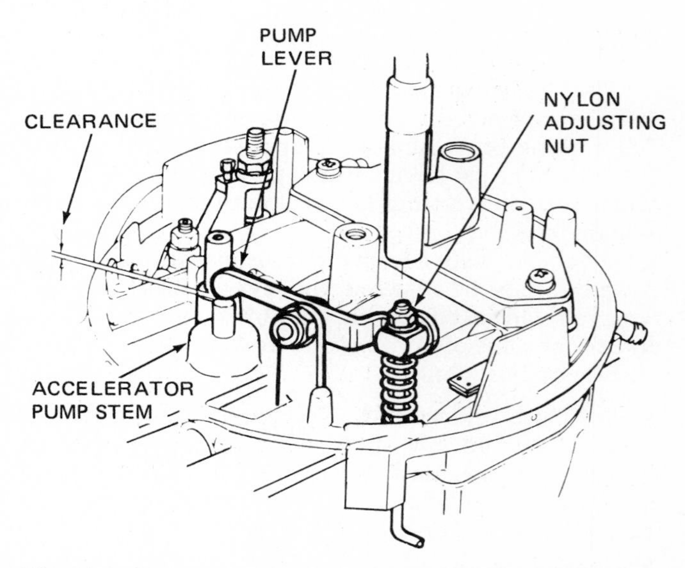

Fig. 12-3 Checking the accelerator pump stem clearance. (*Ford Motor Company*)

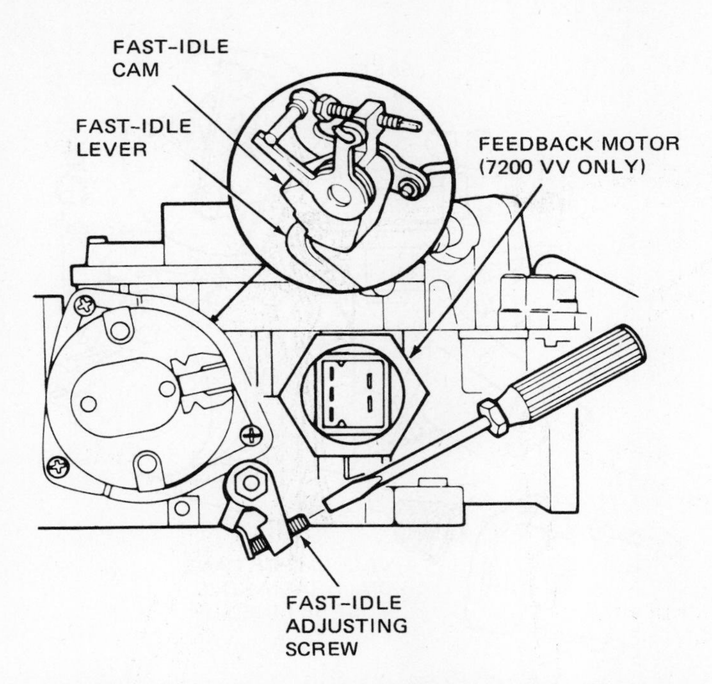

Fig. 12-4 Adjusting the fast-idle speed. (*Ford Motor Company*)

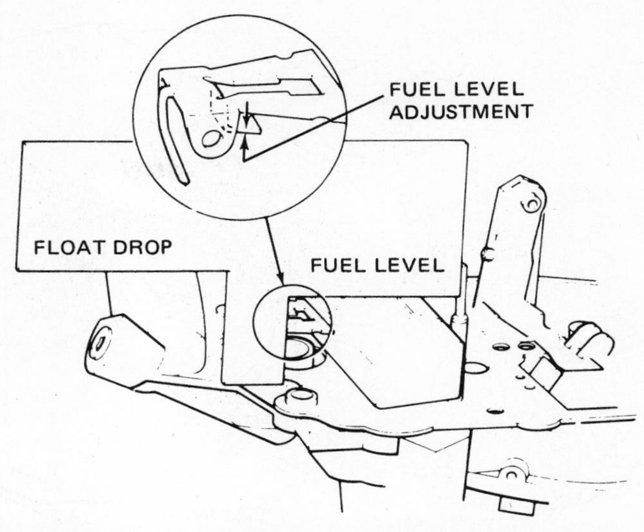

Fig. 12-6 Measuring fuel level. (*Ford Motor Company*)

⚙ 12-8 Float drop Using the gauge as shown in Fig. 12-7, measure the vertical distance between the metal surface of the upper body and the bottom of the float. Adjust by bending the stop tab.

⚙ 12-9 Venturi valve limiter This adjustment is done on the bench. Remove the venturi valve cover and roller bearings. Use a center punch and carefully remove the expansion plug at the rear of the main body on the throttle side of the carburetor.

Use an allen wrench to remove the venturi valve limiter stop screw and torque retention spring (Fig. 12-8). Block the throttle valves wide open. Apply light closing pressure on the venturi valve and check the gap between the valve and the air horn. Adjust, if necessary, as follows.

Hold the venturi valve wide open. Insert an allen wrench into the hole from which the stop screw was removed. This reaches the limiter adjusting screw. Turn it clockwise to increase the gap or counterclockwise to reduce the gap. Remove the allen wrench and recheck the gap.

Reinstall the venturi valve limiter stop screw and turn it clockwise until it touches the valve. Push the venturi valve wide open and check the gap between the valve and the air horn. Turn the stop screw to adjust the gap. Install a new expansion plug in the access hole. Reinstall the venturi valve cover gasket and roller bearings.

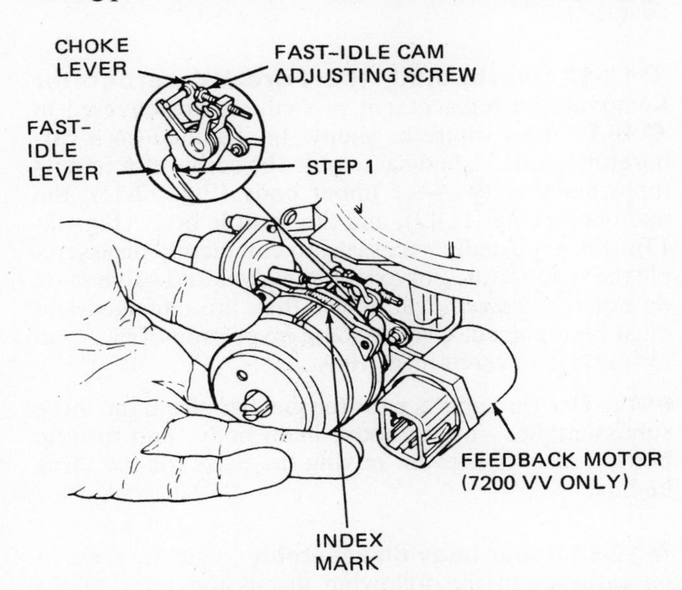

Fig. 12-5 Setting the fast-idle cam. (*Ford Motor Company*)

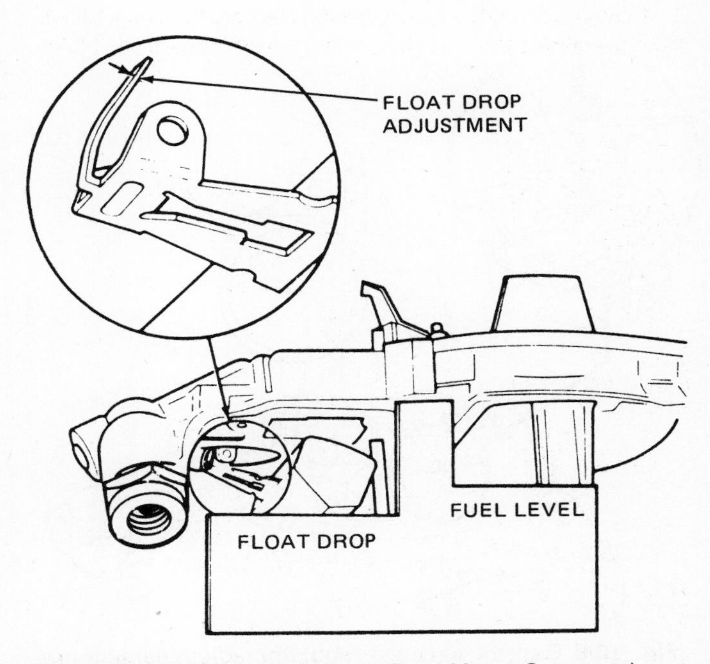

Fig. 12-7 Adjusting float drop. (*Ford Motor Company*)

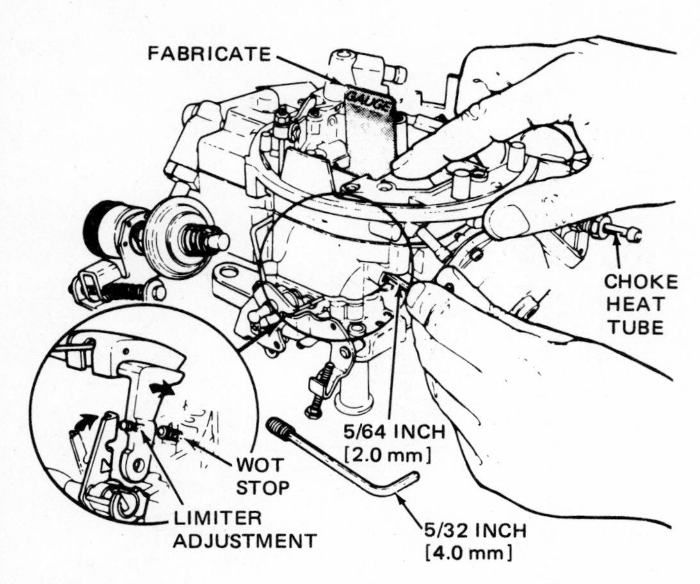

Fig. 12-8 Adjusting the venturi valve limiter. (*Ford Motor Company*)

⚙ 12-10 Control vacuum regulator The cold enrichment metering rod must be checked and adjusted (⚙ 12-3) before this adjustment is made.

Leave the dial indicator in place, but remove the stator cap (Fig. 12-2). Do not reset the dial indicator after removing the stator cap. Press down on the CVR rod until it bottoms (Fig. 12-9). Read the downward travel on the dial indicator. If it is not correct, adjust as follows.

Hold the CVR adjusting nut with a ⅜-inch box wrench. Use a $^{3}/_{32}$-inch allen wrench to turn the CVR rod counterclockwise to increase the travel or clockwise to reduce the travel.

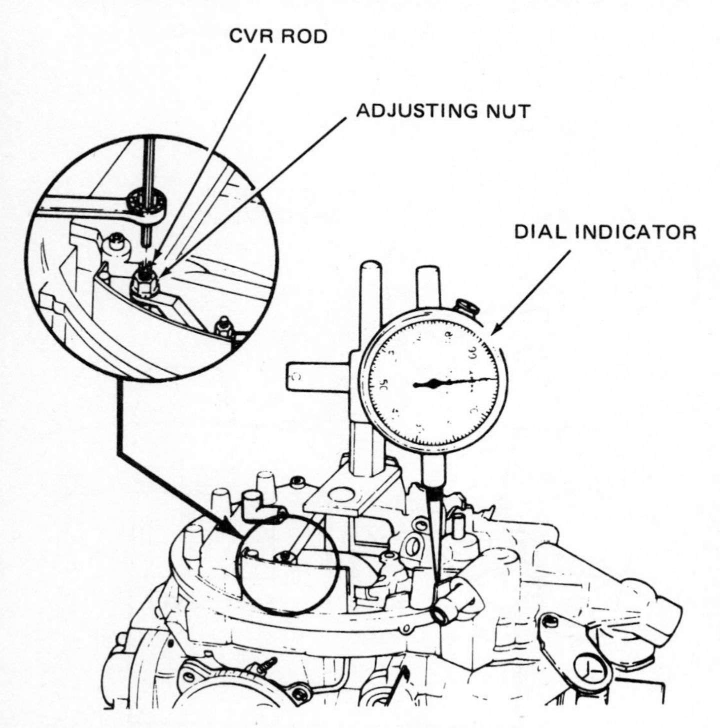

Fig. 12-9 Control vacuum regulator adjustment. (*Ford Motor Company*)

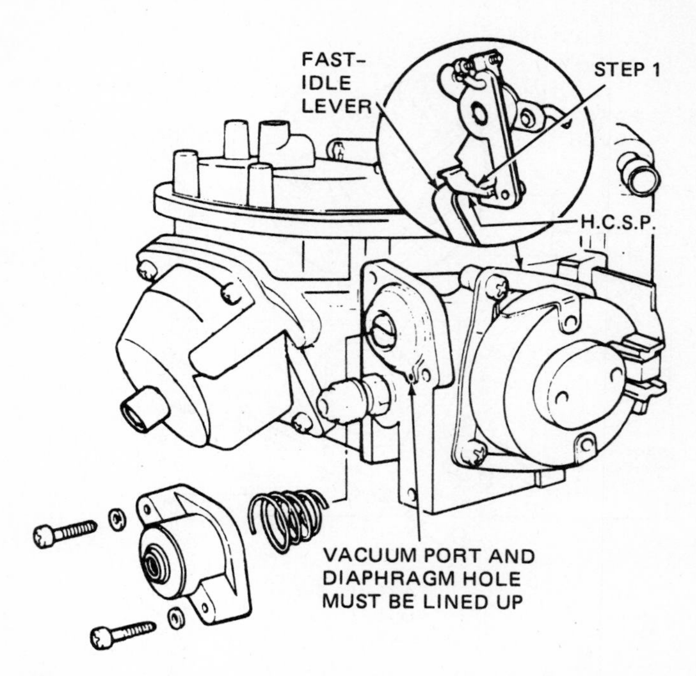

Fig. 12-10 Adjusting the high-cam-speed positioner. (*Ford Motor Company*)

⚙ 12-11 High-cam-speed positioner Put the high-cam-speed positioner (HCSP) in the corner of the specified cam step, counting the highest step as the first (Fig. 12-10). Put the fast-idle lever in the corner of the HCSP. Hold the throttle firmly closed to retain the HCSP and fast-idle lever in position. Remove the diaphragm cover and adjust the diaphragm assembly clockwise until it lightly bottoms on the casting. Then rotate it ½ to 1½ turns, until the vacuum port and diaphragm holes line up. Then install the diaphragm cover.

⚙ 12-12 Overhauling the Ford VV carburetor Removal and replacement procedures are covered in ⚙ 10-7. The carburetor should be installed on a carburetor stand. Then disassemble the carburetor into its three main bodies—the upper body (Fig. 12-11), the main body (Fig. 12-12), and the throttle body (Fig. 12-13). It is not usually necessary to completely disassemble the carburetor. For example, valves and valve seats do not require removal unless they are damaged and must be replaced. Cleaning and inspection of carburetor parts is covered in ⚙ 10-9.

NOTE: Use three parts pans for the parts from the three subassemblies—upper body, main body, and throttle body. This will prevent mixing the parts for the three bodies.

⚙ 12-13 Upper body disassembly The numbers in parentheses in the following discussion refer to the numbered items in Fig. 12-11. For example, items 1

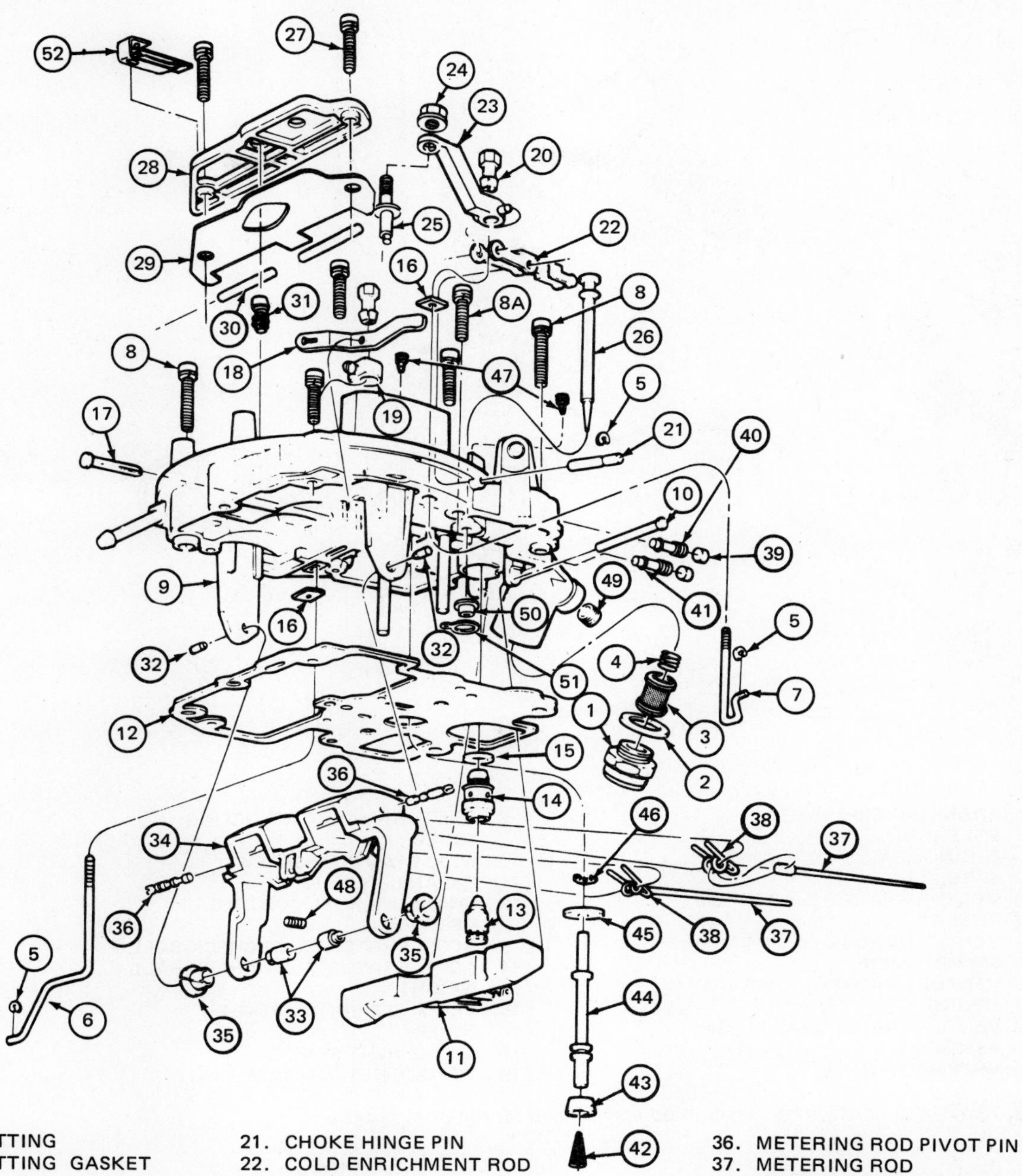

1. FUEL INLET FITTING
2. FUEL-INLET-FITTING GASKET
3. FUEL FILTER
4. FUEL FILTER SPRING
5. RETAINING E-RING
6. ACCELERATOR PUMP ROD
7. CHOKE CONTROL ROD
8. SCREW (2)

8A. SCREW (5)

9. UPPER BODY
10. FLOAT HINGE PIN
11. FLOAT ASSEMBLY
12. FLOAT BOWL GASKET
13. FUEL INLET VALVE
14. FUEL INLET SEAT
15. FUEL INLET SEAT GASKET
16. DUST SEAL
17. PIN
18. ACCELERATOR PUMP LINK
19. ACCELERATOR PUMP SWIVEL
20. NYLON NUT
21. CHOKE HINGE PIN
22. COLD ENRICHMENT ROD LEVER
23. COLD ENRICHMENT ROD SWIVEL
24. CVR ADJUSTING NUT
25. CONTROL VACUUM REGULATOR (CVR)
26. COLD ENRICHMENT ROD
27. SCREW (2)
28. VENTURI VALVE COVER PLATE
29. GASKET
30. ROLLER BEARING
31. VENTURI AIR-BYPASS SCREW AND TORQUE-RETENTION SPRING
32. VENTURI VALVE PIVOT PLUG
33. VENTURI VALVE PIVOT PIN
34. VENTURI VALVE
35. VENTURI VALVE PIVOT-PIN BUSHING
36. METERING ROD PIVOT PIN
37. METERING ROD
38. METERING ROD SPRING
39. CUP PLUG
40. MAIN METERING JET ASSEMBLY
41. O-RING
42. ACCELERATOR-PUMP RETURN SPRING
43. ACCELERATOR-PUMP CUP
44. ACCELERATOR-PUMP PLUNGER
45. INTERNAL VENT VALVE
46. RETAINING E-RING
47. IDLE TRIM SCREW
48. VENTURI VALVE LIMITER ADJUSTING SCREW
49. PIPE PLUG
50. COLD ENRICHMENT ROD SEAL
51. SEAL RETAINER
52. HOT IDLE COMPENSATOR

Fig. 12-11 Upper body of the Ford variable-venturi carburetor in disassembled view. (*Ford Motor Company*)

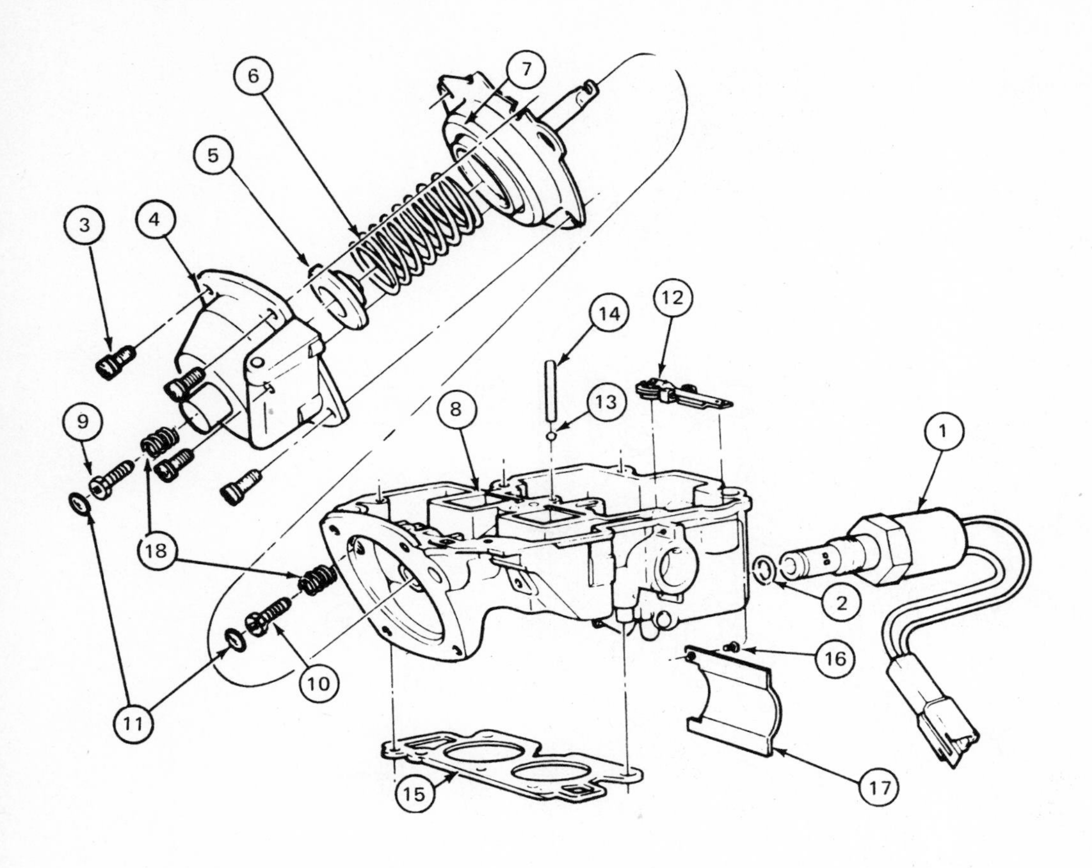

1. CRANKING ENRICHMENT SOLENOID
2. O-RING SEAL
3. SCREW (4)
4. VENTURI-VALVE DIAPHRAGM COVER
5. VENTURI-VALVE DIAPHRAGM SPRING GUIDE
6. VENTURI-VALVE DIAPHRAGM SPRING
7. VENTURI-VALVE DIAPHRAGM ASSEMBLY
8. MAIN BODY
9. VENTURI-VALVE ADJUSTING SCREW
10. WIDE OPEN STOP SCREW
11. EXPANSION PLUG
12. CRANKING FUEL CONTROL ASSEMBLY
13. ACCELERATOR-PUMP CHECK BALL
14. ACCELERATOR-PUMP CHECK BALL WEIGHT
15. THROTTLE BODY GASKET
16. SCREW
17. CHOKE HEAT SHIELD
18. TORQUE RETENTION SPRING

Fig. 12-12 Main body in disassembled view. (*Ford Motor Company*)

and 2 are the fuel inlet fitting and gasket. To disassemble the upper body, remove:

1. Fuel inlet fitting, filter, gasket, and spring (1, 2, 3, 4).
2. E-rings on the accelerator pump rod and choke-control rod and disengage the rods (5, 6, 7).
3. The air-cleaner stud.
4. Seven upper-body screws and upper body (8, 8A). Put it on a clean working surface. Note the position of the two long screws so they can be reinstalled in the same places.
5. Float hinge pin and float assembly (10, 11).
6. Float bowl gasket (12).
7. Fuel inlet valve, seat, and gasket (13, 14, 15).
8. Accelerator pump rod and dust seal (6, 16).
9. Accelerator pump link retaining pin and link (17, 18).
10. Pump swivel and adjusting nut (19, 20).
11. Choke control rod and dust seal by lifting the retainer carefully and sliding the seal out (7, 16).
12. E-ring on choke hinge pin and slide pin out (5, 21).
13. Cold enrichment rod adjusting nut, lever swivel, control vacuum regulator, and adjusting nut as an assembly (20, 22, 23, 24, 25). Disassemble if required.
14. Cold enrichment rod by sliding it out of body and nylon seal (26).
15. Venturi valve cover plate screws, plate gasket, and roller bearings (27, 28, 29, 30).
16. Venturi air bypass screw (30).
17. Use the special tool as shown in Fig. 12-14 to press the tapered plugs out of the venturi-valve pivot pins (31).
18. Push the venturi pivot pins out and slide the venturi valve to the rear until it is free (32, 33).

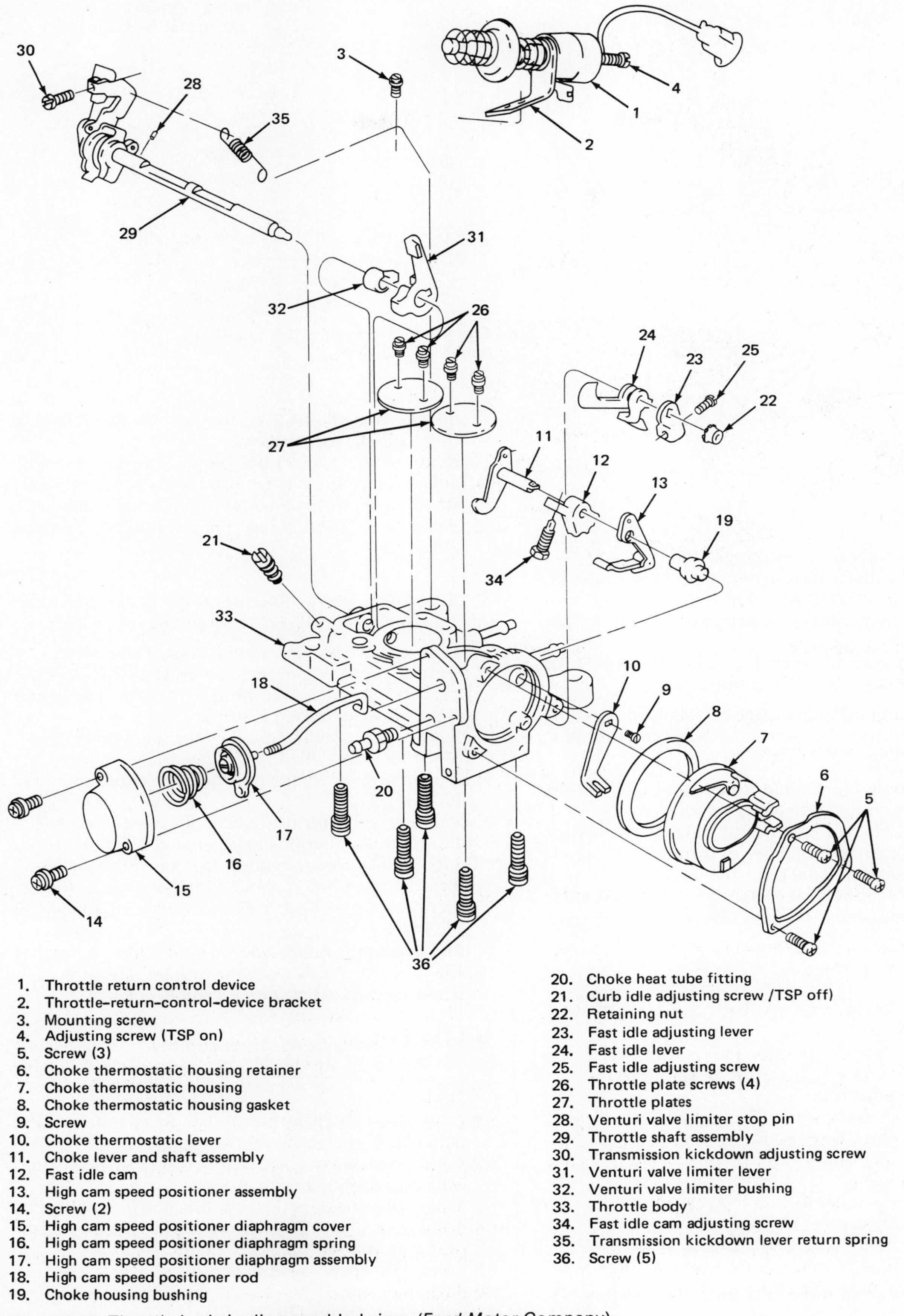

1. Throttle return control device
2. Throttle-return-control-device bracket
3. Mounting screw
4. Adjusting screw (TSP on)
5. Screw (3)
6. Choke thermostatic housing retainer
7. Choke thermostatic housing
8. Choke thermostatic housing gasket
9. Screw
10. Choke thermostatic lever
11. Choke lever and shaft assembly
12. Fast idle cam
13. High cam speed positioner assembly
14. Screw (2)
15. High cam speed positioner diaphragm cover
16. High cam speed positioner diaphragm spring
17. High cam speed positioner diaphragm assembly
18. High cam speed positioner rod
19. Choke housing bushing
20. Choke heat tube fitting
21. Curb idle adjusting screw /TSP off)
22. Retaining nut
23. Fast idle adjusting lever
24. Fast idle lever
25. Fast idle adjusting screw
26. Throttle plate screws (4)
27. Throttle plates
28. Venturi valve limiter stop pin
29. Throttle shaft assembly
30. Transmission kickdown adjusting screw
31. Venturi valve limiter lever
32. Venturi valve limiter bushing
33. Throttle body
34. Fast idle cam adjusting screw
35. Transmission kickdown lever return spring
36. Screw (5)

Fig. 12-13 Throttle body in disassembled view. (*Ford Motor Company*)

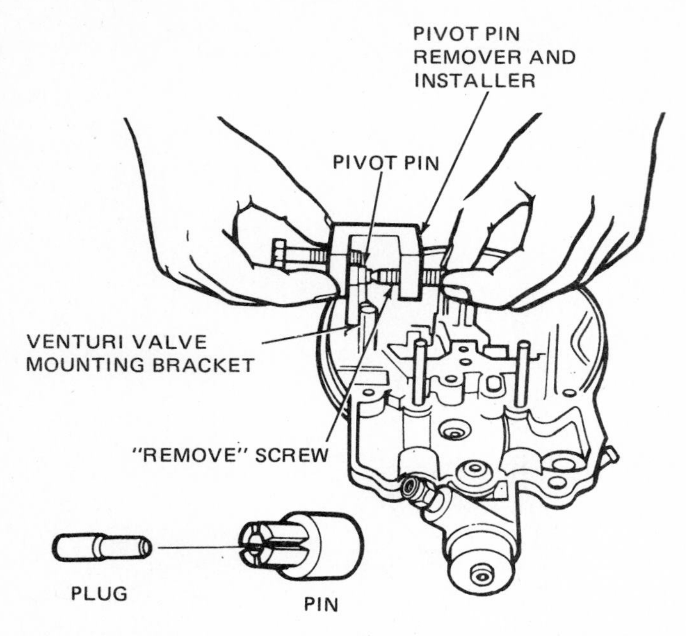

Fig. 12-14 Pressing out the tapered pins. (*Ford Motor Company*)

19. Remove the pivot pin bushings (34).
20. Remove the metering rod pivot pins, metering rods, and springs (35, 36, 37). Tag the rods *throttle* or *choke* so they can be reinstalled in their proper positions.
21. Use the jet plug remover (Fig. 12-15) to remove the cup plugs recessed into the upper body.

NOTE: The following steps must be followed carefully because they affect the overall carburetor calibration and operation.

22. Use the metering-jet adjustment tool (Fig. 12-16) and turn each main metering jet clockwise, counting the number of turns until they bottom in the casting. Record the number of turns to the nearest one-quarter turn. This gives the jet settings.
23. Turn the jet assembly counterclockwise to remove them. Remove the O-rings (39, 40). Tag the jets *throttle* or *choke* side so they can be reinstalled in their original locations.
24. Remove the accelerator-pump plunger assembly and disassemble it if required (41, 42, 43, 44, 45).
25. Remove venturi valve limiter adjusting screw (47).
26. Remove the ⅛-inch pipe plug, if it needs cleaning (48).

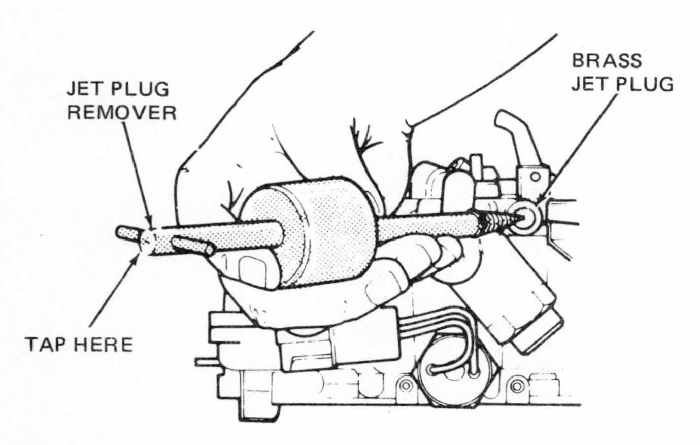

Fig. 12-15 Removing the cup plugs. (*Ford Motor Company*)

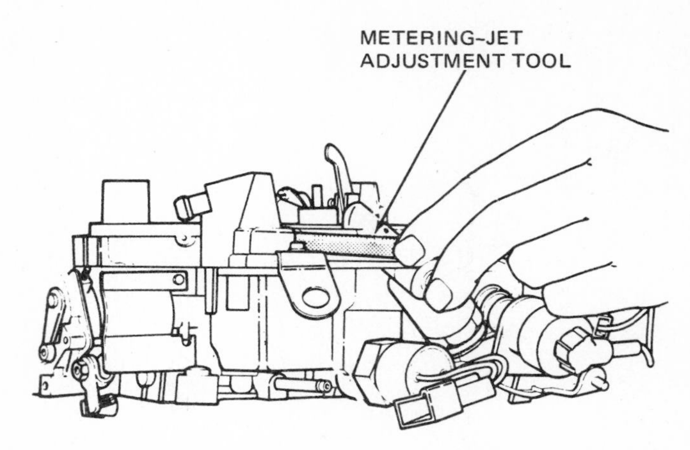

Fig. 12-16 Removing the main jets. (*Ford Motor Company*)

12-14 Main body disassembly To disassemble the main body (Fig. 12-12), remove the following:

1. Cranking enrichment solenoid and O-ring (1, 2).
2. Venturi valve diaphragm cover retaining screws, cover spring guide, and spring (3, 4, 5, 6). Tap lightly on the cover to loosen it. Do not pry it off.
3. Diaphragm by sliding it out of the main body (7).
4. Adjusting screw (9).
5. Limiter stop by center-punching the expansion plug (10, 11).
6. Cranking fuel control assembly if it appears damaged (12). Bend the thermostat up enough to allow the jet plug removal tool to extract the control assembly.
7. Turn the main body upside down and catch the accelerator pump check ball and weight in your hand (13, 14).
8. Five retaining screws and throttle body and gasket (15).
9. Choke heat shield screw and shield (16, 17).

12-15 Throttle body disassembly To disassemble the throttle body (Fig. 12-13), remove the following parts:

1. Throttle-return-control-device assembly and bracket. Unhook the return spring (1, 2, 3, 4, 37).
2. Choke thermostat housing screws, retainer ring, housing, and gasket (6, 7, 8, 9).
3. Choke thermostatic lever screw and lever (10, 11).
4. Choke shaft and lever assembly by sliding it out. Remove fast idle cam (12, 13, 35, 36).
5. High-cam-speed positioner (14).
6. High-cam-speed positioner cover screws, cover, and return spring (15, 16, 17).

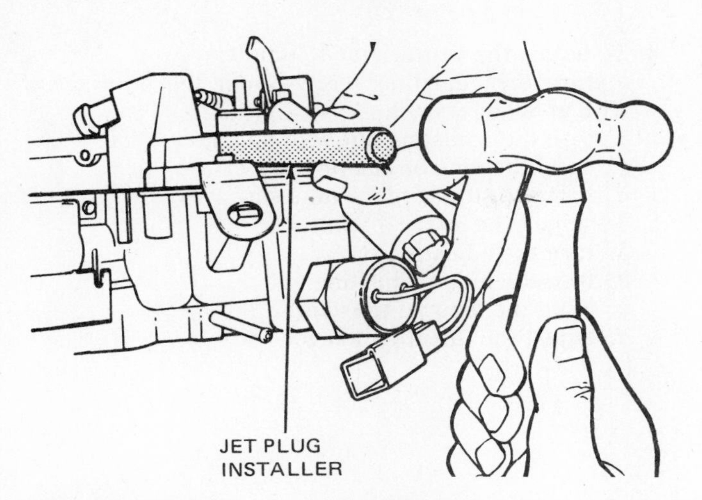

Fig. 12-17 Installing the main-jet plugs. (*Ford Motor Company*)

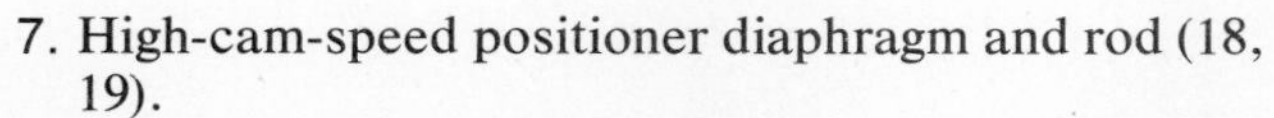

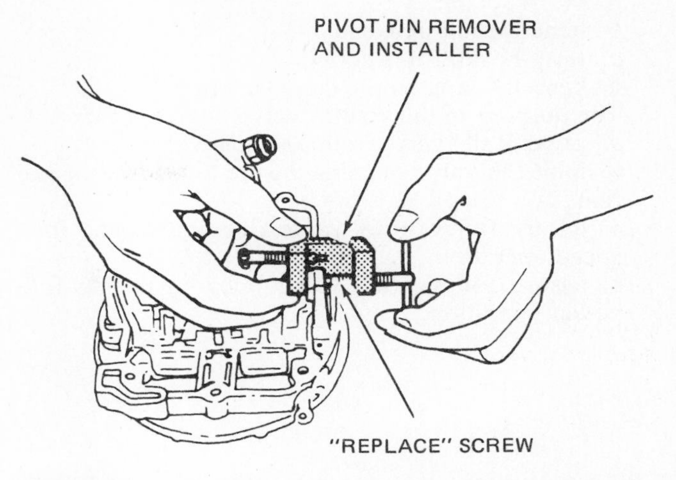

Fig. 12-18 Installing the tapered plugs. (*Ford Motor Company*)

7. High-cam-speed positioner diaphragm and rod (18, 19).
8. Choke housing bushing if it is damaged (20). It is pressed into the casting and staked into place. To remove it, file or grind off the stakes. Then carefully press the bushing out.
9. Choke heat tube fitting (21).
10. Idle-speed adjusting screw (22).
11. Throttle shaft retaining nut, fast-idle adjusting lever, fast-idle lever, and adjusting screw (23, 24, 25, 26).
12. If necessary to remove the throttle plate from the shaft, the staked ends of the screws should be filed off and the old screws discarded (27, 28).
13. To remove the throttle shaft assembly, the limiter lever stop pin must be driven down until it is flush with the shaft (29).
14. Slide the throttle shaft out and remove the transmission kickdown adjusting screw, if necessary (29, 30).
15. If necessary, remove the venturi valve limiter lever and bushing (23, 33).

⚙ 12-16 Carburetor reassembly The reassembly of the carburetor is basically the reverse of the disassembly procedure. Reinstall on the throttle body, main body, and upper body the parts removed. Follow the numbering system of the parts in Figs. 12-11, 12-12, and 12-13 to reinstall the parts. Here are special items to observe:

1. Make the adjustments described in ⚙ 12-2 to 12-11 as the carburetor is reassembled.
2. When putting the throttle body (Fig. 12-14) together, adjust the high-cam-speed positioner diaphragm before installing the spring, cover, and screws (15, 16, 17).
3. On the main body (Fig. 12-13), do not install the venturi valve limiter stop screw and plug (10, 11) until assembly is complete. Refer to the venturi valve limiter adjustment in ⚙ 12-9. Before installing the cranking enrichment solenoid O-ring seal on the solenoid, lubricate it with a mild solution of soapy water.
4. On the upper body (Fig. 12-11), lubricate the O-ring seals with soapy water before installing them. A special jet plug installer is needed to install the metering jets (39). Turn each jet in until it seats. Then back it out the number of turns recorded during disassembly. Use the jet plug installer to drive the plugs (38) into the casting (Fig. 12-17). When installing the tapered plugs into the venturi valve pivot pins, use the special pivot pin tool as shown in Fig. 12-18.

When installing the venturi air bypass screw, turn it in four turns to provide clearance for installing the cover plate (3). When installing the upper body on the main body, hold the pump piston assembly with your finger and guide it into the pump cavity in the main body. Make sure that the venturi valve diaphragm stem engages the venturi valve.

Chapter 12 review questions

Select the *one* correct, best, or most probable answer to each question. Then check your answers against the correct answers given at the end of the book.

1. The Ford VV carburetor has:
   a. one venturi valve
   b. two venturi valves
   c. four venturi valves
   d. none of the above.
2. To adjust the fast-idle speed:
   a. turn the adjusting nut

b. bend the idle-link rod
c. turn the adjusting screw
d. bend the tang on the fast-idle cam.

3. The purpose of the venturi valve limiter is to:
a. prevent the valve from opening wide
b. limit the valve opening during low-speed operation
c. assure full venturi valve opening during low-speed operation
d. assure full venturi valve opening during full-power operation.

4. To adjust the venturi valve limiter:
a. turn the adjusting screw under the stop screw with an allen wrench
b. turn the adjusting nut with an end wrench
c. bend the tang on the limiter
d. turn the stop screw with a screwdriver.

5. To adjust the internal vent:
a. turn the adjusting screw
b. turn the nylon adjusting nut
c. bend the tang on the vent
d. adjust the throttle linkage.

CHAPTER 13

SERVICING A FOUR-BARREL CARBURETOR

Ater studying this chapter, and with proper instruction and equipment, you should be able to:

1. Perform the adjustments on a four-barrel carburetor.
2. Overhaul a four-barrel carburetor.

13-1 Four-barrel carburetors The trend today is toward the smaller engine using a one-barrel or two-barrel carburetor. The four-barrel carburetor is most often used with V-8 engines. The V-8 engine is produced in several versions. The largest and most powerful versions often have four-barrel carburetors. However, the four-cylinder and V-6 engines use the more economical one-barrel and two-barrel carburetors.

The air-fuel distribution pattern for four-barrel carburetors used on V-8 engines is described in ⚙ 4-11. Figure 4-18 illustrates the patterns. The four-barrel carburetor is essentially two 2-barrel fixed-venturi carburetors in one assembly. Each half takes care of four cylinders. In this chapter, a typical four-barrel carburetor used on several models of General Motors vehicles is covered.

⚙ 13-2 Rochester model M4MC carburetor This carburetor (Figs. 4-19 and 13-1) is a four-barrel, fixed-venturi carburetor with a primary and a secondary side, each with two barrels. Each primary barrel has a secondary barrel which comes into operation when full power is demanded. In operation, the carburetor is like two of the two-barrel carburetors described in ⚙ 11-12 and illustrated in Fig. 11-1. The various carburetor systems are also very similar.

The secondary throttle valves are opened by an air valve which senses the flow of air through the primary barrels. When this flow indicates a demand for additional air-fuel mixture for full power, the air valve allows the secondary throttle valves to open. The secondary barrels have baffles which deflect incoming air to increase the vacuum at the main nozzles during full power operation (Fig. 4-19). This increases the flow of fuel from the main nozzles when full power is demanded.

The carburetor on air-conditioned cars has a solenoid. It operates to slightly increase the throttle opening when the air conditioner is switched on. This allows the engine to idle at a slightly higher rpm and take care of the additional power demand of the air conditioner.

This carburetor has an adjustable part throttle (APT). This device (Fig. 4-19) allows the factory to fine-tune the carburetor for low emissions and fuel economy. The APT allows a very accurate setting of the metering rods in the main metering jets. The adjustment feature includes a pin which is pressed into the side of the power piston. This pin extends through a slot in the side of the piston well. When the power piston is down (economy position), the side of the pin rests on a flat surface of the adjustment screw.

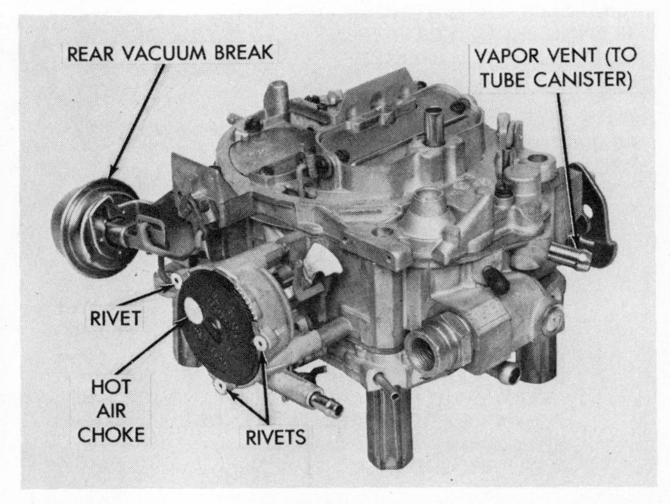

Fig. 13-1 Rochester model M4MC four-barrel carburetor. (*Chevrolet Motor Division of General Motors Corporation*)

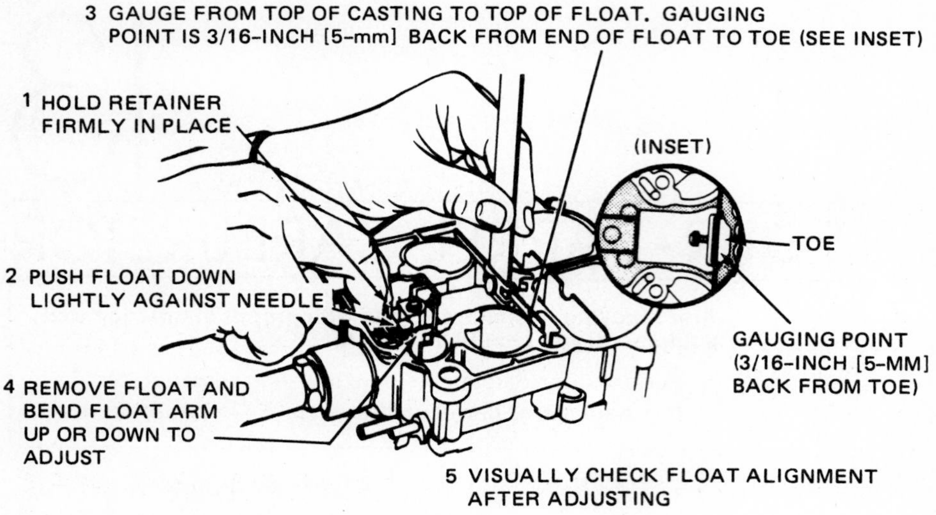

3 GAUGE FROM TOP OF CASTING TO TOP OF FLOAT. GAUGING POINT IS 3/16-INCH [5-mm] BACK FROM END OF FLOAT TO TOE (SEE INSET)

1 HOLD RETAINER FIRMLY IN PLACE

(INSET)

TOE

2 PUSH FLOAT DOWN LIGHTLY AGAINST NEEDLE

GAUGING POINT (3/16-INCH [5-MM] BACK FROM TOE)

4 REMOVE FLOAT AND BEND FLOAT ARM UP OR DOWN TO ADJUST

5 VISUALLY CHECK FLOAT ALIGNMENT AFTER ADJUSTING

Fig. 13-2 Float level adjustment. (*Chevrolet Motor Division of General Motors Corporation*)

If the adjustment screw is turned down (at the factory) it allows the power piston to move farther down and push the metering rods farther into the metering rod jets. This further restricts fuel flow. However, if the adjustment screw is backed out (at the factory) it prevents the metering rods from moving down so far in the rod jets. This sets the minimum amount of leaning out that the system can produce.

NOTE: This is a factory adjustment and should not be changed. If a float bowl replacement is required, the new bowl assembly will include a factory-preset adjustment screw.

The two primary barrels have six systems—float, idle, main metering, power, accelerator pump, and choke. These systems all work in the same general manner as other carburetor systems discussed in previous chapters.

13-3 Carburetor adjustments The carburetor requires the following adjustments which are listed here and described in (Figs. 13-2 to 13-17).

1. Float (Figs. 13-2 and 13-3)
2. Accelerator pump (Fig. 13-4)
3. Choke coil lever (Fig. 13-5)
4. Fast idle (Fig. 13-6)
5. Choke rod (fast-idle cam) (Fig. 13-7)
6. Air valve rod (Fig. 13-8)

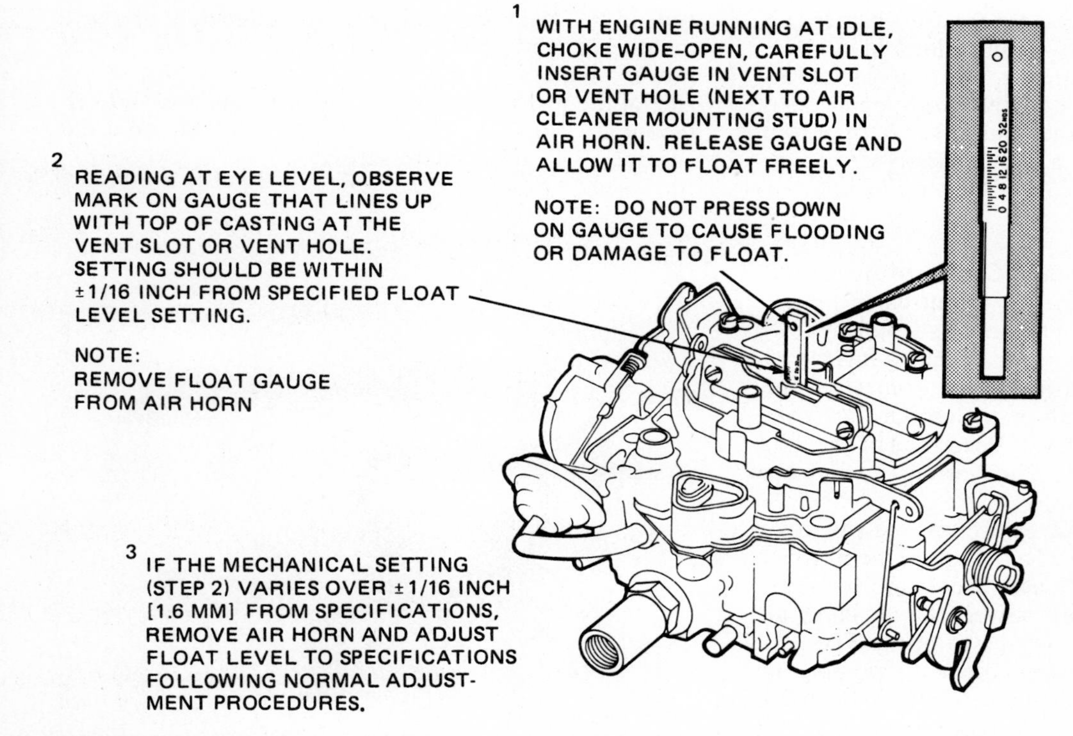

Fig. 13-3 Checking the float level without removing the air horn. (*Chevrolet Motor Division of General Motors Corporation*)

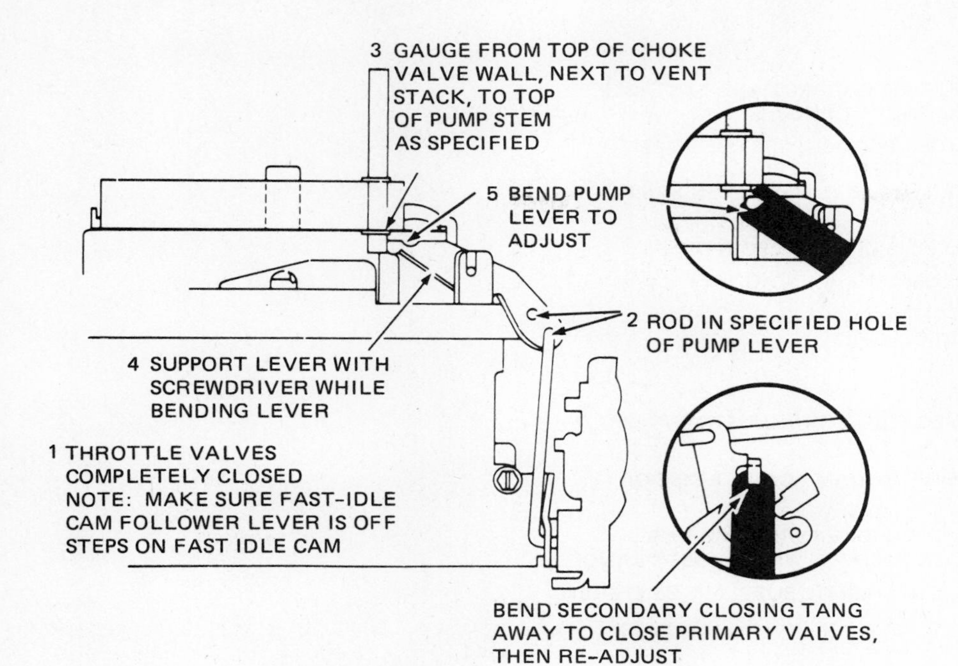

Fig. 13-4 Accelerator pump adjustment. (*Chevrolet Motor Division of General Motors Corporation*)

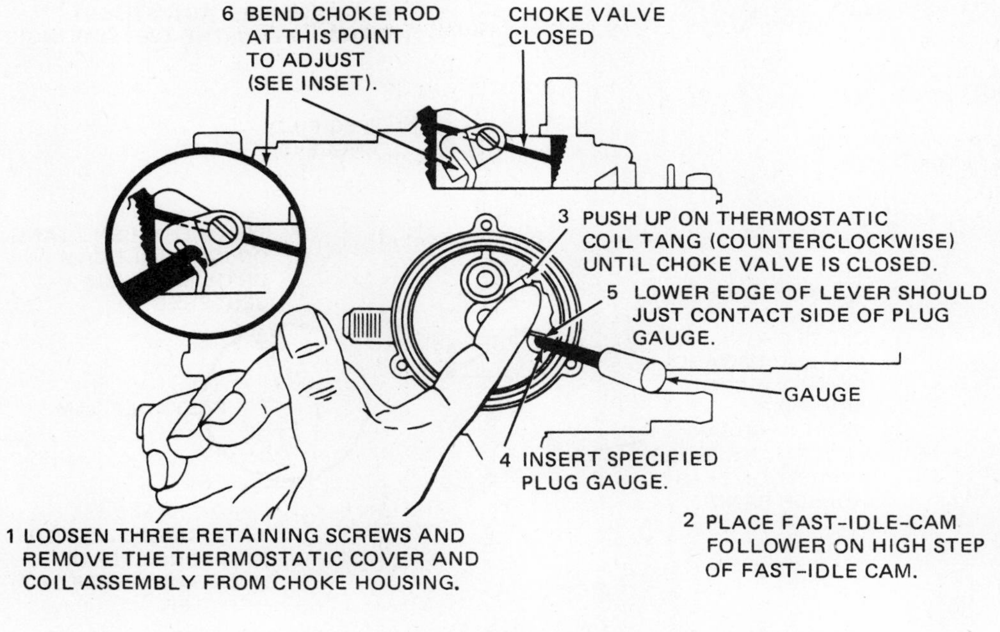

Fig. 13-5 Choke coil lever adjustment. (*Chevrolet Motor Division of General Motors Corporation*)

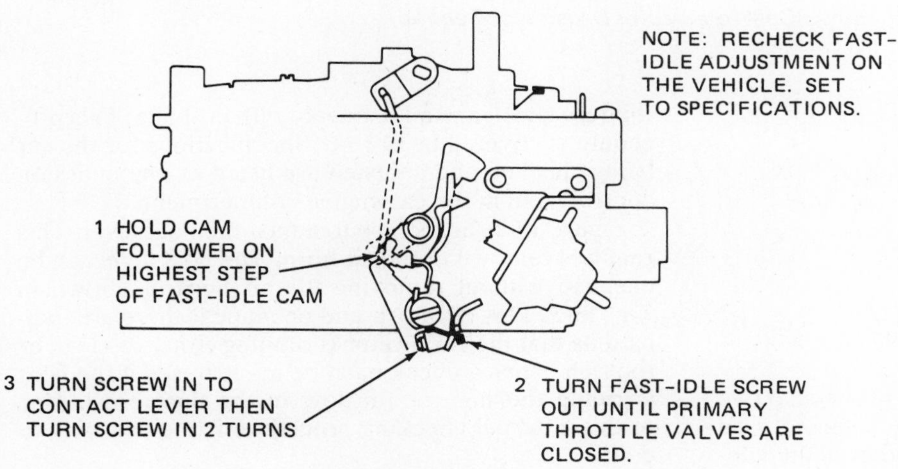

Fig. 13-6 Bench setting the fast-idle adjustment. (*Chevrolet Motor Division of General Motors Corporation*)

FIGURE (A)

1. CHOKE COIL LEVER ADJUSTMENT MUST BE CORRECT AND FAST IDLE ADJUSTMENT MUST BE MADE BEFORE PROCEEDING.
2. USE CHOKE VALVE MEASURING GAUGE. TOOL MAY BE USED WITH CARBURETOR ON OR OFF ENGINE. IF OFF ENGINE, PLACE CARBURETOR ON HOLDING FIXTURE SO THAT IT WILL REMAIN IN SAME POSITION WHEN GAUGE IS IN PLACE.
3. ROTATE DEGREE SCALE UNTIL ZERO (0) IS OPPOSITE POINTER.
4. WITH CHOKE VALVE COMPLETELY CLOSED, PLACE MAGNET SQUARELY ON TOP OF CHOKE VALVE.
5. ROTATE BUBBLE UNTIL IT IS CENTERED.

FIGURE (B)

6. ROTATE SCALE SO THAT DEGREE SPECIFIED FOR ADJUSTMENT IS OPPOSITE POINTER.
7. PLACE CAM FOLLOWER ON SECOND STEP OF CAM AGAINST RISE OF HIGH STEP.
8. CLOSE CHOKE BY PUSHING UPWARD ON CHOKE COIL LEVER OR VACUUM BREAK LEVER TANG (HOLD IN POSITION WITH RUBBER BAND).
9. TO ADJUST, BEND TANG ON FAST-IDLE CAM UNTIL BUBBLE IS CENTERED.
10. REMOVE GAUGE.

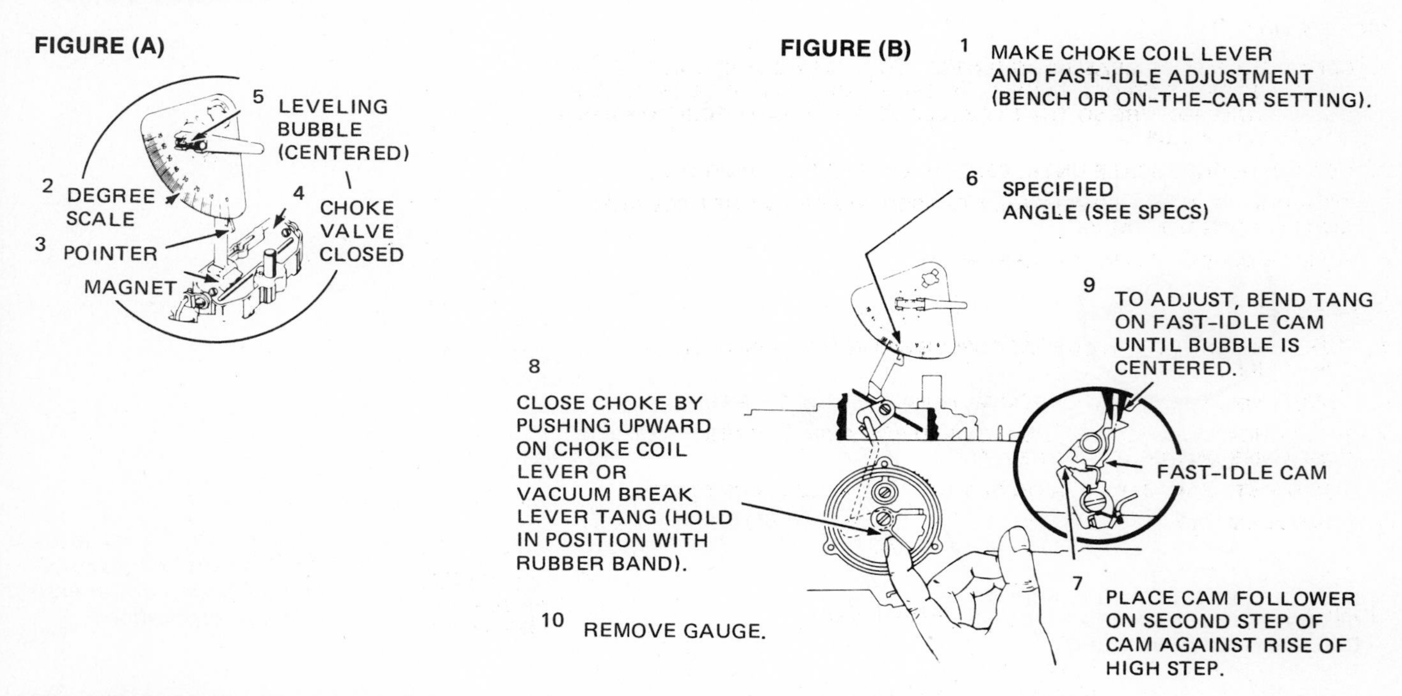

Fig. 13-7 Adjusting the choke rod (fast-idle cam). (*Chevrolet Motor Division of General Motors Corporation*)

7. Vacuum break (Fig. 13-9)
8. Automatic-choke coil (Fig. 13-10)
9. Unloader (Fig. 13-11)
10. Secondary lockout (Fig. 13-12)
11. Secondary closing (Fig. 13-13)
12. Secondary opening (Fig. 13-14)
13. Air valve spring (Fig. 13-15)
14. Idle speed—without solenoid (Fig. 13-16)
15. Idle speed—with solenoid (Fig. 13-17)

In addition to these adjustments, there are also two other checks to be made. One is to the solenoid used in air-conditioned cars (⚙ 13-4). The other is the idle-mixture check and adjustment, with propane. This procedure is covered in ⚙ 10-6. Specifications for the car being checked and adjusted are listed on the emission control label in the car engine compartment.

Figure 13-2 shows how to adjust the float level. This requires removal of the air horn. The float level can be checked without removing the air horn as shown in Fig. 13-3. This check should be made if there are indications that the carburetor is running either too lean or too rich. Then a check must be made to see if the float is riding too high or too low in the float bowl. Use of this external checking procedure is a quick way to do it.

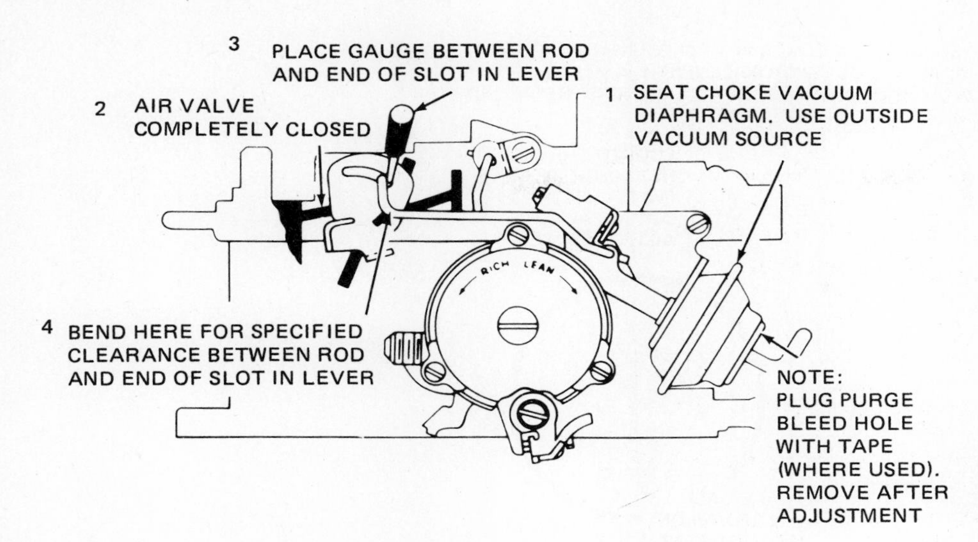

Fig. 13-8 Air valve rod adjustment. (*Chevrolet Motor Division of General Motors Corporation*)

FIGURE (A)

1. USE CHOKE VALVE MEASURING GAUGE. TOOL MAY BE USED WITH CARBURETOR ON OR OFF ENGINE. IF OFF ENGINE, PLACE CARBURETOR ON HOLDING FIXTURE SO THAT IT WILL REMAIN IN SAME POSITION WHEN GAUGE IS IN PLACE.
2. ROTATE DEGREE SCALE UNTIL ZERO (0) IS OPPOSITE POINTER.
3. WITH CHOKE VALVE COMPLETELY CLOSED, PLACE MAGNET SQUARELY ON TOP OF CHOKE VALVE.
4. ROTATE BUBBLE UNTIL IT IS CENTERED.

FIGURE (B)

5. ROTATE SCALE SO THAT DEGREE SPECIFIED FOR ADJUSTMENT IS OPPOSITE POINTER.
6. SEAT CHOKE VACUUM DIAPHRAGM USING VACUUM SOURCE.
7. HOLD CHOKE VALVE TOWARDS CLOSED POSITION, PUSHING COUNTER-CLOCKWISE ON INSIDE COIL LEVER.
8. TO ADJUST, TURN SCREW IN OR OUT UNTIL BUBBLE IS CENTERED.
9. REMOVE GAUGE.

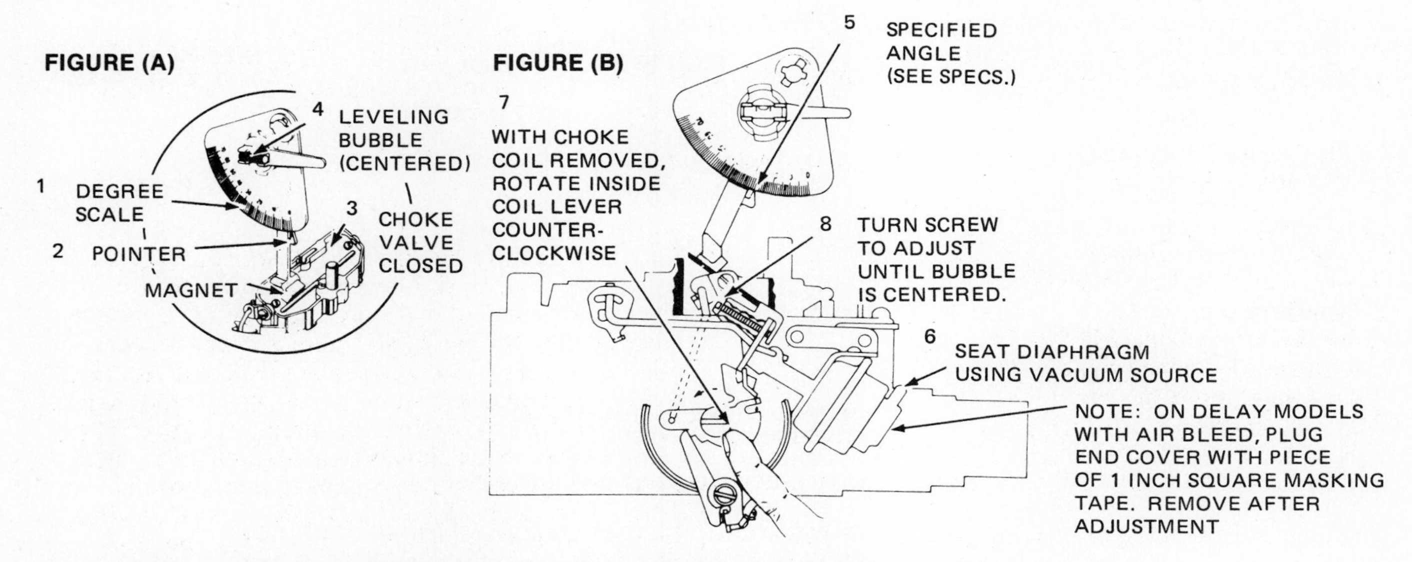

Fig. 13-9 Vacuum-break adjustment. (*Chevrolet Motor Division of General Motors Corporation*)

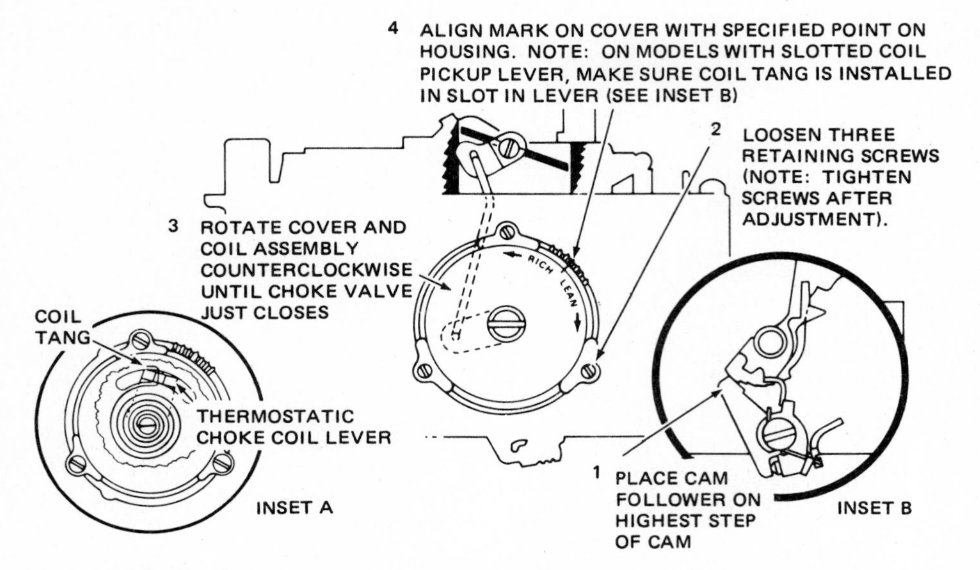

Fig. 13-10 Automatic-choke coil adjustment. (*Chevrolet Motor Division of General Motors Corporation*)

FIGURE (A)

1. USE CHOKE VALVE MEASURING GAUGE. TOOL MAY BE USED WITH CARBURETOR ON OR OFF ENGINE. IF OFF ENGINE, PLACE CARBURETOR ON HOLDING FIXTURE SO THAT IT WILL REMAIN IN SAME POSITION WHEN GAUGE IS IN PLACE.
2. ROTATE DEGREE SCALE UNTIL ZERO (0) IS OPPOSITE POINTER.
3. WITH CHOKE VALVE COMPLETELY CLOSED, PLACE MAGNET SQUARELY ON TOP OF CHOKE VALVE.
4. ROTATE BUBBLE UNTIL IT IS CENTERED.

FIGURE (B)

5. ROTATE SCALE SO THAT DEGREE SPECIFIED FOR ADJUSTMENT IS OPPOSITE POINTER.
6. INSTALL CHOKE THERMOSTATIC COVER AND COIL ASSEMBLY IN HOUSING. ALIGN INDEX MARK WITH SPECIFIED POINT ON HOUSING.
7. HOLD PRIMARY THROTTLE VALVES WIDE OPEN.
8. ON WARM ENGINE, CLOSE CHOKE VALVE BY PUSHING UP ON TANG ON VACUUM BREAK LEVER (HOLD IN POSITION WITH RUBBER BAND).
9. TO ADJUST, BEND TANG ON FAST-IDLE LEVER UNTIL BUBBLE IS CENTERED.
10. REMOVE GAUGE.

FIGURE (A)

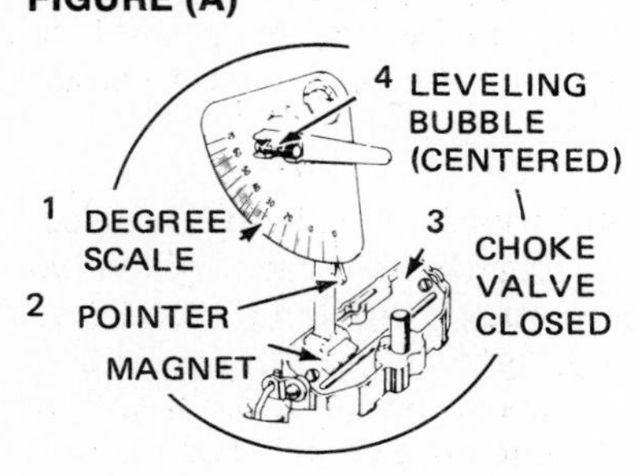

FIGURE (B)

7 HOLD PRIMARY THROTTLE VALVES WIDE OPEN.

5 SPECIFIED ANGLE (SEE SPECS)

6 INSTALL CHOKE THERMOSTATIC COVER AND COIL ASSEMBLY IN HOUSING. ALIGN INDEX MARK WITH SPECIFIED POINT ON HOUSING.

9 BEND TANG TO ADJUST UNTIL BUBBLE IS CENTERED.

8 ON WARM ENGINE, CLOSE CHOKE BY PUSHING UP ON TANG ON VACUUM BREAK LEVER (HOLD IN POSITION WITH RUBBER BAND).

Fig. 13-11 Unloader adjustment. (*Chevrolet Motor Division of General Motors Corporation*)

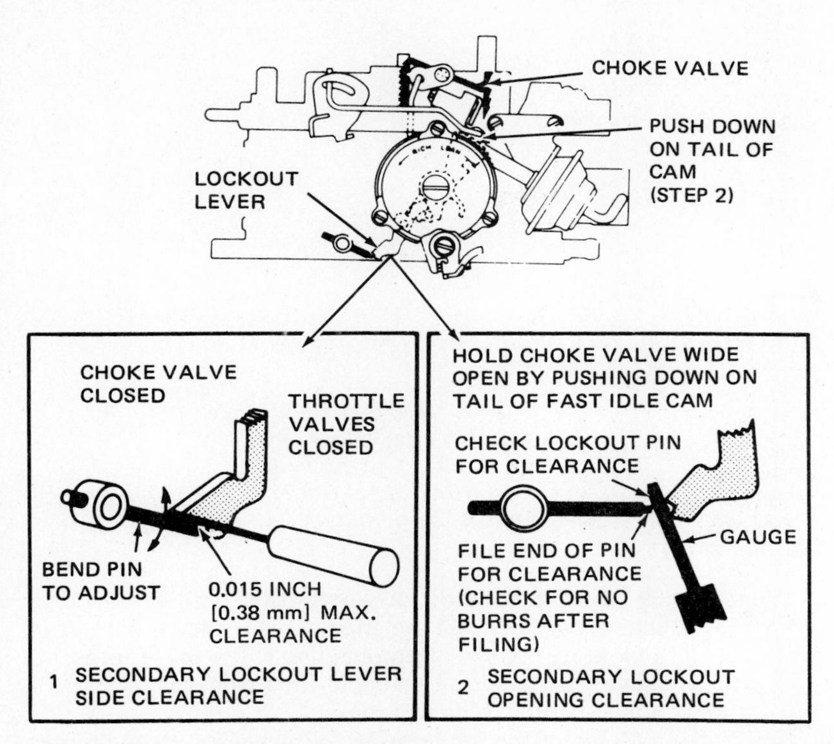

Fig. 13-12 Secondary lockout adjustment. (*Chevrolet Motor Division of General Motors Corporation*)

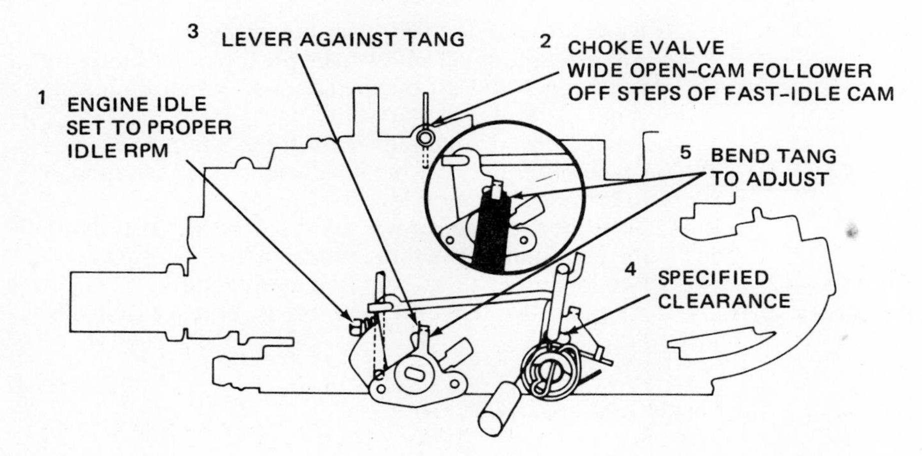

Fig. 13-13 Secondary closing adjustment. (*Chevrolet Motor Division of General Motors Corporation*)

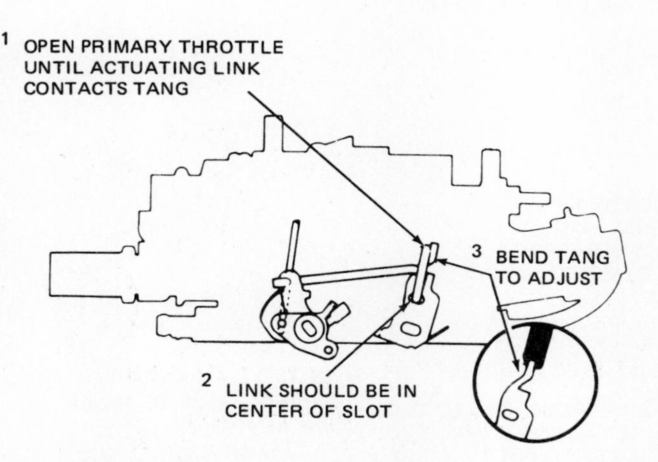

Fig. 13-14 Secondary opening adjustment. (*Chevrolet Motor Division of General Motors Corporation*)

Should disassembly of the carburetor be required, it is removed and reinstalled as discussed in ⚙ 10-7. The parts should be cleaned and inspected as covered in ⚙ 10-9. Hints and cautions on cleanliness and safety are given in ⚙ 8-1.

⚙ 13-4 Checking the solenoid A solenoid is used on air-conditioned cars to slightly increase engine idle speed when the air conditioner is working. To check the solenoid, turn on the ignition but do not start the engine. Then turn the air conditioner on. Open the throttle to allow the solenoid to extend. Then close the throttle. Disconnect the lead at the solenoid. The plunger should drop away from the throttle lever. Reconnect the lead. The plunger should extend out to touch the throttle lever. If it does not, either the solenoid is defective or the feed circuit to the solenoid is faulty.

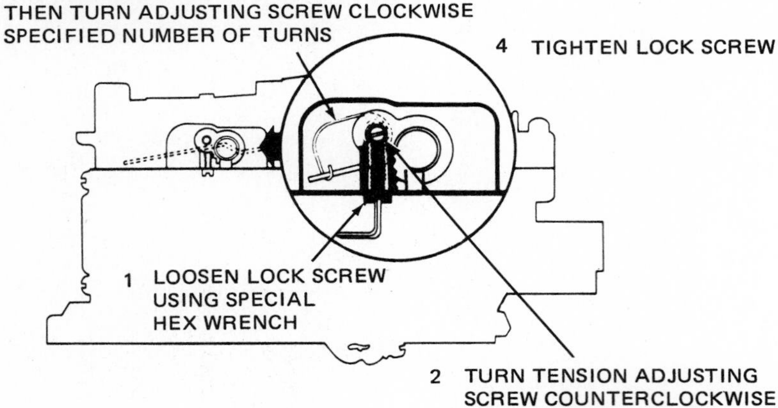

Fig. 13-15 Air valve spring adjustment. (*Chevrolet Motor Division of General Motors Corporation*)

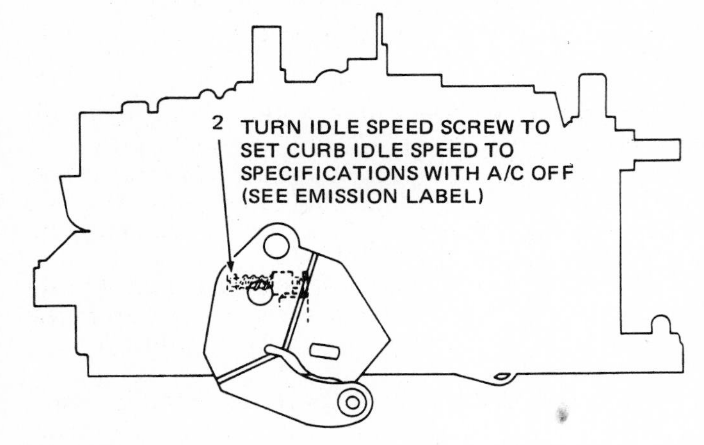

Fig. 13-16 Idle-speed adjustment, without solenoid. (*Chevrolet Motor Division of General Motors Corporation*)

13-5 Carburetor disassembly After removing the carburetor from the engine, place the carburetor in a holding fixture. Remove the solenoid, if present. Never immerse the solenoid in carburetor cleaner or other solvent. This will ruin it.

Remove the air horn as follows. Detach the upper choke lever from the end of the choke shaft. Then rotate the lever to remove the choke rod from the slot in the lever.

1. Detach the choke rod from lower lever.
2. Take out the small screw from the top of the metering rod hanger and lift off the secondary metering rods (Fig. 13-18).
3. To detach the pump lever from the pump rod, use a special tool to drive the small roll pin in just enough to release the pump lever.
4. Remove the nine screws to lift off the air horn. Note the locations of the three different kinds of screws—two long, five short, and two countersunk.
5. Remove the front vacuum-break bracket and dia-

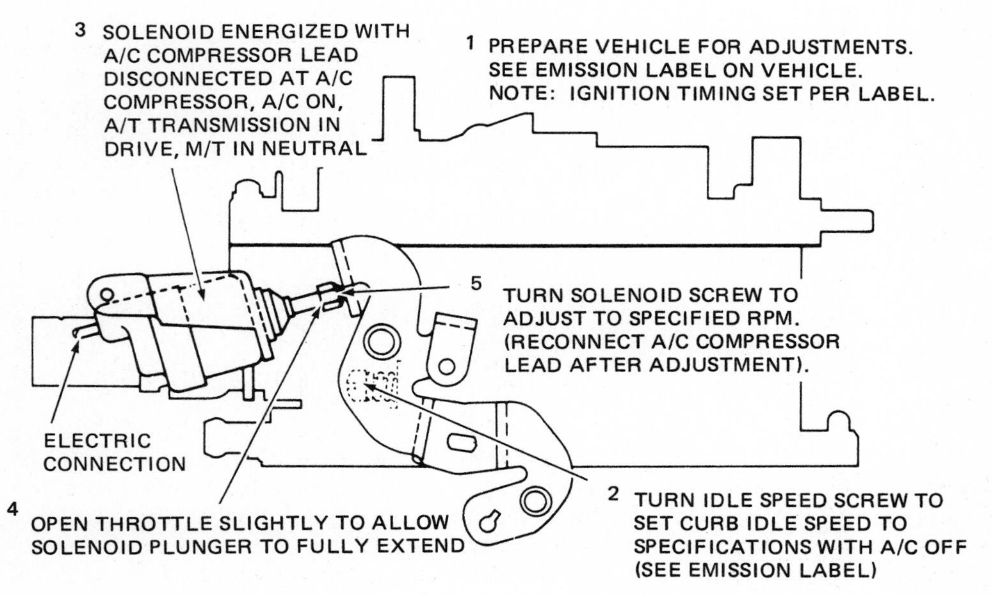

Fig. 13-17 Idle-speed adjustment, with solenoid. (*Chevrolet Motor Division of General Motors Corporation*)

Fig. 13-18 Removing secondary metering rods. (*Chevrolet Motor Division of General Motors Corporation*)

phragm assembly. Do not clean the assembly in carburetor cleaner, since this would ruin it.

6. No further disassembly is required to clean the air horn. The choke valves can be removed if necessary, by filing off the staking on the screws. The air valves and air valve shaft should not be removed.

⚙ 13-6 Float bowl To disassemble the float bowl, remove the following:

1. Air horn gasket, being careful not to distort the main metering rod springs.
2. Pump plunger and spring.
3. Power piston, metering rod, and spring. Depress piston stem and allow it to snap up. Repeat as necessary to work the retainer free. Do not use pliers!
4. Plastic filler block over the float valve.
5. Float assembly, needle, seat, and gasket.
6. Aneroid cavity insert.
7. The primary main metering jets, but only if they are damaged. If the secondary metering jets are damaged, the float bowl must be replaced.
8. Pump discharge check ball retainer and ball.
9. If replacement is required, remove the secondary air baffles.
10. Remove the pump well fill-slot baffle.

⚙ 13-7 Choke disassembly Remove the three attaching screws and retainers. Then pull straight out to remove the cover and coil from housing. On the M4MC hot-air choke, it is not necessary to remove the baffle plate from under the thermostatic coil. This could distort the coil.

If necessary, take off the choke housing by removing the retainer screw. Other parts that can be removed include the secondary throttle valve lockout lever, lower choke lever (from inside bowl), and plastic tube seal.

⚙ 13-8 Throttle body The important point of disassembling the throttle body is the procedure to remove the idle-mixture screws. This requires a punch which is used to shatter or break out the steel plugs covering

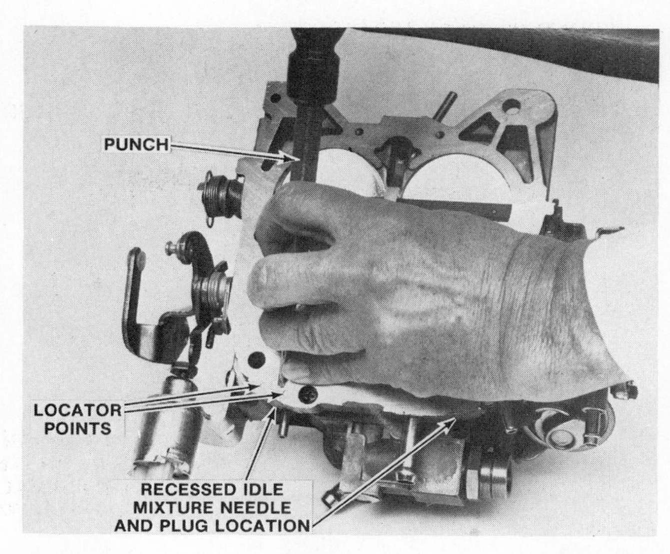

Fig. 13-19 Removing the idle-mixture screw plugs. (*Chevrolet Motor Division of General Motors Corporation*)

the screws (Fig. 13-19). This procedure is covered in ⚙ 11-14.

⚙ 13-9 Throttle body reassembly If the idle-mixture screw has been removed, reinstall it with the washer and spring, using a 3/16-inch thin-wall socket. Turn in to seat, then back it out four turns. This is a preliminary adjustment. Final adjustment must be made on the car (⚙ 10-6).

Install the lower end of the pump rod in the throttle lever. Align the rod tang with the slot in the lever. The rod end should point out toward the throttle lever.

⚙ 13-10 Beginning float bowl reassembly Turn the float bowl upside down. Put a new gasket on the bowl. Lower the throttle body on the bowl and secure it with three screws. Put assembly in a holding fixture, right side up. Install the fuel inlet filter spring, new gasket, and inlet nut. Tighten the nut to specifications.

⚙ 13-11 Choke reassembly Reassemble the choke using new seals. Be sure the steps on the fast-idle cam face downward. The inside thermostatic choke lever is properly aligned when both the inside and outside levers face toward the fuel inlet. A special tool helps to install the choke rod lever (Fig. 13-20). Do not install the choke cover and coil assembly until the inside coil lever is adjusted (Fig. 13-5).

⚙ 13-12 Completing float bowl reassembly On reassembly, install the air baffle in the secondary side of the float bowl with notches toward the top. The top edge of baffle must be flush with the bowl surface. Then install the baffle in the pump well fill slot. Install:

1. Pump discharge check ball and retainer. Tighten retainer.
2. Primary main metering jets (if removed).
3. Aneroid cavity insert.
4. New needle seat assembly, with gasket.

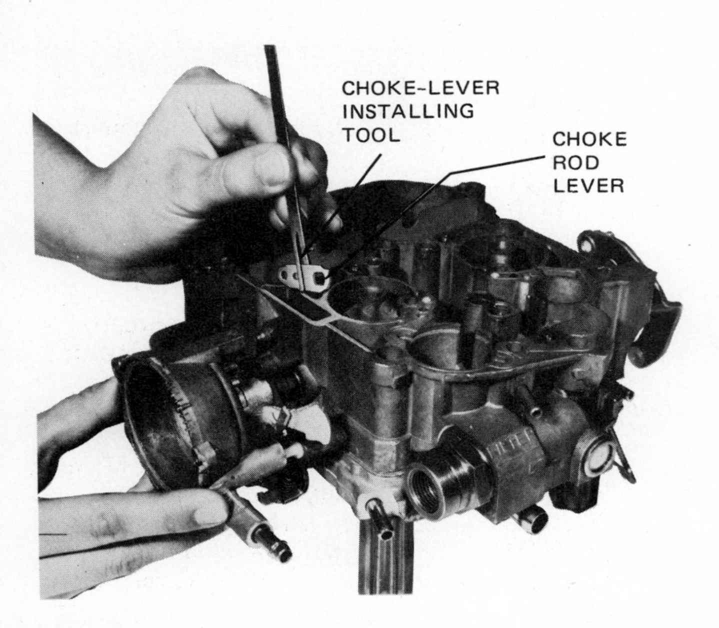

Fig. 13-20 Installing the choke rod lever. (*Chevrolet Motor Division of General Motors Corporation*)

5. Float arm, first bending it upward at notch in the arm before installing it to make installation easier.
6. Needle, by sliding float lever under the needle pull clip. Hook the clip over the edge of the flat on the float arm facing the float pontoon (Fig. 13-21). With the float lever in the pull clip, hold the float assembly at its toe and install the retaining pin from the aneroid side. Do not install the float needle pull clip in the holes in the float arm.
7. Adjust float level (Fig. 13-2).

13-13 Air horn reassembly If they were removed, install the choke shaft, valve and attaching screws. Make sure the choke valve moves freely. Hold

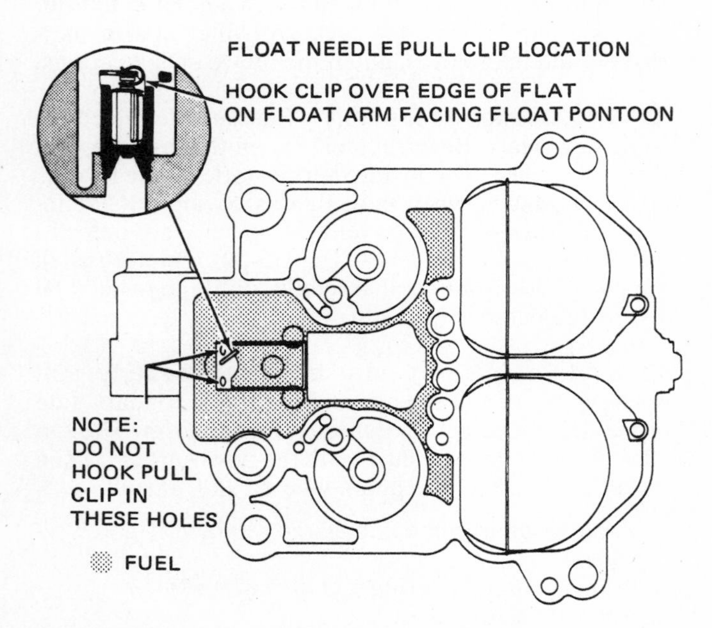

Fig. 13-21 Locating the float-needle-valve pull clip. (*Chevrolet Motor Division of General Motors Corporation*)

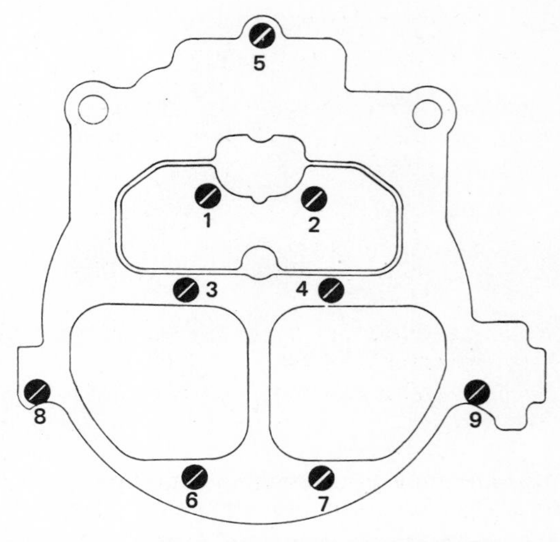

Fig. 13-22 Air horn screw-tightening sequence. (*Chevrolet Motor Division of General Motors Corporation*)

down on the air horn gasket at the pump plunger and carefully lower the air horn onto the float bowl. Make sure the bleed tubes, accelerator well tubes, pullover enrichment tubes (if used), and pump plunger enter the holes in the air horn gasket.

NOTE: If the air horn does not settle down into place easily, something is not lined up. Do not force the air horn down but find out what is hanging. Then secure the air horn with nine screws (two long, five short, and two countersunk). Figure 13-22 shows the screw-tightening sequence.

Install the vacuum-break diaphragm, connecting the diaphragm rod to the slot in the lever in the end of the air valve shaft. The other end of the rod goes into the hole in the diaphragm plunger. Adjust the vacuum break (Fig. 13-9).

Connect the upper end of the pump rod to the pump lever. Align the hole in the pump lever with the hole in the air horn with a special tool. Use a small screwdriver to push the pump lever roll pin back through the casting until the end of the pin is flush. Use care to avoid damaging the pump lever bosses on the casting.

Install the choke rod and lower choke lever and tighten the retaining screw securely. The lever should point to the rear of the carburetor and the number on the rod should face out. The flats on the end of the choke shaft should align with the flats in the choke lever.

The front and rear vacuum-break units, fast-idle cam (choke rod), and inside thermostatic choke-coil lever must be adjusted before the thermostatic coil and cover are installed. These adjustments are shown in Figs. 13-5, 13-7 and 13-9.

Install the thermostatic coil, cover, and gasket. Rotate the cover assembly until the choke valve just closes. The tang on the coil must be in the slot inside the choke coil lever pickup arm. Align the index point on the cover with the specified mark on the housing. Install the three choke cover screws.

Install the hose on the vacuum break. Then, if used, install the solenoid.

Chapter 13 review questions

Select the *one* correct, best, or most probable answer to each question. Then check your answers against the correct answers given at the end of the book.

1. The four-barrel carburetor is essentially:
   a. four 1-barrel carburetors in one assembly
   b. two 2-barrel carburetors, each with two primary barrels
   c. two 2-barrel carburetors, each with one primary and one secondary barrel
   d. two 2-barrel carburetors, each with two secondary barrels.
2. The four-barrel carburetor:
   a. is commonly used on four-, six-, and eight-cylinder engines
   b. has a great future for the new smaller cars
   c. is not normally used on four-cylinder engines
   d. will be on the majority of cars produced in the 1980s.
3. The secondary barrels of the four-barrel carburetor come into operation:
   a. to assure a smooth idle
   b. when the carburetor throttle plates start to open
   c. during most operating modes
   d. when full power is demanded.
4. The adjustable part throttle:
   a. allows the factory to fine-tune the carburetor
   b. allows very accurate setting of the metering rods in the jets
   c. is not adjustable in the field
   d. all of the above.
5. The purpose of the solenoid on the Rochester M4MC carburetor used on air-conditioned cars is to:
   a. turn the air conditioner on when required
   b. increase idle speed when the air conditioner is on
   c. reduce idle speed when the throttle valves are opened
   d. prevent dieseling.

CHAPTER 14

SERVICING GASOLINE FUEL-INJECTION SYSTEMS

After studying this chapter, and with proper instruction and equipment, you should be able to:

1. Describe the four types of gasoline fuel-injection systems used on cars today.
2. Diagnose and service the Bosch K-type fuel-injection system.
3. Diagnose and service the Bosch D-type fuel-injection system.
4. Diagnose and service the Bosch L-type fuel-injection system.
5. Diagnose and service the Cadillac electronic fuel-injection system.
6. Use the fuel-injection analyzer.
7. Perform a quick check of the solenoid-operated injection valve with an oscilloscope.
8. Diagnose and service a digital electronic fuel-injection system.

14-1 Servicing gasoline fuel-injection systems When a car equipped with any type of gasoline fuel-injection system has a problem, first be certain that the ignition system or the engine itself is not causing it. In troubleshooting, the fuel-injection system usually is the last place to look when an engine is not running properly. Basically, any fuel-injection system has the same job as the carburetor it replaces. That is to supply the cylinders with the proper mixture of air and fuel. How the fuel-injection system does this depends on its design.

Today, there are four general types of fuel-injection systems installed on gasoline engines in passenger cars (Fig. 14-1). One type is a mechanical system, which does not use electronics to control the continuous spray of fuel from the injection valves (Fig. 14-1*d*). It was first used by Porsche in 1973. The other three systems are all various types of electronic fuel injection. One of these is the Digital Electronic Fuel-Injection (DEFI) system (Fig. 14-1*c*). It was first used by Cadillac on some 1980 models, and a similar system was used by Ford the same year. In the DEFI system, two injectors are mounted in the throttle body on the intake manifold, and intermittently spray fuel into the airstream passing through. The other two EFI systems both also have intermittent or timed injection. However, they differ in the way the fuel metering is controlled.

The system shown in Fig. 14-1*a* is the Bosch D-type fuel-injection system. It is the most widely used type of electronic fuel injection. In it, the amount of fuel injected is determined primarily by changes in engine speed and intake-manifold pressure (vacuum). This is the system that was first mass-produced in 1968 on the type 3 Volkswagen with the 1600-cc engine. In 1975, a version of it appeared on certain models of Cadillac.

The Bosch L-type electronic fuel injection (Fig. 14-1*b*) was introduced by Porsche and Volkswagen in 1974. In this system, the fuel metering is controlled primarily by engine speed and the quantity of intake air.

Although the fuel-injection systems in use today were designed by Bendix and Bosch, some car manufacturers make their own parts and have further adapted the systems to particular engines. Therefore, all installations of the same type of fuel-injection system are not identical in operation or appearance. For this reason, always try to have the manufacturer's service manual available when it is necessary to troubleshoot or service a fuel-injection system.

14-2 Servicing Bosch K-type continuous injection system This system (Fig. 14-1*d*) has been installed on cars built by Audi, BMW, Mercedes-Benz, Porsche, Saab, Volkswagen, and Volvo. Figure 14-2 shows the system installed on a Volkswagen four-cylinder engine.

Before attempting to check out or service this system using tools or test instruments, always make a complete and thorough visual inspection. First, unplug the leads from the warm-up regulator and auxiliary-air device

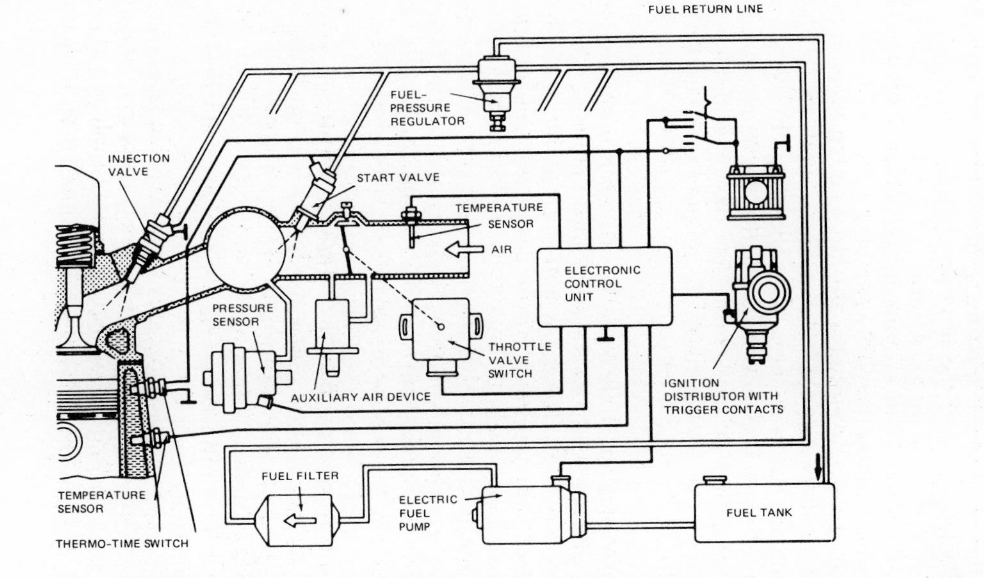

(*a*) BOSCH D-TYPE ELECTRONIC FUEL INJECTION (EFI)

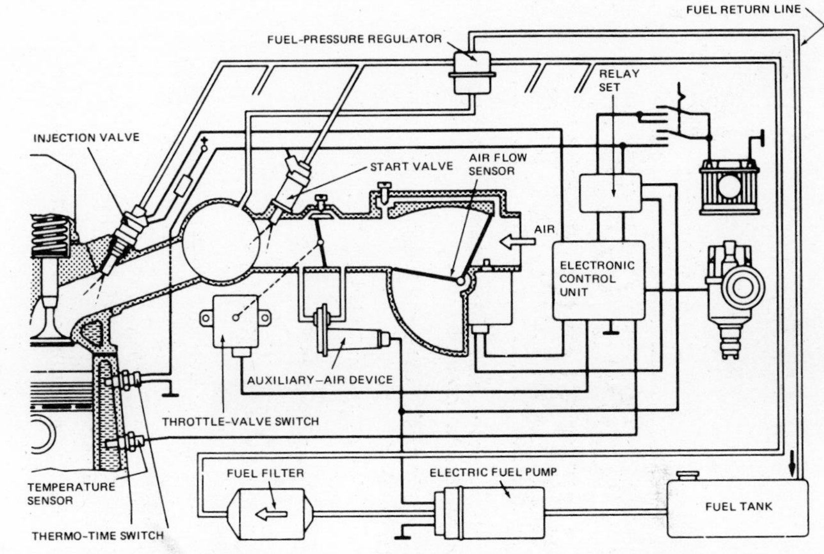

(*b*) BOSCH L-TYPE ELECTRONIC FUEL INJECTION (EFI)

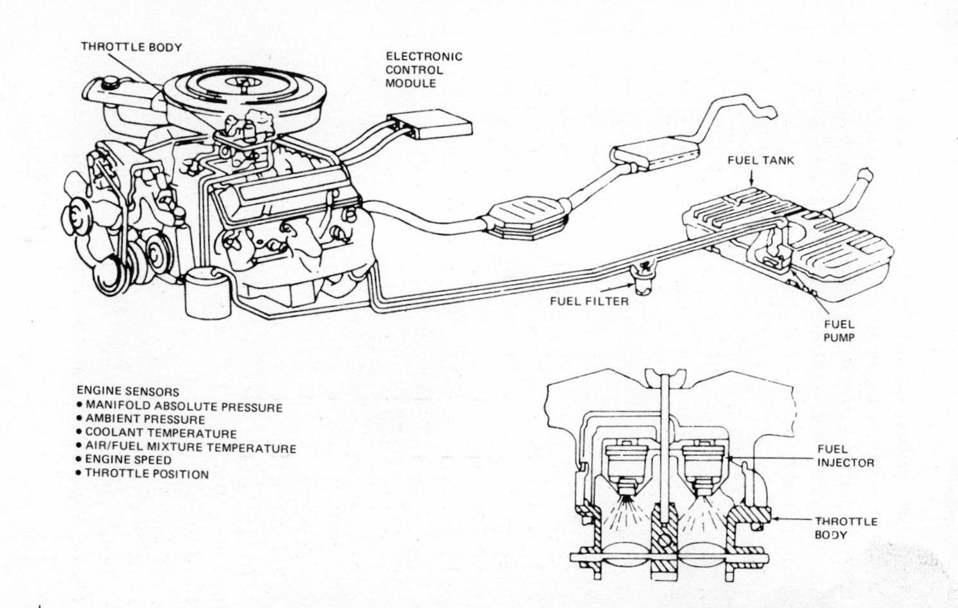

(*c*) DIGITAL ELECTRONIC FUEL INJECTION (DEFI)

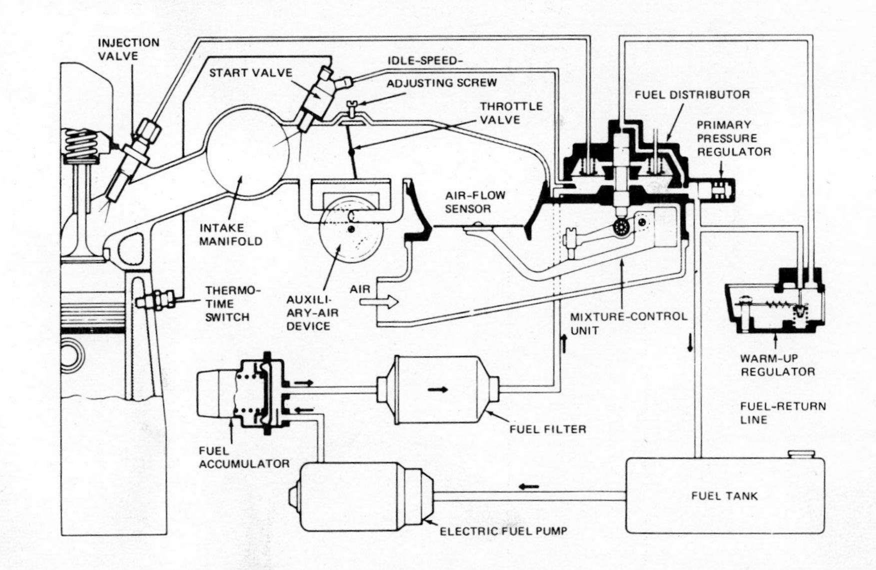

(*d*) BOSCH K-TYPE MECHANICAL CONTINUOUS INJECTION SYSTEM (CIS)

Fig. 14-1 Various types of gasoline fuel-injection systems. (*Robert Bosch Corporation; Cadillac Motor Car Division of General Motors Corporation*)

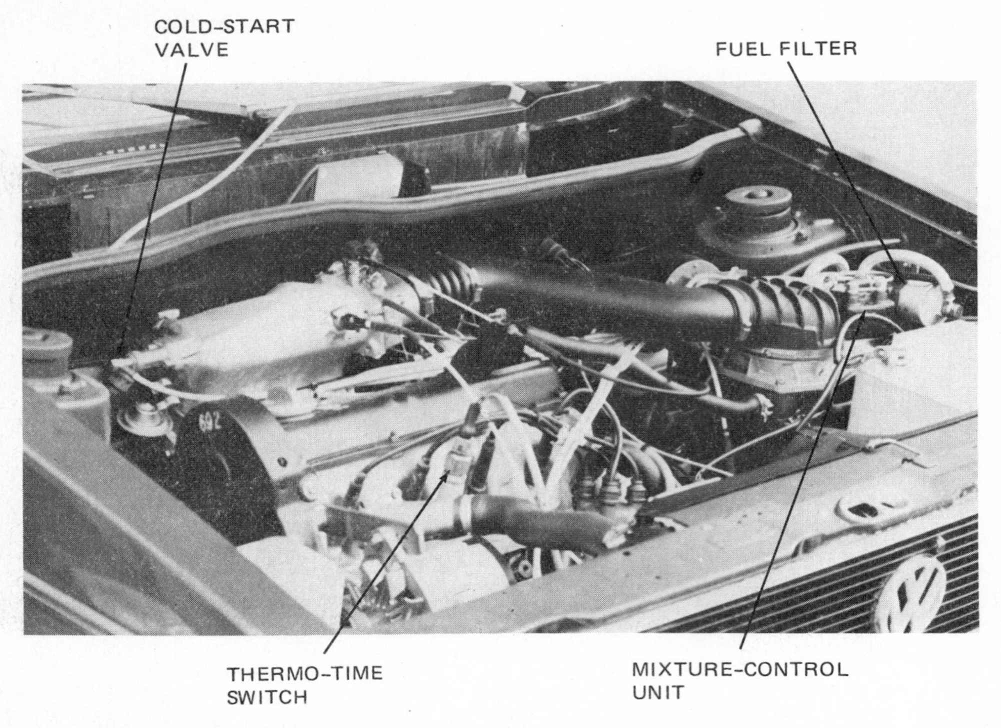

Fig. 14-2 A Bosch K-type continuous-injection system installed on a Volkswagen four-cylinder engine. (*ATW*)

(Fig. 14-1*d*). This prevents the heating elements in these parts from overheating. Now, you are ready to begin the visual inspection.

The two major problems to look for when inspecting a gasoline fuel-injection system are fuel leaks and air leaks. Either of these can cause drivability problems, and possibly engine damage and excessive emissions. A fuel leak is a fire hazard.

1. Fuel leaks Check carefully that there are no fuel leaks from any fuel line connection, from around the electric fuel pump, or from the fuel accumulator. Heavy layers of dirt may make fuel leaks and their source difficult to find. New seals should be installed in any leaking connections, and any defective hoses should be replaced.

Use a box wrench to disconnect the fuel lines from the fuel distributor. Then use a flare nut wrench to disconnect the control pressure connection. Use of regular open-end wrenches on these fittings may damage the adjacent fuel lines.

Some hoses are equipped with screw-type joints. On these, the complete fuel hose assembly, including the nipples and ring connectors, must be replaced. The complete hose assembly is available through the car-dealer's parts department.

Clean all fuel line connections thoroughly before they are opened. Always install new seals when the connections are restored. When installing the mixture control unit, tighten the mounting screws evenly. When detaching or tightening the tubing at the injection valves, use a wrench to hold the hexagonal part stationary.

2. Air leaks No air leaks may occur in the air intake system between the mixture control unit and the engine. Air leaks into the intake manifold cause an excessively lean air-fuel mixture to be delivered to the cylinders. This is because air drawn in through the leak is not metered by the airflow sensor.

Air leaks can occur at several places. These include the connection between the mixture-control unit and the intake manifold, the seal at the flange of the start valve, all hose connections at the intake manifold and at the auxiliary-air device, the seal ring at the injection valves, and the intake manifold support at the cylinder head.

To test the air intake system for leaks, remove one air hose from the auxiliary air regulator. Pressurize this hose with air while holding the throttle open. No air should leak out. Use soapy water on all air hose and manifold connections to detect leaks. Replace any defective hoses or seals.

Figure 14-3 is a trouble-diagnosis chart for the Bosch K-type continuous-fuel-injection system. The chart lists the most frequent complaints and their possible causes, and the checks or corrections to be made. Many of the corrections are self-explanatory, and can be performed quickly and easily. However, others require that you use testers and refer to the manufacturer's service manual for the car you are servicing.

⚙ 14-3 Servicing Bosch D-type electronic fuel injection The Bosch D-type system of electronic fuel injection (Fig. 14-1*a*) is widely used by many manufacturers. These include variations of the system installed by Datsun, Jaguar, Mercedes-Benz, Porsche, Renault, Saab, Volkswagen, Volvo, and Cadillac. Complete di-

COMPLAINT: Engine does not start in cold condition	Engine does not start in warm condition	Engine starts poorly in cold condition	Engine starts poorly in warm condition	Irregular idle during warm-up	Irregular idle with warm engine	Engine backfires into intake manifold	Engine backfires into exhaust system	Engine misses when driven on road	Driving performance unsatisfactory	Engine runs on	Fuel consumption too high	CO concentration at idle too high	CO concentration at idle too low	Idle speed cannot be adjusted	POSSIBLE CAUSE	CHECK OR CORRECTION
●	●														Electric fuel pump not operating	Check pump fuse, pump relay, and pump
●	●							●							Loose contact at electric fuel pump	Check pump wiring
●		●		●											"Cold" control pressure outside tolerance	Test pressure
	●		●		●	●			●				●		"Warm" control pressure too high	Test pressure
	●		●		●		●		●		●	●			"Warm" control pressure too low	Test pressure
			●	●	●						●			●	Auxiliary air valve does not close	Check valve for correct function
●		●		●											Auxiliary air valve does not open	Check valve for correct function
●		●													Cold start valve does not open	Check cold start valve
●		●	●	●	●		●		●	●	●	●			Cold start valve leaking	Check cold start valve
								●							Primary system pressure out of tolerance	Test pressure and adjust with shims
●	●	●	●												Air flow sensor plate stop incorrectly set	Check and reset
●	●	●	●		●				●	●		●			Sensor plate and/or plunger not moving freely	Check for free movement
●	●	●	●	●	●	●			●				●		Leaks in air intake system (false air)	Check air system for leaks
●	●	●	●	●	●		●	●			●	●			Fuel system leakage	Inspect fuel system for leaks
	●		●	●	●					●					Injector leaking, opening pressure low	Check injectors on tester
	●		●		●		●		●		●	●			Idle mixture too rich	Check and adjust CO level
	●		●		●	●			●				●		Idle mixture too lean	Check and adjust CO level
									●						Throttle butterfly does not open completely	Check butterfly and stops in throttle venturi
●		●													Thermo-time switch defective	Test for resistance readings vs. temperature

Fig. 14-3 Trouble diagnosis chart for the Bosch K-type fuel-injection system. (*Robert Bosch Corporation*)

Engine cranks but does not start	Engine starts but then dies	Rough or unstable idle	Idle speed incorrect	Erratic running	Engine misses when driving	Fuel consumption too high	No maximum power	POSSIBLE CAUSE	CHECK OR CORRECTION
●	●	●	●	●	●	●	●	Defect in ignition system	Check battery, distributor, plugs, coil, and timing
●	●	●	●	●	●	●	●	Mechanical defect in engine	Check compression, valve adjustment, and oil pressure
●								Fuel pump not operating	Check pump fuse, pump relay, and pump
●	●							Relay defective; wire to injector open	Test relay, check wiring harness
●	●	●					●	Blockage in fuel system	Check fuel tank, filter, and lines for free flow
	●	●	●	●	●		●	Leaks in air intake system	Check all hoses and connections; eliminate leaks
		●		●	●	●	●	Fuel system pressure incorrect	Test and adjust at pressure regulator
	●				●			Trigger contacts in distributor defective	Replace trigger contacts
●		●				●		Cold start valve defective	Check for spray or leakage
●								Thermo-time switch defective	Test thermo-time switch for correct function
	●		●					Auxiliary air valve not operating correctly	Must be open with cold engine; closed with warm
●	●					●		Temperature sensor II defective	Test for 2–3 kΩ at 68°F [20°C]
●	●	●		●		●	●	Pressure sensor defective	Test with ohmmeter
		●	●				●	Throttle butterfly does not completely close or open	Readjust throttle stops
		●	●	●			●	Throttle valve switch incorrectly adjusted or defective	Adjust as necessary or replace
	●	●	●					Idle speed incorrectly adjusted	Adjust idle speed with bypass screw
		●	●	●	●		●	Defective injection valve	Check valves individually for spray
●	●	●		●	●			Loose connection in wiring harness or system ground	Check and clean all connections
●	●			●	●		●	Control unit defective	Use known good unit to confirm defect

Fig. 14-4 Trouble diagnosis chart for the Bosch D-type electronic fuel-injection system. (*Robert Bosch Corporation*)

agnosis and service procedures for the system as used by Cadillac are given in later sections of this chapter.

Figure 14-4 shows the trouble diagnosis chart for the Bosch D type of electronic fuel injection. This chart is similar to the trouble diagnosis chart shown in ⚙ 14-7, which is for the system used by Cadillac. Because of the complex circuitry in the electronic control module, some types of failure can be pinpointed exactly only by the use of the special electronic-fuel-injection analyzer (⚙ 14-16). In addition, the EFI analyzer can check components on or off the car, and with the engine running or inoperative.

To check the components in the D-type system, a voltmeter, ohmmeter, pressure gauge, and basic hand tools are needed. With these tools, you can locate faults in the system, isolate the component to be tested, and then test the component for proper operation and specification. All electrical measurements are made at the terminals of the large plug at the end of the wiring harness, after it is removed from the electronic control unit.

Most of the components in an electronic fuel-injection system cannot be repaired. After you have determined that a part is defective, it must be replaced.

One of the biggest problems in working with electronic fuel injection is to identify and locate each component of the system on the vehicle. Trouble diagnosis charts in the service manual for each car tell you what to check on that vehicle, where the component is located, and how to check it.

Sometimes, you may not be sure which type of electronic fuel injection is on the engine. One quick way to tell the difference between the Bosch D type and the L

Fig. 14-5 Connecting the scope to the injection valve. (*Autoscan, Inc.*)

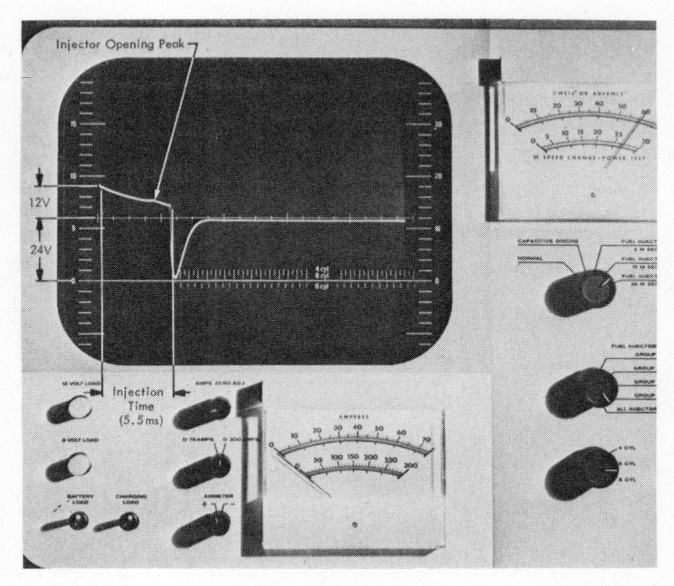

Fig. 14-6 Typical scope pattern showing the operation of the injector. (*Autoscan, Inc.*)

type is by looking at the ignition distributor. The D type requires that the distributor have a speed sensor with a set of trigger contacts (Fig. 6-6). The L type does not require trigger contacts mounted in the ignition distributor.

After performing any service or repair on an electronic fuel-injection system, the ignition timing, idle speed, and idle mixture should be checked. They should be readjusted, if necessary, before returning the car to the customer.

The solenoid-operated injection valve in electronic fuel-injection systems can be checked with an oscilloscope. Special fuel-injection scopes, and special probes for other automotive scopes, are available for this purpose. However, a quick check of the injection valve operation can usually be made with almost any type of shop scope.

To hook up the scope, the proper primary lead is connected to ground. Then connect the fuel-injection probe, or other primary lead, to the voltage feed wire on the injector (Fig. 14-5). The solenoid injection valve may have two wires attached to it. One is a ground wire, and the other is the feed wire. If no pattern appears on the scope screen, touch the primary lead to the other wire at the connector on the injection valve.

Figure 14-6 shows a typical pattern showing operation of the injection valve. Any variation from this basic pattern indicates a fault.

⚙ 14-4 Servicing the Bosch L-type electronic fuel injection This system is found on various models of BMW, Opel, Porsche, Renault, and Volkswagen. It can be checked with a voltmeter, ohmmeter, pressure gauge, and basic hand tools by following the steps in the manufacturer's service manual. As in the D type, the electrical measurements are made at the end of the wiring harness, after it is removed from the electronic control unit. However, some tests may be made at the terminals on the individual components. As with other

Engine cranks but does not start	Engine starts but then dies	Rough or unstable idle	Idle speed incorrect	CO value incorrect	Erratic running	Engine misses when driving	Fuel consumption too high	No maximum power	POSSIBLE CAUSE	CHECK OR CORRECTION
●	●	●	●		●	●	●	●	Defect in ignition system	Check battery, distributor, plugs, coil, and timing
●	●	●	●		●	●	●	●	Mechanical defect in engine	Check compression, valve adjustment, and oil pressure
●	●	●	●	●	●	●		●	Leaks in air intake system (false air)	Check all hoses and connections; eliminate leaks
●	●	●						●	Blockage in fuel system	Check fuel tank, filter, and lines for free flow
●									Relay defective; wire to injector open	Test relay; check wiring harness
●									Fuel pump not operating	Check pump fuse, pump relay, and pump
●	●	●		●	●		●	●	Fuel system pressure incorrect	Check pressure regulator
●									Cold start valve not operating	Test for spray, check wiring and thermo-time switch
●	●	●	●	●	●		●		Cold start valve leaking	Check valve for leakage
●									Thermo-time switch defective	Test for resistance reading vs. temperature
●	●	●	●						Auxiliary air valve not operating correctly	Must be open with cold engine; closed with warm
●	●	●		●			●		Temperature sensor defective	Test for 2–3 kΩ at 68°F [20°C]
●	●	●	●	●	●		●	●	Air flow meter defective	Check pump contacts; test flap for free movement
		●	●					●	Throttle butterfly does not completely close or open	Readjust throttle stops
		●						●	Throttle valve switch defective	Check with ohmmeter and adjust
	●	●	●	●					Idle speed incorrectly adjusted	Adjust idle speed with bypass screw
		●		●	●	●		●	Defective injection valve	Check valves individually for spray
	●	●	●	●	●		●		CO concentration incorrectly set	Readjust CO with screw on air flow meter
●	●	●			●	●			Loose connection in wiring harness or system ground	Check and clean all connections
●	●	●				●		●	Control unit defective	Use known good unit to confirm defect

Fig. 14-7 Trouble diagnosis chart for the Bosch L-type electronic fuel-injection system. (*Robert Bosch Corporation*)

electronic fuel-injection systems, most of the components cannot be repaired. Any component found to be defective must be replaced. A trouble diagnosis chart for the L-type system is shown in Fig. 14-7.

Listed below are seven service cautions that must be followed when working on cars equipped with the Bosch L-type system. This is to prevent damage to electronic devices on the car that might occur by using an improper procedure.

1. Never start the engine without the battery cables connected and tightened.
2. Never jump the battery to start the car.
3. Never remove the cables from the battery with the engine running.
4. Always remove the cables from the battery before charging it.
5. Never remove or attach the wiring harness plug to the electronic control unit with the ignition on.
6. When cranking the engine to check compression, unplug the red cable from the battery to the relays.
7. Before testing the L-type system, make sure that the ignition timing and dwell are within specifications, and that the spark plugs are firing properly.

⚙ 14-5 Servicing the Cadillac electronic fuel-injection system

NOTE: To provide the complete procedure on inspection, diagnosis, and repair of a typical fuel-injection system, this and following sections outline the steps recommended by Cadillac. Figure 6-19 shows the components of the Cadillac system. This is similar to the Bosch D-type system (⚙ 14-3).

Before checking for defects in the electronic fuel-injection system, make sure the engine and especially the ignition system are not causing the complaint. If they are not at fault, make the following visual inspection of the fuel-injection system components.

1. Visually check all wiring harness connections for:
   a. Loose or detached connectors
   b. Broken or detached wires
   c. Terminals not completely seated in connector housings
   d. Partially broken or frayed wires at terminal connections
   e. Excessive corrosion
2. Start the engine. Plug the idle bypass passage on top of the throttle body with a clean shop towel, to make it easier to hear any vacuum leaks. Then visually check all vacuum lines:
   a. To ensure all vacuum lines are securely connected to their proper fittings (Fig. 14-8)
   b. For broken, pinched, or cracked lines
3. Visually check the fuel lines for:
   a. Leakage
   b. Kinks

CAUTION: Do not loosen fuel system fittings until the pressure has been relieved. Fuel may be under high pressure, which could spray it out. This could result in a fire.

To relieve the pressure in the fuel system:

1. Remove the protective cap from the pressure fitting.
2. Loosely install a valve depressor on the fitting.
3. Hold a shop towel or suitable container so any fuel that comes out can be caught without spurting around.
4. Slowly tighten the valve depressor to relieve pressure.
5. Dispose of gasoline or shop towel properly. Put towel outside in a safe place to dry out.
6. Remove the valve depressor and install the protective cap.

⚙ 14-6 Preliminary diagnosis If the cause of the problem is not found during the visual inspection, then a preliminary diagnosis is in order. Cadillac lists nine different types of problems that might occur in cars with electronic fuel injection. If the engine and all other systems are operating properly, these problems may be traced to the electronic fuel-injection system. The problems are:

1. Engine cranks but will not start
2. Hard starting
3. Poor fuel economy
4. Engine stalls after start
5. Rough idle
6. Prolonged fast idle
7. No fast idle
8. Engine hesitates or stumbles on accleration
9. Lack of high-speed performance

After the visual inspection, a fuel-injection analyzer must be used to locate the cause of any of these problems.

During the preliminary diagnosis, keep in mind certain relationships between parts in the carburetor systems and components in the electronic fuel-injection system (Fig. 14-9). These relationships can help you to identify and pinpoint a defective component.

⚙ 14-7 Trouble diagnosis chart The chart that follows on page 221 lists various complaints that can be traced to electronic fuel-injection systems, their possible causes, and the checks or corrections to be made. The information in the chart will shorten the time you need to correct a trouble.

Note that the troubles and possible causes are not listed according to how often they occur. Item 1 (or item *a* under *Possible Cause*) does not necessarily occur more often than item 2 (or item *b*).

⚙ 14-8 Engine will not start As with most other problems, it is necessary to distinguish between fuel problems and electrical problems. To check the fuel pressure, install a pressure gauge and crank the engine. If the fuel pressure is not normal, connect the fuel-injection analyzer (⚙ 14-16), and perform the fuel system tests.

A normal pressure reading of 37 to 42 psi (255 to 290 kPa) while cranking will eliminate the fuel delivery system as a cause of the no-start complaint. In this case the following electrical components should be checked:

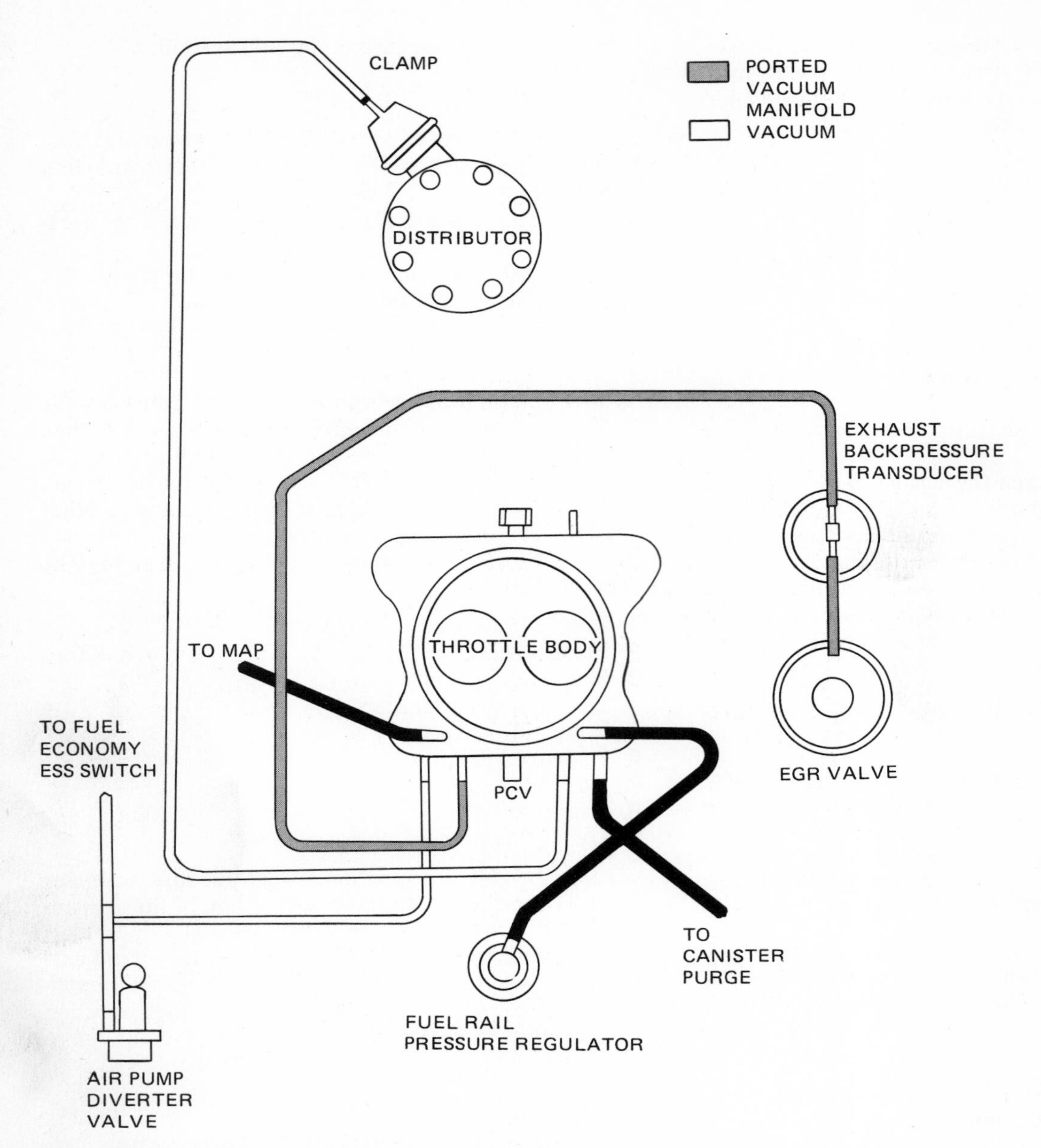

Fig. 14-8 Vacuum hose schematic for the Cadillac electronic fuel-injection system. (*Cadillac Motor Car Division of General Motors Corporation*)

1. Speed sensor The speed sensor (or distributor trigger switch) consists of a pair of switches, located in the distributor housing, that tell the ECU when to activate the injector group and which injector group should be activated. If either of these switches is sticking or does not provide a signal (an "open"), then the ECU will not pulse the injector group. This will prevent the engine from starting.

Proper operation of the speed sensor can be observed on the fuel-injection analyzer.

2. Cranking signal A wire between the starter solenoid and the ECU provides a signal that the engine is being cranked. The ECU uses this signal to close the fuel pump fast-idle valve relay and to provide starting enrichment. Lacking this signal, these functions will not occur and starting will be impaired. Existence of

Carburetor	Electronic Fuel Injection
1. Accelerator pump	1. Throttle position switch
2. Fast-idle cam	2. Electric fast-idle valve
3. Float	3. Fuel-pressure regulator
4. Power valve metering rods	4. Manifold absolute pressure sensor
5. Metering jets and idle fuel system	5. Injection valves, electronic control unit (ECU)

Fig. 14-9 Chart showing the relationship between carburetor components and components in the Cadillac electronic fuel-injection system. (*Cadillac Motor Car Division of General Motors Corporation*)

Electronic Fuel-Injection Trouble Diagnosis Chart

CONDITION	POSSIBLE CAUSE	CHECK OR CORRECTION
1. Engine will not start; fuel pump does not run (⚙ 14-8)	a. Fuse to pump relay terminal blown; cables to pump or pump relay or cables on fuel pump relay defective	Renew fuse; check whether pump relay energizes (switch ignition on and off, and listen for audible action of relay); if necessary, test with voltmeter; check plug on fuel pump for good connection.
	b. No voltage at terminal of pump relay (12 volts), because main relay is not operating or cable is defective	Eliminate open circuit.
	c. Voltage present at terminal of pump relay, but no ground connection	Fuel pump operates for 1 to 2 seconds after ignition is switched on; check with meter (ground is made by control unit); replace ECU.
	d. Open circuit in cable from pump relay to pump plug	Check plug connection; eliminate open circuit.
2. Engine will not start; fuel pump runs (⚙ 14-8)	a. Connection from cable harness to starter terminal defective	Check with special tester
	b. Pressure sensor cable not connected; open circuit	Push on pressure sensor cable; repair.
	c. Open circuit in cable connection at temperature sensor (coolant)	Check cables; if necessary, replace temperature sensor.
	d. No fuel pressure building up (line compressed or pressure regulator defective)	Check pressure with gauge; if necessary replace pressure regulator.
3. Engine starts cold but stalls (⚙ 14-9 and 14-11)	a. Cable connector for triggering contacts not pushed on at ignition distributor; open circuit in cable	If necessary, connect analyzer and locate the fault; replace trigger contacts or cable harness.
	b. Trigger contacts defective	Replace.
	c. Pressure sensor defective	Replace.
	d. See item 1, Engine will not start	
4. Engine cuts out when driving (usually preceded by misfiring) (⚙ 14-12)	a. Trigger contacts have excessive contact resistance or are dirty	Replace trigger contacts.
	b. Plug loose	Check.
	c. No fuel pressure	Check pressure; determine cause
5. Engine runs irregularly, one cylinder not firing; exhaust clear (⚙ 14-14)	a. One injector sticking	Replace injector.
	b. Connection to injector or injector coil defective	Check connections; replace injector; test system with analyzer.
6. Engine misfiring, not caused by ignition system (⚙ 14-14)	c. Loose connections; main ground cable has poor connection to car body	Check connections; tighten ground connection.
7. Engine not reaching full power	a. Fuel pressure low	Check pressure regulator.
	b. Pressure sensor	Replace.
	c. Throttle valve does not open sufficiently	Check throttle valve.
8. Fuel consumption too high (⚙ 14-15)	a. Sensors of ECU not functioning correctly; electrical connections have too high resistance	Test system with analyzer.
	b. Throttle valve switch incorrectly adjusted.	Adjust, using analyzer.
	c. Fuel pressure incorrect	Check pressure regulator; replace if necessary.
9. Engine hunts excessively at idle (between 1000 and 1800 rpm)	a. Hose between auxiliary air regulator and induction manifold detached or defective	Push hose into position or replace.
	b. Throttle valve stop incorrectly adjusted (open too wide)	Readjust throttle valve stop.
	c. Idling speed set too high	Adjust idling speed.
10. Engine misfires when accelerating	Temporary enrichment device in throttle valve switch not functioning; plug incorrectly connected	Check throttle valve switch with analyzer.
11. Too-high idling speed; idling speed cannot be adjusted	a. Idling air system leaks	Check idling air system.
	b. Rubber sealing ring under the injector	Replace rubber sealing ring.
	c. Throttle incorrectly adjusted	Readjust throttle valve.

this signal and continuity of this circuit is checked with the fuel-injection analyzer.

3. Throttle position switch One of the functions of the throttle position switch is to provide a signal for the ECU when wide-open throttle is reached. However, this signal is also used to perform another function. While the possibility of a flooded engine is remote with EFI, the possibility does exist. So, just as with carburetors, a means of clearing a flooded engine is incorporated. If the throttle is held wide open, the throttle-position switch provides a wide-open-throttle (WOT) signal. If the ECU receives this signal along with a cranking signal, it stops pulsing the injectors.

Therefore, if the WOT section of the throttle-position switch is shorted, it will provide this "unload" signal continuously. This results in a no-start condition even though fuel pressure is adequate. Testing of the WOT circuit is accomplished with the fuel-injection analyzer.

4. Coolant sensor Just as on a carbureted engine, extra fuel is required for cold enrichment. The coolant sensor provides coolant temperature information to the ECU so that the pulse width may be lengthened. With a low coolant temperature, the resistance value of the sensor is also low. As the coolant warms up, the sensor increases its resistance in a direct proportion. Should the sensor malfunction so that it provides a high resistance when the engine is cold or is disconnected (has infinite resistance), the ECU cannot provide the additional enrichment. Therefore a no-start condition could occur in cool to cold temperatures. It may sometimes be possible to get around this condition by pumping the accelerator pedal, thereby providing acceleration enrichment (AE) pulses. However, this is not a normal starting procedure. The ability to start a no-start car by using the AE pulses points to a coolant sensor circuit problem. Diagnosis of the coolant sensor is performed using the fuel-injection analyzer.

5. Electronic control unit All of the sensor input information is analyzed by the ECU, which then determines how long to pulse the injectors and when. If all of the input information appears to be generated in its proper sequence, check for an ECU malfunction.

A new ECU should be substituted on a trial basis. The proper procedure for performing this substitution is to connect a new (or known good) ECU of the same part number and try to start the car. If the car will start, reconnect the suspected ECU and again try to start the car. If the car operates properly on both ECUs do not replace the ECU. The problem was probably caused by a poor connection.

If a trial ECU is not available, attempt to start the engine on the *substitute ECU* portion of the fuel-injection analyzer. If the engine will start and run off the analyzer and if the cranking signal, throttle position switch, speed sensor, and coolant sensor circuits are known to be good, the ECU is indicated as being defective.

Next, reconnect the suspected ECU and again try to start the car. If the car now operates properly the problem was probably caused by a poor connection. However, if the car will not start, a new ECU should be tried.

If the car will not run from the fuel-injection analyzer, then the problem is in the car components and was probably misdiagnosed the first time around.

14-9 Engine hard to start when cold A hard-start condition may be considered as a mild case of no-start. Therefore, many of the same components must be checked. First, determine whether the condition exists with the engine hot or cold. Discussed below are the conditions that frequently can make a cold engine hard to start.

1. Fuel system A problem in the fuel system is not the only probable cause of an engine that is hard to start when cold. However, the test is easy to perform and quickly eliminates many of the potential problems. To test the fuel system, install a pressure gauge. Then watch the gauge as the engine is cranked and started. The pressure should rise to 30 to 42 psi (255 to 290 kPa) as soon as the key is turned on, and remain in that range after the engine starts and idles. If there is no fuel pressure, connect the fuel-injection analyzer. Check the operation of the START light. An open in this circuit can also prevent fuel pump operation during cranking. Continue the diagnosis using the fuel-injection analyzer.

2. Coolant and air sensors The coolant and air sensors are responsible for informing the ECU that the engine is cold and needs extra enrichment. If these signals are not transmitted to the ECU, it may not lengthen the pulse width enough to start the car normally. As with the no-start condition, the need for pumping the accelerator pedal and providing AE pulses is a hint that the sensors are malfunctioning. Testing of the sensors is accomplished with the fuel-injection analyzer.

3. Fast-idle valve To accommodate the additional fuel and to provide fast idle, the fast-idle valve controls the size of a bypass passage. On a cold engine (with lots of cold enrichment), the passage is large, providing corresponding fast idle. As the engine runs and begins to warm up, the fast-idle valve heater is electrically heated. This forces the valve against spring pressure toward the closed position. The fast-idle valve will never completely seal the bypass passage. However, this normal opening is calibrated into the idle system.

Electrically, the fast-idle valve heater can be inspected with the fuel-injection analyzer. However, mechanical malfunctions can also exist. The position of the valve is a result of opposing forces. The spring is trying to push the valve up while the plunger on the heater is pushing down.

If the valve is pushed down too far, it can become latched in the seat in the throttle body. This results in no additional air being supplied and too low an idle speed. A quick check for this condition is to wiggle the heating element on a cold engine. If the valve is stuck down, there will not be any spring tension on the plunger and the heater will wiggle. If this condition exists, remove the heater and free the valve. Inspect the heater assembly for cracks, wax leaking out of the brass pellet (white heat-transfer compound around pellet is normal), missing $^{1}/_{16}$-inch-diameter [1.59-mm] push rod, damaged microswitch arm, or missing plunger spring clip. Damage to the valve requires replacement. Make certain that all pieces are removed from the throttle body and are not dropped in the intake manifold.

4. Idle speed and timing Timing on EFI engines is different than on carbureted engines. For proper engine operation, the timing must be set to specifications.

A potential problem on EFI cars is that if the secondary ignition wires are improperly located in the distributor cap even though the rotor is moved to compensate, the speed sensor signals will be significantly out of time. Correct wire location is shown in the shop manual. Each engine has its own specific wire locations.

Two different vacuum spark-advance systems are used. In one system, ported vacuum operates the vacuum spark advance through the thermal-vacuum switch (TVS). On the other engines, the spark advance is supplied with manifold vacuum at all times, and the TVS is eliminated.

Ported vacuum is obtained from a port located above the throttle valves in the throttle body. The vacuum port is not exposed to engine vacuum until the throttle valves are opened. This delays the operation of the vacuum advance until the engine is off idle. A manifold vacuum port located below the throttle valves is exposed to vacuum whenever the engine is running. By using this port as a source, the vacuum advance is operational at idle.

With either system, the vacuum-advance hose should be disconnected and plugged when initial timing is adjusted and reconnected after the idle speed is set.

5. Electronic control unit As with most other EFI problems, the ECU could be malfunctioning. The test procedure when using the fuel-injection analyzer will allow specific signals to be sent to the ECU and the resulting output to be observed on the analyzer. This diagnosis should be performed rather than simply replacing the ECU.

6. Fuel quality In most areas of the country, petroleum refineries blend different grades of gasoline for summer and winter operation. Occasionally the season will change before some gasoline retailers have sold all of the previous seasonal blend. By purchasing this "out-of-season" gas, the owner may experience a hard starting condition temporarily.

⚙ 14-10 Engine hard to start when hot A hard-start condition which develops only when the engine is hot is usually an indication of a low idle speed or a fuel supply problem.

1. Idle speed One characteristic of the electronic fuel-injection system is its ability to provide smooth engine idle over a wide range of idle speeds. While this is desirable, it can sometimes be misleading. The idle speed may be too slow while still quite smooth. Idle speed must be checked and, if necessary, adjusted to 600 to 650 rpm with the engine at normal operating temperature.

2. Fuel system When the engine is shut down, the fuel system acts much like the cooling system. The fuel system absorbs a lot of heat from the surrounding parts without the benefit of coolant (in this case, cool gasoline from the tank) flowing through the system. If enough heat is absorbed, the gasoline will boil, creating vapor pockets in the system. This condition is *vapor lock*.

Like most electric fuel pumps, the chassis-mounted pump is an excellent device for pushing fuel through the system. However, it is not as good at pulling fuel from the tank. The in-tank fuel pump is used to prime the chassis-mounted pump.

Determine the fuel pressure when the hard-starting condition is experienced. Install the fuel pressure gauge. Observe the fuel pressure while cranking. Then return the ignition to the OFF position. If the fuel pressure drops faster than 5 psi [35 kPa] within 2 minutes after the pumps are shut off, follow the procedure using the fuel-injection analyzer.

To check the pressure regulator, pinch off the rubber fuel-return line at the pressure regulator while the pumps are activated. A pressure reading of more than 46 psi [317 kPa] indicates good fuel pumps and a defective regulator. However, if the pressure is less than 46 psi [317 kPa], the pumps must be tested. Using adapters, tee the pressure gauge into the fuel line between the in-tank pump and the chassis-mounted pump.

With the ignition on, the pressure should be at least one psi [6.9 kPa]. This represents the output of the in-tank pump, and is sufficient to prevent vapor lock in this section of the line. If the in-tank pump is functioning properly, inspect the fuel filter and fuel line for restrictions.

However, if the in-tank pressure is less than 1 psi [6.9 kPa], the problem is an electrical or mechanical malfunction of the in-tank pump. Since it is necessary to remove the fuel tank to inspect the pump for mechanical malfunctions, first check the electrical circuit using a 12-volt test light. Normally the pump operates for only one second after the ignition is turned on unless the engine is cranked or running. If the system is OK electrically, remove the fuel tank and replace the in-tank pump. Always retest after a repair has been made, including an attempt to duplicate the original problem.

2. Electrical system It is always easier to locate the cause of a problem if you are working on the system when the problem occurs. Should this happen, the fuel-injection analyzer has a position called MONITOR which allows electrical evaluations to be made at this time.

With the analyzer connected and on monitor, operate the car until the hard-start condition develops. Next, crank the engine and observe the speed sensor and injector lights. An improper speed-sensor signal will indicate replacement of the sensor is required. Malfunctioning injector lights require ECU replacement.

3. Coolant sensor The coolant sensor changes in resistance value with temperature. A hard-starting condition while the engine is hot could be caused by a low-resistance (shorted) signal being sent to the ECU. This results in a long pulse width and rich mixture. An open circuit in the coolant sensor will provide less than normal starting enrichment. Test the coolant sensor with the fuel-injection analyzer.

⚙ 14-11 Engine starts cold but stalls While this condition could also be caused by no fuel pressure, the fact that the engine is cold is a hint that an electrical malfunction is more likely. On a cold engine the coolant sensor and the fast-idle valve provide extra fuel and extra air required for cold operation. The first step in diagnosing this condition should be to eliminate the fuel system as the source of the problem. Install the fuel pressure gauge and watch the fuel pressure as the stall develops. If pressure drops from the fuel rail, install the fuel-injection analyzer. If fuel pressure is maintained, check the electrical signals from the coolant sensor and fast-idle valve and at the EGR system.

1. Temperature sensors Just as on a carbureted engine, extra fuel is required for cold enrichment. The temperature sensors provide temperature information to the ECU so that the pulse width may be lengthened. With a low temperature the resistance value of the sensor is also low. As the engine warms up, the sensor increases its resistance in a direct proportion. Should either sensor malfunction so that it provides a high resistance when the engine is cold or is disconnected (has infinite resistance), the ECU may not provide enough additional fuel to keep the engine running. Sensor resistance can be determined using the fuel-injection analyzer.

High resistance will also result in hard starting and a lack of fast idle.

2. Fast-idle valve With additional fuel being added on signal from the coolant sensor, the additional air is provided by the fast-idle valve. Both electrical and mechanical malfunctions can prevent proper operation of the valve. Electrical analysis of the fast-idle valve circuit is accomplished with the fuel-injection analyzer. The "wiggle test" described for the fast-idle valve in ⚙ 14-9 should be performed to check for mechanical malfunctions such as valve latched in the throttle body.

3. EGR system In normal operation, EGR valve operation is delayed by three separate controls. First, when the engine is cold the EGR solenoid is energized by the ECU, which blocks the source vacuum to the EGR valve. Once the engine warms up, the solenoid is deenergized and the second control is used. This is the use of *ported* vacuum. Ported vacuum is obtained from a port located above the throttle valves in the throttle body. When a vacuum port is so located, it is not exposed to engine vacuum until the throttle valves are opened. Therefore, EGR signal vacuum is delayed until the engine is substantially off idle. These controls are placed on EGR operation for many reasons, one of which is to ensure a smooth idle.

If for some reason EGR operation occurs at idle, a stall could result. This can happen either as a result of a mechanical malfunction that causes the valve to stick open (carbon deposits on valve seat) or if the wrong signal is delivered to the valve (solenoid malfunction or hose connected to manifold vacuum).

The third control is the exhaust-pressure transducer which bleeds off EGR vacuum until the exhaust system pressure reaches the prescribed level. This normally occurs at about 20 to 25 mph [32 to 40 km/h] on a level road. To diagnose the EGR system a visual inspection is necessary along with the solenoid evaluation using the fuel-injection analyzer.

The EFI system incorporates another EGR control which does not have significant effect on the way that the engine performs but could be misdiagnosed. The wide-open-throttle signal from the throttle position switch is used by the ECU to reactivate the EGR solenoid and shut off EGR. With the analyzer operating in the monitor mode (step 20), the EGR solenoid light will come on at wide-open throttle. This is a normal condition.

⚙ 14-12 Engine stalls when hot After determining that the car's ignition system is operating properly, the following components should be checked.

1. Idle speed A common cause of a stall condition is a low idle speed. A 600- to 650-rpm idle is essential. Correct idle speed will eliminate many EFI problems. Idle speed and ignition timing should be set to specifications before continuing.

2. Fuel system Insufficient fuel delivery at the injector is a result of either a lack of an electrical signal at the fuel pumps or an inability of the pumps and pressure regulator to deliver and maintain fuel pressure. To establish the fuel system as the source of the problem, run the engine with the pressure gauge attached and duplicate the stalling condition. If, when the engine stalls, there is still 30 to 40 psi [207 to 276 kPa] pressure in the fuel rail, the fuel system may be assumed to be OK. However, if the fuel pressure is low, then determine whether the problem is caused by the pressure regulator, the in-tank fuel pump, or the chassis-mounted fuel pump.

Since the regulator is easy to check, test it first. Pinch off the rubber fuel-return line at the regulator while the pumps are activated. A pressure reading of over 46 psi [317 kPa] indicates good pumps and a defective regulator. However, if the fuel pressure is below 46 psi [317 kPa] determine whether the in-tank pump or the chassis-mounted pump is at fault. Tee the pressure gauge into the fuel line between the pumps. This allows the gauge to read the output of the in-tank pump only. With both pumps on, the pressure should be at least 1 psi [6.9 kPa]. If not, inspect the feed wire for 12 volts, and the ground wire for good connections. If the wiring is OK, replace the in-tank pump.

However, if the in-tank pump is delivering at least 1 psi [6.9 kPa], the problem lies in the high-pressure side of the system. In this case, again inspect the feed wire for 12 volts and the ground wire for a good connection. If they are OK electrically, inspect the fuel filter and fuel lines between the pump and fuel rail for restrictions. If everything looks OK, assume that the chassis-mounted pump is not capable of supplying enough pressure and replace it.

3. Speed sensor No signal (an open) from the speed sensor will be interpreted by the ECU as an engine-off signal and will normally shut off the fuel pumps. Speed sensor operation may be checked using the analyzer. If the sensor checks good on the analyzer, the possibility still exists that an intermittent problem may be occurring. Operate the engine with the analyzer in MONITOR position. Watch the TRIGGER POINTS lights as the stall develops. If the lights operate properly until the engine stalls, crank the engine immediately and observe the speed sensor lights. If the lights still work properly the ECU is probably stopping of its own accord. In this case the malfunctioning ECU should be checked as outlined in step 5 below.

4. EGR valve An EGR valve sticking open can dilute the air-fuel mixture so much that a stall or no-idle complaint could result. The previously described checks of the EGR system should be performed (❂ 14-11).

5. Electronic control unit Since the ECU is ultimately responsible for actuating the injectors, an internal malfunction of the ECU can cause the engine to stop running. If the analyzer fails to indicate an ECU fault, a new ECU can be substituted on a trial basis. The proper procedure for performing this substitution is to duplicate the complaint and immediately connect a new ECU. If the condition is corrected, continue to try to duplicate the condition until you are sure that the complaint cannot be reproduced. Next, reconnect the suspected ECU and again try to reproduce the complaint. If the car operates properly on both ECUs do not replace the ECU. The problem was probably caused by a poor connection.

❂ 14-13 Engine runs rich with rough idle A rich idle results from too much fuel for the available air. However, with EFI this condition can be caused by either a fuel delivery system problem or as a result of an erroneous signal to the ECU which makes it think that the engine needs more fuel than is actually required.

The following fuel system components could cause this condition.

1. Pressure regulator At idle, engine vacuum is high. The pressure regulator senses this and drops the fuel pressure to approximately 30 psi [207 kPa]. If the regulator allows a higher pressure, a greater quantity of fuel will be injected for any given opening time of the injectors. The high pressure could be caused by a disconnected, plugged, or leaking vacuum hose between the regulator and throttle body. These conditions prevent the vacuum signal from reaching the regulator, which is the same as zero vacuum or wide-open throttle. The regulator will normally raise the fuel pressure under these circumstances.

Inspect the vacuum hose for these conditions prior to regulator testing. Fuel pressure at idle can be determined with the formula: Normal pressure = 39 − ½ manifold vacuum. For example, if the engine is developing 14 inches of vacuum at idle, the normal fuel pressure should be 39 − one-half of 14, or 39 − 7, which is 32 psi (221 kPa). Allow ± 2 psi (14 kPa) tolerance for gauge error.

Operation of the pressure regulator should be checked using the pressure gauge and a hand vacuum pump. Kinks in the fuel line can be located by visual inspection.

2. Injection valve A contaminated fuel system could result in the injector pintle being held off its seat. This allows a steady rather than controlled flow of fuel to the cylinders. To diagnose this condition, install the fuel pressure gauge and build up pressure in the fuel system (either with the analyzer switch or by turning the ignition key on and off several times). If the fuel pressure drops off rapidly and immediately (assuming there are no external leaks), then there are three places for the fuel to go: (1) back through the chassis-mounted pump (because of a leaking check valve), (2) through the pressure regulator and return line (because of a malfunctioning regulator), or (3) through a stuck-open injector.

To isolate the problem, pinch off the rubber return hose and again build up pressure. If the pressure holds, the regulator is malfunctioning. If pressure again drops, leave the clamp on the return hose, build up pressure and immediately clamp off the rubber hose between the chassis-mounted pump and the steel fuel line. If pressure now holds, the check valve in the chassis pump is malfunctioning. If pressure still drops, one injector (or more) is leaking. To determine which injector is leaking, remove the spark plug from each cylinder and examine the firing end for wetness.

A rich rough idle may also be the result of improper information (sensor signals) supplied to the ECU or a malfunction within the ECU. The following sensor signals could cause this condition.

1. MAP sensor The MAP (mean-atmospheric-pressure) sensor reports changes in the intake-manifold vacuum. At idle, engine vacuum is relatively high. If the MAP sensor senses less than the actual vacuum, it

will lengthen the pulse width to compensate, resulting in excessive fuel. To determine if the proper signal is being supplied to the MAP sensor, inspect the hose between the throttle body and ECU for cuts or pinches. If the hose is visually OK, disconnect the hose at both ends and blow through the hose to check for obstructions. If the hose is clear, connect the hose to the ECU and attach a hand vacuum pump to the opposite end.

Evacuate the hose to 20 inches [51 mm] of vacuum and check for leak-down. If a leak is present, plug the hose at the ECU and check again. If a leak is still present, replace the harness. If not, replace the ECU. If the hose is tight and clear, compare the vacuum readings. This is accomplished by teeing into the MAP sensor hose (Fig. 14-8) and observing the vacuum reading. Now reconnect the MAP sensor hose and check actual engine vacuum at the car-harness vacuum port. If the MAP sensor reading is lower than engine vacuum, remove the throttle body and inspect the MAP sensor cavity for leaks.

2. Coolant sensor The coolant sensor increases the pulse width only while the engine coolant is below 180°F [82°C]. With a cold engine the sensor has a low resistance. As coolant temperature increases, so does sensor resistance. Therefore, if the sensor resistance is low (shorted) and the engine is warm, the ECU knows only that the resistance is low and will lengthen the pulse width. This action results in a richer mixture. The coolant sensor may be tested with the fuel-injection analyzer.

3. Air temperature sensor The air temperature sensor provides the ECU with the same type of signal as does the coolant sensor. However, its influence on pulse width is not nearly as great. A shorted air temperature sensor can cause a rich idle. The air temperature sensor can be checked with the fuel-injection analyzer.

4. Electronic control unit Any one of several malfunctions within the ECU could cause a rich idle condition. If an analysis of the EFI system with the analyzer fails to indicate a specific problem, and there are no suspicious components indicated in the sensor tests, a new ECU may be installed on a trial basis. If the new ECU solves the problem, reconnect the original ECU to be sure that the problem was not a poor connection.

An important item to remember on any rich-idle condition is that any driving and testing done under these conditions will usually result in sooty spark plugs. The car must be driven after corrections have been made to eliminate this buildup from the ignition and exhaust systems.

⚙ 14-14 Engine idles rough The most likely cause of a rough idle condition is a low idle speed or improper ignition timing. A 600- to 650-rpm idle speed is essential and will eliminate many EFI problems. Ignition timing should be set prior to checking other EFI components.

Other possible solutions to a rough idle condition are the following.

1. Speed sensor The speed sensor consists of a pair of switches that tell the ECU when to activate the injector group and which injector group should be activated. If either of these switches does not close properly, then the ECU will not pulse any injectors. If one of these switches is sticking, the ECU will pulse the injectors but not for the proper length of time. This will result in a rough idle.

Proper operation of the speed sensor can be observed on the analyzer. The INJECTOR GROUP lights on the analyzer are indications that the injector groups have been activated by the ECU as a result of the speed sensor signals. The TRIGGER POINTS 1 light should blink on at essentially the same time as the INJECTOR GROUP 1 light. This can be more accurately analyzed while the analyzer is operating in the MONITOR position. A sticking speed sensor will be indicated by a SHORT blink or by a blink out of time with the INJECTOR GROUP light. Perform this same evaluation for group 2 lights.

2. EGR system Various controls are used to delay EGR operation to provide smooth idle (⚙ 14-11). Any malfunction in the control system or in the valve itself could result in a rough idle condition. The visual checks and electrical analysis with the analyzer should be performed.

3. MAP sensor Any leak in the MAP sensor hose may be large enough to cause a rough idle. Check the sensor hose as previously described.

4. Fuel pressure regulator A leak in the fuel pressure regulator vacuum hose will result in a higher fuel pressure. It will also lower the vacuum reading to the MAP sensor. The result of these two factors will usually be a rough idle. Again the size of the leak will determine the severity of the condition.

Low fuel pressure will cause a lean idle. This can be checked by installing the pressure gauge and a hand vacuum pump. Compare the results to specifications.

5. Injection valve Plugged or partially plugged injectors can cause a rough idle by causing the mixture in the cylinder to be too lean.

Another condition which can cause a smaller-than-normal injector opening is a high clamping force on the injectors. This could be caused by a bent injector hold-down bracket or fuel rail. It can be determined by a squashed appearance of the grommet between the injector and bracket. If this condition is found, the injector should be replaced only if other factors point to that cylinder as contributing to the condition.

6. Air leaks An air leak at the seal between the injection valve and the intake manifold could cause that cylinder to run lean. One method of isolating a weak cylinder is to plug the idle bypass passage on the throttle body by blocking the opening with something like a credit card. This produces a low idle speed and makes the effect of a weak cylinder more pronounced. The air-conditioning compressor may have to be shut off and the passage blocked slowly to prevent stalling during this check. Disconnect injector wires one at a time, noting speed change, to find the bad cylinder. A short

length of vacuum hose may also be used to listen for leaks at the injectors.

7. Electrical connections Poor electrical connections can easily cause an idle miss (or intermittent miss), resulting in a rough idle. Inspect all electrical connectors, especially the injector connectors and the ground wires at the compressor bracket or fender. Occasionally connections which appear clean and tight may be at fault because of bent terminals in the connector body. The injector connectors and ECU connectors should be carefully inspected for this condition.

8. Electronic control unit Malfunctions within the ECU could cause a rough idle. Analysis of this component should be the same as described previously.

⚙ 14-15 Fuel consumption too high Many of the conditions discussed previously may cause a complaint of excessive fuel consumption. However, the most obvious result of the condition will be a drivability complaint. The degree to which a malfunction exists can determine which symptom is most obvious. Therefore, the solution to a fuel economy complaint can sometimes be found by looking for other abnormal conditions which some owners may "drive around" because of their driving habits or carburetor experience.

Lacking other hints, the following areas should be investigated as a cause of poor fuel economy.

1. Idle speed and timing While both of these items have influence on fuel economy, many times they are overlooked in favor of ECU malfunctions. However, as on carbureted vehicles, timing and idle speed are basic factors in determining proper engine operation.

2. Centrifugal and vacuum spark advance Proper operation of the spark-advance mechanisms is no less critical on cars with EFI. The time necessary to ensure that they are operating correctly will save a great deal of looking for an EFI problem which does not exist.

3. EGR system A sticking EGR valve can be a cause of poor fuel economy. However, this condition will usually be accompanied by a rough idle. Regardless, the security of vacuum hoses, and the proper operation of the EGR controls and valve should be checked on poor fuel economy complaints.

4. Throttle position switch One of several functions of the throttle position switch is to provide an ECU signal that wide-open throttle has been reached. A short circuit in this section of the switch or harness will result in the ECU providing WOT enrichment at all times. Operation of this circuit can be checked with the analyzer.

5. Temperature sensors The cold enrichment provided by the temperature sensors can remain if these components are totally shorted or out of calibration. Analysis of their operation can be made with the analyzer.

6. Electronic control unit The ECU's evaluation of all its sensor inputs determines pulse width. By sending specific information to the ECU and observing what happens as a result, we are able to determine if the ECU is capable of performing normally. This analysis is provided by the analyzer.

With the analyzer hooked up, the digital meter should stabilize at a value which falls within the specification band at zero vacuum for the altitude and the particular ECU being worked on. When the temperature sensor buttons are depressed, an increase in pulse width within the specification band should result.

7. Fuel system Either high fuel pressure or low fuel pressure can cause poor fuel economy. High pressures cause a greater quantity of fuel to be injected for any given injector duration. Low fuel pressures contribute to fuel economy by causing a lean condition, which in turn results in low manifold vacuum and therefore limited operation of the vacuum spark-advance mechanism. Further testing of the fuel system can be performed by the analyzer.

8. Driving habits With today's numerically low drive-axle ratios and low-slip ("tight") torque converters, city fuel economy is probably more closely related to driving habits than previously. If the driver is satisfied with moderate performance, better fuel economy will result. But if the driver insists on the same performance as in previous years, there will be a corresponding decrease in fuel economy.

9. Ignition system As with other ignition system problems, do not overlook the importance of spark plugs and secondary wiring as a source of trouble. Although poor fuel economy may result, secondary ignition problems usually are readily evident.

⚙ 14-16 Using the fuel-injection analyzer Figure 14-10 shows the fuel-injection analyzer used to diagnose troubles in the Cadillac electronic fuel-injection system. The analyzer should be connected as shown in

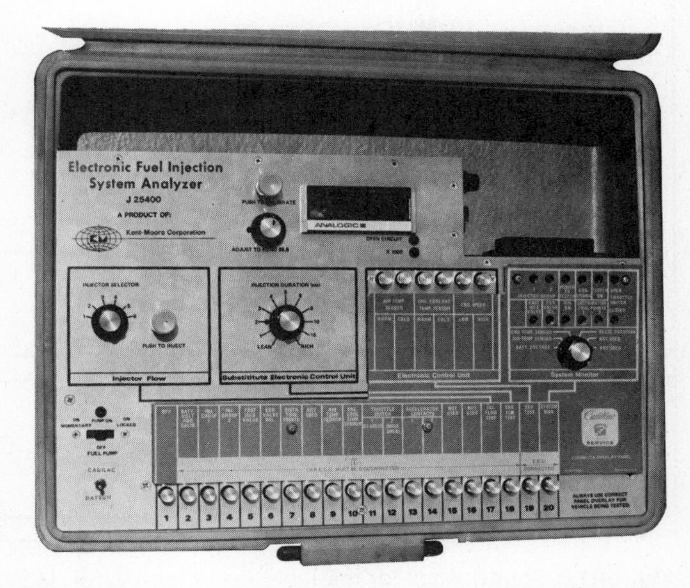

Fig. 14-10 Fuel-injection analyzer used to diagnose troubles in the electronic fuel-injection system. (*ATW*)

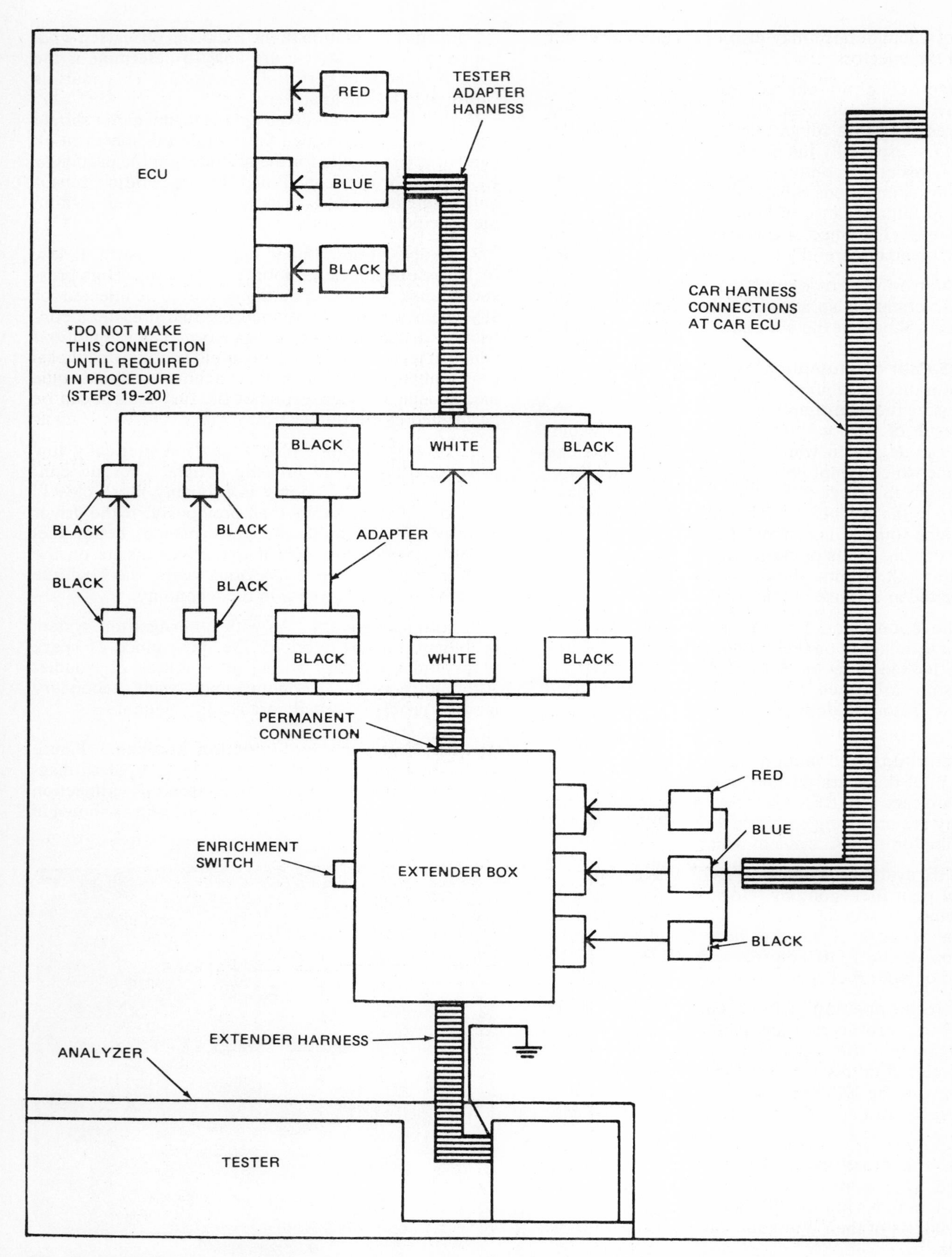

Fig. 14-11 Connections for the fuel-injection analyzer. Male and female references in the drawing are to the plastic connectors and not to the wire terminals. (*Cadillac Motor Car Division of General Motors Corporation*)

Fig. 14-11. This is step 1 in the procedure. All electrical equipment must be turned off during the test. Even leaving a door open so that interior lights are on could interfere with the operation of the analyzer. For each step in the test procedure, a light should come on, or a specified meter reading should be obtained on the analyzer.

The following steps are then required for a complete analysis of the system:

2. Calibrate the analyzer. To do this, depress switch 2, and perform the calibration procedure. There is a separate numbered switch for each step in the test procedure. Each step is described in detail in the EFI diagnosis guide that accompanies the analyzer.
3. Test injectors 1, 2, 7, and 8. Depress analyzer switch 3. Figure 14-12 shows the active electrical circuits in heavy line.
4. Test injectors 3, 4, 5, and 6.
5. Test the fast-idle valve. Depress switch 5.
6. Test the EGR valve solenoid. Depress switch 6.
7. Test distributor trigger points. Depress switch 7.
8. There is no step 8 in the testing procedure used by Cadillac. Therefore, switch 8 on the analyzer is not used.

There are 10 more tests on Cadillacs, using switches 9 to 14 and 16 to 20 on the analyzer. These check:

9. Air temperature sensor
10. Coolant temperature sensor
11. Closed throttle switch
12. Wide-open-throttle switch
13. Accelerator contacts 1
14. Accelerator contacts 2

Steps 15 and 16 are not used on Cadillac systems.

17. Fuel system
18. Substitute electronic control unit. This step inserts a skeleton ECU into the system. It provides manual control of injection pulse widths, which allows the car to be driven for short distances in case the problem is in the ECU or the car. The test is made with switch 18 depressed.
19. ECU functional test. This test is made with switch 19 depressed. It checks the ECU and the ECU harness.

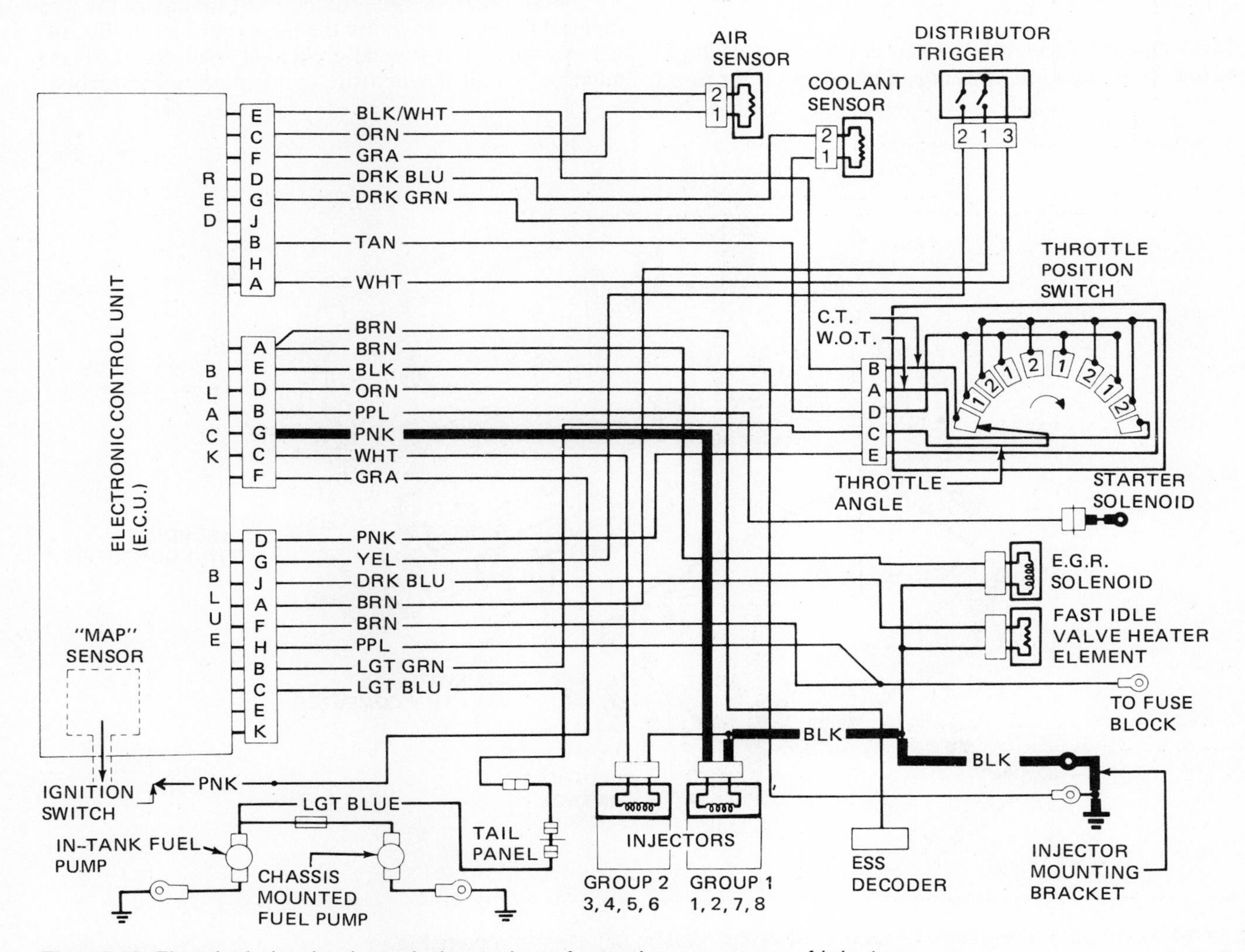

Fig. 14-12 Electrical circuits through the analyzer for testing one group of injectors. (*Cadillac Motor Car Division of General Motors Corporation*)

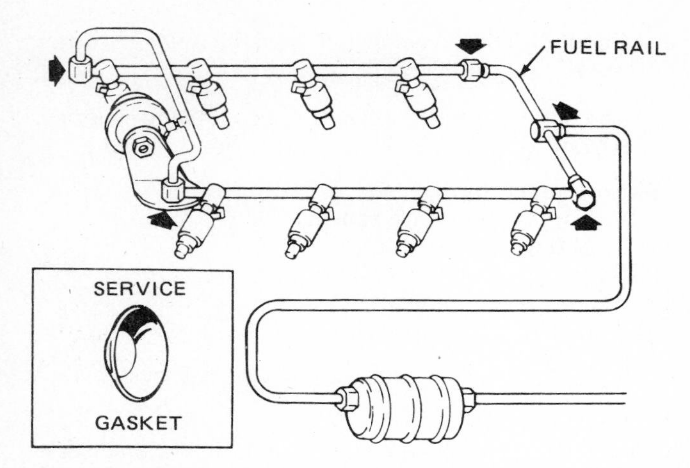

Fig. 14-13 A metal gasket must be installed in each fuel-rail connection any time the fitting is opened and reconnected. (*Cadillac Motor Car Division of General Motors Corporation*)

20. System monitor. This monitors the complete system while the car is being driven. Its purpose is to give a final check to the system after all components have been checked as described above.

14-17 On-car service Various components of the electronic fuel-injection system can be adjusted, or removed and replaced. These are listed below. Consult the manufacturer's service manual for the specific procedures.

1. Fuel-line gaskets (Fig. 14-13)
2. Idle-speed adjustment
3. Throttle position switch adjustment
4. Electronic control unit (Fig. 14-14)
5. Throttle body assembly
6. Throttle position switch
7. Fast-idle valve
8. Fuel rail
9. Injector valve
10. Fuel pressure regulator
11. Fuel filter
12. Fuel filter mounting
13. Fuel pump
14. Temperature sensors
15. Oxygen sensors
16. Hoses

Here are several important points about servicing components and making various adjustments to the Cadillac electronic fuel-injection system.

1. **Fuel line gaskets** Each time that any of the five fuel rail fittings (shown by the heavy arrows in Fig. 14-13) are opened, a special conical-shaped metal gasket must be installed when the connection is reattached.

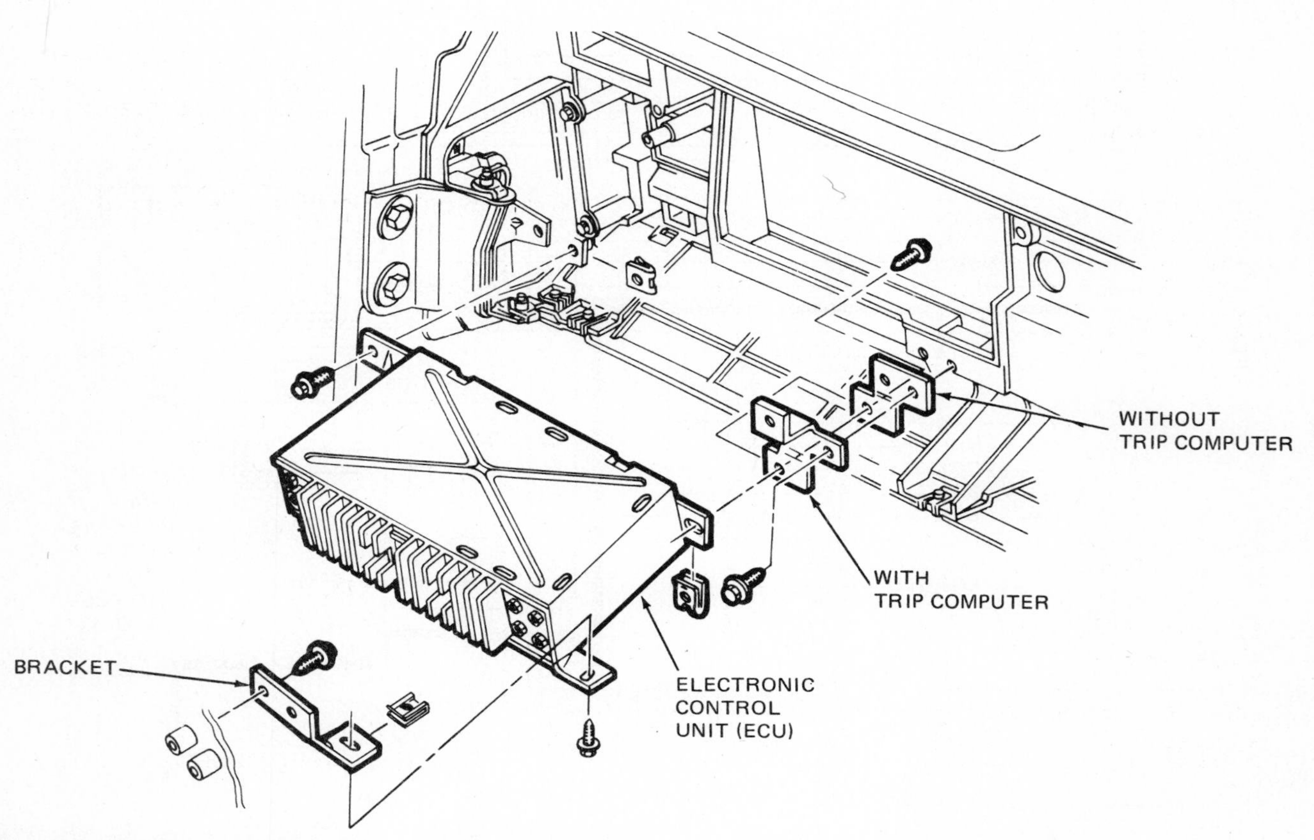

Fig. 14-14 Never open the electronic control unit. It cannot be adjusted or repaired in the field. (*Cadillac Motor Car Division of General Motors Corporation*)

The gasket is not installed during new car assembly. However, to minimize the chances of fuel leaks during service, a new gasket must be installed each time a fitting is separated and reconnected in the field. The gasket cannot be reused.

2. Electronic control unit Never open the ECU (Fig. 14-14) if it is to be used again, or if it must be returned to the parts department in lieu of a core charge. The ECU is a preprogrammed, complex, expensive analog computer. There are no adjustments or repairs that can be made to it in the field.

Both of the fuel pumps are protected by a 10-ampere fuse located in the harness near the ECU. If the fuse blows, replace it only with an AGG-10 or a 3AC-10 fuse of the same physical size. Never install a "slow-blow" or larger-amperage fuse. This could result in damage to the ECU.

When charging the battery on a car equipped with EFI, always make sure that the ignition switch remains off. If the ignition switch is turned to ON, a rapid false triggering of the injectors may occur which will result in a flooded engine. In addition, electrical damage may occur to the ECU.

3. Fuel filter When an engine fails to receive fuel, the problem may be a clogged fuel filter. Check that the filter has been serviced regularly.

4. In-tank fuel pump A filter or strainer is located on the inlet end of the fuel pickup tube in the fuel tank. Sometimes the fuel pump appears to lack capacity and is replaced, when only the filter needs replacement. The filter is a press fit on the end of the fuel-pump assembly, and is available from the dealer's parts department. After installing a new filter, be sure to align it so that it cannot contact the float.

5. Hoses Fuel hoses in the EFI system are made of special materials and cut to certain lengths. When replacing a fuel hose, use only a recommended type of replacement hose. Be sure that it is cut to the proper length. Hose material is designed to withstand the effects of gasoline and high temperatures. Correct hose length allows the new hose to be installed with the same routing so that it can be securely retained.

14-18 Servicing digital electronic fuel injection. In this system (Fig. 14-1*c*), the electronic control module (ECM) performs certain diagnostic and backup system functions. The ECM detects certain system and component problems and identifies them through a coded digital readout. Diagnostic codes appear on the digital display panel normally used as part of the electronic climate control (ECC) system (Fig. 14-15).

When the ECM detects an improper sensor signal, indicating a failure, it takes over the job of the defective sensor. Substitute values from its stored memory replace the missing information from the sensor. If the ECM itself has failed, an analog backup circuit takes over. This allows the car to be driven, but with severely reduced performance. The ability to determine the cause its own trouble makes DEFI a self-diagnosing system.

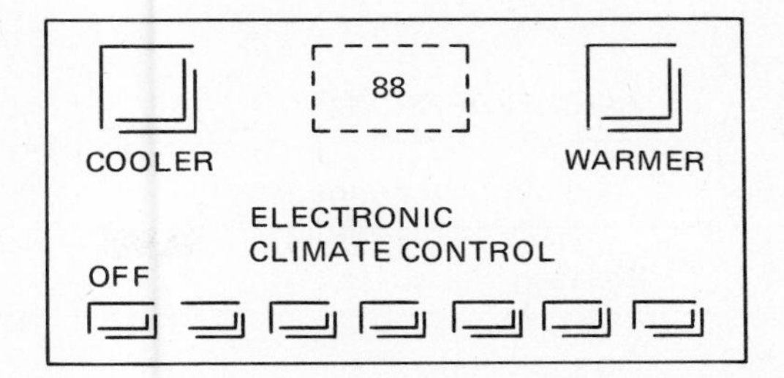

Fig. 14-15 Trouble codes display on the digital display panel on the dash. (*Cadillac Motor Car Division of General Motors Corporation*)

An amber dash-mounted CHECK ENGINE light informs the driver that the ECM has detected a system malfunction or abnormality. These conditions may be related to the various sensors or to the ECM itself. If the fault clears up, the light resets automatically. However, the ECM stores the trouble code associated with the failure until the diagnostic system is cleared.

To read out the stored failure codes, depress the OFF and WARMER buttons simultaneously on the ECC panel until ". . ." appears in the display section. Then the code "88" will appear (Fig. 14-15), which indicates the beginning of the diagnostic readout. Trouble codes stored in the ECM as a result of troubles that have occurred will now display beginning with the lowest numbered code.

In addition, the ECM also has been programmed to analyze the function of several switches on the car. Figures 14-16 and 14-17 show the trouble codes displayed on the DEFI system used by Cadillac. Figure 14-16 shows the codes that indicate trouble with one of the switches, Fig. 14-17 the codes for trouble with the sensors and ECM. Diagnosis charts in the manufacturer's service manual provide the steps to pinpoint and repair the defects.

TROUBLE CODE	SWITCH
71	Brake light switch
72	ISC throttle switch
73	Drive (ADL) switch
74	Backup lamp switch
78	MPG panel "AVG" switch
79	MPG panel "RESET" switch

Fig. 14-16 Trouble codes indicating trouble with various switches on the car. (*Cadillac Motor Car Division of General Motors Corporation*)

TROUBLE CODE	CIRCUIT AFFECTED
00	All diagnostics complete
12	No tach signal
14	Shorted coolant sensor circuit
15	Open coolant sensor circuit
21	Shorted throttle position sensor circuit
22	Open throttle position sensor circuit
28	Shorted idle speed control circuit
29	Idle speed control circuit
30	Idle speed control circuit
31	Shorted MAP sensor circuit
32	Open MAP sensor circuit
33	MAP/BARO sensor correlation
34	MAP hose
35	Shorted BARO sensor circuit
36	Open BARO sensor circuit
37	Shorted MAT sensor circuit
38	Open MAT sensor circuit
55	ECM
56	

Fig. 14-17 Trouble code chart, showing the circuits affected. Consult the pages in the manufacturer's service manual for further diagnosis and repair. (*Cadillac Motor Car Division of General Motors Corporation*)

Chapter 14 review questions

Select the *one* correct, best, or most probable answer to each question. Then check your answers against the correct answers given at the end of the book.

1. Fuel injection for gasoline engines sprays the fuel:
   a. into the combustion chamber
   b. into the precombustion chamber
   c. into the air cleaner
   d. into the intake manifold.
2. In the Digital Electronic Fuel-Injection system, the injectors are mounted:
   a. in the valve port
   b. in the intake manifold
   c. above the throttle valves
   d. below the throttle valves.
3. In the Bosch K-type continuous-fuel-injection system, air leaks into the intake manifold cause:
   a. a rich mixture
   b. a lean mixture
   c. no change in the mixture
   d. the airflow sensor to compensate for the leaks.
4. The Cadillac electronic fuel-injection system is most similar to:
   a. the Bosch type D
   b. the Bosch type L
   c. the Bosch type K
   d. Digital Electronic Fuel Injection.
5. When the electronic control unit has failed, the normal procedure is to:
   a. the Bosch D-type
   b. the Bosch L-type
   c. the Bosch K-type
   d. send the unit to a rebuilder.
6. The opening of the solenoid-operated injection valve can be checked with:
   a. a pressure gauge
   b. a dynamometer
   c. a voltmeter
   d. an oscilloscope.
7. The first step in checking a Cadillac electronic fuel-injection system is to:
   a. connect the fuel-injection analyzer
   b. replace the electronic control unit
   c. make a visual inspection
   d. none of the above.
8. Excessive fuel consumption can be caused by:
   a. high fuel pressure
   b. low fuel pressure
   c. both *a* and *b*
   d. neither *a* nor *b*.
9. During cranking, normal fuel pressure in a Cadillac electronic fuel-injection system should be:
   a. 0 to 10 psi [0 to 69 kPa]
   b. at least 1 psi [6.9 kPa]
   c. between 37 and 42 psi [255 to 290 kPa]
   d. more than 60 psi [414 kPa].
10. A car with electronic fuel injection cannot be started unless the accelerator pedal is pumped several times. This most likely indicates a problem with the:
    a. in-tank fuel pump
    b. chassis-mounted fuel pump
    c. electronic control unit
    d. coolant sensor.

CHAPTER 15

SERVICING DIESEL FUEL-INJECTION SYSTEMS

After studying this chapter, and with proper instruction and equipment, you should be able to:

1. Diagnose troubles in the diesel engine fuel-injection system.
2. Correct leaks in the fuel lines and connections.
3. Service the fuel filter.
4. Locate a misfiring cylinder.
5. Diagnose troubles in the fuel-injection nozzle.
6. Remove, install, and time the fuel-injection pump.
7. Adjust the linkage and idle speed on a diesel engine.
8. Check the fuel pressure at the injection pump.
9. Make a compression test on a diesel engine.

15-1 Servicing diesel fuel-injection systems A variety of diesel fuel-injection systems have been made. In this book, we are interested primarily in the systems found on diesel engines used in automobiles. The basic types are described in Chap. 7. These include the cam-operated in-line plunger pump system and the rotary distributor system. Both are high-pressure systems in which the fuel leaves the pump at high pressure and flows through lines to the fuel nozzles in the cylinders.

Following sections cover troubleshooting the fuel-injection system and servicing fuel lines, filters, air cleaners, and injector nozzles. Servicing the fuel-injection pump is a specialty that requires special tools and training. For this reason, only its removal and reinstallation is discussed. Usually when a fuel-injection pump goes bad, it is replaced with a new or rebuilt pump.

⚙ 15-2 Troubleshooting the fuel system Many of the troubles in a diesel engine can be traced to the fuel system. Troubles such as failure to start during normal cranking, poor idling, lack of power, misfiring, noise, and excessive black smoke can be caused by conditions in the fuel system. The chart that follows lists various engine troubles that could be due to fuel system problems. Conditions such as slow or no cranking or overheating are not listed because these are not caused by troubles in the fuel system.

The troubles and possible causes in the chart are not listed in their order of frequency of occurrence. Item 1 does not necessarily occur more often than item 2. Sections following the chart explain in detail the complaints, possible causes, and corrections.

⚙ 15-3 Engine cranks normally but will not start If the fuel is dirty, it may have clogged the system. This

Diesel Fuel-Injection System Trouble Diagnosis Chart

(See ⚙ 15-3 to 15-10 for detailed explanations of the trouble causes and corrections listed below)

COMPLAINT	POSSIBLE CAUSE	CHECK OR CORRECTION
1. Engine cranks normally but will not start (⚙ 15-3)	a. Incorrect or dirty fuel	Flush system—use correct fuel.
	b. No fuel to nozzles or injection pump	Check for fuel to nozzles.
	c. Plugged fuel return	Check return, clean.
	d. Pump timing off	Retime.
	e. Inoperative glow plugs, incorrect starting procedure, or internal engine problems	
2. Engine starts but stalls on idle (⚙ 15-4)	a. Fuel low in tank	Fill tank.
	b. Incorrect or dirty fuel	Flush system—use correct fuel.

Diesel Fuel-Injection System Trouble Diagnosis Chart

(See ❁ 15-3 to 15-10 for detailed explanations of the trouble causes and corrections listed below)

	c. Limited fuel to nozzles or	Check for fuel to nozzles and to pump.
	injection pump	Check return, clean.
	d. Restricted fuel return	Reset idle.
	e. Idle incorrectly set	Retime.
	f. Pump timing off	Install new pump.
	g. Injection pump trouble	
	h. Internal engine problems	
3. Rough idle, no abnormal noise or smoke (❁ 15-5)	a. Low idle incorrect	Adjust.
	b. Injection line leaks	Fix leaks.
	c. Restricted fuel return	Clear.
	d. Nozzle trouble	Check, repair or replace.
	e. Fuel supply pump problem	Check, replace if necessary.
	f. Uneven fuel distribution to nozzles	Selectively replace nozzles until condition clears up.
	g. Incorrect or dirty fuel	Flush system—use correct fuel.
4. Rough idle with abnormal noise and smoke (❁ 15-6)	a. Injection pump timing off	Retime.
	b. Nozzle trouble	Check in sequence to find defective nozzle.
5. Idle okay but misfires as throttle opens (❁ 15-7)	a. Plugged fuel filter	Replace filter.
	b. Injection pump timing off	Retime.
	c. Incorrect or dirty fuel	Flush system—use correct fuel.
6. Loss of power (❁ 15-8)	a. Incorrect or dirty fuel	Flush system—use correct fuel.
	b. Restricted fuel return	Clear.
	c. Plugged fuel tank vent	Clean.
	d. Restricted fuel supply	Check fuel lines, fuel-supply pump, injection pump.
	e. Plugged fuel filter	
	f. Plugged nozzles	Replace filter.
	g. Internal engine problems, loss of compression, compression leaks.	Selectively test nozzles, replace as necessary.
7. Noise—"rap" from one or more cylinders (❁ 15-9)	a. Air in fuel system	Check for cause and correct.
	b. Gasoline in fuel system	Replace fuel.
	c. Air in high-pressure line	
	d. Nozzle sticking open or with low opening pressure	Replace defective nozzle.
	e. Engine problems	
8. Combustion noise with excessive black smoke (❁ 15-10)	a. Timing off	Reset.
	b. Injection pump trouble	Replace pump.
	c. Internal engine problems	

requires cleaning of the pumps, fuel lines, and nozzles. The wrong fuel can also cause failure to start. Use the recommended fuel.

NOTE: Make sure the injection-pump timing mark aligns with the mark on the adaptor. This is to check that the timing is correct.

To check fuel nozzles for fuel flow, loosen the injection line at a nozzle. Do not disconnect it. Wipe the connection dry. Crank for 5 seconds. Fuel should flow from the loose connection. If it does not, check the fuel solenoid and the fuel supply line to the injection pump. Disconnect the line at the fuel inlet at the injection pump. Connect a hose from this line to a metal container. Crank the engine. If no fuel flows, the trouble is in the fuel supply line or the fuel supply pump. If fuel flows, the trouble is in the injection pump fuel filter or the pump itself. Replace the filter first. If this does not cure the problem, replace the injection pump.

To check for a plugged fuel return line, disconnect the line at the injection pump and connect this line to a metal container. Connect a hose from the injection pump connection to the metal container. Crank the engine. If it starts and runs, the trouble is in the fuel return line.

❁ 15-4 Engine starts but stalls on idle This could be caused by any of the troubles discussed in ❁ 15-3. These include incorrect or dirty fuel, limited fuel to nozzles or injection pump, a restricted fuel return, incorrect pump timing, or defects in the injection pump. Also, stalling on idle after starting could be caused by an incorrectly set idle, or by low fuel in the tank.

❁ 15-5 Rough idle, no abnormal noise or smoke First check if the low idle speed is correctly set. Then look for injection-line leaks. Wipe off the lines and run the engine. If there are leaks, tighten connections or replace lines as necessary to eliminate them. Next, check for a restricted fuel return system (❁ 15-3). Disconnect the fuel return line to see if the engine runs normally.

To check for a defective nozzle, start the engine and then loosen the injection line fitting at each nozzle in turn. This relieves the pressure and prevents normal injection nozzle action. Be careful to avoid spraying of fuel onto hot engine parts. When a nozzle that is good is prevented from operating in this way, the rhythm of the engine will change. The engine will run more roughly. If you find a nozzle that does not change the idle when partly disconnected, then that nozzle was not working and should be replaced.

To check for internal fuel leaks at the fuel nozzle, disconnect the fuel return system from the nozzles on one bank at a time. With the engine running, note the fuel seepage at the nozzles. Replace any nozzle with excessive fuel leakage.

A rough idle can also be caused by a fuel supply problem. The fuel supply pump, line, and fuel filter should be checked. In addition, rough idle can also be caused by dirty fuel or the wrong fuel.

❁ 15-6 Rough idle with abnormal noise and smoke First check the injection pump timing. Then disable the nozzles one at a time, to check their operation. Do this by loosening the injection line connection at the nozzle (❁ 15-5). When you find a nozzle that does not change the noise or smoke when it is disabled, that is the bad nozzle. It should be replaced.

❁ 15-7 Idles OK but misfires as throttle opens This can be caused by a plugged fuel filter, incorrect injection pump timing, or incorrect or dirty fuel. Check and correct these conditions as explained earlier.

❁ 15-8 Loss of power This is a general complaint that could result from many conditions in the engine systems or engine as well as outside the engine. For example, dragging brakes, excessive resistance in the power train, or underinflated tires can produce an impression of low power. In the fuel system, loss of power could result from incorrect or dirty fuel, restricted fuel return, a restricted fuel supply, or plugged nozzles. Previous sections describe how to check for these conditions. Another possible cause is a plugged fuel-tank vent. This would prevent normal fuel flow to the injection pump and engine so that the engine could not produce full power. Also, if the engine overheats, it will lose power.

❁ 15-9 Noise (rap) from one or more cylinders One possible cause of this condition could be air in the fuel system. Air could cause a very uneven flow of fuel to the nozzles. The air expands and contracts with changing pressure, causing too much or too little fuel to feed. To correct the problem, loosen the injection line at each nozzle to allow the air to bleed from the system. Another cause of the noise or rap could be a nozzle sticking open or having a very low opening pressure. Loosening the injection lines at each nozzle in turn will locate a defective nozzle.

❁ 15-10 Combustion noise with excessive black smoke This could be caused by incorrect injection pump timing, or injection pump troubles. The cylinders receive too much fuel, or fuel at the wrong time. It could also be caused by internal engine problems.

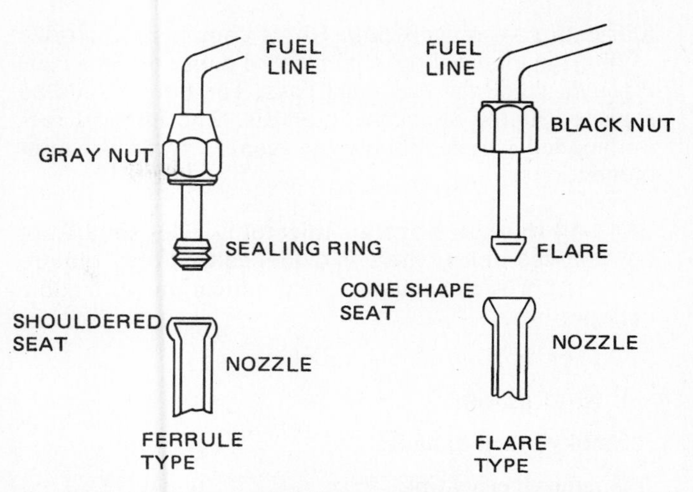

Fig. 15-1 Two types of fittings used to connect the fuel line to the injection nozzle. (*Oldsmobile Division of General Motors Corporation*)

❁ 15-11 Diesel fuel system service procedures After a trouble is located, it must be corrected. We now look at specific servicing operations on the diesel engine fuel-injection system. These include service to the fuel-injection lines, fuel filter, injection nozzles, and the injection and fuel-supply pumps.

❁ 15-12 Fuel lines and fittings The fuel lines are special. They must withstand pressures of several thousand pounds per square inch and must be noncorrosive. If a hydraulic line requires replacement, it should be replaced with the service part recommended by the engine manufacturer.

Figure 9-14 shows various types of fittings, or couplings. Figure 15-1 shows two types of fittings used to connect fuel lines to the injection nozzle. When disconnecting a fuel line, use two wrenches. Hold the fitting with one wrench and turn the coupling nut. If you try to loosen the nut without using the second wrench, you can twist the line and damage it. It is not necessary to use the second wrench when disconnecting the lines from the fuel-injection pump.

Whenever lines are disconnected, the lines, nozzles, and pump fittings must be capped. This prevents dirt from entering the fuel system. Cleanliness is very important when working on the fuel system. A dirt particle almost invisible to your eye can clog an injector.

New hydraulic lines are preformed. Be careful when installing them to avoid twisting or bending them out of shape. If the line that is to be replaced is under other lines, you may have to remove these upper lines.

❁ 15-13 Fuel filter Fuel for the diesel engine must be clean and free of contaminants. Almost invisible specks of dirt can clog the fuel nozzles or damage the injector-pump parts. Therefore, the fuel filter is an essential part of the system. Figure 15-2 shows its location on one engine. It works like the filter used in engine lubricating systems (Chap. 16). It contains a cartridge of filtering material (special pleated paper or fiber mat) through which the fuel must pass. The filter should be replaced at the specified intervals. The interval recommended varies. Follow the manufacturer's recommendations.

15-14 Injector nozzle Injector nozzles should not be removed unless there is evidence that they require servicing or replacement. Usual indications of trouble include:

One or more cylinders knocking

Loss of power

Smoky black exhaust

Engine overheating

Excessive fuel consumption

One way to check for a faulty fuel injector is to run the engine at fast idle. Then loosen the connector at each nozzle in turn, one at a time. Wrap a cloth around the connection before you loosen it to keep fuel from spurting out. If loosening a connector causes the engine speed to drop off, the injector is probably working. If the engine speed remains the same, then the injector is probably not performing properly. It could be clogged so that no fuel flows through. Or the holes could be partly clogged so that the spray is inadequate or does not have the required pattern (Fig. 15-3).

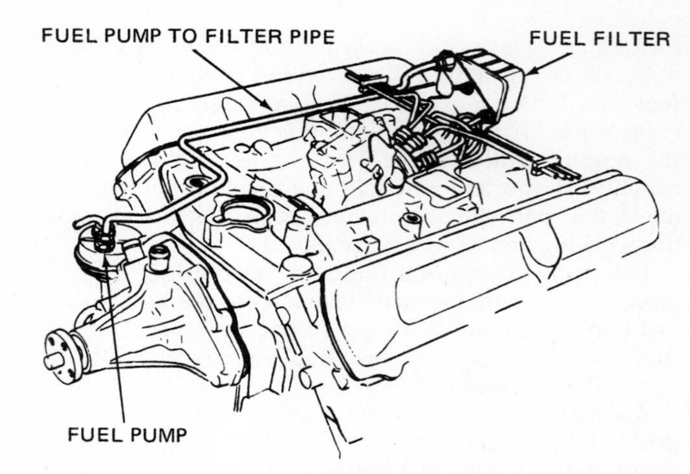

Fig. 15-2 Location of the fuel filter on a V-8 diesel engine. (*Chevrolet Motor Division of General Motors Corporation*)

The fuel nozzle troubleshooting chart below lists possible nozzle troubles and their causes. They are not listed in the order of their frequency of occurrence. Item 1 does not necessarily occur more often than item 2.

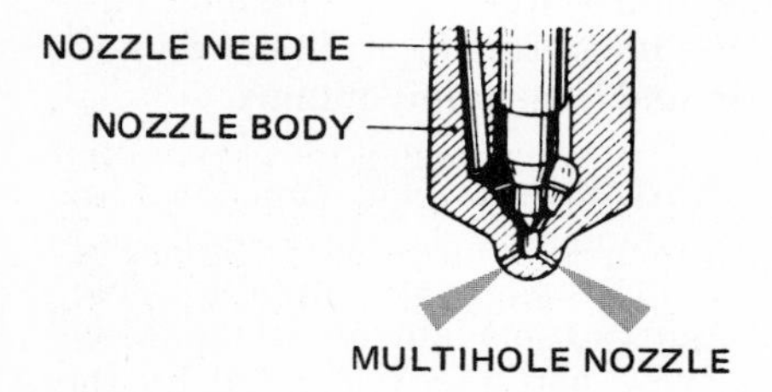

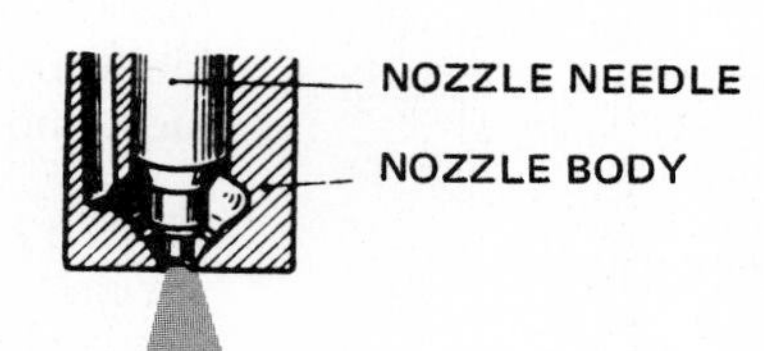

Fig. 15-3 Spray patterns for a multihole nozzle and from a pintle nozzle. (*Robert Bosch Corporation*)

Fuel-Injection Nozzle Trouble Diagnosis Chart

CONDITION	POSSIBLE CAUSE	CHECK OR CORRECTION
1. Spray pattern distorted	a. Excessive carbon on tip of needle valve	Clean nozzle.
	b. Injection holes in nozzle body partially blocked	Clean nozzle.
2. Nozzle drip	a. Nozzle leak because of carbon deposit	Clean nozzle.
	b. Sticking needle valve	Clean nozzle and retorque; replace nozzle body and valve.
3. Nozzle pressure too low	a. Nozzle spring broken	Replace spring and readjust pressure.
	b. Nozzle valve sticking in nozzle body	Clean nozzle.
	c. Pressure adjustment incorrect	Adjust pressure.
4. Nozzle opening pressure too high or too low	a. Adjusting screw out of adjustment	Adjust pressure.
	b. Needle valve seized or corroded	Replace nozzle and valve.
	c. Needle valve dirty or sticky	Clean nozzle.
	d. Nozzle openings clogged with dirt or carbon	Clean nozzle.
5. Nozzle bluing	Faulty installation, tightening, or cooling of nozzle	Replace nozzle valve and body; check cooling system.
6. Excessive fuel leakoff	a. Too much clearance between needle valve and body	Replace.
	b. Foreign matter between faces of nozzle and nozzle holder.	Clean.

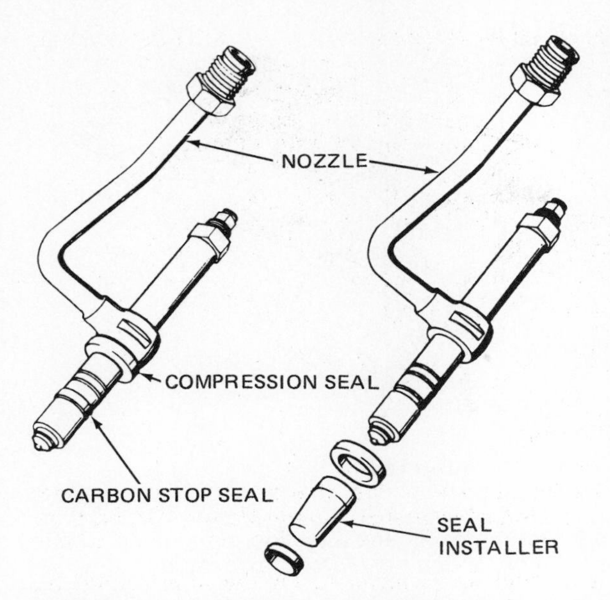

Fig. 15-4 Injector nozzles with seals. (*Chevrolet Motor Division of General Motors Corporation*)

A variety of nozzles are used, but all have a check valve. The check valve opens when pressure is applied so the fuel can flow through. When the pressure drops, the check valve closes to shut off the flow rapidly and completely. To remove an injection nozzle, first remove the fuel return line clamps and return line. Then remove the nozzle hold-down clamp and spacer or other connector arrangement. Remove the nozzle. Cap the nozzle inlet line and the tip of the nozzle. Figure 15-4 shows one installation arrangement which includes a compression seal and a carbon stop seal.

Some manufacturers recommend a spray test of the detached nozzle. This requires a special hydraulic pump which has a pressure gauge (Fig. 15-5). Attach the nozzle to the pump and work the pump. The fuel should spray when the pump pressure reaches the specified value. When the pressure is released the spray should stop abruptly and the nozzle should not drip.

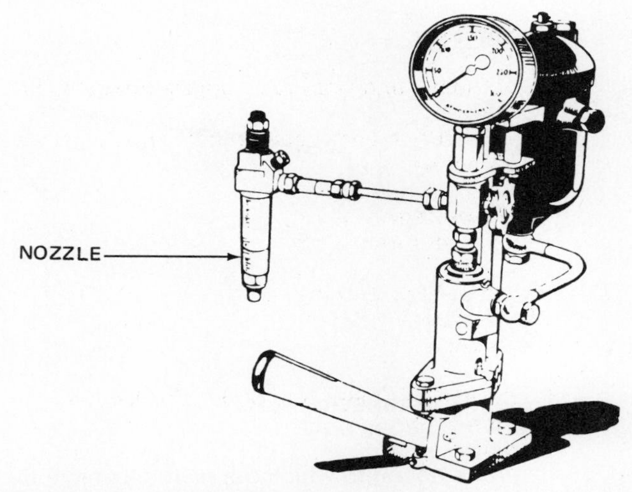

Fig. 15-5 Tester for the fuel-injection nozzle. (*Perkins Engines, Inc.*)

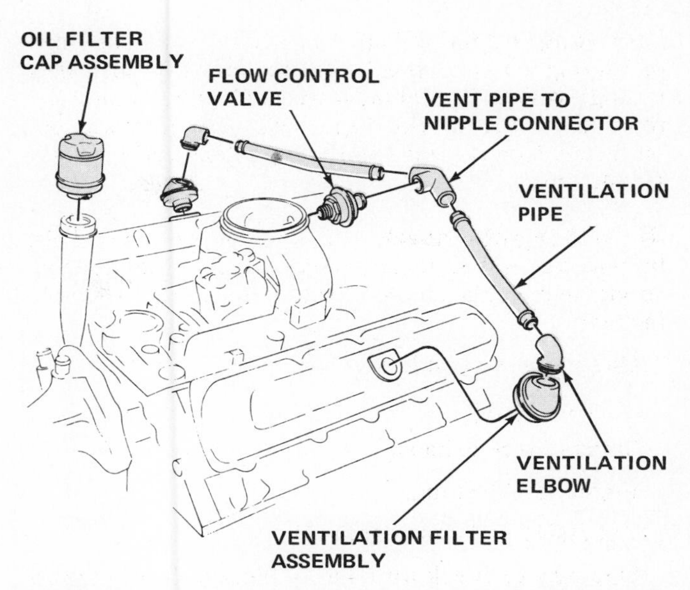

Fig. 15-6 Crankcase ventilating system on an automotive diesel engine. (*Chevrolet Motor Division of General Motors Corporation*)

CAUTION: Direct the spray from the nozzle into a suitable container. Do not allow the spray to hit your skin. The pressure is high enough to force fuel oil through the skin. This can cause serious trouble. You can be seriously injured because the oil could get into the blood stream and produce an infection.

If the nozzle does not work properly, it can be disassembled and cleaned. Some manufacturers recommend replacing a malfunctioning nozzle. If you do disassemble a nozzle, work carefully to avoid damaging the tip or enlarging the holes.

⚙ 15-15 Injection pump service We now cover the replacement of the injection pump used on the General Motors V-8 diesel engine (Fig. 7-8). This is one of several types of injection pump. The procedure that follows is not typical for all pumps. There will be variations in the procedure from one model of diesel engine to another. Follow the special instructions of the engine manufacturer.

The servicing of the fuel-injection pump requires special training and tools. For this reason, General Motors service manuals do not carry disassembly or assembly information on the pump.

To remove the pump, first remove the following:

1. Air cleaner, filters, and pipes from valve covers and air crossover (Fig. 15-6). Cap the intake manifold with special screened covers.
2. Disconnect the throttle rod and return spring (Fig. 15-7).
3. Remove the bell crank and throttle cable (from intake manifold brackets), and push the cable away from the engine.
4. Remove the lines to the fuel filter and then remove the filter and bracket (Fig. 15-2). Disconnect the fuel line at the fuel pump. Disconnect the fuel-return line from the injection pump.

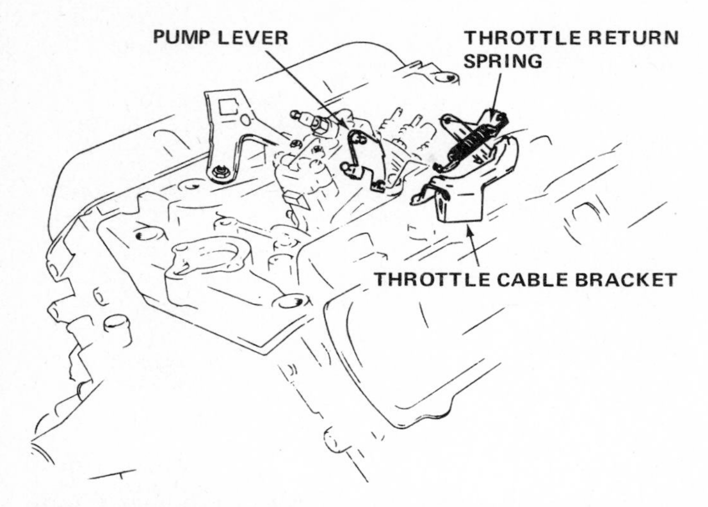

Fig. 15-7 Location of the throttle-return spring. (*Chevrolet Motor Division of General Motors Corporation*)

5. Use two wrenches and disconnect the hydraulic lines from the fuel nozzles. Remove the three nuts attaching the injection pump. This takes a special tool. Then remove the pump and cap all open lines and nozzles.

To install the pump, remove the protective caps. Line up the offset tang on the pump drive shaft with the pump-driven gear and install the pump (Fig. 15-8). Attach it with three nuts and lock washers, lightly run down on the studs. Connect the fuel lines to the nozzles, using two wrenches.

Align the mark on the injector pump (Fig. 15-9) with the line on the adapter and tighten the attaching nuts. Use a ¾-inch-end wrench on the boss at the front of the injection pump to aid in rotating the pump to align the marks.

Install the fuel line from the fuel pump to the fuel filter. Adjust the throttle rod. Install the bell crank and the throttle return spring.

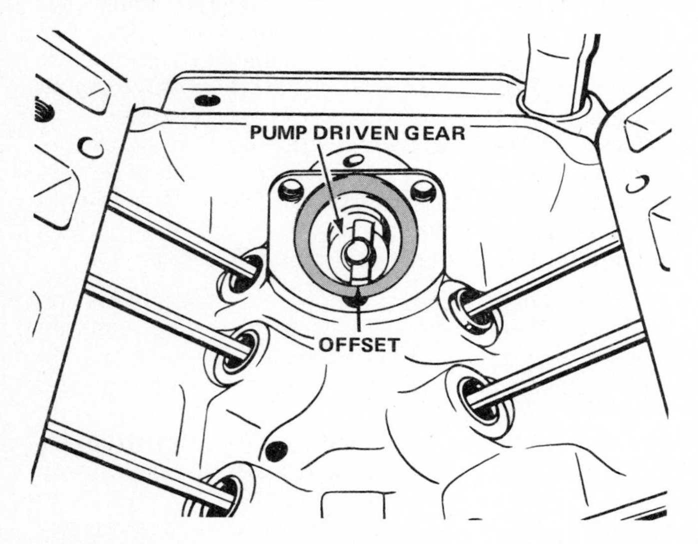

Fig. 15-8 Location of the offset tang on the pump-driven gear. (*Chevrolet Motor Division of General Motors Corporation*)

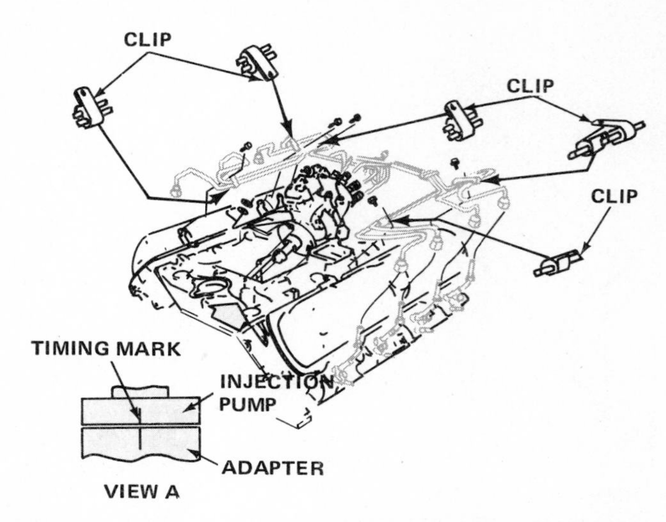

Fig. 15-9 Timing marks and injection pump lines. (*Oldsmobile Division of General Motors Corporation*)

Start the engine and check for fuel leaks. Tighten connectors as necessary. Then install the air crossover (Fig. 15-6) and air cleaner.

15-16 Timing the injection pump The marks on the top of the injection pump adapter and the flange of the injection pump must align (Fig. 15-9). This corrects the timing of the fuel delivery to the nozzles. The adjustment is made with the engine not running. To make the adjustment, loosen the three pump retaining nuts and align the marks. Use a ¾-inch-end wrench on the boss at the front of the pump to help when you rotate the pump into alignment.

15-17 Linkage adjustments Injection pump timing must be correct before the linkage is adjusted. The following may require adjustment (details are in the manufacturer's shop manual):

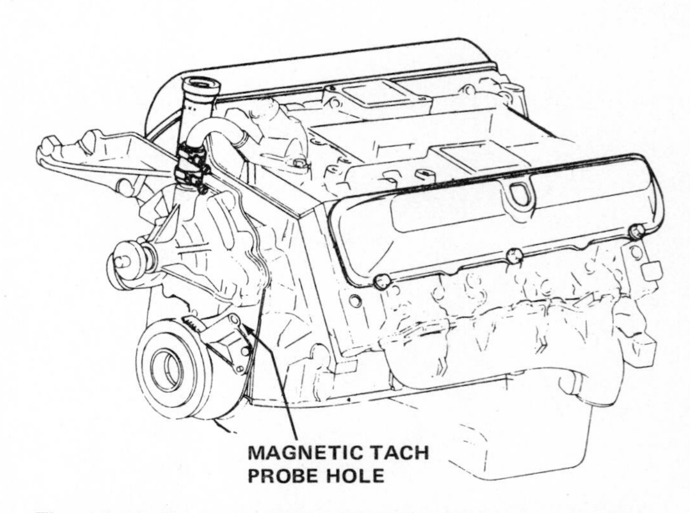

Fig. 15-10 Probe hole into which the magnetic tachometer probe is inserted to check engine rpm. (*Oldsmobile Division of General Motors Corporation*)

1. Throttle rod
2. Transmission throttle valve (TV) or detent cable
3. Transmission vacuum valve.
4. Slow idle speed. This requires a magnetic tachometer inserted in the tach hole (Fig. 15-10). The slow-idle adjustment screw is shown in Fig. 12-10.
5. Fast-idle solenoid. The fast-idle speed is controlled by the plunger in the fast-idle solenoid. It should be checked with the air conditioner on and the compressor wires disconnected.
6. Cruise-control servo relay rod.

Diesel engines, having no electric ignition system, require a different type of tachometer than a gasoline engine. Some tachometers for diesel engines are triggered by a magnet on the crankshaft. Each time that the crankshaft rotates, the magnet swings by a magnetic probe which senses its passing (Fig. 15-10). In this way, the tachometer measures engine speed in revolutions per minute of the crankshaft.

15-18 Checking injection-pump-housing fuel pressure Remove the air cleaner and crossover. Cover the manifold openings with screened covers. From the pump (Fig. 12-10) remove the pressure tap plug. Attach low-pressure gauge by screwing adapter into the tap-plug hole. Install the magnetic pickup tachometer (Fig. 15-10). Check pressure at 1000 rpm (transmission in PARK.) It should be 8 to 12 psi [55 to 83 kPa] with not more than 2 psi [14 kPa] fluctuation. If incorrect, remove the pump for repair.

Reinstall the air crossover after removing the screened covers. Install the air cleaner.

15-19 Diesel engine compression test As on a gasoline engine, sometimes it is necessary to make a compression test. However, the compression test is different for diesel engines. The procedure for one engine is as follows. First, remove the air cleaner and install a manifold cover to keep dirt out of the engine. Disconnect the wire from the fuel shutoff solenoid terminal of the injection pump (Fig. 7-10). This prevents delivery of fuel during the test. Disconnect the glow-plug wires and remove the glow plugs. Screw the compression-tester fitting into the glow-plug hole of the cylinder to be checked. Then crank the engine for at least 12 crankshaft revolutions (six ''puffs'').

Check all cylinders the same way. The lowest compression reading should be not less than 70 percent of the highest. No cylinder should read less than 275 psi [1892 kPa].

In addition to the pressure reached, note the following. If everything is normal, the compression builds up quickly and evenly. If there is leakage past the piston rings, the compression is low on the first strokes but will tend to build up toward normal with later strokes. However, it does not reach normal and the pressure is rapidly lost after cranking stops.

Chapter 15 review questions

Select the *one* correct, best, or most probable answer to each question. Then check your answers against the correct answers given at the end of the book.

1. When a diesel engine cranks normally but will not start, the cause may be:
   a. water in the fuel system
   b. ice in the fuel system
   c. both *a* and *b*
   d. neither *a* nor *b*.
2. A diesel engine that starts but will not idle may have:
   a. no fuel to the injection pump
   b. a low battery
   c. one or more clogged fuel nozzles
   d. ice in the fuel tank.
3. Smoke in the exhaust may be caused by:
   a. sticking nozzles
   b. clogged nozzles
   c. a clogged fuel filter
   d. dirty fuel.
4. To check for a defective nozzle in a diesel engine, start the engine and then:
   a. disconnect the wire to the solenoid
   b. loosen the injection line fitting at the nozzle
   c. disconnect the fuel return line
   d. none of the above.
5. Loss of power may be caused by:
   a. dirty fuel
   b. restricted fuel return line
   c. plugged nozzles
   d. all of the above.

CHAPTER 16

ENGINE LUBRICATING SYSTEMS

After studying this chapter, you should be able to:

1. List the five jobs the lubricating oil does in the engine.
2. Discuss the properties of an engine oil.
3. Define *viscosity*.
4. Explain the service ratings of engine oil.
5. Discuss why oil changes are needed in an engine.
6. Describe the operation of the engine lubricating system.
7. Explain the difference between an oil level indicator and an oil pressure indicator.
8. Discuss the other lubricants required on automotive vehicles.

16-1 Purpose of the lubricating system The lubricating system sends lubricating oil to all engine parts that have relative motion (Fig. 16-1). This includes the bearings that support the crankshaft and camshaft. It also includes the pistons and rings as they slide up and down in the engine cylinders. There are many other moving parts in the engine that require lubrication. The lubricating oil allows the parts to move over each other with relative ease and also reduces wear of the parts to a minimum. The oil reduces the *friction* between the moving parts. In the engine the oil has other jobs to do, as covered in later sections. First, however, let us discuss friction.

16-2 Friction Friction is the resistance to motion between two objects in contact with each other. If you place a book on a table, it will take a certain force to push the book across the tabletop. This force overcomes the friction. The book is a load, or weight. The tabletop is a bearing surface.

If you place a second book on top of the first book, you will have to push harder to move the two books. The more weight you add, the harder you must push. This simple experiment shows that friction, or resistance to motion, increases with the load.

In the engine, the load between the moving bearing surfaces may be more than 1000 psi [6895 kPa]. This means that friction could be quite high. However, the lubricating oil keeps the friction at a relatively low value, as explained in the following sections. Friction can be divided into three classes. These are: (1) dry or mechanical friction, (2) greasy or boundary friction, and (3) viscous or fluid friction.

16-3 Dry friction Dry friction is also called mechanical friction. It is the resistance to motion between two objects in contact with each other, and without any lubricant between them. Here is an example of dry friction. To drag a rough board across a rough floor, a certain pulling force is required. The amount of force depends on the roughness of the surfaces and on the weight of the board.

For example, suppose a pulling force of 10 pounds [4.5 kg] is required to move the board. If you smoothed the floor and the board with sandpaper, then it might take a pull of only 5 pounds [2.3 kg] to move the board. This is an example of dry friction. It is the resistance to motion caused by surface irregularities that catch against each other. Even objects machined to extreme smoothness have microscopic irregularities that cause resistance to relative motion or friction (Fig. 16-2). Without lubrication, smooth hard-metal surfaces with relative motion under load would soon wear. This is because of the dry mechanical friction between the two surfaces.

The tiny irregularities would catch on each other and tear off metal particles, as shown in Fig. 16-2. These particles would then gouge out pits and scratches in the surfaces. Soon the metal surfaces would be very rough, and bigger particles would be broken off. The friction and wear would go up rapidly. In addition, large quantities of heat would be produced by the rubbing and gouging action. Enough heat might result to cause the metal to melt in spots. When this happens, the two surfaces would momentarily weld at the melted spots. Then, there would be an actual joining of the two metal surfaces by small spot welds. With further relative movement, these welds would break, making the surfaces even rougher.

This sort of action actually happens in an engine. For example, under certain conditions small spots on the piston rings in an engine cylinder can weld to the cyl-

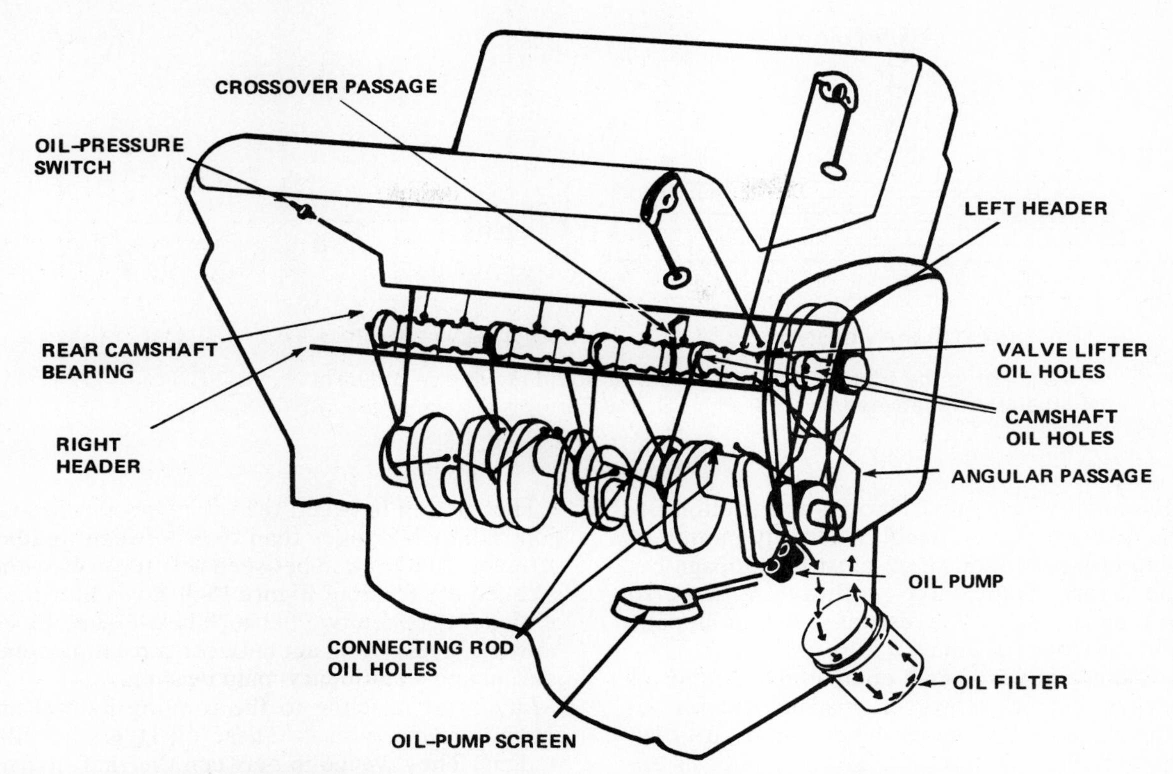

Fig. 16-1 A typical engine lubricating system, shown for an overhead-valve engine with five main bearings. (*Chrysler Corporation*)

inder walls. These welds are pulled apart as the rings continue to move. This leaves gouged-out pits and holes in the rings and cylinder walls.

16-4 Greasy friction Greasy friction is the friction between two surfaces that have been coated with a very thin film of lubricant, such as grease or oil. The surfaces then have what is called borderline or boundary lubrication. The thin coating or film of oil fills the surface irregularities so that they are almost perfectly smooth. Then when the surfaces move, there is less tendency for the irregularities to catch on each other.

When greasy friction exists, the resistance to motion between two surfaces is much less than with dry friction. However, some high spots will still catch, causing some wear as the surfaces move over each other.

When an automobile engine is first started, there may be greasy friction between the moving parts. If the engine has not been operated for a while, most of the lubricating oil will have drained away. Then, when the engine starts, the parts start to move against each other before the lubricating system can get oil to them.

After a short time, oil does flow and the moving parts are adequately lubricated. However, during these first few seconds there is only greasy friction. Excessive wearing of the parts takes place when compared with wear during the viscous friction of full lubrication.

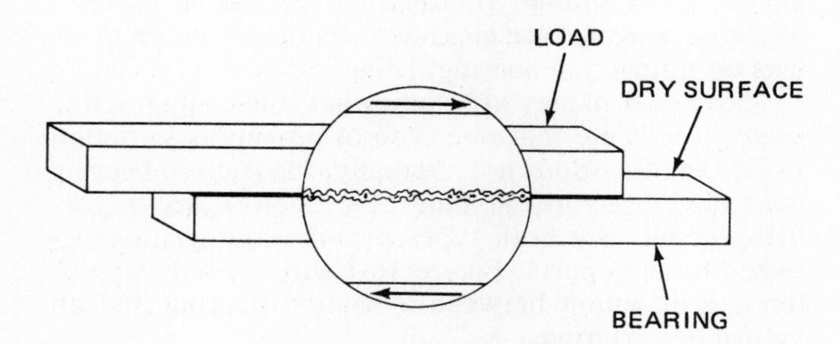

Fig. 16-2 Dry friction. Magnified view of metal-to-metal contact without oil, showing chips shearing off as one surface slides over the other.

16-5 Viscous friction Viscous is a word that refers to the tendency of liquids such as oil and gasoline to resist flowing. For example, a "heavy" oil is more viscous than a "light" oil. The heavy oil flows more slowly. Gasoline is light and has a relatively low viscosity; it flows easily.

Viscous friction is also called fluid friction, because any liquid is a type of fluid. Fluid or viscous friction is the resistance to relative motion between two adjacent layers of liquid. It is the resistance to flow in the fluid itself.

As applied to machines, viscous friction occurs during relative motion between two lubricated surfaces that are flooded with oil (Fig. 16-3). Now, to better understand viscous friction, let us examine Fig. 16-3 closely. It shows in greatly exaggerated view how one object moves across a stationary object. The surfaces of the two objects are separated by layers of lubricating oil. For simplicity, the oil is shown in five layers, A to E. Actually, there would be many more layers of oil.

The nearer a layer is to the stationary layer E, the less the moving layer moves. This is shown in Fig. 16-3 by the shorter and shorter arrows in layers B, C,

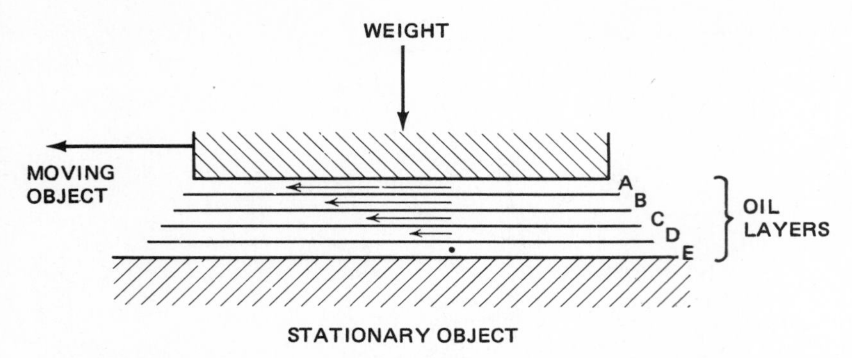

Fig. 16-3 Viscous friction is the friction between layers of liquid moving at different speeds, or in the illustration, between layers A, B, C, D, and E.

and D. Essentially, viscous friction is the friction between the layers in the oil itself. When any motion of the moving object occurs, there must be slippage between the layers. Some force is always required to make the slippage occur. The resistance to this slippage is called the viscous friction.

The resistance is very small compared with that of dry or greasy friction. However, viscous friction becomes important in the engine where there are high pressures between the surfaces. The viscosity of the oil affects the viscous friction. The heavier the oil, the higher the viscosity. Therefore, there is a greater resistance to relative movement between engine parts.

During cold weather, some types of oil get very thick. They can become so viscous in the engine that they prevent cranking of the engine.

⚙ 16-6 Theory of lubrication Several things happen when there is a coating of oil between two surfaces. The two objects in relative motion are held apart by a film or layers of oil. Therefore, there is friction only between the moving layers of oil, instead of between the surfaces of the objects.

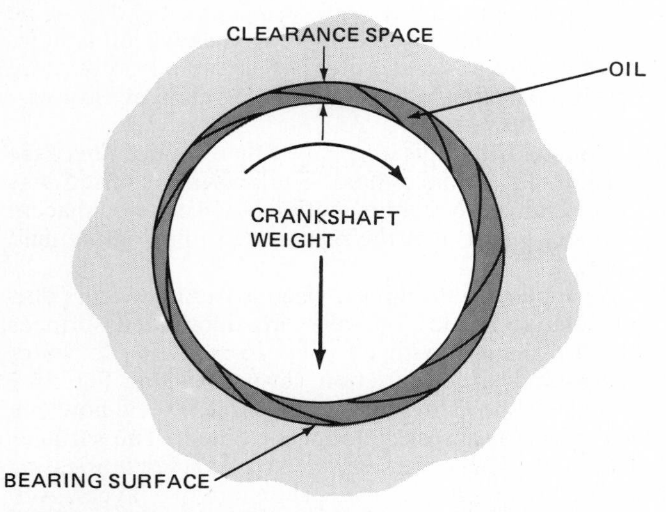

Fig. 16-4 Shaft rotation causes layers of clinging oil to be dragged around with it. Oil moves from the wide space to the narrow space, thereby supporting the shaft weight on an oil film.

The friction between the oil layers, or viscous friction, is much smaller than that between unlubricated surfaces. The friction between two surfaces without oil is called dry friction. Figure 16-3 shows how the layers of oil act between two flat surfaces. Figure 16-4 shows how the layers of oil act between a rotating crankshaft journal and a stationary main bearing.

Layers of oil cling to the rotating journal and are carried around with it. These oil layers act like tiny wedges. They wedge in between the shaft journal and the stationary bearing. The wedging action actually lifts the journal off the bearing so that the shaft weight is supported by the oil.

Figure 16-5 shows how the area of maximum loading, or the high-pressure area between the shaft and the bearing, shifts around with changing shaft speed. When the shaft is at rest (Fig. 16-5*a*), the load is straight down. The lubricating oil is squeezed out from between the shaft and the bearing.

When the shaft starts to rotate (Fig. 16-5*b*), the oil layers wedge between the shaft and the bearing, lifting the shaft off the bearing. In effect, the shaft tries to climb the right side of the bearing because of the viscous friction between the oil layers. However, as shaft speed increases, the wedging action also increases. This transfers the area of maximum pressure toward the left, as shown in Fig. 16-5*c*.

⚙ 16-7 Types of engine bearings Generally, the word bearing means anything that supports a load. In machines, it means anything that supports or confines an object in sliding, rotating, or oscillating motion. Machine bearings are classified as either friction bearings or antifriction bearings (Fig. 16-6).

These two names are somewhat misleading. They seem to indicate that one type of bearing has friction while the other does not. Actually, the friction bearing does have a greater amount of friction, other factors being equal. But both types provide low friction between moving parts. Figure 16-7 shows the basic difference in action between a friction bearing and an antifriction bearing.

In the friction bearing, one object slides over another. The load is supported on layers of oil, as shown earlier in Fig. 16-3. There is sliding friction between the

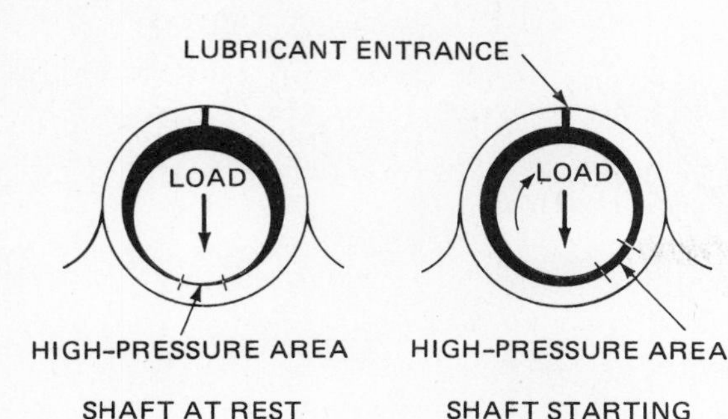

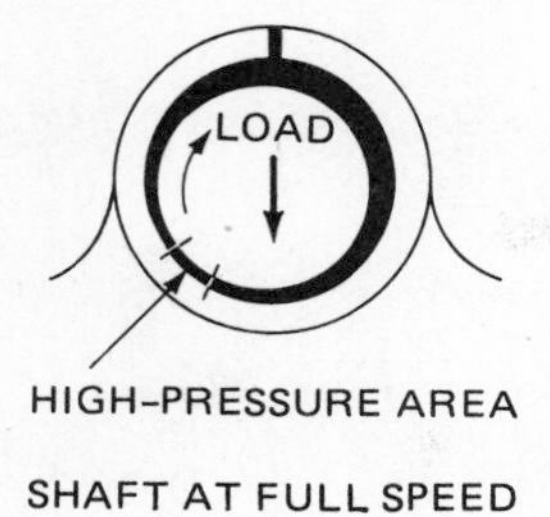

Fig. 16-5 The high-pressure area, or area of maximum loading, varies with shaft speed. The clearance between the shaft and the journal is exaggerated.

two surfaces. In the antifriction bearing, the surfaces are separated by balls or rollers. There is rolling friction between the two surfaces and the balls or rollers (Fig. 16-7).

❂ 16-8 Friction bearings

Friction bearings are also called sleeve bearings and precision-insert bearings. They provide sliding contact between two moving surfaces. When the surfaces are lubricated, the load actually is supported on the layers of oil. In the automobile engine, there are three types of bearing surfaces that are friction bearings. These are illustrated in Fig. 16-8 and are called *journal bearings, guide bearings,* and *thrust bearings*.

The journal type of friction bearing can be symbolized by two hands holding a turning shaft (Fig. 16-8*a*). The hands support the turning shaft in the same way that the bearings support a shaft journal in an engine.

There are several bearings of this type in the engine (Fig. 16-9). The crankshaft main bearings, connecting-rod bearings, camshaft bearings, and piston pin bearings are all the journal type of friction bearing. Some of these bearings are split into an upper half and a lower half. Others are of the bushing or one-piece type.

The bearing surface between the cylinder wall and the piston and piston rings is a guide type of friction bearing (Fig. 16-8*b*). The cylinder wall "guides" the piston up and down in its path.

There is one main bearing in the engine that has thrust faces (Fig. 16-8*c*). The thrust faces hold the crankshaft in position so that it does not shift endways too much as it rotates. Therefore, the thrust bearing takes the end thrust of the crankshaft as it attempts to move backward and forward in the cylinder block.

❂ 16-9 Friction-bearing lubrication

In the automobile engine, the friction bearings and the lubricating system are designed to permit a constant flow of lubricating oil across the bearing surfaces. Oil enters the clearance space between the bearing and the journal (Fig. 16-4). Then the oil passes across the bearing face and drains back into the oil reservoir or crankcase at the bottom of the engine.

Many sleeve bearings have oil grooves (Fig. 16-6) that help spread the oil across the face of the bearing. The grooves also serve as oil reservoirs to hold some oil for initial lubrication just after the engine is started.

Figure 16-10 shows a typical bearing half with its various parts named. Note the annular and distributing grooves that are cut in the bearing face. Oil enters from the oil hole and moves around in the annular groove to the distributing grooves. There the oil is picked up by the rotating shaft journal and carried around so that oil is distributed over the entire face of the bearing.

❂ 16-10 Antifriction bearings

Figure 16-11 shows the three types of antifriction bearings. They are ball, roller, and tapered-roller bearings. The ball bearing (Fig. 16-11*a*) has an inner and an outer race in which symmetrical grooves have been cut. Balls roll in these two race grooves. The balls are held apart by a spacer. When one of the races is held stationary and the other rotates, the balls roll in the two races, permitting low-friction rotation.

Roller bearings (Fig. 16-11*b* and *c*) are similar to ball bearings. However, either plain or tapered rollers are used instead of balls. The rollers roll between the inner

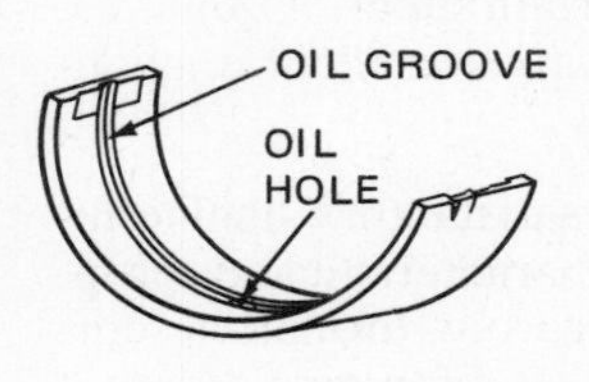

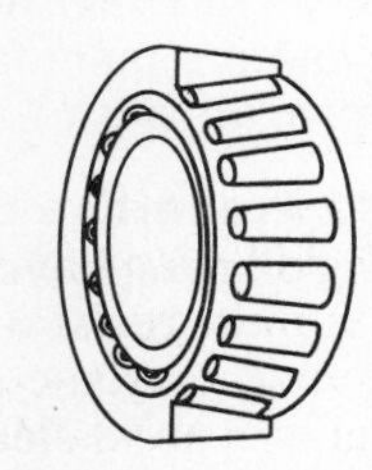

Fig. 16-6 Types of bearings.

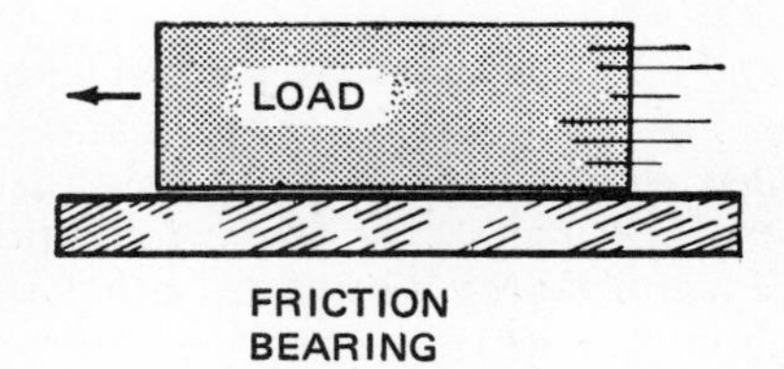

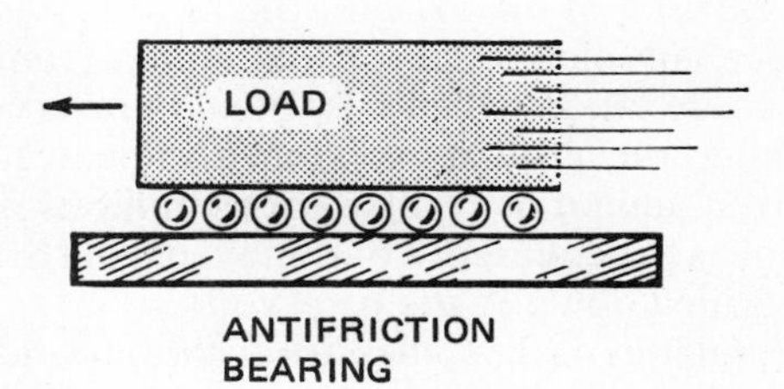

Fig. 16-7 The basic difference in action between a friction bearing and an antifriction bearing.

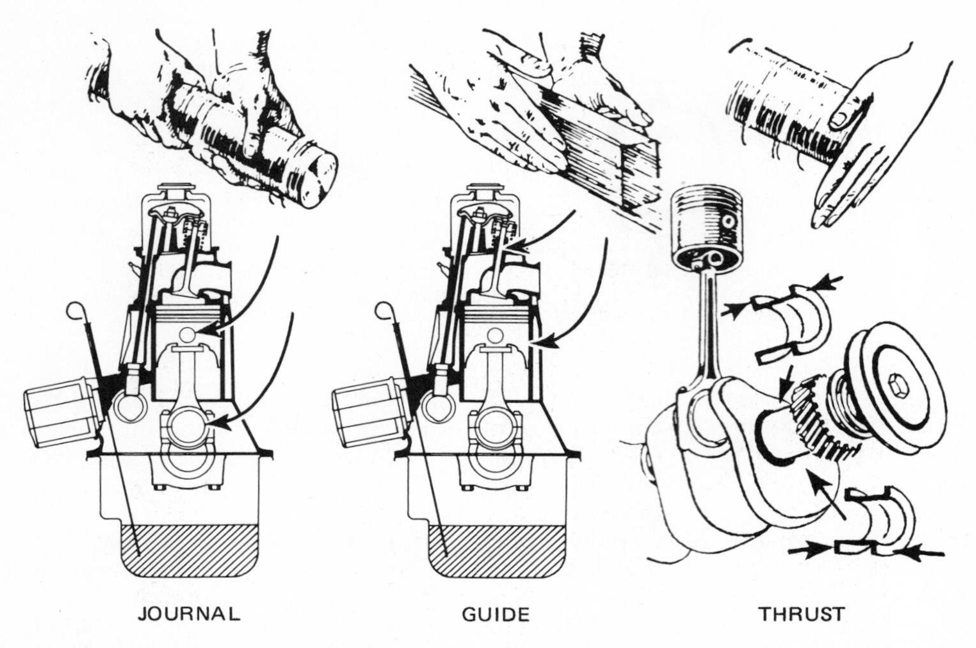

Fig. 16-8 Three types of friction-bearing surfaces in an engine.

and outer races. In the ball bearing, there is only a spot contact between each ball and the races. In the roller bearing, there is line contact between the rollers and the races.

Ball and roller bearings provide rolling friction (Fig. 16-11) rather than sliding friction as in sleeve bearings (Fig. 16-8). The action of the oil in ball and roller bearings is very similar to that in sleeve bearings. However, in the antifriction bearings, the layers of oil are carried around with the balls or rollers. The oil wedges between the balls or rollers and the inner and outer races. Therefore, the balls or rollers float on layers of oil so that there is no actual metal-to-metal contact.

Antifriction bearings may be lubricated with grease instead of oil. Essentially, grease is oil mixed with a solidifying agent known as *soap*. The soap does not directly lubricate the balls or rollers. The job of the soap is to hold the oil in the bearing so that it gets adequate lubrication.

Some ball and roller bearings are packed with grease when they are manufactured. Then the bearing is permanently sealed with metal shields to hold the grease in place. These bearings never require additional lubrication. The trapped grease provides adequate lubrication throughout the life of the bearing.

⚙ 16-11 Source of oil Engine oil, as well as gasoline and various automobile lubricants, comes from petroleum, or crude oil. Petroleum is found in reservoirs, or pools, under the ground. Evidence indicates it was formed from animal or plant sources millions of years ago. The oil is "recovered," or removed from the earth, by wells drilled down to the reservoirs.

The petroleum as it comes from the ground is not usable for lubricating purposes. It must first be refined. This refining process separates the petroleum into various parts or constituents. In the basic refining process, the petroleum is heated in an enclosed chamber, or still. As the petroleum temperature increases, the more volatile parts evaporate first. The vapors are led from the enclosed chamber through tubing to cooler chambers, where they condense. These more volatile parts of the petroleum form gasoline.

As the petroleum is heated to higher and higher temperatures, the less and less volatile parts form engine oil and various heavier products, including tar. Properties of gasoline are described in Chap. 2. Properties of engine oils are discussed in ⚙ 16-12. Properties of other automotive lubricants are described in ⚙ 16-14.

NOTE: Liquefied petroleum gas is also obtained from petroleum. LPG is the most volatile fraction (or part) of the petroleum, tending to turn to vapor at atmospheric pressure and relatively low temperature.

In recent years, synthetic oils have been developed. These are oils made by chemical processes and do not necessarily come from petroleum. Some oil manufacturers claim that these synthetic oils have superior lubricating properties. Actually, there are several types of synthetic oils. The type most widely used at present is produced from organic acids and alcohols (from plants of various types). A second type is produced from coal and crude oil. Although tests have shown that these synthetics do have certain superior qualities, no automotive manufacturer has given them unqualified approval yet.

⚙ 16-12 Properties of oil A satisfactory engine lubricating oil must have certain characteristics, or properties. It must resist oxidation, carbon formation, corrosion, rust, extreme pressures, and foaming. Also, it must act as a good cleaning agent, pour at low temperatures, and have good viscosity (ability to flow) at very high and low temperatures.

No oil, by itself, has all these properties. Lubricating-

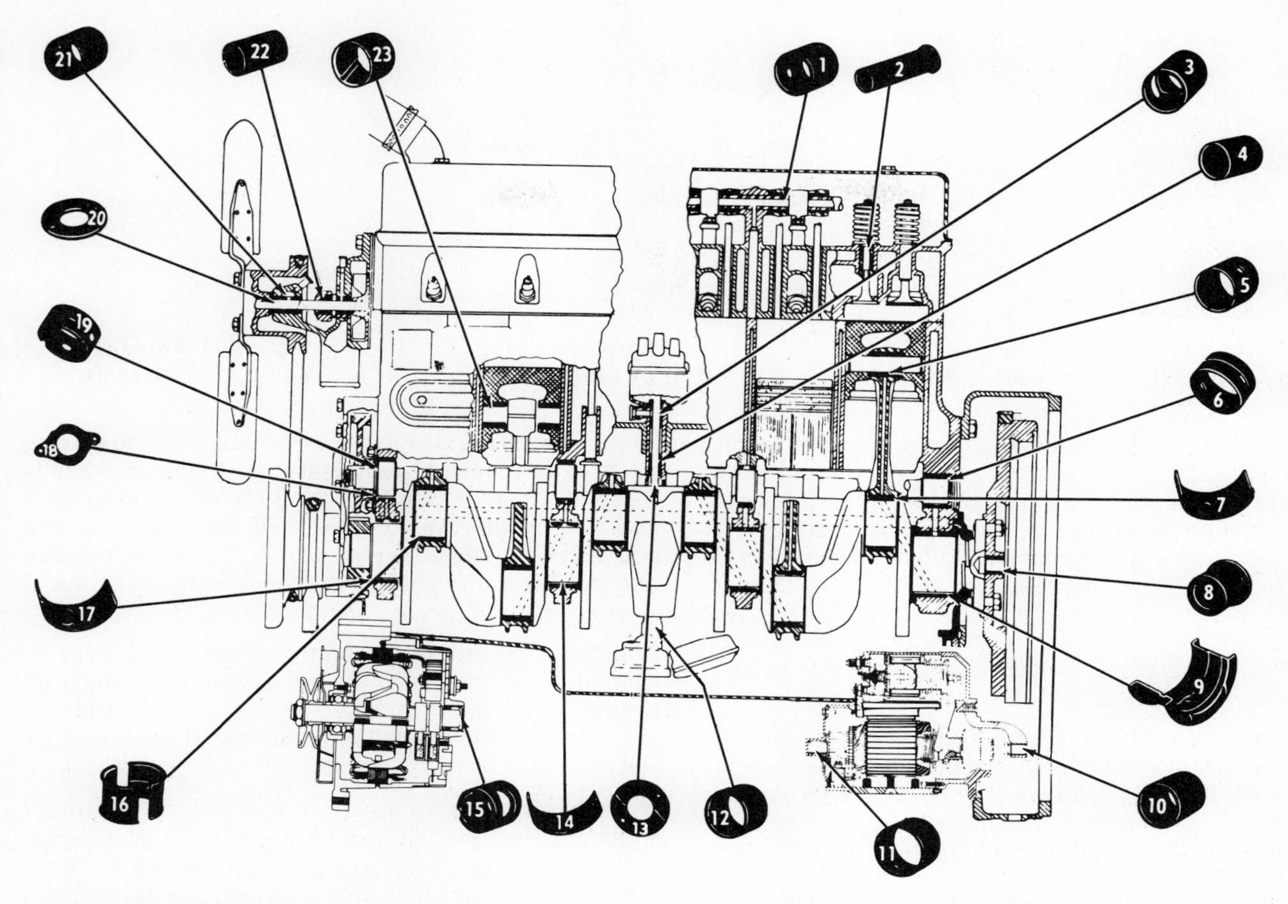

1. ROCKER-ARM BUSHING
2. VALVE-GUIDE BUSHING
3. DISTRIBUTOR BUSHING, UPPER
4. DISTRIBUTOR BUSHING, LOWER
5. PISTON-PIN BUSHING
6. CAMSHAFT BUSHINGS
7. CONNECTING-ROD BEARING
8. CLUTCH PILOT BUSHING
9. CRANKSHAFT THRUST BEARING
10. STARTING-MOTOR BUSHING, DRIVE END
11. STARTING-MOTOR BUSHING, COMMUTATOR END
12. OIL-PUMP BUSHING
13. DISTRIBUTOR THRUST PLATE
14. INTERMEDIATE MAIN BEARING
15. ALTERNATOR BEARING
16. CONNECTING-ROD BEARING, FLOATING TYPE
17. FRONT MAIN BEARING
18. CAMSHAFT THRUST PLATE
19. CAMSHAFT BUSHING
20. FAN THRUST PLATE
21. WATER-PUMP BUSHING, FRONT
22. WATER-PUMP BUSHING, REAR
23. PISTON-PIN BUSHING

Fig. 16-9 Bearings and bushings used in typical engine. (*Johnson Bronze Company*)

oil manufacturers therefore put several additives into the oil during the manufacturing process. An engine oil may have any or all of the following additives:

Viscosity index improver

Pour-point depressants

Oxidation inhibitors

Corrosion and rust inhibitors

Foam inhibitors

Detergent dispersants

Extreme-pressure compounds

Antifriction modifiers

Graphite and molybdenum compounds

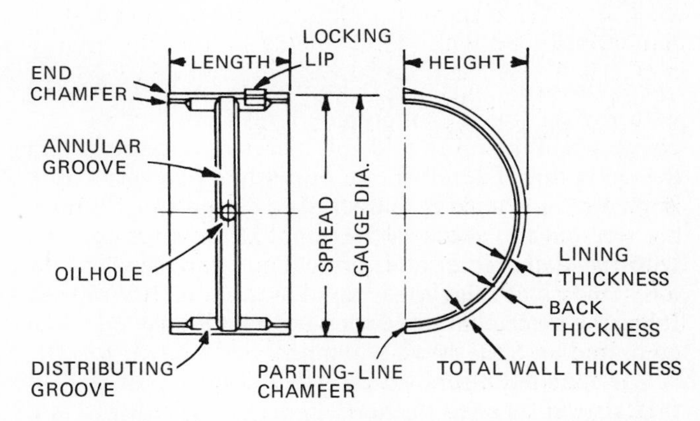

Fig. 16-10 Typical sleeve-type bearing half with its parts named. Many bearings do not have annular and distributing grooves. (*Federal-Mogul Corporation*)

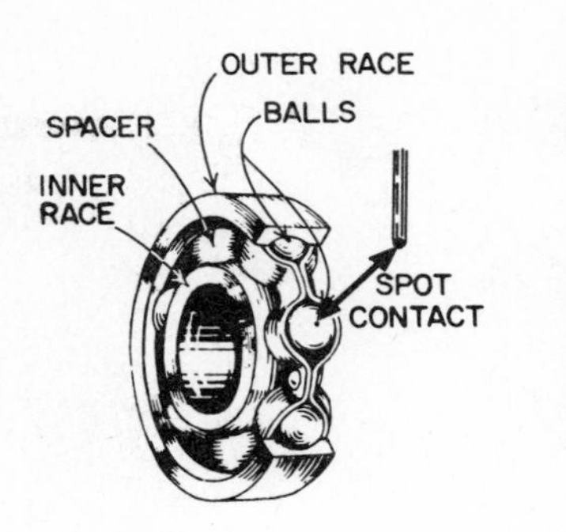

BALL BEARING

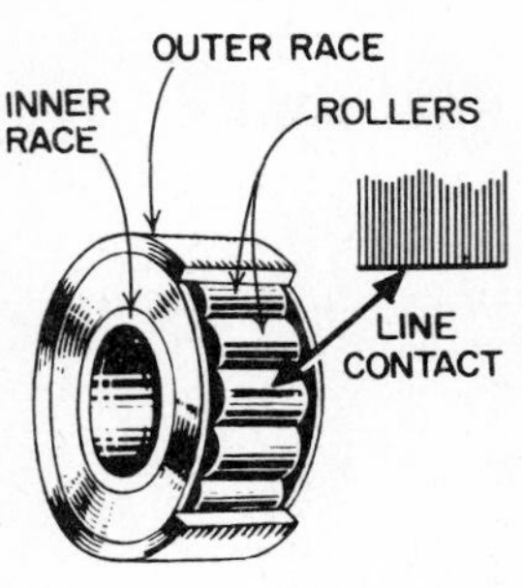

ROLLER BEARING

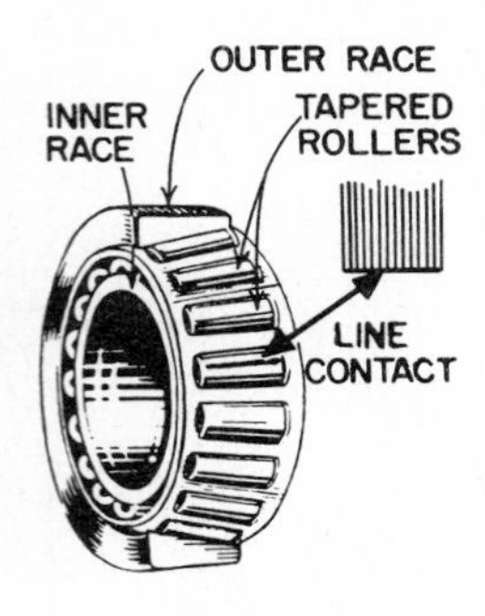

TAPERED ROLLER BEARING

Fig. 16-11 Three types of antifriction bearings.

1. Viscosity Viscosity is the most important property of lubricating oil. Viscosity refers to the tendency of oil to resist flowing. In a bearing and journal, layers of oil adhere to the bearing and journal surfaces. These layers must move or slip over each other. The viscosity of the oil determines the ease with which this slipping can take place. Temperature influences viscosity. Increasing temperature reduces viscosity; it makes the oil thin out. Decreasing temperature causes oil viscosity to increase. The oil thickens.

2. Viscosity grades Viscosity of oil is determined by a viscosimeter. This device determines the length of time required for a certain amount of oil to flow through an opening of a definite size at a specified temperature. Oils with the lower numbers are of lower viscosity. They are thinner.

The Society of Automotive Engineers (SAE) rates oil viscosity in two different ways: for winter use and for other than winter use. Winter-grade oils are tested at 0 and 212°F [−18 and 100°C]. There are three grades: SAE 5W, SAE 10W and SAE 20W, the W indicating winter grade. For other than winter use, the grades are SAE 20, SAE 30, SAE 40, and SAE 50 (all without the W suffix.) All these grades are called single-viscosity oils.

Some oils have multiple ratings, which means they are equivalent in viscosity to several single-rating oils (see below).

3. Viscosity index improver When oil is cold, it is thicker and runs more slowly than when it is hot. A cold engine is harder to start because the oil is more viscous. To overcome this problem, chemists developed viscosity index (VI) improvers. When blended with the oil, these compounds tend to reduce oil viscosity when the oil is cold and increase viscosity when the oil is hot. Therefore, an oil with a viscosity index improver in it makes cold starting easier and yet does not thin out too much as the temperature goes up. Oils with these characteristics are called *multiple-viscosity oils*. One oil of this type is rated as an SAE 10W-40 oil. It has the same characteristics as an SAE 10W oil when cold, and an SAE 40 oil when hot.

Car manufacturers specify the viscosity of the oil that should be used in their engines. Figure 16-12 is a chart showing how outside temperature affects the viscosity requirements of the engine. For example, a 5W-20 oil is recommended if the temperature will be 20°F [−8°C] or less. This oil is good for starting and driving in these low outside temperatures. If the weather is going to warm up, a different oil should be used such as a 10W-30 or 10W-40. These oils will hold their viscosities in the higher temperatures. They will not thin out too much and yet are thin enough for easy initial starting and cold-engine operation.

4. Pour-point depressants Pour-point depressants depress, or lower, the temperature at which oil becomes too thick to flow. Therefore, this additive keeps the oil fluid at low temperatures for cold-weather starts.

5. Resistance to carbon formation Cylinder walls, pistons and piston rings operate at high temperatures. These temperatures are high enough to cause the oil to break down and form carbon. The less carbon in the engine cylinders, the better. Therefore oil chemists regulate the refining process to make sure that lubricating oil has good resistance to carbon formation.

6. Oxidation inhibitors When oil is heated and then stirred (as happens in the crankcase), the oxygen in the air tends to combine, or oxidize, with the oil. As oil oxidizes, various harmful substances can form, including some that are tarlike and others that are like

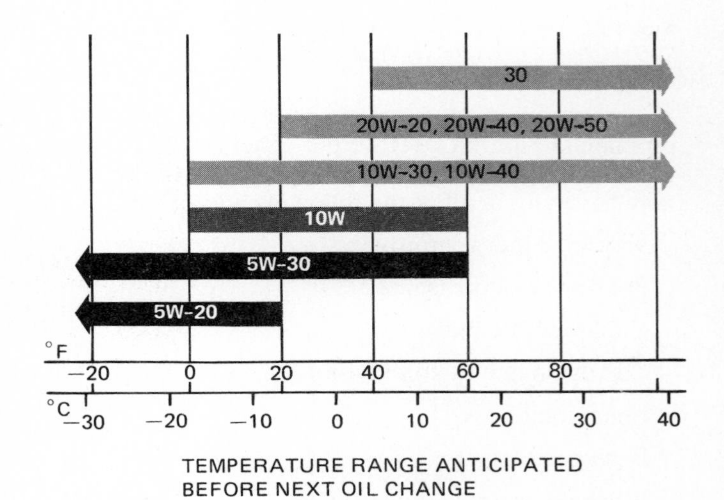

Fig. 16-12 Selecting engine oil based on outside temperature. (*Chevrolet Motor Division of General Motors Corporation*)

varnish. To prevent this, additives are put into the oil to prevent oxidation.

7. Corrosion and rust inhibitors At high temperatures, especially if there is excessive blowby, acids may form in the oil which can corrode engine bearings and other parts. Corrosion inhibitors are added to the oil to prevent this corrosion. Also, rust inhibitors are added. These displace water from metal surfaces so that oil coats the surfaces. Rust inhibitors also neutralize combustion acids.

8. Foam inhibitors The churning action in the engine crankcase also causes the engine oil to foam. As the oil foams, it tends to overflow, or to be lost through the positive crankcase ventilating system. In addition, the foaming oil does not provide normal lubrication of bearings and other moving parts. Foaming oil in hydraulic valve lifters causes them to function poorly and noisily, wear rapidly, and possibly break. To prevent foaming, foam inhibitors are mixed with the oil.

9. Detergent dispersants Despite the air filters for the intake air and for the PCV system, dirt does get into the engine crankcase. In addition, as the engine runs, the combustion processes leave deposits of carbon on piston rings, valves, and other parts. Also, some oil oxidation may take place, resulting in still other deposits. Then, too, metal wear in the engine puts particles of metal into the oil. As a result, deposits tend to build up in and on engine parts. The deposits reduce engine performance and speed up wear of parts.

To prevent or slow the formation of these deposits, engine oils contain a detergent additive. A detergent is similar to soap. When you wash your hands with soap, the soap surrounds and loosens the particles of dirt. You can then rinse off the dirt with water. Similarly, the detergent in the oil loosens the particles of carbon, gum, and dirt on engine parts and carries them away. Some of the particles drop to the bottom of the crankcase, where they are drained away when the oil is changed. Other particles are trapped in the oil filter.

To prevent the detached small particles from clotting and to keep them finely divided a dispersant is added to the oil. Without the dispersant, the particles would tend to collect and form larger particles. These larger particles might then block the oil filter and reduce its effectiveness. They could also build up in oil passages and plug them. This would deprive bearings and other engine parts of oil. The dispersant prevents this. It greatly increases the amount of particles, or contaminants, the oil can carry and still function effectively.

Lubricating-oil manufacturers now place more emphasis on the dispersant qualities of an additive than on its detergent qualities. If the contaminants can be kept suspended in the oil as small particles, they will not be deposited on engine parts. There will be less need for detergent action.

10. Extreme-pressure agents In automotive engines, the lubricating oil is subjected to very high pressures in the bearings and valve train. To prevent the oil from squeezing out, extreme-pressure agents are added to the oil. These agents react chemically with metal surfaces to form very strong, slippery films, which may be only about one molecule thick. Therefore, these additives supplement the oil by providing protection during periods of extreme pressure during which the oil itself is likely to be squeezed out.

11. Antifriction modifiers Certain chemicals when added to oil in effect tend to plate the metal surfaces that the oil is lubricating. This layer allows the oil to pass through the space between the parts more easily. As a result, the friction in the engine is reduced, and fuel consumption is decreased. When added to the oil, a friction modifier dissolves completely and becomes part of the oil itself.

12. Graphite and molybdenum modifiers Some oils now have small particles of molybdenum (''moly'') held in suspension in the oil. These particles are very slippery, much more slippery than the oil they are in. As a result, the amount of engine power required to overcome the internal friction of the engine is reduced. One theory is that the small particles actually support part of the load, thereby reducing the friction between moving metal surfaces. In the oil, the small particles do not dissolve. They are held in suspension to prevent their settling out while the engine is stopped and the oil is stored.

⚙16-13 Service ratings of oil Lubricating oil is rated by viscosity grade (SAE 10W-30, for example). Lubricating oil is also rated in another way by the American Petroleum Institute (API), as to its service designation. This means the oil is rated according to the type of service for which it is best suited. There are six API service ratings for gasoline engine lubricating oils: SA, SB, SC, SD, SE, and SF. There are four API service ratings for diesel engine lubricating oils: CA, CB, CC, and CD. The oils differ in their properties and in the additives they contain.

1. SA oil SA oil is used in utility gasoline and diesel engines that are operated under mild conditions so that protection by additives is not required. This oil may have pour-point depressants and foam inhibitors.

2. SB oil SB oil is used in gasoline engines that are operated under such mild conditions that only minimum protection by additives is required. Oils designed for this service have been used since the 1930s. They provide only antiscuff capability and resistance to oil oxidation and bearing corrosion.

3. SC oil SC oil is for service typical of gasoline engines in the 1964 to 1967 models of passenger cars and trucks. It is intended primarily for use in passenger cars. This oil provides control of high- and low-temperature engine deposits, wear, rust, and corrosion.

4. SD oil SD oil is for service typical of gasoline engines in passenger cars and trucks beginning with 1968 models. This oil provides more protection from high- and low-temperature engine deposits, wear, rust, and corrosion than SC oil.

5. SE oil SE oil is for service typical of gasoline engines in passenger cars and some trucks beginning

with 1972 (and some 1971) models. This oil provides more protection against oil oxidation, high-temperature engine deposits, rust, and corrosion than do oils with the SC and SD ratings.

6. SF oil SF oil is for service typical of gasoline engines in passenger cars and some trucks beginning with 1981 (and some 1980) models. This oil provides more protection than do SE oils against sludge, varnish, wear, oil-screen plugging, and engine deposits.

Notice that this is an open-end series. When the car manufacturers and oil producers see the need for other types of oil they can bring out SG and SH service-rated oils. SA and SB oils are not recommended for use in automobile engines. These are nondetergent oils. Detergent oils are required in modern automotive engines.

Diesel engine oils must have different properties than oils for gasoline engines. The CA, CB, CC, and CD ratings indicate oils for increasingly severe diesel-engine operation. For example, CA oil is for light-duty service. CD oil is for severe-duty service typical of turbocharged high-output diesel engines operating on fuel oil with a high sulfur content.

Modern high-speed automotive diesel engines require a special "combination" type of lubricating oil. For example, General Motors specifies that the oil for their 1978 model V-8 passenger-car diesel engine should be oil which has the service designation SE/CD marked on the can. For 1979 and later models, the can must be marked SE/CC. Do *not* use oil labeled only SE or CD. These oils could cause engine damage. General Motors states that a single-viscosity grade oil such as SAE 20W or SAE 30 is better for their V-8 automotive diesel engines than multiviscosity oils, for sustained high-speed driving.

NOTE: The viscosity grade and the service rating of an oil are different. A high-viscosity oil is not necessarily a "heavy-duty" oil. Viscosity grade refers to the thickness of the oil. Thickness is not a measure of heavy-duty quality. An oil has two ratings, viscosity and service. Therefore, an SAE 30 oil may be an SC, SD, or SE oil (for automotive engines). Likewise, an oil of any other viscosity grade can have any one of the service ratings.

16-14 Automotive lubricants In addition to engine oil, many other lubricants are required for the automobile. Wherever one part slides on or rotates in another part, some kind of lubricant is used to protect the parts from undue wear. The steering system, axles, differential, transmission, brakes, alternator, ignition distributor, all require special types of lubricant.

1. Gear lubricants The gears in manual transmissions and differentials must be lubricated with special heavy oils that resist oil-film puncture and thereby prevent actual metal-to-metal contact between the moving gear teeth. The oil must flow readily even at low temperatures so that it does not "channel" as the gears begin to rotate. Channeling of the oil takes place if the oil is so thick that the teeth cut out channels in the oil and the oil does not readily flow to fill the channels.

The lubricant used in hypoid-gear differentials is subjected to very severe service. Hypoid gears have teeth that not only roll over one another but also slide over each other. This combined rolling and sliding action puts additional pressure on the lubricant. So that the lubricant will stand up under this service, it is specially compounded. It contains certain added chemicals that enable it to withstand much greater pressure than oil alone could withstand. Such lubricants are called extreme-pressure, or EP, lubricants. There are two classifications of these lubricants: (1) strong extreme-pressure lubricants for use on heavy-duty applications, and (2) mild extreme-pressure lubricants for use on applications with less severe requirements.

2. Grease Essentially, lubricating grease is oil to which certain thickening agents have been added. The oil furnishes the lubricating action. The thickening agents simply function to hold the oil in place so that it does not run away. The thickening agents are usually called *soap*. This is not the kind of soap used in washing, but is any one of several metallic compounds. The type of soap used depends on the service required of the grease. This is also true of the viscosity grade (or thickness) or the oil that goes into the grease. For some services, a relatively light oil is used. For others, a heavy oil is used.

a. Aluminum grease Aluminum grease contains a thickening agent made up of aluminum compounds. This grease has good adhesive properties and is widely used for chassis lubrication. Although it will not stand extreme temperatures, it is highly resistant to moisture and is therefore valuable for lubricating springs and other chassis parts subjected to road splash.

b. Soda grease Soda grease contains a thickening agent made up of sodium compounds that gives the grease a thick, fibrous appearance, even though it contains no actual fiber. This grease is often called fibrous grease, or fiber grease. It is less resistant to moisture than some other greases, but it is very adhesive and clings tightly to rotating parts. Therefore, it is valuable for rotating parts, such as wheel bearings and universal joints.

c. Calcium grease Calcium grease uses calcium compounds as a thickening agent. It is often known as *cup grease* and is used in lubricating parts supplied with grease cups. It has a tendency to separate into liquid oil and solid soap at high temperatures.

d. Mixed greases Each of the various greases mentioned above has special valuable characteristics. Mixed greases are blends of these different greases. This blending produces greases that can better meet the requirements of certain specific applications.

e. Other greases There are other special greases. Brake grease is specified for the moving parts in the drum brake mechanisms. Distributor breaker cam grease is specified for the cam in ignition distributors. Speedometer cable lubricant is another special lubricant.

3. Automatic-transmission fluid There are several types of automatic-transmission fluid. Each model

of automatic transmission has special lubricating requirements. Most automative automatic transmissions require either type F or Dexron® II automatic-transmission fluid.

4. Power-steering fluid This is another special fluid that meets the special needs of the power-steering unit.

5. Other fluids Several other fluids are used in automobiles. These include the antifreeze (ethylene glycol) and brake fluid. In service work, a variety of fluids are used to clean or loosen parts. Carburetor cleaner is one example. Manifold heat control valve solvent, used to loosen up the valve if it gets stuck, is another.

NOTE: The lubricant and fluid makers have tried to make the technician's job as easy as possible by supplying the specific lubricant or fluid needed for each service. Service manuals contain specific lubrication charts for each model of automobile. These charts indicate the type of lubricant or fluid to use at every place needing lubrication. They also indicate the intervals at which these services should be supplied.

16-15 Purpose of the lubricating oil We normally think of lubricating oil in the engine as a fluid that reduces wear and friction between adjacent surfaces (Fig. 16-13). However, the lubricating oil that circulates through the engine performs other jobs (Fig. 16-6). The engine lubricating oil must:

1. Lubricate moving parts to minimize wear and power loss from friction.
2. Remove heat from engine parts by acting as a cooling agent.
3. Absorb shocks between bearings and other engine parts, thereby reducing engine noise and extending engine life.
4. Form a good seal between piston rings and cylinder walls.
5. Act as a cleaning agent.

1. Minimizing wear and power loss from friction Friction encountered in the engine is normally viscous friction. This is the fluid friction between adjacent moving layers of oil. If the lubricating system does not function properly, sufficient oil will not be supplied to moving parts. Greasy or even dry friction could result between moving surfaces.

This would cause considerable power loss since much power must be used to overcome these types of friction. At worst, major damage would occur to engine parts as greasy or dry friction developed. Bearings would wear rapidly. The heat resulting from greasy or dry friction would cause bearing failure, so that connecting rods and other parts could be broken. Insufficient lubrication of cylinder walls would cause rapid wear and scuffing of walls, rings, and pistons. A properly operating engine lubricating system supplies all moving parts with enough oil so that only viscous friction exists in the engine.

2. Removing heat from engine parts The engine oil circulates through the engine lubricating system. All bearings and moving parts are bathed in streams of oil. In addition to lubricating parts, the oil absorbs heat from engine parts and carries this heat back into the oil pan. The oil pan absorbs heat from the oil, transferring it to the surrounding air. Therefore, the oil acts as a cooling agent.

3. Absorbing shocks between bearings and other engine parts As the piston approaches the end of the compression stroke, the air-fuel mixture in the cylinder is ignited. Pressure in the cylinder suddenly increases many times. A load of as much as 4000 pounds (1814 kg) is suddenly placed on the top of a 3-inch (76.2-mm) piston. This sudden increase in pressure causes the piston to thrust down hard through the piston-pin bearing, connecting rod, and connecting-rod bearing.

There is always some space or clearance between bearings and journals. This space is filled with oil. When the load on the piston suddenly increases, the layers of oil between bearings and journals act as cushions. They must resist penetration, or "squeezing out." A film of oil must remain between the adjacent metal surfaces. In absorbing and cushioning the hammerlike effect of the sudden loads, the oil quiets the engine and reduces wear of parts.

4. Forming a seal between piston rings and cylinder walls Piston rings must form a gas-tight seal with the cylinder walls. The lubricating oil that is delivered to the cylinder walls helps the piston rings to accomplish this. The oil film on the cylinder walls makes up for microscopic unevenness in the fit between the rings and walls. The film fills in any gaps through which gas might escape. The oil film also lubricates the rings, so that they move easily in the ring grooves and on the cylinder walls.

5. Acting as a cleaning agent As the oil circulates, it tends to wash off and carry away dirt, particles of carbon, and other foreign matter. The oil picks up this material and carries it back to the crankcase. There, larger particles drop to the bottom of the oil pan. Smaller particles are filtered out by the oil filter.

16-16 Types of lubricating systems Two types of lubricating systems have been used on four-cycle engines. They are the splash system and the pressure-feed system. Two-cycle engines require a different kind of lubrication. The Wankel engine also requires a special lubricating system.

1. Splash In the splash lubricating system, oil is splashed from the oil pan into the lower part of the crankcase. Usually, the connecting rod has a dipper that dips into the crankcase oil each time the piston reaches BDC. This splashes the oil (Fig. 16-14). Some small engines also use oil slingers driven by the camshaft. These are gearlike parts that throw oil from the oil pan up onto moving engine parts. The splash system is used on most small four-cycle engines for power lawn mowers and similar applications.

2. Pressure feed Today the automobile engine is

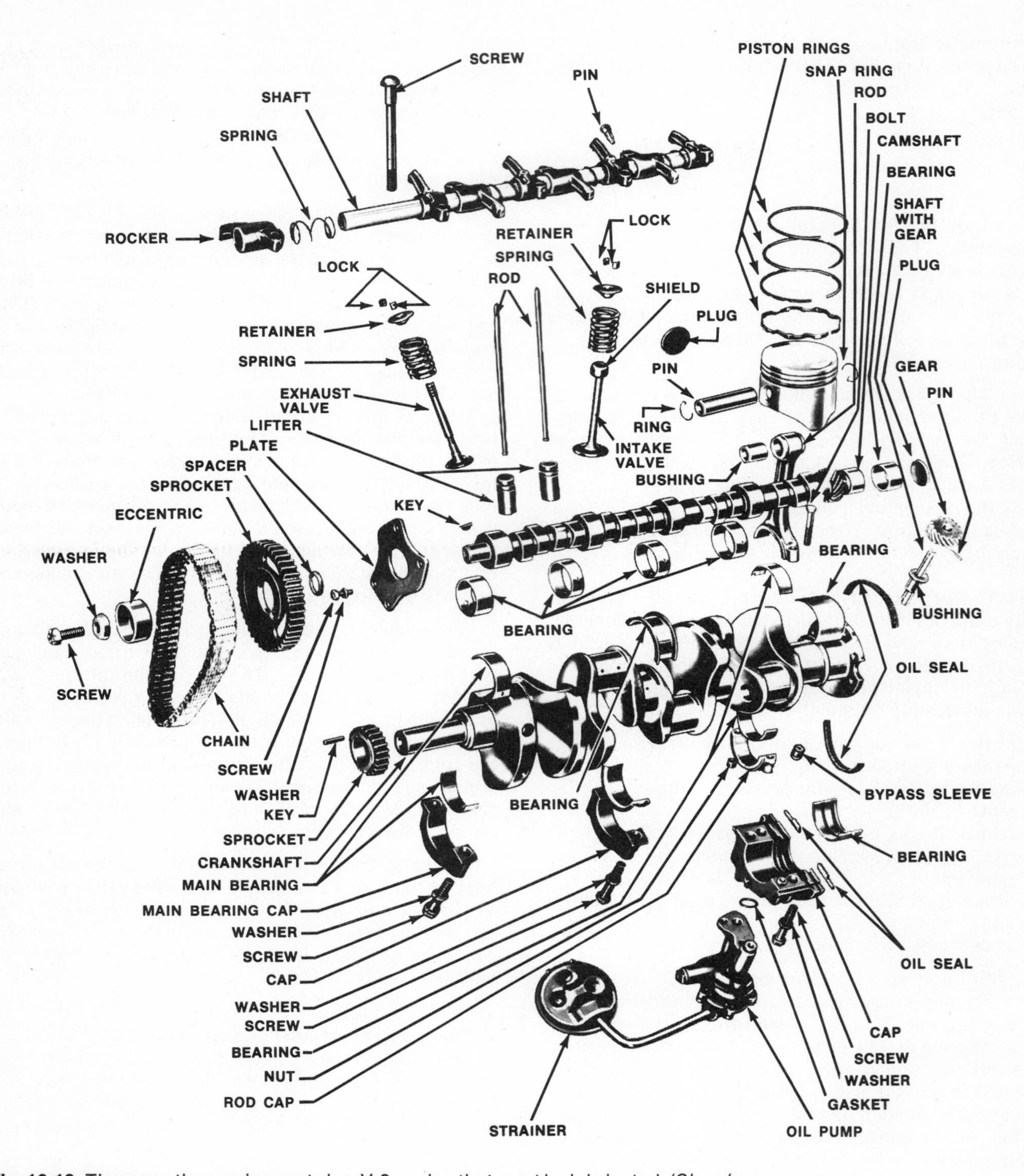

Fig. 16-13 These are the moving parts in a V-8 engine that must be lubricated. (*Chrysler Corporation*)

lubricated by a pressurized, or pressure-feed, lubricating system (Fig. 16-15). In the pressure-feed lubricating system, many of the engine parts are lubricated by oil that is fed to them under pressure from the oil pump (Figs. 16-15 and 16-16). The oil from the pump passes through an oil filter. It then enters an oil line (or a drilled header, channel, or gallery, as it is variously called). From the oil line, the oil flows to the main bearings, camshaft bearings, and hydraulic lifters. The main bearings have oil-feed holes or grooves that feed oil into drilled passages in the crankhaft. The oil flows through these passages to the connecting-rod bearings.

In overhead-valve engines, oil is fed under pressure to the valve mechanisms in the head. For example, some engines have the rocker arms mounted on hollow shafts (Fig. 16-13). These shafts feed oil to the rocker

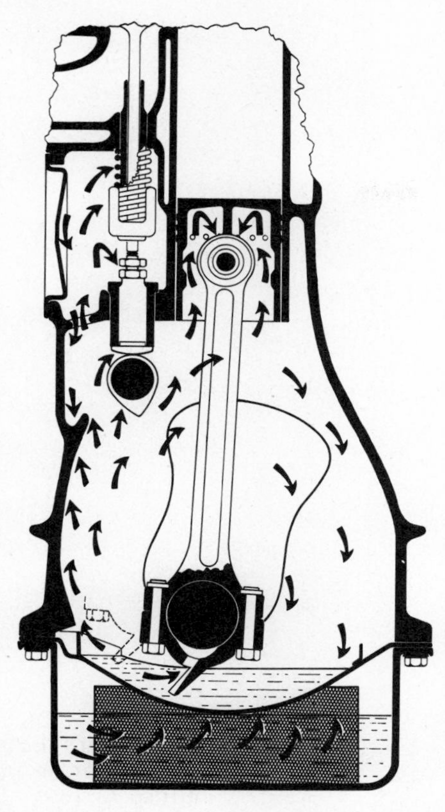

Fig. 16-14 Splash lubricating system, used on older automotive engines and on small engines.

arms. Some engines with independently mounted rocker arms have hollow mounting studs. These studs feed oil from an oil gallery in the head to the rocker-arm ball pivots. On other engines, the oil flows up through hollow push rods to lubricate the valve stems and other valve train parts (Fig. 16-16). The oil spills off the rocker arms and provides lubrication for the valve stems and push rod and valve-stem tips. Therefore, all valve mechanism parts are adequately lubricated.

Cylinder walls are lubricated by splashing oil thrown off from the connecting-rod bearings. Some engines have oil spit holes or grooves in the connecting rods that index or align with drilled holes in the crankpin journals with each revolution. As this happens a stream of oil is spit or thrown onto the cylinder walls (Fig. 16-17).

On many V-type engines, the oil spit holes or grooves are arranged so that they deliver their jets of oil into opposing cylinders in the other cylinder bank. For example, the oil spit holes in the connecting rods in the right bank lubricate the cylinder walls in the left bank. When oil spit holes are not used, the engine relies on oil that flows through the side clearance between the side of the connecting rod and the crankshaft for cylinder wall lubrication.

In many engines, the piston pins are lubricated with oil scraped off the cylinder walls by the piston rings. The pistons have grooves, holes, or slots to feed oil from the oil-ring groove on the piston to the piston pin bores.

Oil passages in the block and head permit circulation of oil to bearings and moving parts (Fig. 16-18). Some

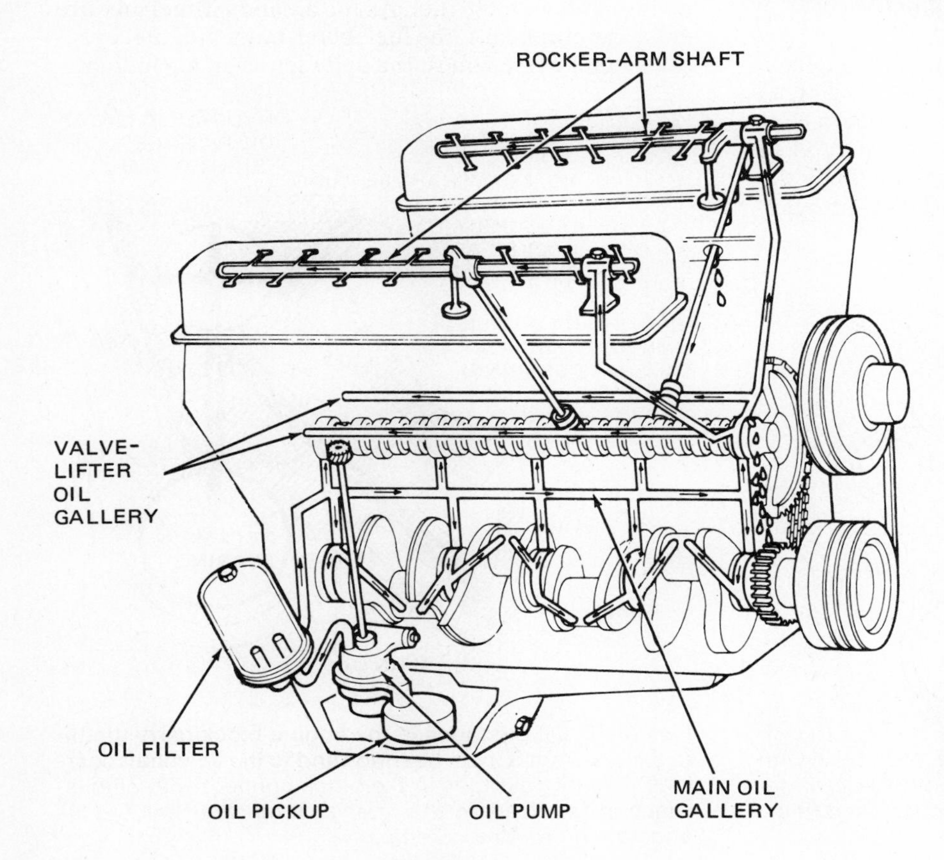

Fig. 16-15 Lubricating system for an overhead-valve V-8 engine with five main bearings. (*Chrysler Corporation*)

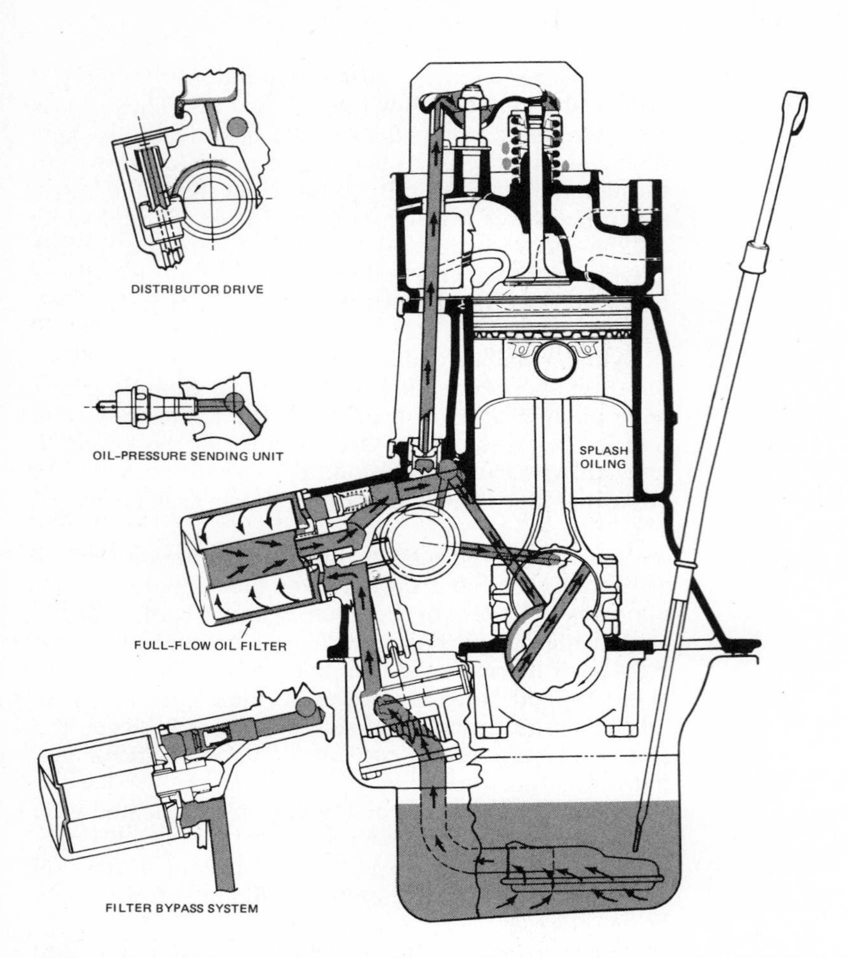

Fig. 16-16 Lubricating system for an in-line engine. Arrows show the flow of oil to the moving parts in the engine. (*Chrysler Corporation*)

older engines had holes drilled in the connecting rod to carry oil up to the piston pin and lubricate it.

16-17 Two-cycle engines In two-cycle engines, the air-fuel mixture passes through the crankcase on its way from the carburetor to the engine cylinders. For this reason it is not possible to maintain a reservoir of oil in the crankcase. The oil would be picked up by the passing air-fuel mixture, carried to the engine cylinders, and burned.

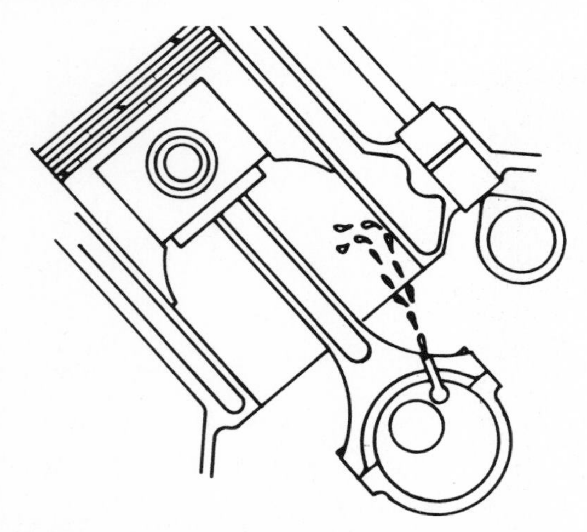

Fig. 16-17 When a hole in the connecting rod aligns with a hole in the crankpin, oil is sprayed onto the cylinder wall as shown. The action lubricates the piston and rings. (*Ford Motor Company*)

To provide lubrication of two-cycle engine parts, the oil is mixed with the fuel. As the air and oil-fuel mixture enter the crankcase, the fuel, being more volatile, evaporates and passes on to the cylinder as an air-fuel mix-

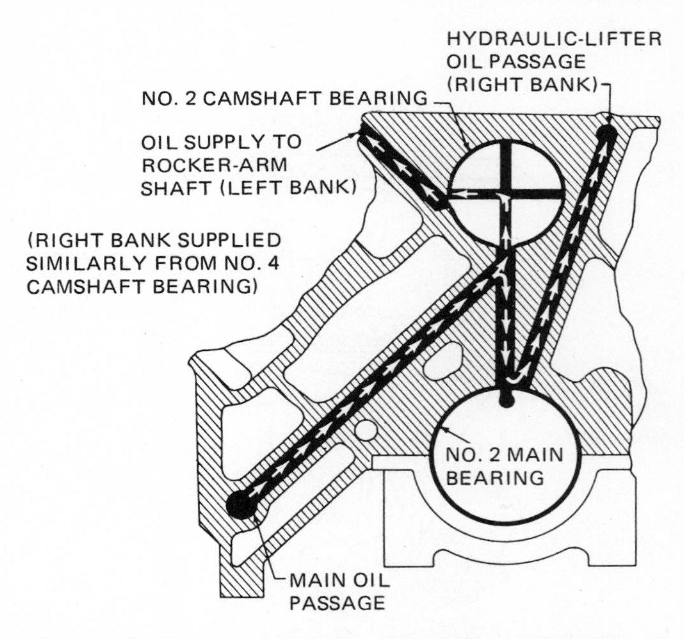

Fig. 16-18 Oil passages in the engine block carry the oil to the crankshaft main bearings and to the camshaft bearings. The oil passages in the block connect with oil passages in the head so that the valve mechanisms are lubricated. (*Ford Motor Company*)

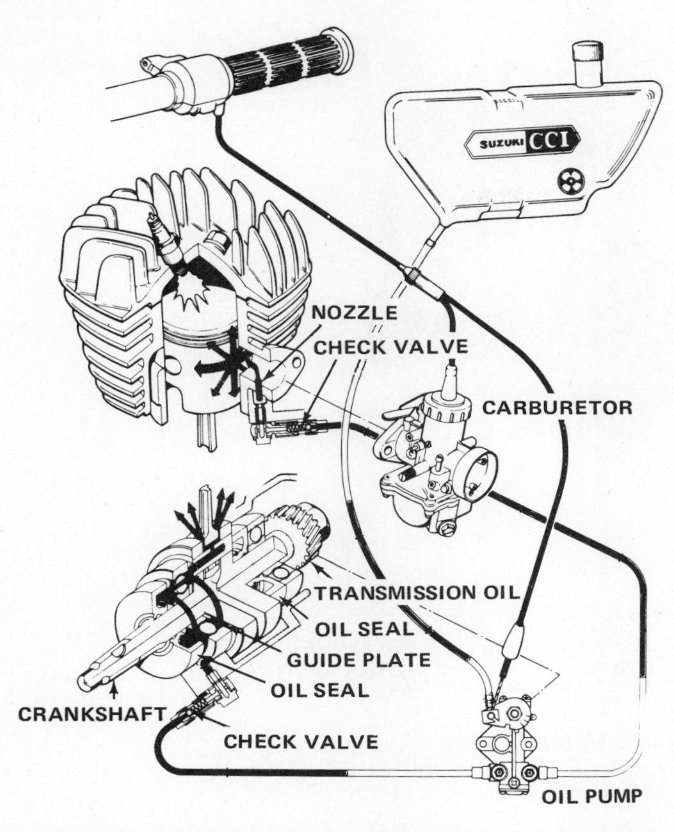

Fig. 16-19 Lubricating system for a one-cylinder two-cycle motorcycle engine. (*Suzuki Motor Company, Ltd.*)

ture. Some of the oil is carried along with the air-fuel mixture and is burned. But enough oil is left behind to keep the moving engine parts coated with oil and adequately lubricated.

Some two-cycle engines, for example many used in motorcycles, have an oil feed to the carburetor (Fig. 16-19). The oil is not mixed with the gasoline for these engines. Instead, the oil enters the air-fuel mixture in the carburetor. The system meters the oil so that the proper amount is added to assure adequate lubrication of the engine under all operating conditions. This is important for motorcycle engines which operate at greatly varying speeds. A standard gasoline-oil mix such as used in power lawn mowers, for example, would not meet the requirements. The system shown in Fig. 16-19, which is typical for many motorcycle engines, varies the amount of oil entering the air-fuel mixture. It provides more oil at high speeds, satisfying operating requirements.

Through a separate circuit, this type of system may supply lubricating oil to the engine bearings and other components in the lower part of the engine.

16-18 Overhead-camshaft engine lubrication In the overhead-camshaft engine, additional lubrication must be furnished to the cylinder head for the camshaft bearings. Usually there is an oil gallery running the length of the cylinder head. It supplies oil to the camshaft bearings. The gallery also supplies oil to the hydraulic valve lifters and other valve-train parts.

16-19 Wankel engine lubrication Figure 16-20 shows the lubricating system for the Mazda Wankel

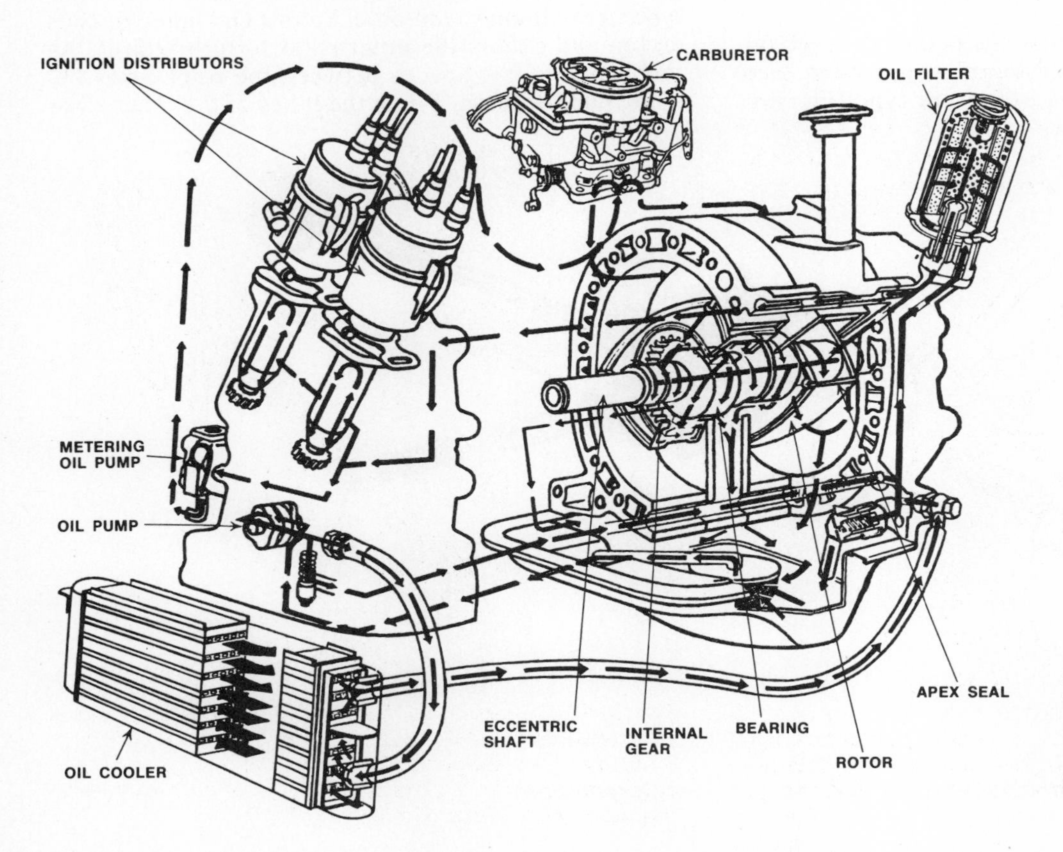

Fig. 16-20 Lubricating system for the Mazda Wankel engine. Arrows show the flow of oil in the engine. (*Toyo Kogyo Company, Ltd.*)

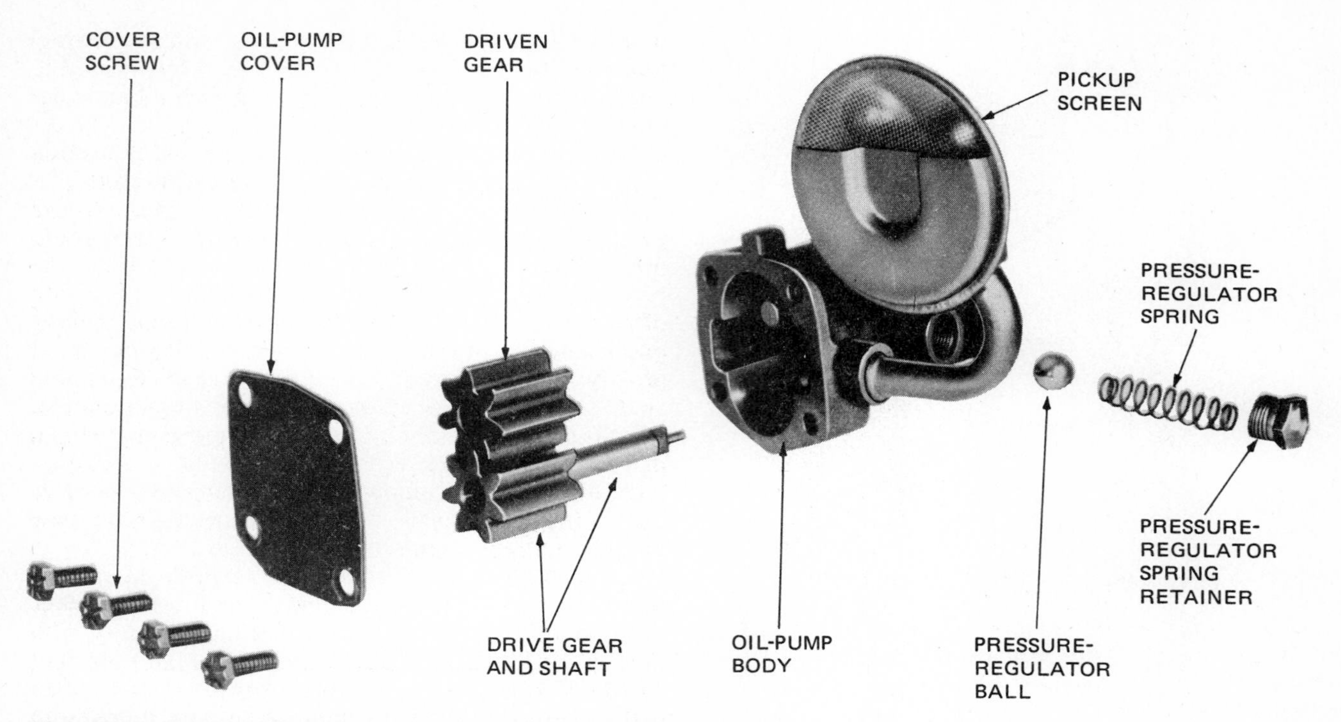

Fig. 16-21 Disassembled view of a gear-type oil pump. (*Pontiac Motor Division of General Motors Corporation*)

engine. The oil pump is at the bottom of the engine (to the lower left). The arrows show the flow of oil from the oil pump to the engine. Note that the system uses an oil cooler. The oil cooler helps to remove excess heat from the engine.

⚙ 16-20 Oil pumps The two general types of oil pumps used in pressure-feed lubricating systems are the gear type (Fig. 16-21) and the rotor type (Fig. 16-22). The gear-type pump uses a pair of meshing gears. As the gears rotate, the spaces between the gear teeth open by moving apart. This draws oil in from the oil inlet. The gear teeth seal to the housing and carry the oil around it to the oil outlet. Then, as the teeth mesh, the oil is forced out through the oil outlet. The rotor-type uses an inner and outer rotor. The inner rotor is driven and causes the outer rotor to turn with it. As this happens, the spaces between the rotor lobes become filled with oil. When the lobes of the inner rotor

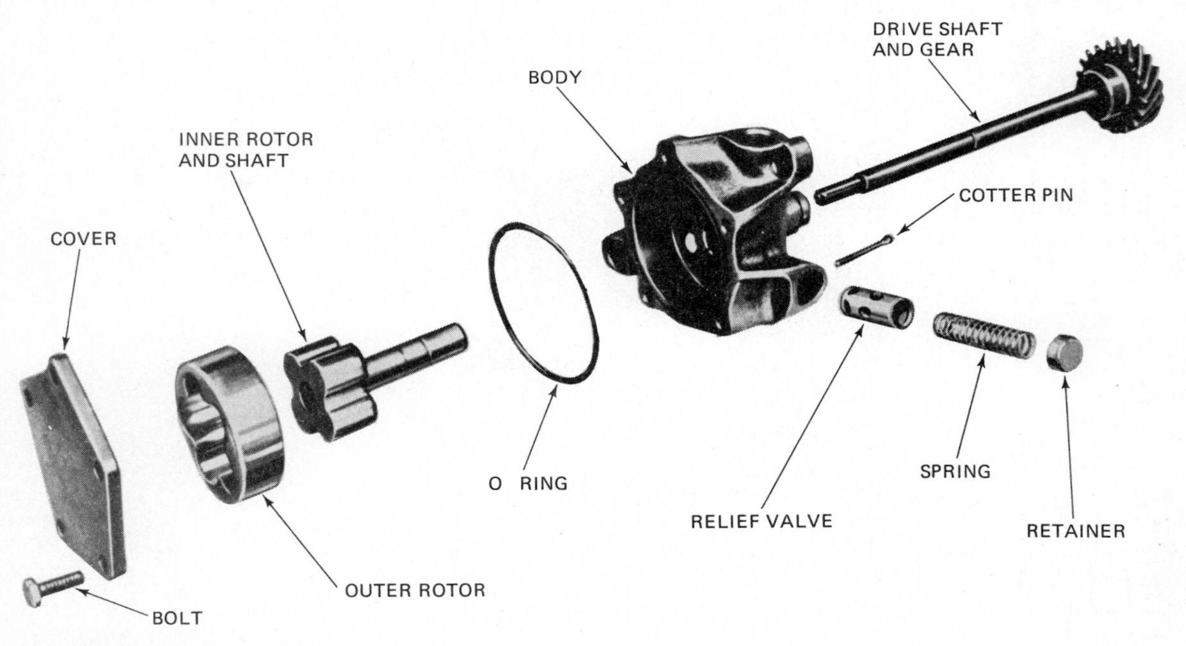

Fig. 16-22 Disassembled view of a rotor-type oil pump. (*Chrysler Corporation*)

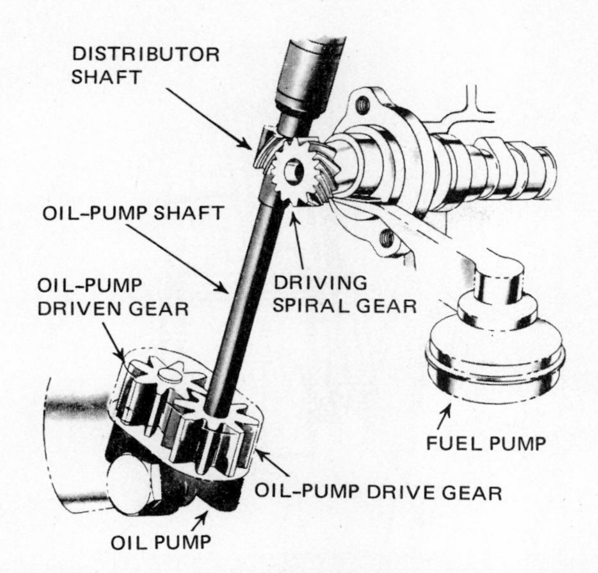

Fig. 16-23 Oil pump, ignition distributor, and fuel pump drives. The oil pump is the gear type. A gear on the end of the camshaft drives the ignition distributor. An extension of the distributor shaft drives the oil pump. The fuel pump is driven by an eccentric on the camshaft. (*Buick Motor Division of General Motors Corporation*)

move into the spaces of the outer rotor, oil is squeezed out through the outlet.

Oil pumps are usually driven from the engine camshaft from the same spiral gear that drives the ignition distributor (Fig. 16-23). In most engines, the oil intake for the oil pump is a rigidly mounted screen assembly. It is mounted to the oil pump or engine block and positioned just above the bottom of the pan. The screen prevents large dirt and metal particles from entering the pump, where they could damage or jam the pump gears or rotor.

⚙ 16-21 Relief valves To keep the oil pump from building up too much pressure, a relief valve is included in the lubricating system. The valve has a spring-loaded ball (Fig. 16-21) or spring-loaded plunger (Fig. 16-22). When the pressure reaches the preset valve, the ball or plunger moves against its spring. This action opens a port through which oil can flow back to the oil pump inlet. Enough of the oil flows past the relief valve to prevent excessive pressure.

The oil pump can normally deliver much more oil than the engine requires. This is a safety factor that ensures delivery of enough oil under extreme operating conditions and high speeds.

⚙ 16-22 Oil coolers Some engine lubricating systems have an oil cooler (Fig. 16-20). Oil coolers are used on almost all automotive air-cooled engines and on Mazda Wankel engines (⚙ 16-19). One type of oil cooler is a small heat exchanger mounted on the side of the engine block. Oil and engine coolant circulate through separate circuits in the heat exchanger. As the coolant circulates, it picks up heat. The heat is carried by the coolant to the radiator, where the heat is transferred to the cooler air flowing through. This process helps cool the oil and keeps it at its normal operating temperature.

Another type of oil cooler uses a small section of the cooling-system radiator so that a separate heat exchanger is not required. The oil cooler for air-cooled engines consists of a small heat exchanger much like the radiator used in the liquid-cooled engine cooling system. Air flows through the oil cooler to remove heat from the oil.

⚙ 16-23 Oil filters The oil filter is the engine's main protection against dirt. The oil from the oil pump must first pass through an oil filter (Fig. 16-1) before the oil goes up to the engine. The filter removes particles of carbon and dirt so they do not get into the engine and damage engine bearings and other parts. The filter contains a filtering element made of pleated paper or fibrous material. The oil passes through the filter, and the paper or fibers trap the dirt particles.

Figure 16-24 is a cutaway view of an oil filter. The filter element is housed in a disposable can which is thrown away when the element becomes clogged with dirt. The filter has a bypass valve. If the filter element becomes so clogged that all the oil needed by the engine cannot pass through the filter, the increased pressure from the oil pump causes the valve to open. Now, some of the oil from the pump can bypass the filter and go directly to the engine to prevent oil starvation. However, this is unfiltered oil. The filter should be changed before this happens.

The filter shown in Fig. 16-24 is called a *single-stage* filter. It uses folded or pleated paper as the filtering element. Dirty oil enters the filter around the outside of the filtering element as shown by the heavy black arrows. Then the oil is forced through the pleated pa-

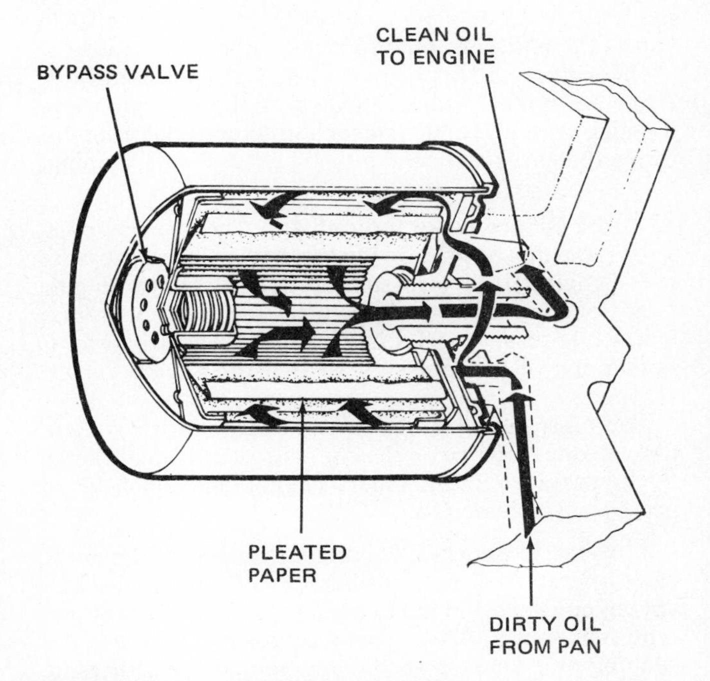

Fig. 16-24 Cutaway view of an oil filter with a built-in bypass valve. (*Chrysler Corporation*)

per, which filters out the dirt and contaminants by trapping them in the paper. The paper has fine pores through which the oil can pass but the dirt particles are too large to pass through these tiny pores.

Cleaned oil flows into the hollow center of the filter element. From there, the clean oil flows on to the engine bearings and other surfaces requiring lubrication.

There is a reason for sending the dirty oil from the outside to the inside of the filtering element. This way, the filtering element has its largest surface exposed to the dirty oil. Therefore, the filter can be used for a longer time before being changed.

Some filters have what is called a *depth-type* filtering element. These filters do not use a pleated-paper element. Instead, the filtering is done by a different type of material, usually a fiber such as cotton, wood, rayon, and nylon. The dirty oil passes through the packed fiber, and the dirt is trapped in it.

Another type of filter that is sold in stores today is the *dual filter*. This filter has two different kinds of filtering material. One claim for the dual filter is that it can be used longer and does not need to be changed so often.

The oil filter should be replaced before it stops working properly. Car manufacturers recommend that oil filters be replaced periodically, as explained in Chapter 17. The usual recommendation is that the filter be replaced at the first oil change and then every other oil change after that.

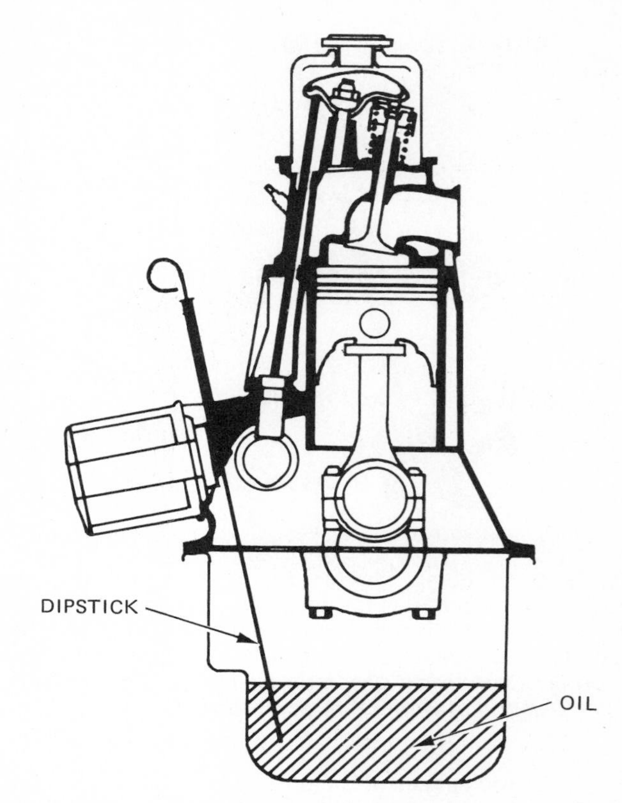

Fig. 16-25 Location of the oil level indicator or dipstick in the engine.

16-24 Oil level indicator A dipstick is used to check the level of the oil in the oil pan (Fig. 16-25). To use the dipstick, allow the engine to sit 1 minute after running. This allows the oil to drain back to the pan. Then pull the dipstick out, wipe it off, and put it back in place. Then pull it out again so that you can check the level of the oil shown on the dipstick. Figure 16-26 shows the markings on different engine oil dipsticks.

On engines with a positive crankcase ventilating system, the dipstick tube is sealed at the top when the dipstick is in place. This keeps unfiltered air from entering the crankcase, and blowby gases from escaping.

16-25 Oil pressure indicators Cars are equipped with some means of showing the driver the oil pressure in the engine. If the pressure drops too low, the engine is not being properly lubricated. Continued operation at low oil pressure will ruin the engine. The driver must be warned of low pressure so that the engine can be stopped.

There are two general types of oil pressure indicators. In one, a gauge on the car instrument panel shows the oil pressure. In the other, a light comes on if the oil pressure drops too low.

1. Electric gauge There are two types of electrically operated oil pressure gauges. They are the balancing-coil type and the bimetal thermostat type. Each type makes use of two separate units. These are the sending unit located on the engine and the indicating unit or gauge which is located on the instrument panel (Fig. 16-27).

The engine unit consists of a variable resistance and a movable sliding contact. The movable contact moves from one end of the resistance to the other as oil pressure against a diaphragm varies. As pressure increases, the diaphragm moves inward. This causes the contact to move along the resistance. More resistance is placed in the circuit between the engine and the indicating unit. This reduces the amount of current that can flow in the circuit.

The indicating unit consists of two magnetic coils that balance each other. When the oil pressure is low, the engine unit has very little resistance. Most of the current flowing through the left coil flows through the low resistance to ground. This provides maximum current flow through the left coil. Therefore, the left coil is stronger magnetically. It pulls the armature around so the pointer swings to the left to indicate a low oil pressure.

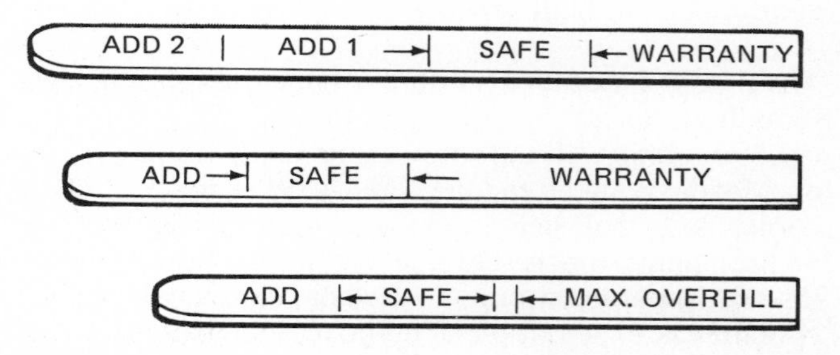

Fig. 16-26 Typical markings on an engine oil dipstick. (*Chrysler Corporation*)

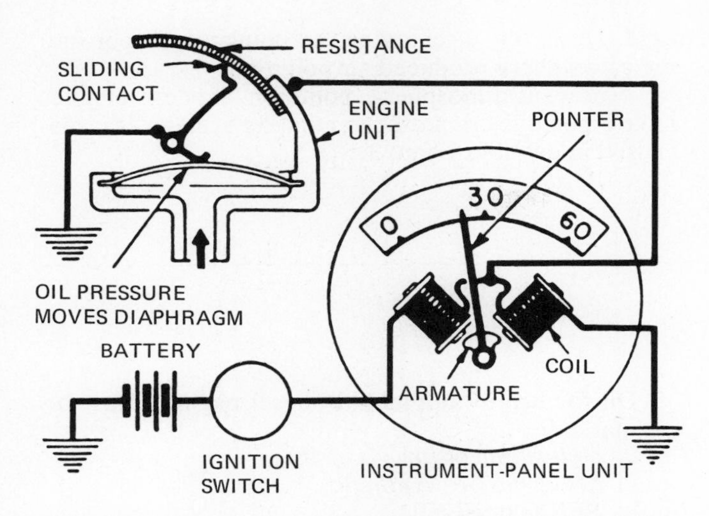

Fig. 16-27 Electric circuit of a magnetic or balancing-coil type of oil pressure gauge, which has a variable resistance in the engine sending unit.

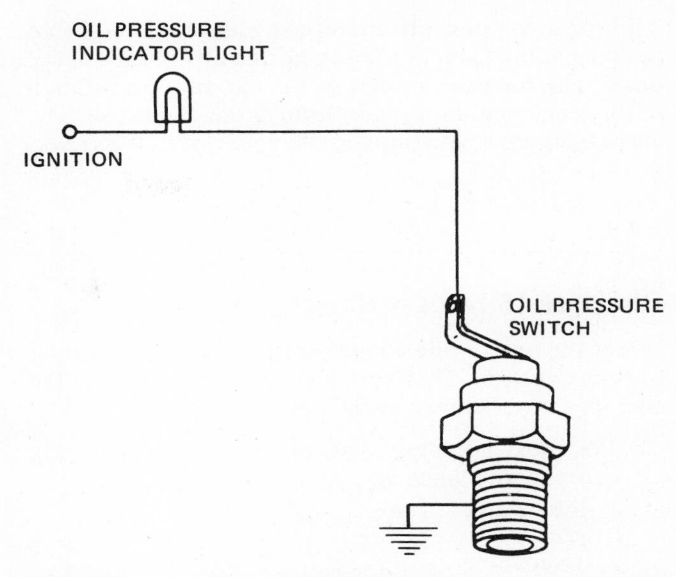

Fig. 16-28 Schematic wiring diagram for an engine oil pressure light. (*ATW*)

When the oil pressure goes up, the resistance of the engine unit increases. Now less current can flow through the engine unit. This allows more current to flow through the right coil. It becomes stronger magnetically and pulls the armature around. The pointer swings to the right to indicate increased oil pressure.

The bimetal thermostat type of oil pressure indicator is similar to the balancing-coil type. However, there is a thermostat in the dash unit. The thermostat is linked to the pointer. As the engine-unit resistance changes, the changing current flowing through the heater coil surrounding the thermostat changes. This causes the thermostat to bend varying amounts to indicate the oil pressure.

2. Indicator light Instead of a gauge, most automobiles have an oil pressure indicator light (Fig. 16-28). The light comes on when the ignition is turned on and the oil pressure is low. Normally, after the engine has started and oil pressure has built up, the light goes off. If it does not, then the engine should be shut off. The lubricating system should be checked at once to find the cause of the low oil pressure.

Just as in the oil pressure gauge system, there are two parts to the oil pressure light system. These parts are the oil-pressure switch and the indicator light. On the engine, the oil pressure switch connects into one of the engine oil gallerys or headers (Fig. 16-29).

The switch is a two-position switch. It is either closed or open, since there are only two positions for the electrical contacts in it. With the engine off, or with low oil pressure, a small spring holds the contacts closed. After the engine starts and the oil pressure gets high enough (usually 10 to 15 psi [69 to 102 kPa]) it overcomes the spring and the contacts are opened.

The indicator light on the instrument panel is electrically connected to the oil pressure switch on the engine. The light and the switch are connected in series to the battery through the ignition switch. When the ignition switch is turned on, the indicator light comes on. It stays on until the engine starts and the oil pressure builds up enough to open the contacts in the switch.

⚙ 16-26 Crankcase ventilation

Air must circulate through the crankcase when the engine is running (⚙ 3-21). This removes the water and liquid gasoline that appear in the crankcase when the engine is cold. Also, it removes blowby gases from the crankcase.

Unless the water, liquid gasoline, and blowby gases are removed from the crankcase, there will be trouble. Sludge and acids will form. Sludge can clog oil filters and lines and starve the lubricating system. Acids corrode metal parts. Either of these troublemakers can ruin the engine.

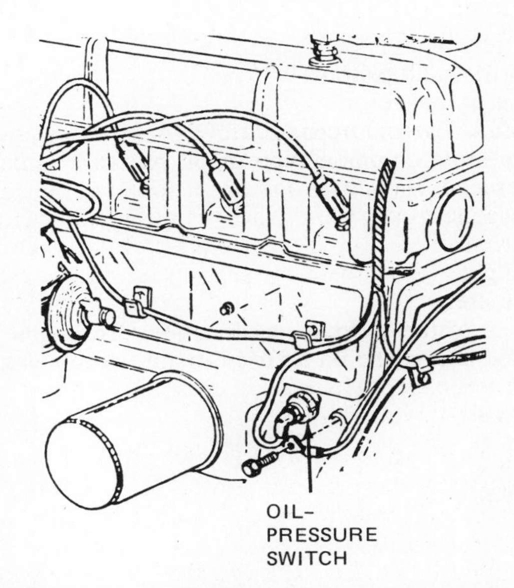

Fig. 16-29 Typical oil-pressure switch location on a six-cylinder engine. (*Ford Motor Company*)

In older engines, the crankcase was ventilated by an opening at the front of the engine and a vent tube at the back. The forward motion of the car and the rotation of the crankshaft moved air through the crankcase. The air passing through removed the vapor from the crankcase. However, discharging the crankcase vapor into the atmosphere produced air pollution.

To prevent atmospheric pollution, modern engines have a positive crankcase ventilating system. It is discussed in detail in Chap. 3.

Chapter 16 review questions

Select the *one* correct, best, or most probable answer to each question. Then check your answers against the correct answers given at the end of the book.

1. The most common type of friction in a running automotive engine is:
   a. dry
   b. greasy
   c. viscous
   d. mechanical.
2. Other factors being equal, the bearing having the least friction is the:
   a. friction bearing
   b. antifriction bearing
   c. sleeve bearing
   d. journal bearing.
3. Almost all bearings used in automotive engines are:
   a. friction bearings
   b. antifriction bearings
   c. guide bearings
   d. thrust bearings.
4. In addition to lubricating engine parts and acting as a cooling agent, the lubricating oil must:
   a. improve carburetion, aid fuel pump, and seal
   b. increase clearances, cool engine, and clean
   c. cool engine, improve combustion, and seal bearings
   d. absorb shocks, seal, and clean.
5. The two types of oil pumps in automotive engines are:
   a. gear and piston
   b. rotor and piston
   c. gear and rotor
   d. full flow and bypass.
6. The most common type of oil pressure indicator uses a
   a. gauge
   b. tube
   c. light
   d. meter.
7. None of the service classifications for engine oils listed below should be used in automobile engines today *except*:
   a. SB
   b. CB
   c. SA
   d. SE.
8. Oil for use in automotive diesel engines must be marked:
   a. *high sulfur* or *low viscosity*
   b. *heavy duty* or *synthetic*
   c. SE/CC or SE/CD
   d. none of the above.
9. Two types of engine lubricating systems are:
   a. pressure feed and force feed
   b. pressure and splash
   c. oil pump and pressure
   d. splash and nozzle.
10. The purpose of the relief valve in the pressure lubricating system is to:
    a. ensure minimum pressure
    b. prevent excessive pressure
    c. prevent insufficient lubrication
    d. ensure adequate oil circulation.
11. Two types of oil filters used in automotive engines are:
    a. single stage and depth type
    b. open and closed
    c. low pressure and high pressure
    d. full flow and flow-through.
12. The purpose of crankcase ventilation is to:
    a. remove liquid gasoline and water
    b. remove vaporized water and gasoline
    c. cool the oil
    d. supply oxygen to the crankcase.
13. Viscosity can be defined as:
    a. ease of flow and fluidity
    b. foaming and flowing
    c. resistance of flow
    d. body and penetration.
14. The substance added to the oil which helps keep the engine clean is called a:
    a. detergent
    b. soap
    c. grease
    d. thickening agent.
15. Most of the dilution of the oil in the crankcase takes place during:
    a. high-speed operation
    b. long trips
    c. engine overheating
    d. engine warm-up.

CHAPTER 17

LUBRICATING SYSTEM SERVICE

After studying this chapter, and with proper instruction and equipment, you should be able to:

1. Locate excess bearing clearance with a bearing prelubricator.
2. Check engine oil pressure with an oil pressure gauge.
3. Describe the formation of sludge in an engine.
4. Diagnose a complaint of excessive oil consumption.
5. Change the oil and oil filter.
6. Perform a chassis lubrication.
7. Replace the oil pan.
8. Overhaul the oil pump.
9. Diagnose and repair a defective oil pressure indicator.

17-1 Lubricating system test instruments The lubricating system is an integral part of the engine. This means that any test of the lubricating system involves making a test on the engine. Some lubricating system testers, such as the bearing prelubricator, require that the engine is not running. Other testers, such as the oil pressure gauge, are used only on a running engine. Various lubricating system checks, the troubles encountered in the lubricating system, and the problems with lubricating oils are discussed in following sections.

17-2 Bearing prelubricator The bearing prelubricator for pressurizing the lubricating system is shown in use in Fig. 17-1. It has two uses. One is to detect excessively worn bearings such as in an older engine. The other use is to pressurize the lubricating system in a rebuilt engine to prelubricate the bearings before starting. The bearing prelubricator includes a pressure tank partly filled with engine oil, with fittings and hose for attaching the tank to a source of compressed air and to the engine lubrication system. The prelubricator is connected to the outlet line of the oil pump or to any point in the system where oil pressure can be applied.

With the oil pan removed so that the main and connecting-rod bearings can be seen, air pressure is applied to the oil in the prelubricator tank. This forces oil into the engine oil galleries. Then as the oil circulates through the engine and runs out, any leak, restriction, or obstruction can easily be detected by the amount of oil running out (Fig. 17-1). In addition, a worn bearing will be disclosed, since it permits the escape of a steady stream of oil around the ends of the bearing.

One prelubricator manufacturer specifies that with SAE 30 oil and 25 psi [172 kPa] of air pressure, a good bearing should leak 20 to 150 drops of oil per minute. This indicates that the bearing clearance is satisfactory. Less than 20 drops of oil per minute indicates a tight bearing or an obstruction in the oil passage. More than 150 drops per minute indicates a worn bearing.

NOTE: When an oil-passage hole in the crankshaft indexes with an oil-passage hole in a bearing, too much oil will be fed to the bearings. Then, an excessive

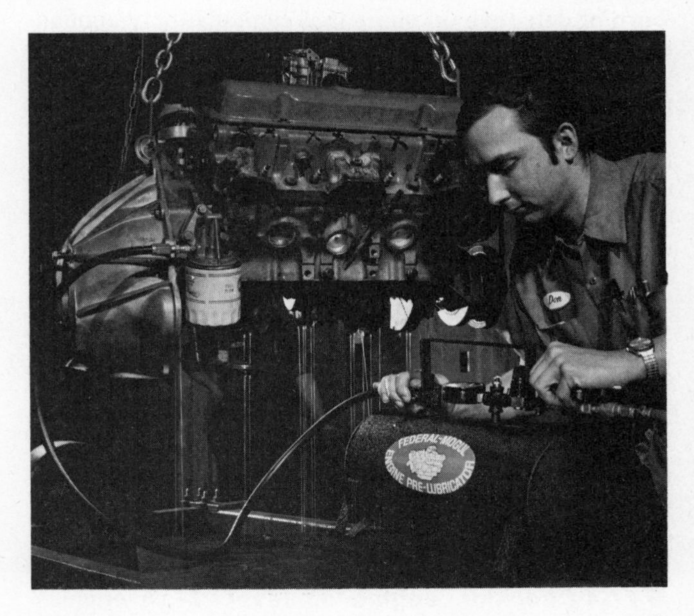

Fig. 17-1 Using an engine prelubricator to make an oil leakage test. (*Federal-Mogul Corporation*)

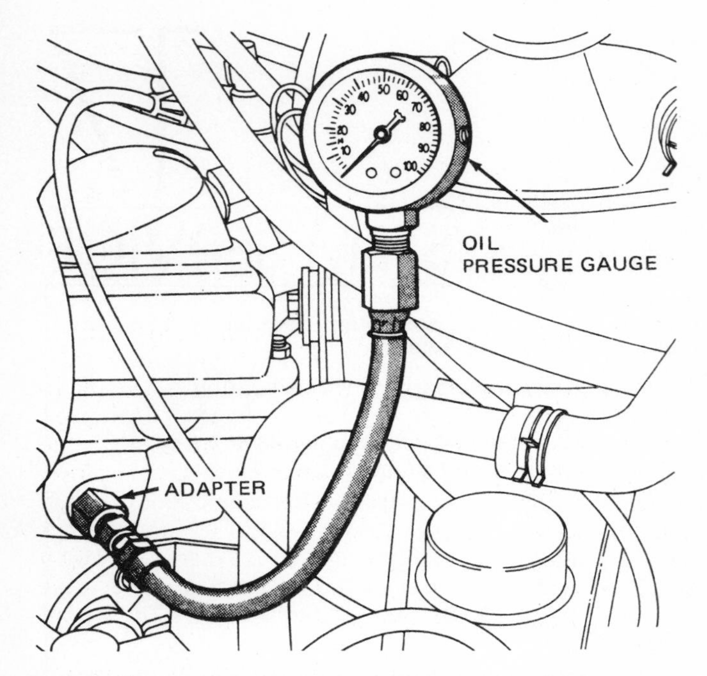

Fig. 17-2 An oil pressure gauge is used to check the oil pressure in a running engine. (*Chrysler Corporation*)

amount of oil will stream out as though the bearings were worn. Rotate the crankshaft slightly to move the holes out of index before making the test. Bearings that have annular grooves (Fig. 16-10) cannot be tested by making a bearing leakage test.

⚙ 17-3 Oil pressure tester Sometimes you need to know the actual oil pressure in the engine. For example, in a car with an oil pressure indicator light, the light might stay on after the engine starts. The cause of this condition could be a defective oil pressure switch in the block. The condition could also be caused by a worn or defective oil pump that cannot supply enough oil to the engine.

A quick and accurate way to check out this problem is to use an oil pressure gauge (Fig. 17-2). To use the gauge, disconnect the wire from the oil-pressure switch and remove the switch. Then install an adaptor for the oil pressure gauge and connect the gauge. Start the engine and note the oil pressure on the gauge with the engine idling.

The manufacturer of one engine specifies that the idling oil pressure should be between 4 and 8 psi [28 to 55 kPa]. Then, with the engine warmed up, the oil temperature at 176°F [80°C], and the engine running at 2000 rpm, the minimum pressure should be 28 psi [193 kPa].

The oil pressure tester can also be used to check the operation of the oil pressure switch and dash unit in a car that has an oil pressure gauge in the dash instead of an indicator light.

⚙ 17-4 Checking engine oil level Most engines have a dipstick type of oil level indicator (⚙ 16-24). The dipstick can be withdrawn, wiped clean, reinserted, and again withdrawn so that the oil level on the dipstick can be seen.

The dipstick is marked to indicate the proper oil level (Fig. 16-27). If the oil level is low, the oil should be drained and the crankcase refilled with clean oil. The oil should be changed at regular intervals (⚙ 17-6). ⚙ 17-6 also explains how to change the oil.

⚙ 17-5 Sludge formation Sludge is a thick, creamy, black substance that often forms in the crankcase. It clogs oil screens and oil lines, preventing normal circulation of lubricating oil to engine parts. This can result in engine failure from oil starvation.

1. How sludge forms Water collects in the crankcase in two ways: First, water is formed as a product of combustion (⚙ 2-1). Second, the crankcase ventilating system (⚙ 3-21) carries air with moisture in it through the crankcase. If the engine parts are cold, the water condenses and drops into the crankcase. There, it is churned up with the lubricating oil by the crankshaft. The crankshaft acts much like a giant eggbeater. It whips the oil and water into the thick, black, mayonnaise-like "goo" called sludge. The black color comes from dirt and carbon. Figure 17-3 shows an oil pickup screen clogged with sludge.

2. Why sludge forms If a car is driven for long distances each time it is started, the water that collects in the crankcase while the engine is cold quickly evaporates. The crankcase ventilating system then removes the water vapor. No sludge will form. However, if the engine is operated when cold most of the time, then sludge will form. For example, the home-to-store-to-home sort of driving, each trip being only short distance, is sludge-forming. When a car is used for short-trip start-and-stop driving, the engine never has a chance to warm up enough to get rid of the water that collects in the crankcase. The water accumulates and forms sludge.

3. Preventing sludge formation To prevent sludge formation, a car must be driven long enough for the engine to heat up and get rid of the water in the crankcase. This means trips of 12 miles [19 km] or more in winter (but shorter in summer). If trips of this length

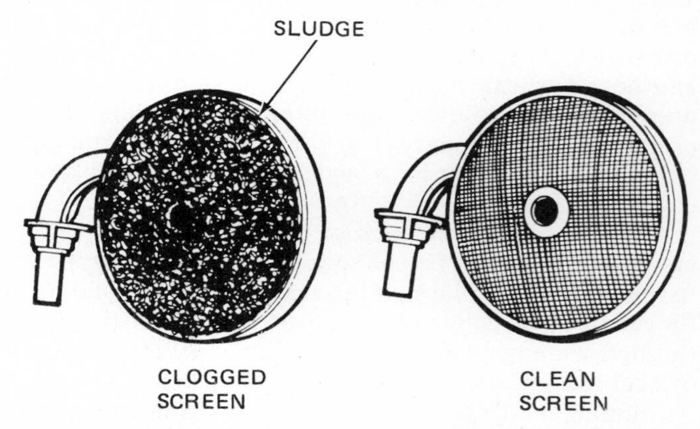

Fig. 17-3 An oil pickup screen clogged with sludge, compared with a clean screen. (*Chrysler Corporation*)

are impractical, then the oil must be changed frequently.

⚙ 17-6 Oil changes From the day that fresh oil is put into the engine crankcase, it begins to lose its effectiveness as an engine lubricant. This gradual loss of effectiveness is largely a result of the depletion or "wearing out" of the additives. For example, the antioxidant additive becomes used up, allowing gum and varnish to form. The corrosion and rust inhibitors are gradually depleted, allowing corrosion and rust formation to begin.

In addition, during engine operation, carbon tends to form in the combustion chamber. Some of this carbon gets into the oil. Also, the air that enters the engine air cleaner carries a certain amount of dust. Even though the air filter is operating efficiently, it will not remove all the dust. Also, the engine releases fine metal particles as it wears.

All these substances tend to circulate with the oil. As the mileage increases, the oil accumulates more and more of these contaminants. Even though the engine has an oil filter, some of these contaminants remain in the oil. Finally, the oil is so loaded with contaminants that it is not safe to use. Unless the oil is drained and clean oil is put in, engine wear will increase rapidly.

Different automotive manufacturers have different recommendations on how often engine oil should be changed. For example, Chrysler recommended for their 1977–1978 cars that the oil should be changed every 6 months or 7500 miles [12,000 km], whichever comes first. In 1979, Chrysler liberalized this recommendation to every 12 months or 7500 miles [12,000 km]. For severe service conditions such as extended periods of idling and short-trip operation, dusty driving conditions, and trailer towing, the oil should be changed at least twice as often.

Ford has different recommendations for their various engines. For example, on their 1979 in-line engines, they recommend changing the oil *and oil filter* every 10,000 miles [16,000 km] or 12 months, whichever comes first. For their V-8 engines, the recommendation is to change the engine oil every 7500 miles [12,000 km] or 12 months, whichever comes first. The oil filter should be replaced on the first oil change and then every other oil change thereafter. For severe service, change the oil twice as often.

General Motors recommends oil changes every 7500 miles [12,000 km] or 12 months, whichever comes first. For severe service, change the oil twice as often. On the General Motors V-8 diesel engine, the recommendation is to change the oil every 3000 miles [4800 km].

If you look back to the recommendations for previous years, you will notice that oil changes were more frequent. For example, Chrysler at one time recommended changing the oil every 4000 miles [6400 km]. The reason that oil change intervals have been increased is because engine oil and oil filters have been improved.

The procedure of changing the engine oil and oil filter is covered in ⚙ 17-10 and 17-11. Always check the manufacturer's specifications on oil-change intervals before deciding when to change the oil and the oil filter.

⚙ 17-7 Oil consumption Oil is lost from the engine in three ways: by burning in the combustion chambers, by leakage in liquid form (Fig. 17-1), and by passing out of the crankcase as a mist through the PCV system (⚙ 3-21). Two main factors affect oil consumption, *engine speed* and the *amount that engine parts have worn*.

High speed produces high temperature. This, in turn, lowers the viscosity of the oil. Now it can more readily work past the piston rings into the combustion chamber, where it is burned. In addition, the high speed exerts a centrifugal effect on oil feeding through the crankshaft to the connecting-rod journals. Therefore, more oil is fed to the bearings and thrown on the cylinder walls. Also, at high speeds the oil-control rings cannot function as effectively. Crankcase ventilation (⚙ 3-21) causes more air to pass through the crankcase at high speed. This causes oil to be lost in the form of mist.

As engine parts wear, oil consumption increases. Worn bearings tend to throw more oil onto the cylinder walls. Tapered and worn cylinder walls prevent normal oil-ring action. The rings cannot change shape rapidly enough to conform with the worn cylinder walls as the rings move up and down. More oil gets into the combustion chamber, where the oil burns and fouls spark plugs, valves, rings, and pistons. Carbon formation worsens the condition, since it further reduces the effectiveness of the oil rings. The burning of oil in the combustion chamber usually produces blue smoke from the tail pipe.

Worn intake valve guides also increase oil consumption. Oil leaks past the valve stems and is pulled into the combustion chamber along with the air-fuel mixture every time the intake valves open. Worn exhaust valve guides can also cause high oil consumption. In this case, the oil is burned as the hot exhaust gases hit it when the exhaust valve opens. Installation of new valves guides, reaming of guides and installation of valves with oversize stems, or replacement of valve-stem seals will reduce oil consumption from these causes.

NOTE: See ⚙ 17-8 for a discussion of oil consumption and the checks to make to find the cause.

⚙ 17-8 Diagnosing troubles in the lubricating system Most troubles in the lubricating system are directly related to other engine problems. The most common lubricating system troubles are discussed below.

1. Excessive oil consumption Most lubricating system troubles produce excessive oil consumption. The cause of this problem is not always easy to determine. Oil can be lost from the engine in three ways. These are by burning in the combustion chamber and exhaust manifold, by liquid oil leaking out, and by oil vapor and mist flowing through the PCV system into the cylinders.

Complaints of excessive oil consumption must be

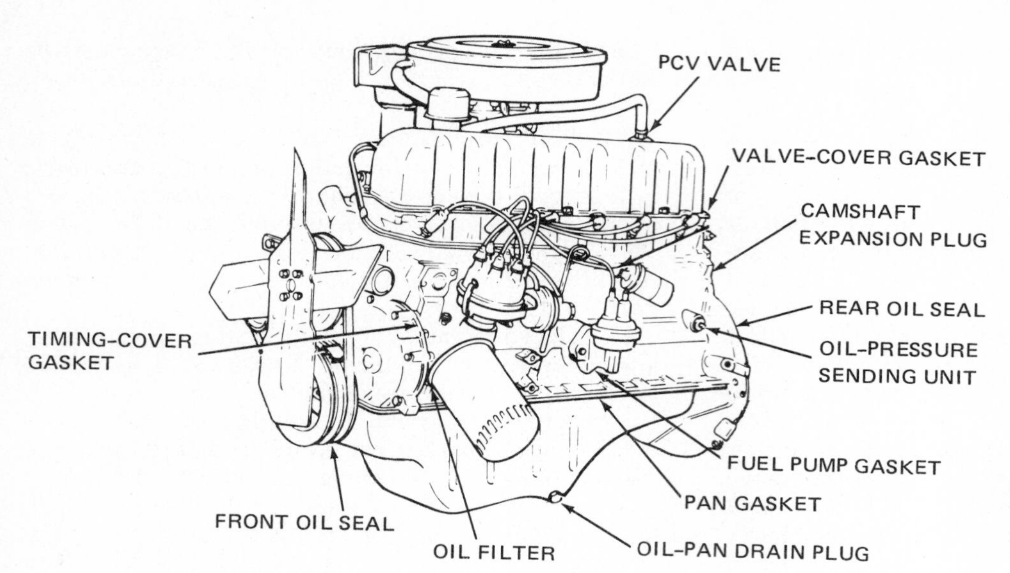

Fig. 17-4 Places where oil may be lost by leakage from an engine. (*Ford Motor Company*)

verified. The need to frequently add oil to maintain the proper oil level in the crankcase is a good indicator. However, the actual amount of oil consumption can be accurately checked by filling the crankcase to the correct level with oil. Next, operate the car for about 500 miles [805 km]. Then measure the additional oil that must be added to bring the oil back to the original level.

Figure 17-4 shows some of the places where oil may leak out of an engine. External leaks often can be seen by looking at the seals around the oil pan, valve cover and timing-gear housing, and at oil line and oil filter connections. The presence of an excessive amount of oil indicates leakage.

The burning of oil in the combustion chamber usually produces blue smoke from the tail pipe. Oil can enter the combustion chamber in three ways. These are through the clearance between the intake-valve guides and valve stems, around the piston rings, and through the PCV system, as discussed earlier.

a. *Intake valve guides* Oil can enter the combustion chamber through clearance caused by wear between the valve guides and the valve stems (Fig. 17-5). When clearance is excessive, oil is drawn into the combustion chamber on each intake stroke. The appearance of the underside of a valve provides a clue to the condition of its stem and the guide.

When the underside of the valve has an excessive amount of carbon, the valve guide and possibly the valve stem are excessively worn. Some of the oil that passes around the valve remains on the underside, forming carbon deposits. When this condition is found, it is usually necessary to install valve seals or a new valve guide. The condition may be helped by knurling the valve guide. A new valve may also be required.

Excessive clearance between the exhaust valve and its guide allows oil to enter the exhaust manifold. In some engines, more oil may be consumed in this way than through a worn intake valve guide.

b. *Rings and cylinder walls* Probably the most common cause of excessive oil consumption in an old engine is wear. It allows oil to pass between the piston rings and cylinder walls into the combustion chamber. This condition also is called *oil pumping*. It results from worn or stuck piston rings.

In addition, when the bearings are worn, an excessive amount of oil is thrown onto the cylinder walls. The piston rings may not be able to handle all of the oil. Then, they allow too much oil to work up into the combustion chamber.

Failure of the rings to control the oil on the cylinder walls is not limited to old or worn engines. Sometimes new piston rings fail to seat. This also allows excess oil to get into the combustion chamber.

c. *Speed* Another possible cause of excessive consumption is engine speed. High speed produces high oil temperatures and thin oil. This combination causes more oil to be thrown onto the cylinder walls. The piston rings, moving at high speed, cannot function effectively. Therefore, more oil slips past the rings into the combustion chamber.

High engine speed increases the churning effect as the crankshaft strikes the oil falling into the pan. More oil vapor and mist are created in the crankcase. Then more oil is lost as the oil vapor and mist are

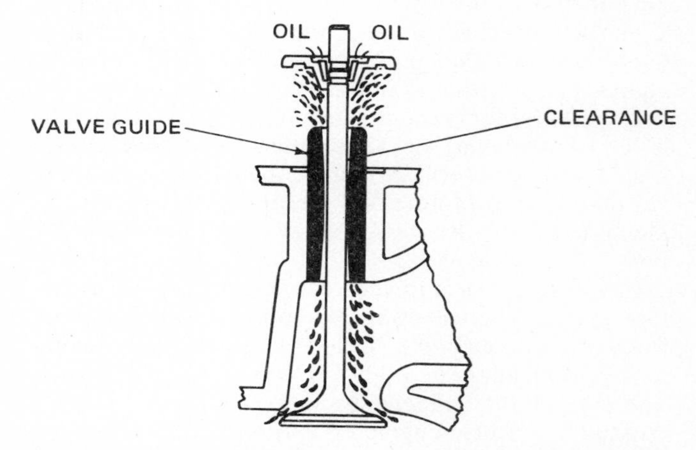

Fig. 17-5 Excessive oil consumption can be caused by oil leaking into the combustion chamber through worn valve guides. (*Perfect Circle Division of Dana Corporation*)

carried through the PCV system. Tests have shown that an engine uses much more than twice as much oil at 60 miles per hour [96 km/h] as it does at 30 mph [48 km/h]

2. Low oil pressure Low oil pressure can result from a weak relief-valve spring in the oil pump, a worn oil pump, a broken or cracked oil line, obstructions in the oil lines, insufficient or thin oil, or bearings so badly worn that they can pass more oil than the oil pump is capable of delivering. A defective oil pressure indicator may show low oil pressure.

3. Excessive oil pressure Excessive oil pressure may result from a stuck relief valve in the oil pump, an excessively strong relief-valve spring, a clogged oil line, or thick oil. A defective oil pressure indicator may show high oil pressure.

4. Oil dilution When the car is used for short trips with sufficient time between runs to allow the engine to cool, the engine is operating most of the time on warm-up. Under this condition, the oil is subject to dilution by unburned gasoline seeping down into the crankcase past the piston rings. In addition, water collects in the crankcase. This happens because the engine does not operate long enough at temperatures high enough to evaporate the water.

Water and gasoline change the lubricating properties of the oil by forming sludge. The result is that engine parts wear more rapidly. When this type of engine operation occurs, the oil should be changed at half the normal time or mileage interval. This is to remove the sludge and diluted oil.

⚙ 17-9 Lubricating system service There are certain lubricating system services that are included during other engine service and repair jobs. For example, the oil pan is removed and cleaned during the replacement of crankshaft bearings or piston rings.

When the crankshaft is removed, you should clean out the oil passages in it. Also, the oil passages in the cylinder block should be cleaned out. The following sections discuss how to service the lubricating system by changing the oil and filter, cleaning the oil pan, and servicing the oil pump and oil pressure indicators. Checking the oil level was covered in ⚙ 17-4.

⚙ 17-10 Changing oil The reasons for changing the oil, and the recommended intervals for changing it, were discussed in ⚙ 17-6. It is the using up of the oil additives, and the accumulation of contaminants, that makes oil changes necessary. To change the oil you need as many quarts or liters of oil as listed in the manufacturer's specifications to replace the oil you drain from the engine. The oil must be of the correct type and grade for the engine. Proceed as follows.

1. If the engine is cold, run it for about 5 minutes to warm up the oil. Then shut off the engine.

CAUTION: Operate the engine only in a well-ventilated area. Exhaust gas contains deadly carbon monoxide which can kill in a closed area.

Fig. 17-6 Engine oil is changed by placing the container under the oil-pan drain hole and then removing the drain plug.

2. Raise the car on a lift (Fig. 17-6). Or, you can use a jack to raise the front end of the car. Then place stands under the front end and let the car down onto the stands.

CAUTION: Do not go under the car with only the jack holding the front end up. If the jack slips, the car could fall and injure you.

3. With a wrench or socket, loosen the drain plug slightly until you can turn it with your fingers (Fig. 17-7).
4. Place the drain pan under the plug to catch the oil. Be sure the pan is properly placed to catch the oil and that it will hold the amount of oil that is in the crankcase. The oil will flow outward at first rather than straight down.
5. Remove the plug with your fingers and let the oil drain into the pan (Fig. 17-8). While removing the

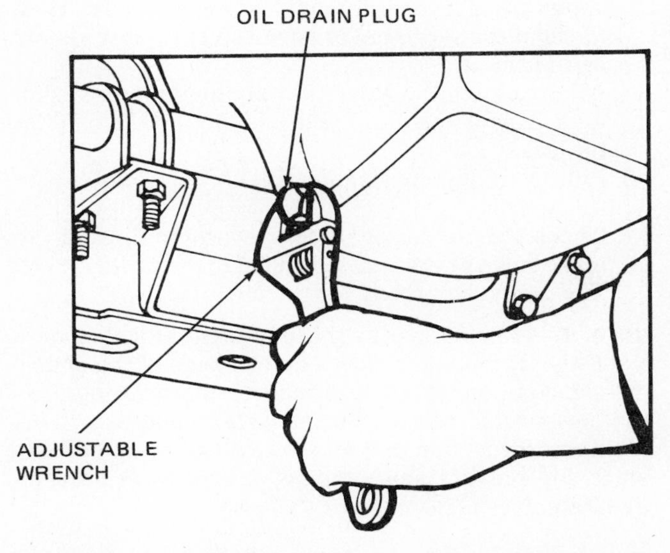

Fig. 17-7 Removing the drain plug from the oil pan. (*Chrysler Corporation*)

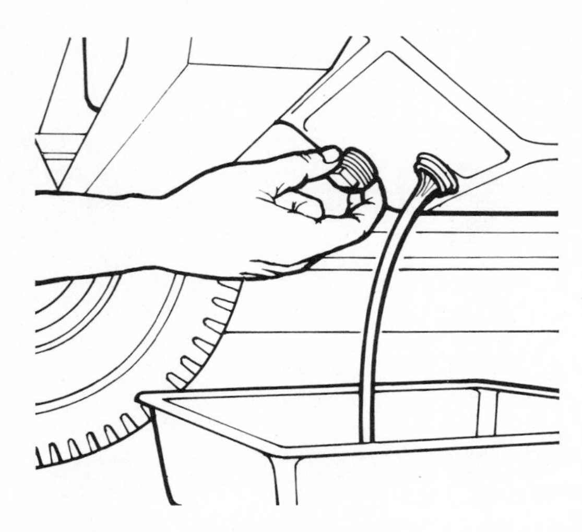

Fig. 17-8 Draining the oil from the engine. (*Chrysler Corporation*)

plug, keep your arm above the plug or else the oil will run down your arm.

CAUTION: If the engine is hot, the oil can be hot enough to burn you.

6. While the oil is draining, look at the drain plug gasket to make sure it is in good condition. After the oil has drained, start the plug into the threads in the drain hole. Be careful that you do not cross-thread it. Then tighten the plug securely, but be careful not to overtighten it.
7. If the filter does not need to be changed, lower the car. Remove the oil filler cap. Fill the crankcase to the required level with the proper type and grade of engine oil. Start the engine.

NOTE: If the oil filter needs to be changed, see ⚙ 17-11.

8. Watch the oil indicator light or gauge on the instrument panel. It should glow or the gauge should show low pressure for only a few seconds. Then the light should go out or normal oil pressure should be indicated.
9. After running the engine for about 5 minutes, shut it off. Check the drain plug and around the oil filter for leaks.
10. Fill out a new lubrication sticker and attach it to the vehicle.
11. Check the crankcase oil level with the dipstick, to make sure the crankcase is filled to the proper level. The car is then ready for the customer.

NOTE: In some shops, every time a car is raised on a hoist, the technician makes a quick check of the under-the-car components. These include the steering linkage, suspension parts, exhaust system, and tires. This is a quick inspection that may detect trouble at an early stage and prevent more expensive repairs or an accident later.

⚙ 17-11 Changing oil filter According to the manufacturer's recommendations, the oil filter should be changed the first time the oil is changed in a new engine (Fig. 17-9). After that, the filter should be changed every other time the oil is changed, on most engines. For some engines, the manufacturer specifies that the oil filter should be changed every time the oil is changed. Refer to the manufacturer's specifications for the recommended oil-change interval.

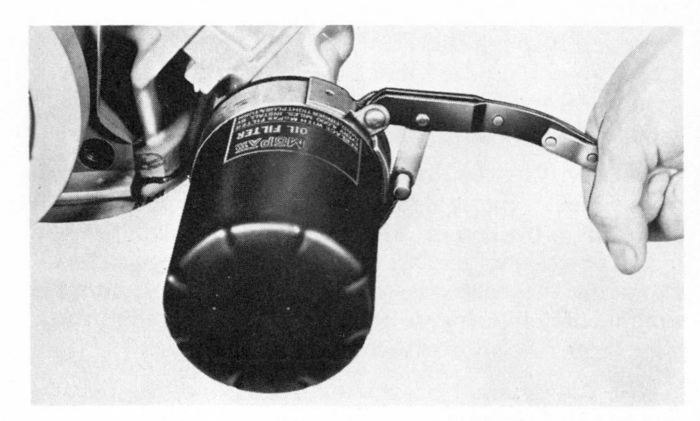

Fig. 17-9 Replacing an oil filter. (*Chrysler Corporation*)

You can tell whether or not a filter is working by feeling it after the engine has been running. If the filter is hot, it has been passing oil. If it is only warm, it may be clogged and not allowing oil to filter through. On automotive engines, the filter element and container are replaced as a single disposable unit. Some heavy-duty oil filters have a separate cartridge type of filter element. The container remains on the vehicle while the filter element inside is replaced (⚙ 17-12). Here is the way to replace the integral filter and container.

1. After draining the oil and installing the drain plug (item 6 in ⚙ 17-10), move the drain pan under the oil filter. Using the oil-filter wrench (Fig. 17-9), loosen the filter one or two turns counterclockwise. Do not remove it until oil from the filter drains into the drain pan.
2. Then unscrew the filter and put it in the drain pan, gasket-end facing up. Make sure that the old filter gasket is not on the engine. Clean any sludge out of the filter mounting recess on the engine. Clean the engine oil-filter gasket surface.
3. Compare the new filter with the old filter. If the new filter is a different size, make sure it will have sufficient clearance from the frame and suspension after it is installed.
4. Check the gasket end of the new filter. It should be the same size as on the old filter. The threads for attaching the filter to the engine must be the same for both filters.
5. Coat the face of the gasket on the new filter with clean oil. If the filter mounts in a position that allows it to be filled with oil before it is installed, fill it with fresh oil. Make sure the gasket is properly positioned on the filter. Then hand-tighten the filter until the gasket makes contact with the engine gasket surface. Be careful to avoid cross-threading the filter into the engine.
6. Now tighten the filter by hand an additional half

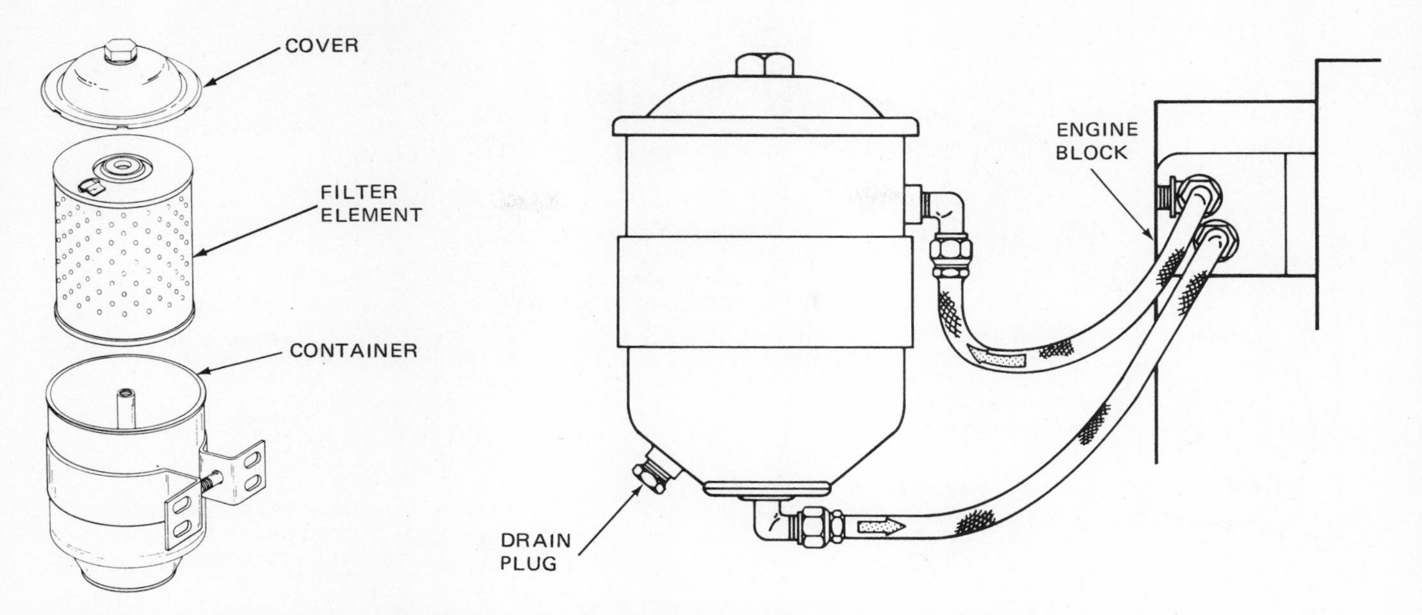

Fig. 17-10 An oil filter using a replaceable or cartridge type of filter element. (*Kohler Company*)

turn, or the amount specified, to make sure it is securely tightened. Do not use the wrench to tighten the filter, unless it is specified. Wipe the filter and mounting area clean.

7. Fill the crankcase with oil, as explained in item 7 in ⚙ 17-10. Start the engine and watch the oil indicator light or gauge for oil pressure. If the light stays on or the pressure gauge shows no pressure after 30 seconds, stop the engine and check for the cause.
8. If oil pressure is OK, run the engine for a few minutes. Shut it off and check around the oil filter and drain plug for leaks.
9. On the lubrication sticker, write the odometer reading and the date on which the oil filter was changed. As a final check, use the dipstick to make sure the crankcase is full. Add oil if necessary.

⚙ 17-12 Heavy-duty filter On heavy-duty engines, the filter element is a separate cartridge that can be removed from the container (Fig. 17-10). To do this, oil lines may have to be disconnected. The procedure is as follows.

Remove the drain plug (if present) from the bottom of the filter housing. Take the cover off by loosening the center bolt or clamp. Lift out the element. If the housing has no drain plug, remove the oil or sediment with a pump. Wipe the inside of the housing with a clean cloth. Be sure that no trace of lint or dirt remains. Install the new filter element.

Install the drain plug and cover, using a new gasket. Start the engine, and check for leaks around the cover. Note whether the oil pressure has changed. With a new element, which passes oil more easily, oil pressure may be lower. Check the level of oil in the crankcase, and add oil if necessary. On larger engines, the new element may soak up as much as a quart of oil. It may be necessary to add oil.

⚙ 17-13 Automatic engine-oil refiller Some heavy-duty trucks have a device that dumps additional oil from a reservoir into the engine crankcase when the oil level falls below a safe level. On one engine, the reservoir holds 2.5 gallons [9.5 L] of oil. The refiller is mounted close to the engine so that the oil is kept warm. When the oil level in the crankcase drops below a safe level, oil from the refiller flows down into the crankcase and restores the correct crankcase oil level.

⚙ 17-14 Chassis lubrication In addition to changing the oil and filter, the technician may also check the lubrication sticker to see if a chassis lubrication is due. Various chassis parts should be checked periodically. Some require lubrication. Others do not. Figures 17-11 and 17-12 provide typical examples of the engine and chassis lubrication points.

The lubricants, the amount, and how often they should be used are included in the manufacturer's service manual. Many owner's manuals also may contain this information.

⚙ 17-15 Oil pan service Whenever the oil pan is removed from the engine, the pan should be cleaned to remove the accumulated deposits and sludge. Then the pan should be inspected for dents, cracks, and damage around the drain hole and on the gasket and seal surfaces.

1. Removing oil pan Oil pan removal varies on different cars because of the interference of various other parts. On many cars, the steering idler arm or other steering linkage must be detached. In such cases, carefully note how the linkage is attached and also the number of shims (when used) so that the linkage can be correctly reattached. In addition, certain other parts may require removal. For example, on some engines the exhaust crossover pipe, starting motor, and fly-

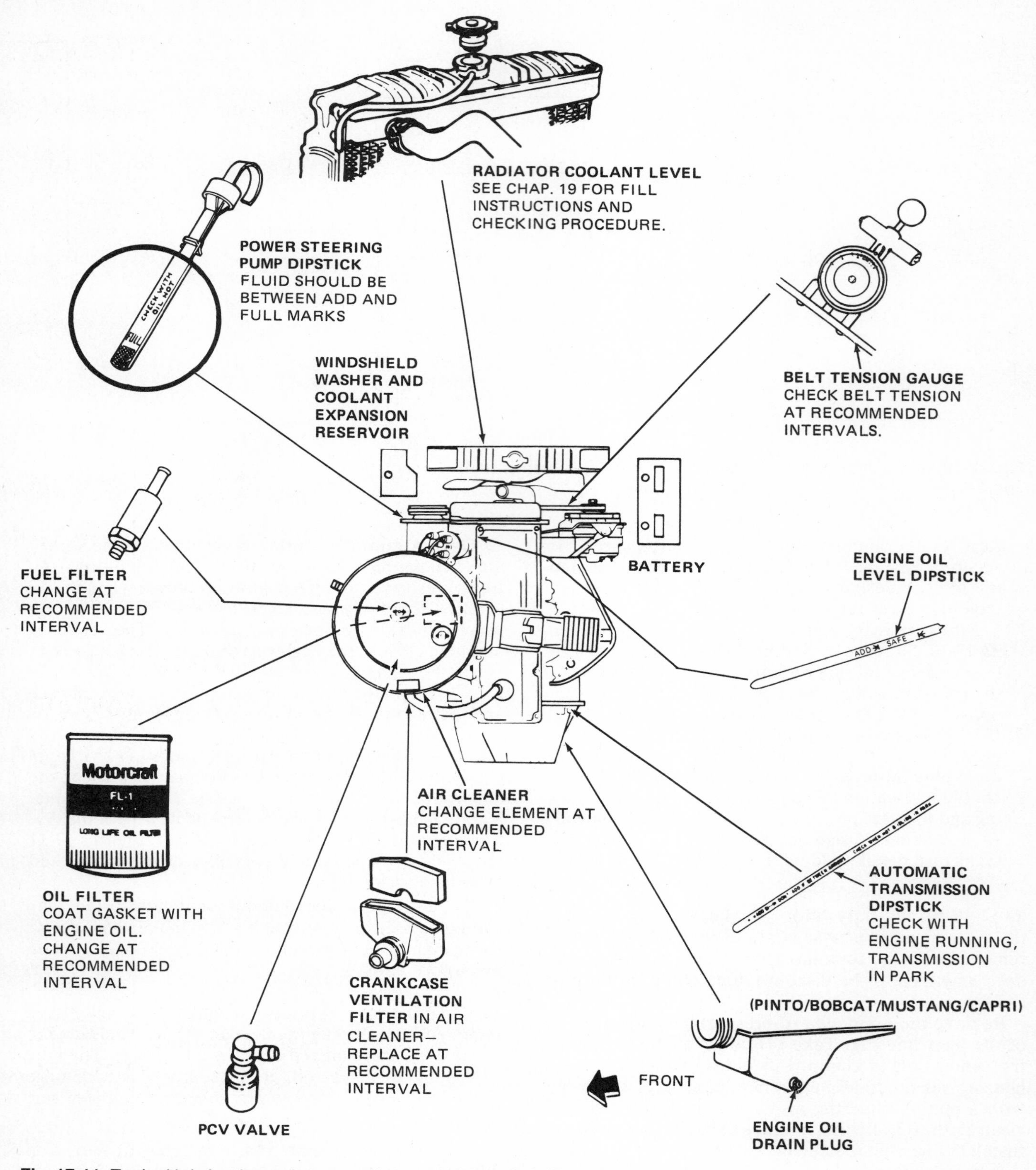

Fig. 17-11 Typical lubrication points on and around the engine. (*Ford Motor Company*)

wheel housing cover must be removed. On other cars, it is necessary to remove engine mounting bolts and to raise the front end of the engine. On some engines the clutch housing dustcover should be removed to prevent damage to the oil pan gaskets.

With the preliminaries out of the way, the drain plug should be removed so that the oil can drain out. Then the attaching bolts should be taken off so that the oil pan can be removed. To prevent the pan from dropping, steady it while the last two bolts are being taken out.

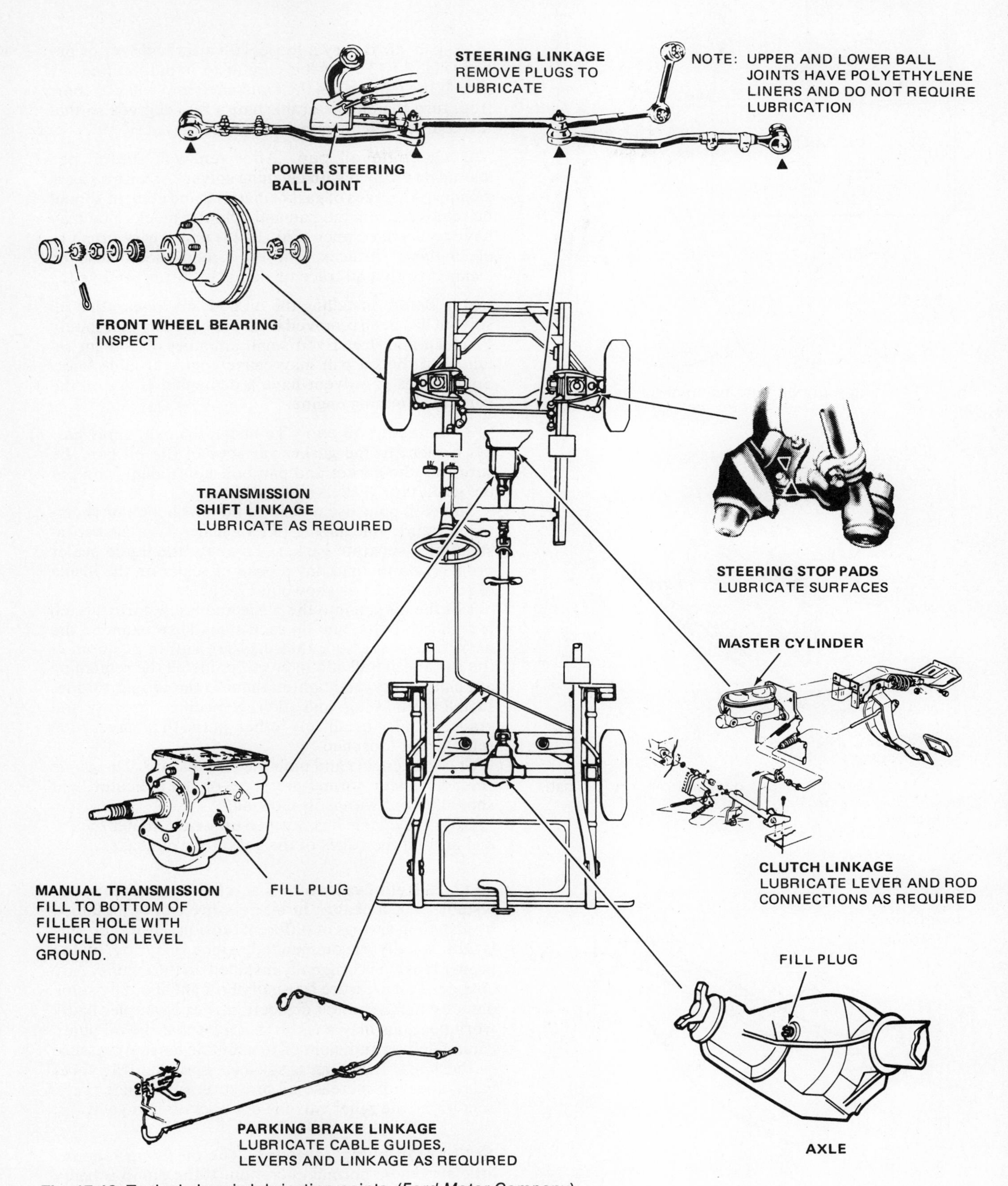

Fig. 17-12 Typical chassis lubrication points. (*Ford Motor Company*)

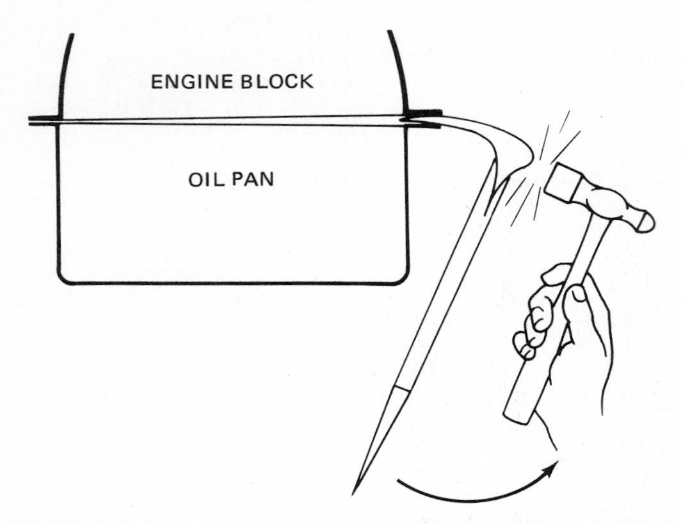

Fig. 17-13 Using a pry bar and hammer to break the oil pan loose.

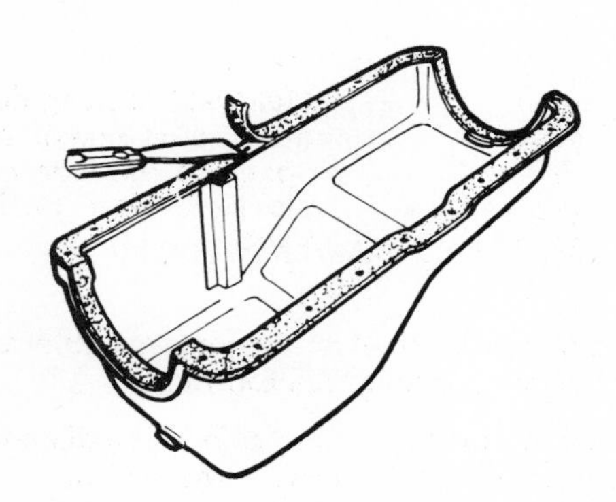

Fig. 17-14 Scraping the old gasket from the oil pan. (*Chrysler Corporation*)

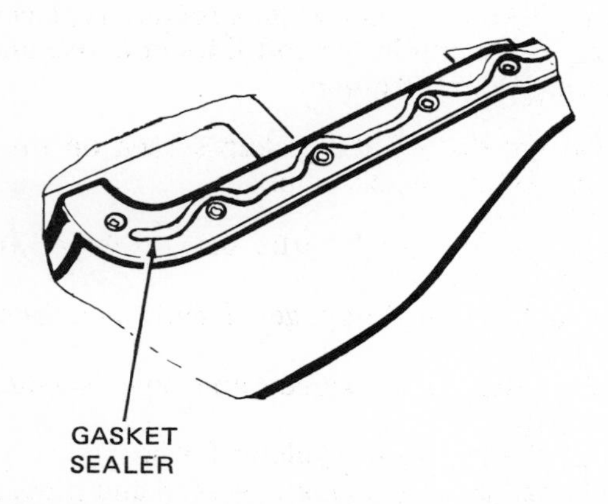

Fig. 17-15 Some pan gaskets use a gasket made with a bead of special gasket sealer, instead of a cork-type gasket. (*Chrysler Corporation*)

If the pan sticks, pry it loose with a screwdriver or pry bar (Fig. 17-13). But be careful to avoid denting the pan. If the pan strikes the crankshaft and will not come free, turn the engine crankshaft a few degrees so that the counterweights move out of the way.

2. Cleaning oil pan After removal, the oil pan should be cleaned with cleaning solvent or with a steam cleaner. All traces of gasket material and cement should be removed from the pan and cylinder block. You may have to use a scraper (Fig. 17-14) and a wire brush to clean these surfaces. The oil screen should also be cleaned so that all trace of sludge or dirt is removed.

NOTE: Before installing the oil pan, make sure that all solvent has been removed from the pan. The pan should be clean and dry. Even small amounts of solvent retained in the oil pan may cause engine trouble later. Some types of solvent have a damaging effect on the parts in a running engine.

3. Installing oil pan To install the pan, apply gasket cement to the gasket surfaces of the oil pan. Be sure that the gasket and pan bolt holes align, and put the gasket (or gaskets) into position.

Some oil pans use a bead of liquid silicone or room-temperature vulcanizing (RTV) sealer for a gasket instead of a separate gasket. To apply the liquid sealer properly on the pan, lay a bead of sealer on the inside of the bolt holes as shown in Fig. 17-15.

Lift the oil pan into the place and temporarily attach it with two bolts, one on each side. Then examine the gaskets to make sure that they are still in position. If the gaskets are all still in position, install the remaining attaching bolts, and tighten them to the proper torque. Install the oil plug, and add the specified quantity and grade of oil. Install any other parts that have been removed or loosened.

Start the engine and note the oil pressure. The gauge should register normal pressure, or the indicator light should go out within 30 seconds.

After the engine has warmed up, check under the car and around the edges of the gasket for oil leaks.

⚙ 17-16 Relief valves Most relief valves are not adjustable, but a change in oil pressure can be obtained by installing springs of different tension. However, this is not usually recommended, since a spring of the proper tension is originally installed on the engine. Any change of oil pressure is usually brought about by some defect which requires correction. For example, badly worn bearings may pass so much oil that the oil pump cannot deliver sufficient oil to maintain normal pressure in the lines. Installing a stronger spring in the relief valve does not increase oil pressure, since under such conditions the relief valve is not operating anyway.

⚙ 17-17 Oil pump removal The oil pump requires little service in normal operation. If the pump is badly worn, it will not maintain oil pressure and should be removed for repair or replacement.

An oil pump should not be serviced or replaced unless oil pressure cannot be maintained. Before blaming

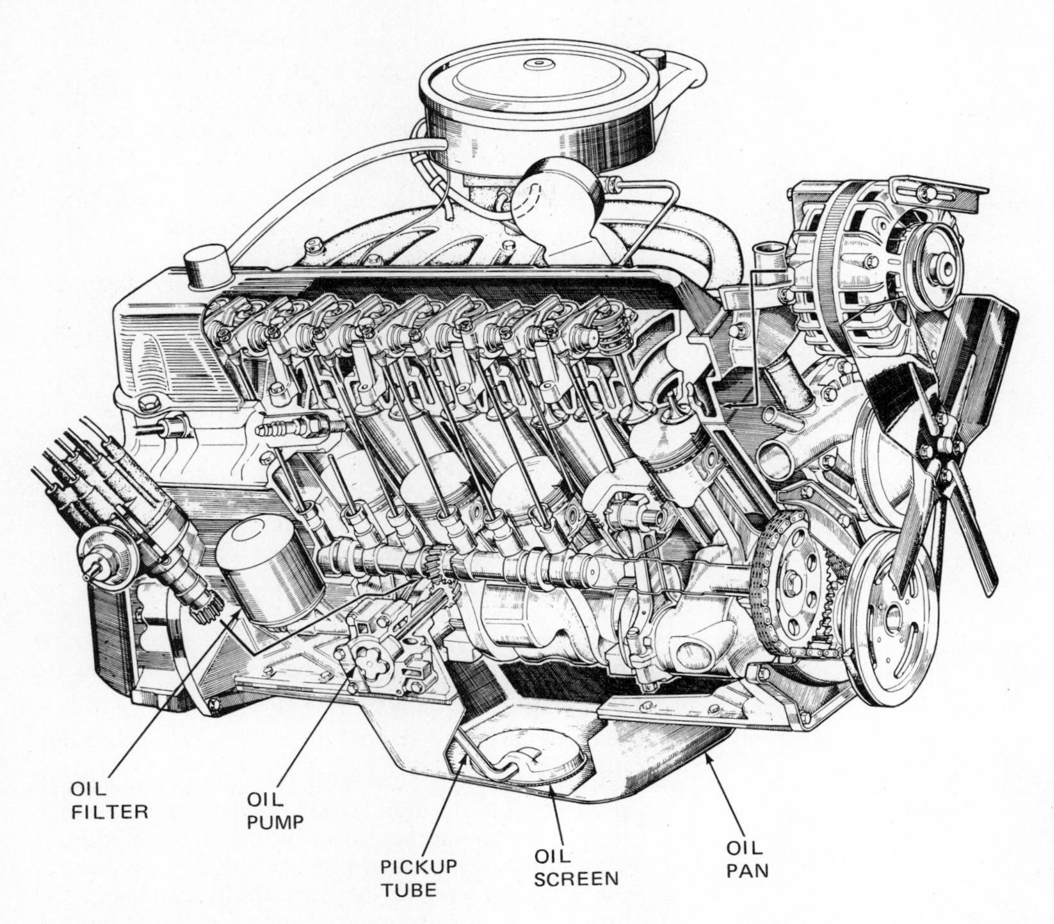

Fig. 17-16 An engine that has a rotor-type oil pump bolted to the side of the block. (*Chrysler Corporation*)

the pump, be sure the crankcase oil level is correct. Check that the oil pressure problem is not the result of a defective oil pressure indicator, or some engine condition. For example, worn engine bearings may pass so much oil that the pump cannot deliver enough oil to maintain pressure.

In some engines, the pump is mounted above the oil pan on the side of the block (Fig. 17-16). On these engines, it is not necessary to drain the oil or remove the pan to replace the pump. If the oil pump is located in the crankcase, the oil must be drained and the oil pan removed as explained in ۞ 17-15. You may have to remove the bolts from the front engine mounts and raise the engine about 2 inches [51 mm]. Put small wood blocks between the engine and the mount brackets to hold the engine up. Do not raise the engine by using a jack under the crankshaft pulley.

Remove the oil pump attaching bolts, the oil pump, the gasket, and intermediate drive shaft, if used. See ۞ 17-18 for repair procedures.

۞ 17-18 Servicing oil pumps There are two types of oil pump, gear and rotor, as described in ۞ 16-20 and illustrated in Figs. 16-22 and 16-23. Oil pumps for some late-model engines are serviced by complete replacement. If the pump is defective, it is discarded and a new pump is installed. Always refer to the manufacturer's shop manual covering the engine you are working on to see if disassembly is recommended. If so, make sure that spare parts that might be required for repair are available before you disassemble a pump. Typical servicing procedures follow.

1. Gear-type pump Figure 17-17 is a disassembled view of a gear pump. To service the pump:

1. Remove the pump cover attaching screws, cover, and gasket. Mark the gears so you can put them back into the pump body with the same teeth indexing.
2. Remove the gears and shaft.
3. Remove the pressure regulator retaining pin, spring, and valve.
4. If the pickup screen and pipe require replacement, clamp the pump in the soft jaws of a vise and pull the pipe from the pump.

NOTE: Do not disturb the pickup screen on the pipe. This is serviced as an assembly.

5. Wash and clean all parts, using solvent and air hose.
6. Inspect the pump body, gears, and cover for cracks and wear.
7. Inspect the pickup screen and pipe assembly for damage.
8. Check the pressure regulator for fit.
9. If you removed the pickup screen and pipe, a new screen-and-pipe assembly must be installed. If you reinstalled the old assembly, air would leak around the pipe where it fits into the pump body and the

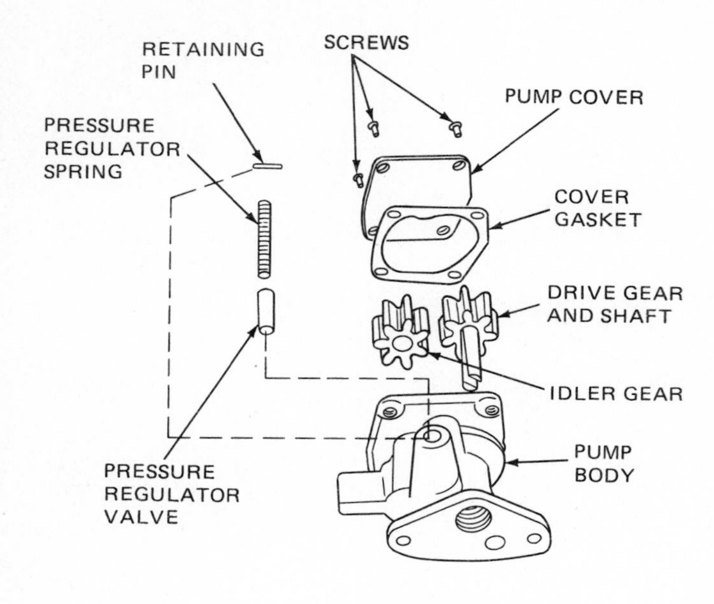

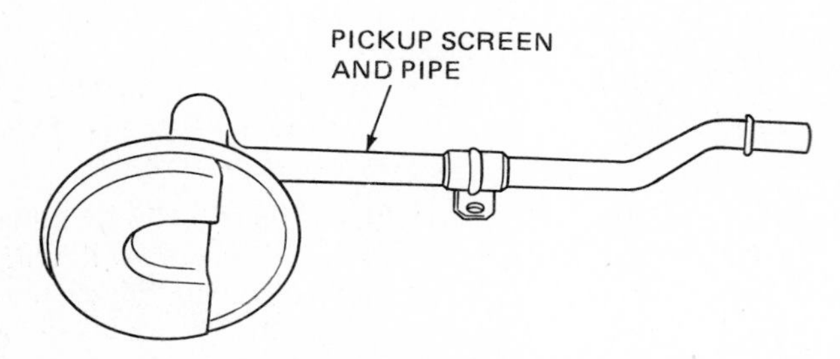

Fig. 17-17 A disassembled gear-type oil pump. (*Chevrolet Motor Division of General Motors Corporation*)

oil pressure would be very low. To install the new screen and pipe, clamp the pump body in a vise with soft jaws. Apply sealer to the end of the pipe. Then tap the pipe into place with a plastic hammer. The pickup screen must be parallel to the bottom of the oil pan when the oil pump is installed. Work carefully so you do not twist, shear, or collapse the pipe.

10. Install the pressure regulator, spring, and pin.
11. Install the gears. The smooth side of the idler gear goes toward the pump cover. Install the pump cover with a new gasket, and torque the attaching screws to specifications. Turn the drive shaft by hand to check for smooth operation.
12. When reinstalling the pump in the engine, align the oil-pump drive-shaft slot with the tang on the distributor shaft. Put the pump into position and attach it with the attaching screws.

NOTE: The pump should go into place easily. If it does not, recheck the alignment of the slot in the drive shaft and the tang on the distributor shaft.

2. Rotor-type pump Figure 17-18 is a disassembled view of a rotor-type oil pump. To disassemble it, remove the pump cover and seal ring. Support the drive gear and press the shaft out of the gear. Remove rotors and shaft. Then remove the pressure relief valve plug, gasket, spring, and valve.

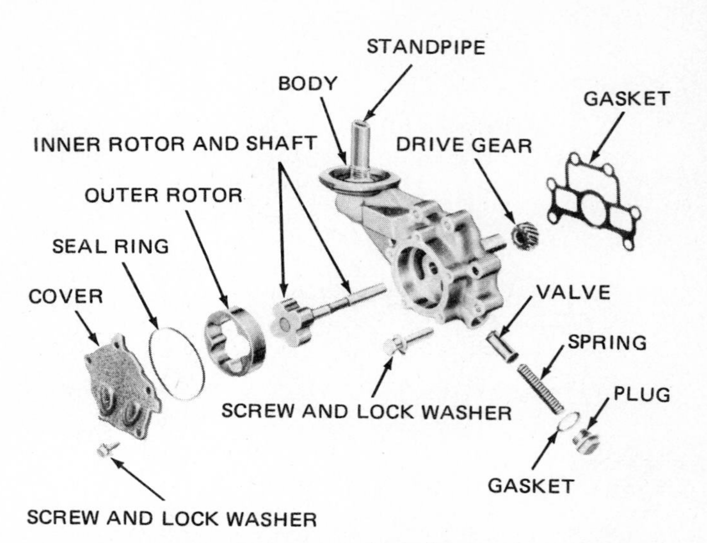

Fig. 17-18 A disassembled rotor-type oil pump. (*Chrysler Corporation*)

NOTE: Individual parts are not serviced separately. If the pump cover, rotors, pump body, or other parts are worn or damaged, install a new pump.

Clean all parts and inspect them for wear or damage.

1. Lay a straightedge across the pump cover and check whether a 0.0015-inch [0.038-mm] thickness gauge can pass between the cover and the straightedge. If it can, the cover is worn and a new pump is required.
2. Use a micrometer to measure the thickness of the rotors. Compare the measurements with the factory specifications. If the rotors are worn, replace the pump.
3. When reassembling the pump, measure the clearance between the inner and outer rotors, and between the outer rotor and pump body, with thickness gauges. Also, lay a straightedge across the pump body and use a thickness gauge to check the clearance between the rotors and straightedge. If any clearance exceeds factory specifications, replace the pump.
4. Use new oil-seal rings between the pump body and cover when installing the cover. Tighten the cover bolts to specifications.
5. Prime the oil pump before installing it by filling the rotor cavity with engine oil.
6. Install the pump on the engine, tightening the attaching bolts to specifications.

NOTE: Some oil pumps require priming before they are installed because they cannot start oil circulation in the engine if there is only air in the pump cavity. Engine oil can be used to prime the pump or, if this is difficult, the pump may be primed by packing it with petroleum jelly (Fig. 17-19).

⚙ 17-19 Checking oil pressure indicators Oil pressure indicators (⚙ 16-27) require very little service. Defects in either the dash unit or the engine unit usually require replacement of the defective unit.

1. Indicator light check A typical oil pressure in-

dicator light system is shown in Fig. 16-29. The light should glow if the oil pressure is low.

1. Turn the ignition switch on but do not start the engine. The light should glow.
2. Start the engine. The light should go out. If it does not, check for a defective oil-pressure switch or low oil pressure.
3. If the light does not glow when the ignition switch is turned on, short the terminal of the oil-pressure switch to ground. If the light now comes on, the switch is defective. If the light does not come on, the light bulb is burned out or the wire from the bulb to the ignition switch or oil-pressure switch is defective.

2. **Oil pressure gauge** Operation of this gauge is described in ❂ 16-25. To check the gauge, install a T-fitting between the engine oil-pressure sending unit and its block fitting. Then attach a pressure gauge to the fitting (Fig. 17-2). Start the engine. If the engine oil pressure is normal on the test gauge, the sending unit on the dash unit is defective. Replace the sending unit and try again. If this does not cure the trouble, the dash unit is defective. Replace it.

❂ 17-20 Chemical cleaners When valves and piston rings have become so coated with carbon and other deposits that they cease to operate properly, it may be necessary to overhaul the engine. Regular oil changes tend to remove the impurities held in suspension in the oil before they have a chance to settle on engine parts.

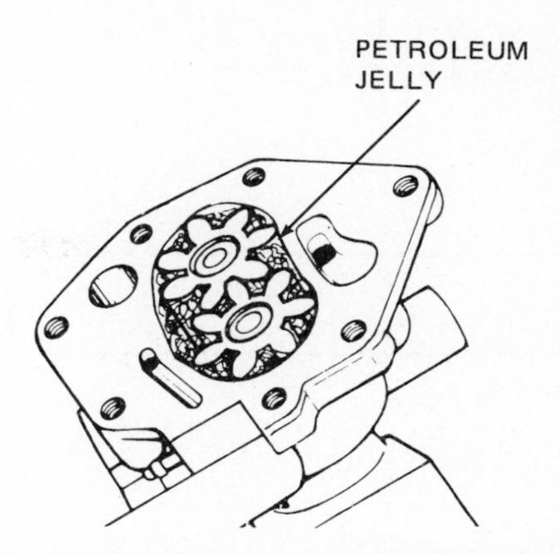

Fig. 17-19 Some oil pumps must be packed with petroleum jelly before they will prime themselves and start pumping. (*Chrysler Corporation*)

Some technicians suggest the introduction of chemical compounds into the engine oil and through the intake manifold as an aid in freeing sticking valves and rings. When engine trouble is a result of carbon deposits on valves and rings, and when these parts are not excessively worn or damaged, improved engine performance may be obtained without engine overhaul.

NOTE: Manufacturers caution that some fuel and oil additives must not be used in engines equipped with a catalytic converter. Use of these chemical compounds in catalytic-converter-equipped vehicles may damage the catalyst and render it inoperative.

Chapter 17 review questions

Select the *one* correct, best, or most probable answer to each question. Then check your answers against the correct answers given at the end of the book.

1. Common causes of excessive oil consumption include:
   a. heavy oil and tight bearings
   b. high speed and worn engine parts
   c. short trips and cold weather
   d. frequent oil changes and weak valve springs.
2. Oil is lost from the engine in three ways, by passing as a mist through the PCV system, by leaking in liquid form, and by:
   a. evaporating
   b. burning in the combustion chambers
   c. condensing
   d. leaking into the transmission.
3. Oil can enter the combustion chambers in two ways—around the valve stems and:
   a. past the float bowl needle
   b. past the manifold gaskets
   c. through the intake manifold
   d. past the piston rings.
4. Sludge forms more rapidly in engine oil with:
   a. high-speed driving
   b. slow-speed driving
   c. short-trip driving
   d. long-trip driving.
5. Oil-filter elements, according to usual recommendations, should be replaced every:
   a. 2000 miles [3219 km]
   b. every oil change
   c. every other oil change
   d. every 2 years.

CHAPTER 18

ENGINE COOLING SYSTEMS

After studying this chapter, you should be able to:

1. Describe the operation of the two types of engine cooling systems.
2. Explain the operation of the water pump.
3. Define *variable-speed fan.*
4. Discuss the flow of coolant through the two types of automotive radiators.
5. Explain why cars with automatic transmissions are equipped with a transmission-oil cooler.
6. Discuss how the heater works.
7. Explain the operation of the thermostat.
8. Name the two valves used in the radiator pressure cap.
9. Define *antifreeze*.

18-1 Purpose of the cooling system The purpose of the cooling system is to keep the engine at its most efficient operating temperature at all speeds and under all operating conditions. During the combustion of the air-fuel mixture in the engine cylinders, temperatures of 4000°F [2200°C] or higher may be reached by the burning gases. Some of this heat is absorbed by the cylinder walls, cylinder head, and pistons. They, in turn, must be provided with some means of cooling so that they will not get too hot.

Cylinder wall temperature must not go higher than about 400 to 500°F [205 to 260°C]. Temperatures higher than this cause the lubricating-oil film to break down and lose its lubricating properties. However, the engine operates best at temperatures as close to the limits imposed by oil properties as possible. Removing too much heat through the cylinder walls and head lowers the thermal efficiency of the engine. Cooling systems are designed to remove about one-third (30 to 35 percent) of the heat produced in the combustion chambers by the burning of the air-fuel mixture.

The engine is very inefficient while cold. Therefore the cooling system includes devices that prevent normal cooling action during engine warm-up. These devices allow the engine parts to reach their normal operating temperatures more quickly. This shortens the inefficient cold operating time. When the engine reaches its normal operating temperature, the cooling system begins to function. The cooling system cools the engine rapidly when it is hot, and slowly or not at all when the engine is cold or warming up.

Two general types of cooling systems are used on automobile engines. They are air cooling and liquid cooling. Most automotive engines are liquid-cooled. Most engines for airplanes, snowmobiles, motorcycles, power lawn mowers, and chain saws are air-cooled.

⚙ 18-2 Air-cooled engines In air-cooled engines, the cylinders are semi-independent. They are not grouped in a block. There are metal fins on the heads and cylinders to help dissipate the heat from the engine. Shrouds and fans are used on some air-cooled engines to improve the air circulation around the cylinders and heads. Figure 18-1 illustrates a six-cylinder air-cooled engine made by Chevrolet. Figure 18-2 shows the flow of air around the cylinder of a small air-cooled engine.

⚙ 18-3 Liquid-cooled engines In operation, the engine cooling system is really a temperature-regulating system. It maintains the engine temperature within certain limits, neither too hot nor too cold. A cold engine is inefficient. Combustion of the air-fuel mixture is incomplete. The engine may run roughly, have excessive fuel consumption and exhaust emissions, and contaminate the crankcase with excessive blowby and liquid fuel dripping into it. The job of the cooling system is to get the engine up to normal operating temperature as quickly as possible and then maintain it at that temperature. The cooling system should not allow overheating or overcooling.

In the liquid-cooling system (Fig. 18-3), a liquid is circulated around the cylinders to absorb heat from the cylinder walls. The liquid is water to which an antifreeze solution (⚙ 18-19) is added to prevent freezing in cold weather. The mixture is called the coolant. The coolant is heated as it passes through the engine. Then the hot coolant flows through a radiator in which the heat in the coolant is passed on to air that is flowing through the radiator. The water passage, radiator size, and other details are so designed as to maintain the cylinder walls, head, pistons, and other working parts at efficient, but not excessive, temperatures.

After the coolant has been heated by passing through

Fig. 18-1 Sectional view from the top of a six-cylinder overhead-valve air-cooled engine. This type of engine is sometimes referred to as a pancake engine. (*Chevrolet Motor Division of General Motors Corporation*)

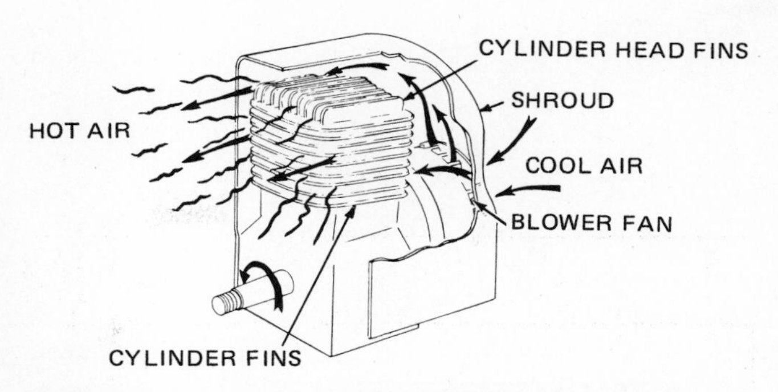

Fig. 18-2 Circulation of air around the cooling fins of an air-cooled engine.

the engine, it flows through the radiator where it loses heat. The cooled coolant then flows back through the engine. This circulation of the coolant continually removes heat from the engine. The coolant is kept in circulation by the water pump.

18-4 Water jackets Just as we might put on a sweater or a jacket to keep warm on a cool day, so water jackets are placed around the engine cylinders. There is this difference: Water jackets are designed to keep the cylinders cool. The water jackets are open spaces between the outside wall of the cylinder and the

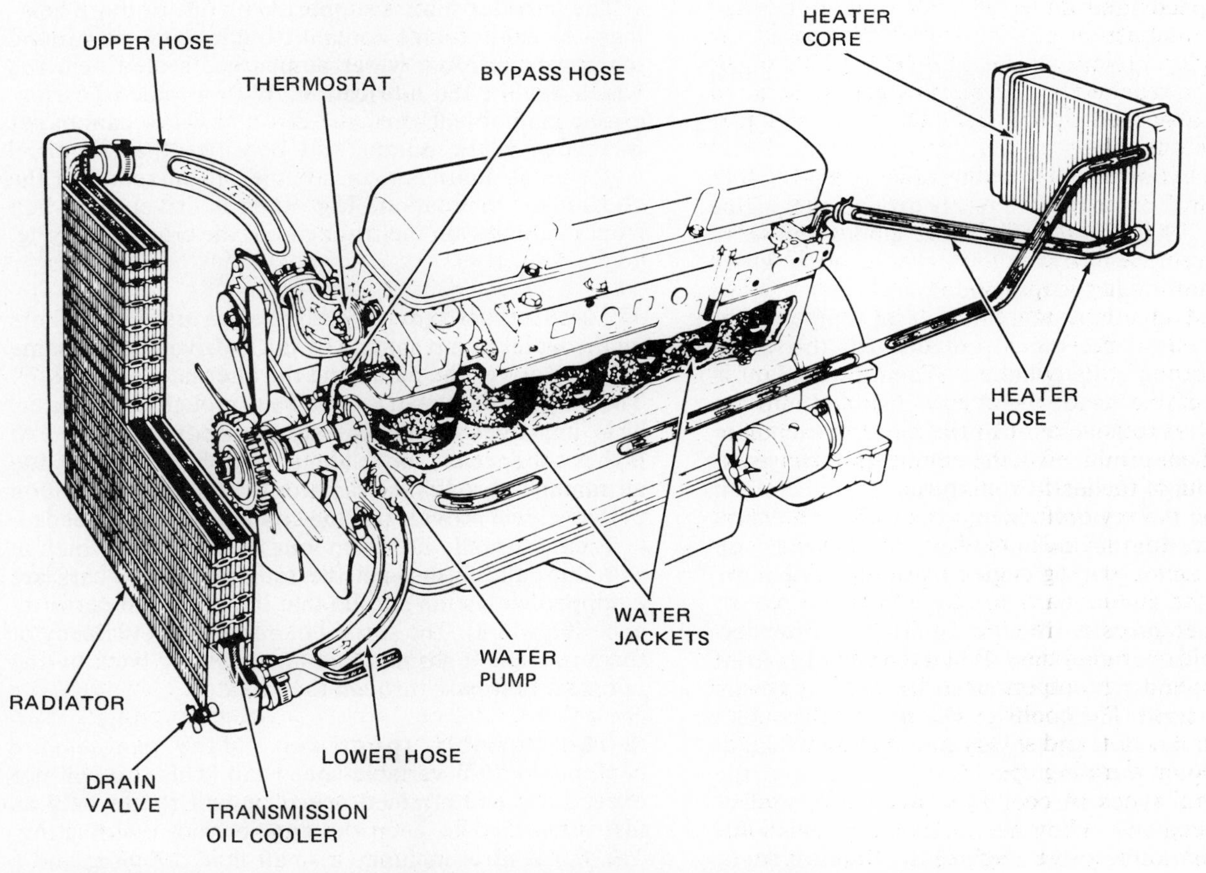

Fig. 18-3 Operation of the liquid cooling system on an engine. The small arrows show the direction of coolant flow. (*Chrysler Corporation*)

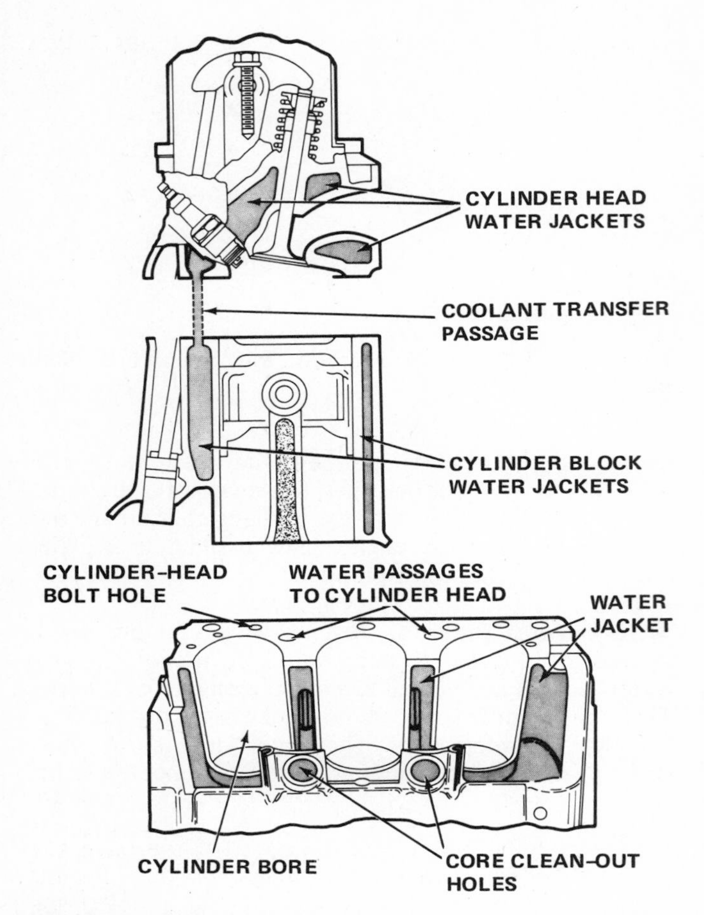

Fig. 18-4 Water jackets in cylinder head and cylinder block.

inside of the cylinder block and head (Fig. 18-4). Note that the coolant can circulate freely around the engine "hot spots." These include the valve guides and valve seats, and the upper parts of the cylinder walls where the pistons and rings slide up and down.

When the engine is running at normal temperature, coolant flows into the block and through the water jackets surrounding the cylinders. Then the coolant is forced through the head gasket openings and into the cylinder head water jackets. In the head, the coolant flows around the combustion chambers and valve seats, picking up additional heat. From the heads, the coolant flows through the upper hose into the radiator. There, the temperature of the coolant is lowered, and the coolant is drawn again into the engine by the water pump.

⚙ 18-5 Water pumps Water pumps are impeller-type centrifugal pumps. They are attached to the front end of the cylinder block between the block and the radiator, as shown in Fig. 18-3. The pump (Figs. 18-5 and 18-6) can circulate up to 7500 gallons [28,390 L] of coolant per hour between the water jackets and the radiator. The pump consists of a housing, with a coolant inlet and outlet, and an impeller. The impeller is a flat plate mounted on the pump shaft with a series of flat or curved blades, or vanes.

When the impeller rotates, the coolant between the

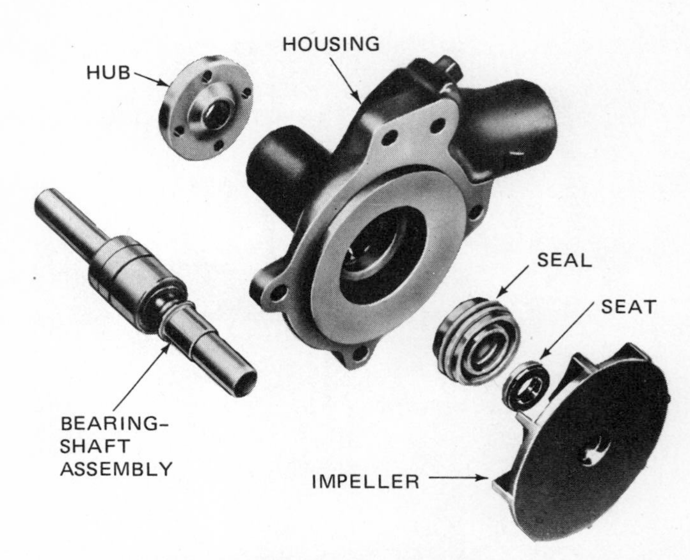

Fig. 18-5 Disassembled water pump. (*Pontiac Motor Division of General Motors Corporation*)

blades is thrown outward by centrifugal force. Then the coolant is forced through the pump outlet and into the cylinder block. The pump inlet is connected by a hose to the bottom of the radiator. Coolant from the radiator is drawn into the pump to replace the coolant forced through the outlet. The arrows in Fig. 18-3 show how the coolant flows through the cooling system.

The impeller shaft is supported on one or more bearings. A seal prevents coolant from leaking out around the bearings. Most water pumps use sealed bearings which never need lubrication. With a sealed bearing, grease cannot leak out, and dirt and water cannot get in. Older water pumps had bearings that required water-pump lubricant, or soluble oil, mixed with the coolant for lubrication. The pump is driven by a belt from the pulley on the front end of the crankshaft (Fig. 18-6).

⚙ 18-6 Engine fan The engine fan usually mounts on the water pump shaft. The fan is driven by the same belt that drives the pump and the alternator (Fig. 18-7). The purpose of the fan is to pull air through the radiator. This improves cooling at slow speeds and idle. At higher car speeds (above about 40 mph [64 km/h]), the air rammed through the radiator by the forward motion of the vehicle provides all the cooling air that is needed.

The fan usually has from four to six blades which in rotating pull air through the radiator. Some cars are equipped with a fan shroud that improves fan performance (Fig. 18-8). The shroud increases the efficiency of the fan, since it assures that all air pulled back by the fan must first pass through the radiator.

⚙ 18-7 Variable-speed fan Many engines are equipped with a variable-speed fan. This fan will not exceed a predetermined speed or will rotate only as fast as needed to keep the engine from overheating. The fan control includes a small fluid coupling and a thermostatic device.

One type is shown in Figs. 18-9 and 18-10. It has a

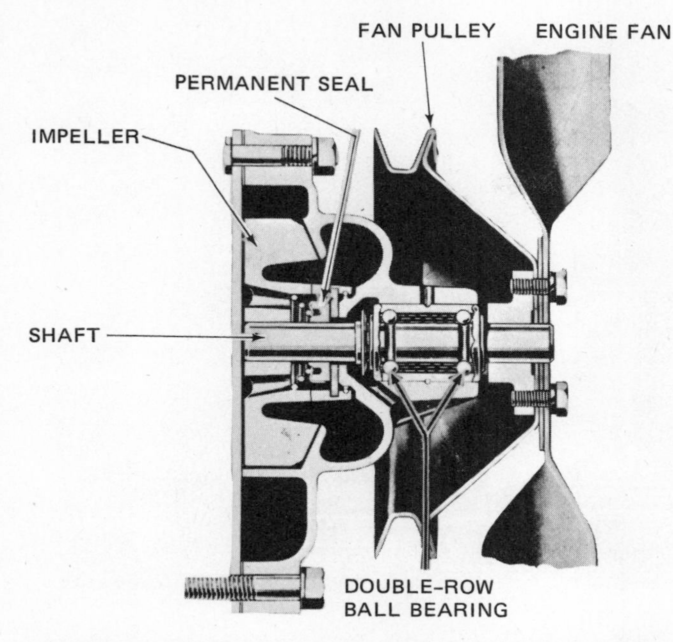

Fig. 18-6 Sectional view of a water pump. Note the double-row ball bearings which support the shaft. The fan and pulley are mounted on the shaft.

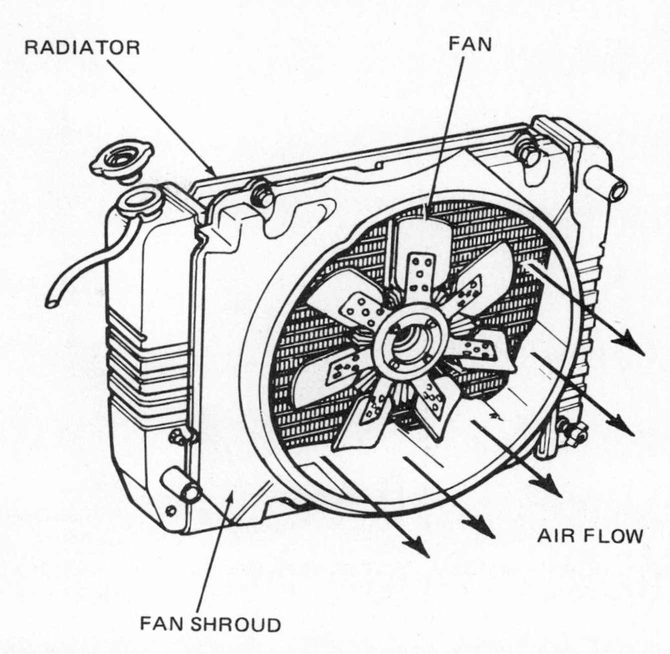

Fig. 18-8 A fan shroud is attached to the radiator to improve fan performance. (*Chrysler Corporation*)

small fluid coupling that is partly filled with a special silicone oil. When the temperature of the air coming through the radiator rises, the heat causes the bimetal coil spring on the drive to expand. As the coil expands, it slightly turns the shaft in the center of the drive. This allows more oil to enter the fluid coupling, which now begins to rotate the fan. But when the engine is cold, the fluid coupling slips, and the fan coasts. This reduces noise and saves engine power, thereby reducing fuel consumption.

Another type of temperature-controlled fan uses a flat bimetal-strip spring instead of a coil spring. The two outer ends of the strip are attached to the drive face. The center of the strip is attached to the control piston in the center of the drive.

When the engine temperature rises, the air flowing through the radiator gets hotter. The hot air causes the metal strip to expand and bow outward. As the strip bows, it pulls the control piston outward. This allows more oil to flow into the fluid coupling. The fan then speeds up for improved cooling.

As the engine temperature drops, the thermostatic strip straightens, forcing the control piston in. This causes oil to leave the fluid coupling. With the drive uncoupled, the fan slows to its normal coast speed.

Figure 18-11 shows how the fan bolts to the drive, and then attaches by the drive bolts through the pulley to the hub of the water pump. The pulley, driven by a belt from the crankshaft, turns while the engine is running. Because the flange on the drive shaft is bolted to

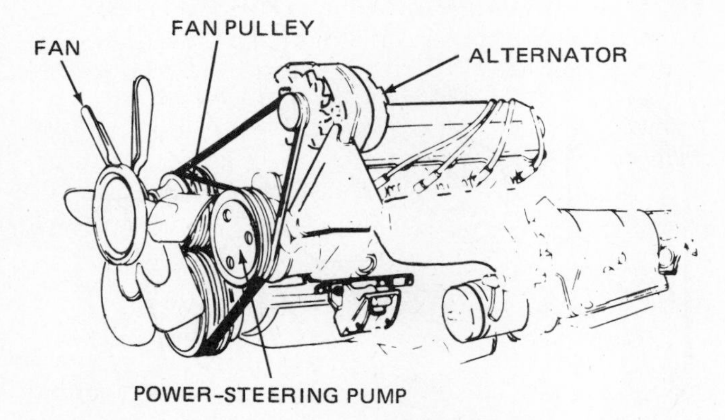

Fig. 18-7 Engine fan driven by two belts, which also drive the water pump, alternator, and power-steering pump. (*Buick Motor Division of General Motors Corporation*)

Fig. 18-9 Installation of a fan-drive clutch. The clutch is positioned between the fan hub and the water pump shaft. (*Ford Motor Company*)

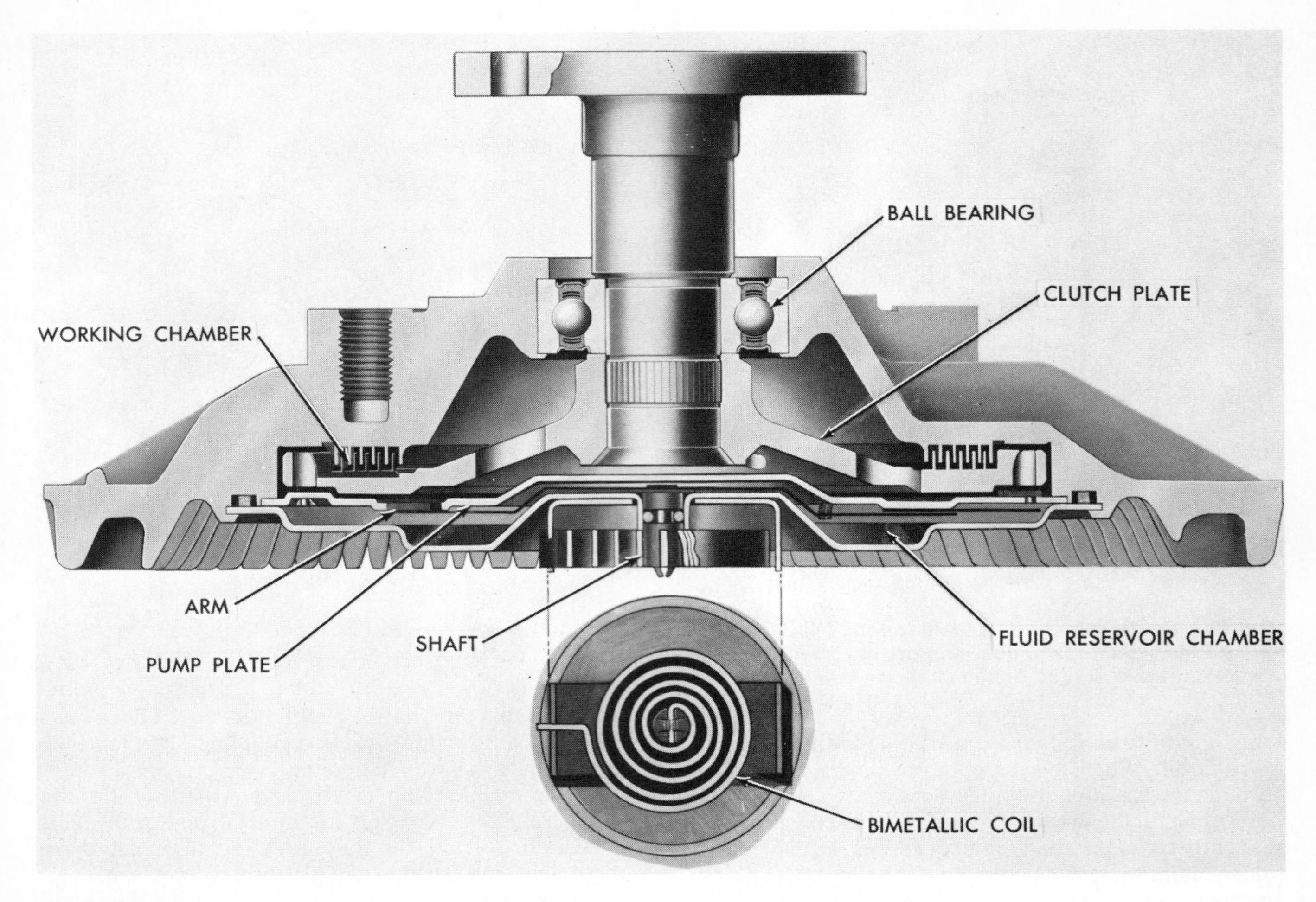

Fig. 18-10 Sectional view of a fan clutch drive. (*Chevrolet Motor Division of General Motors Corporation*)

the pulley, the shaft also turns. But there is no direct mechanical connection through the fluid coupling in the drive. The fluid coupling must fill with oil before the outside housing of the drive rotates. Since the fan bolts to the drive housing, the fan can be driven at normal speed only after the fluid coupling locks up.

⚙ 18-8 Flexible-blade fan Another way to reduce the power needed to drive the fan, and to reduce fan

Fig. 18-11 Mounting of a fixed-blade fan to the drive, and the drive to the hub of the water pump. (*Ford Motor Company*)

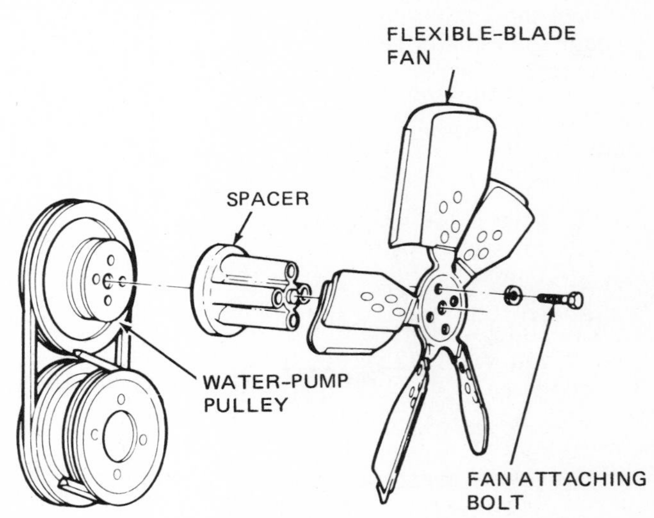

Fig. 18-12 Mounting of a flexible-blade fan to the spacer for the water pump pulley. (*Ford Motor Company*)

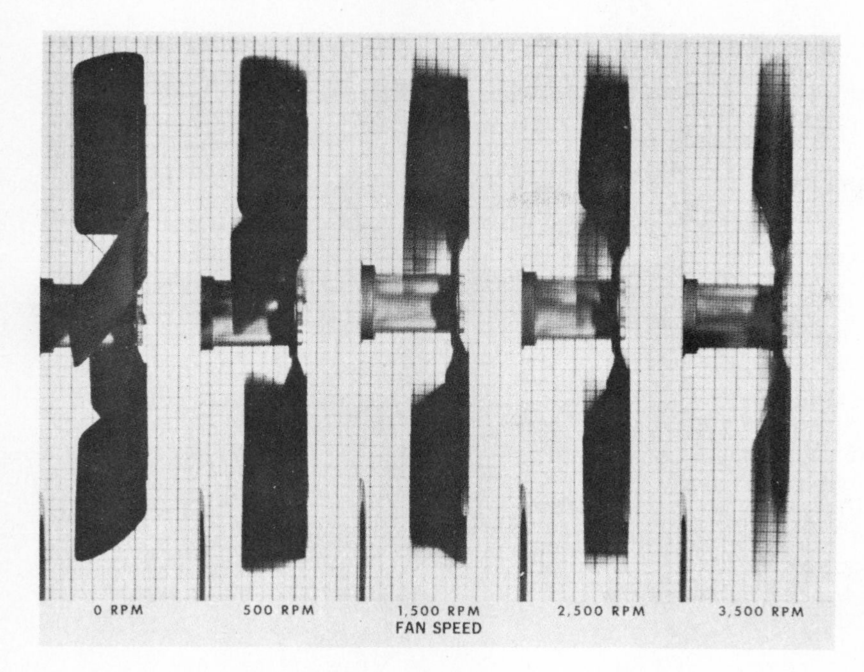

Fig. 18-13 Change of pitch of fan blades as fan speed increases.

noise, is to use a fan with flexible plastic blades. Figure 18-12 shows the installation of this type of fan. In operation, the pitch of the blades decreases as fan speed increases. This is a result of a flattening effect on the blades caused by centrifugal force. Each blade therefore takes a smaller bite of air per revolution. Airflow is reduced because of the shallower pitch. This, in turn, saves power and lowers the noise level. Figure 18-13 shows how the fan blades change pitch as speed increases.

In Fig. 18-12, note the use of the spacer between the water pump pulley and the fan. Various sizes and shapes of spacers are used to properly position the fan. For example, on radiators without a shroud, one recommendation is to position the fan in the center of the radiator with the blades two inches (51 mm) in back of it. The distance between the blades and the radiator allows the engine to shift in its mounts without the blades striking the radiator fins.

CAUTION: Fan blades can break and fly off. Whenever. the engine is running, never stand directly in line with the rotating fan. Also, avoid allowing your clothing, hands, or tools to get near the fan blades or fan belt.

18-9 Drive belts A fan belt, or drive belt, uses friction, tension, and proper fit in the pulley groove to turn the water pump and fan. Most drive belts are V-shaped. When the belt is properly tightened, it wedges into the pulley groove. Friction between the sides of the belt and the sides of the groove transmits power through the belt from one pulley to the other. Power is *not* transmitted through the bottom of a V-belt.

The V-belt provides a large surface area on its sides in contact with the pulley groove. Because of this, considerable power can be transmitted by a V-belt. The wedging action of the belt as it curves into the pulley groove aids in preventing belt slippage. Figure 18-7 shows a V-belt in place on an engine. The belt transmits power from the crankshaft pulley to the water pump pulley, and to the alternator pulley.

NOTE: Some engines use two V-belts with double pulleys, which are pulleys that have two belt grooves. The added belt provides the power needed to drive the alternator and water pump. These belts are a matched set. If one belt must be replaced, then replace both belts at the same time. Otherwise the new belt will carry most of the load. It will wear rapidly, and may slip. The reason is that the remaining old belt will have stretched slightly. Therefore it will be looser, after the new belt is installed, and will carry less of the load.

18-10 Electric fan for transverse engines Engines mounted crosswise at the front (and driving the front wheels) often use another method of driving the engine fan. Figure 18-14 shows a typical arrangement. The radiator is located at the front, as usual. The fan is driven by an electric motor. A thermostatic switch turns the motor on only when it is needed. For example, in one engine, the switch turns the motor on when the coolant reaches 193 to 207°F [89 to 97°C]. It turns the motor off if the coolant temperature drops below these figures. On cars with air conditioning, the thermostatic switch is bypassed and the motor runs all the time the air conditioner is turned on.

The advantage of the electric fan is less power drain on the engine and less fan noise. Also, there is no fan belt to inspect, adjust, or replace. This means less cooling system maintenance.

Another way to drive the fan when the engine is mounted transversely is to have the drive belt "turn the corner." In this arrangement, the radiator is in its normal position facing the front of the car. Idler pulleys are positioned so that the belt is turned 90°. The crankshaft pulley still drives the belt, but the fan pulley is at a 90° angle to the crankshaft pulley.

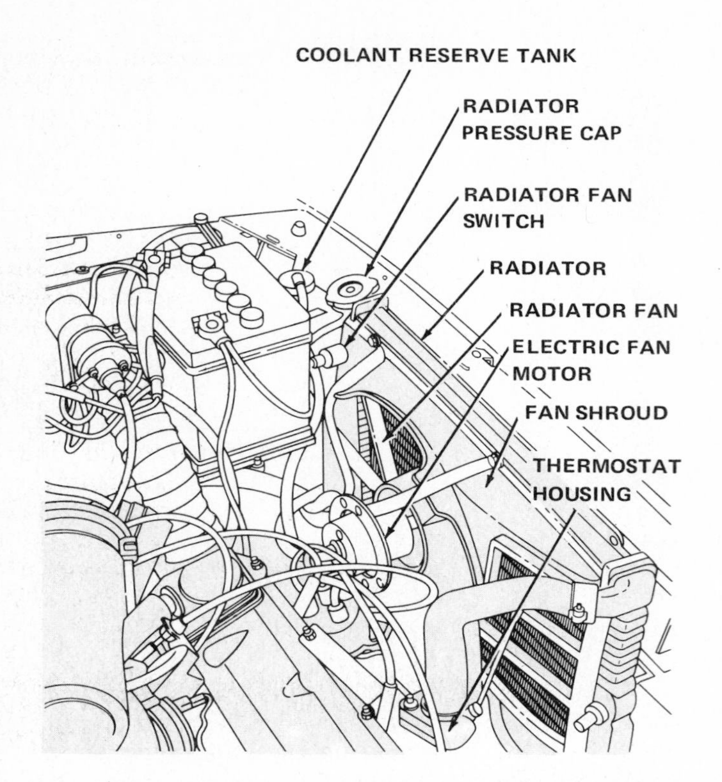

Fig. 18-14 A typical arrangement with the engine mounted transversely. The engine fan is driven by an electric motor. (*Chrysler Corporation*)

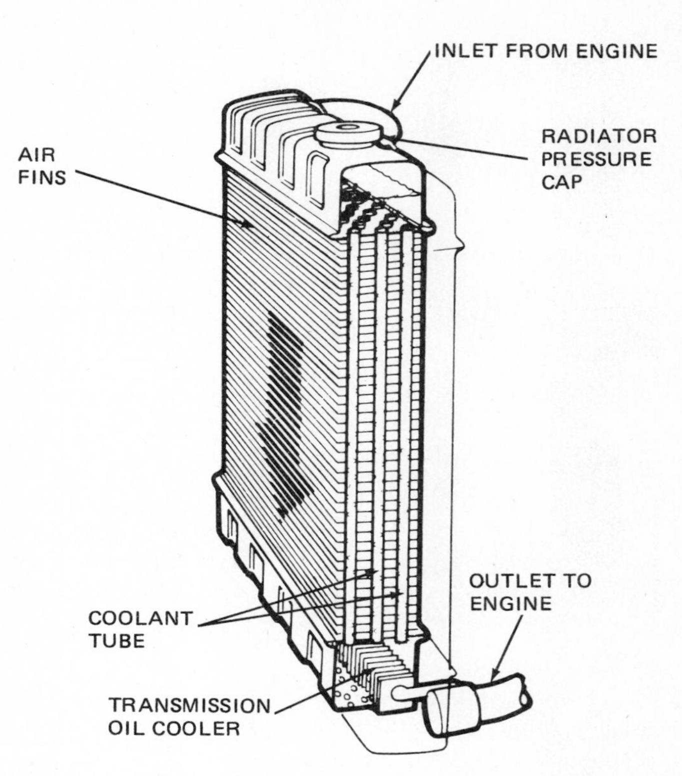

Fig. 18-15 Construction of a down-flow type of tube-and-fin radiator. (*Chrysler Corporation*)

18-11 Radiator In the cooling system, the radiator is a heat exchanger that removes heat from coolant passing through it. The radiator holds a large volume of coolant in close contact with a large volume of air so that heat will transfer from the coolant to the air. Figure 18-8 shows the airflow through a radiator with a shroud. The radiator core is divided into two separate and intricate compartments. Coolant passes through one, and air passes through the other. Several types of radiator cores have been used. The two most common are the tube-and-fin and the ribbon-cellular types.

A tube-and-fin radiator (Fig. 18-15) consists of a series of tubes extending from the top to the bottom of the radiator. The tubes run from the upper inlet tank to the lower outlet tank. Fins are placed around the outside of the tubes to improve heat transfer. Air passes between the fins. As the air passes by, it absorbs heat from the fins which have, in turn, absorbed heat from the coolant.

In a typical radiator, there are five fins per inch [25.4 mm]. Radiators used in cars that have factory-installed air conditioning have seven fins per inch [25.4 mm]. This provides the additional cooling surface required to handle the additional heat load imposed by air conditioning.

The ribbon cellular radiator core (Fig. 18-16) is made up of many narrow coolant passages formed by pairs of thin sheet-metal ribbons. These ribbons are soldered together along their edges which run from the upper to the lower tank. The sealed outside edges of the coolant passages form the front and back surfaces of the radiator core. The coolant passages are separated by open fins of metal ribbon. These provide air passages between the coolant passages. Air moves through these passages from front to back, absorbing heat from the fins. The fins, in turn, absorb heat from the coolant moving through the coolant passages. As a result, the temperature of the coolant is lowered.

On every radiator a small inlet tank is provided at the top or side of the radiator into which hot coolant is delivered from the engine. Another small tank, the outlet tank, is located on the other side or at the bottom of the radiator. This tank receives the coolant after it

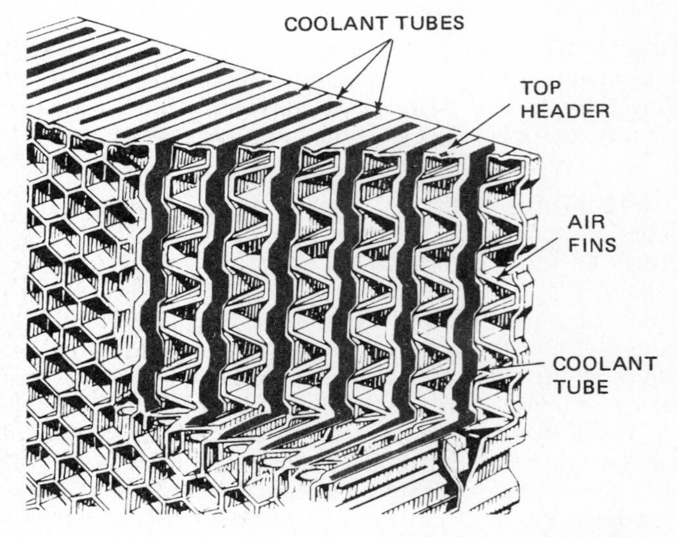

Fig. 18-16 Construction of a ribbon-cellular radiator core.

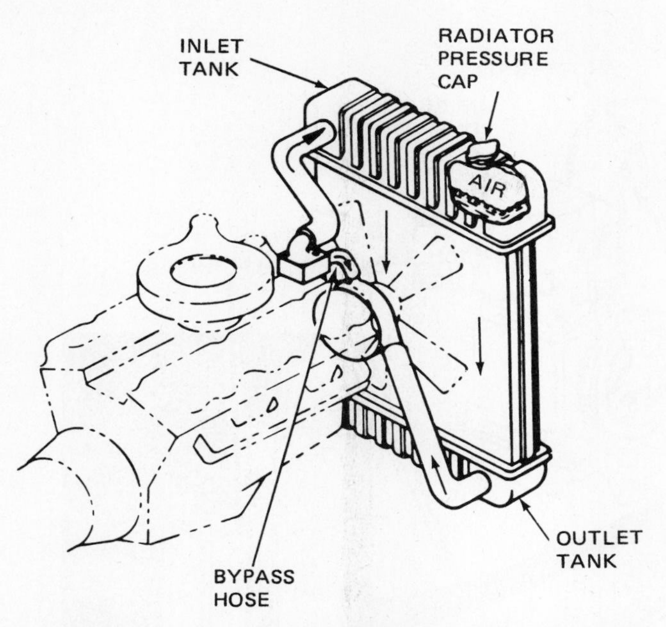

Fig. 18-17 Cooling system using a down-flow radiator. *(Harrison Radiator Division of General Motors Corporation)*

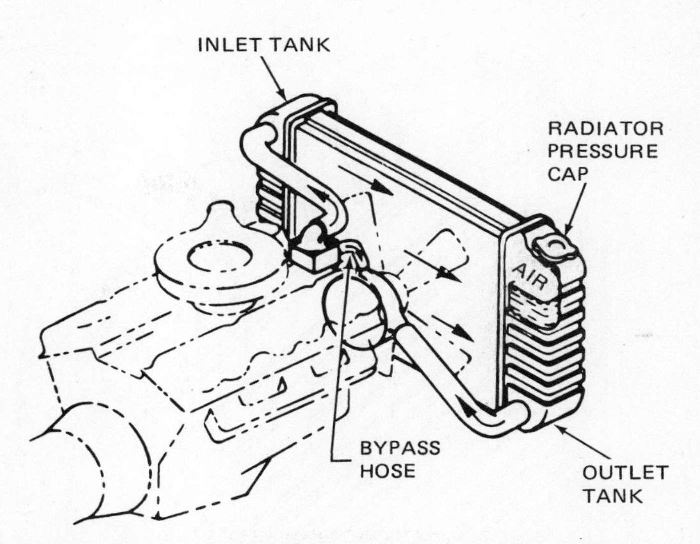

Fig. 18-18 Cooling system using a cross-flow radiator. *(Harrison Radiator Division of General Motors Corporation)*

has passed through the radiator core. A filler cap (called a radiator cap) placed on one of the tanks can be removed to add coolant to the cooling system. Coolant can be lost because of evaporation or leakage.

Radiators can be classified in another way, according to the direction that the coolant flows through them. In some, the coolant flows from top to bottom (down-flow type) as shown in Fig. 18-17. Until about 1970, most automotive radiators were of this type.

In the down-flow radiator, the top tank has a radiator filler neck, sealed by a radiator pressure cap (⚙ 18-18). The cap is removed to add coolant to the cooling system on many cars. On others, the coolant is added to the expansion tank (⚙ 18-12). On cars with automatic transmission, the bottom or outlet tank has an oil cooler (⚙ 18-15). Some radiators also have a drain valve in the bottom tank.

Most late-model cars have a cross-flow radiator (Fig. 18-18). In this type, the coolant flows horizontally from the inlet tank on one side to the outlet tank on the other side. Basically, the cross-flow radiator is a down-flow radiator turned on its side. This allows the car body to be designed with a lower hood line.

In the cross-flow radiator, the outlet tank has the filler neck and sealing pressure cap. The outlet tank also contains the transmission oil cooler.

On any radiator, the inlet tank (above or to one side) serves two purposes. It provides a reserve supply of coolant. It also provides a place where the coolant can be separated from any air that might be circulating in the cooling system (Figs. 18-17 and 18-18).

⚙ 18-12 Expansion tanks Many cooling systems have a separate clear plastic coolant reservoir or expansion tank (Fig. 18-19). The expansion tank is partly filled with coolant and is connected to the overflow tube from the radiator filler neck. The coolant in the engine expands as the engine heats up. Instead of dripping out the overflow tube onto the street and being lost from the cooling system completely, the coolant flows into the expansion tank.

When the engine cools, a vacuum is created in the cooling system. (More about pressure and vacuum in the cooling system is discussed later.) The vacuum siphons some of the coolant back into the radiator from the expansion tank. In effect, a cooling system with an expansion tank is a closed system. Coolant can flow back and forth between the radiator and the expansion tank. This occurs as the coolant expands and contracts from heating and cooling. Under normal conditions, no coolant is lost.

An advantage to the use of an expansion tank is that it eliminates almost all air bubbles from the cooling

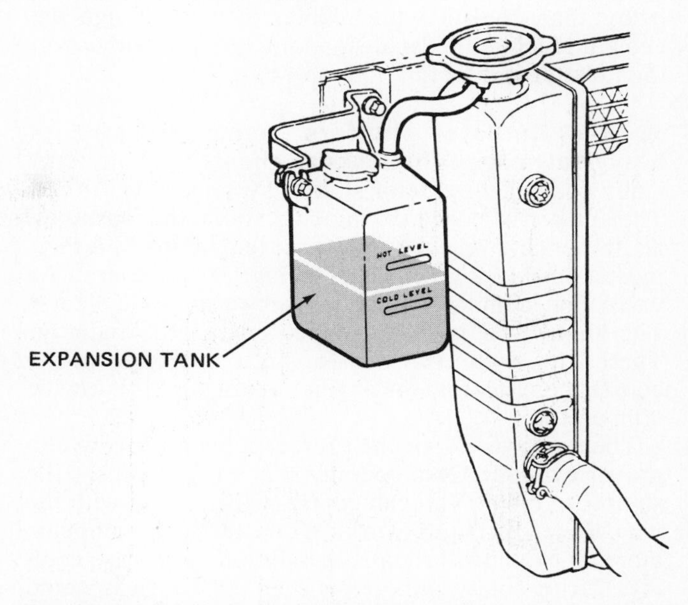

Fig. 18-19 Cooling system using an expansion tank. *(Ford Motor Company)*

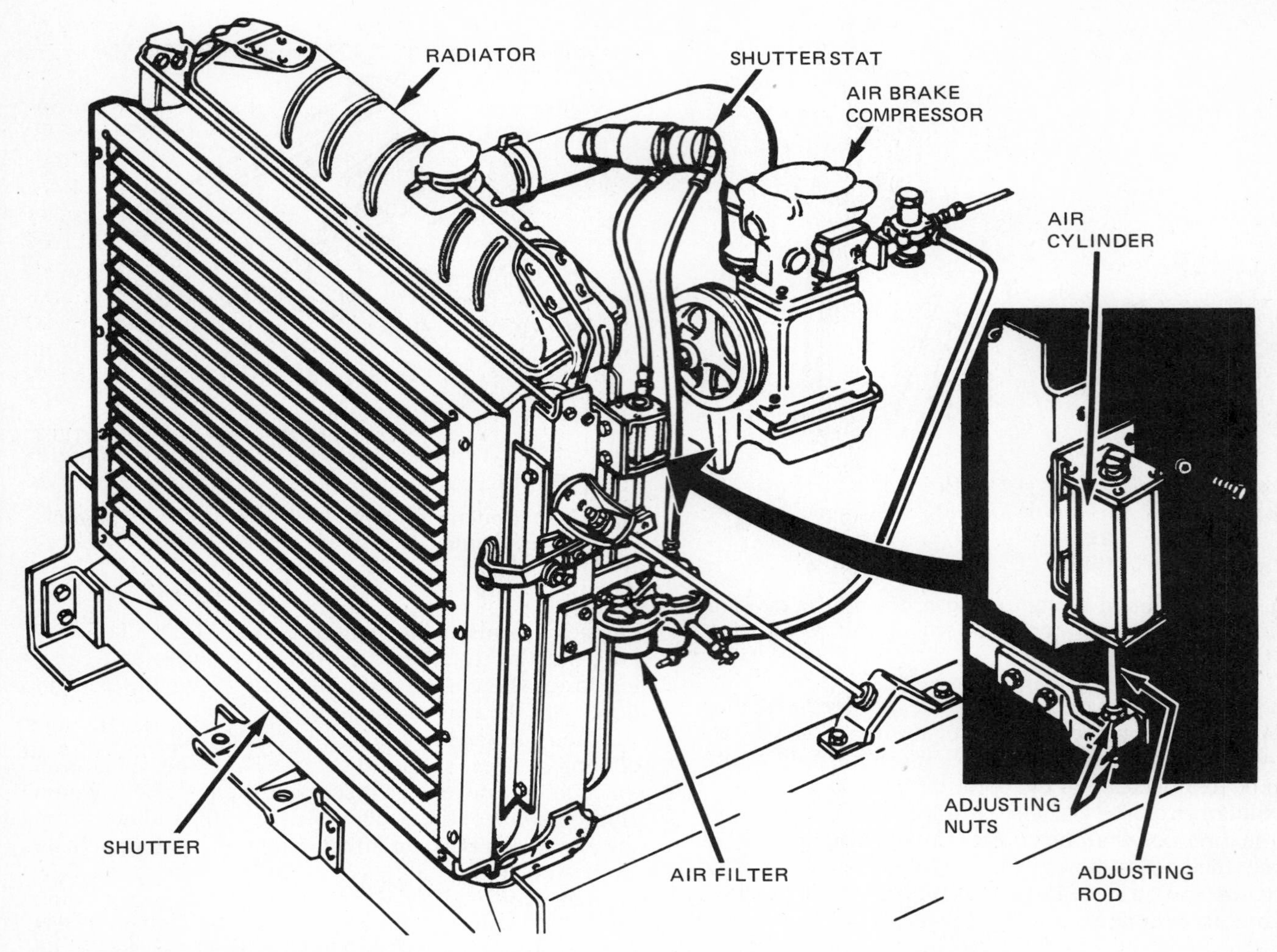

Fig. 18-20 Automatically controlled radiator shutter system used on trucks. (*Ford Motor Company*)

system. Coolant without bubbles absorbs heat much better than coolant with bubbles in it. Although the coolant level in the expansion tank goes up and down, the radiator is kept full.

18-13 Radiator shutters Some radiators for heavy-duty engines in trucks and buses have automatically controlled radiator shutter systems (Fig. 18-20). The shutter is closed during cold starts and warm-up. As the engine reaches operating temperature, a thermostat (called a *shutterstat* in Fig. 18-20) operates a valve that admits compressed air to an air cylinder. The shutterstat is in the upper line to the radiator. There, the shutterstat senses the temperature of the coolant coming from the engine. In effect, this is engine temperature.

The compressed air then forces a piston to move in the air cylinder. This operates a lever that causes the shutter to open. You can compare this action with the way Venetian blinds work. When the shutter opens, more air can flow through the radiator to increase cooling. During cold-weather operation, too much air could overcool the engine. However, as the engine temperature drops, the shutterstat causes the shutter to close. The compressed air that operates the radiator shutters comes from the air-brake compressor (Fig. 18-20).

18-14 Radiator grilles Radiator grilles, which add to the streamlined appearance of the car, may place an added heat load on the cooling system. Some grilles tend to restrict the flow of air through the radiator. However, the cooling system is designed with enough capacity to meet all cooling requirements adequately.

In addition to cooling the engine, the cooling system must cool the automatic-transmission fluid. It must also eliminate the heat from the air-conditioner condenser and any other auxiliary oil cooler mounted near the radiator. To do this, the grille must allow enough air to reach the radiator.

However, not all car body designs use a grille. Some body styles have slots or ducts to allow air through the body into the engine compartment. Other cars have a wide-open type of grille that almost fills the entire space between the headlights.

18-15 Transmission oil coolers Cars with automatic transmissions are equipped with an oil cooler for the automatic-transmission fluid. The oil cooler is

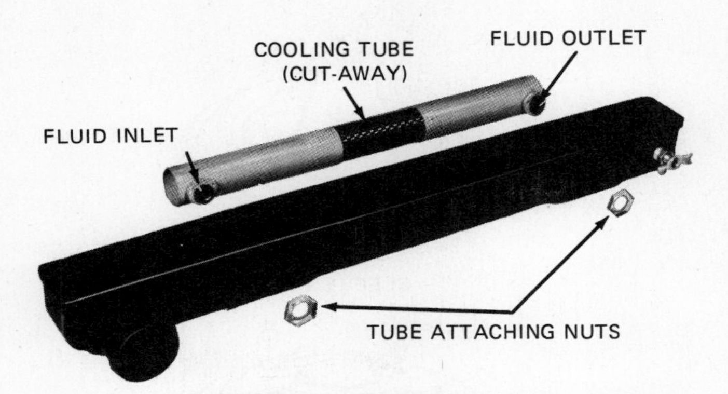

Fig. 18-21 Outlet tank of the radiator and the oil-cooler tube, showing how the tube fits into the lower tank. (*Chrysler Corporation*)

needed because it is possible for the oil in automatic transmissions to overheat. Overheating reduces transmission performance and can damage the transmission.

The transmission oil cooler is usually a tube located in the outlet (bottom or side) tank of the radiator (Fig. 18-21). Figures 18-3 and 18-15 show the location of the oil cooler in the outlet tank of a cross-flow radiator and a down-flow radiator. The oil-cooler tube, being immersed in the lower-temperature coolant, is cooled by the coolant. This cools the transmission fluid that is passing through. Attached to the outlet tank of the radiator are two metal tubes called the transmission-cooler lines (Fig. 18-22). They carry the fluid between the transmission and the oil cooler.

Figure 18-22 shows the use of an auxiliary oil cooler for the automatic transmission fluid. Cars that are factory-equipped with a trailer-towing package often have an auxiliary oil cooler. It is mounted in front of the radiator and air-conditioner condenser. The auxiliary oil cooler is connected in series with the radiator oil cooler. When both oil coolers are used, the hot oil flows from the transmission through the radiator oil cooler. From there, the oil flows through the auxiliary oil cooler. Then the cool oil returns to the transmission. The small arrows in Fig. 18-22 show the flow of oil through this system.

The auxiliary oil cooler transfers heat from the oil to the cooler air passing through. This is similar to the operation of the radiator, which was discussed earlier. Both types of transmission oil coolers place a greater load on the engine cooling system. The radiator oil cooler transfers heat to the coolant. The auxiliary oil cooler, mounted in front of the radiator, raises the temperature of the air before it reaches the radiator.

18-16 Car heater Today most cars are equipped with a hot-coolant type of car heater. The basic part of this device is the heater core (Fig. 18-3), which might be considered a secondary radiator. It transfers heat

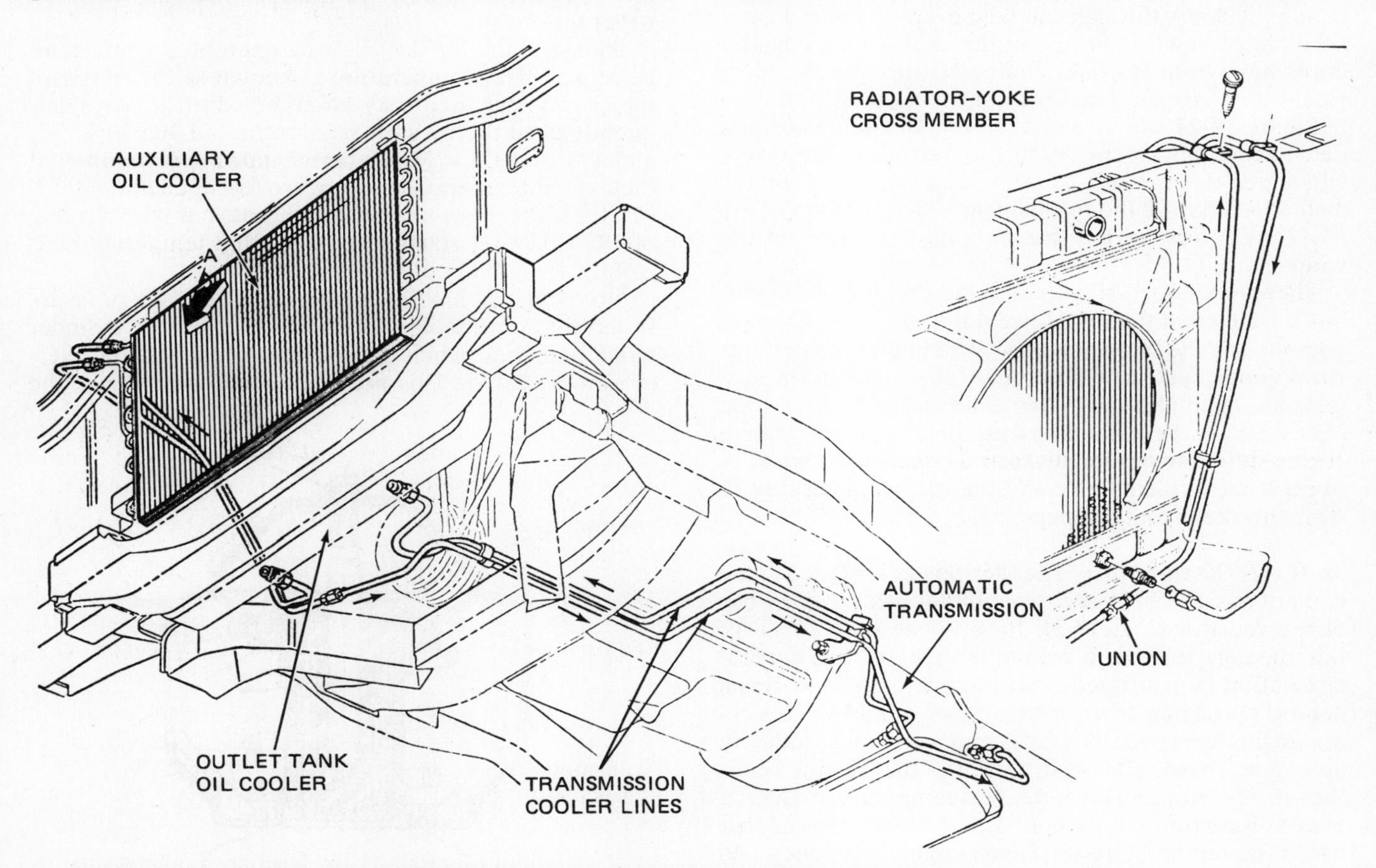

Fig. 18-22 An auxiliary oil cooler for the automatic-transmission fluid, which mounts in series with the oil-cooler tube in the outlet tank of the radiator. (*Chrysler Corporation*)

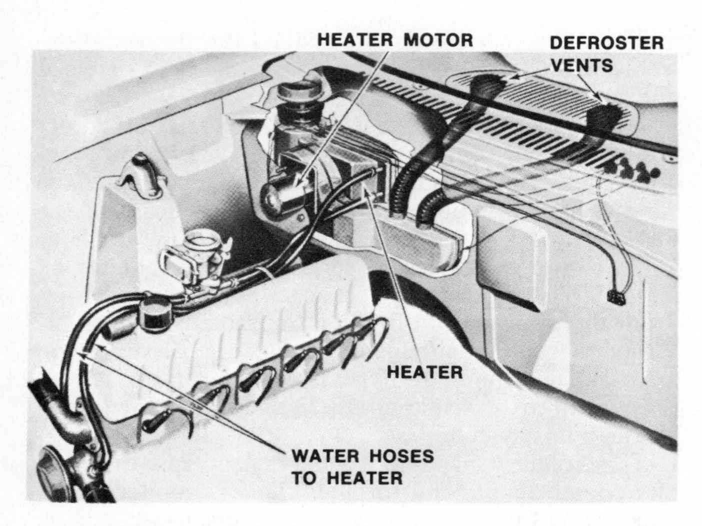

Fig. 18-23 Car heater system. Hot coolant from the engine cooling system circulates through a small radiator called the *heater core.* The fan blows air through the heater core.

from the cooling system to the passenger compartment, instead of to the air passing through the main radiator.

The car heater gets its heat from the top (or hottest part) of the engine. Hot coolant from the engine flows through the heater core. A fan, driven by a small electric motor, blows air through the core. The air absorbs heat as it flows through the heater core. Some heater cores have small inlet and outlet tanks. Other heater cores have both the inlet and outlet tube on the same side of the core, and no tanks.

Figure 18-23 shows another view of the heater and defroster arrangement. With the defroster, the driver can adjust the controls to send heated air up along the inside surface of the windshield. The heated air removes any mist or ice that may have formed on the windshield.

There are some cars which have this type of heater but no air conditioner. However, most cars now are equipped at the factory with a combination heater–air-conditioner system. Some of these combination systems are manually controlled. The driver decides what is to operate. In automatic systems, the driver sets a control at a desired temperature. Then the system takes over. It either heats or cools the air, as necessary to maintain the preset temperature.

⚙ 18-17 Thermostat The thermostat is placed in the coolant passage between the cylinder head and the top of the radiator (Fig. 18-3). Its purpose is to close off this passage when the engine is cold so that coolant circulation is restricted, causing the engine to reach normal operating temperature more quickly. This reduces the formation of acids, moisture, and sludge in an engine. Also, after warm up, the thermostat keeps the engine running at a higher temperature than it would, for example, without a thermostat. The higher operating temperature improves engine efficiency and reduces exhaust emissions.

The thermostat consists of a thermostatic device and a valve (Fig. 18-24). Various valve arrangements and

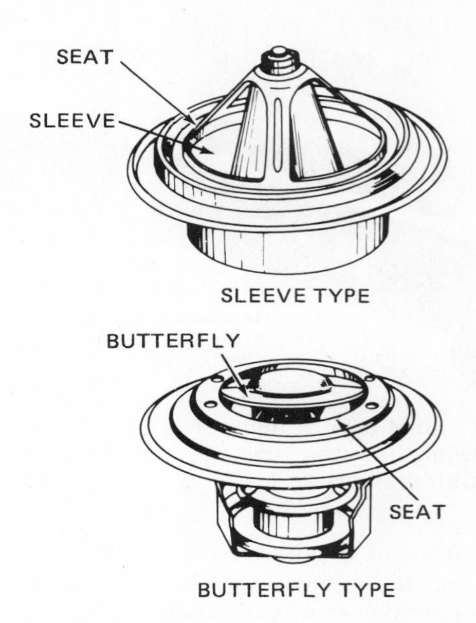

Fig. 18-24 Two types of thermostats for engine cooling systems. (*Chrysler Corporation*)

thermostatic devices have been used. Most thermostats today are operated by a wax pellet, which expands with increasing temperature to open a valve. The sleeve and the butterfly thermostats shown in Fig. 18-24 both use the wax pellet. Figure 18-25 is a sectional view of a wax pellet thermostat.

Thermostats are designed to open at specific temperatures. This temperature is known as the *rating* of the thermostat, and may be stamped on it. Two frequently used thermostats have ratings of 185°F [85°C] and 195°F [91°C]. Most thermostats *begin* to open at their rated temperature. They are full open about 20°F [11°C] higher. For example, a thermostat with a rating of 195°F [91°C] starts to open at that temperature. It is fully open at about 215°F [102°C].

Most engines have a small coolant-bypass passage. It permits some coolant to circulate within the cylinder block and head when the engine is cold and the thermostat is closed. This provides equal warming of the

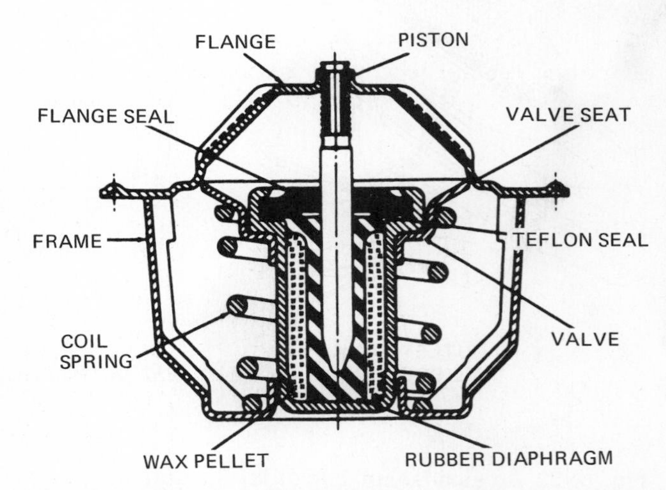

Fig. 18-25 Sectional view of a wax pellet thermostat. (*Chevrolet Motor Division of General Motors Corporation*)

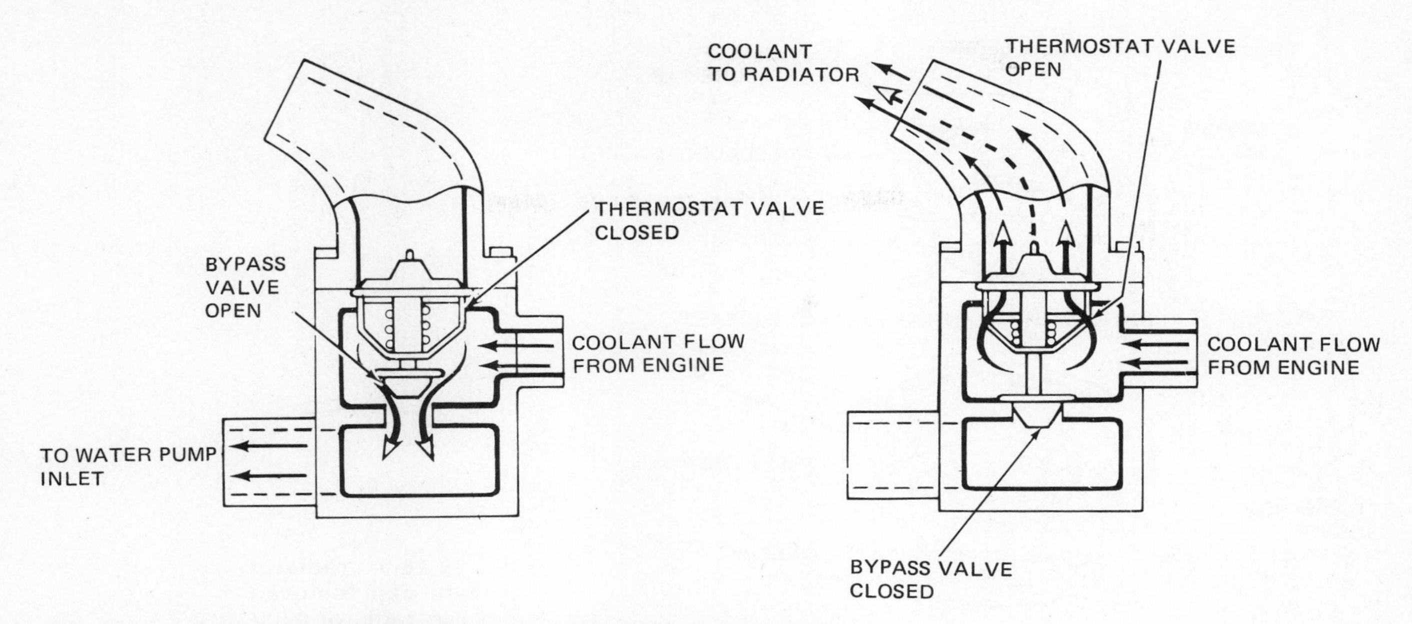

Fig. 18-26 Operation of the blocking-bypass thermostat. (*Chrysler Corporation*)

cylinders and prevents hot spots. When the engine warms up, the bypass must close or become restricted. Otherwise, the coolant would continue to circulate within the engine itself, and too little would go to the radiator for cooling.

The bypass passage may be an internal passage, or an external bypass hose. In Figs. 18-3 and 18-18, notice the small external bypass hose at the top of the water pump.

One internal bypass system uses a small, spring-loaded valve located in back of the water pump. The valve is forced open by coolant pressure from the pump when the thermostat is closed. As the thermostat opens, the coolant pressure drops within the engine and the bypass valve closes.

Another internal bypass system has a blocking bypass thermostat (Fig. 18-26). This thermostat operates like those already described, but it also has a secondary or bypass valve. When the thermostat valve is closed, the circulation to the radiator is shut off. However, the bypass valve is open, permitting coolant to circulate through the bypass. As the thermostat valve opens, permitting coolant to flow to the radiator, the bypass valve closes. This blocks off the engine bypass passage.

⚙ 18-18 Radiator pressure cap The cooling systems on automobile engines today are sealed and pressurized by a radiator pressure cap (Figs. 18-27 and 18-28). There are two advantages to sealing and pressurizing the cooling system. First, the increased pressure raises the boiling point of the coolant. This increases the efficiency of the cooling system. Second, sealing the cooling system reduces coolant losses from evaporation or from surge losses. Surge losses could occur during heavy braking when the coolant surges up the radiator filler neck. However, the sealed pressure cap prevents this loss.

At normal atmospheric pressure, water boils at 212°F [100°C]. If the air pressure is increased, the temperature at which water boils is also increased. For example, if the pressure is raised to 15 psi [103 kPa] over atmospheric pressure, the boiling point is raised to about 260°F [127°C]. Every 1-psi [7-kPa] increase in pressure raises the boiling point of water about 3¼°F [1.8°C]. This is the principle upon which the pressurized cooling system works.

Here is what happens in the pressurized cooling system. As the pressure goes up, the boiling point goes up. Therefore, the coolant can be safely run at a temperature higher than 212°F [100°C] without boiling. The higher the coolant temperature, the greater the difference between it and the outside air temperature. This difference in temperatures is what causes the cooling system to work. The hotter the coolant, the faster the heat moves from the radiator to the cooler passing air. This means that the pressurized, sealed cooling system can take heat away from the engine faster. Therefore, the cooling system works more efficiently when the coolant is under higher pressure.

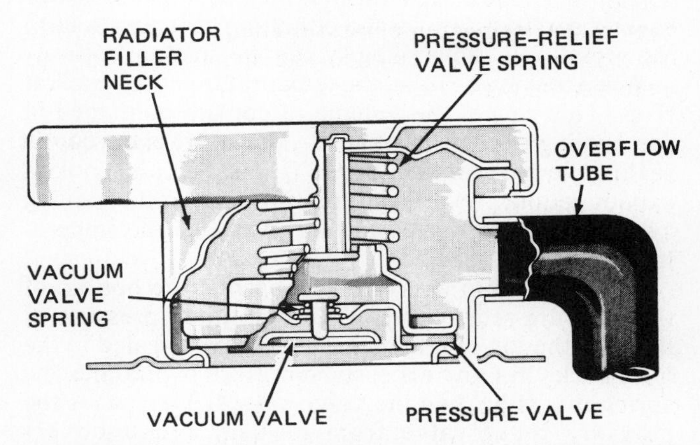

Fig. 18-27 Cutaway view of a radiator pressure cap, showing the pressure valve and the vacuum valve. (*Ford Motor Company*)

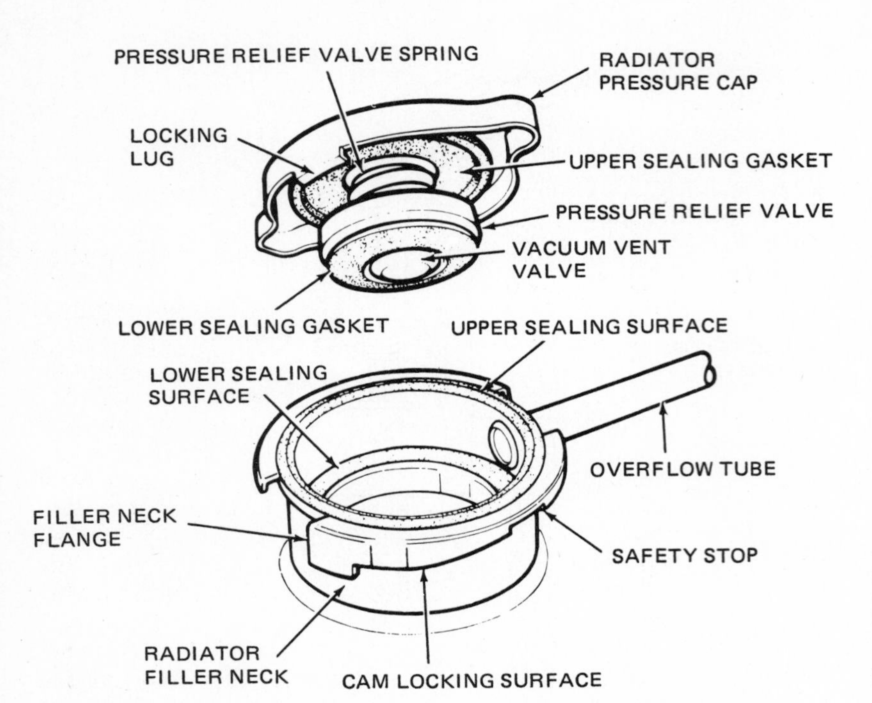

Fig. 18-28 A radiator pressure cap removed from the radiator filler neck. (*Chrysler Corporation*)

However, the cooling system can be pressurized too much. If the pressure in the system gets too high it can damage the radiator and blow off the hoses. To prevent this, the radiator cap has a pressure-relief valve (Figs. 18-27 and 18-28). When the pressure gets too high, it raises the valve so that the excess pressure can escape into the overflow tube. Here is how the pressure relief valve works.

The radiator pressure cap fits over the radiator filler neck and seals tightly around *two* sealing surfaces. Both the upper sealing surface and the lower sealing surface are shown in Fig. 18-28. Notice the position of the overflow tube opening into the filler neck, between the two sealing surfaces.

The upper seal is tight when you place the cap in the filler neck and then turn the cap to lock it in place. As you install the cap, the lower seal also is made tight. This shuts off the overflow-tube opening. Locking the cap compresses the pressure relief valve spring (Fig. 18-28). It applies its rated pressure against the lower sealing surface. This is the condition that exists until the pressure in the cooling system begins to increase.

When the engine runs, the temperature of the coolant rises. This causes the volume of coolant contained in the cooling system to expand. But a pressure cap is sealing the radiator filler neck. Trapped air and coolant cannot get out. As a result, the pressure in the cooling system rises more with further increases in temperature.

Cooling system pressure acts against the bottom of the pressure-relief valve. Up to the rated pressure of the cap, the lower sealing gasket remains sealed in the filler neck. But any excessive pressure overcomes the spring that is holding the valve closed. This raises the pressure relief valve from its seat and uncovers the opening into the overflow tube. Air, steam, or coolant now can pass through the overflow tube to the ground, or into the expansion tank. When the pressure against the pressure-relief valve drops to the rating of the cap, the spring will again close the valve.

The radiator pressure cap also has a vacuum vent valve (Figs. 18-27 and 18-28). This valve protects the system from developing a vacuum that could collapse the radiator. When the engine is shut off and cools, the coolant volume is reduced. Cold coolant takes up less space than hot coolant. As the temperature of the coolant drops, a vacuum develops in the cooling system. To prevent excessive vacuum from developing, the vacuum valve opens to allow outside air or coolant from the expansion tank to flow into the cooling system. This relieves the vacuum that could otherwise cause outside air pressure to collapse the radiator.

There are two types of radiator pressure cap, the constant-pressure type (with a spring) and the pressure vent type (which uses a small weight). In the constant-pressure type, the spring holds the valve closed (Fig. 18-27). As the pressure in the system increases, it reaches a value high enough to overcome the spring and open the pressure valve. This is the type of cap usually found on General Motors and Ford cars.

The pressure vent cap does not have a spring to hold the vacuum valve closed. Instead, a small weight on the valve holds it open. Under normal operating conditions, the cooling system is not pressurized. However, when a liquid surge hits the bottom of the cap, or when the vapor flow is heavy, the valve is forced closed. Now the system is sealed. Pressure begins to build up.

The advantage claimed for the pressure vent cap is that the load is reduced on all parts of the cooling system. Hoses, connections, seams, and the water pump all last longer. This type of cap may be found on AMC and Chrysler cars.

Both types of radiator cap are removed and installed in the same way. When the cap is placed on the filler neck, the locking lug on the cap fits under the filler-neck flange (Fig. 18-28). As the cap is turned, the cam locking surface of the flange tightens the cap. It also preloads the pressure relief valve spring.

To remove the cap, press down and slowly turn the cap back to the safety stop. In this position, any steam or boiling coolant can escape through the overflow tube. This is a safety device to allow the pressure to be relieved. Then the cap can be removed safely from a hot engine.

Several types of safety caps are also used on radiators. The safety cap has a button or lever on top of the cap. By pressing the button, or lifting the lever, the pressure is relieved. This eliminates the chance that steam or boiling coolant could scald you when the cap is removed.

CAUTION: In cooling systems with an expansion tank, the radiator pressure cap is more or less permanently installed. Manufacturers' service manuals warn you never to remove the cap except for major service, such as flushing out the system. The cap should not be removed just to check the coolant level or to add coolant. Instead, coolant level is checked at the expansion tank. Many tanks are marked to indicate normal hot and cold coolant levels. If coolant is needed, it is poured into the expansion tank.

❁ 18-19 Antifreeze solution Water freezes at 32°F [0°C]. If water freezes in the engine cooling system, it stops coolant circulation. Some parts of the engine will overheat. This could seriously damage the engine. Worse, however, is the fact that water expands when it freezes. Water freezing in the cylinder block or cylinder head could expand enough to crack the block or head. Water freezing in the radiator could split the radiator seams. In either case, there is serious damage. A cracked block or head cannot be repaired satisfactorily. A split radiator is hard to repair.

To prevent freezing of the water in the cooling system, antifreeze is added to form the coolant. The most commonly used antifreeze is ethylene glycol, although an alcohol-base antifreeze has been used in the past. A mixture of half water and half ethylene glycol will not freeze above −34°F [−36.7°C]. This is 34°F below zero and it seldom gets that cold in the United States, except in Alaska. A higher concentration of antifreeze will prevent freezing of the coolant at temperatures as low as −84°F [−64.4°C].

Some antifreeze compounds have been sold that plug small leaks in the cooling system. These antifreeze compounds contain tiny plastic beads or inorganic fibers which circulate with the coolant. When a leak develops, the beads or fibers jam in the leak and plug it. This is the same action provided by adding "stop-leak" or "sealer" to the cooling system in an emergency. However, if the leak is too large, no chemical can stop it. Also, a chemical cannot stop leaks in hoses, cylinder-head gaskets, or water-pump seals. The only permanent repair for a leak in a cooling-system component is to repair or replace the component. One problem blamed on "sealing" antifreezes and some types of stop leak is their tendency to plug the heater core.

Corrosion protection is also built into antifreeze solutions. Compounds are added that fight corrosion inside engine water jackets and the radiator. Corrosion shortens the life of metal parts. Also, corrosion forms an insulating layer which reduces the amount of heat transferred from the metal parts to the coolant. In engines with severe corrosion, it is possible for the coolant to maintain normal temperature. But at the same time, the cylinder head and cylinders may be overheating.

Some antifreeze manufacturers add a foam inhibitor to the ethylene glycol. Air does not conduct heat as well as coolant. Any air in the cooling system may cause excess foaming of the coolant as it is whipped up by the water pump impeller. A foam inhibitor tends to reduce this problem.

Small cans of "rust inhibitor" are widely sold. However, the contents should not be added to new antifreeze. Certain types may contaminate the antifreeze, which already has the proper antirust agent in it. Rust inhibitor usually is added to the cooling system when water is used as the coolant.

In addition to having rust and foam inhibitor added, most antifreezes are colored with a dye, such as red, green, or yellow. The dye allows the antifreeze to serve as a leak indicator. The distinctive coloring makes a leak easier to locate.

Antifreeze solutions also serve a purpose during hot-weather operation. They raise the boiling point of the coolant so it does not boil away in hot weather. Also, they continue to fight corrosion.

When the rust and foam inhibitors are used up, the coolant becomes rust-colored. Then the cooling system should be serviced. Car manufacturers recommend that the cooling system be drained, flushed out, and refilled with a fresh mixture of water and antifreeze periodically. One recommendation is that this be done every 2 years. Another is that it be done every year, preferably in the late fall, just before freezing weather sets in. The procedure is covered in Chap. 19.

When adding antifreeze to an engine that has an aluminim block, cylinder head, or radiator, add only the recommended type of antifreeze. Some commercial antifreeze compounds are not safe for use in engines with aluminum components.

NOTE: Automotive cooling systems should never be filled with water only. Indicator lights do not come on until well above the boiling point of water (❁ 18-20). Therefore, plain water could boil, even though the indicator light does not come on. Severe engine damage could occur before the driver is aware of the problem. Coolant will not boil until a higher temperature is reached.

❁ 18-20 Temperature indicators The driver should know the temperature of the coolant in the cooling system at all times. For this reason, a temperature-indicating light or gauge is installed in the instrument panel or dash of the car. An abnormal heat rise is a warning of abnormal conditions in the engine. The indicator warns the driver to stop the engine before serious damage is done. Temperature gauges are of two

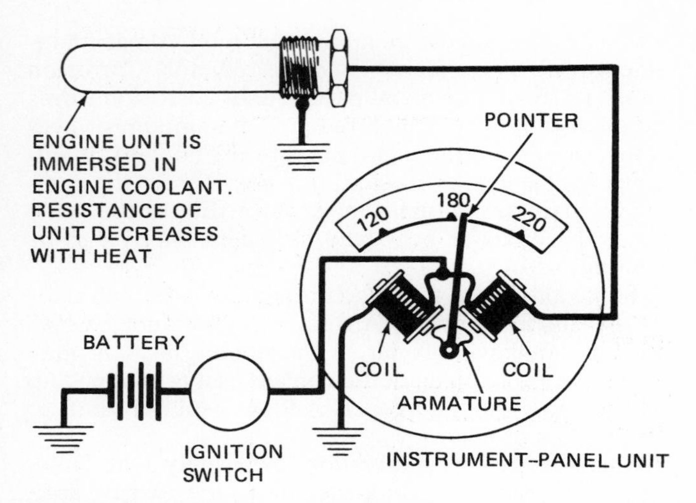

Fig. 18-29 A magnetic or balancing-coil type of temperature gauge system.

general types: the balancing-coil (magnetic) type and the bimetal-thermostat (thermal) type. Instead of a gauge, many cars use one or two temperature-indicator lights.

1. Balancing-coil gauge The balancing-coil type of oil pressure gauge (⚙ 16-25), fuel gauge (⚙ 13-9), and temperature gauge all operate in a similar manner. The instrument panel units are very similar. Each consists of two coils and an armature to which a pointer is attached. Two coils of wire are placed at an angle of 90° to each other (Fig. 18-29). This unit is located in the dash in front of the driver.

An engine unit, or sending unit, that changes resistance with temperature is placed in the engine so that the end of the unit is in the coolant. The resistance of the engine unit drops as coolant temperature goes up. At higher temperatures, it allows more current to flow through the right coil. Then when the engine is cold, it allows less current. Therefore, the pointer or needle in the gauge moves to the right or left, as the magnetic field around the right coil varies in strength.

With the engine cold, current flows through the ignition switch to the coils in the dash unit. The same amount of current always passes through the left coil, as long as the ignition switch is closed. When the engine unit is cold, it allows only a small current to flow through the right coil of the dash unit. Now, the left coil has more magnetism than the right coil. Therefore, the armature between the two coils is pulled to the left. The pointer is attached to the armature, and moves left with it to indicate that the engine is cold.

As the engine warms up, the engine unit passes more current. More current flows through the right coil of the dash unit. This creates a stronger magnetic field around the right coil. Therefore the pointer swings right to indicate a higher coolant temperature.

2. Bimetal-thermostat gauge The bimetal-thermostat type of temperature gauge is similar to the balancing-coil type except for the use of a bimetal thermostat in the dash unit (Fig. 18-30). This thermostat is linked to the pointer. As the engine unit warms up and

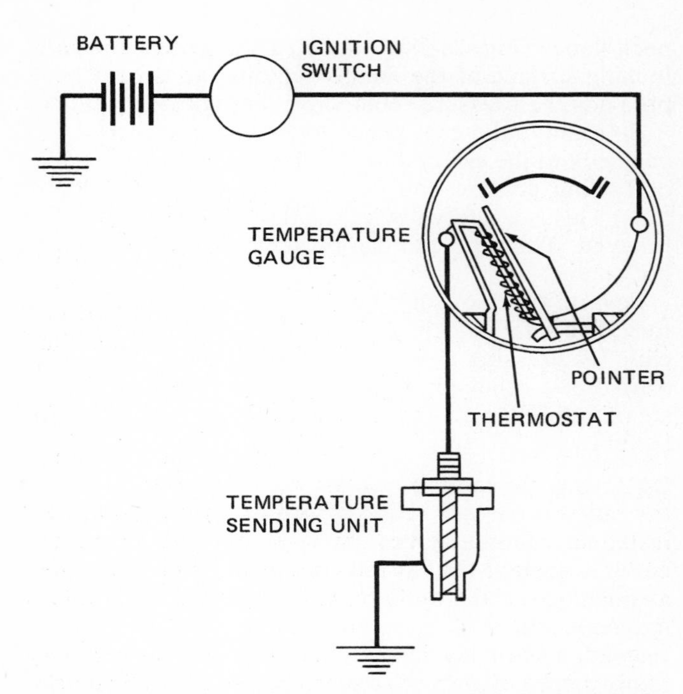

Fig. 18-30 A thermostatic or thermal type of temperature gauge. (*Chrysler Corporation*)

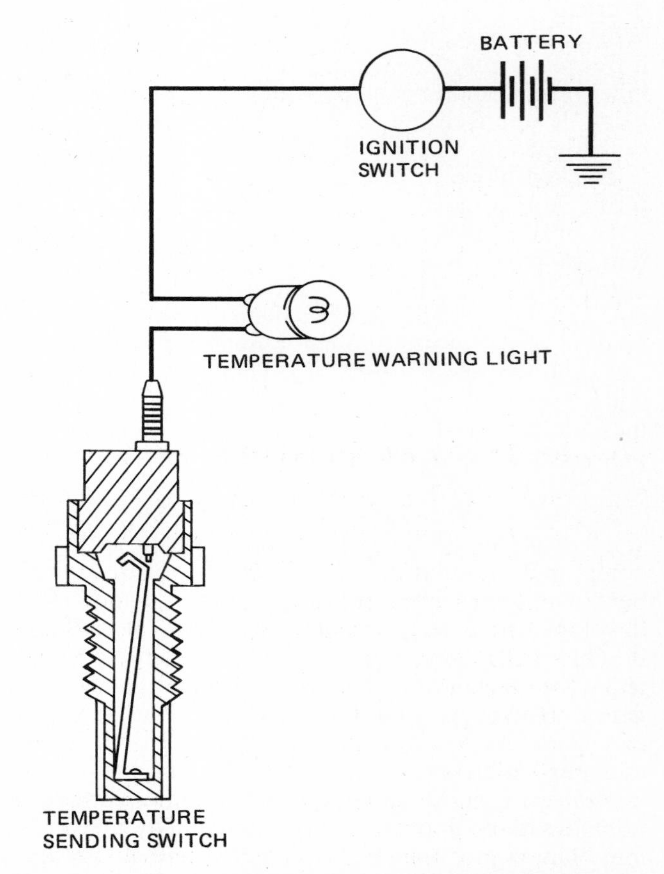

Fig. 18-31 Temperature warning light which indicates engine overheating. (*ATW*)

passes more current, the thermostat heats up and bends. This causes the pointer to swing to the right to indicate that the engine temperature is rising.

3. Indicator light The temperature indicator light system has two units. One is the light on the instrument panel and the other is the coolant temperature switch or sending unit (Fig. 18-31). The sending unit is mounted on the engine so that the end of the unit is in the coolant. When the temperature goes too high, the sending unit connects the light bulb to the battery. Then the indicator light comes on to signal the driver that the engine is overheating.

The indicator light warns of an overheating condition at about 5 to 10°F [2.8 to 5.6°C] below the coolant boiling point. A "prove-out" circuit is incorporated in the system. When the ignition switch is turned from OFF to RUN, the light should come on, proving that the system is working. If the light does not come on, either the bulb is burned out or the sending unit or connecting wire is defective. The light will go out normally after the engine starts.

Another indicator light system is shown in Fig. 18-32. This system uses two light bulbs which can become connected to the battery through the sending unit when the ignition switch is turned ON. When the ignition switch is first turned on, to start a cold engine, the sending-unit thermostatic blade is in the proper position to connect the COLD light to the battery. It comes on.

The COLD light, which appears in blue on the instrument panel, remains on until the engine approaches operating temperature. As this happens, the thermostatic blade in the sending unit is bent by the increasing temperature. The blade therefore moves off the cold terminal, disconnecting the COLD light so that it goes out. If the engine should overheat, the thermostat will warp further so that it moves under the hot terminal. This connects the HOT bulb to the battery so that it glows and appears in red on the instrument panel. This is a signal to the driver that the engine has overheated and should be stopped before damage results.

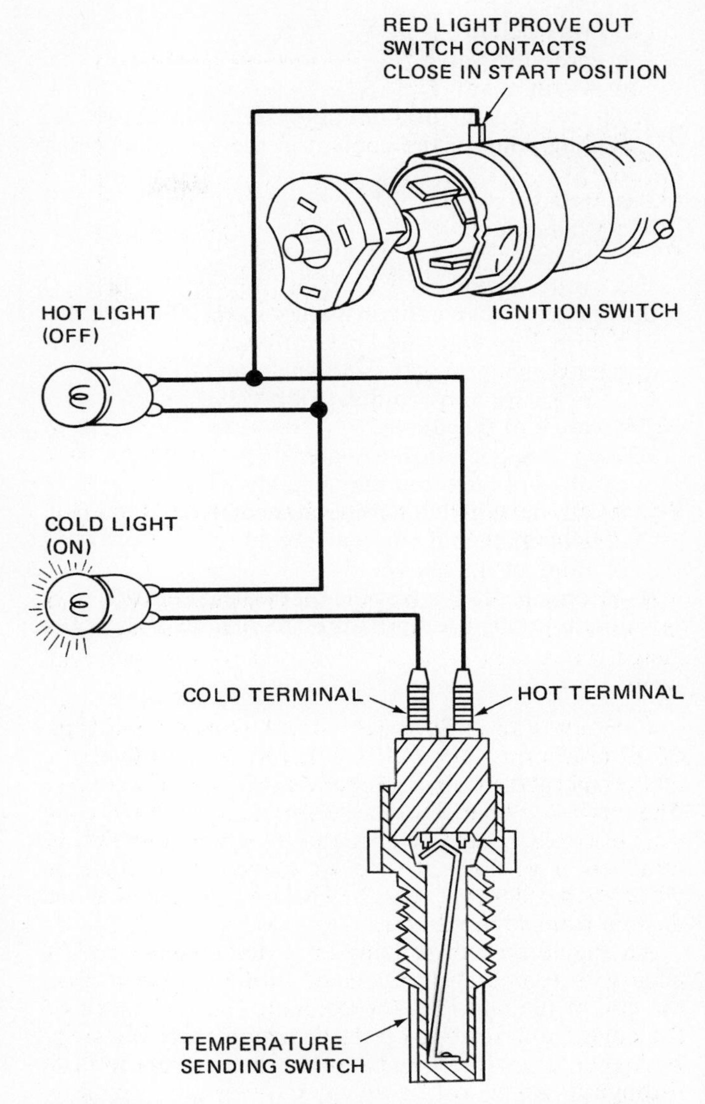

Fig. 18-32 Temperature-indicating system using COLD and HOT indicating lights. (*Ford Motor Company*)

Chapter 18 review questions

Select the *one* correct, best, or most probable answer to each question. Then check your answers against the correct answers given at the end of the book.

1. The pump part that rotates to cause coolant circulation between the radiator and engine water jackets is called the:
   a. impeller
   b. fan
   c. body
   d. bypass.
2. Which of these statements is (are) true about a liquid cooling system?
   I. A thermostat that is installed backwards will cause the engine to run at a lower than normal temperature.
   II. A radiator pressure cap that is *not* fully seated will cause the coolant to boil over at a higher than normal temperature.
   a. I only
   b. II only
   c. both I and II
   d. neither I nor II.
3. In normal operation, coolant in the down-flow radiator circulates:
   a. from top to bottom
   b. from bottom to top
   c. in a circular path in the radiator
   d. none of the above.
4. The part of the cooling system thermostat that functions to open and close the valve is called the:
   a. seater

b. wax pellet
c. pressure valve
d. vacuum valve.

5. The device in the cooling system that raises the boiling point of the coolant in the system is called the:
   a. pressure cap
   b. vacuum valve
   c. radiator
   d. water jacket.
6. A pressure cap contains two valves. They are the:
   a. pressure valve and blowoff valve
   b. atmospheric valve and vacuum valve
   c. pressure valve and vacuum valve
   d. none of the above.
7. Two types of antifreeze are:
   a. alcohol-base and ethylene glycol
   b. ethylene glycol and permanent
   c. isooctane and ethylene glycol
   d. none of the above.
8. Mechanic X says that the thermostat controls maximum coolant temperature. Mechanic Y says that an engine that overheats will be repaired by replacing the thermostat with one of a lower temperature. Who is right?
   a. X only
   b. Y only
   c. both X and Y
   d. neither X nor Y.
9. What is the main purpose of the water pump bypass hose in the engine cooling system?
   a. to reduce pressure at the water pump outlet during high engine speeds
   b. to allow coolant flow within the engine when the thermostat is closed
   c. to prevent air pockets in the water pump housing
   d. to prevent collapse of the lower radiator hose.
10. The percent of the heat produced in the combustion chambers that the cooling system removes from the engine is
    a. 30 to 35 percent
    b. 50 to 60 percent
    c. 10 to 80 percent
    d. 85 to 90 percent.

COOLING SYSTEM SERVICE

After studying this chapter, and with proper instruction and equipment, you should be able to:

1. Diagnose cooling system troubles using the pressure tester and the cooling system analyzer.
2. Test and adjust antifreeze strength.
3. Check the thermostat.
4. Replace and adjust the fan belt.
5. Clean and flush the cooling system.
6. Locate and repair leaks in the cooling system.
7. Replace the water pump.
8. Replace an expansion core plug

19-1 Working safely on the cooling system There are several safety hazards you must watch for when working on engines and the cooling system:

1. Keep you hand away from the moving fan! When the engine is running, the fan is turning so fast it is a blur. But it can mangle your hand and cut off fingers if your hand should get into the fan.
2. Never stand in a direct line with the fan. A fan blade could break off and fly out from the engine compartment. Anyone standing in line with the fan could be injured or killed. Before starting the engine, examine the fan for cracked or loose blades. If you find any damage, the fan must be replaced.
3. Electrically operated and thermostatically controlled fans may not be shut off when the ignition switch is turned to OFF. Because the coolant-temperature switch turns the fan motor on and off solely on the basis of coolant temperature, the fan may start and run even with the engine stopped. When working near an electric fan, disconnect the lead to the fan motor to avoid injury should the fan start.
4. Keep the fingers away from the moving belt and pulleys! Your fingers could be pinched and cut off if they are caught between the belt and a pulley.
5. Never attempt to remove the radiator cap from the cooling system of an engine that is near or above its normal operating temperature. Releasing the pressure may cause instant boiling of the coolant. Boiling coolant and steam spurting from the filler neck can cause scalding and burns. Allow an engine to cool before attempting to remove the cap.

⚙ 19-2 Cooling system trouble diagnosis Two common complaints about the cooling system are that the engine is overheating and that the cooling system leaks. If the engine is slow to warm up, this could also be blamed on the cooling system. The possible causes of these complaints are discussed below.

The troubleshooting chart on page 290 lists (1) various cooling system troubles, (2) possible causes of these various troubles, and (3) checks or corrections to be made.

There are several quick checks that can be made when certain types of troubles are reported. These quick checks often immediately indicate the cause of trouble. Sometimes it may be necessary to use special testing instruments to find the cause (as explained later). Often, the first step is to refill the cooling system. Quick checks to be made, and causes and corrections of various cooling troubles, are described below.

NOTE: If a cooling system analyzer (Fig. 19-1) is available, it can pinpoint many troubles in the cooling system. There is more about this type of analyzer in ⚙ 19-7.

⚙ 19-3 Causes of loss of coolant Many leaks can be spotted easily for two reasons. First, the cooling system requires frequent refilling. Second, the point of the leak usually can be found at the top of a telltale stain. Dye is added to most antifreeze to make leak detection easier.

There are two types of coolant leaks, external leaks and internal leaks. External leaks are those where the coolant can drip onto the ground. These can be seen. Typical leak points are from hose and hose connections, heater core, radiator core, and expansion core plugs (freeze plugs) in the block and head.

Internal leaks occur when the coolant leaks from the cooling system into some other part of the engine.

Cooling System Trouble Diagnosis Chart

CONDITION	POSSIBLE CAUSE	CHECK OR CORRECTION
1. Loss of coolant (⚙ 19-3)	a. Pressure cap and gasket defective	Inspect. Wash gasket and test. Replace only if cap will not hold pressure specified.
	b. Leakage	Pressure-test system.
	c. External leakage	Inspect hose, hose connections, radiator, edges of cooling system, gaskets, core plugs, drain plugs, oil-cooler lines, water pump, heater system components. Repair or replace as required.
	d. Internal leakage	Check torque of head bolts; retorque if necessary. Disassemble engine as necessary. Check for cracked intake manifold, blown head gasket, warped head or block gasket surfaces, cracked cylinder head or engine block.
2. Engine overheating (⚙ 19-4)	a. Low coolant level	Fill as required. Check for coolant loss.
	b. Loose fan belts	Adjust.
	c. Pressure cap defective	Test. Replace if necessary.
	d. Radiator or air conditioner condenser obstructed	Remove bugs, leaves, and debris.
	e. Closed thermostat	Test. Replace if necessary.
	f. Fan-drive clutch defective	Test. Replace if necessary.
	g. Ignition faulty	Check timing and advance. Adjust as required.
	h. Temperature gauge or HOT light defective.	Check electrical circuits. Repair as required.
	i. Inadequate coolant flow	Check water pump and block for blockage.
	j. Exhaust system restricted	Check for restrictions.
3. Engine fails to reach normal operating temperature; slow warm-up (⚙ 19-5)	a. Open or missing thermostat	Test. Replace or install as necessary.
	b. Defective temperature gauge or COLD light	Check electrical circuits. Repair as required.

These leaks cannot be seen. However, having to add coolant frequently is a clue that an internal leak may exist.

Internal leaks can severely damage the engine. The coolant may contaminate the oil and cause rust. A coolant leak into the combustion chamber while the engine is stopped may fill the combustion chamber. Then when the engine is cranked, the upward moving piston could cause the head, piston, or cylinder to crack or the connecting rod to bend.

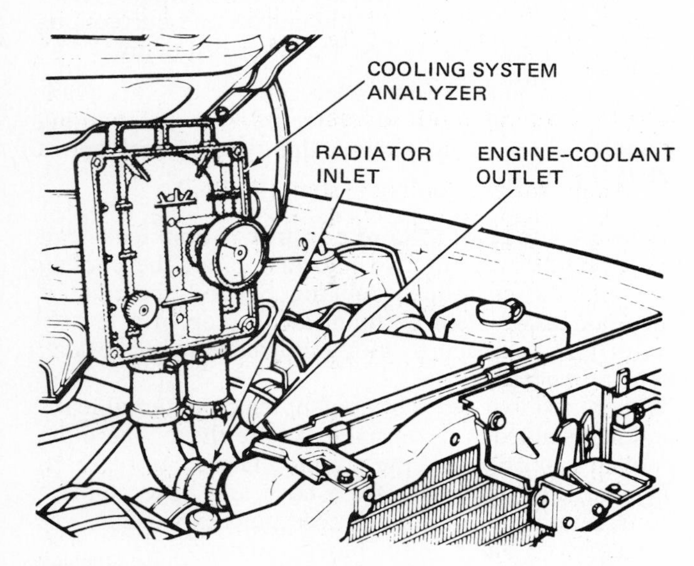

Fig. 19-1 Using a cooling system analyzer to check the volume of coolant flow. (*Ford Motor Company*)

If coolant leaks from a gasket (cylinder head, water pump), the gasket may require replacement. Attaching bolts should be tightened to the correct torque. If the leak is from the radiator, it should be removed and either repaired or replaced. Oil in the coolant indicates leakage of the transmission oil cooler in the outlet tank of the radiator (Fig. 18-22). If the leak is at a hose connection, the hose connection should be tightened. If a hose is leaking, the hose should be replaced. Pressure testing the cooling system to locate leaks is covered in ⚙ 19-14.

⚙ 19-4 Causes of engine overheating The driver may notice that the red light stays on or the temperature gauge registers in the overheating zone. Also, the driver may complain that the engine boiled over. Possible causes of engine overheating include:

1. Low coolant level caused by leakage of coolant from the system.
2. Accumulation of rust and scale in the system which prevents normal circulation of coolant. Antifreeze compounds contain additives which tend to prevent the formation of rust and corrosion.
3. Collapsed hoses which prevent normal coolant circulation.
4. Defective thermostat which does not open normally, blocking circulation of coolant. If the engine overheats without the radiator becoming normally warm, and if the fan belt is properly tightened, then the thermostat is probably at fault. Sometimes on new cars, grains of sand from the sand core for the engine block or head may lodge in the thermostat, prevent-

ing it from opening. A thermostat that is installed backwards usually cannot open, and will also cause overheating. Many thermostats are marked with an arrow to indicate the up side.

5. Defective water pump which does not circulate enough coolant throught the engine. A quick check of water pump operation can be made by installing a clear plastic hose in place of the upper radiator hose and running the engine. Then you can see how much coolant is circulating, and if any air is in it. A more accurate test of water pump capacity is made by installing a cooling system analyzer in place of the upper radiator hose (Fig. 19-1).

   One cause of water pump bearing failure is an overtight drive belt. A fan belt always should be tightened correctly with a belt-tension gauge (⚙ 19-16). Bearing failure usually makes the water pump noisy. A quick check of the bearing can be made with the fan belt off by grasping the tips of the fan blades and trying to move the fan in and out (Fig. 19-2). Any movement, or a rough and grinding feeling as the fan is slowly turned, indicates a defective bearing. Drops of coolant leaking from the water pump ventilation hole indicates a leaking water pump seal. The ventilation hole is below and behind the water pump pulley (Fig. 19-2).
6. A loose or worn fan belt will not drive the water pump fast enough. The belt should be tightened or replaced. Where a pair of belts is used, both belts should be replaced at the same time, not just the one that appears most worn. When you replace only one belt, all the load is on the new belt. It will wear rapidly. When both belts are replaced with a new matched pair, then each belt will carry half the load.
7. Overheating may be caused by afterboil. This may occur when the coolant starts to boil after the engine has been turned off, for example, after a long hard drive. The engine has so much heat in it that, when the water pump stops circulating coolant, it starts to boil.
8. Boiling can occur if the coolant is frozen. This hinders or stops its circulation. Then the coolant in the engine around the combustion chamber and cylinders becomes so hot that the coolant boils. Freezing of coolant in the radiator, cylinder block, or head may crack the block or head and open up seams in the radiator. Operating an engine in which the coolant is frozen may cause serious engine damage.

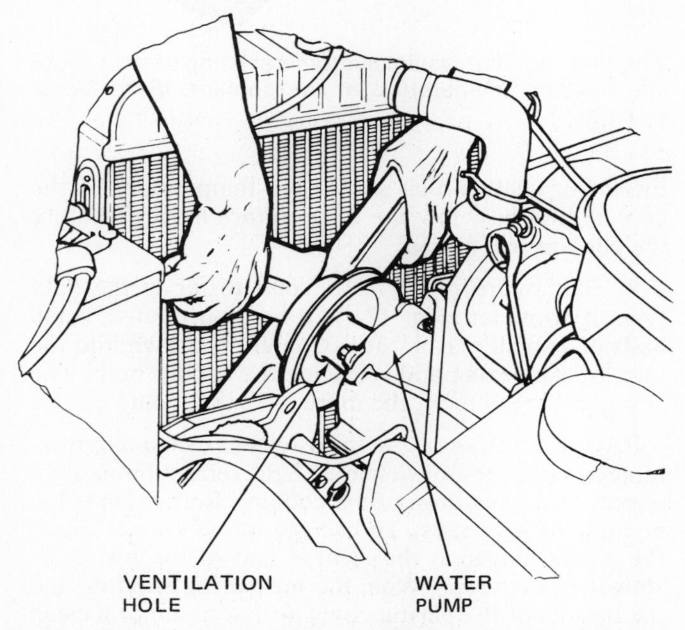

Fig. 19-2 Checking the water pump bearing. (*Chrysler Corporation*)

NOTE: There are other causes of engine overheating which have nothing to do with conditions in the cooling system. High-altitude operation, insufficient oil, overloading of the engine, hot-climate operation, improper ignition timing, long periods of slow-speed or idling operation, improperly operating emission controls—any of these can cause overheating of the engine.

⚙ 19-5 Causes of slow warm-up The most likely cause of slow engine warm-up is a thermostat that is stuck open (⚙ 18-26). This allows the coolant to circulate between the engine and the radiator even though the engine is cold. Therefore the engine has to run longer to reach normal operating temperature. As a result, engine wear is greater because the engine operates cold for a longer time.

Another possible cause of slow warm-up is that the thermostat has been removed. Never remove a thermostat and leave it out of the cooling system. This does not improve coolant circulation. It does delay warm-up and increases engine wear and sludge formation.

A quick check for a missing or stuck-open thermostat can be made by squeezing the upper hose immediately after starting a cold engine. No coolant flow through the upper hose should be felt. If you feel movement, the thermostat is missing or open.

In the winter, the driver's complaint often is that it takes a long time for the car heater to start delivering heat. If you hear this complaint, suspect a defective or missing thermostat. How to test thermostats is covered in ⚙ 19-10.

⚙ 19-6 Cooling system test Cooling system tests include:

1. Analyzing the cooling system (⚙ 19-7)
2. Checking coolant level (⚙ 19-8)
3. Checking antifreeze strength (⚙ 19-9)
4. Testing the thermostat (⚙ 19-10)
5. Checking the hose and hose connections (⚙ 19-11)
6. Testing the water pump (⚙ 19-12)
7. Checking for exhaust gas leakage into the system (⚙ 19-13)
8. Pressure testing the system (⚙ 19-14) and cap (⚙ 19-15)
9. Checking the fan belt or belts for wear and tension (⚙ 19-16)
10. Checking the cooling system for rust and scale (⚙ 19-17)

These are covered in detail in following sections.

⚙ 19-7 Analyzing the cooling system One difficulty in trying to diagnose certain problems is that you cannot see what actually is happening in the cooling system. For example, eroded or broken blades on the water pump impeller may greatly reduce the capacity of the pump. This will cause the engine to overheat. However, there is no way to see this condition.

You could eliminate all other likely causes, and then remove the water pump for inspection. But this is a slow and costly way to diagnose the problem. Another condition that you cannot see is the amount of air and foaming in the coolant. Both of these reduce the capacity of the coolant to transfer heat.

Use of a cooling system analyzer, such as shown in Fig. 19-1 simplifies the analysis. This type of analyzer connects into the cooling system so that the coolant flows through it. The connections are made between the coolant outlet housing on the engine and the upper radiator hose.

Most cooling system analyzers have a flow meter, so that the pump capacity can be checked. Also, most analyzers have a clear section so that you may visually inspect the coolant passing through for color, flow, bubbles, and foaming. Some analyzers have a temperature gauge. By watching the gauge and the coolant flow you can see exactly when the thermostat starts to open and when it is fully open.

When using a cooling system analyzer, follow the manufacturer's instructions for the connections and testing procedures.

⚙ 19-8 Checking coolant level On cooling systems using an expansion tank, it is not necessary to remove the radiator cap to check the coolant level. Car manufacturers warn against removing the radiator cap except for major service. (See caution 5 in ⚙ 19-1). The coolant level can be checked by looking at the expansion tank. It is plastic so that you can see the level of the coolant (Fig. 18-19).

On cooling systems without an expansion tank, remove the radiator cap to check the coolant level.

Never remove the radiator cap from a running engine. Use care when removing a radiator pressure cap. The engine may be hot. Cover the cap with a folded cloth to protect your hand. Then turn the cap slowly to the safety stop and step back. Any pressure in the system will be released through the overflow tube. (See caution 5 in ⚙ 19-1.)

⚙ 19-9 Testing antifreeze strength The amount of antifreeze in the coolant must be great enough to protect against freezing at the lowest temperatures that might occur. The strength of the antifreeze can be checked with any of three testers: the float hydrometer, the ball hydrometer, and the refractometer.

1. Float hydrometer The float-type hydrometer is shown in Fig. 19-3. The higher the float rises in the coolant, the higher the percentage of antifreeze in the coolant. To use the hydrometer, put the rubber tube into the coolant. Then squeeze and release the rubber bulb. Note how high the float rises in the coolant. Check

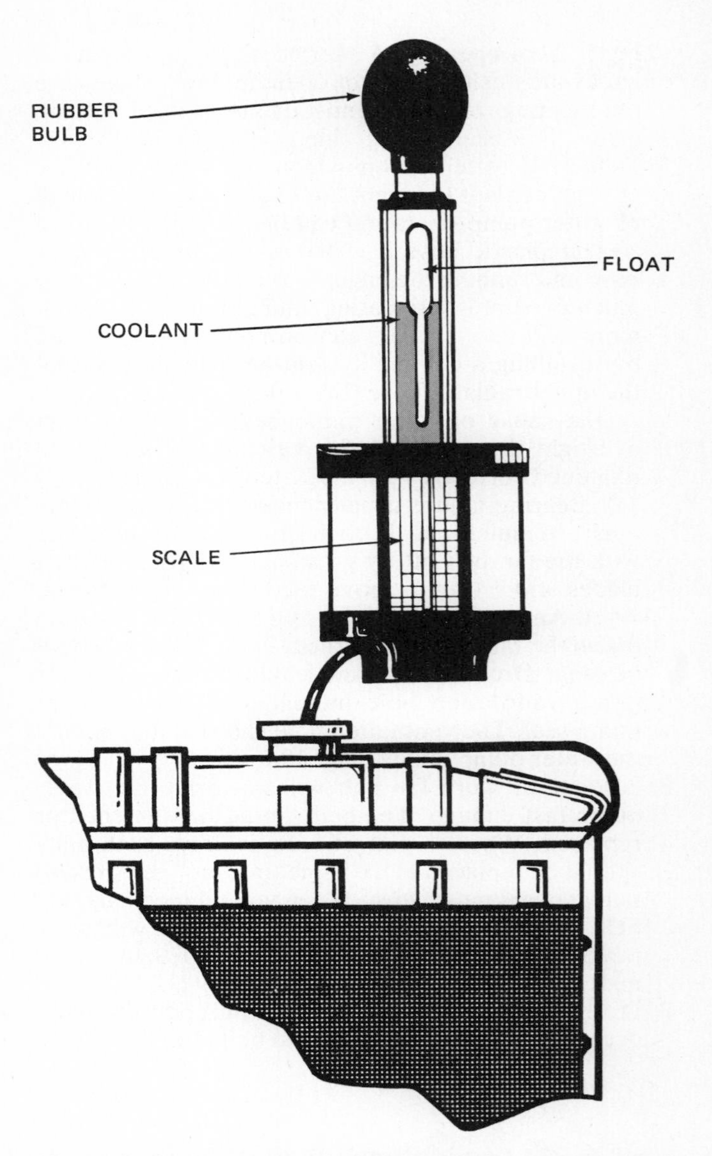

Fig. 19-3 Cooling system hydrometer being used to check the freezing temperature of the coolant. (*Ford Motor Company*)

the lower scale which shows the temperature of the coolant and how low the temperature must go before the coolant will freeze.

2. Ball hydrometer A second tester is the ball-type hydrometer (Fig. 19-4). It has four or five small balls in a small plastic tube. Coolant is drawn into the tube by squeezing and releasing the rubber bulb. The stronger the solution, the more balls that float.

3. Refractometer A third tester, called a refractometer, uses the principle of light refraction as light passes through a drop of the coolant. Refraction is the bending of light rays. To use the refractometer, open the plastic cover at the slanted end of the tester, as shown in Fig. 19-5. Wipe the measuring window and the bottom of the plastic cover with a tissue or a clean cloth.

Close the plastic cover. Release the tip of the tube

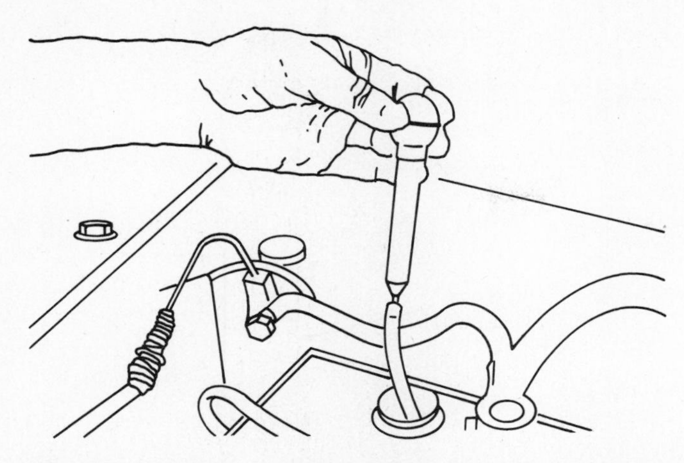

Fig. 19-4 Ball-type hydrometers are available which will check antifreeze strength.

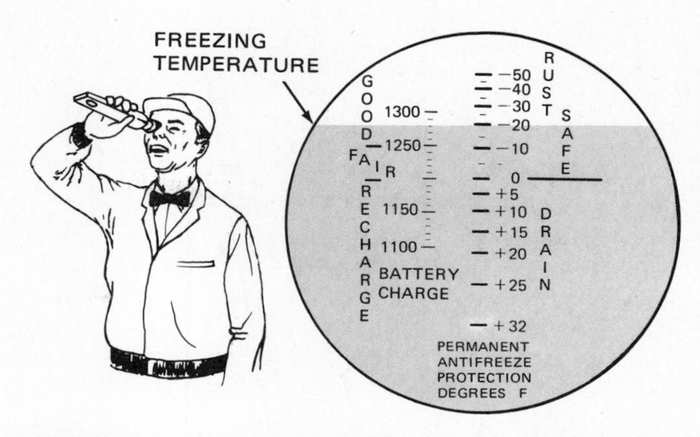

Fig. 19-6 Reading the refractometer. The freezing temperature of the coolant is shown by the line that separates the clear area (top) from the unclear area below. *(Pontiac Motor Division of General Motors Corporation)*

from the tester housing. Insert the tip of the tube into the radiator filler neck. Be sure the end of the tube is below the level of the coolant. Now press and release the bulb. This draws a small sample of coolant into the tube. Bend the tube around so that the tip can be inserted into the cover plate opening. Squeeze the bulb to put a few drops of coolant into the measuring window.

Now point the tester toward the light and look into the eyepiece, as shown in Fig. 19-6. The freezing temperature of the coolant is where the dividing line between light and dark (the edge of the shadow) crosses the scale. Readings on the lower half of the scale indicate solutions without sufficient antifreeze.

NOTE: A refractometer can also be used to check battery electrolyte to determine the battery state of charge. This is the scale shown on the left of Fig. 19-6.

CAUTION: COOLANT IS POISONOUS! It can cause serious illness and even death if it is swallowed! Never place the hydrometer or refractometer tube in your mouth. Always wash your hands thoroughly after getting coolant on them.

19-10 Testing the thermostat Different car manufacturers have different testing procedures for checking thermostats. Chevrolet recommends suspending the thermostat in a solution of one-third antifreeze and two-thirds water. The solution should be heated to 25°F [14°C] above the temperature stamped on the thermostat. The thermostat should open. Then submerge the thermostat in the same solution with the temperature at 10°F [5.5°C] below the temperature stamped on the thermostat. The thermostat should close completely. If it does not open and close during the test, it is defective.

Plymouth recommends testing the thermostat in the engine cooling system. With the cooling system filled to the proper level, warm the engine by driving the car for about 10 minutes. Remove the radiator cap, observing the safety cautions. Insert a thermometer into the coolant. Idle the engine with the hood raised. The coolant temperature should remain steady at no lower than 8°F [4.4°C] below the thermostat opening temperature.

Ford recommends immersing the thermostat in boiling water, as shown in Fig. 19-7. If the thermostat does not open, it is defective. If the problem is slow warm-up, the thermostat may be leaking. Hold the thermostat up to the light to see if the valve is closing completely. If there is a gap between the valve and the valve seat with the thermostat at room temperature, replace the thermostat.

19-11 Checking the hose and hose connections To check the condition of the radiator hose, squeeze the hose (Fig. 19-8). It should not collapse easily when squeezed. The appearance of the hose and hose connections usually indicates their condition (Fig. 19-9). Hose that is soft, hard, rotted, or swollen should be replaced. The hose must be in good condition and connections should be properly tightened to avoid leaks.

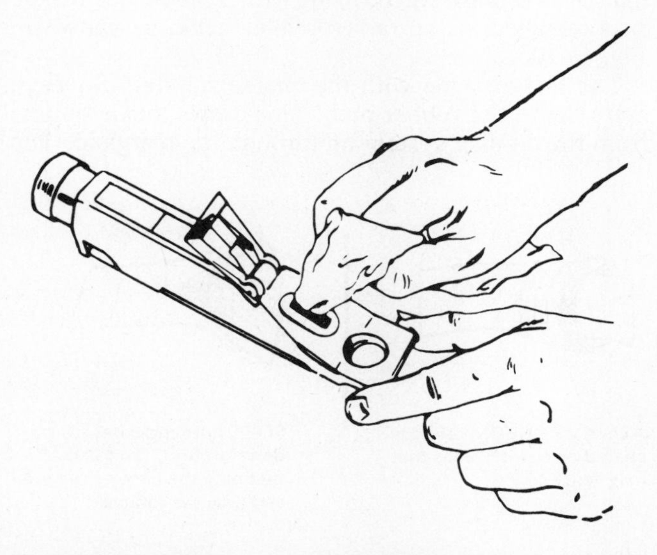

Fig. 19-5 Cleaning the refractometer measuring window. *(Pontiac Motor Division of General Motors Corporation)*

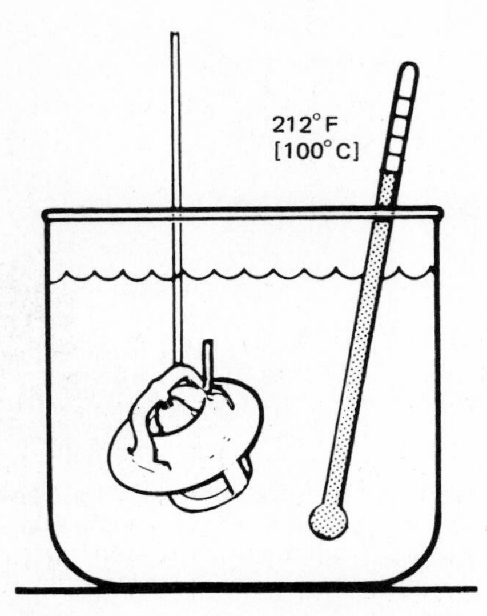

Fig. 19-7 Testing a cooling system thermostat. (*Ford Motor Company*)

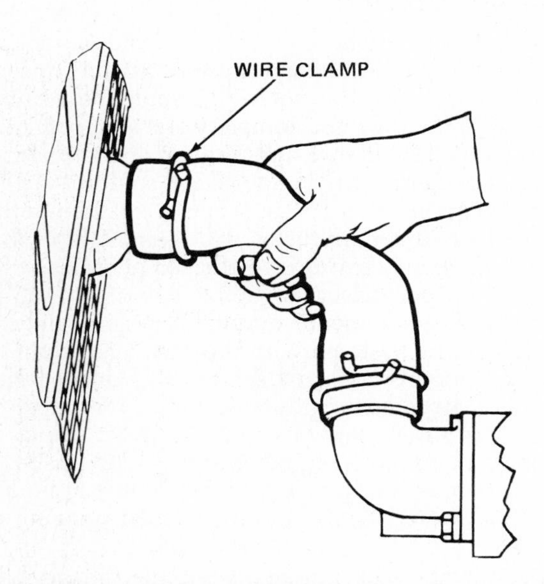

Fig. 19-8 To check the condition of a radiator hose, squeeze the hose. (*Chrysler Corporation*)

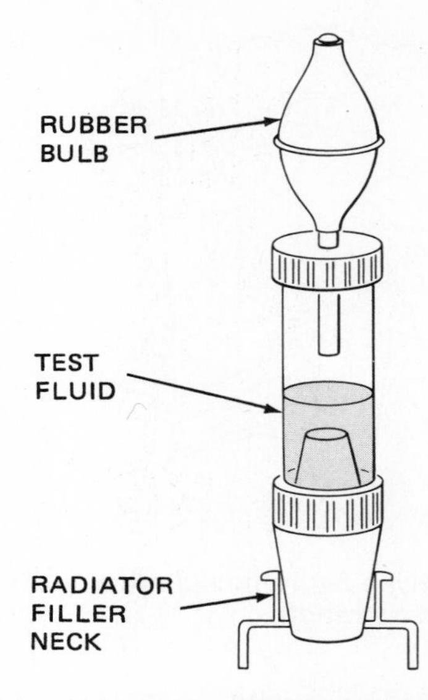

Fig. 19-10 Checking for exhaust gas leakage into the cooling system with a Bloc-Chek tester.

⚙ 19-12 Testing the water pump Checking the water pump bearing and seal as described in ⚙ 19-4. To check the water pump capacity, substitute a clear plastic pipe for the upper radiator hose. Then, when the engine is running, you can see how much coolant is circulating. A more accurate test of the water pump can be made using a cooling system analyzer (Fig. 19-1), as discussed in ⚙ 19-7. Always be careful when checking and tightening fan belts. Overtightening the fan belt can cause the water pump bearing to fail.

⚙ 19-13 Checking for exhaust gas leakage into the system A defective cylinder head gasket may allow exhaust gas to leak into the cooling system. This is very damaging. Strong acids can form as the gas unites with the water in the coolant. These acids corrode the radiator and other cooling system parts. A test for exhaust-gas leakage can be made with a Bloc-Chek tester. It is installed in the radiator filler neck, as shown in Fig. 19-10.

The test is made with the engine running. Squeeze and release the rubber bulb. This draws an air sample from the cooling system up through the test fluid. The

HARD hose can fail. Tightening hose clamps will not seal the connection or stop leaks.

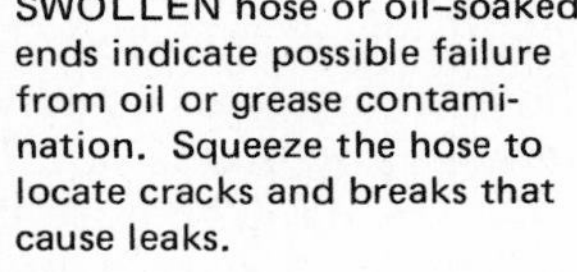

SWOLLEN hose or oil-soaked ends indicate possible failure from oil or grease contamination. Squeeze the hose to locate cracks and breaks that cause leaks.

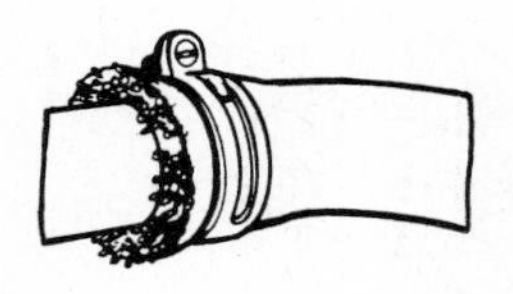

ALWAYS CHECK hose for chafed or burned areas that may fail.

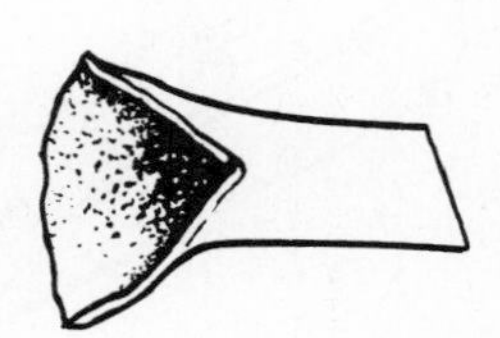

SOFT hose indicates inside deterioration. This can contaminate the cooling system and clog the radiator.

Fig. 19-9 Conditions that indicate failure of a radiator hose.

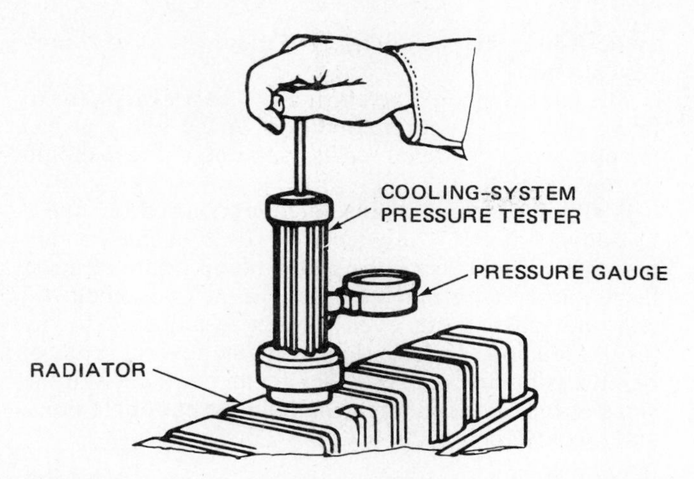

Fig. 19-11 Using a cooling system pressure tester to check the cooling system for leaks. (*Chrysler Corporation*)

test fluid is ordinarily blue. However, if combustion gas is leaking into the cooling system, the test fluid will change to a yellow color. If a leak is indicated, the exact location can be found by removing one spark plug wire at a time and retesting. When a leaking cylinder is firing, the liquid will change to yellow. When nonleaking cylinders only are firing, the liquid will remain blue.

Undetected combustion leaks in the valve areas can cause cracked valve seats and cylinder heads. The coolant is forced away from the cracked area during heavy acceleration by the leakage of combustion gases through the leak. This causes excessive heat buildup. When acceleration stops, the diverted coolant rushes back to the overheated area. The sudden cooling of the area can crack the head and valve seat.

19-14 Pressure testing the cooling system To pressure test the cooling system, apply pressure with the cooling system pressure tester as shown in Fig. 19-11. The tester usually will cause the coolant to drip or spray from the leak.

To use the tester, remove the radiator cap and fill the radiator until the coolant level is about ½ inch [13 mm] below the bottom of the filler neck. Wipe the neck sealing surface and attach the tester. Then operate the pump to apply a pressure that does not exceed 3 psi [21 kPa] above the manufacturer's specifications. If the pressure holds steady, the system is not leaking. If the pressure drops, there are leaks. Look for coolant leaks at hose connections, hose, engine expansion plugs, water pump and cylinder head gaskets, water pump shaft seal, and radiator.

If no external leaks are visible, remove the tester and start the engine. Run the engine until operating temperature is reached. Reattach the tester, apply a pressure of 15 psi [103 kPa], and increase engine speed to about 3000 rpm. If the needle of the pressure gauge fluctuates with engine speed, it indicates an exhaust gas leak, probably through a cylinder head gasket. On a V-type engine, you can determine which bank is at fault by grounding the spark plugs in one bank.

If the needle does not fluctuate, sharply accelerate

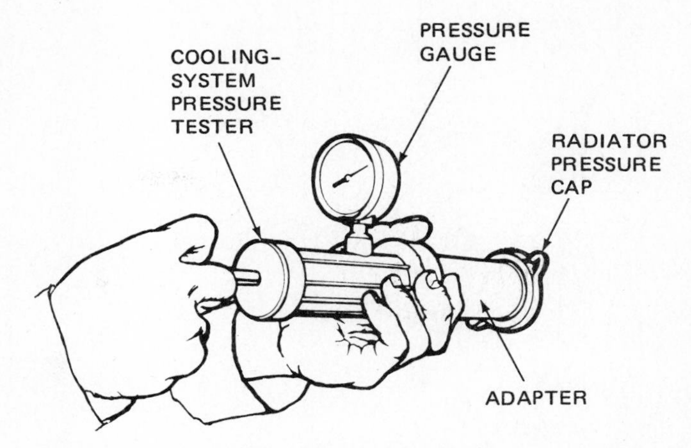

Fig. 19-12 Using the cooling system pressure tester to check a radiator pressure cap. (*Chrysler Corporation*)

the engine several times and check for abnormal discharge of liquid through the tail pipe. This would indicate a cracked block or head, or a defective head gasket.

19-15 Pressure testing the radiator cap The cooling system pressure tester can be used to check the radiator pressure cap. An adapter is attached to the tester pump so it will fit the cap (Fig. 19-12). Then the pump is operated to apply the rated pressure against the cap. If the cap will not hold its rated pressure, it should be discarded.

19-16 Testing the fan belt Fan belts, or drive belts, should be checked for wear and tension. Most wear occurs on the underside of the belt. To check a belt, be sure the engine is off and will not be cranked. Then twist the belt with your fingers. Check for small cracks, grease, glazing, and tears or splits (Fig. 19-13).

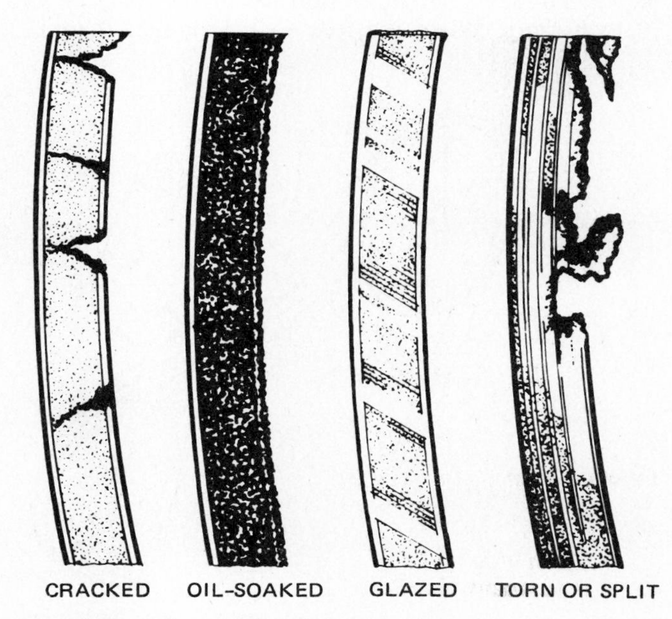

Fig. 19-13 Conditions to look for when inspecting a drive belt. (*Chrysler Corporation*)

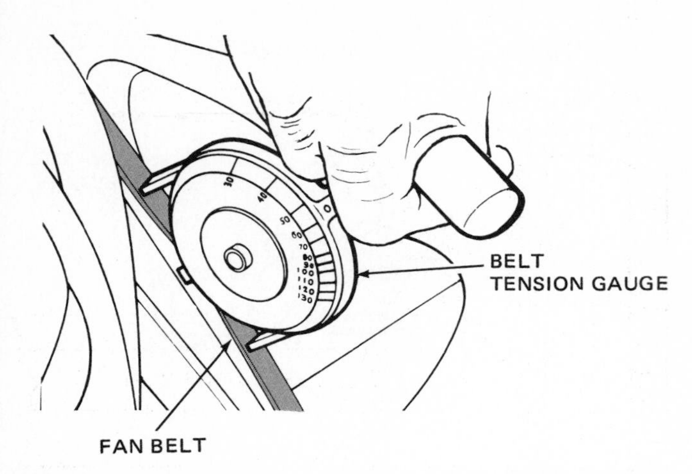

Fig. 19-14 Using a belt tension gauge to check the tension of a drive belt. (*Chrysler Corporation*)

Small cracks will enlarge as the belt is flexed. Grease rots rubber and makes the side slick so that the belt slips easily. A high-pitched squeal is the typical sound of a loose and slipping belt. Glazed belts result from slippage. Large tears or splits in a belt allow it to be tossed from the pulley easily.

On cars with a set of two fan belts, if one is worn and requires replacement, then both should be replaced. They come in matched sets. If only one is replaced, the new belt will take most of the wear because it is not stretched like the old belt. This means the new belt will wear rapidly while the old belt may slip, overheat, and also wear rapidly.

Use a belt tension gauge (Fig. 19-14) to check and adjust the fan belt tension. When you do not have a gauge, or if space does not allow use of a gauge, you can make a quick check of belt tension. To do this, press in the middle of a free span, as shown in Fig. 19-15. When the free span is less than 12 inches [305 mm] between pulleys, belt deflection should be ⅛ to ¼ inch [3 to 6 mm]. When the free span is longer than 12 inches [305 mm], belt deflection should be ¼ to ½ inch [6 to 13 mm].

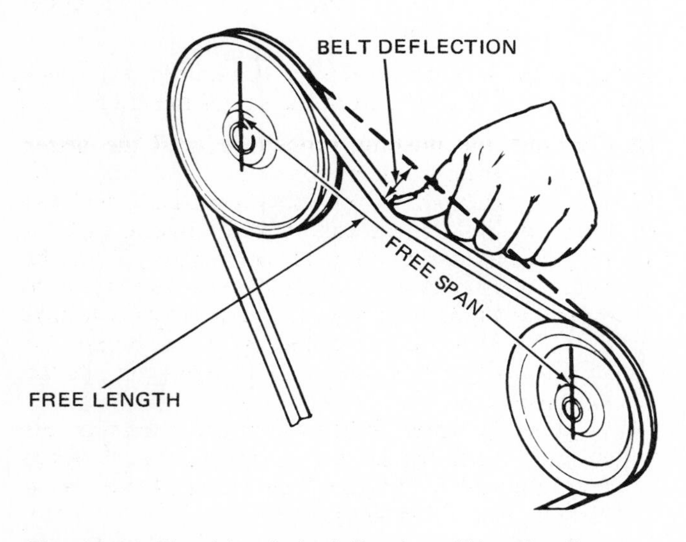

Fig. 19-15 Checking belt deflection. (*Chrysler Corporation*)

The fan belt should be checked at least every year to make sure it is in good condition. A fan belt that has become worn or frayed, or has separated plies, should be replaced.

A slipping belt can cause engine overheating and a run-down battery. These troubles result because a slipping belt cannot drive the water pump and alternator fast enough for normal operation. Sometimes a belt will slip and make noise even after it is adjusted to the proper tension. To help this problem, several types of belt dressing are available which can be applied to the sides of the belt. Belt dressing helps to eliminate noise and increase belt friction.

⚙ 19-17 Testing the system for accumulation of rust and scale Although ethylene glycol is called a "permanent" antifreeze, it really is not. However, it is much more permanent than the previously used alcohol-based antifreeze. The alcohol boiled away quickly when used in an engine in hot weather.

To prolong the life of the old-type alcohol-based antifreeze, it was drained and stored in a closed container during the summer months. The cooling system was refilled with plain water. This caused the rapid formation of rust and scale. However, with pressurized cooling systems and the use of a permanent "antifreeze and summer coolant" year round, less rust collects in the cooling system.

Some scale may accumulate from minerals in the water. This is the reason why many manufacturers recommend periodic cleaning of the cooling system. The modern cooling system is sensitive to accumulations of scale and rust. There is no accurate way to determine the actual amount of deposit buildup in the cooling system. However, an approximation can be made if you know the original capacity of the system. This specification is given in most auto repair manuals and in the manufacturers' service manuals.

⚙ 19-18 Cleaning the cooling system The original additives in antifreeze to fight rust and corrosion break down and are ineffective after 1 to 2 years. This is because of the continual exposure to the heat in the cooling system. This also explains why even a permanent antifreeze must be replaced periodically.

When the antifreeze has severely deteriorated, its color may be used as an indicator of its condition. After the additives in the antifreeze break down, rust begins to form rapidly. Therefore, a rust-colored antifreeze is an indication that cooling system service is needed.

A quick check of the conditions inside the cooling system can be made. Remove the cap and wipe the inside of the radiator filler neck with your finger. If you find any oil, grease, rust, or scale, the cooling system should be cleaned. Reverse flushing also may be needed.

The cooling system should be cleaned periodically to remove rust, scale, grease, oil, and any acids formed by exhaust gas leakage into the coolant. Recommendations vary. For example, Chevrolet recommends that

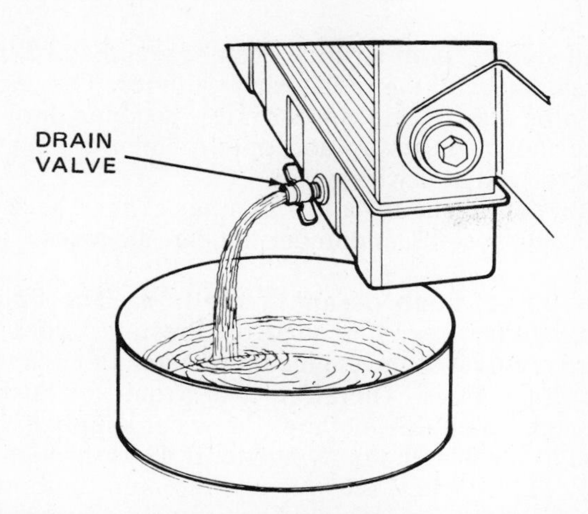

Fig. 19-16 To service the cooling system, open the radiator drain valve. (*Chrysler Corporation*)

the cooling system be drained and flushed with plain water every 2 years. Then the cooling system should be filled with a solution of fresh antifreeze and water.

Here is the general procedure for cleaning the cooling system. It begins with the use of a chemical cooling-system cleaner. This is used before reverse flushing to break loose the rust and scale deposits in the cooling system.

1. Drain the cooling system (Fig. 19-16). Remove the thermostat and reinstall the thermostat housing. Close the radiator drain valve. Pour the cooling system cleaner into the radiator. Then fill the cooling system with water to about 3 inches [76 mm] below the top of the overflow pipe in the radiator.

NOTE: Several types of chemical cooling-system cleaner are available. Some are liquid. Others are powder. Follow the instructions on the container.

2. Run the engine at fast idle for about 20 minutes. Stop the engine if the water begins to boil.
3. Stop the engine and wait for it to cool. Then open the valve to drain out the water and chemical cleaner.

NOTE: Some cooling systems do not have a separate drain valve. On these, you must loosen or remove the lower hose on the radiator to drain the cooling system.

4. Direct a stream of water from a hose into the radiator to flush out any loosened rust and scale.
5. Remove the lower hose from the radiator.
6. Use a water hose or compressed air to clean dirt, bugs, and other trash from the fins of the radiator. Blow from the engine side. In many cars that can be done only after removal of the fan shroud.
7. If possible, straighten any bent fins. Bent fins reduce the cooling capacity of the radiator. Be very careful when doing this. The fins on a radiator are very sharp and can easily cut you.
8. Reverse-flush the radiator and engine block. Each of these operations is done separately. In reverse flushing, water is forced through the radiator or engine in the direction opposite to normal flow.

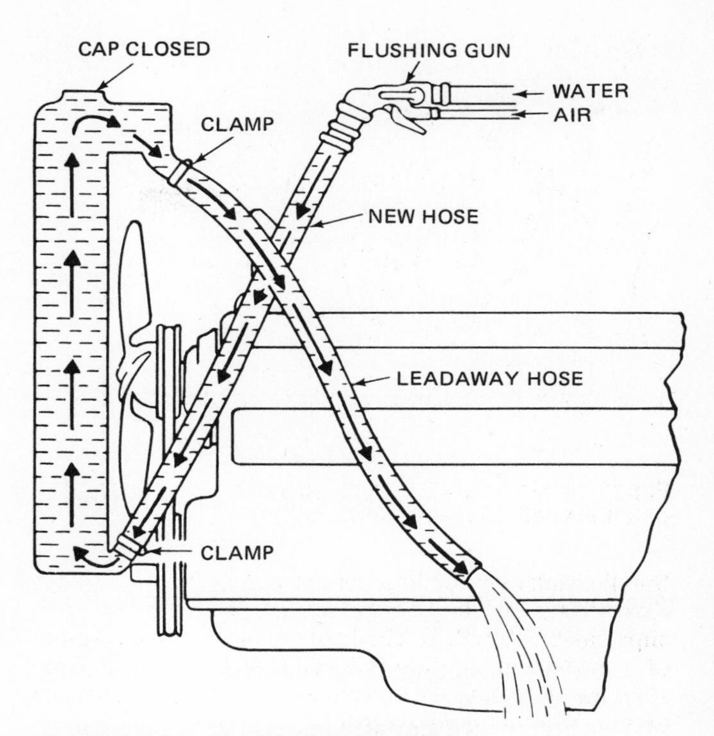

Fig. 19-17 Reverse flushing a radiator.

This gets behind the scale and rust to loosen it so that it can be flushed away.

9. To reverse-flush the radiator, remove the upper and lower hoses from the radiator. Then install the radiator cap.
10. Connect a long leadaway hose to the top of the radiator as shown in Fig. 19-17. Connect a new hose to the outlet at the bottom of the radiator. Connect the water hose of the flushing gun to a water outlet, and the air hose to an air line. Insert the nozzle of the flushing gun into the hose from the radiator outlet.
11. Turn on the water. When the radiator is full, use short blasts of compressed air to force water through the radiator. Allow the radiator to fill between blasts of air.

NOTE: Apply the air gradually. A radiator will stand only 20 psi [138 kPa] before the seams may leak.

12. Continue the flushing procedure until the water from the leadaway hose runs clear.
13. To reverse-flush the engine, first disconnect the heater hose from the engine, as shown in Fig. 19-18. Cap the connection at the engine. With the radiator hose removed, attach a leadaway hose to the water pump inlet. Connect a length of new hose to the coolant outlet at the top of the engine, as shown in Fig. 19-18. Insert the flushing gun into the top hose.
14. Turn on the water. When the engine water jackets are full, use short blasts of compressed air to force water through the engine. Never exceed an air pressure of 20 psi [138 kPa].

NOTE: Excessive air pressure can blow out the freeze plugs (also called expansion core plugs and water jacket

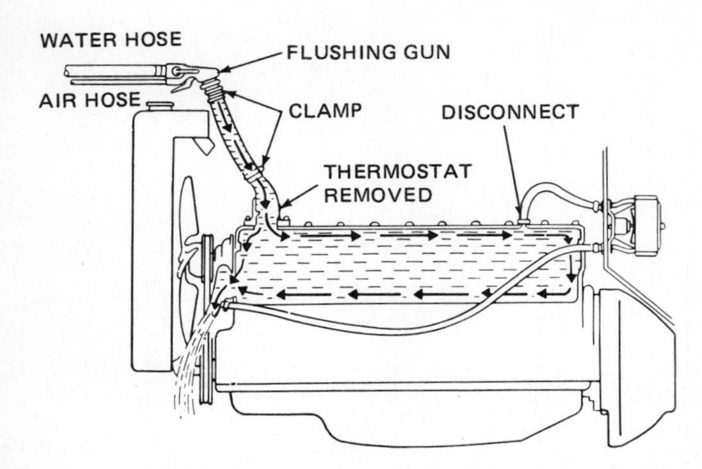

Fig. 19-18 Reverse flushing the water jackets in the engine.

plugs) in the cylinder head and block. Servicing these plugs is covered in ⚙ 19-23.

15. Continue flushing until the water from the leadaway hose runs clear.
16. The heater core may also be reverse-flushed. However, a high air pressure must not be used. The core can be damaged by excessive air pressure.
17. Install the thermostat and the upper and lower radiator hoses. Use new hoses if the old hoses are worn or damaged. Make sure the hoses are properly installed on the inlet and outlet tubes of the radiator. These tubes have a ridge, or raised bead, to help seal the connection. Then install the hose clamps, and tighten them if necessary.
18. Add enough antifreeze to protect against freezing at the lowest temperature expected. Then fill the system with water. Because the water is cold, the thermostat will close and prevent quick filling of the cooling system. With the thermostat closed, air is trapped in back of it, as shown in Fig. 19-19. The thermostat has a small hole that permits air to leak out slowly. If an air lock occurs, you may have to wait and refill the radiator a few times. The engine can be started and run for a few seconds until the thermostat heats up and opens. Then completely fill the radiator with water.
19. Run the engine for a few minutes. Then check the cooling system, and under the car, for leaks.

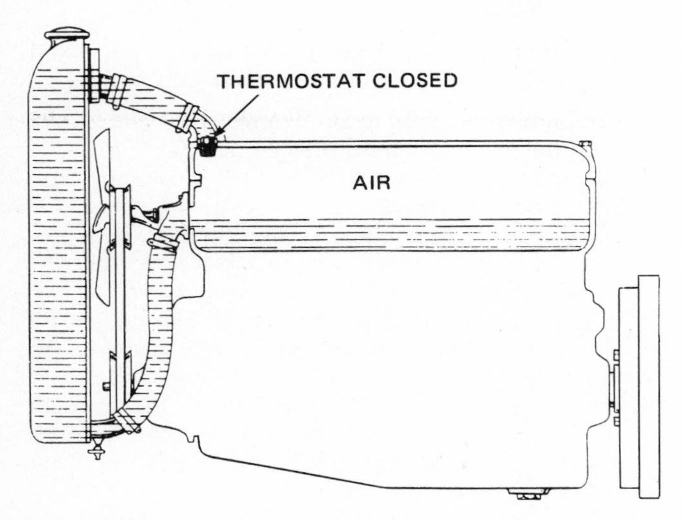

Fig. 19-19 Air trapped in back of a closed thermostat as the engine cooling system is filled.

⚙ 19-19 Fast-flush cleaning method The following procedure, recommended by Prestone, does not require removal of the engine thermostat and cylinder-block drain plugs. Therefore it is a quicker flushing procedure. With this method, the water supply is connected to the heater supply nipple. It flows through the engine water jacket, the water pump, and the bottom of the radiator. The water exhausts mainly through the filler neck of the top radiator tank. The specific directions follow.

1. Set the heater temperature control (1) to high. If the car has a vacuum-operated heater valve, start the engine and run it at idle during flushing. Be sure to turn it off before shutting off the water.
2. Open the radiator drain valve (2).
3. Remove the radiator cap, and install a leadaway hose or a deflection elbow (3).
4. Remove the hose from the heater supply nipple at the engine block (4). Point the hose down so that it will drain. (If the nipple is hard to get at, remove the heater supply hose at the heater. Connect the water supply to flow into this hose—not into the heater. Attach a short piece of hose at the heater, and point it down (5). Never connect the water supply directly to the heater! This could damage the heater.)
5. After connecting the water supply either to the heater supply nipple (4) or to the heater supply hose, turn on the water. A flushing gun can also be used, as noted in the flushing instructions given earlier. Avoid excessive air pressure if air is used. During the last minute, squeeze the upper radiator hose to remove any trapped liquid.
6. Turn off the water. Reconnect the heater supply hose, and disconnect the deflection elbow. Allow enough water to drain so there is room in the cooling system for the antifreeze. Then close the radiator drain valve. Add the antifreeze and adjust its strength.

⚙ 19-20 Fast-flush-and-fill machine This is a machine that flushes, tests, and refills the cooling system in about 10 minutes. It is not necessary to remove the thermostat, open the drains at the radiator and engine block, or remove and replace cooling-system hoses when using the machine. Here is the procedure:

1. Open the car hood, and remove the radiator cap. Attach a plastic tube in the radiator filler neck. This is the leadaway hose through which the system is flushed.
2. Set the car temperature control to high.
3. Locate the heater inlet hose, which is connected to the top of the engine. Apply temporary clamps on the two sides of the place you will cut. Cut the hose,

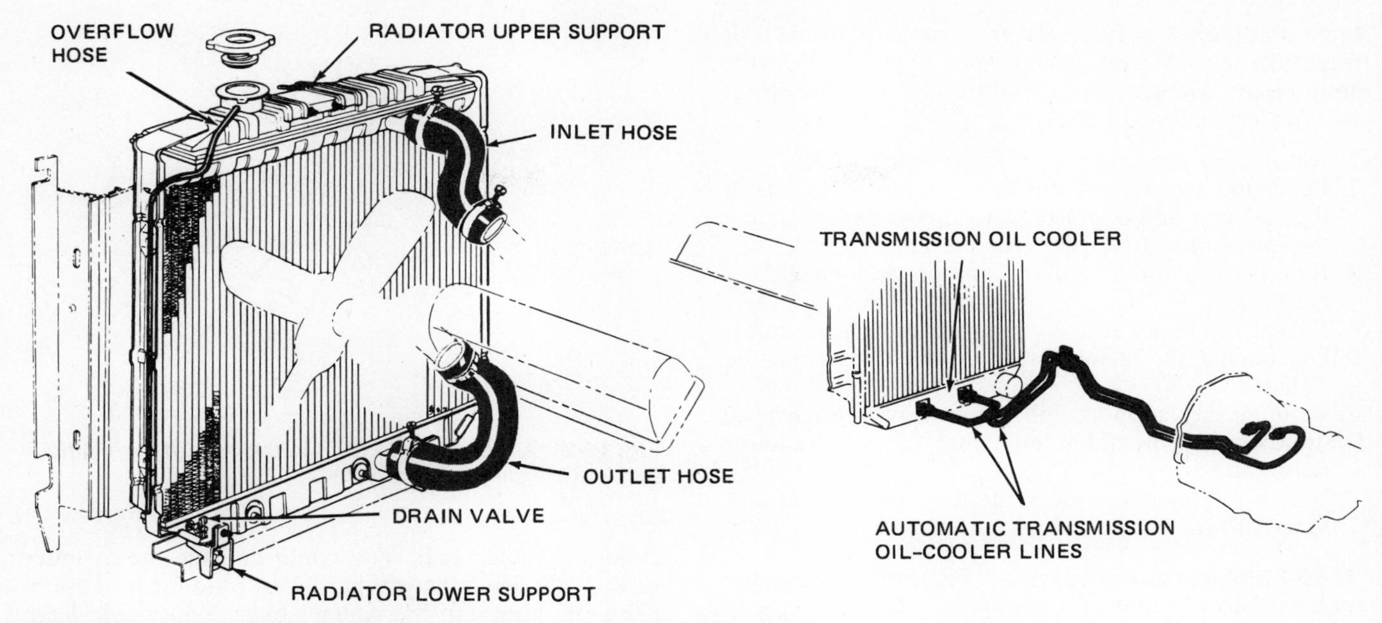

Fig. 19-20 Radiator installation on a V-8 engine. (*Ford Motor Company*)

and install the T-fitting with permanent clamps. The T-fitting now becomes a permanent part of the cooling system. Tighten the clamps enough to make leakproof joints. Remove the temporary clamps.

4. Attach the hoses from the flush-and-fill machine to the T-fitting. You are now ready to start the machine. The machine must be connected to an electric outlet and a water source.
5. Turn the selector switch on the machine to FLUSH. Operate the machine for 4 minutes, or until the flush water coming out the plastic tube is clean. This is a reverse-flushing operation.
6. While the machine is working, pour the required amount of antifreeze into the machine reservoir.
7. When flushing is finished, turn the selector switch off. Disconnect the plastic tube, and reinstall the radiator cap. Operate the pressure-test valve on the machine, and note the gauge reading. This tells you whether or not the system is tight. If the indication is that there are leaks, inspect the connections, hoses, radiator, water pump, and engine gaskets for signs of leakage.
8. If the system has no leaks, remove the radiator cap. Reinstall the plastic tube on the radiator filler neck. Turn the selector switch to FILL so the machine will pump the antifreeze into the system. When the machine reservoir is empty, turn off the machine. Remove the plastic tube from the radiator filler neck, and reinstall the radiator cap.
9. Disconnect the coupling of the machine hose. Cap the T-fitting. The job is now complete.

✿ 19-21 Locating and repairing radiator leaks

Figure 19-20 shows the radiator and its connections for a V-8 engine. Leaks from a radiator usually are easy to find. Telltale marks form below the leak because of the dye in the antifreeze.

A cooling system pressure tester can be used to find leaks (✿ 19-14). When the cooling system is pressurized, coolant may spray or drip from any small hole or open seam. Also, on copper-core radiators, a greenish-white corrosion may form on any leaking seam between the tanks and the core. Automatic-transmission fluid may leak from the cooler lines connected into the outlet tank of the radiator (Fig. 19-20).

An accurate way to locate radiator leaks is to remove the radiator from the car and drain out all the coolant. Then close the openings at top and bottom, and immerse the radiator in water. Air bubbles will escape from the radiator through any leaks. In a radiator repair shop, air pressure may be applied to the radiator while it is immersed in a special tank. This will locate almost all leaks in the radiator. However, some cooling system leaks may occur only while the engine is cold. Other leaks may occur only after the engine is at normal operating temperature.

Small leaks sometimes can be repaired without removing the radiator from the car, but not often. Liquid stop-leak and radiator seal compounds, when poured into the radiator, may seep through the leaks. The chemicals harden on contact with the air, sealing off any small openings. The best way to repair radiator leaks is to solder them.

If there are several leaks in the radiator core, it may not be worthwhile to repair it. You might stop one leak temporarily, but other leaks may soon develop. Heavy deposits of rust and scale that restrict coolant flow through the core also may make the repair of a radiator impractical. A radiator may be "recored" by removing the tanks and installing them on a new core. However, a new core is expensive. When the price of labor to change the tanks from the old radiator to the new core is considered, it often is cheaper to install a complete new radiator.

Radiator repair and recoring are specialty jobs usually done by a radiator repair shop. This means that in most automotive service shops, a radiator in need of repair is removed from the car and sent to the radiator shop. Here is a general procedure to follow when removing a radiator

1. Disconnect the negative cable from the battery.
2. Drain the cooling system by opening the drains in the radiator and engine block, or by disconnecting the lower hose from the radiator.
3. Remove the upper radiator support and shroud, as necessary.
4. Detach the upper and lower hoses from the radiator.
5. Disconnect the transmission cooler lines from the radiator.
6. Remove any support bolts, horns, wiring harness, or other equipment that might interfere with radiator removal.
7. With all other parts out of the way and the radiator loose, lift it straight up and out of the car.

⚙ 19-22 Water pump service Figure 19-2 shows the water pump installed on an engine. The water pump requires no service in normal operation. Older pumps required lubrication through grease fittings or through the use of soluble oil mixed with the coolant. This was known as "water pump lubricant."

Today water pumps have sealed bearings which require no lubrication. If the pump develops noise, leaks, or becomes otherwise defective, it is replaced with a new or remanufactured pump. The procedures vary for replacing the water pump. A typical procedure follows.

To remove the water pump, drain the cooling system. Remove the inlet hose and the heater hose from the pump. Remove the fan belt. Then remove the attaching bolts and take off the pump. On some engines, be sure to pull straight out to avoid damaging the impeller and shaft.

In normal operation, the impeller blades may wear away. This may result from abrasive action of sand in the cooling system. Rusting away of the blades may result from use of old antifreeze without active rust and corrosion inhibitors, or from the use of plain water in the cooling system.

Before installing a new water pump, check the impeller size of the new pump against the old pump. Some water pumps look alike and will bolt to more than one engine. However, impellers of different diameters are installed on the pump shaft to change the pump output. Damage to the cooling system or overheating may result from installation of a pump with the wrong impeller.

Install the water pump on the engine, following backward the steps you used to remove it. Then adjust the fan belt to the proper tension (⚙ 19-16).

⚙ 19-23 Expansion core plugs A leaking core plug (Fig. 19-21) must be replaced. To do this, place the pointed end of a pry bar against the center of the plug. Tap the end of the bar with a hammer until the point goes through the plug. Then press the pry bar to one side to pop the plug out.

Do not drive the pry bar or drill past the plug. On some engines the plug is only about ⅓ inch [9.5 mm]

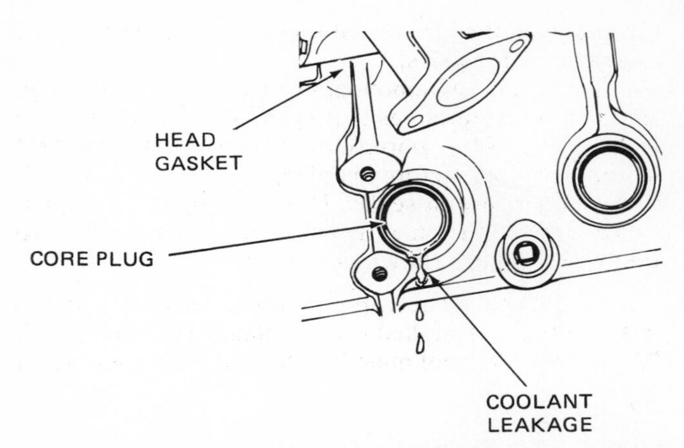

Fig. 19-21 A leaking core plug in the side of the cylinder block. (*Chrysler Corporation*)

from a cylinder wall. You could damage the cylinder wall if you drive the pry bar or drill too far in. Do not drive the plug into the water jacket. You would have trouble getting it out. Left in, it could block coolant circulation.

Another method of core plug removal is to drill a small hole in the center of the plug. Then pry the plug out. Special tools also are available to remove core plugs. However, removal and installation of certain core plugs on some engines is very difficult to do with

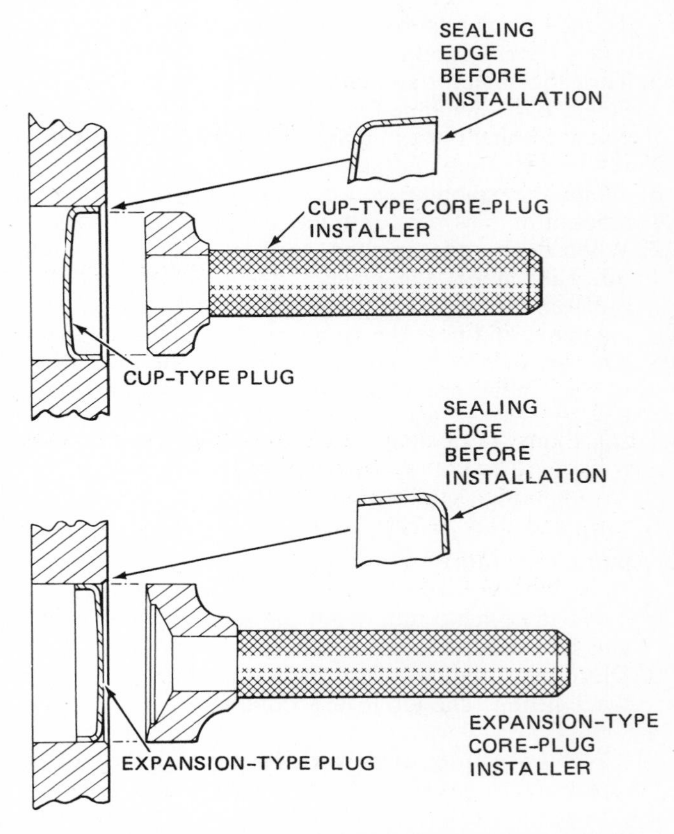

Fig. 19-22 Two types of expansion core plugs and the installation tool required for each type. (*Ford Motor Company*)

the engine in the car, or without removing parts of the engine or accessories.

After the plug is removed, inspect the bore for rough spots, nicks, or grooves that might allow a new plug to leak. If necessary, bore out the hole to take the next larger size plug. Before installing the new plug, coat it with water-resistant sealer. Use the proper installation tool and install the plug. Installation procedures for two types of plugs are shown in Fig. 19-22.

1. Cup type The cup-type plug, shown at the top of Fig. 19-22, is installed with the flanged edge outward. The proper size tool must be used. It must not contact the flange. All driving must be against the internal cup. The flange must be driven in until it is below the chamfered edge of the bore.

2. Expansion type The expansion-type plug, shown at the bottom of Fig. 19-22, is installed with the flanged edge inward. The proper tool must be used. The crowned center part must not be touched when the plug is driven in. Instead, the tool must drive against the outer part of the plug. The plug must be driven in until the center of the crown is below the chamfered edge of the bore.

Chapter 19 review questions

Select the *one* correct, best, or most probable answer to each question. Then check your answers against the correct answers given at the end of the book.

1. Accumulation of rust and scale in the engine cooling system causes:
   a. slow warm-up
   b. reduced heating capacity
   c. overheating
   d. rough idle.
2. If the thermostat is stuck closed, the engine will:
   a. warm up slowly
   b. overheat
   c. fail to start
   d. idle roughly.
3. If the thermostat is stuck open, the engine will:
   a. warm up slowly
   b. overheat
   c. fail to start
   d. stall.
4. The strength of the antifreeze solution in the cooling system is checked by a special:
   a. micrometer
   b. hydrometer
   c. barometer
   d. thermometer.
5. Exhaust gas leakage into the cooling system is most likely to be caused by a defective:
   a. cylinder head gasket
   b. manifold gasket
   c. water pump
   d. radiator hose.
6. Air will be drawn into the cooling system if there are leaks at any point between the:
   a. water pump and jackets
   b. radiator and water pump
   c. thermostat and radiator
   d. radiator cap and expansion tank.
7. When the coolant boils after the engine has been turned off and a long drive, the condition is known as:
   a. overheating
   b. hard running
   c. clogged radiator
   d. afterboiling.
8. When reverse flushing the radiator, the flushing gun is connected to the:
   a. upper tank
   b. pump inlet
   c. intake manifold
   d. lower tank.
9. When reverse-flushing the engine water jackets, the flushing gun is connected to the:
   a. upper tank
   b. lower tank
   c. thermostat housing
   d. pump inlet.
10. Troubles of the engine related directly to the cooling system include:
   a. hard starting and slow warm-up
   b. slow warm-up and overheating
   c. slow cranking and warm-up
   d. none of the above.

GLOSSARY

This glossary of automotive words and phrases provides a ready reference for the automotive technician. The definitions may differ somewhat from those given in a standard dictionary. They are not intended to be all-inclusive but to cover only what specifically applies to the automotive service field.

ABDC Abbreviation for *after bottom dead center;* any position of the piston between bottom dead center and top dead center, on the upward stroke.

absolute pressure A pressure measured on a scale having as its zero point the complete absence of pressure (known as a *perfect vacuum*). Atmospheric pressure on the absolute scale is 14.7 psi [101.35 kPa] or 29.92 inches [760 mm] of mercury (Hg).

accelerator A foot-operated pedal linked to the throttle valve in the carburetor; used to control the flow of fuel to the engine.

accelerator pump In the carburetor, a small pump (linked to the accelerator) which momentarily enriches the air-fuel mixture when the accelerator is depressed at low speed.

additive A substance added to gasoline or oil to improve some property of the gasoline or oil.

advance To make faster or to push ahead. To adjust ignition timing so that the spark plug fires earlier or more degrees before TDC; this effect is produced by moving the distributor (initial advance); by centrifugal or vacuum devices in accordance with engine speed and intake manifold vacuum; or electronically.

afterboil Boiling of fuel in the carburetor or of coolant in the engine immediately after the engine is stopped.

afterrun The condition in which an engine continues to run after the ignition is turned off. Sometimes referred to as *dieseling* or *run-on*.

AIR Abbreviation for air-injection reactor, part of a system of exhaust emission control. See *air-injection system*.

air bleed An opening into a fuel passage through which air can pass, or "bleed," into the fuel as it moves through the passage.

air cleaner A device mounted on or connected to the engine air intake for filtering dirt and dust out of air entering the engine.

air-cooled engine An engine that is cooled by the passage of air around the cylinders, and not by the passage of a liquid through water jackets.

air filter A filter that removes dirt and dust particles from air passing through it.

air-fuel mixture The air and fuel traveling to, and in, the combustion chamber and available for burning.

air-fuel ratio The proportions (by weight) of air and fuel supplied for combustion; an average mixture is 15 parts of air to 1 part of gasoline.

air horn In the carburetor, a tubular passage on the atmospheric side of the venturi through which the incoming air must pass, and which contains the choke valve.

air-injection system An exhaust-emission control system; injects air at low pressure into the exhaust manifold or thermal reactor to complete the combustion of unburned hydrocarbons and carbon monoxide in the exhaust gas.

air line A hose, pipe, or tube through which air passes.

air nozzle In an air-injection system, the tube through which air is delivered to the exhaust gas.

air pollution Contamination of the air by natural and manufactured pollutants, such as smoke, gases, and dust.

air pressure Atmospheric pressure; also the pressure produced by an air pump or by compression of air in a cylinder.

air pump Any device for compressing air. In the air-injection system of exhaust emission control, an engine-driven (belt-driven) pump incorporating a rotor and vanes.

alcohol A colorless, volatile liquid which in some forms can be used as a fuel for engines.

ambient temperature The temperature of the outside air surrounding an object, such as a car.

aneroid A device with a sealed bellows that provides mechanical movement by extending or retracting in response to varying atmospheric pressure.

antibackfire valve A valve used in the air-injection system to prevent backfiring in the exhaust system during deceleration.

antidieseling solenoid See *idle-stop solenoid.*

antifreeze A chemical, usually ethylene glycol, that is added to the engine coolant to raise the coolant boiling temperature and lower its freezing temperature.

antifriction bearing Name given to almost any type of ball, roller, or tapered-roller bearing.

anti-icing system A heating system designed to prevent the formation of ice on a surface or in a passage of a carburetor.

antiknock compound An additive put into gasoline to suppress spark knock or detonation—usually a lead compound (which becomes an air pollutant in the engine exhaust gas).

antipercolator A vent in the carburetor that opens to release fuel vapors when the throttle is closed; prevents fuel from being pushed out through the fuel nozzle by pressure buildup.

antisiphon system A small passage designed into a carburetor to prevent fuel from siphoning from the float bowl into the engine.

ATDC Abbreviation for *after top dead center;* any position of the piston between top dead center and bottom dead center, on the downward stroke.

atmosphere The mass of air that surrounds the earth.

atmospheric pressure The weight of the atmosphere per unit area. Atmospheric pressure at sea level is 14.7 psi absolute [101.35 kPa]; it decreases as altitude increases.

atom The smallest particle into which an element can be divided.

atomization The spraying of a liquid through a nozzle so that the liquid is broken into a very fine mist.

automatic choke A carburetor choke that positions the choke valve automatically in accordance with carburetor needs.

automotive air pollution Evaporated and unburned fuel and other undesirable by-products of combustion which escape from a motor vehicle into the atmosphere; mainly carbon monoxide (CO), hydrocarbons (HC), nitrogen oxides (NO_x), sulfur oxides (SO_x), and particulates.

automotive emissions See *automotive air pollution.*

backfire Noise made by the explosion of air-fuel mixture in the intake or exhaust system, usually during cranking and deceleration.

backfire-suppressor valve An antibackfire valve used in the air-injection system of exhaust emission control.

back pressure Pressure in the exhaust manifold of a running engine; affects volumetric efficiency.

balanced carburetor Carburetor in which the float bowl is vented into the air horn to partially compensate for the effects of a clogged air filter.

balancing-coil gauge An indicating device (fuel supply, oil pressure, engine temperature) that contains a pair of magnetic coils in the instrument-panel unit; a magnetic gauge.

ball check valve A valve consisting of a ball and a seat. Fluid can pass in one direction only; flow in the other direction is checked by the ball sealing tightly on the seat.

barrel Term sometimes applied to the cylinders in an engine; used in referring to the number of throttle bores in a carburetor.

BDC Abbreviation for *bottom dead center.*

bearing groove A channel cut in the surface of a bearing to distribute oil.

bearing oil clearance The space purposely provided between a shaft and a bearing through which lubricating oil can flow.

bearing prelubricator A special tank, attached to an air line; it supplies oil to the engine lubricating system at a predetermined and maintained pressure when the engine is not operating.

bellows A device, usually metal, that can lengthen or shorten much like an accordion.

bimetal A thermostatic element made up of two metals with different heat expansion rates. Temperature changes produce a bending or distortion of the element.

Bloc-Chek A special measuring device that, when inserted in the radiator filler neck of a running engine, can detect the leakage of exhaust gas into the cooling system.

blowby Leakage of compressed air-fuel mixture and burned exhaust gases past the piston rings into the crankcase.

blower Term for a supercharger or a two-stroke diesel engine intake air compressor. Also, the fan motor in a heater or air-conditioning system.

boiling Conversion from the liquid to the vapor state, which takes place throughout the liquid. The conversion is accompanied by bubbling as vapor rises from below the surface.

boiling point The temperature at which a liquid begins to boil.

borderline lubrication Type of poor lubrication resulting from greasy friction; moving parts are coated with a very thin film of lubricant.

bottom dead center (BDC) The piston position at the lower limit of its travel in the cylinder, such that the cylinder volume is at its maximum.

breather On engines without emission control devices, the opening that allows air to circulate through the crankcase and thereby provides crankcase ventilation.

British thermal unit (Btu) A measure of heat quantity. The amount of heat necessary to raise the temperature of 1 pound of liquid water by 1°F.

BTDC Abbreviation for *before top dead center;* any position of the piston between bottom dead center and top dead center, on the upward stroke.

Btu See *British thermal unit.*

butane A type of liquefied petroleum gas that is liquid below 32°F [0°C] at atmospheric pressure.

butterfly valve A pivoted flat plate used to regulate the flow of air, as in a carburetor.

bypass A separate passage which permits a liquid, gas, or electric current to take a path other than that normally used.

bypass valve In an oil filter, a valve that opens when the filter has clogged to allow oil to reach the engine.

cam A rotating lobe or eccentric which can be used with a cam follower to change rotary motion to reciprocating motion.

camshaft The shaft in the engine with cam lobes used mainly for operating the valves. It is driven by gears, or by sprockets and a toothed belt or chain from the crankshaft.

camshaft gear A gear, driven by the crankshaft, that is mounted on and rotates the camshaft; it is twice as large as the crankshaft gear.

canister A special container in an evaporative control system that contains charcoal to trap vapors from the fuel system.

capacity The ability to perform or to hold.

carbon (C) A black deposit left on engine parts such as pistons, rings, and valves by the combustion of fuel, and which inhibits their action.

carbon canister See *charcoal canister.*

carbon dioxide (CO_2) A colorless, odorless gas which results from complete combustion; usually considered harmless. The gas absorbed from air by plants in photosynthesis; also used to carbonate beverages.

carbon monoxide (CO) A colorless, odorless, tasteless, poisonous gas which results from incomplete combustion. A pollutant contained in engine exhaust gas.

carburetion The actions that take place in the carburetor: converting liquid fuel to vapor and mixing it with air to form a combustible mixture.

carburetor The device in an engine fuel system which mixes fuel with air in correct proportions and delivers this mixture to the intake manifold.

carburetor insulator A spacer, or insulator, used to prevent excess engine heat from reaching the carburetor.

carburetor kickdown Moderate depressing of the accelerator pedal to change the engagement of the choke fast-idle-speed screw from the high step to a lower step of the cam.

carcinogen A substance or agent that produces or incites cancer. Also, *carcinogenic,* tending to produce or incite cancer.

catalyst A substance that can speed or slow a chemical reaction between substances without itself being consumed by the reaction. In the catalytic converter, platinum and palladium are the active catalysts.

catalytic converter A muffler-like device in the exhaust system of an engine that converts harmful exhaust gases into harmless gases by promoting a chemical reaction between a catalyst and the pollutants.

cc Abbreviation for *cubic centimeter.*

Celsius A thermometer scale (formerly called *centigrade*) on which water boils at 100° and freezes at 0°. The formula $°C = 5/9\,(°F - 32)$ converts Fahrenheit readings to Celsius readings.

centigrade See *Celsius.*

centimeter (cm) A unit of linear measure in the metric system; 1 centimeter equals approximately 0.390 inches.

centrifugal force The force acting on a rotating body which tends to move it outward and away from the center of rotation. The force increases as rotational speed increases.

ceramic A type of material made from various minerals by baking or firing at high temperatures; can be used as an electric insulator, a filter element, or a catalyst substrate in a catalytic converter.

ceramic filter A filter for gasoline or other liquid that uses porous ceramic as the filter element.

cetane number An indicator of the ignition quality of diesel fuel. A high-cetane fuel ignites more easily (at a lower temperature) than a low-cetane fuel.

change of state Transformation of a substance from solid to liquid, or from liquid to vapor, or vice versa.

charcoal canister A container filled with activated charcoal used to trap gasoline vapor from the fuel tank and carburetor while the engine is off; also called *carbon canister.*

check valve A valve that opens to permit the passage of air or fluid in one direction only, or operates to prevent (check) some undesirable action.

chemical reaction The formation of one or more new substances when two or more substances are brought together.

choke In the carburetor, a plate or valve that is closed while a cold engine is started; it restricts or "chokes off" the airflow through the air horn, producing a partial vacuum that causes greater fuel delivery and a richer air-fuel mixture. Operates automatically on most cars. See also *electric-assist choke*.

CID Abbreviation for *cubic-inch displacement*.

clearance The space between two moving parts, or between a moving and a stationary part, such as a journal and a bearing. The bearing clearance is filled with lubricating oil when the mechanism is running.

closed-crankcase ventilation system A system in which the crankcase vapors (blowby gases) are discharged into the engine intake system and pass through to the engine cylinders rather than being discharged into the air.

cm See *centimeter*.

CO See *carbon monoxide*.

CO_2 See *carbon dioxide*.

combustion Burning; fire produced by the proper combination of fuel, heat, and oxygen. In the engine, the rapid burning of the air-fuel mixture in the combustion chamber.

combustion chamber The space between the top of the piston at TDC and the cylinder head, in which the air-fuel mixture is burned.

compound vortex-controlled combustion engine A type of stratified-charge engine built by Honda, and known as a CVCC engine.

compression Reducing the volume of a gas by squeezing it into a smaller space. Increasing the pressure reduces the volume and increases the density and temperature of the gas.

compression ignition The ignition of fuel solely by the heat generated when air is compressed in the cylinder; the method of ignition in a diesel engine.

compression pressure The pressure in the combustion chamber at the end of the compression stroke.

compression ratio The volume of the cylinder and combustion chamber when the piston is at BDC divided by the volume when the piston is at TDC.

compression tester An instrument for testing the amount of pressure, or compression, developed in an engine cylinder during cranking.

condensation A change of state during which a gas turns to liquid, usually because of temperature or pressure changes. Also, moisture from the air deposited on a cool surface.

coolant The liquid in the cooling system, which is a mixture of about 50 percent antifreeze and 50 percent water, used to carry heat out of the engine.

cooling system The system that removes heat from the engine by the forced circulation of coolant, and thereby prevents engine overheating. It includes the water jackets, water pump, radiator, and thermostat.

core In a radiator, the coolant passages surrounded by fins through which air flows to carry away heat.

corrosion Chemical action, usually by an acid, that eats away (decomposes) a metal.

crankcase The part of the engine that surrounds the crankshaft; usually the lower section of the cylinder block.

crankcase breather The opening or tube that allows air to enter and leave the crankcase, thereby providing crankcase ventilation.

crankcase dilution Dilution of the lubricating oil in the oil pan by gasoline; caused by gasoline condensing from the blowby in a cold engine and seeping down the cylinder walls.

crankcase emissions Pollution emitted into the atmosphere from any portion of the engine crankcase ventilating or lubricating system.

crankcase ventilation The circulation of air through the crankcase of a running engine to remove water, blowby, and other vapors; prevents oil dilution, contamination, sludge formation, and pressure buildup.

cross-flow radiator A radiator in which the coolant flows horizontally from the inlet tank on one side of the radiator, through the individual coolant passages, to the outlet tank on the opposite side of the radiator.

cubic centimeter (cm^3 or cc) A unit of volume in the metric system; equal to approximately 0.034 fluid ounces.

cubic-inch displacement The cylinder volume swept out by the pistons of an engine as they move from BDC to TDC, measured in cubic inches.

cutout In a running engine, the temporary complete loss of power, usually at irregular intervals; worse during heavy acceleration.

CVCC See *compound vortex-controlled combustion*.

cycle Any series of events which repeat continuously. In an engine, the four (or two) piston strokes that together produce power.

cylinder A circular tube-like opening in an engine cylinder block or casting in which a piston moves up and down.

cylinder block The basic framework of the engine, in and on which the other engine parts are attached. It includes the engine cylinders and the upper part of the crankcase.

cylinder compression tester See *compression tester*.

cylinder head The part of the engine that covers and encloses the cylinders. It contains cooling fins or water jackets, the combustion chambers, and, on overhead-valve engines, the valves.

cylinder leakage tester A testing device that forces compressed air into the cylinder through the spark-plug hole when the valves are closed and the piston is at TDC on the compression stroke. The percentage of compressed air that leaks out is measured, and the source of the leak accurately pinpoints the defective part.

dashpot A device on the carburetor that prevents the throttle valve from closing too suddenly.

deceleration A decrease in velocity or speed. Also, coasting of the car or engine to idle speed from a higher speed with the accelerator at or near the idle position.

deceleration valve A device used in conjunction with the dual-diaphragm vacuum-advance unit to advance the timing under deceleration conditions.

deenergize The removal of the current from a device to allow it to return to its normal "at rest" position.

detergent A chemical added to engine oil; helps keep internal parts of the engine clean by preventing the accumulation of deposits.

detonation Commonly called spark knock or ping. In the combustion chamber, an uncontrolled second explosion (after the spark occurs at the spark plug), with spontaneous combustion of the remaining compressed air-fuel mixture resulting in a pinging noise.

die-out The condition in which an engine stalls without movement of the accelerator pedal.

diesel cycle An engine operating cycle in which air is compressed, and fuel oil is injected into the compressed air at the end of the compression stroke. The heat produced by the compression ignites the fuel oil, eliminating the need for spark plugs or a separate ignition system.

diesel engine An engine operating on the diesel cycle and burning oil instead of gasoline.

dieseling A condition in which a spark-ignition engine continues to run after the ignition is off. Caused by carbon deposits or hot spots in the combustion chamber glowing sufficiently to furnish heat for combustion.

dipstick See *oil-level indicator*.

dispersant A chemical added to oil to prevent dirt and impurities from clinging together in lumps that could clog the engine lubricating system.

displacement In an engine, the total volume of air-fuel mixture an engine is theoretically capable of drawing into all cylinders during one operating cycle. Also, the volume swept out by the piston in moving from one end of a stroke to the other.

diverter valve In the air-injection system of exhaust emission control, a valve that diverts air-pump output into the air cleaner or the atmosphere during deceleration; prevents backfiring and popping in the exhaust system.

DOHC See *double-overhead-camshaft engine*.

double-overhead-camshaft engine An engine with two camshafts in each cylinder head to actuate the valves; one camshaft operates the intake valves, and the other operates the exhaust valves.

downdraft carburetor A carburetor in which the air horn is so arranged that the air flows down through it on its way to the intake manifold.

down-flow radiator A radiator in which the coolant enters the radiator at the top, and loses heat as it flows down through passages to the bottom of the radiator.

drivability The general operation of an automobile, usually rated from good to poor; based on characteristics of concern to the average driver, such as smoothness of idle, even acceleration, ease of starting, quick warm-up, and no tendency to overheat at idle.

dry sump A type of engine-lubricating system in which the oil supply is carried in a separate oil tank instead of in the engine oil pan.

dual carburetors Two carburetors mounted on one engine.

dual quad A carburetion system that uses two 4-barrel carburetors.

duct A tube or channel used to convey air or liquid from one place to another. In emission systems, a tube on an air cleaner that has a vacuum motor mounted on it to help regulate the temperature of the intake air.

dwell meter A precision electric instrument used to measure the cam angle, or dwell, or the number of degrees the distributor shaft or cam rotates while the distributor points are closed.

dynamometer A device for measuring the power output, or brake horsepower, of an engine. An engine dynamometer measures the power output at the flywheel; a chassis dynamometer measures the power output at the drive wheels.

eccentric A disk or offset section (of a shaft, for example) used to convert rotary to reciprocating motion. Sometimes called a cam.

ECU See *electronic control unit*.

efficiency The ratio between the power of an effect and the power expended to produce the effect; the ratio between an actual result and the theoretically possible result.

EGR system Abbreviation for *exhaust gas recirculation system*.

electric-assist choke A choke in which a small electric heating element warms the choke spring, causing it to release more quickly. This reduces exhaust emissions during the start-up of a cold engine.

electric system In an automobile, the system that electrically cranks the engine for starting; furnishes high-voltage sparks to the engine cylinders to fire the compressed air-fuel charges, lights the lights, and powers the heater motor, radio, and other accessories. Consists, in part, of the starting motor, wiring, battery, alternator, regulator, ignition distributor, and ignition coil.

electronic control unit A solid-state device that receives information from sensors and is programmed to operate various circuits and systems based on that information.

electronic fuel-injection system A system that injects fuel into a spark-ignition engine, and that includes an electronic control unit to time and meter the fuel flow.

emission control Any device or modification added onto or designed into a motor vehicle for the purpose of reducing air-polluting emissions.

emitter An engine with measurable exhaust emissions; sometimes preceded by the word *high* or *low* to indicate the degree of emission.

energize To activate; to cause movement or action.

energy The capacity or ability to do work. Usually measured in work units of foot-pounds [newton-meters], but also expressed in heat-energy units (Btu's [joules]).

engine A machine that converts heat energy into mechanical energy. A device that burns fuel to produce mechanical power; sometimes referred to as a power plant.

engine speed See *rpm*.

engine tuneup A procedure for inspecting, testing, and adjusting an engine, and replacing any worn parts, to restore the engine to its best performance.

Environmental Protection Agency The independent agency of the United States government that sets standards and coordinates activities related to automotive emissions and the environment.

EPA Abbreviation for *Environmental Protection Agency*.

ethyl See *tetraethyl lead*.

ethylene glycol Chemical name of a widely used type of antifreeze.

evaporation The transforming of a liquid to its gaseous state.

evaporative control system A system which prevents the escape of fuel vapors from the fuel tank or carburetor to the atmosphere while the engine is off. The vapors are stored in a charcoal canister or in the engine crankcase until the engine is started.

exhaust emissions Pollutants emitted into the atmosphere through any opening downstream of the exhaust ports of an engine.

exhaust gas The burned and unburned gases that remain (from the air-fuel mixture) after combustion.

exhaust-gas analyzer A device for sensing the amounts of air pollutants in the exhaust gas from an engine. The analyzers used in automotive shops check HC and CO; those used in testing laboratories can also check NO_x.

exhaust-gas recirculation (EGR) system An NO_x control system that recycles a small part of the inert exhaust gas back through the intake manifold to lower the combustion temperature.

exhaust manifold A device with several passages through which exhaust gases leave the engine combustion chambers and enter the exhaust piping system.

exhaust pipe The pipe connecting the exhaust manifold with the muffler.

exhaust system The system through which exhaust gases leave the vehicle. Consists of the exhaust manifold, exhaust pipe, muffler, tail pipe, and resonator (if used).

expansion plug A slightly dished plug that is used to seal core passages in the cylinder block and cylinder head. When driven into place, it is flattened and expanded to fit tightly.

expansion tank A tank at the top of an automobile radiator which provides room for heated coolant to expand and to give off any air that may be trapped in the coolant. Also, a similar device used in some fuel tanks to prevent fuel from spilling out of the tank through expansion.

extreme-pressure lubricant A special lubricant for use in hypoid-gear differentials; needed because of the heavy wiping loads imposed on the gear teeth.

fan The bladed device on the front of the engine that rotates to draw cooling air through the radiator or around the engine cylinders; an air blower such as the heater fan or the air-conditioning blower.

fast flushing A method of cleaning the cooling system; uses a special machine to circulate the cleaning solution.

fast-idle cam A mechanism on the carburetor, connected to the automatic choke, that holds the throttle valve slightly open when the engine is cold; causes the engine to idle at a higher rpm as long as the choke is applied.

filter A device through which air, gases, or liquids are passed to remove impurities.

fins On a radiator or heat exchanger, thin metal projections over which cooling air flows to remove heat from hot liquid flowing through internal passages. On an air-cooled engine, thin metal projections on the cylinder and head which greatly increase the area of the heat-dissipating surfaces and help cool the engine.

flat spot Lack of normal acceleration or response to throttle opening; implies no loss of power, but also no increase in power.

float bowl In a carburetor, the reservoir from which fuel is metered into the passing air, and in which the float is located.

float level The float position at which the needle valve closes the fuel inlet to the carburetor to prevent further delivery of fuel.

float system In the carburetor, the system that controls the entry of fuel and the fuel level in the float bowl.

flooded The condition in which the engine cylinders

receive "raw" or liquid fuel, or an air-fuel mixture too rich to burn.

fluid Any liquid or gas.

four-barrel carburetor A carburetor with four throttle valves. In effect, two 2-barrel carburetors in a single assembly.

four-cycle See *four-stroke cycle*.

four-stroke cycle The four piston strokes—intake, compression, power, and exhaust—that make up the complete cycle of events in the four-stroke-cycle engine. Also called *four-cycle* and *four-stroke*.

friction The resistance to motion between two bodies in contact with each other.

friction bearing A bearing in which there is sliding contact between the moving surfaces. Sleeve bearings, such as those used in connecting rods, are friction bearings.

friction horsepower The power used up by an engine in overcoming its own internal friction; usually increases as engine speed increases.

fuel Any combustible substance. In a spark-ignition engine, the fuel (gasoline) is burned, and the heat of combustion expands the resulting gases, which force the piston downward and rotate the crankshaft.

fuel decel valve A device which supplies additional air-fuel mixture to the intake manifold during deceleration to control exhaust-gas hydrocarbon emissions.

fuel filter A device located in the fuel line, ahead of the float bowl; removes dirt and other contaminants from fuel passing through.

fuel gauge A gauge that indicates the amount of fuel in the fuel tank.

fuel-injection system A system which delivers fuel under pressure into the combustion chamber, intake manifold, or valve ports. Replaces the carburetor on a spark-ignition engine.

fuel line The pipe or tubes through which fuel flows from the fuel tank to the carburetor or fuel injectors.

fuel mixture See *air-fuel mixture*.

fuel nozzle The tube in the carburetor through which gasoline feeds from the float bowl into the passing air. In a fuel-injection system, the tube that delivers fuel into the compressed air or the passing airstream.

fuel pump The electric or mechanical device in the fuel system which draws fuel from the fuel tank and forces it to the carburetor or fuel-injection system.

fuel system In an automobile, the system that delivers the fuel and air to the engine cylinders. Consists of the fuel tank and lines, gauge, fuel pump, intake manifold, and carburetor or fuel-injection system.

fuel tank The storage tank for fuel on the vehicle.

fuel-vapor recovery system See *vapor recovery system*.

full-flow oil filter An oil filter designed so that all the oil from the oil pump flows through it before reaching the bearings.

full throttle Wide-open throttle position, with the accelerator pressed all the way down to the floorboard.

gallery A passageway inside a wall or casting. The main oil gallery within the block supplies lubrication to all parts of the engine.

gasket A layer of material, usually made of cork or metal, or both, placed between two parts to make a tight seal.

gasket cement A liquid adhesive material, or sealer, used to install gaskets; sometimes a layer of gasket cement is used as the gasket.

gasoline A liquid blend of hydrocarbons obtained from crude oil; used as the fuel in most automobile engines.

gauge pressure A pressure reading on a scale which ignores atmospheric pressure. Atmospheric pressure of 14.7 psi absolute is equivalent to 0 psi gauge.

gear-type pump A pump in which a pair of rotating gears mesh to force oil (or other liquid) from between the teeth to the pump outlet.

glow plug A plug-type heater containing a coil of resistance wire that is heated by a low-voltage current to help ignite fuel sprayed into the intake manifold; used as a cold-starting aid in diesel engines.

governor A device that controls, or governs, another device, usually on the basis of speed or load—for example, the governor used in certain automatic transmissions to control gear shifting in relation to car speed.

gram (g) A measurement of mass and weight in the metric system; 1 ounce equals 28.33 grams.

grams per mile Unit of measurement for the amount (weight) of pollutants emitted into the atmosphere with the vehicle exhaust gases. Antipollution laws set maximum limits for each exhaust pollutant in grams per mile.

grease Lubricating oil to which thickening agents have been added.

greasy friction The friction between two solids coated with a thin film of oil.

gulp valve In the air-injection system, a type of antibackfire valve which allows a sudden intake of fresh air through the intake manifold during deceleration; prevents backfiring and popping in the exhaust system.

hard start, cold Excessive cranking times or numerous false starts while starting a cold engine.

hard start, hot Excessive cranking times while starting a hot engine.

HC Abbreviation for *hydrocarbon*.

head See *cylinder head*.

header A special exhaust manifold made of separate tubes.

heat A form of energy released by the burning of fuel; in an engine, heat energy is converted to mechanical energy.

heat-control valve In the engine, a thermostatically operated valve in the exhaust manifold; diverts heat to the intake manifold to warm it before the engine reaches normal operating temperature.

heated-air system A system in which a thermostatically controlled air cleaner supplies hot air from a stove around the exhaust manifold to the engine during warm-up; improves cold-engine operation.

heater core A small radiator, mounted under the dash, through which hot coolant circulates. When heat is needed in the passenger compartment, a fan is turned on to circulate air through the hot core.

heat exchanger A device in which heat is transferred from one fluid, across a tube or other solid surface, to another fluid.

heat of compression The temperature rise in a gas (such as air or an air-fuel mixture) as it is compressed.

hemispherical combustion chamber A combustion chamber resembling a hemisphere, or one-half of a round ball.

hesitation A lack of response to initial throttle opening, occurring when driving from a standstill or accelerating from any speed.

high altitude Classification of engine certified by the Environmental Protection Agency to comply with federal emission specifications for vehicles to be used above 4000 feet [1219 m].

high compression Term used to refer to the increased compression ratios of modern automotive engines as compared with engines built in the past.

high-speed system In the carburetor, the system that supplies fuel to the engine at speeds above about 25 mph [40 km/h]. Also called the *main metering system*.

horsepower A measure of mechanical power, or the rate at which work is done. 1 horsepower equals 33,000 foot-pounds of work per minute; it is the power necessary to raise 33,000 pounds a distance of 1 foot in 1 minute.

H_2O Chemical symbol for hydrogen oxide, commonly known as water.

hot-idle compensator A thermostatically controlled carburetor valve that opens whenever inlet air temperatures are high. Allows additional air to discharge below the throttle plates at engine idle to improve idle stability and prevent overly rich air-fuel mixtures.

hot soak A condition that may arise when an engine is stopped for a prolonged period after a hard, hot run. Heat transferred from the engine evaporates fuel out of the carburetor, so that the carburetor needs priming before the engine will start and run smoothly. A longer cranking period is required.

humidity A measure of the amount of water vapor in the air.

hydraulic lifter A valve lifter that uses oil pressure from the engine lubricating system to keep it in constant contact with the cam lobe and with the valve stem, push rod, or rocker arm. As the hydraulic lifter automatically adjusts to any variation in valve stem length, valve noise is reduced.

hydraulic pressure Pressure exerted through the medium of a liquid.

hydraulics The use of a liquid under pressure to transfer force or motion, or to increase an applied force.

hydraulic valve lifter See *hydraulic lifter*.

hydrocarbon (HC) A compound containing only carbon and hydrogen atoms, usually derived from fossil fuels such as petroleum, natural gas, and coal; an agent in the formation of photochemical smog. Gasoline is a blend of liquid hydrocarbons refined from crude oil.

hydrogen (H) A colorless, odorless, highly flammable gas whose combustion produces water; the simplest and lightest element.

IC See *internal combustion engine*.

idle Engine speed when the accelerator pedal is fully released, and there is no load on the engine.

idle limiter A device that controls the maximum richness of the idle air-fuel mixture in the carburetor; also aids in preventing overly rich idle adjustments. Limiters are of two types: the external plastic cap type, installed on the head of the idle-mixture screw, and the internal needle type, located in the idle passages of the carburetor.

idle limiter cap A plastic cap placed over the head of the idle-mixture screw to limit its travel and prevent the idle mixture from being set too rich.

idle mixture The air-fuel mixture supplied to the engine during idling.

idle-mixture screw The adjustment screw (on some carburetors) that can be moved in or out to lean or enrich the idle mixture.

idle port The opening into the throttle body through which the idle system in the carburetor discharges fuel.

idle speed The speed, or rpm, at which the engine runs without load when the accelerator pedal is released.

idle-stop solenoid An electrically operated two-position plunger used to provide a predetermined throttle-valve opening at idle.

idle system In the carburetor, the passages through which fuel is fed when the engine is idling.

idle vent An opening from an enclosed chamber through which air can pass under idle conditions.

ignition The action of the spark in starting the burning of the compressed air-fuel mixture in the combustion chamber of a spark-ignition engine. In a diesel engine, the start of the burning of fuel after its temperature has been raised by the heat of compression.

ignition advance The moving forward, in time, of the ignition spark relative to the piston position. TDC or 1° ATDC is considered advanced as compared with 2° ATDC.

ignition lag In a diesel engine, the delay in time between the injection of fuel and the start of combustion.

ignition system In the engine, the system that furnishes high-voltage sparks to the cylinders to fire the compressed air-fuel mixture. Consists of the battery, ignition coil, ignition distributor, ignition switch, wiring, and spark plugs.

ignition temperature The lowest temperature at which a fuel will begin to burn.

ignition timing The delivery of the spark from the coil to the spark plug at the proper time for the power stroke, relative to the piston position.

impeller A rotating finned disk; used in centrifugal pumps, such as water pumps, and in torque converters.

indicator A device used to make some condition known by use of a light or a dial and pointer; for example, the temperature indicator or oil pressure indicator.

inertia Property of an object that causes it to resist any change in its speed or direction of travel.

infrared analyzer A test instrument used to measure very small quantities of HC and CO in exhaust gas. See *exhaust gas analyzer*.

injector The tube or nozzle through which fuel is injected into the intake airstream or the combustion chamber.

insulator A poor conductor of electricity or heat.

intake manifold A device with several passages through which air or the air-fuel mixture flows from the air intake or throttle body to the ports in the cylinder head or cylinder block.

internal-combustion engine An engine in which the fuel is burned inside the engine itself rather than in a separate device (as in a steam engine).

jet A calibrated passage in the carburetor through which fuel flows.

journal The part of a rotating shaft which turns in a bearing.

kg/cm^2 Abbreviation for kilograms per square centimeter, a metric engineering term for the measurement of pressure; 1 kilogram per square centimeter equals 4.22 pounds per square inch.

kilogram (kg) In the metric system, a unit of weight and mass; approximately equal to 2.2 pounds.

kilometer (km) In the metric system, a unit of linear measure; equal to 0.621 miles.

kilowatt (kW) 1000 watts; a unit of power, equal to about 1.34 horsepower.

kinetic energy The energy of motion; the energy stored in a moving body through its momentum; for example, the kinetic energy stored in a rotating flywheel.

knock A heavy metallic engine sound which varies with engine speed; usually caused by a loose or worn bearing. Name also used for detonation, pinging, and spark knock. See *detonation*.

kPa Abbreviation for kilopascals, the metric measurement of pressure; 1 kilopascal equals 0.145 pounds per square inch.

kW Abbreviation for *kilowatt*.

leaded gasoline Gasoline to which small amounts of tetraethyl lead are added to improve engine performance and reduce detonation.

lean mixture An air-fuel mixture that has a relatively high proportion of air and a relatively low proportion of fuel. An air-fuel ratio of 16:1 indicates a lean mixture, compared with an air-fuel ratio of 13:1.

lifter See *valve lifter*.

linkage An assembly of rods or links used to transmit motion.

liquefied petroleum gas A hydrocarbon suitable for use as an engine fuel, obtained from petroleum and natural gas; it is a vapor at atmospheric pressure but becomes a liquid under sufficient pressure. Butane and propane are the liquefied gases most frequently used in automotive engines.

liquid-cooled engine An engine that is cooled by the circulation of liquid coolant around the cylinders.

liter (L) In the metric system, a measure of volume; approximately equal to 0.26 gallons (U.S.), or about 61 cubic inches (33.8 fluid ounces, or 1 quart 1.8 ounces). Used as a metric measure of engine cylinder displacement.

loading An overrich air-fuel mixture that causes hard starting or rough engine operation and the emission of black smoke from the tail pipe.

low-lead fuel Gasoline which is low in tetraethyl lead, containing not more than 0.5 gram per gallon.

low-speed system The system in the carburetor that supplies fuel to the air passing through during low-speed, part-throttle operation.

LPG Abbreviation for *liquefied petroleum gas*.

lubricant Any material, usually a petroleum product

such as grease or oil, that is placed between two moving parts to reduce friction.

lubricating system The system in the engine that supplies engine parts with lubricating oil to prevent contact between any two moving metal surfaces.

lugging Low-speed, full-throttle engine operation in which the engine is heavily loaded and overworked; usually caused by failure of the driver to shift to a lower gear when necessary.

main bearings In the engine, the bearings that support the crankshaft.

main jet The fuel nozzle, or jet, in the carburetor that supplies fuel when the throttle is partially to fully open.

manifold A device with several inlet or outlet passageways through which a gas or liquid is gathered or distributed. See exhaust manifold, intake manifold, and manifold gauge set.

manifold heat control See *heat control valve*.

manifold vacuum The vacuum in the intake manifold that develops as a result of the vacuum in the cylinders during their intake strokes.

matter Anything that has weight and occupies space.

mechanical efficiency In an engine, the ratio between brake horsepower and indicated horsepower.

meter A unit of linear measure in the metric system, equal to 39.37 inches. Also, the name given to any test instrument that measures a property of a substance passing through it, as an ammeter measures electric current. Also, any device that measures and controls the flow of a substance passing through it, as a carburetor jet meters fuel flow.

metering rod and jet A device consisting of a small, movable, cone-shaped rod and a jet; increases or decreases fuel flow according to engine throttle opening, engine load, or a combination of both.

millimeter (mm) In the metric system, a unit of linear measure, approximately equal to 0.039 inch.

misfire In an engine, a failure to ignite the air-fuel mixture in one or more cylinders without stalling; may be intermittent or continuous.

miss See *misfire*.

mm See *millimeter*.

mode Term used to designate a particular set of operating characteristics.

modification An alteration; a change from the original.

moisture Humidity, dampness, wetness, or very small drops of water.

molecule The smallest particle into which a substance can be divided and still retain the properties of that substance.

MON Abbreviation for *motor octane number*.

motor octane number (MON) Laboratory octane rating of a fuel established on single-cylinder, variable-compression-ratio engines.

motor vehicle A vehicle propelled by a means other than muscle power, usually mounted on rubber tires, which does not run on rails or tracks.

mph Abbreviation for miles per hour, a measure of speed.

muffler In the engine exhaust system, a device through which the exhaust gases must pass to reduce the exhaust noise.

multiple-viscosity oil An engine oil which has a low viscosity when cold for easier cranking and a higher viscosity when hot to provide adequate engine lubrication.

needle bearing An antifriction bearing of the roller type, in which the rollers are very small in diameter (needle-sized).

needle valve A small, tapered, needle-pointed valve which can move into or out of a valve seat to close or open the passage through the seat. Used to control the fuel level in the carburetor float bowl.

neoprene A synthetic rubber that is not affected by the various chemicals that are harmful to natural rubber.

nitrogen (N) A colorless, tasteless, odorless gas that constitutes 78 percent of the atmosphere by volume and is a part of all living things.

nitrogen oxides (NO_x) Any chemical compound of nitrogen and oxygen. Nitrogen oxides result from high temperature and pressure in the combustion chambers of automobile engines and other power plants during the combustion process. When combined with hydrocarbons in the presence of sunlight, nitrogen oxides form smog. Nitrogen oxides are a basic air pollutant; automotive exhaust emission levels of nitrogen oxides are controlled by law.

no start Engine will not start.

nonleaded gasoline See *unleaded gasoline*.

NO_x control system Any device or system used to reduce the amount of NO_x produced by an engine.

nozzle The opening, or jet, through which fuel or air passes as it is discharged.

octane number The number used to indicate the octane rating of a gasoline.

octane rating A measure of the antiknock properties of a fuel. The higher the octane rating, the more resistant the fuel is to spark knock or detonation.

octane requirement The minimum-octane-number fuel required to enable a vehicle to operate without detonation.

OHC See *overhead-camshaft engine*.

OHV See *overhead-valve engine*.

oil A liquid lubricant made from crude oil and used to provide lubrication between moving parts. In a diesel engine, oil is used for fuel.

oil cooler A small radiator that lowers the temperature of oil flowing through it.

oil dilution Thinning of oil in the crankcase by liquid fuel.

oil filter A filter which removes impurities from the engine oil passing through it.

oil-level indicator The indicator that is removed and inspected to check the level of oil in the crankcase of an engine or compressor. Usually called the dipstick.

oil pan The detachable thin steel or plastic cover bolted to the crankcase, which encloses the crankcase and acts as an oil reservoir.

oil pressure indicator A gauge that indicates to the driver the oil pressure in the engine lubricating system.

oil pump In the lubricating system, the device that delivers oil from the oil pan to the moving engine parts.

oil pumping Leakage of oil past the piston rings and into the combustion chamber; usually the result of defective rings or worn cylinder walls.

oil seal A seal placed around a rotating shaft or other moving part to prevent leakage of oil.

oil seal and shield Two devices used to control oil leakage past the valve stem and guide, and into the ports or combustion chamber of an engine.

oil separator A device for separating oil from air or from another liquid. Used with some engine crankcase emission control systems.

oil strainer A wire-mesh screen placed at the inlet end of the oil pump pickup tube; prevents dirt and other large particles from entering the oil pump.

open system A crankcase emission control system which draws air through the oil-filter cap, and does not include a tube from the crankcase to the air cleaner.

orifice A small calibrated hole in a line carrying a liquid or gas.

orifice spark-advance control A system used on some engines to aid in the control of nitrogen oxides. Consists of a valve which delays the change in vacuum to the distributor vacuum-advance unit between idle and part-throttle.

O ring A type of sealing ring, made of rubber or other similar material; in use, the O ring is compressed into a groove to provide the sealing action.

overflow Spilling of the excess of a substance; also, running or spilling over the sides of a container, usually because of overfilling.

overflow tank See *expansion tank*.

overhead-camshaft (OHC) engine An engine in which the camshaft is mounted over the cylinder head instead of inside the cylinder block.

overhead-valve (OHV) engine An engine in which the valves are mounted in the cylinder head above the combustion chamber, instead of in the cylinder block; in this type of engine, the camshaft is usually mounted in the cylinder block, and the valves are actuated by push rods.

overheat To heat excessively; also, to become excessively hot.

oxidation The combining of a material with oxygen. Rusting is slow oxidation, and combustion is rapid oxidation.

oxidation catalyst In a catalytic converter, a substance that promotes the combustion of exhaust-gas hydrocarbons and carbon monoxide at a lower temperature.

oxides of nitrogen See *nitrogen oxides*.

oxygen (O) A colorless, tasteless, odorless, gaseous element which makes up about 21 percent of air. Capable of combining rapidly with all elements except the inert gases in the oxidation process called *burning*. Combines very slowly with many metals in the oxidation process called rusting.

pan See *oil pan*.

particle A very small piece of metal, dirt, or other impurity which may be contained in the air, fuel, or lubricating oil used in an engine.

particulates Small particles of lead occurring as solid matter in the exhaust gas.

passage A small hole or gallery in an assembly or casting, through which air, coolant, fuel, or oil flows.

PCV Abbreviation for *positive crankcase ventilation*.

PCV valve The valve that controls the flow of crankcase vapors in accordance with ventilation requirements for different engine speeds and loads.

percolation The condition in which a bowl vent fails to open when the engine is turned off, and pressure in the float bowl forces liquid fuel through the main jets into the manifold.

petroleum The crude oil from which gasoline, lubricating oil, and other such products are refined.

photochemical smog Smog caused by hydrocarbons and nitrogen oxides reacting photochemically in the atmosphere. The reactions take place under low wind velocity, bright sunlight, and an inversion layer in which the air mass is trapped (as between the ocean and mountains in Los Angeles). Can cause eye and lung irritation.

ping Engine spark knock that occurs primarily during acceleration. Usually associated with acceleration or lugging at relatively low speeds, especially with a manual transmission. However, it may occur in higher speed ranges under heavy-load conditions. Caused by too much advance of ignition timing or by low-octane fuel.

piston displacement The cylinder volume displaced by the piston as it moves from the bottom to the top of the cylinder during one complete stroke.

plastic gasket compound A plastic paste which can be squeezed out of a tube to make a gasket in any shape.

Plastigage A plastic material available in strips of various diameters; used to measure crankshaft main bearing clearance and connecting-rod bearing clearances.

plunger A sliding reciprocating piece driven by an auxiliary power source, having the motion of a ram or piston.

pollutant Any substance that adds to the pollution of the atmosphere. In a vehicle, any such substance in the exhaust gas from the engine or which has evaporated from the fuel tank or carburetor.

pollution Any gas or substance in the air which makes it less fit to breathe. Also, "noise pollution" is the name applied to excessive noise from machinery, vehicles, or engines.

polyurethane A synthetic substance used in filtration materials; normally associated with the filtering of carburetor inlet air.

poor gas mileage Excessive fuel consumption; may be caused by the driver, vehicle, or operating conditions.

pop-back Condition in which the air-fuel mixture is ignited in the intake manifold. Because combustion takes place outside the combustion chamber, the combustion may "pop back" through the air intake.

port In the engine, the opening in which the valve operates and through which air, air-fuel mixture, or burned gases pass; the valve port.

ported vacuum switch A coolant-temperature-sensing vacuum control valve used in distributor and EGR vacuum systems. Sometimes called the *vacuum control valve* or *coolant override valve*.

positive crankcase ventilation (PCV) A crankcase ventilation system; uses intake manifold vacuum to return crankcase vapors and blowby gases to the intake manifold to be burned, thereby preventing their escape into the atmosphere.

pour point The lowest temperature at which an oil will flow.

power The rate at which work is done. A common power unit is the horsepower, which is equal to 33,000 foot-pounds per minute.

power piston In some carburetors, a vacuum-operated piston that allows additional fuel to flow at wide-open throttle; permits delivery of a richer air-fuel mixture to the engine.

power plant The engine or power source of a vehicle.

power train The mechanisms that carry the rotary motion developed in the engine to the car wheels; these include the clutch, transmission, drive shaft, differential, and axles.

ppm Abbreviation for *parts per million;* the unit used in measuring the level of hydrocarbons in exhaust gas with an exhaust-gas analyzer; 1 part per million is 1 drop in 16 gallons.

precombustion chamber In some engines, a separate small combustion chamber into which the fuel is injected and where combustion begins.

preignition Ignition of the air-fuel mixture in the combustion chamber, by some unwanted means, before the spark occurs at the spark plug.

premium gasoline The best or highest-octane gasoline available to the motorist.

pressure Force per unit area, or force divided by area. Usually measured in pounds per square inch (psi) and kilopascals (kPa).

pressure cap A radiator cap with valves which causes the cooling system to operate under pressure at a higher and more efficient temperature without boiling of the coolant.

pressure-feed oil system A type of lubricating system that makes use of an oil pump to force oil to various engine parts.

pressure relief valve A valve in the oil line that opens to relieve excessive pressure.

pressure tester An instrument that clamps in the radiator filler neck; used to pressure-test the cooling system for leaks.

pressurize To apply more than atmospheric pressure to a gas or liquid.

PROCO Short for *programmed combustion;* a type of stratified charge engine developed by Ford.

programmed combustion See *PROCO*.

progressive linkage A linkage arrangement used with multiple-carburetor installations to progressively open the secondary carburetors.

propane A type of LPG that is liquid below −44°F [−42°C] at atmospheric pressure and sometimes is used as an engine fuel.

psi Abbreviation for *pounds per square inch,* a measurement of pressure.

psig Abbreviation for *pounds per square inch gauge* pressure.

quad carburetor A four-barrel carburetor.

quench The removal of heat during combustion from the end gas or outside layers of air-fuel mixture by the cooler metallic surfaces of the combustion chamber; this reduces the tendency for detonation to occur.

quench area The area of the combustion chamber near the cylinder walls which tends to cool (quench) combustion through the effect of the nearby cool water jackets.

races The metal rings on which ball or roller bearings rotate.

radiator In the cooling system, the device that removes heat from coolant passing through it; takes hot coolant from the engine and returns the coolant to the engine at a lower temperature.

radiator pressure cap See *pressure cap.*

radiator shutter system An engine-temperature control system used mostly on trucks that regulates the amount of air flowing through the radiator by use of a shutter system.

ram-air cleaner An air cleaner for high-performance cars that opens an air scoop on the hood to provide a ram effect when the throttle is wide open.

refractometer An instrument used to measure the specific gravity of a liquid such as battery electrolyte or engine coolant; gives a reading that is already adjusted for the temperature of the liquid being tested.

relative humidity The actual moisture content of the air, as a percentage of the total moisture that the air can hold at a given temperature. For example, if the air contains three-fourths of the moisture it can hold at its existing temperature, then its relative humidity is 75 percent.

research octane number A number used to describe the octane rating of a gasoline. See also *motor octane number.*

resonator A device in the exhaust system that reduces certain exhaust noises.

retard To make slower or hold back. To adjust ignition timing so the spark plug fires later or fewer degrees before TDC; the opposite of advance.

reverse flushing A method of cleaning a radiator or engine cooling system by flushing it in the direction opposite to normal coolant flow.

ribbon cellular radiator core A type of radiator core consisting of ribbons of metal soldered together along their edges.

rich mixture An air-fuel mixture that has a relatively high proportion of fuel and a relatively low proportion of air. An air-fuel ratio of 13:1 indicates a rich mixture, compared with an air-fuel ratio of 16:1.

roadability The steering and handling qualities of a vehicle while it is being driven on the road.

road-draft tube A method of removing the fumes and pressure from the engine crankcase; used prior to crankcase emission control systems. The tube, which was connected into the crankcase and suspended slightly above the ground, depended on venturi action to create a partial vacuum as the vehicle moved. The method was ineffective below about 20 mph [32 km/h].

road load The power required to hold a constant vehicle speed on a level road.

RON Abbreviation for *research octane number.*

room temperature 68 to 72°F [20 to 22°C].

rotor oil pump A type of oil pump in which a pair of rotors, one inside the other, produce the pressure required to circulate oil to engine parts.

rough idle Unsteady, uneven, or erratic engine idle, which may make the car shake.

rpm Abbreviation for *revolutions per minute;* a measure of rotational speed, usually of the engine crankshaft.

run-on See *dieseling.*

SA Designation for lubricating oil that is acceptable for use in engines operated under the mildest conditions.

SAE Abbreviation for *Society of Automotive Engineers.* Used to indicate a grade or weight of oil measured according to Society of Automotive Engineers standards.

sag The condition in which the engine responds initially, then flattens out or slows down, and then recovers; may cause engine stall if severe enough.

SB Designation for lubricating oil that is acceptable for minimum-duty engines operated under mild conditions.

SC Designation for lubricating oil that meets requirements for use in the gasoline engines in 1964 to 1967 passenger cars and trucks.

scale The accumulation of rust and minerals (from the water) within the cooling system. Also, a series of graduations used to designate specific values.

scavenging The displacement of exhaust gas from the combustion chamber by fresh air or mixture.

scored Scratched or grooved; a cylinder wall may be scored by abrasive particles moved up and down by the piston rings.

screens Pieces of fine-mesh metal fabric; used to prevent solid particles from circulating through any liquid or vapor system and damaging moving parts.

scuffing A type of wear in which there is a transfer of material between parts moving against each other; shows up as pits or grooves in the mating surfaces.

SD Designation for lubricating oil that meets requirements for use in the gasoline engines for 1968 to 1971 passenger cars and in some trucks for those years.

SDV Abbreviation for *spark-delay valve;* a calibrated restrictor in the vacuum-advance hose which delays the vacuum spark advance.

SE Designation for lubricating oil that meets requirements for use in gasoline engines in 1972 and later cars, and in certain 1971 passenger cars and trucks.

seal A material, shaped around a shaft, used to close off the operating compartment of the shaft, preventing oil leakage.

sealer A thick, tacky compound, usually spread with a brush, which may be used as a gasket or sealant, to seal small openings or fill surface irregularities.

seat The surface upon which another part rests, such as a valve seat. Also, to wear into a good fit; for example, new piston rings seat after a few miles of driving.

secondary air Air that is pumped to thermal reactors, catalytic converters, exhaust manifolds, or the cylinder head exhaust ports to promote the chemical reactions that reduce exhaust gas pollutants.

sediment The accumulation of matter which settles to the bottom of a liquid.

sensor Any device that receives and reacts to a signal, such as a change in voltage, temperature, or pressure.

service rating A designation that indicates the type of service for which an engine lubricating oil is best suited. See SA, SB, SC, SD, and SE.

SF Designation for lubricating oil that meets requirements for use in gasoline engines in 1981 and later passenger cars and trucks and in certain 1980 models.

shroud A hood placed around an engine fan to improve fan action.

side clearance The clearance between the sides of moving parts when the sides do not serve as load-carrying surfaces.

single-overhead-camshaft (SOHC) engine An engine in which a single camshaft is mounted over each cylinder head, instead of inside the cylinder block.

sintered bronze Tiny particles of bronze pressed tightly together so that they form a solid piece which is highly porous and often used as a filter for fuel.

sludge Black, soft deposits throughout the interior of the engine, caused by dirt, oil, and water being whipped together by moving parts; sludge is very viscous and tends to reduce lubrication.

sluggish The condition in which the engine delivers limited power under load or at high speed, and will not accelerate as fast as normal, loses too much speed on hills, or has a lower top speed than normal.

smog A term coined from the words *smoke* and *fog*. First applied to the fog-like layer that hangs in the air under certain atmospheric conditions; now generally used to describe any condition of dirty air and/or fumes or smoke. Smog is compounded from smoke, moisture, and numerous chemicals that are produced by combustion.

smoke Small gasborne or airborne particles, exclusive of water vapor, that result from combustion, such particles being emitted by an engine into the atmosphere in sufficient quantity to be observable.

smoke in exhaust A visible blue or black substance often present in the automotive exhaust. A blue color indicates excessive oil in the combustion chamber; black indicates excessive fuel in the air-fuel mixture.

solenoid An electrically operated magnetic device used to mechanically operate some other device through movement of an iron core placed inside a coil. When current flows through the coil, the core attempts to center itself in the coil, thereby exerting a strong force on anything connected to the core.

splash-feed oil system A type of engine lubricating system in which oil is splashed onto moving engine parts.

spray cone A pattern formed when a material is atomized under pressure, narrow at the base and wider as it projects.

spring A device that changes shape under stress or pressure, but returns to its original shape when the stress or pressure is removed; the component of the automotive suspension system that absorbs road shocks by flexing and twisting.

squeak A high-pitched noise of short duration.

squeal A continuous high-pitched noise.

squish The action in some combustion chambers in which the last part of the compressed air-fuel mixture is pushed, or squirted, out of a decreasing space between the piston and cylinder head.

stacks Term for short, tubular carburetor intake pipes; also, for short, individual exhaust pipes.

stalls The condition in which an engine quits running, at idle or while driving.

standpipe assembly See vapor-liquid separator.

stratified charge In a spark-ignition engine, an air-fuel charge with a small layer or pocket of rich air-fuel mixture; the rich mixture is ignited first, after which ignition spreads to the leaner mixture filling the rest of the combustion chamber. The diesel engine is a stratified-charge engine.

stroke In an engine cylinder, the distance that the piston moves in traveling from BDC to TDC or from TDC to BDC.

stumble A severe sudden loss of engine power.

substance Any matter or material; may be a solid, a liquid, or a gas.

substrate In a catalytic converter, the supporting structure to which the catalyst is applied; usually made of ceramic. Two types of substrate used in catalytic converters are the monolithic or one-piece substrate, and the bead- or pellet-type substrate.

sulfur oxides (SO_x) Acids that can form in small amounts as the result of a reaction between hot exhaust gas and the catalyst in a catalytic converter.

supercharger In the intake system of the engine, a device that pressurizes the intake air or the air-fuel mixture. This increases the amount of mixture delivered to the cylinders, which increases the engine power. If the supercharger is driven by the engine exhaust gas, it is called a turbocharger.

surface ignition Ignition of the air-fuel mixture in the combustion chamber by hot metal surfaces or heated particles of carbon.

surge To occur suddenly to an excessive or abnormal value. The condition in which the engine speed increases and decreases slightly under constant throttle operation.

***S/V* ratio** The ratio of the surface area *S* of the combustion chamber to its volume *V*, with the piston at TDC. Often used as a comparative indicator of hydrocarbon emission levels from an engine.

switch A device that opens and closes an electric circuit.

synthetic oil An artificial oil that is manufactured; not a natural mineral oil made from petroleum.

tachometer A device for measuring the speed of an engine in revolutions per minute (rpm).

tank unit The part of the fuel-indicating system that is mounted in the fuel tank.

TDC Abbreviation for *top dead center*.

TEL Abbreviation for *tetraethyl lead*.

temperature The measure of heat intensity in degrees. Temperature is not a measure of heat quantity.

temperature gauge A gauge that indicates to the driver the temperature of the coolant in the engine cooling system.

temperature indicator See *temperature gauge*.

temperature-sending unit A device, in contact with the engine coolant, whose electric resistance changes as the coolant temperature increases or decreases; these changes control the movement of the indicator needle of the temperature gauge.

tetraethyl lead A chemical which, when added to engine fuel, increases its octane rating, or reduces its tendency to detonate. Also called *ethyl* and TEL.

thermal Of or pertaining to heat.

thermal efficiency Ratio of the energy output of an engine to the energy in the fuel required to produce that output.

thermistor A heat-sensing device with a negative temperature coefficient of resistance; as its temperature increases, its electric resistance decreases. Used as the sensing device for engine temperature-indicating devices.

thermometer An instrument which measures heat intensity (temperature) by the thermal expansion of a liquid.

thermostat A device for the automatic regulation of temperature; usually contains a temperature-sensitive element that expands or contracts to open or close off the flow of air, a gas, or a liquid.

thermostatic gauge An indicating device (for fuel quantity, oil pressure, engine temperature) that contains a thermostatic blade or blades.

thermostatic vacuum switch A temperature-sensing device extending into the coolant; connects full manifold vacuum to the distributor when the coolant overheats. The resultant spark advance causes an increase in engine rpm, which lowers the coolant temperature.

thermostatically controlled air cleaner An air cleaner in which a thermostat controls the preheating of intake air.

throat A venturi in a carburetor. See *venturi*.

throttle A disk valve in the carburetor base or throttle body that pivots in response to accelerator pedal position; allows the driver to regulate the volume of air or air-fuel mixture entering the intake manifold, thereby controlling the engine speed. Also called the *throttle plate* or *throttle valve*.

throttle-return check See *dashpot*.

throttle solenoid positioner An electric solenoid which holds the throttle valve open in the hot-idle position, and then permits the throttle valve to close completely when the ignition is turned off to prevent dieseling.

throttle valve A round disk valve in the throttle body of the carburetor; can be turned to admit more or less air, thereby controlling engine speed.

timing In an engine, delivery of the ignition spark or operation of the valves (in relation to the piston) for the power stroke. See *ignition timing* and *valve timing*.

timing chain A chain that is driven by a sprocket on the crankshaft and that drives the sprocket on the camshaft.

timing gear A gear on the crankshaft; drives the camshaft by meshing with a gear on its end.

timing light A stroboscopic light that is connected to the secondary circuit of the ignition system and flashes each time the No. 1 spark plug fires; directing these flashes of light at the whirling timing marks makes the marks appear to stand still, thereby allowing timing to be set by aligning the moving and stationary marks.

top dead center (TDC) The piston position when the piston has reached the upper limit of its travel in the cylinder and the center line of the connecting rod is parallel to the cylinder walls.

transducer Any device which converts an input signal of one form into an output signal of a different form. For example, the automobile horn converts an electric signal to sound.

transistor A semiconductor device that can be used as an electronic (solid-state) switch, and can operate on low voltage. Used to replace the contact points in electronic ignition systems.

transmission-controlled spark (TCS) system An NO_x exhaust emission control system; makes use of the transmission gear position to allow distributor vacuum advance in high gear only.

transmission oil cooler A small finned tube, either mounted separately or as part of the engine radiator, which cools the transmission fluid.

transmission-regulated spark (TRS) system A Ford exhaust emission control system, similar to the General Motors transmission-controlled spark system;

allows distributor vacuum advance in high gear only.

TRS See *transmission-regulated spark system.*

tube-and-fin radiator core A type of radiator core consisting of tubes to which cooling fins are attached; coolant flows through the tubes between the upper and lower radiator tanks.

tuned intake system An engine air intake system in which the manifold has the proper length and volume to produce an air ramming or supercharging effect.

tuneup A procedure for inspecting, testing, and adjusting an engine, and replacing any worn parts, to restore the engine to its best performance.

turbocharger A supercharger driven by the engine exhaust gas.

turbulence The state of being violently mixed or swirled. In the engine, the rapid swirling motion imparted to the air-fuel mixture entering a cylinder.

TVS Abbreviation for *thermostatic vacuum switch.*

two-barrel carburetor A carburetor with two throttle valves.

two cycle Short for *two-stroke cycle.*

two-stroke cycle The two piston strokes during which fuel intake, compression, combustion, and exhaust take place in a two-stroke-cycle engine.

unleaded gasoline Gasoline to which no lead compounds have been intentionally added; gasoline that contains 0.05 g or less of lead per gallon. Required by law to be used in 1975 and later vehicles equipped with catalytic converters.

unloader A device linked to the throttle valve; opens the choke valve when the throttle is moved to the wide-open position.

upshift To shift a transmission into a higher gear.

vacuum A pressure less than atmospheric pressure; a negative pressure. Vacuum can be measured in pounds per square inch, but is usually measured in inches or millimeters of mercury (Hg); a reading of 30 inches [762 mm] Hg would indicate a perfect vacuum.

vacuum advance The advancing or retarding of ignition timing by changes in intake manifold vacuum. Also, the unit on the ignition distributor that uses intake manifold vacuum working on a diaphragm to adjust ignition timing.

vacuum-advance control Any type of NO_x emission control system designed to allow vacuum advance only during certain modes of engine and vehicle operation.

vacuum-advance solenoid An electrically operated two-position valve which allows or denies intake manifold vacuum to the distributor vacuum-advance unit.

vacuum control temperature-sensing valve A valve that connects manifold vacuum to the distributor advance mechanism under hot-idle conditions.

vacuum gauge In automotive engine service, a device that measures intake manifold vacuum and thereby indicates actions of engine components.

vacuum motor A small motor, powered by intake manifold vacuum; used for jobs such as raising and lowering headlight doors.

vacuum pump A mechanical device used to evacuate or pump out a system.

vacuum switch A switch that closes or opens its contacts in response to changing vacuum conditions.

valve Any device that can be opened or closed to allow or stop the flow of a liquid or gas. There are many different types.

valve-in-head engine See *overhead-valve (OHV) engine.*

valve seat The surface against which a valve comes to rest to provide a seal against leakage.

valve timing The timing of the opening and closing of the valves in relation to the piston position.

vane A flat, extended surface that is moved around an axis by or in a fluid. Part of the internal revolving portion of an air supply pump.

vapor A gas; any substance in the gaseous state, as distinguished from the liquid or solid state.

vapor-fuel separator Same as *vapor-liquid separator.*

vapor-liquid separator A device in the evaporative emission control system; prevents liquid fuel from traveling to the engine through the charcoal canister vapor line.

vapor lock A condition in the fuel system in which gasoline vaporizes in the fuel line or fuel pump; bubbles of gasoline vapor restrict or prevent fuel delivery to the carburetor, causing slow, hot starts, no starting, or reduced power.

vapor-recovery system An evaporative emission control system that recovers gasoline vapor escaping from the fuel tank.

vapor-return line A line from the fuel pump to the fuel tank; allows vapor that has formed in the fuel pump to return to the fuel tank.

vapor-saver system Same as *vapor-recovery system.*

vapor separator A device used on cars equipped with air conditioning to prevent vapor lock by sending gasoline vapors back to the fuel tank through a separate line.

vaporization A change of state from liquid to vapor or gas by evaporation or boiling; a general term including both evaporation and boiling. In the carburetor, breaking gasoline into fine particles and mixing it with incoming air.

variable-venturi carburetor A carburetor in which the size of the venturi changes according to engine speed and load.

vehicle vapor recovery See *vapor-recovery system.*

V engine See *V-type engine*.

vent An opening through which air can leave an enclosed chamber.

ventilation The circulating of fresh air through any space, to replace impure air. The basis of crankcase ventilation systems.

venturi In the carburetor, a narrowed passageway or restriction which increases the velocity of air moving through it; produces the vacuum responsible for the discharge of fuel from the main nozzle.

VI Abbreviation for *viscosity index*.

viscosity The resistance to flow exhibited by a liquid. A thick oil has greater viscosity than a thin oil.

viscosity index A number indicating how much the viscosity of an oil changes with heat.

viscosity rating An indicator of the viscosity of engine oil. There are separate ratings for winter driving and for summer driving. The winter grades are SAE 5W, SAE 10W, and SAE 20W. The summer grades are SAE 20, SAE 30, SAE 40, and SAE 50. Many oils have multiple viscosity ratings, as, for example, SAE 10W-30.

viscous Thick; tending to resist flowing.

viscous friction The friction between layers of a liquid.

volatile Evaporating readily. For example, Refrigerant-12 is volatile (evaporates quickly) at room temperature.

volatility A measure of the ease with which a liquid vaporizes; has a direct relationship to the flammability of a fuel.

volumetric efficiency The ratio of the amount of air-fuel mixture that actually enters an engine cylinder to the theoretical amount that could enter under ideal conditions.

V-type engine An engine with two banks, or rows, of cylinders, set on an angle to form a V.

VV carburetor See *variable-venturi carburetor*.

VVR Abbreviation for vehicle vapor recovery. See *vapor-recovery system*.

water jacket The space around cylinders and valves that is hollow so that coolant can flow through to provide cooling.

water pump In the cooling system, the device that circulates coolant between the engine water jackets and the radiator.

wedge combustion chamber A combustion chamber resembling a wedge in shape.

work The changing of the position of an object against an opposing force; measured in foot-pounds, newton-meters, or joules. The product of a force and the distance through which it acts.

WOT Abbreviation for *wide-open throttle*.

INDEX

D

E

F

G

H

ANSWERS TO REVIEW QUESTIONS

The answers to the chapter review questions are given here. If you want to figure your grade on any quiz, divide the number of questions in the quiz into 100. This gives you the value of each question. For instance, suppose there are 10 questions: 10 goes into a hundred 10 times. Each correct answer, therefore, gives you 10 points. If you answered 8 correct out of the 10, then your grade would be 80 (8×10).

If you are not satisfied with the grade you make on a test, restudy the chapter and retake the test. This review will help you remember the important facts.

Remember, when you take a course in school, you can pass and graduate even though you make a grade of less than 100. But in the automotive shop, you must score 100 percent all the time. If you make 1 error out of 100 service jobs, for example, your average would be 99. In school that is a fine average. But in the automotive shop that one job you erred on could cause such serious trouble (a ruined engine or a wrecked car) that it would outweigh all the good jobs you performed. Therefore, always proceed carefully in performing any service job and make sure you know exactly what you are supposed to do and how you are to do it.

CHAPTER 1

1. *b* 2. *c* 3. *b* 4. *c* 5. *c*
6. *c* 7. *d* 8. *a* 9. *b* 10. *b*

CHAPTER 2

1. *a* 2. *b* 3. *d* 4. *c* 5. *b*
6. *c* 7. *d* 8. *d* 9. *c* 10. *a*

CHAPTER 3

1. *a* 2. *a* 3. *b* 4. *b* 5. *b*
6. *c* 7. *b* 8. *b* 9. *a* 10. *a*

CHAPTER 4

1. *d* 2. *c* 3. *c* 4. *b* 5. *b*
6. *b* 7. *c* 8. *a* 9. *a* 10. *c*
11. *b* 12. *b* 13. *a* 14. *a* 15. *c*
16. *c* 17. *b* 18. *c* 19. *b* 20. *a*
21. *c* 22. *a* 23. *a* 24. *b* 25. *c*

CHAPTER 5

1. *d* 2. *c* 3. *c* 4. *b* 5. *b*
6. *b* 7. *a* 8. *c* 9. *c* 10. *d*

CHAPTER 6

1. *b* 2. *c* 3. *b* 4. *a* 5. *a*
6. *c* 7. *a* 8. *b* 9. *a* 10. *b*

CHAPTER 7

1. *a* 2. *b* 3. *c* 4. *c* 5. *b*
6. *c* 7. *c* 8. *a* 9. *c* 10. *a*

CHAPTER 8

1. *a* 2. *a* 3. *c* 4. *c* 5. *a*
6. *b* 7. *b* 8. *a* 9. *b* 10. *d*
11. *c* 12. *d* 13. *b* 14. *a* 15. *c*
16. *b* 17. *b* 18. *a* 19. *a* 20. *b*
21. *b* 22. *d* 23. *d* 24. *b* 25. *d*

CHAPTER 9

1. *a* 2. *b* 3. *b* 4. *b* 5. *d*
6. *b* 7. *d* 8. *a* 9. *c* 10. *a*
11. *d* 12. *c* 13. *b* 14. *c* 15. *b*
16. *c* 17. *b* 18. *c* 19. *b* 20. *c*

CHAPTER 10

1. *b* 2. *d* 3. *c* 4. *d* 5. *c*
6. *b* 7. *d* 8. *b* 9. *d* 10. *d*

CHAPTER 11

1. *d* 2. *a* 3. *b* 4. *c* 5. *a*
6. *c* 7. *c* 8. *b* 9. *b* 10. *b*

CHAPTER 12

1. *b* 2. *c* 3. *d* 4. *a* 5. *b*

CHAPTER 13

1. *c* 2. *c* 3. *d* 4. *d* 5. *b*

CHAPTER 14

1. *d* 2. *c* 3. *b* 4. *a* 5. *a*
6. *d* 7. *c* 8. *a* 9. *c* 10. *d*

CHAPTER 15

1. *c* 2. *c* 3. *a* 4. *b* 5. *d*

CHAPTER 16

1. *c* 2. *b* 3. *a* 4. *d* 5. *c*
6. *c* 7. *d* 8. *c* 9. *b* 10. *b*
11. *a* 12. *b* 13. *c* 14. *a* 15. *d*

CHAPTER 17

1. *c* 2. *b* 3. *d* 4. *c* 5. *c*

CHAPTER 18

1. *a* 2. *d* 3. *a* 4. *b* 5. *a*
6. *c* 7. *a* 8. *d* 9. *b* 10. *a*

CHAPTER 19

1. *c* 2. *b* 3. *a* 4. *b* 5. *a*
6. *b* 7. *d* 8. *d* 9. *c* 10. *b*